Grundlehren der mathematischen Wissenschaften 246

A Series of Comprehensive Studies in Mathematics

M. A. Naimark
A. I. Štern

Theory of Group Representations

Translated by Elizabeth Hewitt
Translation Editor Edwin Hewitt

With 3 Figures

Springer-Verlag
New York Heidelberg Berlin

AMS Subject Classification (1980): 20Cxx

Library of Congress Cataloging in Publication Data
Naĭmark, M. A. (Mark Aronovich)
  Theory of group representations.

  (Grundlehren der mathematischen Wissenschaften;
246 — Comprehensive Studies in Mathematics; 246)
  Translation of: Teoriia predstavleniĭ grupp.
  Bibliography: p.
  Includes index.
  1. Representations of groups.   I. Title.   II. Shtern, Aleksandr Isaakovich,
1941–   III. Series: Grundlehren der mathematischen Wissenschaften; 246.
QA171.N3213   512′.2   81-23274
                        AACR2
Title of the Russian Original Edition: Teoriya predstavleniĭ grupp. Publisher: Nauka,
Moskow, 1975

ISBN 0-387-90602-9   Springer-Verlag New York   Heidelberg   Berlin
ISBN 3-540-90602-9   Springer-Verlag Berlin   Heidelberg   New York

Preface

Author's Preface to the Russian Edition

This book is written for advanced students, for predoctoral graduate students, and for professional scientists—mathematicians, physicists, and chemists—who desire to study the foundations of the theory of finite-dimensional representations of groups.

We suppose that the reader is familiar with linear algebra, with elementary mathematical analysis, and with the theory of analytic functions. All else that is needed for reading this book is set down in the book where it is needed or is provided for by references to standard texts.

The first two chapters are devoted to the algebraic aspects of the theory of representations and to representations of finite groups. Later chapters take up the principal facts about representations of topological groups, as well as the theory of Lie groups and Lie algebras and their representations.

We have arranged our material to help the reader to master first the easier parts of the theory and later the more difficult. In the author's opinion, however, it is algebra that lies at the heart of the whole theory.

To keep the size of the book within reasonable bounds, we have limited ourselves to finite-dimensional representations. The author intends to devote another volume to a more general theory, which includes infinite-dimensional representations.

The author expresses his deep gratitude to A. I. Štern, who has contributed enormously to the work not only in an editorial capacity, but indeed as a co-author. Dr. Štern wrote Chapters VIII–XI, Sections 2 and 3 of Chapter IV, Sections 4 and 5 of Chapter II, and 2.10 of Chapter I.

The author is also deeply obligated to Professor A. A. Kirillov, who read the book in manuscript and made a series of valuable suggestions.

May 1975   M. A. NAĬMARK

Editor's Preface to the English Edition

Mark Aronovič Naĭmark, Professor at the Steklov Institute of The Akademija Nauk in Moscow, USSR, was born on 5 December 1909 and died on 30 December 1978. He was one of the outstanding mathematicians of his day, a leader in the Soviet school of functional analysis, operator algebras, and representations of topological groups. He was also a gifted and productive writer, the author of several classical monographs. Perhaps the most renowned of these is the treatise Naĭmark [1]. (Throughout the book, numbers in square brackets refer to the bibliography, found on pages 560–563.)

The present book was Naĭmark's *Schwanengesang*. In his last years, Professor Naĭmark was seriously ill, a burden which he bore with grace and unfailing good humor. Much of this book was dictated to his wife. In his preface to the Russian edition, Professor Naĭmark describes his collaboration with A. I. Štern in writing the book.

This English edition is not the first translation of the book. A French translation, under the title *Théorie des Représentations des Groupes*, was published by Éditions Mir (Moscow, 1979). The French translation appears to be a word-for-word translation of the Russian, including a faithful transcription of misprints. The French translation lists A. I. Štern as co-author.

A few words about the present edition may be in order. The English translation was made by Elizabeth Hewitt. The translation was edited, and a careful comparison with both the Russian and French editions was made, by Edwin Hewitt. Misprints and trifling errors have been corrected without mention. Here and there actual improvements presented themselves, and these too have been made, identified by editor's footnotes. The editor feels certain that Professor Naĭmark would have welcomed these small changes. Naturally the editor makes no claim of authorship: the book is Naĭmark's and Štern's.

Seattle, Washington   EDWIN HEWITT
Charlottesville, Virginia   ELIZABETH HEWITT

December 1981

Contents

Chapter V
Finite-Dimensional Representations of Connected Solvable
Groups; the Theorem of Lie

Chapter VI
Finite-Dimensional Representations of the Full Linear Group

Chapter VII
Finite-Dimensional Representations of the Complex
Classical Groups

Chapter VIII
Covering Spaces and Simply Connected Groups

Chapter IX
Basic Concepts of Lie Groups and Lie Algebras

Chapter X
Lie Algebras

Chapter XI
Lie Groups

Chapter XII
Finite-Dimensional Irreducible Representations of Semisimple
Lie Groups

Bibliography

Chapter I

Algebraic Foundations of Representation Theory

In this chapter we set forth those concepts and propositions of representation theory which have a purely algebraic character and consequently do not utilize topological or analytical facts. Strictly speaking, it would be appropriate to add the adjective "algebraic" to each concept introduced in Chapter I, for example, algebraic group, algebraic isomorphism, algebraic equivalence, etc. However, for the sake of brevity we will assume "algebraic" in Chapter I and use it in other chapters only in cases where some doubt might arise.

§1. Fundamental Concepts of Group Theory

1.1. Definition of a Group

A nonempty set G is called a *group* if a *product* $g_1 g_2$ is defined for every two elements $g_1, g_2 \in G$, for which the following conditions hold[1]:

(a) $g_1 g_2 \in G$, for all $g_1, g_2 \in G$;

(b) $(g_1 g_2) g_3 = g_1 (g_2 g_3)$, for all $g_1, g_2, g_3 \in G$;

(c) there exists a unique element e in G such that $eg = ge = g$ for all $g \in G$; e is called the *identity element* of the group G;

(d) for every element $g \in G$ there exists a unique element, designated g^{-1}, for which $gg^{-1} = g^{-1}g = e$; the element g^{-1} is called the *inverse* of g. It is evident that g is the inverse of g^{-1}, so that $(g^{-1})^{-1} = g$.

A group G is called *commutative* (or *abelian*) if $g_1 g_2 = g_2 g_1$ for every $g_1, g_2 \in G$ and *noncommutative* in the opposite case. In the case of a commutative group we often write $g_1 + g_2$ instead of $g_1 g_2$ and 0 for the identity element of G. When writing the product this way, we say that the group is given in *additive notation*.

---

[1] Actually, these conditions can be relaxed. For example, it is sufficient in condition (c) to only require the existence of an identity element. Its uniqueness then follows. In fact, if e and e' are identity elements, then $e'e = e'$ and $e'e = e$, and therefore $e' = e$. (For more details see for example, Kuroš [1]). However, we will not need a minimal list of axioms that define a group.

A group is called *finite* if the number of its elements is finite; in the opposite case it is called *infinite*. The number of elements of a finite group G is called its *order* and is denoted by $|G|$. A finite group G, consisting of elements $g_1, \ldots, g_m$, $m = |G|$, may be defined by its multiplication table:

	g_1	g_2	$\cdots$	g_m
g_1	$g_1 g_1$	$g_1 g_2$	$\cdots$	$g_1 g_m$
g_2	$g_2 g_1$	$g_2 g_2$	$\cdots$	$g_2 g_m$
$\cdot$				
g_m	$g_m g_1$	$g_m g_2$	$\cdots$	$g_m g_m$,

in which the product $g_j g_k$ is the intersection of the j-th row and the k-th column. This table is called a *Cayley table* of the group G.

Examples

1. The set $\mathbf{R}^1$ of all real numbers[2] becomes a group if we define multiplication as addition of real numbers. This group is called the *additive group of real numbers*. The identity element of this group is the number zero and the inverse element to the number x is the number $-x$. The *additive group* $\mathbf{C}^1$ *of complex numbers* is defined analogously.

2. The set of $\mathbf{R}_0^1$ of all nonzero real numbers forms a group if we define multiplication as ordinary multiplication of numbers. This group is called the *multiplicative group of real numbers*. The number 1 is the identity of this group and the inverse of the number x is the number $1/x$. The *multiplicative group* $\mathbf{C}_0^1$ *of complex numbers* is defined analogously.

3. The set $G_0 = \{1, i, -1, -i\}$ with ordinary multiplication is a group. The Cayley table of this group takes the following form:

	1	i	-1	$-i$
1	1	i	-1	$-i$
i	i	-1	$-i$	1
-1	-1	$-i$	1	i
$-i$	$-i$	1	i	-1

4. Let X be a linear space, G_X the set of all one-to-one linear operators on X that map X *onto* X. We define multiplication in G_X as multiplication of operators. Then G_X is a group. The identity element here is the identity operator 1 (that is, the operator such that $1x = x$ for all $x \in X$) and the inverse element of the operator A is the inverse operator A^{-1}. If X is finite-dimensional ($\dim X = n < \infty$), then for a fixed basis in X the operators

---

[2] As a rule, the symbol $\mathbf{R}^1$ will denote the *additive group* of real numbers and the letter $\mathbf{R}$ alone will be used when the set of real numbers is regarded as a *field*.

$A \in G_X$ are described by non-singular (that is, with determinant $\neq 0$) square matrices of order n.

5. The set of all complex matrices of order n with nonzero determinant is a group if multiplication is defined as multiplication of matrices; this group is usually denoted as $GL(n, \mathbf{C})$. Its identity element is the identity matrix and the inverse element of the matrix a is the inverse matrix a^{-1}. The group $GL(n, \mathbf{R})$ of all real matrices of order n with nonzero determinant is defined similarily. If $n \geqslant 2$ these groups are noncommutative.

6. Let $SL(n, \mathbf{C})$ be the set of all complex matrices of order n with determinant equal to 1. We define a product in $SL(n, \mathbf{C})$ as the product of matrices. Then $SL(n, \mathbf{C})$ is a group, since in multiplication of matrices, determinants also multiply. The group $SL(n, \mathbf{R})$ of all real matrices of order n with determinant equal to 1 is defined similarly.

7. Let G'_0 be the set of all the rotations of a square $ABCD$ around its center O, carrying the square onto itself. There are four different[3] rotations: the rotation α_0 through the angle 0, the rotation α_1 through 90°, the rotation α_2 through 180° and the rotation α_3 through 270° (all counterclockwise). These carry the point A to A, B, C, D, respectively.

The *product* $\alpha\beta$ of two rotations α, β is defined as the result of first applying the rotation β and then the rotation α. It is easy to show that under the definition of the product, G'_0 is a group of order 4.

8. The set $\mathbf{N}$ of all integers is a group if we define multiplication in $\mathbf{N}$ as addition of integers. This group is called *the (additive) group of integers.*

9. Let $\mathbf{N}_p$ be the set of all integers that are multiples of a fixed positive integer p, that is, $\mathbf{N}_p = \{np, n \in \mathbf{N}\}$. Again, we define multiplication in $\mathbf{N}_p$ as addition. It is clear that $\mathbf{N}_p$ is a group.

10. Let Ω_p be the set of all p^{th} roots of unity where p is a fixed positive integer. It is known that Ω_p consists of the numbers $e^{i2\pi k/p}$, $k = 0, 1, \ldots, p - 1$. Let the product of numbers in Ω_p be their ordinary product. It is then evident that Ω_p is a group. Note that $\Omega_4 = G_0$ (see example 3).

The groups described in examples 1–3 and 7–10 are commutative. The groups of examples 3 and 7 are finite and of order 4; in example 10 the group is finite of order p. The groups of examples 1, 2, 4–6, 8, and 9 are infinite.

1.2. Subgroups; Cosets

A nonempty set $H \subset G$ is called a *subgroup* of a group G if $g_1 g_2^{-1} \in H$ for all g_1, g_2 in H. For $g_1 = g_2$ we get $e \in H$ and therefore if $g_1, g_2 \in H$, it follows that also $g_1^{-1} \in H$ and $g_1 g_2 \in H$. Consequently, with the definition of multiplication that we have in G, the set H is also a group.

Thus, $\mathbf{R}^1$, $\mathbf{R}_0^1$, $GL(n, \mathbf{R})$ and $\mathbf{N}$ are all subgroups of $\mathbf{C}^1$, $\mathbf{C}_0^1$, $GL(n, \mathbf{C})$ and $\mathbf{R}^1$ respectively (see examples 1, 2, 5, 8 of 1.1); $SL(n, \mathbf{C})$ and $SL(n, \mathbf{R})$ are

---

[3] Two rotations are not considered different if they lead to the same position of the square.

subgroups of $GL(n, \mathbf{C})$ and $GL(n, \mathbf{R})$. $SL(n, \mathbf{R})$ is a subgroup of $SL(n, \mathbf{C})$ (see examples 5 and 6 of 1.1). It is evident that the original group G, as well as the subset $\{e\}$, consisting solely of the identity element of the group G, are subgroups. They are called *trivial subgroups* of group G. All other subgroups of G (if any exist) are called *nontrivial subgroups*.

It is also evident that the intersection of any set of subgroups of G is also a subgroup of G; specifically, the intersection of all subgroups containing given subset S of G is a subgroup. It is the smallest subgroup containing S and is written $G(S)$.

I. *Let H be the set of all finite products of elements $g_i \in S$ and their inverses g_i^{-1}; then $G(S) = H$.*

Proof. Plainly, H is a subgroup containing S. On the other hand, all subgroups containing S contain H; consequently $G(S) = H$ by virtue of the minimality of $G(S)$. ☐

Suppose that S consists of a single element g_0. The subgroup $G(g_0)$ is called *cyclic*. Plainly $G(g_0)$ contains all possible powers g_0^n, $n = 0, \pm 1, \pm 2, \ldots, $; some of these powers may coincide. If all powers g_0^n are distinct, then g_0 is called an *element of infinite order* and $G(g_0)$ a *cyclic group of infinite order*. If two powers g^m and g^l are equal, $g_0^l = g_0^m$, $(m > l)$, then $g_0^{m-l} = e$. In this case g_0 is called an *element of finite order*. The smallest positive integer p for which $g_0^p = e$ is called the *order of the element g*. If g_0 has finite order p, the group $G(g_0)$ consists of the elements $e, g_0, g_0^2, \ldots, g^{p-1}$, which are all distinct; $G(g_0)$ is then called a *cyclic group of order p*. A group G is called *cyclic* if there exists an element $g_0 \in G$ such that $G = G(g_0)$.

Let H be a subgroup of a group G; any set Hg_0 (by which we mean the set of all elements hg_0 for all $h \in H$) is called a *right coset of the subgroup H in the group G*. *Left cosets* are defined in like manner. Each element of a coset is called a *representative* of this coset; a coset containing a representative g is written $\{g\}$ or $\tilde{g}$. If g is a representative of a coset Hg_0 then $Hg = Hg_0$. To see this note that $g = hg_0$ for some $h \in H$ and therefore $Hg = H(hg_0) = (Hh)g_0 = Hg_0$. From this it follows that *distinct right (left) cosets are disjoint*. Besides this, every element $g \in G$ belongs to some right coset, namely, the set Hg. We infer the following.

II. *The group G falls into pairwise disjoint cosets.*

It is clear that the analogous proposition applies to left cosets.

The set of all right cosets of the subgroup H of G, each regarded as a single element, is called the *factor space* or the *space of cosets of the subgroup H in G*. It is denoted by G/H. The number of elements in G/H, if it is finite, is called the *index of the subgroup H in G* and is usually written as $|G/H|$.

If the group G is finite, then H is also finite and the number of elements of each coset Hg is the same and coincides with $|H|$. (Note that $H = He$ is a coset.) From this and from II we conclude that $|G| = |H| \, |G/H|$. This equality can be restated as follows.

III. *The order of a subgroup of a finite group is a divisor of the order of the group.*

Examples

1. Let $G = \mathbf{R}^1$, $H = \mathbf{N}$ (see examples 1 and 8 of 1.1). Then each set $\tilde{g} \in \mathbf{R}^1/\mathbf{N}$ has the form $g_\alpha = \{n + \alpha, \alpha \in \mathbf{N}\}$, where α is a number determined by the coset. Note that $0 \leqslant \alpha < 1$; α is called the *fractional part of the number* $n + \alpha$. In this way, an element of $\mathbf{R}^1/\mathbf{N}$ is uniquely defined by the fractional part of all of its representatives.

2. Let $G = \mathbf{N}$, $H = \mathbf{N}_2$ (see examples 8 and 9 in 1.1); that is, H is the subgroup of $\mathbf{N}$ consisting of all even integers. Then $\mathbf{N}/H$ consists of two elements: $\tilde{g}_0 = H$, $\tilde{g}_1 = \{1 + h, h \in H\}$. In other words, $\tilde{g}_0$ is the set of all even integers and $\tilde{g}_1$ is the set of all odd integers.

3. Let $G = \mathbf{N}$, $H = \mathbf{N}_p$. Then $\mathbf{N}/H$ consists of p sets $\tilde{g}_0, \tilde{g}_1, \ldots, \tilde{g}_{p-1}$, where $\tilde{g}_k = \{k + h, h \in H\}$, $k = 0, 1, \ldots, p - 1$. These sets are called *sets of residues modulo p*.

1.3. Normal Divisor; Factor Group

A subgroup H of a group G is called a *normal divisor of the group* G if for every element $g \in G$, we have

$$gH = Hg. \tag{1.3.1}$$

That is, for every $g \in G$, $h_1 \in H$ there exists an element $h_2 \in H$ for which $gh_1 = h_2 g$. It is evident that relation (1.3.1) indicates that every left coset gH coincides with the right coset Hg. If our group G is commutative, then every subgroup of G is a normal divisor of G. Many groups admit that are not normal divisors. The group G itself and the identity subgroup $\{e\}$ are normal divisors of G. They are called *trivial normal divisors*. All other normal divisors, if any, are called *nontrivial normal divisors in G*. The group G is called *simple* if it has no nontrivial normal divisors. If H is a normal divisor of G, we can define multiplication in the factor space G/H as follows: the *product* $(Hg_1)(Hg_2)$ of cosets Hg_1, Hg_2 is defined as the coset Hg_1g_2. This definition does not depend upon the selection of representatives of sets Hg_1, Hg_2. Indeed, if $g'_1 \in Hg_1$, $g'_2 \, Hg_2$, then $g'_1 = h_1g_1$, $g'_2 = h_2g_2$ and by virtue of (1.3.1) $g_1h_2 = h'_2g_1$ for some $h'_2 \in H$. From this we have $g'_1g'_2 = (h_1g_1)(h_2g_2) = h_1(g_1h_2)g_2 = h_1h'_2g_1g_2$; consequently, $Hg'_1g'_2 = Hh_1h'_2g_1g_2 = Hg_1g_2$. Also, it is easy to verify that the product so defined satisfies conditions

(a) through (c) in 1.1. The identity element $\tilde{e}$ in G/H is the coset $\tilde{e} = H$. Therefore G/H is a group. It is called the *factor group of the group G by the normal divisor H* and is written G/H, as before.

Examples and Exercises

1. Let $G = GL(n, \mathbf{C})$, $H = SL(n, \mathbf{C})$ (see example 5 in 1.1). Then H is a normal divisor of G. In fact, for all $g \in G$, $h \in H$ we have $\det(ghg^{-1}) = \det g \cdot \det h \cdot \det g^{-1} = \det h = 1$ and therefore $ghg^{-1} \in H$, so that (1.3.1) holds.

Let us find the factor group G/H. If g_1 and g_2 belong to the same set $\tilde{g} \in G/H$, then $g_2 = hg_1$ and therefore $\det g_2 = \det h \cdot \det g_1 = \det g_1$. Conversely, if $\det g_2 = \det g_1$, then $\det(g_2 g_1^{-1}) = 1$ and therefore $g_2 g_1^{-1} \in H$, that is, g_2 and g_1 belong to the same set $\tilde{g} \in G/H$. Thus, set $\tilde{g}$ is defined uniquely by the number $\lambda_{\tilde{g}} = \det g$, where $g \in \tilde{g}$. From the definition of product in G/H it follows that $\lambda_{\tilde{g}_1 \tilde{g}_2} = \det(g_1 g_2) = \det g_1 \cdot \det g_2$, where $g_1 \in \tilde{g}_1$, $g_2 \in \tilde{g}_2$, that is $\lambda_{\tilde{g}_1 \tilde{g}_2} = \lambda_{\tilde{g}_1} \lambda_{\tilde{g}_2}$. Thus, the mapping $\tilde{g} \to \lambda_{\tilde{g}}$ is a one-to-one mapping of the group $G/H = GL(n, \mathbf{C})/SL(n, \mathbf{C})$ onto the group $\mathbf{C}_0^1$ (see example 2 in 1.1). Likewise, the mapping $g \to \lambda_{\tilde{g}} = \det g$, where $g \in \tilde{g} \in GL(n, \mathbf{R})/SL(n, \mathbf{R})$, is a one-to-one mapping of $GL(n, \mathbf{R})/SL(n, \mathbf{R})$ onto $\mathbf{R}_0^1$ satisfying the condition $\lambda_{\tilde{g}_1 \tilde{g}_2} = \lambda_{\tilde{g}_1} \lambda_{\tilde{g}_2}$.

2. Let K_2 be the set of all matrices

$$k = \left\| \begin{matrix} \lambda^{-1} & 0 \\ \mu & \lambda \end{matrix} \right\|$$

and Z_2 the set of all matrices

$$z = \left\| \begin{matrix} 1 & 0 \\ \mu & 1 \end{matrix} \right\|$$

where λ and μ are complex numbers such that $\lambda \neq 0$. Then K_2 is a subgroup of $SL(2, \mathbf{C})$ and Z_2 is a normal divisor of K_2. Prove that:

(1) Z_2 is a normal divisor of K_2;
(2) every element $\tilde{k} \in K_2/Z_2$ is uniquely defined by the number $\lambda_{\tilde{k}} = \lambda$, where

$$k = \left\| \begin{matrix} \lambda^{-1} & 0 \\ \mu & \lambda \end{matrix} \right\| \in \tilde{k};$$

(3) the mapping $\tilde{k} \to \lambda_{\tilde{k}}$ is a one-to-one mapping of the group K_2/Z_2 onto the group $\mathbf{C}_0^1$ (see example 2 in 1.1).

3. Let $G = \mathbf{R}^1$, $H = \mathbf{N}$ (see examples 1 and 8 in 1.1). Since $\mathbf{R}^1$ is commutative, $\mathbf{N}$ is a normal divisor of $\mathbf{R}^1$; $\mathbf{R}^1/\mathbf{N}$ consists of the sets $\tilde{g} = \{n + \alpha, n \in \mathbf{N}\}$,

$0 \leqslant \alpha < 1$ (see example 1 in 1.2). By the definition of multiplication in $\mathbf{R}^1/\mathbf{N}$, we have

$$\tilde{g}_\alpha \tilde{g}_\beta = \{n + \alpha + \beta, n \in \mathbf{N}\} = \{n + \gamma, n \in \mathbf{N}\} = \tilde{g}_\gamma,$$

where $\gamma = [\alpha + \beta]$.[4]

4. Let $G = \mathbf{N}$, $H = \mathbf{N}_p$ (see examples 8 and 9 in 1.1); $\mathbf{N}$ is commutative and therefore $\mathbf{N}_p$ is a normal divisor of $\mathbf{N}$ and $\mathbf{N}/\mathbf{N}_p = \{g_0, g_1, \ldots, g_{p-1}\}$ (see example 3 in 1.2). From the definition of multiplication in $\mathbf{N}/\mathbf{N}_p$, we see that $\tilde{g}_k \tilde{g}_l = \{k + l + pn, n \in \mathbf{N}\} = \{m + pn, n \in \mathbf{N}\} = \tilde{g}_m$. Observe that m is the remainder upon division of the number $k + l$ by p. The group $\mathbf{N}/\mathbf{N}_p$ is called the *group of residues modulo p*.

1.4. The Center

Let G be a group. The set of all elements of G that commute with every element of G is called the *center of the group* G. It is written $Z(G)$. Thus an element g_0 of G belongs to $Z(G)$ if and only if $gg_0 = g_0 g$ for all $g \in G$.

Plainly, $Z(G)$ is a *subgroup* of the group G. If $g_1, g_2 \in Z(G)$, then $g_1 g = gg_1$ and $g_2 g = gg_2$ for all $g \in G$. From this we obtain $g^{-1}g_2^{-1} = g_2^{-1}g^{-1}$. Since g^{-1} runs through all of G as g runs through G, we infer that $g_2^{-1} \in Z(G)$. For all $g \in G$, we also have $g_1 g_2^{-1} = g_1 gg_2^{-1} = gg_1 g_2^{-1}$; consequently, $g_1 g_2^{-1}$ belongs to $Z(G)$.

Elements of $Z(G)$ commute with every element of G; in particular, they commute with each other. That is, $Z(G)$ is a *commutative group*. If the group G itself is a commutative, then $Z(G) = G$. Also $Z(G)$ is a *normal divisor* of G, since for every $g \in G$ we have $gZ(G) = Z(G)g$.

Examples and Exercises

1. Let $G = GL(n, \mathbf{C})$ (see example 5 of 1.1). Let us find $Z(G)$. If $z \in Z(G)$, then

$$gz = zg \quad \text{for all } g \in G. \tag{1.4.1}$$

In (1.4.1), we choose

$$g = \begin{Vmatrix} \lambda_1 & & & \\ & \lambda_2 & & \\ & & \ddots & \\ & & & \lambda_n \end{Vmatrix},$$

where $\lambda_j \neq 0, j = 1, \ldots, n$, the λ_j are all distinct and all elements not on the

---

[4] For numbers α and β in the interval $[0, 1)$, define $[\alpha + \beta]$ as $\alpha + \beta$ if $\alpha + \beta < 1$ and as $\alpha + \beta - 1$ if $\alpha + \beta \geqslant 1$.

main diagonal are zero. From (1.4.1) we find $\lambda_j z_{jk} = \lambda_k z_{jk}$ and therefore, $z_{jk} = 0$ for $j \neq k$. That is, z is a diagonal matrix. We next choose in (1.4.1)

$$
g = \left\| \begin{matrix} 0 & 1 & & & \\ 1 & 0 & & & \\ & & 1 & & \\ & & & \cdot & \\ & & & & \cdot \\ & & & & & 1 \end{matrix} \right\|,
$$

where all places not marked are occupied by zeroes. Multiplying out, we find that $z_{11} = z_{22}$. Analogously we can prove that $z_{jj} = z_{kk}$ for all $j, k = 1, \ldots, n$. We have proved that $z = \lambda 1$ where $\lambda = z_{jj}, j = 1, \ldots, n$. In other words, $Z(GL(n, \mathbf{C}))$ consists of the matrices $\lambda 1, \lambda \in \mathbf{C}_0^1$. Similarly $Z(GL(n, \mathbf{R}))$ consists of all matrices $\lambda 1, \lambda \in \mathbf{R}_0^1$.

2. Find the factor groups $GL(n, \mathbf{C})/Z(GL(n, \mathbf{C}))$ and $GL(n, \mathbf{R})/Z(GL(n, \mathbf{R}))$.

1.5. Mappings

Let X and Y be two arbitrary sets. A *mapping* f of X into Y is any rule or correspondence which assigns a fixed point $f(x)$ in Y to every point x in X. The point $y = f(x)$ is called the *image* of the point x under the mapping f. Given a subset M of X, we call the set of all points $f(x)$ for x in M the *image* of M under the mapping f. We write this set as $f(M)$. For a subset N of Y, the set of all $x \in X$ for which $f(x) \in N$ is called the *inverse image of the set N* under the mapping f. It is written $f^{-1}(N)$. If the image $f(X)$ of the set X is the entire set Y, then f is said to be a mapping of X *onto* Y.

A mapping f of the set X onto Y is called *one-to-one* if the inverse image of every point $y \in f(X)$ consists of a single point. The mapping of X onto X in which every image coincides with itself is called the *identity mapping of X onto X*.

Let f be a mapping of X into Y and ϕ a mapping of Y into Z. Consider the mapping of X into Z obtained by first applying f and then ϕ to every image $f(x)$. We write this mapping as ϕf. Thus $\phi f(x)$ is defined as $\phi(f(x))$ for all x in X. The mapping ϕf is called the *product of ϕ and f*.[5] Note that ϕf maps X into Z.

Suppose that $Z = X$ and that ϕf is the identity mapping of X onto itself. Then ϕ is called the *inverse of f* and is written f^{-1}. The identities $(\phi f)(x) = x$ and $(f\phi f)(x) = f(x)$ show that ϕ is a mapping of Y onto X and also that $f\phi$ is the identity mapping of Y onto itself. Thus f is the inverse ϕ^{-1} of ϕ. Conversely, if f is one-to-one, we can define f^{-1} as the mapping of $f(X)$ onto X such that $f^{-1}(f(x))$ is the unique element of X whose image under f is $f(x)$.

I. *The inverse mapping f^{-1} of f exists if and only if the mapping f is one-to-one.*

---

[5] The product ϕf is also written $\phi \circ f$ and is also called the *composition of f and ϕ*.

The necessity of this condition is obvious. Conversely, if the mapping f is one-to-one, we assign to each point y in $f(X)$ its inverse image under f. This gives us the inverse mapping f^{-1}.

Let f be a mapping of X into Y and let Z be a subset of X; the mapping ϕ of the set Z into Y defined by the formula $\phi(z) = f(z)$ for all $z \in Z$, is called the *restriction* of the mapping f to Z. It is written $f|_Z$.

1.6. Homomorphisms and Isomorphisms of Groups

A mapping f of a group G into a group G' is called a *homomorphism of G into G'* if

$$f(g_1 g_2) = f(g_1)f(g_2) \text{ for all } g_1, g_2 \in G. \tag{1.6.1}$$

The condition (1.6.1) means that if f maps g_1 to g_1' and g_2 to g_2', the f also maps $g_1 g_2$ to $g_1' g_2'$. The inverse image $f^{-1}(e')$ of the identity element e' of G' is called the *kernel of the homomorphism f*. It is written Ker f. If the image of G is the whole group G' (that is, $f(G) = G'$), then f is called a *homomorphism of G onto G'*.

I. *If f is a homomorphism of a group G into a group G', then*

(1) $f(e)$ *is the identity element in G'*;
(2) $f(g^{-1}) = (f(g))^{-1}$;
(3) f *maps every subgroup of G onto a subgroup of G'*;
(4) f *maps every normal divisor of G onto a normal divisor of the group $f(G)$*;
(5) Ker f *is a normal divisor of G*.

Proof. (1) We set $\hat{e} = f(e)$ and write the identity element in G as e'. Then, by virtue of (1.6.1)

$$\hat{e}\hat{e} = f(e)f(e) = f(ee) = f(e) = \hat{e} = \hat{e}e'.$$

If we multiply both sides of this equality on the left by $\hat{e}^{-1}$, we get $\hat{e} = e'$. This proves (1). Assertion (2) follows from the relations $f(g)f(g^{-1}) = f(g^{-1})f(g) = f(gg^{-1}) = f(e) = e'$ and (3) from the relations

$$f(h_1)f(h_2)^{-1} = f(h_1 h_2^{-1}) \in f(H).$$

Let H be a normal divisor of G and let $g' \in f(G)$; then $g' = f(g)$ for some $g \in G$ and therefore we have

$$g'f(H) = f(g)f(H) = f(gH) = f(Hg) = f(H)f(g) = f(H)g'.$$

This proves (4).

Let us write Ker $f = H$; then $f(H) = e'$. If h_1, h_2 are in H, then $f(h_1) = e'$, $f(h_2) = e'$ and therefore $f(h_1 h_2^{-1}) = f(h_1)f(h_2)^{-1} = e'$. That is, $h_1 h_2^{-1}$ belongs to H; consequently H is a subgroup of G.

For all $g \in G$, we have

$$f(gHg^{-1}) = f(g)f(H)f(g^{-1}) = f(g)e[f(g)]^{-1} = e'.$$

It follows that $gHg^{-1} \subset H$ and $gH \subset Hg$. Replacing g by g^{-1} and multiplying both sides of the resulting equality on the left and right by g, we get $Hg \supset gH$; consequently $Hg = gH$ and (5) is proved. $\square$

We can construct a simple example of homomorphism as follows. Let H be a normal divisor of a group G. We map every g in G onto the set $\tilde{g} \in G/H$ that contains g. It follows from the definition of multiplication in G/H that the mapping $g \to \tilde{g}$ is a homomorphism of G onto the group G/H. This is called the *canonical homomorphism* (also the *natural mapping*) of G onto the *factor group* G/H. A one-to-one homomorphism is called an *isomorphism*. It is clear that all of Proposition I is valid for isomorphisms.

II. *A homomorphism f is an isomorphism if and only if* Ker $f = \{e\}$.

Proof. If f is an isomorphism, then Ker $f = \{e\}$ since f is one-to-one. Conversely, if Ker $f = \{e\}$ and $f(g_1) = f(g_2)$, then $f(g_1 g_2^{-1}) = f(g_1)f(g_2)^{-1} = e$ and therefore $g_1 g_2^{-1} = e$, $g_1 = g_2$. Consequently, f is one-to-one. $\square$

A mapping g of a group G into group G' is called an *antihomomorphism* if

$$f(g_1 g_2) = f(g_2)f(g_1), \quad \text{for all } g_1, g_2 \in G,$$

and an *antiisomorphism* if in addition f is one-to-one. If $f(G) = G'$, then f is called an *antihomomorphism* (or *antiisomorphism*) of G onto G'. The set $f^{-1}(e)$ is called the *kernel* of the antihomomorphism f.

It is easy to verify Propositions I and II for antihomomorphisms and antiisomorphisms. Two groups G, G' are called *isomorphic* if there is an isomorphism of G onto G'. In many cases it makes no difference which of the two isomorphic groups G and G' one considers. One then identifies the elements of G with their images under the isomorphism of G onto G'.

III. *If f is a homomorphism of a group G onto a group G' and H is the kernel of this homomorphism, then*:

(1)   *the group G' is isomorphic to the factor group G/H;*
(2)   *$f = \psi\phi$, where ϕ is the canonical homomorphism of the group G onto the group G/H and ψ is an isomorphism of G/H onto G'.*

Proof. We define the mapping ψ of G/H onto G' by setting

$$\psi(\tilde{g}) = f(g) \tag{1.6.2}$$

for $\tilde{g} \in G/H$ and $g \in \tilde{g}$. This definition does not depend on the choice of the representative $g \in \tilde{g}$. Indeed, if g_1 is an element of $\tilde{g}$, then $g_1 = gh$ for some $h \in H$, and we have $f(g_1) = f(gh) = f(g)f(h) = f(g)e' = f(g)$.

Let $\tilde{g}_1, \tilde{g}_2$ be elements of G/H and $g_1 \in \tilde{g}_1, g_2 \in \tilde{g}_2$. By virtue of (1.6.2) we have

$$\psi(\tilde{g}_1\tilde{g}_2) = f(g_1g_2) = f(g_1)f(g_2) = \psi(\tilde{g}_1)\psi(\tilde{g}_2),$$
$$\psi(G/H) = f(G) = G'.$$

Consequently, ψ is a homomorphism of G/H onto G'. Let us find Ker ψ. If $\tilde{g} \in$ Ker ψ, then $\psi(\tilde{g}) = e'$. From this we have $f(g) = e'$ for all $g \in \tilde{g}$, by virtue of (1.6.2). This means that $g \in H$, and consequently we have $\tilde{g} = H = \tilde{e}$. Similarly, Ker $\psi = \{\tilde{e}\}$, so that by virtue of II ψ is an isomorphism of G/H onto G'. This proves assertion (1).

In order to prove (2) we note that by the definition of the canonical homomorphism $\phi(g) = \tilde{g}$ and therefore (1.6.2) can be written in the form

$$\psi(\phi(g)) = f(g)$$

so that $\psi\phi = f$. $\square$

An isomorphism of a group G onto itself is called an *automorphism of G*. We have as an example of an automorphism of group G the mapping

$$g \rightarrow f(g) = g_0^{-1}gg_0, \tag{1.6.3}$$

where g_0 is a fixed element of group G. We have

$$f(g_1)f(g_2) = g_0^{-1}g_1g_0g_0^{-1}g_2g_0 = g_0^{-1}g_1g_2g_0 = f(g_1g_2),$$

and if $g_0^{-1}g_1g_0 = g_0^{-1}g_2g_0$, then $g_1 = g_2$.

Automorphisms of group G of the type (1.6.3) are called *inner* and all other automorphisms are called *outer*.

Examples and Exercises

1. The mapping $f: g \rightarrow \det g$ is a homomorphism of the group $GL(n, \mathbf{C})$ onto the group $\mathbf{C}^1$ (see examples 2 and 5 in 1.1), since $\det(g_1g_2) = \det g_1 \det g_2$ and Ker $f = SL(n, \mathbf{C})$. The factor group $\tilde{G} = GL(n, \mathbf{C})/SL(n, \mathbf{C})$ is isomorphic to $\mathbf{C}_0^1$ and an isomorphism ψ of the group $\tilde{G}$ onto $\mathbf{C}_0^1$ can be described by the formula $\psi: \tilde{g} \rightarrow \lambda_g = \det g$ for $g \in \tilde{g}$ (see example 1 in 1.3). An analogous statement holds for $GL(n, \mathbf{R})$, $SL(n, \mathbf{R})$ and $\mathbf{R}_0^1$. Verify the equality $f = \psi\phi$ for these examples.

2. The groups G_0 and G_0' (see examples 3 and 7 in 1.1) are isomorphic. The mapping

$$f:1 \to \alpha_0, \quad i \to \alpha_1, \quad -1 \to \alpha_2, \quad -i \to \alpha_3$$

is obviously an isomorphism.

3. The groups G_X when dim $X = n$ and $GL(n, \mathbf{C})$ (see examples 4 and 5 in 1.1) are isomorphic. To construct an isomorphism, we assign to each operator $A \in G$ its matrix in some fixed basis of X.

4. The interval $\mathcal{T}^1 = [0, 1)$ is a group if we consider $[\alpha + \beta]$ the product in $\mathcal{T}^1$ of the numbers $\alpha + \beta \in \mathcal{T}^1$. It is called the *one-dimensional torus*. We can also think of $\mathcal{T}^1$ as the interval $[0, 1]$ with 0 and 1 identified. The correspondence $\tilde{g}_\alpha = \{n + \alpha, n \in N\} \to \alpha$ is an isomorphism of the group $\mathbf{R}^1/\mathbf{N}$ onto $\mathcal{T}^1$ (see example 3 in 1.3); consequently, $\mathbf{R}^1/\mathbf{N}$ and $\mathcal{T}^1$ are isomorphic.

5. The groups Ω_p and $\mathbf{N}/\mathbf{N}_p$ (examples 10 in 1.1 and 4 in 1.3) are isomorphic. The mapping

$$f:\tilde{g}_k = \{k + np, n \in \mathbf{N}\} \to e^{ik\pi/p}, \quad k = 0, 1, \ldots, p - 1$$

is an isomorphism.

6. Let g be a matrix of order n. Let $\bar{g}$ denote the matrix whose elements are the complex conjugates of the elements of g and let g' denote the transposed matrix of g, so that $\bar{g} = (\bar{g}_{jk})$ and $g'_{jk} = g_{kj}$. Prove that the mappings

$$g \to \bar{g} \quad \text{and} \quad g \to g'^{-1}$$

are outer automorphisms of the groups $GL(n, \mathbf{C})$ and $SL(n, \mathbf{C})$.

1.7. Transformation Groups

Let X be any set. A one-to-one mapping of X onto itself is called a *transformation of X*. The result of applying the transformation of g to an element $x \in X$ is denoted by xg or gx so that the transformation g is written either as $x \to xg$ or $x \to gx$.

In the first case, g is called a *right transformation* and in the second, a *left transformation*. Actually, right and left transformations coincide and differ only in their notation. Later on it will prove useful to employ both notations.

If X is finite and consists of n elements, then a transformation of the set X is called a *permutation*. In this case, it is convenient to write the elements of the set X as $1, 2, \ldots, n$. Then the permutation is written by the expression

$$\begin{pmatrix} 1 & 2 & \cdots & n \\ k_1 & k_2 & \cdots & k_n \end{pmatrix},$$

where $(k_1, k_2, \ldots, k_n)$ is some arrangement of the numbers $1, 2, \ldots, n$. This expression means that $1, 2, \ldots, n$ are mapped into $k_1, k_2, \ldots, k_n$, respectively.

The transformation obtained by first applying g_1 and then g_2, i.e.,

$$x(g_1 g_2) = (x g_1) g_2 \qquad (1.7.1a)$$

is called the *product $g_1 g_2$ of the right transformations g_1 and g_2.*

The transformation obtained by first applying g_2 and then g_1, i.e.,

$$(g_1 g_2) x = g_1 (g_2 x) \qquad (1.7.1b)$$

is called the *product $g_1 g_2$ of left transformations.*

It is easy to see that the set of all right transformations of the set X, which we will write as $G_r(X)$, is a group. The identity element e in $G_r(X)$ is the identity transformation, that is, the mapping $x \to x$ which leaves every element $x \in X$ fixed. The inverse transformation of g is g^{-1}, defined by $x'g^{-1} = x$, if $xg = x'$.

The group $G_l(X)$ of all left transformations of the set X is defined analogously. It is clear that the identity mapping $g \to g$ is an antiisomorphism of the group $G_r(X)$ onto the group $G_l(X)$ as well as of the group $G_l(X)$ onto the group $G_r(X)$. If X is finite and consists of n elements, $G_l(X)$ is the group of all permutations of n elements, and $G_l(X)$ is the group of all permutations of n elements. It is called a *symmetric group* and is written S_n. Its order is equal to the number of permutations of n elements and consequently is equal to $n!$. In the future the term *transformation* will mean right transformation and we will write $G(X)$ instead of $G_r(X)$.

Any subgroup G of the group $G(X)$ is called a *(right) transformation group of the set X.* The pair (X, G) is called a *space X with transformation group G;* the elements $x \in X$ are called *points* of the space X.

The space X with a left transformation group is defined similarly.

Consider a space X with transformation group G, say (X, G). A set $O_x = \{xg, g \in G\}$, where x is fixed and g runs through all of G, is called a *trajectory* or *orbit* relative to G.

I. *Two points $x_1, x_2 \in X$ belong to the same orbit if and only if $x_2 = x_1 g$ for a certain $g \in G$.*

If $x_2 = x_1 g$, then x_2 and x_1 belong to the orbit containing x_1. Conversely, if x_2 and x_1 belong to the same orbit, then $x_1 = xg_1$, $x_2 = xg_2$ for certain $x \in X$, $g_1 g_2 \in G$ and so $x_2 = x_1 g_1^{-1} g_2$. $\square$

From I we infer that *the space X is the union of pairwise disjoint orbits.*

The space X is called *transitive* or *homogeneous* relative to G, if X consists of one orbit. This means that for every pair $x_1, x_2 \in X$ there exists an element $g \in G$ such that $x_2 = x_1 g$.

In the general case we will consider the restriction of every transformation g to a fixed orbit O_x; the set of these restrictions is again written as G. Then O_x is a transitive space with transformation group G and the entire space falls into pairwise disjoint transitive spaces, each with transformation group

G. Every group G can be represented as a transformation group in the following manner. We take $X = G$. With each element g_0 of G we associate the mapping $\hat{g}_0$ of G onto itself defined by

$$g \rightarrow g\hat{g}_0 = gg_0. \tag{1.7.2}$$

The transformation (1.7.2) is called a *right translation on* G. The set of all right translations is denoted by $\hat{G}$.

From the relations

$$g(\hat{g}_1\hat{g}_2) = (g\hat{g}_1)\hat{g}_2 = (gg_1)g_2 = g(g_1g_2) = g(\widehat{g_1g_2})$$

we infer that $\hat{G}$ is a group and that the correspondence $g \rightarrow \hat{g}$ is a homomorphism of G onto $\hat{G}$. Clearly this homomorphism is one-to-one, and so the correspondence $g \rightarrow \hat{g}$ is an isomorphism of G onto $\hat{G}$. To summarize:

II. *Every group G is isomorphic to the group of all right translations on G.*

The isomorphism $g \rightarrow \hat{g}$ is called the *right regular representation*[6] of the group G. The *left regular representation* of group G is constructed similarly. For each element $g_0 \in G$, we define the transformation $\breve{g}_0$ of G onto itself by the formula

$$g \rightarrow \breve{g}_0g = g_0g. \tag{1.7.3}$$

This transformation is called a *left translation* on G. The set of all left translations on G we will write as $\breve{G}$. From the relations

$$(\breve{g}_1\breve{g}_2)g = \breve{g}_1(\breve{g}_2g) = g_1(g_2g) = (g_1g_2)g = (g_1g_2)g$$

and the fact that the mapping $g \rightarrow \breve{g}$ is one-to-one, we infer that $\breve{G}$ is a group and that the correspondence $g \rightarrow \hat{g}$ is an isomorphism of G onto $\breve{G}$. This isomorphism is called the *left regular representation* of the group G. If the group G is finite and consists of n elements, then $\breve{g}$ is a permutation of the finite set G. Therefore $\breve{G}$ is a subgroup of the group S_n. From this and from II we conclude the following.

III. *Every finite group G is isomorphic to a certain subgroup of the symmetric group S_n, where $n = |G|$.*

The set of all automorphisms of a group G is an important example of a transformation group. Here $X = G$ and the automorphisms of G serve as our transformations. If f_1 and f_2 are two automorphisms of G, then $f_1f_2^{-1}$ is clearly also an automorphism of G. Consequently, the set of all

---

[6] Below (see, for example, 1.3, chapter II) this term will be used in a different sense.

automorphisms of G is a group. It is called the *automorphism group* of group G and is written $A(G)$. The set of all inner automorphisms

$$a_{g_0} : g \to g_0^{-1} g g_0 \tag{1.7.4}$$

is a subgroup of the group $A(G)$. We will write this set as $A_i(G)$. Observe that

$$g(a_{g_1} a_{g_2}) = (g a_{g_1}) a_{g_2} = g_2^{-1}(g_1^{-1} g g_1) g_2 = (g_1 g_2)^{-1} g(g_1 g_2) = g a_{g_1 g_2}.$$

Thus the correspondence $g \to a_g$ is a homomorphism of G into the group $A(G)$. Consequently, $A_i(G)$ is a subgroup *of* $A(G)$ by virtue of Proposition I in 1.6; $A_i(G)$ is called the *inner automorphism group* of G. Each orbit in G relative to $A_i(G)$ consists of elements in the form $g^{-1} g_0 g$, where g_0 is fixed and g runs through all of the group G. Two elements g_1 and g_2 belong to the same orbit if and only if

$$g_2 = g^{-1} g_1 g \tag{1.7.5}$$

for a certain $g \in G$ (see I). Elements g_1 and g_2 that satisfy condition (1.7.5) for some $g \in G$ are called *conjugate*. The set of all elements of G that are conjugate to a fixed element (and consequently conjugate among themselves) is called a *conjugacy class*. From the above we conclude that *orbits in G relative to $A_i(G)$ are the conjugacy classes of G.*

Let us find the kernel Ker a of the homomorphism $g \to a_g$. An element g_0 of G belongs to Ker a if and only if a_{g_0} is the identity mapping, that is, $g a_{g_0} = g$ for all $g \in G$. By virtue of (1.7.4), this means that $g_0^{-1} g g_0 = g$, that is, $g g_0 = g_0 g$ for all $g \in G$. *The kernel of the homomorphism $g \to a_g$ therefore coincides with the center $Z(G)$ of G.*

We note also the following.

IV. *A subgroup H of a group G is a normal divisor of G if and only if H is invariant under all inner automorphisms a_g of G.*

In fact, if H is a normal divisor, then $Hg = gH$ and that means that $g^{-1} Hg = H$, that is, H is invariant under all a_g. Conversely, if H is invariant under all a_g, then $g^{-1} Hg \subset H$ for all $g \in G$. Inserting g^{-1} here instead of g and multiplying the resulting inclusion $gHg^{-1} \subset H$ on the left by g^{-1} and on the right by g, we obtain $H \subset g^{-1} Hg$. Thus $H = g^{-1} Hg$, $gH = Hg$, and H is a normal divisor of the group G. $\square$

For this reason, normal divisors are sometimes called *invariant subgroups*.

Examples and Exercises

1. Let X be a circle and Γ^1 the set of all rotations of X. Two rotaions are not considered different if they produce the same new position of the original circle. Clearly, each rotation $\gamma \in \Gamma^1$ is prescribed by the angle ϕ

through which γ rotates the circle X. We may take $0 \leqslant \phi < 2\pi$. The angle ϕ corresponding to the product $\gamma = \gamma_1 \gamma_2$ is defined by the condition $\phi_1 + \phi_2 = \phi + 2n\pi$ where $n \in \mathbf{N}$ and $0 \leqslant \phi < 2\pi$. The transformation group Γ^1 is called the *group of rotations of the circle*. It is clear that the circle X is homogenous with respect to Γ^1. The group Γ^1 is isomorphic to the torus $\mathscr{T}$ (see example 4 in 1.6) under an isomorphism f, defined by the formula

$$f: \phi \to \alpha = \frac{1}{2\pi} \phi.$$

2. Let $X = (-\infty, +\infty)$, let S be the set of all left transformations

$$s: x \to sx = ax + b \tag{1.7.6}$$

of the space $(-\infty, +\infty)$, where a,b are real numbers and $a > 0$; it is easy to see that S is a group. Two subgroups of S are: the group S^1 of all transformations $s_b: x \to s_b x = x + b$ and the group S^2 of all transformations $s_a: x \to s_a x = ax$. One can show that S^1 is a normal divisor of S. The group S is called *the group of linear transformations of the line* (or, more precisely, the group of linear transformations of the line that preserve direction). The subgroup S^1 is *the group of translations of the line* and S^2 is *the group of dilations of the line*. The space X is homogenous with respect to S^1, hence also with respect to S.

Exercise. Find the factor group S/S^1.

3. Let Π^1 be the complex plane with an adjoined point at infinity. Let F^1 be the set of all right transformations $f: z \to z' = fz$ defined by

$$\begin{aligned} z' &= (az + b)/(cz + d) && \text{for } z \neq \infty \text{ and } z \neq -d/c, \\ z' &= \infty && \text{for } z = -d/c, \\ z' &= a/c && \text{for } z = \infty. \end{aligned}$$

The complex numbers a, b, c, d are arbitrary except for the requirement $ad - bc \neq 0$. It is easy to check that F^1 is a group and that Π^1 is homogenous with respect to F^1; F^1 is called *the group of fractional-linear transformations of the plane* Π^1.

4. We will use $\mathbf{C}^n$ to indicate the set of all sequences $x = (x_1, x_2, \ldots, x_n)$ where all x_j belong to $\mathbf{C}^1$. Let G^n denote the set of all left linear transformations $g: x \to gx$ of the space $X = \mathbf{C}^n$, defined by

$$(gx)_j = \sum_{k=1}^{n} g_{jk} x_k. \tag{1.7.7}$$

We also require that $\det(g_{ik}) \neq 0$. It is easy to see that G^n is a group isomorphic to the group $GL(n, \mathbf{C})$ and that $\mathbf{C}^n \setminus \{0\}$ is homogenous with respect to $\mathbf{C}^n$.

5. A permutation $s \in S_n$ (S_n is a symmetric group) is called a *cycle* if it has the form

$$\begin{pmatrix} k_1 & k_2 & \cdots & k_{l-1} & k_l \\ k_2 & k_3 & \cdots & k_l & k_1 \end{pmatrix} \tag{1.7.8}$$

where all $k_1, \ldots, k_l$ are distinct, that is, if s changes k_1 into k_2, k_2 into $k_3, \ldots$, k_{l-1} into k_l and k_l into k_1 and leaves all other numbers between $1, 2, \ldots, n$ in place. The numbers $k_1, k_2, \ldots, k_l$ are called *elements of the cycle* and the number l is called the *length of the cycle*. The cycle shown in (1.7.8) is abbreviated as $(k_1, k_2, \ldots, k_l)$. A cycle (k_1, k_2) with length 2 is called a *transposition* of the elements k_1, k_2. Prove the following.

(a) Each permutation $s \in S_n$ is a product of transpositions. The number of transpositions for a given s is either always even or always odd, regardless of the factorization used. The permutation s is called *even* if it is a product of an even number of transpositions and *odd* if it is a product of an odd number of transpositions.

(b) The set P_n of all even permutations $s \in S_n$ is a normal divisor of S_n. (Find the factor group S_n/P_n.) The group P_n is called an *alternating group*.

(c) Every permutation $s \in S_n$ can be uniquely written as a product of cycles with no common elements (order plays no role in the product, since cycles with no common elements commute). Two permutations $s, s' \in S_n$ are conjugate if and only if the set of lengths of the cycles making up the permutations are the same for s and s'.

6. Write a Cayley table for the group S_3. Also find all the sets of conjugate elements.

7. Find all the sets of conjugate elements in the groups $GL(n, \mathbf{C})$ and $SL(n, \mathbf{C})$.

1.8. The Canonical Realization of a Homogenous Space

There is a simple way to construct a homogenous space, which consists of the following. Let G be a group, and H a subgroup of G. For the sake of brevity we will introduce the symbol $\tilde{G}$ for G/H. For each element g_0 of G, we define the transformation $\bar{g}_0$ of $\tilde{G}$ by the formula[7]

$$\{g\}\bar{g}_0 = \{gg_0\} \tag{1.8.1}$$

where $\{g\}$, $\{gg_0\}$ are right cosets of H in G with representatives g and gg_0 (see 1.2). It is easy to see that the mapping $g \to \bar{g}$ is a homomorphism of G into

---

[7] The mapping $\bar{g}_0$ of (1.8.1) is a one-to-one mapping of $\tilde{G}$ onto itself, that is, a transformation of $\tilde{G}$. To see this, suppose that $\{g_1\}\bar{g}_0 = \{g_2\}\bar{g}_0$, that is, $\{g_1g_0\} = \{g_2g_0\}$. Then $g_1g_0 = hg_2g_0$ for some $h \in H$. From this we have $g_1 = hg_2$, that is, $\{g_1\} = \{g_2\}$. Further, if $\{g_1\}$ is an arbitrary element of $\tilde{G}$, then when $g = g_1g_0^{-1}$ we get $\{g\}\bar{g}_0 = \{gg_0\} = \{g_1g_0^{-1}g_0\} = \{g_1\}$, so that $\bar{g}_0$ maps $\tilde{G}$ onto $\tilde{G}$.

the group $G_n(\tilde{G})$; consequently, the image $\bar{G}$ of G under this mapping is a group (see I, 1.6). Thus we obtain the space $\tilde{G}$ with a transformation group $\bar{G}$. This space is homogenous with respect to $\bar{G}$. In fact, for two points $\{g_1\}, \{g_2\}$ we write $g_0 = g_1^{-1}g_2$. Then

$$\{g_1\}\bar{g}_0 = \{g_1 g_0\} = \{g_1 g_1^{-1} g_2\} = \{g_2\},$$

so that $\tilde{G} = G/H$ is a *homogenous space with transformation group* $\bar{G}$. If $H = \{e\}$, then $\tilde{G} = G$ and we arrive at the right regular representation $g \to \hat{g}$ of the group G. Let H_0 be the kernel of the homomorphism $g \to \bar{g}$. By virtue of I in 1.6, H_0 is a normal divisor of the group G. By definition, H_0 consists exactly of those elements $g_0 \in G$ for which $\bar{g}_0$ is the identity transformation, that is, for which $Hgg_0 = Hg$ for all $g \in G$. This is equivalent to the relation

$$gg_0 g^{-1} \in H, \quad \text{for all } g \in G. \tag{1.8.2}$$

Setting $g = e$ in (1.8.2), we conclude that $g_0 \in H$; consequently

$$H_0 \subset H. \tag{1.8.3}$$

If $g_1 \in H_0$ and $g_2 \in H_0$, then by virtue of (1.8.2) we have $gg_1 g^{-1} \in H$ and $gg_2 g^{-1} \in H$, for all $g \in G$. Consequently, we have

$$gg_1 g_2^{-1} g^{-1} = (gg_1 g^{-1})(gg_2 g^{-1})^{-1} \in H,$$

and therefore $g_1 g_2^{-1} \in H_0$. This means that H_0 is a subgroup of H. Thus, *the kernel of the homomorphism $g \to \bar{g}$ is a subgroup of H which is a normal divisor of G.* Conversely, let H_1 be a subgroup of the subgroup H which is a normal divisor of G. Then we have $gH_1 g^{-1} \subset H_1 \subset H$ for all $g \in G$ and therefore $H_1 \subset H_0$. We have the following.

I. *The kernel of the homomorphism $g \to \bar{g}$ is the largest subgroup of H which is a normal divisor of G.*

II. *The homomorphism $g \to \bar{g}$ is an isomorphism if and only if the group G contains no subgroup distinct from $\{e\}$ that is a normal divisor of G.*

We will now show that the above method of constructing a homogeneous space is actually universal. Let X be an arbitrary homogeneous space with transformation group G and let x_0 be a fixed point in the space X. Let H denote the set of all transformations h in the group G that leave x_0 fixed, that is,

$$x_0 h = x_0. \tag{1.8.4}$$

It is clear that H is a subgroup of G. It is called the *stationary group* of the point x_0. Since X is homogeneous, each point $x \in X$ is obtained from x_0 by applying a certain transformation $g \in G$: $x = x_0 g$. If g_1, g_2 are two such transformations, so that $x = x_0 g_1$ and $x = x_0 g_2$ then $x_0 g_2 = x_0 g_1$ and consequently $x_0 g_2 g_1^{-1} = x_0$. This means that $g_2 g_1^{-1} \in H$, that is, $g_2 \in H g_1$, therefore g_2 and g_1 belong to the same right coset of H in G. Conversely, if g_1 and g_2 belong to the same right coset of H, then $g_2 = h g_1$ for some $h \in H$; (1.8.4) implies that $x_0 g_2 = x_0 h g_1 = x_0 g_1$. Thus g_1 and g_2 transform x_0 into the same point x. Thus we have established a one-to-one correspondence $f: x \to \tilde{g}$ between points $x \in X$ and right cosets of H in G, that is, points g of the factor space $\tilde{G} = G/H$. The right coset $\tilde{g}$ corresponding to a point x consists exactly of those group elements g for which

$$x_0 g = x. \tag{1.8.5}$$

Let us see what happens to the transformation g under the mapping f. Let a point $x \in X$ map into the set $\{g\}$ under f, so that $x_0 g = x$. Then $x g_0 = x_0 g g_0$ and thus the point $x g_0$ maps into the set $\{g g_0\} = \{g\} \bar{g}_0$; therefore *under the mapping f, the transformation g becomes the transformation $\bar{g}_0$ of the space* $\tilde{G} = G/H$.

Let H_1 be a subgroup of the group H that is a normal divisor of the original group G and let g_1 be an element of H_1. In view of Proposition I, we have $g_1 = e$, that is, $\{g g_1\} = \{g\}$ for all $g \in G$. Applying both parts of this relation to the element x_0 and setting $x_0 g = x$, we find that $x g_1 = x$ for all $x \in X$. This means that $g_1 = e$, since G is a transformation group. Thus $H_1 = \{e\}$, and the correspondence $g \to \bar{g}$ is an isomorphism by virtue of II. The following proposition is proved in the same manner.

III. *Let X be a homogeneous space with transformation group G; let H be the stationary group of a fixed point $x_0 \in X$ and let f be the mapping which assigns to each point $x \in X$ the right coset $\tilde{g} = \{g\} \in G/H$ of all elements $g \in G$ for which $x_0 g = x$. Then f is a one-to-one mapping of X onto G/H and in this mapping every transformation g_0 is mapped onto the transformation $\bar{g}_0$ defined by the formula*

$$\{g\} \bar{g}_0 = \{g g_0\} \tag{1.8.6}$$

so that

$$f\{xg\} = f(x)\bar{g}. \tag{1.8.7}$$

The resulting correspondence $g \to \bar{g}$ is an isomorphism of the group G onto the group $\bar{G}$ of all transformations $\bar{g}$.

Usually points $x \in X$ are identified with the corresponding sets $f(x) \in X$. Then, by virtue of Proposition III, X coincides with G/H, transformations g

with the transformations $\bar{g}$ and the group G with the group G. This identity is
called *the canonical model* of the homogeneous space in question.

The preceding reasoning leads to the following definition.

Two homogeneous spaces, each with a transformation group, say (X, G)
and (X', G'), are called *equivalent* if there exist:

(a) an isomorphism $\phi : g \to g'$ of the group G onto the group G';
(b) a one-to-one mapping $f : x \to x'$ of the space X onto X' such that if
     $x \to x'$ then $xg \to x'g'$. That is,

$$f(xg) = f(x)\phi(g). \tag{1.8.8}$$

It is not difficult to verify that the equivalence just defined satisfies all the
axioms for an equivalence relation.

Proposition III means that *every homogeneous space (X, G) is equivalent
to some canonical model.*

IV. *Two homogeneous spaces (X, G), (X', G') are equivalent if and only if
there exist:*

(1) *an isomorphism $\phi : g \to g'$ of the group G onto the group G';*
(2) *points $x_0 \in X$ and $x_0' \in X'$ such that ϕ maps the stationary group H
     of the point x_0 onto the stationary group H' of the point x_0'.*

Proof. Let (X, G) and (X', G') be equivalent and let ϕ and f be as in (a) and
(b). Let x_0 be a fixed point in X; we write $x_0' = f(x_0)$. Then $x_0' \in X'$. Let H
and H' be the stationary groups of the points x_0 and x_0', respectively. Then ϕ
maps H onto H'. In fact, if $g \in H$, then $x_0 g = x_0$. This and (1.8.8) show that
$f(x_0) = f(x_0 g) = f(x_0)\phi(g)$; that is, $x_0' = x_0'\phi(g)$. We infer that $\phi(g) \in H'$.
Conversely, if $\phi(g) \in H'$, then $x_0' = x_0'(\phi)g$, that is, $f(x_0) = f(x_0)\phi(g) =
f(x_0 g)$. Therefore $x_0 = x_0 g$ since f is one-to-one, i.e., $g \in H$. Thus ϕ, x_0, x_0'
satisfy conditions (1) and (2).

Conversely, let ϕ, x_0, x_0' satisfy conditions (1) and (2). We will construct
a mapping f of the space X onto X' by setting

$$f(x) = f(x_0 g) = x_0'\phi(g) \tag{1.8.9}$$

when $x = x_0 g$. This definition does not depend on the selection of the element
$g \in G$ for which $x = x_0 g$. If $x = x_0 g_1$, then $g_1 = hg$, where $h \in H$. Therefore

$$x_0'\phi(g_1) = x_0'\phi(hg) = x_0'\phi(h)\phi(g) = x_0'\phi(g),$$

since $\phi(h) \in H$. It is also easy to show that f is a one-to-one mapping of X
onto X'. It follows from (1.8.9) that condition (1.8.8) is satisfied; therefore
(X, G) and (X', G') are equivalent. $\square$

Remark. A canonical realization analogous to III occurs also for homogeneous spaces with a group G of left transformations. In this case G/H

denotes the space of left cosets of H, where H is a stationary group defined in exactly the same way as above; the transformation g_0 in this realization becomes the transformation

$$\bar{g}_0 : \{g\} \to \{g_0 g\}.$$

Examples and Exercises

1. Let Π^2 be the set of all pairs $z = \{z_1, z_2\}$, $z_1, z_2 \in \Pi^1$ and let F^2 be the set of all transformations $f : z \to z' = \{z'_1, z'_2\}$ where

$$
\begin{aligned}
z'_i &= (az_i + c)/(bz_i + d), && \text{for } z_i \neq \infty \text{ and } z_i \neq -d/b, \\
z'_i &= a/b, && \text{for } z_i = \infty && (1.8.10) \\
z'_i &= \infty, && \text{for } z_i = -d/b,\ i = 1, 2;\ a/d - bc \neq 0.
\end{aligned}
$$

We can easily verify that F^2 is a group, so that Π^2 is a homogeneous space with transformation group F^2. Find all orbits in Π^2 relative to F^2.

2. Let Π'^2 be the set of all pairs $z = \{z_1, z_2\}$, $z_1, z_2 \in \Pi^1$, such that $z_1 \neq z_2$, and let F^2 consist of the restrictions of the transformations f (see (1.8.10); we will again write them as f) on Π'^2. Further, let Π'^2 be the set of all pairs $x = (z, \zeta)$, $z, \zeta \in \Pi^1$, $\zeta \neq \infty$ and let $\tilde{F}^2$ be the set of all transformations $f : x \to x' = \{z', \zeta'\}$, where

$$
\begin{aligned}
z' &= (az + c)/(bz + d), && \text{for } z \neq \infty, z \neq -d/c, \\
z' &= a/b, && \text{for } z = \infty, \\
z' &= \infty, && \text{for } z = -d/b, \\
\zeta' &= (bz + d)^2 \zeta + b(bz + d).
\end{aligned}
$$

Prove that:

(1) Π'^2 is homogeneous with respect to F^2 and $\tilde{\Pi}^2$ is homogeneous with respect to $\tilde{F}^2$;
(2) (Π'^2, F^2) and $(\tilde{\Pi}'^2, \tilde{F}^2)$ are equivalent.

Find an isomorphism ϕ and a mapping f which transform (Π'^2, F^2) into $(\tilde{\Pi}^2, \tilde{F}^2)$. Hint. Consider the mapping $\phi : f \to \tilde{f}$, where f and $\tilde{f}$ correspond to the same a, b, c, d. Consider the stationary groups of points $z_0 \in \Pi'^2$, $x_0 \in \Pi'^2$, where $z_0 = (0, \infty)$, $x_0 = (0, 0)$.

1.9. The Direct Product of Groups

The *direct product* of the groups $G_1, \ldots, G_n$ is the set of all sequences $g = \{g_1, \ldots, g_n\}$, $g_1 \in G_1, \ldots, g_n \in G_n$, in which multiplication is defined by the formula

$$
\begin{aligned}
gg' &= \{g_1 g'_1, g_2 g'_2, \ldots, g_n g'_n\} \\
&\text{for } g = \{g_1, \ldots, g_n\}, \qquad g' = \{g'_1, \ldots, g'_n\}.
\end{aligned}
\qquad (1.9.1)
$$

We easily verify that G is a group under this definition of multiplication. The identity element of G is $e = \{e_1, \ldots, e_n\}$ where $e_1, \ldots, e_n$ are the identity elements of $G_1, \ldots, G_n$. The direct product of the groups $G_1, \ldots, G_n$ is denoted by $G_1 \times G_2 \times \cdots \times G_n$. The mapping $g_1 \to \{g_1, e_2, \ldots, e_n\}$ is an isomorphism of the group G_1 into $G = G_1 \times \cdots \times G_n$; therefore g_1 is often identified with $\{g_1, e_2, \ldots, e_n\}$. We can then consider G_1 to be a subgroup of G. Similarly we can regard $G_2, \ldots, G_n$ as subgroups of G. In this identification each element $g \in G$ is represented uniquely by

$$g = g_1 g_2, \ldots, g_n, \qquad g_1 \in G_1, \ldots, g_n \in G_n. \tag{1.9.2}$$

Note that

$$g_j g_k = g_k g_j, \qquad j \neq k. \tag{1.9.3}$$

That is, every $g_j \in G_j$ commutes with every $g_k \in G_k$ for $j \neq k$. Conversely, if a certain group G admits subgroups $G_1, \ldots, G_n$ which satisfy conditions (1.9.2) and (1.9.3), then G is called *the direct product of its subgroups* $G_1, \ldots, G_n$ and is again written as

$$G = G_1 \times \cdots \times G_n. \tag{1.9.4}$$

Thus, (1.9.4) has two different meanings, which coincide under the identifications indicated above.

If $G = G \times \cdots \times G_n$, where $G_1, \ldots, G_n$ are subgroups of G, it is evident that each of the subgroups $G_1, G_2, \ldots, G_n$ is a normal divisor of G.

Examples

1. Let $G_1 = G_2 = \cdots = G_n = \mathbf{C}^1$ (see example 1 in 1.1); then $G_1 \times \cdots \times G_n$ is the set of all sequences $x = \{x_1, x_2, \ldots, x_n\}$ of complex numbers $x_1, \ldots, x_n$, where the product of the two sequences $\{x_1, x_2, \ldots, x_n\}$ and $\{x'_1, x'_2, \ldots, x'_n\}$ is $\{x_1 + x'_1, x_2 + x'_2, \ldots, x_n + x'_n\}$. This group $G_1 \times \cdots \times G_n$ is called *the complex n-dimensional vector group* and is denoted by $\mathbf{C}^n$. We define *the real n-dimensional vector group* $\mathbf{R}^n$ in the same way. The only difference here is that in the case of $\mathbf{R}^n$, x_j and x'_j are real. The group $\mathbf{C}^n$ is isomorphic to the group $\mathbf{R}^{2n}$. (Prove this.)

2. Let $G_1 = G_2 = \cdots = G_n = \mathbf{C}_0^1$ (see example 2 in 1.1). Then $G_1 \times \cdots \times G_n$ is the set of all sequences $x = \{x_1, x_2, \ldots, x_n\}$, $x_j \in \mathbf{C}_0^1$, where the product of the two sequences $\{x_1, \ldots, x_n\}$ and $\{x'_1, \ldots, x'_n\}$ is $\{x_1 x'_1, \ldots, x_n x'_n\}$. In this case, the group $G_1 \times \cdots \times G_n$ is written $\mathbf{C}_0^n$. The group $\mathbf{R}_0^n$ is defined in the same manner.

3. Let $G_1 = G_2 = \cdots = G_n = \mathscr{T}^1$ (see example 4 in 1.6); then $G_1 \times G_2 \times \cdots \times G_n$ consists of all sequences $\{x_1, x_2, \ldots, x_n\}$, $x_1, \ldots, x_n \in [0, 1)$ and the product of the two sequences $\{x_1, \ldots, x_n\}$, $\{x'_1, \ldots, x'_n\}$ is

$\{[x_1 + x'_1], \ldots, [x_n + x'_n]\}$. In this case $G_1 \times \cdots \times G_n$ is called *the n-dimensional torus* and is written $\mathcal{T}^n$.

§2. Fundamental Concepts and the Simplest Propositions of Representation Theory

2.1. The Definition of a Representation

Let G be a group and let X be a complex linear space $\neq \{0\}$. Consider a mapping T of G into the set of all linear operators carrying X into itself, written $g \to T(g)$, with the following properties:

(1) $T(e) = 1$, where 1 is the identity operator in X;
(2) $T(g_1 g_2) = T(g_1)T(g_2)$ for all $g_1, g_2 \in G$.

Then T is called a *representation of G in the space X*. The space X is called *the representation space* and the operators $T(g)$ *representation operators*. From properties (1) and (2) it follows that $T(g^{-1})T(g) = T(g^{-1}g) = T(e) = 1$ and analogously, $T(g)T(g^{-1}) = 1$. Therefore, *every operator $T(g)$ is a one-to-one mapping of X onto X and*

$$T(g^{-1}) = (T(g))^{-1}. \qquad (2.1.1)$$

Property (2) thus means that *a representation T is a homomorphism of the group G into the group G_X* (that is, into the group of all one-to-one linear operators carrying X onto X; see example 4 in 1.1). This property may be regarded as another definition of a representation (see I in 1.6).

Let T be a representation of a group G in a linear space X and let H be a subgroup of G. Considering all the operators $T(g)$, $g = h \in H$, we obtain a representation $T|_H$ of the group H. It is called *the restriction of the representation T to the subgroup H*.

A subspace $M \subset X$ is said to be *invariant under a representation T* if it is invariant under all operators $T(g)$ of this representation. Let a subspace $M \subset X$ be invariant under a representation T of the group G in X. Considering the operators $T(g)$ only on M, we obtain a representation of the group G in M; it is called *the restriction of the representation T to M* and is denoted by $T|_M$. Furthermore, the operators $T(g)$ generate operators $\tilde{T}(g)$ in the factor space $\tilde{X} = X/M$ and we are easily convinced that the correspondence $g \to \tilde{T}(g)$ satisfies conditions (1) and (2) of the definition of a representation. This correspondence accordingly gives a representation (which we will write as $\tilde{T}$) of G in X/M. The representation $\tilde{T}$ is called the representation *generated* by T in the factor space X/M.

A representation is called *finite-dimensional* if the representation space X is finite-dimensional and *infinite-dimensional* in the opposite case. If X is finite-dimensional, then its dimension dim X is called *the dimension* (also *degree*) *of the representation* T and is denoted by n_T. Let T be finite-dimensional and $n_T = n$. Choosing a basis $e_1, e_2, \ldots, e_n$ in X, we can describe the operators $T(g)$ by matrices of order n:

$$t(g) = \begin{Vmatrix} t_{11}(g) & \cdots & t_{1n}(g) \\ \vdots & & \vdots \\ t_{n1}(g) & \cdots & t_{nn}(g) \end{Vmatrix}. \tag{2.1.2}$$

This means that

$$T(g)e_k = \sum_{j=1}^{n} t_{jk}(g)e_j. \tag{2.1.3}$$

Conditions (1) and (2) then take the form

$$t(e) = 1, \qquad t(g_1 g_2) = t(g_1)t(g_1), \tag{2.1.4}$$

or, in more detail

$$t_{jk}(e) = \begin{cases} 1, & \text{for } j = k, \\ 0, & \text{for } j \neq k \end{cases} \tag{2.1.5}$$

and

$$t_{jk}(g_1 g_2) = \sum_{s=1}^{n} t_{js}(g_1)t_{sk}(g_2). \tag{2.1.6}$$

The matrix $t(g)$ is called a *matrix of the representation* T and the functions $t_{jk}(g)$ are called *matrix elements* of the representation in the basis $e_1, \ldots, e_n$. Conversely, suppose that a matrix function $g \to t(g)$ of order n is defined on the group G in such a way that the conditions (2.1.4) hold. For each $g \in G$ we define an operator $T(g)$ in the coordinate space $\mathbf{C}^n$ with matrix $T(g)$ by the rule

$$T(g) : x_j \to x'_j = \sum_{k=1}^{n} t_{jk}(g)x_k \tag{2.1.7}$$

for $\{x_1, x_2, \ldots, x_n\} \in \mathbf{C}^n$. It follows from (2.1.4) that the correspondence $g \to T(g)$ is a representation of the group G. Hence we can also view a finite-dimensional representation as a matrix function $g \to T(g)$ satisfying the conditions of (2.1.4). If the group G itself consists of matrices g of a fixed order, for example, $GL(n, \mathbf{C})$, $SL(n, \mathbf{C})$ (example 5 and 6 in 1.1), then one of the simplest representations is obtained when $T(g) = g$. This representation is called *the self-representation*.

A representation in a space X is called *irreducible* if X admits no subspace except for $\{0\}$ and X itself that is invariant under all operators of the representation. A representation that is not irreducible is called *reducible*.

It is clear that a one-dimensional representation is irreducible; below we will see that there are groups admitting multi-dimensional and even infinite-dimensional irreducible representations.

The simplest of all representations is obtained if we set $T(g) = 1$ for all $g \in G$. It is called an *identity representation*. Every subspace $M \subset X$ is invariant under the identity representation. Hence the identity representation is irreducible only when it is one-dimensional.

I. *If T is a representation in a finite-dimensional space X, then X contains a nonzero subspace on which the restriction of T is irreducible.*

Proof. If T itself is irreducible, then the proof is trivial: $M = X$. Let T be reducible. That is, there is a subspace M of X such that $\{0\} \subsetneqq M \subsetneqq X$ and such that M is invariant under T. If the restriction of T to M is irreducible, we are done. Otherwise consider a subspace M_1 of M such that $\{0\} \subsetneqq M_1 \subsetneqq M$ and such that M_1 is invariant under T. Continue by induction, obtaining invariant subspaces $M \supsetneqq M_1 \supsetneqq M_2 \cdots \supsetneqq M_l \supsetneqq \{0\}$. Since $\dim(X) > \dim(M) > \dim(M_1) > \cdots > 1$, the process must stop. When it does, we have a proper subspace M_l on which T is irreducible.   □

Examples and Exercises

1. Let $G = \mathbf{R}^1$ (see example 1 in 1.1) for every complex number k; the function $\alpha \to e^{k\alpha}$ on $\mathbf{R}^1$ satisfies the conditions (2.1.4) and therefore defines a one-dimensional representation of $\mathbf{R}^1$.

Analogously, the function $z = x + iy \to e^{ik_1 x} e^{ik_2 y}$ on $\mathbf{C}^1$ (see example 1 in 1.1) is a one-dimensional representation of the group G for all pairs of complex numbers k_1 and k_2.

2. Prove that the following functions are one-dimensional representations of the groups indicated:

   (a) $a \to e^{k(\ln|a|)}$ (sign $a)^\varepsilon$ on $\mathbf{R}_0^1$ for every complex number k and $\varepsilon = 0$ or 1.

   (b) $z \to e^{k(\ln|z|)}$ (arg $z)^n$ on $\mathbf{C}_0^1$ for every complex number k and integer n. (See example 2 in 1.1.)

3. The function $\phi \to e^{in\phi}$ on the group Γ^1 of rotations of the circle (see ex. 1 in 1.6) is a one-dimensional representation of the group Γ^1 for each integer n.

4. Prove that the self-representations of the groups $GL(n, \mathbf{C})$ $SL(n, \mathbf{C})$, $GL(n, \mathbf{R})$ and $SL(n, \mathbf{R})$ are irreducible.

5. Prove that: (a) the matrix function $\alpha \to \begin{pmatrix} 1 & 0 \\ \alpha & 1 \end{pmatrix}$ on $\mathbf{R}$ describes a two-dimensional reducible representation T of the group $\mathbf{R}^1$; (b) there is exactly one one-dimensional subspace M invariant under T. Find this subspace M and prove that: (c) the restriction of T to M is the identity representation; (d) the representation $\tilde{T}$ in the factor space with respect to M, generated by the representation T, is the identity representation.

2.2. Equivalence

Two representations T, S of a group G in spaces X and Y are called *equivalent* (written $T \sim S$) if there is a one-to-one linear operator A carrying S onto Y and satisfying the condition

$$AT(g) = S(g)A, \quad \text{for all } g \in G. \tag{2.2.1}$$

It is possible that $Y = X$, and in this case we speak of the equivalence of representations in the same space. Condition (2.2.1) shows that $AT(g)x = S(g)Ax$ for all $x \in X$ and $g \in G$. That is, if A maps x into y (i.e., $Ax = y$), then A also maps $T(g)x$ into $S(g)y$ (i.e., $AT(g)x = S(g)y$). Condition (2.2.1) can also be written as follows:

$$T(g) = A^{-1}S(g)A, \quad \text{for all } g \in G. \tag{2.2.2}$$

The notion of equivalence just introduced plainly satisfies all the axioms for an equivalence relation.

I. *Let X be finite-dimensional. Representations S and T on X and Y are equivalent representations if and only if $n_S = n_T$ and under a proper choice of bases in X and Y, the matrix elements of the representations S and T coincide.*

Proof. Let $e_1, \ldots, e_n$ be any basis in X. Write $f_j = Ae_j, j = 1, \ldots, n$. The elements $f_1, \ldots, f_n$ are a basis in Y. Apply the operator A to both sides of (2.1.3) and use condition (2.2.1) to obtain

$$S(g)f_k = S(g)Ae_k = AT(g)e_k = \sum_{j=1}^{n} t_{jk}(g)Ae_j = \sum_{j=1}^{n} t_{jk}(g)f_j.$$

Conversely, suppose that $n_T = n_S = n$ and that $e_1, \ldots, e_n, f_1, \ldots, f_n$ are bases in X and Y respectively, such that the matrix elements of T and S coincide in these bases. Writing $A(c_1e_1 + \cdots + c_ne_n) = c_1f_1 + \cdots + c_nf_n$ (for all complex numbers $c, \ldots, c_n$), we obtain a one-to-one operator A, carrying X onto Y and satisfying condition (2.2.1). $\square$

Equivalent one-dimensional representations are defined by the same numerical function $t(g)$.

II. *If $S \sim T$ and T is irreducible, then S is also irreducible.*

The proof follows directly from the definitions of irreducibility and equivalence. We leave the details to the reader.

Lemma 1 (Schur's lemma). *Let T and S be irreducible representations of a group G in spaces X and Y respectively. Let A be an operator carrying X*

into Y, satisfying the condition

$$AT(g) = S(g)A, \quad \text{for all } g \in G. \tag{2.2.3}$$

Then either A is a linear isomorphism of X onto Y and consequently $T \sim S$, or $A = 0$.

Proof. We write $L = AX$; then L is a subspace of Y. It is invariant under all $S(g)$ since (2.2.3) implies that $S(g)Ax = AT(g)x \in AX = L$ for all $g \in G$ and $x \in X$. As S is irreducible, either $L = (0)$ or $L = Y$. In the first case $A = 0$. On the other hand, suppose that $L = Y$, i.e., A maps X onto Y. We will prove that A is one-to-one. We write $M = \{x : Ax = 0\}$; it suffices to prove that $M = (0)$. Observe that M is invariant under T. Indeed, if we have $x \in M$, then $Ax = 0$. Again by virtue of (2.2.3) we have $AT(g)x = S(g)Ax = A(g)0 = 0$, so that $T(g)x \in M$. Since T is irreducible, we conclude that either $M = (0)$ or $M = X$. But the second case is not possible, since it implies that $Y = L = AX = \{0\}$. $\square$

Lemma 2. *Let T be a finite-dimensional irreducible representation of a group G in a space X. Then every linear operator B on the space X that commutes with all operators $T(g)$, $g \in G$, has the form $B = \lambda 1$, where λ is a complex number.*

Proof. By hypothesis, we have

$$BT(g) = T(g)B, \quad \text{for all } g \in G. \tag{2.2.4}$$

Since B is a linear operator on a finite-dimensional space, B has at least one eigenvalue λ. We write $A = B - \lambda 1$; thus A is not a one-to-one mapping on X. From (2.2.4) it follows that $AT(g) = T(g)A$ for all $g \in G$, that is, A satisfies the condition of (2.2.3) for all $S(g) = T(g)$, $Y = X$. Since A is not one-to-one, lemma 1 implies that $A = 0$, which is to say that $B - \lambda 1 = 0$, $B = \lambda 1$.

We will offer a second proof of lemma 2, which does not use lemma 1. Let λ be an eigenvalue of the operator B. We write $L = \{x : x \in X : Bx = \lambda x\}$. The space L is not $\{0\}$, since λ is an eigenvalue of B. Also L is invariant under T. In fact, if $x \in L$, then $Bx = \lambda x$. Thus we have $BT(g)x = T(g)Bx = Tg(\lambda x) = T(g)x$, so that $T(g)x \; L$. Since T is irreducible, we conclude that $L = X$. That is, we have $Bx = \lambda x$ for all $x \in X$, $B = \lambda 1$. $\square$

Corollary. *An irreducible finite-dimensional representation of a commutative group is one-dimensional.*

Proof. Let T be an irreducible representation of a commutative group G in a finite-dimensional space X. For all g_0, $g \in G$, we have $T(g_0)T(g) = T(g_0 g) = T(g g_0) = T(g)T(g_0)$, so that every operator $T(g_0)$ is commutative with all $T(g)$. Lemma 2 shows that $T(g_0) = \lambda(g_0)1$. Every subspace of X is invariant under all operators $T(g) = \lambda(g)1$. If dim $X > 1$, this contradicts the irreducibility of the representation T. $\square$

Thus an irreducible finite-dimensional representation of a commutative group is one-dimensional and consequently is defined by a numerical function $t(g)$, which satisfies the conditions

$$t(e) = 1, \qquad t(g_1 g_2) = t(g_1)t(g_2). \tag{2.2.5}$$

Any numerical function on a commutative group G which satisfies the conditions of (2.2.5) is called a *character* of the group G.

Examples and Exercises

1. Prove that every finite-dimensional representation of a commutative group contains a one-dimensional invariant subspace.

2. Let $G = S_3$. Define a representation $g \to T(g)$ of G in $\mathbf{C}^3$ by the following rule: if $g = \begin{pmatrix} 1 & 2 & 3 \\ i & j & k \end{pmatrix}$ $((i, j, k)$ is a permutation of the numbers 1, 2, 3) then $T(g)e_1 = e_i$, $T(g)e_2 = e_j$ and $T(g)e_3 = e_k$. Find the irreducible subrepresentations of the representation T.

3. Prove that every irreducible representation of the cyclic group G of order n generated by an element a has the form $T_m(a^k) = e^{2\pi mki/n}$, where $m = 0, 1, \dots, n - 1$. (The operator $T_m(a^k)$ is the operator of multiplication by the number $e^{2\pi mki/n}$ in the one-dimensional complex vector space $\mathbf{C}$.)

2.3. Adjoint Representations

Let X and Y be complex linear spaces. A numerical function $\{x, y\} \to (x, y)$ on $X \times Y$ is called a *bilinear form* on (x, y) if it satisfies the following conditions[8]:

(1) $(\alpha x, y) = \alpha(x, y)$,
(2) $(x, \alpha y) = \bar{\alpha}(x, y)$,
(3) $(x_1 + x_2, y) = (x_1, y) + (x_2, y)$,
(4) $(x, y_1 + y_2) = (x, y_1) + (x, y_2)$,

for all $x, x_1, x_2 \in X$, $y, y_1, y_2 \in Y$ and for all complex numbers α. If $Y = X$, we speak of a bilinear form on X. Let (x, y) be a bilinear form on $X \times Y$. A vector $x \in X$ is called *orthogonal to a vector* $y \in Y$ *with respect to* (x, y) if $(x, y) = 0$. We write $x \perp y$. Two sets $E \subset X$ and $E_1 \subset Y$ are called *orthogonal with respect to* (x, y) and are written $E \perp E_1$, if every vector of E is

---

[8] Here we find it useful to deviate from the generally accepted terminology in which the function (x, y) is called a bilinear form if it satisfies conditions (1), (3) and (4), but in place of condition (2), satisfies (2') $(x, \alpha y) = \alpha(x, y)$. In the book of N. Bourbaki [1], a function (x, y) satisfying conditions (1)–(4) is called a *sesquilinear form*.

orthogonal to every vector of E_1. If E is a subset of X, the set of all $y \in Y$ orthogonal to E is called *the orthogonal complement of E in Y* with respect to (x, y) and is denoted by $E^{\perp}_{(x,y)}$, or simply $E^{\perp}$, if there is no question about what form the symbol $\perp$ refers to. The orthogonal complement $E^{\perp}_1$ to $E_1 \subset Y$ in X with respect to (x, y) is defined similarly. Note that $E^{\perp}$ and $E^{\perp}_1$ are linear subspaces of Y and X respectively.

The pair (X, Y), on which the bilinear form (x, y) is defined, is said to be *in duality with respect to the form (x, y)* if it satisfies in addition to (1)–(4) the following conditions:

(5)   if $(x_0, y) = 0$, for all $y \in Y$, then $x_0 = 0$;
(6)   if $(x, y_0) = 0$, for all $x \in X$, then $y_0 = 0$.

Condition (5) shows that $Y^{\perp}_{(x,y)} = (0)$ and condition (6) that $X^{\perp}_{(x,y)} = (0)$. If (x, y) is a bilinear form on the pair (X, Y), then the function $(y, x)_1 = \overline{(x, y)}$ is a bilinear form on the pair (Y, X).

From this we conclude the following.

I. *If the pair (X, Y) is in duality with respect to the form (x, y) then the pair (Y, X) is in duality with respect to the form $(y, x)_1 = \overline{(x, y)}$.*

Consider as an example the case of finite-dimensional spaces X and Y. Let dim $X = m$, dim $Y = n$ and let $e_1, \ldots, e_m$ and $f_1, \ldots, f_n$ be bases in X and Y respectively. Then for $x \in X$, $y \in Y$, we have

$$x = \sum_{j=1}^{m} \xi_j e_j, \qquad y = \sum_{k=1}^{n} \eta_k f_k, \tag{2.3.1}$$

where ξ_j and η_k are complex numbers. Using this and properties (1)–(4), we write

$$(x, y) = \sum_{j=1}^{m} \sum_{k=1}^{n} a_{jk} \xi_k \overline{\eta}_k, \tag{2.3.2}$$

where $a_{jk} = (e_j, f_k)$. Conversely, for any complex numbers a_{jk} the formulas (2.3.1) and (2.3.2) define a bilinear form on the pair (X, Y). The numbers

$$a_{jk} = (e_j, f_k) \tag{2.3.3}$$

are called *coefficients of the form (x, y) with respect to the bases $e_1, \ldots, e_m$; $f_1, \ldots, f_n$.*

We will now find conditions under which the pair (X, Y) is in duality with respect to a form (2.3.2).

We write $y_0 = \sum_{k=1}^{m} \eta_k^0 f_k$. Use condition (6) to show that the equality

$$\sum_{j=1}^{m} \left(\sum_{k=1}^{n} a_{jk} \overline{\eta}_k^0 \right) \xi_j = 0, \quad \text{for all } \xi_j \tag{2.3.4}$$

implies that $\eta_k^0 = 0$, $k = 1, \ldots, n$. Next, (2.3.4) is equivalent to the system of homogeneous equations

$$\sum_{k=0}^{n} a_{jk}\bar{\eta}_k^0 = 0, \qquad j = 1, \ldots, m. \tag{2.3.5}$$

Thus condition (6) shows that the system (2.3.5) has only the trivial solution. For this to hold, it is necessary and sufficient that the rank of the matrix.

$$a = \begin{Vmatrix} a_{11} & \cdots & a_{1n} \\ a_{21} & \cdots & a_{2n} \\ \cdots\cdots\cdots\cdots \\ a_{m1} & \cdots & a_{mn} \end{Vmatrix} \tag{2.3.6}$$

be not less than n. In particular, we must have $m \geq n$. Switching the roles of x and y and applying condition (5), we conclude that also $n \geq m$. Thus $n = m$ and $\det a \neq 0$. In this case we can simplify expression (2.3.2) for the form (x, y). We choose a new basis in Y, $f'_1, f'_2, \ldots, f'_n$, by the formulas $f'_k = \sum_{j=1}^{n} b_{jk}f_j$, where

$$\bar{b} = \begin{Vmatrix} \bar{b}_{11} & \cdots & \bar{b}_{1n} \\ \cdots\cdots\cdots\cdots \\ \bar{b}_{n1} & \cdots & \bar{b}_{nn} \end{Vmatrix}$$

is the inverse matrix of a (b exists since $\det a \neq 0$). Then the coefficients of the form (x, y) with respect to the bases $e_1, e_2, \ldots, e_n; f'_1, \ldots, f'_n$ are

$$a'_{jk} = (e_j, f'_k) = \left(e_j, \sum_{v=1}^{n} b_{vk}f_v\right) = \sum_{v=1}^{n} (e_j, f_v)\bar{b}_{vk}$$

$$= \sum_{v=1}^{n} a_{jv}\bar{b}_{vk} = \begin{cases} 1, & \text{for } j = k, \\ 0, & \text{for } j \neq k, \end{cases}$$

Therefore $(x, y) = \sum_{j=1}^{n} \xi_j\bar{\eta}'_j$, where the η'_j are coordinates of the vector y in the basis $f'_1, \ldots, f'_k$. We have proved the following proposition.

II. *Two finite-dimensional spaces X, Y can be in duality if and only if $\dim X = \dim Y$. In this case, the form (2.3.2) ($n = m = \dim X = \dim Y$) defines a duality of the pair (X, Y) if and only if the determinant of this form is nonzero. Under a proper choice of bases $e_1, \ldots, e_n, f_1, \ldots, f_n$ the expression (2.3.2) for the form (x, y) becomes*

$$(x, y) = \sum_{k=1}^{n} \xi_k\bar{\eta}_k, \tag{2.3.7}$$

where $\xi_1, \ldots, \xi_n$ and $\eta_1, \ldots, \eta_n$ *are coordinates of the vectors* x *and* y *in these bases.*

We return to the general case. Let X, Y be linear spaces in duality with respect to the form (x, y) and let T, S be representations of a group G in X and Y respectively. The representation S is called *adjoint to the representation* T *with respect to* (x, y) if

$$(T(g)x, S(g)y) = (x, y), \quad \text{for all } g \in G, x \in X, y \in Y. \tag{2.3.8}$$

III. *Condition* (2.3.8) *is equivalent to the condition*

$$(T(g^{-1})x, y) = (x, S(g)y), \quad \text{for all } g \in G, x \in X, y \in Y. \tag{2.3.9}$$

To prove this it suffices to replace x in (2.3.8) by $T(g^{-1})x$ and to note that $T(g)$ is a one-to-one mapping of X onto itself.

IV. *If* X *and* Y *are in duality with respect to the form* (x, y), *if* T *is a representation in* X, *and if there is a representation in* Y *that is adjoint to* T *with respect to* (x, y), *then it is unique.*

Proof. Let S and S' be representations in Y both adjoint to T with respect to (x, y). By virtue of (2.3.9) we have $(T(g^{-1})x, y) = (x, S(g)y)$ and $(T(g^{-1})x, y) = (x, S'(g)y)$, which means that $0 = (x, (S(g) - S'(g))y)$ for all $g \in G$, $x \in X$, $y \in Y$. From this and condition (6) on page 29 we find $(S(g) - S'(g))y = 0$ for all $y \in Y$, $g \in G$, so that $S(g) = S'(g)$ and $S = S'$. $\square$

We draw another inference from I.

V. *If* S *is adjoint to* T *with respect to* (x, y), *then* T *is adjoint to* S *with respect to* $(y, x)_1 = \overline{(x, y)}$.

Let us examine more closely the case of finite-dimensional X, Y. Let the representation S in a space Y be adjoint to a representation T in a space X with respect to the form (x, y). Choose bases $e_1, \ldots, e_n$ and $f_1, \ldots, f_n$ in X and Y respectively, (where $n = \dim X = \dim Y$) such that condition (2.3.7) is satisfied (see II) and consequently

$$(e_j, f_k) = \begin{cases} 1, & \text{for } j = k, \\ 0, & \text{for } j \neq k. \end{cases} \tag{2.3.10}$$

Let $t_{jk}(g)$, $s_{jk}(g)$ be the matrix elements of T and S in the bases $e_1, \ldots, e_n$ and $f_1, \ldots, f_n$ respectively, i.e.,

$$T(g)e_j = \sum_{v=1}^{n} t_{vj}(g)e_v, \quad S(g)f_k = \sum_{\mu=1}^{n} s_{\mu k}(g)f_\mu \tag{2.3.11}$$

(see (2.1.3)). The condition (2.3.9) is equivalent to the system of relations

$$(T(g^{-1})e_j, f_k) = (e_j, S(g)f_k), \qquad j, k = 1, \ldots, n, \qquad (2.3.12)$$

which with (2.3.10) and (2.3.11) lead to the relations

$$t_{kj}(g^{-1}) = \overline{s_{jk}(g)}, \qquad j, k = 1, 2, \ldots, n, \qquad n = \dim X = \dim Y. \quad (2.3.13)$$

We have proved the following.

VI. *Two finite-dimensional representations T and S in spaces X and Y respectively are mutually adjoint with respect to the form* (x, y) *if and only if (with the proper choice of bases in X and Y) their matrix elements are connected by the relations* (2.3.13).

From VI. we conclude the following.

VII. *If X and Y are finite-dimensional spaces in duality with respect to the form* (x, y), *and if a representation T of the group G is defined in X, then there is a unique representation S of G in Y adjoint to the representation T.*

Proof. We have proved in IV the uniqueness of S; we now prove its existence. We write $n = \dim X = \dim Y$ and choose bases $e_1, \ldots, e_n$ and $f_1, \ldots, f_n$ in X and Y respectively, satisfying (2.3.7). Let $t_{jk}(g)$ be the matrix elements of T in the basis $e_1, \ldots, e_n$ and let $t(g)$ be the corresponding matrix. We define the operator $S(g)$ in Y by its matrix elements $s_{jk}(g)$ in the basis $f_1, \ldots, f_n$:

$$s_{jk}(g) = \overline{t_{kj}(g^{-1})}, \qquad j, k = 1, \ldots, n. \qquad (2.3.14)$$

Let $s(g)$ be the matrix of the operator $S(g)$. The relations (2.3.14) show that $s(g) = t(g^{-1})^*$ where $*$ denotes the adjoint matrix.

We check that $g \to S(g)$ is a representation of G. We have

$$s(e) = t(e)^* = 1^* = 1,$$
$$s(g_1 g_2) = t((g_1 g_2)^{-1})^* = t(g_2^{-1} g_1^{-1})^* = (t(g_2^{-1})t(g_1^{-1}))^* = s(g_1)s(g_2).$$

From this we have

$$S(e) = 1, \qquad S(g_1 g_2) = S(g_1)S(g_2).$$

Finally, the relations (2.3.14) are equivalent to (2.3.13); consequently S is adjoint to T with respect to (x, y). $\square$

VIII. *Let the finite-dimensional representations T and S be mutually adjoint. Then T is irreducible if and only if S is irreducible.*

Proof. Let X and Y be the spaces of the representations T and S, so that X and Y are in duality with respect to a form (x, y) and dim $X = $ dim Y. Let M be a subspace of Y, invariant under S. Write $L = M^{\perp}$. Clearly, L is a subspace of X, and in addition L is invariant under T. Indeed, if $X \in L$, then $(x, y) = 0$ for all $y \in M$ and therefore $(x, S(g^{-1})y) = 0$ for all $y \in M$, since, according to the hypothesis, M is invariant under S. For all $y \in M$, (2.3.9) shows that

$$(T(g)x, y) = (x, S(g^{-1})y) = 0,$$

so that $T(g)x \in L$.

Suppose that T is irreducible; then either $L = \{0\}$ or $L = X$. In the first case condition (5) of the definition of duality is fulfilled for X and M (see page 29). Clearly, condition (6) is also satisfied, so that X and M are in duality with respect to (x, y). By virtue of II, this cannot happen if dim $M <$ dim $Y = $ dim X, i.e., if $M \neq Y$. In the second case, if $y \in M$, then $(x, y) = 0$ for all $x \in X$. From this and condition (6) on page 29, we conclude that $y = 0$, so that $M = \{0\}$. Thus, if T is irreducible, the subspaces $M = \{0\}$ and $M = Y$ are the only subspaces of Y invariant under S. That is, S is irreducible. Switching the roles of T and S in the preceding discussion, we conclude that T is irreducible if S is. □

Remark. Suppose that we have a form (x, y) that satisfies conditions (1), (3), (4) and instead of (2), the following condition (2) $(x, \alpha y) = \alpha(x, y)$ (see the footnote to page 28). Just as was done above, we can define duality with respect to the form (x, y). A representation S in the space X is called *contragredient* to a representation T in the space Y with respect to such a form (x, y) if $(T(g)x, S(g)y) = (x, y)$ for all $g \in G$, $x \in X$ and $y \in Y$. All the propositions of the present section remain valid for a contragredient representation. We replace formulas (2.3.2), (2.3.7), (2.3.9) and (2.3.13) by the following:

$$(x, y) = \sum_{j=1}^{m} \sum_{k=1}^{n} a_{jk}\xi_j\eta_k, \tag{2.3.2'}$$

$$(x, y) = \sum_{k=1}^{n} \xi_k\eta_k, \tag{2.3.7'}$$

$$(T(g)x, y) = (x, S(g^{-1})y), \tag{2.3.9'}$$

$$t_{kj}(g) = s_{jk}(g^{-1}). \tag{2.3.13'}$$

All proofs of the present section can be carried over almost word for word for contragredient representations.

Examples

1. Let T be a finite-dimensional representation of a group G in a vector space X. Let S be a representation of the group G in a space Y, adjoint to the representation T. Show that a vector subspace $L \subset X$ is invariant under T if and only if the subspace $M = L^{\perp} \subset Y$ is invariant under S. [Hint. See the proof of VIII.]

2. Prove that any n-dimensional representation of a commutative group admits an $(n - 1)$-dimensional invariant subspace.

3. Find the representation adjoint to a one-dimensional representation T of a group G.

4. Let T be a one-dimensional representation of a group G and let T^* be its adjoint. Find a necessary and sufficient condition that T and T^* be equivalent.

5. Let G be a matrix group and let T be its self-representation. For the following groups, determine whether or not T and T^* are equivalent.
(a) $GL(n, \mathbf{R})$; (b) $SL(n, \mathbf{R})$; (c) $U(n)$; (d) $O(n)$; (e) $SU(n)$; (f) $SO(n)$.

2.4. The Direct Sum of Representations

Let G be a group. Let $X_1, X_2, \ldots, X_m$ be linear spaces, and let

$$X = X_1 + \cdots + X_m$$

be their direct sum, so that every vector x of X is represented uniquely in the form $x = x_1 + \cdots + x_m$, where $x_k \in X_k$ (see Kuroš [1]). Suppose that we have a representation T^k in each X_k. We define a linear operator $T(g)$ on X by

$$T(g)(x_1 + \cdots + x_m) = T^1(g)x_1 + \cdots + T^m(g)x_m. \tag{2.4.1}$$

It is clear that

$$T(e) = 1, \qquad T(g_1 g_2) = T(g_1)T(g_2).$$

Hence the correspondence $g \to T(g)$ is a representation of G in the space X. This representation is called the *direct sum of the representations* T^1, $\ldots, T^m$ and is denoted by $T^1 + \cdots + T^m$. Clearly, every X_k is a subspace of X invariant under T. The restriction of T to X_k is T^k.

Suppose in particular that all X_k's are finite-dimensional. Let $e_1^k, \ldots, e_{n_k}^k$ be a basis in X_k, and let $t_{jl}^k(g)$, $(j, l = 1, \ldots, n_k)$ be the matrix elements of the representation T^k in this basis. Write the corresponding matrix as $t^k(g)$. The union of the bases e_1^k is a basis in the direct sum $X_1 + \cdots + X_m$. We see from (2.4.1) and (2.1.3) that

$$T(g)e_j^k = T^k(g)e_j^k = \sum_{\nu=1}^{n} t_{\nu j}^k(g)e_\nu^k.$$

Arrange the bases $e_1^k, \ldots, e_{n_k}^k$ in increasing order of the index k. The matrix of the operator $T(g)$ in this basis in X is

$$
t(g) = \begin{Vmatrix}
t'_{11}(g) & \cdots & t'_{1n_1}(g) & 0 & \cdots & 0 & \cdots & 0 & \cdots & 0 \\
\vdots & & \vdots & \vdots & & \vdots & & \vdots & & \vdots \\
t'_{n_1 1}(g) & \cdots & t'_{n_1 n_1}(g) & 0 & \cdots & 0 & \cdots & 0 & \cdots & 0 \\
0 & \cdots & 0 & t^2_{11}(g) & \cdots & t^2_{1n_2}(g) & \cdots & 0 & \cdots & 0 \\
\vdots & & \vdots & \vdots & & \vdots & & \vdots & & \vdots \\
0 & \cdots & 0 & t^2_{n_2 1}(g) & \cdots & t^2_{n_2 n_2}(g) & \cdots & 0 & \cdots & 0 \\
\vdots & & \vdots & \vdots & & \vdots & & \vdots & & \vdots \\
0 & \cdots & 0 & 0 & \cdots & 0 & \cdots & t^m_{11}(g) & \cdots & t^m_{1n_m}(g) \\
\vdots & & \vdots & \vdots & & \vdots & & \vdots & & \vdots \\
0 & \cdots & 0 & 0 & \cdots & 0 & \cdots & t^m_{n_m 1}(g) & \cdots & t^m_{n_m n_m}(g)
\end{Vmatrix}.
$$

$$(2.4.2)$$

We abbreviate (2.4.2) as

$$
t(g) = \begin{Vmatrix}
t^1(g) & & & \\
& t^2(g) & & \\
& & \ddots & \\
& & & t^m(g)
\end{Vmatrix},
$$

$$(2.4.3)$$

where all entries not specified are zeroes. We summarize as follows.

I. *Let T be the direct sum of representations $T, \ldots, T^m$ in the finite-dimensional spaces $X_1, \ldots, X_m$. Let $e_1^k, \ldots, e_{n_k}^k$ be a basis in X^k and write the union of these bases as above. The matrix of the operator $T(g)$ in this basis has the quasidiagonal form (2.4.3) where the matrices $t^k(g)$ of the representations X^k in the bases $e_1^k, \ldots, e_{n_k}^k$ are arranged along the diagonal.*

We return to the general case. A representation T in a space X is called *completely reducible* if T is the direct sum of a finite number of irreducible representations. In this case we also say that *T factors into a direct sum of irreducible representations.* We do not exclude the possibility that this direct sum consists of only one term since an irreducible representation is also considered completely reducible. A representation T is said to be *divisible by an irreducible representation T^1*, written $T = nT^1$, if T is the direct sum of n representations each of which is equivalent to the irreducible representation T^1. More generally, if a completely reducible representation T is the direct sum of irreducible representations, among which there are n_1 representations equivalent to T_1, n_2 representations equivalent to $T_2, \ldots, n_p$ equivalent to T^p, we write

$$T = n_1 T^1 + n_2 T^2 + \cdots + n_p T^p.$$

We say that T^j enters or is contained in T with the multiplicity n_j $(j = 1, \ldots, p)$.

We can easily give an example of a reducible but not completely reducible (actually two-dimensional) representation. Consider the representation $\alpha \to T(\alpha)$ of the group $\mathbf{R}^1$ (see example 1 in 2.1) in the space $X = \mathbf{C}^2$. Its matrix is the following:

$$t(\alpha) = \begin{Vmatrix} 1 & 0 \\ \alpha & 1 \end{Vmatrix}, \qquad \alpha \in \mathbf{R}^1. \tag{2.4.4}$$

Conditions (1) and (2) for a representation are satisfied here since

$$t(0) = \begin{Vmatrix} 1 & 0 \\ 0 & 1 \end{Vmatrix},$$

$$t(\alpha_1)t(\alpha_2) = \begin{Vmatrix} 1 & 0 \\ \alpha_1 & 1 \end{Vmatrix} \begin{Vmatrix} 1 & 0 \\ \alpha_2 & 1 \end{Vmatrix} = \begin{Vmatrix} 1 & 0 \\ \alpha_1 + \alpha_2 & 1 \end{Vmatrix} = t(\alpha_1 + \alpha_2).$$

Since this representation is two-dimensional, its proper invariant subspaces M are one-dimensional. Let $x = \begin{Vmatrix} \xi_1 \\ \xi_2 \end{Vmatrix}$ be any vector in M. Since M is invariant, we have $t(\alpha)x = \lambda(\alpha)x$, where $\lambda(\alpha)$ is a complex-valued function. Then (2.4.4) implies that

$$\begin{Vmatrix} \xi_1 \\ \alpha\xi_1 + \xi_2 \end{Vmatrix} = \lambda(\alpha) \begin{Vmatrix} \xi_1 \\ \xi_2 \end{Vmatrix}. \tag{2.4.5}$$

The identity (2.4.5) means that

$$\xi_1 = \lambda(\alpha)\xi_1, \qquad \alpha\xi_1 + \xi_2 = \lambda(\alpha)\xi_2. \tag{2.4.6}$$

From the first equality in (2.4.6) we infer that either $\xi_1 = 0$ or $\lambda(\alpha) = 1$. But the second case leads back to the first, since the second equality in (2.4.6) gives $\alpha\xi_1 + \xi_2 = \xi_2$ and this means that $\alpha\xi_1 = 0$ for all α and thus $\xi_1 = 0$. Thus $\left\{ M = \begin{Vmatrix} 0 \\ \xi \end{Vmatrix} \; \xi \in \mathbf{C}^1 \right\}$ is the unique nontrivial invariant subspace of our representation. Therefore this representation cannot be a direct sum of irreducible representations.

In many cases the following proposition is useful for actual factorization of a completely reducible representation.

II. *Let T be a representation of a group G in a space X and let $M_1, M_2, M_3, \ldots$ be a sequence of subspaces in X having the following properties:*

(1)   *every M_k is invariant under T;*
(2)   *the restrictions T^k of the representation T to M_k are irreducible.*

We can choose a subsequence $M_{k_1}, M_{k_2}, \ldots$ *of* $M_1, M_2, M_3, \ldots$ *such that:*

(a) $M_{k_1}, M_{k_2}, \ldots$ *are linearly independent;*
(b) $\sum_j M_{k_j} = \sum_k M_k.$

If we have $\sum_k M_k = X,$ *then also* $\sum_j M_{k_j} = X.$

Proof. We construct M_{k_j} by induction. Let $k_1 = 1$, so that $M_{k_1} = M_1$. Suppose that linearly independent subspaces $M_{k_1}, \ldots, M_{k_n}$ have been constructed and that they satisfy the condition

$$M_{k_1} + \cdots + M_{k_n} = M_1 + M_2 + \cdots + M_{k_n}. \qquad (2.4.7)$$

Consider the subspace $M = (M_1 + \cdots + M_{k_n}) \cap M$ for an arbitrary $k > k_n$. It is invariant under T and is contained in M_k. By virtue of (2), either $M = (0)$ or $M = M_k$. If the first case is true for some $k > k_n$, then we take k_{n+1} to be the least of these k. If the second case holds for $k > k_n$, then we have

$$M_k \subset M_{k_1} + \cdots + M_{k_n}, \quad \text{for } k > k_n,$$

and by virtue of (2.4.7), $M_{k_1} + \cdots + M_{k_n} = \sum_k M_k.$ $\square$

Examples and Exercises

   1. Prove that the representation T of exercise 2 on page 28 is the direct sum of two irreducible representations.

   2. Prove that a representation T is the direct sum of representations $T^{(1)}$ and $T^{(2)}$ if and only if the adjoint representation T^* is the direct sum of the representations $T^{(1)*}$ and $T^{(2)*}$.

   3. Prove that a finite-dimensional representation of a finite commutative group is the direct sum of one-dimensional representations.

2.5. The Semidirect Sum (Linkage) of Representations

A representation T of a group G in a space X is called a *semidirect sum* (or *linkage*) of representations T^k in spaces X_k, $k = 1, \ldots, m$, and we write $T = T^1 \to T^2 \to T^3 \to \cdots \to T^m$, if X contains a sequence of T-invariant subspaces, $M_1 \subset M_2 \subset \cdots \subset M_m = X$, with the following properties. First, the restriction of T to M_1 is equivalent to T^1. Second, the representation $\tilde{T}^2$ generated by T on M_2/M_1 (see 2.1) is equivalent to T^2. Third, the representation $\tilde{T}^3$ generated by the representation T on M_3/M_2 is equivalent to T^3, and so on. Finally, the representation $\tilde{T}^m$ generated by the representation T on X/M_{m-1} is equivalent to T^m.

   Plainly if T is a direct sum of representations $T^1, \ldots, T^m$ (i.e., $T = T^1 + \cdots + T^m$), then T is also a linkage of these representations (i.e., $T = T^1 \to T^2 \to \cdots \to T^m$). One can consider $M_1 = X_1$, $M_2 = X_1 + X_2 \ldots, M_m = X_1 + \cdots + X_m = X$ as the subspaces $M_1, M_2, \ldots, M_m$. The converse, however, does not hold.

To prove this, it suffices to consider the representation T given in the example at the end of 2.4. This representation can be described by the formula

$$T(\alpha)\left\|\begin{matrix}\xi_1\\\xi_2\end{matrix}\right\| = \left\|\begin{matrix}\xi_1\\\alpha\xi_1 + \xi_2\end{matrix}\right\|. \tag{2.5.1}$$

As shown above, X admits only one proper subspace invariant under T:

$$M = \left\{\left\|\begin{matrix}0\\\xi_2\end{matrix}\right\|, \xi_2 \in \mathbf{C}_1\right\}. \quad \text{For } \xi = \left\|\begin{matrix}0\\\xi_2\end{matrix}\right\| \in M, \text{ we have } T\xi = \xi.$$

Thus, $T(\alpha) \equiv 1$ on M. Every vector $\tilde{\xi}$ in the factor space $\mathbf{C}^2/M$ is a set of the following form: all $\left\|\begin{matrix}\xi_1\\\xi_2\end{matrix}\right\|$ where ξ_1 is fixed and ξ_2 runs through all of $\mathbf{C}^1$. It follows from (2.5.1) that the representation $\tilde{T}$ generated in $\mathbf{C}^2/M$ by the representation T also has the form $T(\alpha) \equiv 1$. Thus the representation T is the linkage of two identity representations of the group $\mathbf{R}^1$, but it is not the direct sum of these representations.

Consider a linkage of finite-dimensional representations $T^1, \ldots, T^m$. Let $e_1^k, \ldots, e_{n_k}^k$ be a basis in X_k and $t_{j\nu}^k(g)$ be the matrix elements of the representation T^k in this basis ($k = 1, \ldots, m$). By definition, the representation $\tilde{T}^k$ generated by T in $\tilde{X}_k = M_k/M_{k-1}$ ($k = 1, \ldots, m$, $\tilde{X}_1 = M_1$) is equivalent to the representation T^k. Thus (see I in 2.2), we can choose a basis $\tilde{f}_1^k, \ldots, \tilde{f}_{n_k}^k$ in $\tilde{X}_k$ such that the matrix elements of T^k in this basis coincide with $t_{j\nu}^k(g)$. Let $f_1^k, \ldots, f_{n_k}^k$ be representatives of the classes $\tilde{f}_1^k$, $\ldots, \tilde{f}_{n_k}^k$ in M_k for $k \geq 2$. Let $f_1 = \tilde{f}_1^1, \ldots, f_{n_1}^1 = \tilde{f}_{n_1}^1$. Then all f_ν^k, $\nu = 1$, $\ldots, n_k$; $k = 1, \ldots, m$ form a basis in X. To see this, let $x \in X$ and let $\tilde{x}$ be the image of x under the canonical mapping of X onto X/M_{m-1}. Then we have $x \in \tilde{x}$. Since $\tilde{f}_1^m, \ldots, \tilde{f}_{n_m}^m$ is a basis in X/M_{m-1}, we have $\tilde{x} = \alpha_1^m \tilde{f}_1^m + \cdots + \alpha_{n_m}^m \tilde{f}_{n_m}^m$ for some complex numbers $\alpha_1^m, \ldots, \alpha_{n_m}^m$. From this we have $x = \alpha_1^m f_1^m + \cdots + \alpha_{n_m}^m f_{n_m}^m + y_{m-1}$, where $y_{m-1} \in M_{m-1}$. Applying the above reasoning to y_{m-1} and M_{m-1}/M_{m-2} in place of x and X/M_{m-1} we conclude that $y_{m-1} = \alpha_1^{m-1} f_1^{m-1} + \cdots + \alpha_{n_{m-1}}^{m-1} f_{n_{m-1}}^{m-1} + y_{m-2}$ where $y_{m-2} \in M_{m-2}$. Repeating this reasoning, after a finite number of steps we obtain

$$x = \sum_{k=1}^{m} \sum_{j=1}^{n_k} \alpha_j^k f_j^k.$$

It remains to prove that f_j^k, $j = 1, \ldots, n_k$; $k = 1, \ldots, m$ are linearly independent. Suppose that

$$\sum_{k=1}^{m} \sum_{j=1}^{n_k} \alpha_j^k f_j^k = 0. \tag{2.5.2}$$

We must prove that all $\alpha_j^k = 0$.

Applying the canonical mapping $X \to X/M_{m-1}$ to both sides of (2.5.2), we obtain

$$\sum_{j=1}^{n_m} \alpha_j^m \tilde{f}_j^m = 0. \tag{2.5.3}$$

Since $\tilde{f}_j^m$, $j = 1, \ldots, n_m$ is a basis in X/M_{m-1}, (2.5.3) implies that $\alpha_j^m = 0$, $j = 1, \ldots, n_m$. Therefore (2.5.2) can be rewritten as follows.

$$\sum_{k=1}^{m-1} \sum_{j=1}^{n_k} \alpha_j^k f_j^k = 0. \tag{2.5.4}$$

Applying the canonical mapping $M_{m-1} \to M_{m-1}/M_{m-2}$ to (2.5.4) and repeating the above reasoning, we conclude that $\alpha_j^{m-1} = 0$, $j = 1, \ldots, n_{m-1}$. After a finite number of such steps we see that all $\alpha_j^k = 0$.

Thus the set of all f_ν^k forms a basis in X. We arrange the f_ν^k in the following way:

$$f_1^1, \ldots, f_{n_1}^1; f_1^2, \ldots, f_{n_2}^2; \ldots; f_1^m, f_2^m, \ldots, f_{n_m}^m. \tag{2.5.5}$$

We can find the matrix $t(g)$ of the operator $T(g)$ in this basis. By definition, the matrix elements of the restriction of T to M_1 in the basis $f_1^1, \ldots, f_{n_1}^1$ coincide with $t_{jk}^1(g)$, so that

$$T(g)f_k^1 = \sum_{j=1}^{n_1} t_{jk}^1(g)f_j^1. \tag{2.5.6}$$

Also the matrix elements of the representation T^2 in the basis $\tilde{f}_1^2, \ldots, \tilde{f}_{n_2}^2$ coincide with $t_{jk}^2(g)$, so that

$$\tilde{T}(g)\tilde{f}_k^2 = \sum_{j=1}^{n_2} t_{jk}^2(g)\tilde{f}_j^2.$$

From this we have

$$T(g)f_k^2 = \sum_{j=1}^{n_2} t_{jk}^2(g)f_j^2 + y_1, \qquad y_1 \in M_1. \tag{2.5.7}$$

Repeating this reasoning we conclude that

$$T(g)f_k^\nu = \sum_{j=1}^{n_\nu} t_{jk}^\nu(g)f_j^\nu + y_{\nu-1},$$

where

$$y_{\nu-1} \in M_{\nu-1}, \qquad \nu = 1, \ldots, m, \qquad y_0 = 0. \tag{2.5.8}$$

Hence the matrix $t(g)$ of the operator $T(g)$ in the basis (2.5.5) is

$$
t(g) =
\begin{Vmatrix}
t^1_{11}(g) & \cdots & t^1_{1n_1}(g) & 0 & \cdots & 0 & \cdots & 0 & \cdots & 0 \\
& & & & & & & & & \vdots \\
t^1_{n_1 1}(g) & \cdots & t^1_{n_1 n_1}(g) & 0 & \cdots & 0 & \cdots & 0 & \cdots & 0 \\
* & \cdots & & t^2_{11}(g) & \cdots & t^2_{1n_2}(g) & \cdots & 0 & \cdots & 0 \\
& & & & & & & & & \vdots \\
* & \cdots & * & t^2_{n_2 1}(g) & \cdots & t^2_{n_2 n_2}(g) & \cdots & 0 & \cdots & 0 \\
& & & & & & & & & \vdots \\
* & \cdots & * & * & \cdots & * & \cdots & t^m_{11}(g) & \cdots & t^m_{1n_m}(g) \\
& & & & & & & & & \vdots \\
* & \cdots & * & * & \cdots & * & \cdots & t^m_{n_m 1}(g) & \cdots & t^m_{n_m n_m}(g)
\end{Vmatrix}
$$

$$(2.5.9)$$

The symbols $*$ denote certain unspecified matrix elements. We rewrite (2.5.9) as

$$
t(g) =
\begin{Vmatrix}
t^1(g) & & & \\
* & t^2(g) & & \\
& & \vdots & \\
* & * & \cdots & t^m(g)
\end{Vmatrix},
$$

$$(2.5.10)$$

where $t^k(g)$ is the matrix of the representation T^k in basis $f^k_1, \ldots, f^k_{n_k}$; the asterisks denote certain matrices (dependent on g) and all places not marked are occupied by zeroes.

We have proved the following.

I. *If $T = T^1 \to T^2 \to \cdots \to T^m$, where $T^1, \ldots, T^m$ are finite-dimensional, then we can choose a basis in the space of the representation T such that the matrix of T in this basis has the form (2.5.10) where $t^k(g)$ is the matrix of the representation T^k, $k = 1, \ldots, m$. The asterisks denote certain matrices and all places not marked are occupied by zeroes.*

In the case of a direct sum of representations we can choose a basis so that all matrices marked by asterisks are zero (see I in 2.4). For linkages of representations, this is not possible (see the preceding example.)

Remark. We can also define the linkage $T^1 \leftarrow T^2 \leftarrow \cdots \leftarrow T^m$ by analogy with the definition of the linkage $T^1 \to T^2 \to \cdots \to T^m$ already described. Suppose that the space M of the representation T contains subspaces $M_m \subset M_{m-1} \subset \cdots \subset M_1 = M$, each invariant under T, such that the restriction of the representation T to M_m is equivalent to T_m, while the representation $\tilde{T}^k$, generated by the representation T in M_k/M_{k+1}, $k = 1, \ldots, n - 1$, is equivalent to T^k. Plainly, this linkage differs from the linkage

$T^1 \to T^2 \to \cdots \to T^m$ only in the numbering of the invariant spaces. We suggest that the reader prove the following. If $T = T^1 \leftarrow T^2 \leftarrow \cdots \leftarrow T^m$, then with the proper choice of a basis, the matrix $t(g)$ of the representation T has the following form:

$$\begin{Vmatrix} t^1(g) & * & \cdots & * \\ & t^2(g) & \cdots & * \\ & & \vdots & \\ \cdots & \cdots & \cdots & t^m(g) \end{Vmatrix}, \tag{2.5.11}$$

where $t^1(g), t^2(g), \ldots, t^m(g)$ are matrices of the representations $T^1, T^2, \ldots, T^m$.

Examples and Exercises

1. Prove that the linkage of representations $\begin{Vmatrix} T^{(1)}(g) & A(g) \\ 0 & T^{(2)}(g) \end{Vmatrix}$ is factorable into a direct sum if and only if there is a linear operator Y from the space of the representation $T^{(2)}$ into the space of the representation $T^{(1)}$, such that for a certain number λ the equality $T^{(1)}(g)Y - YT^{(2)}(g) = \lambda A(g)$ holds for all $g \in G$.

2. Show that a finite-dimensional representation of a finite group is never a non-trivial linkage of one-dimensional representations.

3. Show that a linkage of finite-dimensional, irreducible, inequivalent representations $T^{(1)}$, $T^{(2)}$ is in general *not* factorable into a direct sum of representations $T^{(1)}$ and $T^{(2)}$.

2.6. The Tensor Product of Finite-Dimensional Representations

Let T^1, T^2 be representations of a group G in finite-dimensional spaces X_1, X_2. A representation T in the space $X = X_1 \otimes X_2$ (see Zamansky [1]) is called a *tensor product $T^1 \otimes T^2$ of the representations* T^1, T^2 if the following holds. The operators $T(g)$ of this tensor product operate on vectors $x_1 \otimes x_2$, $x_1 \in X_1, x_2 \in X_2$ by the identity

$$T(g)(x_1 \otimes x_2) = T^1(g)x_1 \otimes T^2(g)x_2. \tag{2.6.1}$$

The formula (2.6.1) uniquely defines the operator $T(g)$ throughout the entire space $X = X_1 \otimes X_2$, and the correspondence $g \to T(g)$ is a representation since we have

$$T(e)(x_1 \otimes x_2) = T^1(e)x_1 \otimes T^2(e)x_2 = x_1 \otimes x_2, \tag{2.6.2a}$$

$$\begin{aligned} T(g_1 g_2)(x_1 \otimes x_2) &= T^1(g_1 g_2)x_1 \otimes T^2(g_1 g_2)x_2 \\ &= T^1(g_1)T^1(g_2)x_1 \otimes T^2(g_1)T^2(g_2)x_2 \\ &= T(g_1)(T^1(g_2)x_1 \otimes T^2(g_2)x_2) \\ &= T(g_1)T(g_2)(x_1 \otimes x_2), \end{aligned} \tag{2.6.2b}$$

We infer from (2.6.2a) and (2.6.2b) that $T(e) = 1$, $T(g_1 g_2) = T(g_1) T(g_2)$. Let $n_1 = \dim X_1$, $n_2 = \dim X_2$, and let $e_1^1, \ldots, e_n^1$ and $e_1^2, \ldots, e_n^2$ be bases in X_1 and X_2. Setting

$$e_{jk} = e_j^1 \otimes e_k^2, \qquad (2.6.3)$$

we obtain a basis $e_{jk}, j = 1, \ldots, n_1, k = 1, \ldots, n_2$ in $X_1 \otimes X_2$. The elements of this basis are described by the compound index that consists of the pair of numbers j, k. Therefore, the matrix elements of the representation $T^1 \otimes T^2$ must be described by two compound indices, that is, by a quadruple of numbers j, k, μ, ν. We denote the matrix elements of the representation $T = T^1 \otimes T^2$ in the basis $e_{jk} = e_j^1 \otimes e_k^2$ by the symbol $t_{\mu\nu jk}(g)$. To compute these functions, use (2.6.1) and (2.6.2) to write

$$T(g)e_{jk} = T^1(g)e_j^1 \otimes T^2(g)e_k^2 = \left(\sum_{\mu=1}^{n_1} t_{\mu j}^1(g) e_\mu^1 \right) \otimes \left(\sum_{\nu=1}^{n_2} t_{\nu k}^2(g) e_\nu^2 \right)$$

$$= \sum_{\mu=1}^{n_1} \sum_{\nu=1}^{n_2} t_{\mu j}^1(g) t_{\nu k}^2(g) (e_\mu^1 \otimes e_\nu^2) = \sum_{\mu=1}^{n_1} \sum_{\nu=1}^{n_2} t_{\mu j}^1(g) t_{\nu k}^2(g) e_{\mu\nu}$$

and therefore

$$t_{\mu\nu jk}(g) = t_{\mu j}^1(g) t_{\nu k}^2(g). \qquad (2.6.4)$$

The tensor product of any finite number of finite-dimensional representations is defined similarly. A representation T in the space $X_1 \otimes X_2 \otimes \cdots \otimes X_m$ is called the *tensor product* $T^1 \otimes T^2 \otimes \cdots \otimes T^m$ of *finite-dimensional representations* $T^1, \ldots, T^m$ of the group G in the spaces $X_1, \ldots, X_m$ if the operators are described by the following formula.

$$T(g)(x_1 \otimes \cdots \otimes x_m) = T^1(g)x_1 \otimes \cdots \otimes T^m(g)x_m \qquad (2.6.5)$$

for the special vectors $x_1 \otimes x_2 \otimes \cdots \otimes x_m$. Repeating previous calculations we find the following.

I. *Let* $T^1, \ldots, T^m$ *be representations of a group* G *in the finite-dimensional spaces* $X_1, \ldots, X_m$; *let* $n_1, \ldots, n_m$ *be the dimensions of these spaces. Let* f_1^k, $\ldots, f_{n_k}^k$ *be a basis in* X_k *and let* $t_{\nu j}^k(g)$ *be the matrix elements of the representation* T^k *in this basis* $(k = 1, \ldots, m)$. *Then the matrix elements* $t_{\nu_1 \nu_2 \cdots \nu_m j_1 j_2 \cdots j_m}(g)$ *of the tensor product* $T = T^1 \otimes \cdots \otimes T^m$ *of the representations* $T^1, \ldots, T^m$ *are defined by the following formula:*

$$t_{\nu_1 \nu_2 \cdots \nu_m j_1 j_2 \cdots j_m}(g) = t_{\nu_1 j_1}^1(g) t_{\nu_2 j_2}^2(g) \cdots t_{\nu_m j_m}^m(g). \qquad (2.6.6)$$

Examples and Exercises

1. Let χ_1, χ_2 be characters of G. Show that $(\chi_1 \otimes \chi_2)(g) = \chi_1(g)\chi_2(g)$.

2. Let $T^{(1)}, T^{(2)}$ be finite-dimensional representations of G in the spaces X and Y respectively. Let L be the linear space of all linear mappings A

of the linear space Y' (conjugate to Y) into X. We define the representation $g \to S(g)$ of the group G in L by the formula $S(g)A = T^{(1)}(g)A(T^{(2)}(g))'$ for all $A \in L$. Prove that the representation S is equivalent to the tensor product $T^{(1)} \otimes T^{(2)}$.

3. Prove that the identity representation of G is a subrepresentation of the tensor product of finite-dimensional irreducible representations $T^{(1)}$ and $T^{(2)}$ of G if and only if $T^{(1)}$ and $T^{(2)}$ are mutually conjugate.

2.7. Finite-Dimensional Representations of a Direct Product of Groups

Let $G = G_1 \times G_2$ be the direct product of the groups G_1 and G_2. Let T^1 be a finite-dimensional representation of G_1 in the space X_1 and let T^2 be a finite-dimensional representation of G_2 in the space X_2. Using these two representations, we construct a representation T of G in the space $X = X_1 \otimes X_2$ by defining

$$T(g)(x_1 \otimes x_2) = T^1(g_1)x_1 \otimes T^2(g_2)x_2, \quad \text{for } g = g_1 \times g_2. \quad (2.7.1)$$

Plainly the correspondence $g \to T(g)$ is a representation of G. It is denoted by $T^1_{G_1} \otimes T^2_{G_2}$. Consider the case $G_2 = G_1$, so that T^1 and T^2 are representations of the same group G. The group G contains a subgroup G_0, consisting of all $g = g_1 \times g_1$, $g_1 \in G_1$, which is clearly isomorphic to G_1. Therefore the restriction of the representation $T^1_{G_1} \otimes T^2_{G_1}$ to G_0 is a representation of G_1. On the other hand, from (2.7.1) we have

Comparison with (2.6.1) shows the following.

I. *The restriction of the represention $T^1_{G_1} \otimes T^2_{G_2}$ of the group $G \times G$ to the group G_0 of all $g_1 \times g_1$, $g_1 \in G$ coincides with the tensor product $T^1 \otimes T^2$ of the representations T^1, T^2 of G.*

Proposition I also holds for the direct product $G^n = G \times \cdots \times G$ of an arbitrary finite number of replicas of G. In this case the set of all $g_1 \times g_1 \cdots \times g_1, g_1 \in G_1$ plays the role of the group G.

II. *Every one-dimensional representation*

$$g_1 \times g_2 \times \cdots \times g_n \to f(g_1, g_2, \ldots, g_n),$$
$$g_1 \in G_1, \ldots, g_n \in G_n \quad (2.7.3)$$

of the direct product $G_1 \times G_2 \times \cdots \times G_n$ of the groups $G_1, \ldots, G_n$ is described by the formula

$$f(g_1, g_2, \ldots, g_n) = f_1(g_1)f_2(g_2) \cdots f_n(g_n), \quad (2.7.4)$$

where

$$g_j \to f_j(g_j), \qquad g_j \in G_j, \qquad j = 1, 2, \ldots, n, \tag{2.7.5}$$

are one-dimensional representations of the groups $G_1, G_2, \ldots, G_n$. Conversely, any function $f(g_1, \ldots, g_n)$ of the form (2.7.4), where $g_j \to f_j(g_j)$, $g_j \in G_j$ are one-dimensional representations of the groups $G_1, \ldots, G_n$, is a one-dimensional representation of G. The representation f is unitary[9] if and only if every representation f_j, $j = 1, \ldots, n$ is unitary.

Proof. We write

$$f_1(g_1) = f(g_1, e_2, \ldots, e_n),$$
$$f_2(g_2) = f(e_1, g_2, e_3, \ldots, e_n),$$
$$\vdots$$
$$f_n(g_n) = f(e_1, e_2, \ldots, e_{n-1}, g_n), \tag{2.7.6}$$

where e_j is the identity element in G_j. We then have

$$f_1(e_1) = f(e_1, e_2, \ldots, e_n) = 1$$

and for $g_1, g_1' \in G_1$

$$\begin{aligned}
f_1(g_1 g_1') &= f(gg_1', e_2, \ldots, e_n) \\
&= f(g_1, e_2, \ldots, e_n) f(g_1', e_2, \ldots, e_n) \\
&= f_1(g_1) f_1(g_1').
\end{aligned}$$

Therefore $g_1 \to f_1(g_1)$ is a one-dimensional representation of G. We prove similarly that every mapping $g_j \to f_j(g_j)$, $g_j \in G_j$ is a one-dimensional representation of G_j. We also have

$$\begin{aligned}
f_1(g_1) & f_2(g_2) \cdots f_n(g_n) \\
&= f(g_1, e_2, \ldots, e_n) f(e_1, g_2, \ldots, e_n) \cdots f(e_1, \ldots, e_{n-1}, g_n) \\
&= f(g_1, g_2, \ldots, g_n).
\end{aligned}$$

This proves the first assertion of II. The converse is obvious. The representation (2.7.3) is unitary if and only if $|f(g_1, \ldots, g_n)| = 1$. From (2.7.6) it follows that

$$|f_1(g_1)| = 1, \ldots, |f_n(g_n)| = 1. \tag{2.7.7}$$

The converse assertion follows from (2.7.4) and (2.7.7). □

[9] For the definition of a unitary representation see (2.8) *infra*.

III. *An arbitrary irreducible finite-dimensional representation T of the group $G = G_1 \times G_2$ is equivalent to the tensor product of certain irreducible representations T_1 and T_2 of G_1 and G_2 respectively. Also, T_1 and T_2 are subrepresentations of the restrictions of the representation T to the groups $G_1 \approx G_1 \times \{e_2\}$ and $G_2 \approx \{e_1\} \times G_2$ respectively, where e_1, e_2 are the identity elements of G_1 and G_2.*

Proof. Let T_1 and T_2 be representations of G_1 and G_2; the reader can easily verify that the representation $T = T_1 \otimes T_2$ of $G = G_1 \times G_2$ is irreducible if and only if T_1 and T_2 are irreducible. We will show that any irreducible representation T of $G_1 \times G_2$ in a (finite-dimensional) space E is equivalent to a representation of the form $T_1 \times T_2$, where T_1 and T_2 are irreducible representations of G_1 and G_2 respectively. Let S_1 and S_2 be representations of G_1 and G_2 respectively in a space E, defined by the formulas $S_1(g_1) = T((g_1, e_2))$ and $S_2(g_2) = T((e_1, g_2))$. Let T_1 be the restriction of the representation S_1 to a certain subspace E_1 invariant under S_1 and on which S_1 is irreducible. Let s be a nonzero vector in the subspace E_1. Let E_2 denote the subspace of E formed by finite linear combinations of vectors of the form $S_2(g_2)x$, $g_2 \in G_2$. Let T_2 be the restriction of the representation S_2 to the subspace E_2. We define a mapping A of the tensor product $E_1 \otimes E_2$ into E as follows. Let $x_1 \otimes x_2$ belong to $E_1 \otimes E_2$; then x_1 and x_2 can be represented in the following form:

$$x_1 = \sum_{p=1}^{m} \lambda_p T_1(g_1^{(p)})x, \qquad x_2 = \sum_{q=1}^{n} \mu_q T_2(g_2^{(q)})x, \quad \text{where } g_1^{(p)} \in G_1, \, g_2^{(q)} \in G_2,$$

and λ_p, μ_q are certain numbers. We set

$$A(x_1 \otimes x_2) = \sum_{p=1}^{m} \sum_{q=1}^{n} \lambda_p \mu_q T((g_1^{(p)}, g_2^{(q)}))x.$$

Since $(g_1, e_2)(e_1, g_2) = (e_1, g_2)(g_1, e_2) = (g_1, g_2)$ in $G_1 \times G_2$, we have

$$S_1(g_1)S_2(g_2)y = S_2(g_2)S_1(g_1)y = T((g_1, g_2))y \qquad (2.7.8)$$

for all $g_1 \in G_1$, $g_2 \in G_2$, $y \in E$. Using the equalities (2.7.8) and the definition of T_1 and T_2, the reader can check that the mapping A is defined uniquely and can be extended linearly to a mapping of $E_1 \otimes E_2$ into E, where $A(T_1(g_1)x_1 \otimes T_2(g_2)x_2) = T(g_1, g_2)A(x_1 \otimes x_2)$ for all $x_1 \in E_1$, $x_2 \in E_2$, $g_1 \in G_1$, $g_2 \in G_2$. The operator A demonstrates the equivalence of the representations T and $T_1 \otimes T_2$. Since T is irreducible, T_2 is also irreducible. $\quad\square$

2.8. Unitary Representations

A bilinear form (x_1, x_2) on a linear space X is called *Hermitian* if $(x_2, x_1) = \overline{(x_1, x_2)}$. An Hermitian bilinear form (x_1, x_2) on X is called *nonnegative* if $(x, x) \geqslant 0$ for all $x \in X$ and *positive* if $(x, x) > 0$ for $x \neq 0$. A positive bilinear form on X is also called a *scalar product on X*. A linear space X with a scalar product is called a *pre-Hilbert space*. A finite-dimensional pre-Hilbert space is called a *Euclidean space*. If X is a pre-Hilbert space, (x, y) will always denote (unless the contrary is stipulated) the scalar product on X. Orthogonality in X is orthogonality relative to $(x, y):(x, y) = 0$. A finite-dimensional pre-Hilbert space is dual to itself by the scalar product (x, y) (see 2.3). A linear operator A on a pre-Hilbert space X is called *unitary* if A is a one-to-one mapping of X onto X and

$$(Ax, Ay) = (x, y), \quad \text{for all } x, y \in X. \tag{2.8.1}$$

A representation T of G in a pre-Hilbert space X is called *unitary* if all operators of the representation are unitary. Since operators of a representation in X are by definition one-to-one mappings of X onto X (see 2.1), *a representation T is unitary if and only if*

$$(T(g)x, T(g)y) = (x, y), \quad \text{for all } g \in G, x, y \in X. \tag{2.8.2}$$

Comparing (2.8.2) with (2.3.8) we infer the following.

I. *If a representation T in a pre-Hilbert space X is unitary then it is conjugate with itself relative to (x, y).*

Repeating the proof of (2.3.9), we find that condition (2.8.2) is equivalent to the following condition.

$$(T(g^{-1})x, y) = (x, T(g)y), \quad \text{for all } g \in G, \ x, y \in X. \tag{2.8.3}$$

II. *If T is a unitary representation of G in a pre-Hilbert space X and M is invariant under T, then $M^{\perp}$ is also invariant under T.*

Proof. Let y belong to $M^{\perp}$. For all $x \in M$ we have $T(g^{-1})x \in M$, since M is invariant under T; that is, we have $y \perp T(g^{-1})x$. By (2.8.3) we obtain

$$(x, T(g)y) = (T(g^{-1})x, y) = 0.$$

We have shown that $T(g)y \perp M$, $T(g)y \in M^{\perp}$. This means that $M^{\perp}$ is invariant under T. $\square$

$$T(g)(x_1 \otimes x_2) = T^1(g_1)x_1 \otimes T^2(g_1)x_2 \quad \text{for } g = g_1 \times g_1 \in G_0. \tag{2.7.2}$$

If X is a Euclidean space and dim $X = n$, then we can choose a basis $e_1, \ldots, e_n$ in X for which

$$(e_j, e_k) = \begin{cases} 1, & \text{for } j = k \\ 0, & \text{for } j \neq k. \end{cases} \tag{2.8.4}$$

Such a basis is called *orthonormal*. Let $t_{jk}(g)$ be the matrix elements of a representation T in this basis. The representation T is unitary if and only if

$$(T(g)e_j, T(g)e_k) = (e_j, e_k) = \begin{cases} 1, & \text{for } j = k \\ 0, & \text{for } j \neq k. \end{cases} \tag{2.8.5}$$

By virtue of (2.1.3) and (2.8.4) we have

$$(T(g)e_j, T(g)e_k) = \left(\sum_{v=1}^{n} t_{vj}(g)e_v, \sum_{\mu=1}^{n} t_{\mu k}(g)e_\mu \right)$$

$$= \sum_{v,\mu=1}^{n} t_{vj}(g)\overline{t_{\mu k}(g)}(e_v, e_\mu)$$

$$= \sum_{v=1}^{n} t_{vj}(g)\overline{t_{vk}(g)}.$$

Consequently, (2.8.2) is equivalent to the conditions

$$\sum_{v=1}^{n} t_{vj}(g)\overline{t_{vk}(g)} = \begin{cases} 1, & \text{for } j = k \\ 0, & \text{for } j \neq k. \end{cases} \tag{2.8.6}$$

Condition (2.8.3) is equivalent to

$$t^*(g)t(g) = 1, \tag{2.8.7}$$

where $t^*(g)$ is the Hermitian conjugate of $t(g)$ (see 2.3). A matrix t for which $t^*t = 1$ is called *unitary*. We summarize as follows.

III. *The matrix of a finite-dimensional unitary representation in an orthonormal basis is unitary.*

The relations (2.8.6) mean that

$$t^*(g) = (t(g))^{-1}. \tag{2.8.8}$$

On the other hand, we have $T(g^{-1}) = (T(g))^{-1}$ and so $(t(g))^{-1} = t(g^{-1})$. Therefore we can rewrite (2.8.8) in the forms

$$t(g^{-1}) = t^*(g), \tag{2.8.9}$$

which in matrix elements gives

$$t_{jk}(g^{-1}) = \overline{t_{kj}(g)}. \tag{2.8.10}$$

IV. *Let T be a unitary representation in the Euclidean space X. Let M be a subspace of X invariant under T, and let P be the orthogonal projection* [10] *of X onto M. Then P commutes with all T(g), g ∈ G.*

Proof. Let x belong to X. Then we have $Px \in M$, and so also $T(g)Px \in M$, since M is invariant. It follows that $PT(g) \, Px = T(g)Px$ for all $x \in X$. That is, we have

$$PT(g)P = T(g)P. \qquad (2.8.11)$$

Replacing g in (2.8.11) by g^{-1} and noting that $T(g^{-1}) = (T(g))^{-1} = T^*(g)$, we obtain $PT^*(g)P = T^*(g)P$. But then we have $(PT^*(g)P)^* = (T^*(g)P)^*$, that is, $PT(g)P = PT(g)$. A comparison of the last equality with (2.8.11) gives us the required result: $T(g)P = PT(g)$. □

V. *A unitary representation T in the Euclidean space X is irreducible if and only if every operator A ∈ L(X) commuting with all T(g) is a multiple of the identity operator: $A = \lambda 1$.*

Proof. Let T be irreducible and let Z in $L(X)$ commute with all $T(g)$. Then we have $A = \lambda 1$ by virtue of lemma 2 in 2.2. Conversely, suppose that every operator A in $L(X)$ commuting with all $T(g)$ is a multiple of the identity operator. Let M be a subspace of X invariant under T. Let P be the orthogonal projection in X mapping X onto M. Then P commutes with all $T(g)$ by virtue of IV and our hypothesis implies that $P = \lambda 1$. This is possible only for $\lambda = 0$ and $\lambda = 1$, that is, when $P = 0$ or $P = 1$, which is to say $M = 0$ or $M = X$. Therefore T is irreducible. □

Two representations T, S of G in pre-Hilbert spaces X, Y are called *unitarily equivalent* if there exists a linear operator A from X into Y that satisfies the following conditions:

$$A \text{ is a one-to-one mapping of } X \text{ onto } Y; \qquad (2.8.12)$$

$$(Ax_1, Ax_2) = (x_1, x_2), \quad \text{for all } x_1, x_2 \in X; \qquad (2.8.13)$$

$$AT(g) = S(g)A, \quad \text{for all } g \in G. \qquad (2.8.14)$$

Thus, a unitary equivalence differs from an ordinary equivalence in that condition (2.8.13) is added. This condition specifies that A be an *isometry*.

VI. *If two unitary representations T and S of a group G in Euclidean spaces are equivalent, then they are also unitarily equivalent.*

---

[10] That is, P is a projection and $P^* = P$. Recall that a projection is an operator such that $P^2 = P$. It is a *projection onto M* if $PX = M$.

Proof. Let A be a linear operator carrying X onto Y and satisfying conditions (2.8.12) and (2.8.14). We then have $A = UB$, where B is a one-to-one Hermitian operator carrying X onto X and U is an isometry of X onto Y (see Gantmaher [1]), so that

$$(Ux_1, Ux_2) = (x_1, x_2). \tag{2.8.15}$$

Substituting $A = UB$ in (2.8.14) we obtain

$$UBT(g) = S(g)UB. \tag{2.8.16}$$

Replacing g in (2.8.16) by g^{-1} and recalling that $T(g^{-1}) = T^*(g)$, we conclude that $UBT^*(g) = S^*(g)UB$. Therefore, we have $(UBT^*(g))^* = (S^*(g)UB)^*$, or

$$T(g)BU^{-1} = BU^*S(g). \tag{2.8.17}$$

From (2.8.16) and (2.8.17) we obtain

$$T(g)B^2 = T(g)BU^*UB = BU^*S(g)UB = BU^*UBT(g) = B^2T(g).$$

Thus B^2 and so also B commute with all of the operators $T(g)$. We can now rewrite (2.8.11) in the form

$$UT(g)B = S(g)UB. \tag{2.8.18}$$

Multiplying both sides of (2.8.18) on the right by B^{-1}, we obtain $UT(g) = S(g)U$, which together with (2.8.15) shows that T and S are unitarily equivalent.   □

VII. *Two unitary representations T and S of G in Euclidean spaces X and Y are equivalent if and only if there are orthonormal bases in X and Y with respect to which the matrix elements of the representations T and S coincide.*

Proof. If T and S are equivalent, then by VI they are also unitarily equivalent. Let A be an operator from X onto Y satisfying conditions (2.8.13) and (2.8.14) and let $e_1, \ldots, e_n$ be any orthonormal basis in X. We set $f_1 = Ae_1, \ldots, f_n = Ae_n$; then in view of (2.8.12) and (2.8.13) $f_1, \ldots, f_n$ is an orthonormal basis in Y. Using condition (2.8.14) and repeating part of the proof of Proposition I in 2.2, we conclude that the matrix elements of the representations T and S in the bases $e_1, \ldots, e_n$ and $f_1, \ldots, f_n$ coincide. The converse assertion follows from I in 2.2.   □

We shall also need the following simple fact.

VIII. *If X is a Euclidean space and M is a subspace of X distinct from X, then $M^\perp \neq \{0\}$.*

Proof. Let dim $X = n$, dim $M = m$. The inequality $m < n$ is obvious. Let $e_1, \ldots, e_n$ be a basis in X and let $f_1, \ldots, f_m$ be a basis in M. We will find an element $x = \alpha_1 e_1 + \cdots + \alpha_n e_n \neq 0$ that belongs to $M^\perp$. For this to occur it suffices that $(x, f_k) = 0$, $k = 1, \ldots, m$, or more explicitly:

$$\alpha_1(e_1, f_k) + \cdots + \alpha_n(e_n, f_k) = 0, \qquad k = 1, \ldots, m. \qquad (2.8.19)$$

But (2.8.19) consists of m homogeneous equations in the unknowns $\alpha_1, \ldots,$ and $m < n$. By elementary linear algebra (2.8.19) has nontrivial solution $\alpha_1^0, \ldots, \alpha_1^0$. We set $x = \alpha_1^0 e_1 + \cdots + \alpha_n^0 e_n$ and so obtain a nonzero vector x in $M^\perp$. $\square$

A representation T in a space X is said to be *equivalent to a unitary representation* if there is a scalar product (s, y) in X under which T is unitary.

IX. *If a finite-dimensional representation is equivalent to a unitary representation, then it is completely reducible.*

Proof. Let T be a representation in the space X and let (x, y) be a scalar product under which T is unitary. By I in 2.1, X admits a subspace M_1, $M_1 \neq \{0\}$, invariant under T, on which the restriction of T is irreducible. If $M_1 = X$, we are through. If $M_1 \neq X$, then $M_1^\perp$ is not $\{0\}$ by virtue of IV and $M_1^\perp$ is invariant under T by II. Applying I in 2.1 to the restriction of T of $M_1^\perp$, we obtain a subspace M_2 of $M_1^\perp$ that is invariant under T, on which the restriction of T is irreducible. By construction we have $M_2 \subset M_1^\perp$ and therefore $M_2 \perp M_1$. It is clear that $M_1 + M_2$ is invariant under T. If $M_1 + M_2 = X$, we are through. If $M_1 + M_2 \neq X$, we apply the preceding argument to $(M_1 + M_2)^\perp$ instead of $M^\perp$. Since X is finite-dimensional, a finite number of such steps produces the equality $X = M_1 + \cdots + M_k$, where $M_1, \ldots, M_k$ are mutually orthogonal subspaces all invariant under T an on each of which the restriction of T is irreducible. $\square$

Note. In the preceding proof we established only the *existence* of a decomposition of the representation under consideration into irreducible part. The *actual determination* of such factorizations can present serious difficulties. It is one of the important problems of the theory of finite-dimensional representations.

2.9. The Character of a Finite-Dimensional Representation

Recall that the sum of the diagonal elements of a matrix $a = (a_{ij})$, $i, j = 1, \ldots, n$ is called its *trace* and is written tr a:

$$\text{tr } a = a_{11} + a_{22} + \cdots + a_{nn}. \qquad (2.9.1)$$

I. *The trace* tr *a has the following properties*:

(1) $\text{tr}(\alpha a) = \alpha \, \text{tr}(a)$,
(2) $\text{tr}(a + b) = \text{tr } a + \text{tr } b$,
(3) $\text{tr } 1 = n$,   (2.9.2)
(4) $\text{tr}(ab) = \text{tr}(ba)$,
(5) $\text{tr}(b^{-1}ab) = \text{tr } a$,

if b^{-1} *exists. Here* α *is a number, a, b are matrices of the same order and n is the order of the unit matrix 1.*

Proof. Properties (1)–(3) are obvious; property (4) follows from the equalities

$$\text{tr}(ab) = \sum_{j=1}^{n} (ab)_{jj} = \sum_{j,k=1}^{n} a_{jk}b_{kj}$$

$$= \sum_{j,k=1}^{n} a_{kj}b_{jk} = \sum_{j=1}^{n} (ba)_{jj} = \text{tr}(ba),$$ (2.9.3)

where n is the order of the matrices a and b. Property (5) follows from (4). Indeed we have $\text{tr}(b^{-1}ab) = \text{tr}(bb^{-1}a) = \text{tr } a$. $\square$

II. *The trace of a matrix is equal to the sum of all its eigenvalues, in which the eigenvalue* λ *is repeated k times if* λ *is a zero of multiplicity k in the characteristic equation of this matrix.*

Proof. We know (see, for example Gel'fand[1]) that we can represent the matrix a in the form $a = b^{-1}a_1b$, where a_1 is the so-called *Jordan canonical form* of the matrix a:

$$a_1 = \begin{Vmatrix} \lambda_1 & 1 & & & & & & \\ & \ddots & 1 & & & & & \\ & & \lambda_1 & \ddots & & & & \\ & & & \ddots & 1 & & & \\ & & & & \lambda_r & \ddots & & \\ & & & & & \ddots & 1 & \\ & & & & & & \lambda_r \end{Vmatrix},$$

All eigenvalues of the matrix a are found on the diagonal, each one repeated according to its multiplicity as defined above. All entries in a_1 are zeroes except for those labelled. To prove our assertion, use (5) and the form of a_1. $\square$

Let X be a finite-dimensional linear space, $n = \dim X$, and $e_1, \ldots, e_n$ a basis in X. Every linear operator A in X is determined by its matrix $a = (a_{jk})$ in this basis. The trace of the matrix of A is called the *trace* tr A *of the operator* A. This definition is independent of our choice of a basis. In fact, when there is a change from the basis $e_1, \ldots, e_n$ to another basis $f_1, \ldots, f_n$, the matrix of the operator A changes into the matrix $b^{-1}ab$, where b is the matrix of the operator carrying the basis $e_1, \ldots, e_n$ onto $f_1, \ldots, f_n$. By (5) we have $\operatorname{tr}(a) = \operatorname{tr}(b^{-1}ab)$.

III. *Let A be a linear operator in a finite-dimensional space X and M a subspace of X invariant under A. Let P be a projection (not necessarily orthogonal) of X onto M and let A_M be the restriction of the operator A to M. Then we have*

$$\operatorname{tr}(A_M) = \operatorname{tr}(AP). \tag{2.9.4}$$

Proof. We set

$$N = (1 - P)X. \tag{2.9.5}$$

Let $e_1, \ldots, e_m$ be a basis in M and $e_{m+1}, \ldots, e_n$ be a basis in N. Since P is a projection, $e_1, \ldots, e_n$ is a basis in X. From (2.9.5) and the definition of a projection, we have

$$APe_k = \begin{cases} Ae_k = A_M e_k, & \text{for } k = 1, \ldots, m \\ 0, & \text{for } k = m + 1, \ldots, n. \end{cases} \tag{2.9.6}$$

Therefore the matrix a of the operator AP in the basis $e_1, \ldots, e_n$ has the form

$$a = \left\| \begin{matrix} a_m & 0 \\ \cdot & 0 \end{matrix} \right\|, \tag{2.9.7}$$

where a_m is the matrix of the operator A_M in the basis $e_1, \ldots, e_m$. From this we get (2.9.4). $\square$

Let T be a finite-dimensional representation of a group G. The complex-valued function $g \to \operatorname{tr}(T(g))$ on G is called the *character of* T and is written $\chi_T(g)$. We have

$$\chi_T(g) = \operatorname{tr}(T(g)). \tag{2.9.8}$$

Where no ambiguity arises, we write $\chi(g)$ instead of $\chi_T(g)$.

By (2.9.1) and (2.9.2) we have

$$\chi_T(g) = t_{11}(g) + t_{22}(g) + \cdots + t_{nn}(g), \qquad (2.9.9)$$

where $t_{jk}(g)$ are matrix elements of a representation T in any basis and n is the dimension of the representation. If G is commutative and T is one-dimensional, then $\chi_T(g)$ is a character of G (see (2.2.5)).

IV. *Characters of a finite-dimensional representation T of a group G enjoy the following properties*:

(a) *characters of equivalent representations coincide*;
(b) *characters are constant on conjugacy classes in G*;
(c) *if T and S are conjugate, then $\chi_S(g) = \chi_T(g^{-1})$*
(d) *if T is unitary, then $\chi_T(g^{-1}) = \overline{\chi_T(g)}$*;
(e) *the character of a linkage (in particular of a direct sum) of a finite number of representations is equal to the sum of the characters of these representations*;
(f) *the character of the tensor product of a finite number of representations is equal to the product of the characters of these representations*.

Proof. Suppose that representations T and S in the spaces S and Y are equivalent. By I in 2.2 we can select bases in X and Y such that the matrix elements of T and S coincide in these bases. Property (a) follows from this and (2.9.9).[11]

If g_2 and g_1 are conjugate elements in G, then $g_2 = g^{-1}g_1g$ for some $g \in G$ (see (1.7.5)). From this and from property (5) we conclude that

$$\chi(g_2) = \chi(g^{-1}g_1g) = \operatorname{tr}(T(g^{-1}g_1g))$$
$$= \operatorname{tr}((T(g))^{-1}T(g_1)T(g)) = \operatorname{tr}(T(g_1)) = \chi(g_1).$$

If S and T are conjugate representations of G, then there are bases such that

$$s_{jk}(g) = \overline{t_{kj}(g^{-1})}$$

(see (2.3.13)). Thus we have

$$\chi_S(g) = s_{11}(g) + \cdots + s_{nn}(g) = \overline{t_{11}(g^{-1})} + \cdots + \overline{t_{nn}(g^{-1})} = \overline{\chi_T(g^{-1})}.$$

Property (d) now follows from (c). For, if T is unitary, it is conjugate with itself (see I in 2.8) and therefore $\chi_T(g) = \overline{\chi_T(g^{-1})}$. It follows that

$$\chi_T(g^{-1}) = \overline{\chi_T(g)}.$$

---

[11] We will see *infra* that the converse holds in a number of cases.

Now suppose that $T = T^1 \to T^2 \to \cdots \to T^m$. We can choose a basis in which the matrix $t(g)$ of the representation T has the form

$$
\begin{Vmatrix}
t^1_{11}(g) & \cdots & t^1_{1n_1}(g) & 0 & \cdots & 0 & 0 & \cdots & 0 \\
\hdotsfor{9} \\
t^1_{n_11}(g) & \cdots & t^1_{n_1n_1}(g) & 0 & & 0 & 0 & \cdots & 0 \\
* & \cdots & * & & \hdotsfor{5} \\
& & & & \vdots & & & & \\
* & \cdots & * & * & & * & t^m_{11}(g) & \cdots & t^m_{1n_m}(g) \\
* & \cdots & * & * & & \cdot & t^m_{n_m1}(g) & \cdots & t^m_{n_mn_m}(g)
\end{Vmatrix},
$$

where the t^k_{jv} are matrix elements of the representation T^k, $k = 1, \ldots, n$ (see (2.5.9)). We thus obtain

$$\chi_T(g) = t^1_{11}(g) + \cdots + t^1_{n_1n_1}(g) + t^2_{11}(g) + \cdots + t^2_{n_2n_2}(g) \cdots + t^m_{11}(g) + \cdots + t^m_{n_mn_m}(g)$$
$$= \chi_{T^1}(g) + \chi_{T^2}(g) + \cdots + \chi_{T^m}(g). \tag{2.9.10}$$

In particular (2.9.10) holds if $T = T^1 \dotplus \cdots \dotplus T^m$. Finally, let $T = T^1 \otimes T^2$, let X_1, X_2 be the spaces of the representations T^1, T^2 and let $e^1_1, \cdots, e^1_{n_1}$, $e^2_1, \cdots, e^2_{n_2}$ be bases in X_1, X_2. Then we have

$$t_{\mu v jk}(g) = t^1_{\mu j}(g) t^2_{vk}(g)$$

as matrix elements of the representation T in the basis $e_{jk} = e^1_j \otimes e^2_k$ (see (2.6.4)). The diagonal elements of this matrix are

$$t_{jkjk}(g) = t^1_{jj}(g) t^2_{kk}(g)$$

and so

$$\chi_{T^1 \otimes T^2}(g) = \sum_{j=1}^{n_1} \sum_{k=1}^{n_2} t^1_{jj}(g) t^2_{kk}(g) = \sum_{j=1}^{n_1} t^1_{jj}(g) \sum_{k=1}^{n_2} t^2_{kk}(g) = \chi_{T^1}(g) \chi_{T^2}(g).$$

Dealing similarly with formula (2.6.6), we conclude that

$$\chi_{T^1 \otimes \cdots \otimes T^m}(g) = \chi_{T^1}(g) \cdots \chi_{T^m}(g). \tag{2.9.11}$$

$\square$

Examples and Exercises

   1. Find the irreducible representations of the groups S_3, S_4, P_3, P_4 and compare the characters of these representations.

   2. Find inequivalent representations of a certain infinite group that have the same characters.

2.10. Intertwining Operators

Let T, S be representations of a group G in spaces X and Y. A linear operator A carrying X into Y is intertwining for T and S if

$$AT(g) = S(g)A, \quad \text{for all } g \in G. \tag{2.10.1}$$

If $X = Y$ and $T = S$, formula (2.10.1) assumes the form

$$AT(g) = T(g)A, \quad \text{for all } g \in G. \tag{2.10.2}$$

In this case, we say that A commutes with the representation T (cf. 2.2.3).

Intertwining operators generalize equivalent representations. Indeed, if A is a linear isomorphism of X onto Y and intertwines S and T, the representations T and S are equivalent (see 2.2).

The set of all linear operators A intertwining T and S is denoted by $\text{Hom}(T, S)$ or $\text{Hom}_G(T, S)$.

I. *The set $\text{Hom}(T, S)$ is a linear subspace of the space of linear operators carrying X into Y.*

In fact, if A_1, A_2 belong to $\text{Hom}(T, S)$ then we also have $(A_1 + A_2)T(g) = A_1 T(g) + A_2 T(g) = S(g)A_1 + S(g)A_2 = S(g)(A_1 + A_2)$ for all $g \in G$. That is, $A_1 + A_2$ is in $\text{Hom}(T, S)$. If $A \in \text{Hom}(T, S)$ and λ is a number, we have

$$(\lambda A)T(g) = \lambda(AT(g)) = \lambda(S(g)A) = S(g)(\lambda A), \quad \text{for all } g \in G.$$

Therefore, λA belongs to $\text{Hom}(T, S)$.

The dimension of the linear space $\text{Hom}(T, S)$ is called the *intertwining number* of the representations T and S.

II. *For a single representation T, the set $\text{Hom}(T, T)$ is a subalgebra (see 2.1, chapter II) of the algebra $L(X)$ of all linear operators on X.*

As already proved, $\text{Hom}(T, T)$ is a linear space in $L(X)$. For A, B in $\text{Hom}(T, T)$ we have

$$ABT(g) = AT(g)B = T(g)AB, \quad \text{for all } g \in G.$$

Therefore AB belongs to $\text{Hom}(T, T)$.

III. *Let $A \in \text{Hom}(T, S)$. Let L be the kernel of the operator $A: L = \{x : x \in X, Ax = 0\}$. Let M be the image of $A: M = \{y : y = Ax \text{ for some } x \in X\}$. Then L is invariant under T and M is invariant under S.*

If $x \in L$, that is, $Ax = 0$, then we have $AT(g)x = S(g)Ax = S(g)0 = 0$ for all $g \in G$, which is to say that $T(g)x \in L$ for all $g \in G$. If $y \in M$, that is, $y = Ax$ for some $x \in X$, then $S(g)y = S(g)Ax = A(T(g)x)$ so that $S(g)y \in M$ for all $g \in G$. $\square$

IV. *Let A, L, M be as in* III. *The representation $\tilde{T}$, generated by the representation T in the factor space X/L, is equivalent to the restriction $S|_M$ of the representation S to M.*

Proof. From the operator A we obtain the one-to-one operator $\tilde{A}$, carrying X/L onto M by the formula $A(x + L) = Ax$, $x \in X$. Thus, if $\tilde{T}$ is the representation generated by T in the factor space X/L, then $\tilde{A}\tilde{T}(g)(x + L) = \tilde{A}(T(g)x + L) = AT(g)x = S(g)Ax = S(g)\tilde{A}(x + L)$ for all $x \in X$ and $g \in G$. But $\tilde{A}(x + L)$ belongs to M, and therefore $S(g)\tilde{A}(x + L) = S(g)|_M\tilde{A}(x + L)$, where $S(g)|_M$ is the restriction of the operator $S(g)$ to M. Therefore we have $\tilde{A}\tilde{T}(g) = S(g)|_M\tilde{A}$ for all $g \in G$ and so $\tilde{T} \sim S|_M$. $\square$

V. *Let T be an irreducible finite-dimensional representation of G in the linear space X and let S be a finite-dimensional representation of G in the space Y, where S is a multiple of an irreducible representation $T_1: S = nT_1$. Then we have*

$$\dim \operatorname{Hom}(T, S) = \begin{cases} 0, & \text{if } T \nsim T_1, \\ n, & \text{if } T \sim T_1. \end{cases} \qquad (2.10.3)$$

Proof. Let $Y = Y_1 \dotplus \cdots \dotplus Y_n$, where every subspace $Y_i \subset Y$ is invariant under S and the restriction of S to subspace Y_i is equivalent to T_1. Let S_i denote this restriction. Let P_k be the linear operator in Y that assigns the element $y_i \in Y_i$ to every element $y = y_1 + \cdots + y_n$ ($y_i \in Y_i$, $i = 1, \ldots, n$). That is, we define $P_i(y) = y_i$. By the definition of a direct sum of representations, P_i commutes with the representation S, that is, $P_iS(g) = S(g)P_i$ for all $g \in G$. It follows that if $A \in \operatorname{Hom}(T, S)$, then

$$P_iAT(g) = P_i(S(g)A) = S(g)P_iA \quad \text{for all } g \in G. \qquad (2.10.4)$$

Hence the operator P_iA intertwines T and S. Since $P_iA(X) \subset Y_i$, the operator P_iA (considered a linear operator carrying X into $Y_i = P_iY$) intertwines T and S_i. But S_i is equivalent to T_1 and therefore $P_iA = 0$ when $T \nsim T_1$ (to see this use Schur's lemma). Then we have $Ax = \sum_{i=1}^{n} P_iAx = 0$ for all x X, and so $A = 0$.

Suppose now that $T \sim T_1$. Then S_i is equivalent to T. Let B_i be a linear isomorphism of S onto Y_i which effects the equivalence of T and S_i:

$$B_iT(g) = S_i(g)B_i, \quad \text{for all } g \in G. \qquad (2.10.5)$$

Since B_i is an isomorphism of X onto Y_i (see the definition of equivalence) the operator B_i^{-1} exists. Multiplying both sides of (2.10.5) on the left and right by B_i^{-1}, we obtain

$$T(g)B_i^{-1} = B_i^{-1}S_i(g), \quad \text{for all } g \in G. \tag{2.10.6}$$

From (2.10.4) and (2.10.6) we conclude that

$$P_iAB_i^{-1}S_i(g) = P_iAT(g)B_i^{-1} = S_i(g)P_iAB_i^{-1}.$$

Thus the operator $P_iAB_i^{-1}$ commutes with the irreducible finite-dimensional representation S_i of G. Lemma 2 in 2.2 shows that

$$P_iAB_i^{-1} = \lambda_i 1, \quad \text{where } \lambda \text{ is a number.}$$

Hence we have $P_iA = \lambda B_i$, where all B_i are considered as operators carrying X into Y. Since $y = P_i y + \quad + P_n y$ for all $y \in Y$ (by the definition of P_i) we have

$$A = \sum_{i=1}^{n} P_iA = \sum_{i=1}^{n} \lambda_i B_i.$$

Thus every operator A that intertwines T and S is a linear combination of the B_i's. Conversely, every operator $A = \sum_{i=1}^{n} \lambda_i B_i$ intertwines T and S. In fact, by I it suffices to show that B_i intertwines T and S. The equalities (2.10.4) are equivalent to the equalities $B_i T(g) = S(g)B_i$ for all $g \in G$, since the image of X under the operator B_i is contained in Y_i. Thus we have $A \in \text{Hom}(T, S)$ if and only if $A = \sum_{i=1}^{n} \lambda_i B_i$. Since every operator B_i is an isomorphism of X onto Y_i, we find that $\sum_{i=1}^{n} \lambda_i B_i = 0$ if and only if $\lambda_i = 0$ for all $i = 1, \ldots, n$. Therefore the correspondence $A = \sum_{i=1}^{n} \lambda_i B_i \leftrightarrow (\lambda_1, \ldots, \lambda_n)$ is an isomorphism of the linear spaces $\text{Hom}(T, S)$ and C^n. This proves that dim $\text{Hom}(T, S) = n$. $\square$

VI. *Let T be an irreducible finite-dimensional representation of the group G in the linear space X. Let S be a finite-dimensional representation of G which is a direct sum: $S = n_1 T^1 \dotplus \cdots \dotplus n_p T^p$, where $T^1, \ldots, T^p$ are pairwise inequivalent irreducible representations of G. Then we have*

$$\dim \text{Hom}(T, S) = \begin{cases} n_k, & \text{if } T \text{ is equivalent to } T^k, \\ 0, & \text{if } T \text{ is not equivalent to } T^1, \ldots, T^p. \end{cases} \tag{2.10.7}$$

That is, the dimension of the space $\text{Hom}(T, S)$ of intertwining operators is equal to the multiplicity of the occurrence in S of the irreducible representation T.

Proof. Let Y be the space of the representation S and let $Y = Y_1 + \cdots + Y_p$, where Y_k is the space of the representation $n_k T^k$. Let P_k be the linear operator on Y, defined by $P_k(y) = P_k(y_1 + \cdots + y_p) = y_k$, where each $y_i \in Y_i$. It is clear that $P_1 + \cdots + P_p = 1$. Just as in the proof of V we see that if A is in $\mathrm{Hom}(T, S)$, then $P_k A \in \mathrm{Hom}(T, S)$ and if S^k is the restriction of S to $Y_{k,,}$, then $P_k A \in \mathrm{Hom}(T, S^k)$. Thus we have

$$A = P_1 A + \cdots + P_p A, \qquad (2.10.8)$$

where $P_k A \in \mathrm{Hom}(T, S^k)$. We verify easily that if A is a linear operator from X into Y such that $P_k A \in \mathrm{Hom}(T, S^k)$ then $A \in \mathrm{Hom}(T, S)$. From V we have

$$\dim \mathrm{Hom}(T, S^k) = 0, \quad \text{if } T \nsim T^k,$$
$$\dim \mathrm{Hom}(T, S^k) = n_k, \quad \text{if } T \sim T^k.$$

Therefore, if $T \nsim T^1, \ldots, T \nsim T^p$, then for all $A \in \mathrm{Hom}(T, S)$ we have $P_k A = 0$, and consequently $A = 0$. That is, $\mathrm{Hom}(T, S) = \{0\}$. On the other hand, if $T \sim T^k$, then $T \nsim T^j$, where $j \neq k$. We accordingly have $P_j A = 0$ for $j \neq k$ and all $A \in \mathrm{Hom}(T, S)$. The equality (2.10.8) shows that $A = P^k A$. Thus $\mathrm{Hom}(T, S)$ is isomorphic to $\mathrm{Hom}(T, S^k)$, which has (by V) dimension n_k. $\square$

Examples and Exercises

Let T, S be irreducible finite-dimensional representations of G. The following conditions are equivalent: (a) $\mathrm{Hom}(T, S) \neq \{0\}$; (b) $\dim \mathrm{Hom}(T, S) = 1$; (c) $T \sim S$.

2. If T is a finite-dimensional representation of G, then for some irreducible representations S, S_1 of G, the relations $\mathrm{Hom}(T, S) \neq \{0\}$, $\mathrm{Hom}(S_1, T) \neq \{0\}$ hold.

3. Let $T = m_1 T^1 + \cdots + m_p T^p$, $S = n_1 T^1 + \cdots + n_p T^p$ be decompositions of finite-dimensional representations T and S into direct sums of irreducible representations of G. Then we have

$$\dim \mathrm{Hom}(T, S) = m_1 n_1 + \cdots + m_p n_p.$$

4. Let X, X' be linear spaces, let Y, Y' be linear spaces in duality with X, X' with respect to bilinear forms (x, y) and $(x', y')'$. Let T, T' be representations of G in X, X' respectively. Let S, S' be representations conjugate to T, T' respectively.

(a) Let A be a linear operator carrying X into X'. Let B be a linear carrying Y' into Y such that $(Ax, y')' = (x, By')$ for all $x \in X$, $y' \in Y$. Under these conditions the operator A intertwines T and T' if and only if the operator B intertwines S' and S.

(b) If X and X' are finite-dimensional, then the linear spaces

$$\text{Hom}(T, T') \quad \text{and} \quad \text{Hom}(S', S)$$

are isomorphic.

5. Let T, S be finite-dimensional representations of the group G. Generally speaking, we have dim $\text{Hom}(T, S) \neq$ dim $\text{Hom}(S, T)$. [Hint. Consider linkages of irreducible representations for either T or S.]

Chapter II

Representations of Finite Groups[12]

§1. Basic Propositions of the Theory of Representations of Finite Groups

1.1. The Invariant Mean on a Finite Group

Let G be a group of order N. Let $g_1 = e, g_2, \ldots, g_N$ be all its elements, each listed exactly once. We will consider numerical functions $f(g)$ on G, that is, sequences of N numbers $f(g_1), \ldots, f(g_N)$. For such a function f and an element h of G, let f_h be the function on G such that

$$f_h(g) = f(gh), \quad \text{for all } g \in G. \tag{1.1.1a}$$

The function f_h is called *the right translation of the function f by h*. Similarly, the function f^h, defined by the formula

$$f^h(g) = f(hg) \tag{1.1.1b}$$

is called *the left translation of the function f by h*.

A principal method for the study of representations of a finite group G is the use of an invariant mean on G. The arithmetic mean of the values of the numerical function f on the group G of the order N is called the *invariant mean* of the function f on G. It is denoted by $M(f)$, so that we have

$$M(f) = \frac{1}{N} \sum_{k=1}^{n} f(g_k), \tag{1.1.2a}$$

or, in abbreviated form

$$M(f) = \frac{1}{N} \sum_{g} f(g). \tag{1.1.2b}$$

We sometimes write $M(f(g))$ instead of $M(f)$.

[12] Throughout this chapter we deal only with finite groups and their finite-dimensional representations, although we do not always state this explicitly.

I. *The invariant mean $M(f)$ has the following properties:*

(1) $M(1) = 1$, *where* 1 *on the left is the function* $f \equiv 1$ *on* G;
(2) $M(\bar{f}) = \overline{M(f)}$;
(3) $M(f) \geqslant 0$ *if* $f \geqslant 0$ *and* $M(f) > 0$ *if* $f \geqslant 0$ *and* $f \neq 0$;
(4) $M(f_1 + f_2) = M(f_1) + M(f_2)$;
(5) $M(\alpha f) = \alpha M(f)$, *where* α *is a number*;
(6) $M(f_h) = M(f)$ *and* $M(f^h) = M(f)$;
(7) $M(f(g^{-1})) = M(f(g))$.

Proof. Properties (1)–(5) follow at once from the definition (1.1.2). To prove property (6) it suffices to note that for each $h \in G$, the set $g_1 h, g_2 h, \ldots, g_N h$ coincides with G. That is, $g_1 h, g_2 h, \ldots, g_N h$ are exactly the elements $g_1, \ldots, g_N$, perhaps written in a different order. Therefore we have

$$M(f_h) = \frac{1}{N} \sum_{k=1}^{N} f_h(g_k) = \frac{1}{N} \sum_{k=1}^{N} f(g_k h) = \frac{1}{N} \sum_{k=1}^{N} f(g_k) = M(f).$$

We show similarly that $M(f^h) = M(f)$. By virtue of this property of invariance of $M(f)$ properties (1), (3), (4) and (5) assert that M is a *mean value* (or *mean*) and property (6) that M is an *invariant mean*. The invariant means for vector- and operator-valued functions on a finite group are defined analogously. Let $x = x(g)$ be a vector function on the group $G = \{g_1, g_2, \ldots, g_N\}$ with values in a fixed linear space X. If X is finite-dimensional, then $(x)_j$ denotes the j-th coordinate of the vector x relative to some fixed basis in X. The vector

$$M(x) = \frac{1}{N} \sum_{k=1}^{N} x(g_k). \tag{1.1.3}$$

is called the *invariant mean* of the vector function x on the group G. Sometimes we write $M(x(g))$ instead of $M(x)$.   □

II. *The invariant mean $M(x)$ has the following properties:*

(a) $M(c) = c$, *where* c *on the left is the function* $x(g) \equiv c$ *and* c *on the right is simply the vector* $c \in X$;
(b) $M(x_1 + x_2) = M(x_1) + M(x_2)$, *where* $x_1(g), x_2(g) \in X$;
(c) $M(\alpha x) = \alpha M(x)$, *where* α *is a number*;
(d) $M(Ax) = AM(x)$, *where* A *is a linear operator carrying* X *into* X;
(e) $M(x_h) = M(x)$ *and* $M(x^h) = M(x)$, *where* $x_h(g) = x(gh)$ *and* $x^h(g) = x(hg)$;
(f') $\operatorname{tr}(M(A)) = M(\operatorname{tr}(A))$, *if* $A(g) \in L(X)$ *and* X *is finite-dimensional; taken with respect to the same basis in* X.

All these properties follow from the definition (1.1.3). Property (e) is proved in the same way as property (6) in I; property (d) follows from the equalities

$$M(Ax) = \frac{1}{N} \sum_{k=1}^{N} Ax(g_k) = A\left(\frac{1}{N} \sum_{k=1}^{N} x(g_k)\right) = AM(x).$$

We will now consider the invariant mean of an operator function. In the future, X, Y, Z will denote linear spaces, $L(X)$ the set of all linear operators carrying X into X, and $L(X, Y)$ the set of all linear operators carrying X into Y. If X and Y are finite-dimensional, $e_1, \ldots, e_n$ and $f_1, \ldots, f_m$ are bases in X and Y and $A \in L(X, Y)$, then $(A)_{jk}$ will denote the matrix elements of the operator A with respect to these bases. Let $A = A(g)$ be an operator function on the group $G = \{g_1, g_2, \ldots, g_N\}$ with values lying in $L(X, Y)$. The operator

$$M(A) = \frac{1}{N} \sum_{k=1}^{N} A(g_k) \tag{1.1.4}$$

is called the *invariant mean of the operator function* A on the group G. We sometimes write $M(A(g))$ instead of $M(A)$.

III. *The invariant mean* $M(A)$ *of the operator function* $A = A(g)$, $A(g) \in L(X)$ *has the following properties*:

(a') $M(A) = A$, *if* $A \in L(X, Y)$ *does not depend on* g;
(b') $M(A_1 + A_2) = M(A_1) + M(A_2)$, *where* $A_1(g), A_2(g) \in L(X, Y)$;
(c') $M(\alpha A) = \alpha M(A)$, *where* α *is a number*;
(d') $M(BA) = BM(A)$, $M(AC) = M(A)C$, *if* $B \in L(Y, Z)$, $C \in L(Z, X)$ *and* B *and* C *do not depend on* g, $A(g) \in L(X, Y)$;
(e') $M(A_h) = M(A)$ *and* $M(A^h) = M(A)$, *where* $A_h(g) = A(gh)$ *and* $A^h(g) = A(hg)$;
(f') $\text{tr}(M(A)) = M(\text{tr}(A))$, *if* $A(g) \in L(X)$ *and* X *is finite-dimensional*;
(g') $(M(A))_{jk} = M((A)_{jk})$, *where the matrix elements on the left and right are computed in the same bases in* X *and* Y.

Proof. Properties (a')–(d') follow directly from (1.1.4) and the linearity of the operators B and $A(g)$. Properties (e') and (g') can be proved in the same way as (6) in I and (f) in II. Finally, property (g') is obtained from the equalities

$$\text{tr}(M(A)) = \text{tr}\left(\frac{1}{N} \sum_{k=1}^{N} A(g_k)\right) = \frac{1}{N} \sum_{i=1}^{N} \text{tr}(A(g_k)) = M(\text{tr } A)$$

(see (1) and (2) in 2.9, chapter I). $\square$

1.2. Complete Reducibility of the Representations of a Finite Group[13]

Theorem 1. *Every representation of a finite group is equivalent to a unitary representation.*

Proof. Let T be a representation of a finite group in a finitedimensional space X and let $(x, y)_1$ be any scalar product in X, for example, $(x, y)_1 =$

---

[13] Recall that in this chapter we consider only finite-dimensional representations.

$\sum_{j=1}^{n} \xi_j \bar{\eta}_j$, $n = \dim X$, where ξ_j, η_j are coordinates of the vectors $x, y \in X$ in a fixed but arbitrary basis in X. We define the form (x, y) in X, setting

$$f(g) = (T(g)x, T(g)y)_1, \tag{1.2.1}$$
$$(x, y) = M(f) = M((T(g)x, T(g)y)_1). \tag{1.2.2}$$

The form (x, y) is a scalar product in X. By properties (1), (4), (5) of the mean $M(f)$, the linearity of the operator $T(g)$ and the bilinearity of the form $(x, y)_1$, (x, y) is bilinear. Since $(x, y)_1$ is Hermitian, (2) I in 1.1 shows that

$$(y, x) = M((T(g)y, T(g)x)_1) = \overline{M((T(g)x, T(g)y)_1)} = \overline{(x, y)}.$$

That is, (x, y) is also Hermitian. Finally, we have $(T(g)x, T(g)x_1 \geqslant 0$, since $(x, y)_1$ is a scalar product. From this and (3), I in 1.1 we infer that

$$(x, x) = M((T(g)x, T(g)x)_1) \geqslant 0.$$

Equality holds if and only if $(T(g)x, T(g)x)_1 = 0$ for all $g \in G$. Setting $g = e$, we obtain $(x, x)_1 = 0$; consequently (x, x) vanishes if and only if $x = 0$. That is, (x, y) is a scalar product on X. To complete the proof use (6), I in 1.1, (1.2.1) and (1.2.2) to show

$$(T(h)x, T(h)y) = M(T(g)T(h)x, T(g)T(h)y) = M(T(gh)x, T(gh)y)$$
$$= M(f_h) = M(f) = (x, y)$$

for all $h \in G$. Thus T is unitary with respect to (x, y). $\square$

Having noted that every finite-dimensional representation equivalent to a unitary representation is completely reducible, (see V in 2.8, chapter I) we conclude the following.

Theorem 2. *Every representation of a finite group is completely reducible.*

The fundamental problems of the theory of representations of finite groups are:

(1) finding all irreducible finite-dimensional representations of a given finite group;
(2) decomposing a given finite-dimensional representation of a finite group into its irreducible components.

We will see in 1.8 *infra* that we can solve the second problem if we have solved the first problem. The solution for the first problem is known only for a small number of finite groups.[14]

---

[14] Editor's note. See for example the interesting survey by C. W. Curtis [1*].

1.3. The Space $L^2(G)$; the Regular Representation

We will use $L^2(G)$ to denote the set of all numerical functions $f(g)$ on the finite group $G = \{g_1, \ldots, g_N\}$. Each $f(g)$ is simply a sequence of N complex numbers $f(g_1), f(g_2), \ldots, f(g_N)$. The operations of addition and multiplication by a complex number are defined coordinatewise:

$$(f + g)(h) = f(h) + g(h) \text{ and } (\alpha g)(h) = \alpha(g(h)).$$

Thus $L^2(G)$ is an N-dimensional linear space. Finally, we define the bilinear form (f_1, f_2) in $L^2(G)$ by the formula

$$(f_1, f_2) = M(f_1, \bar{f}_2) = \frac{1}{N} \sum_{k=1}^{N} f_1(g_k)\overline{f_2(g_k)}. \tag{1.3.1}$$

It is easy to see that the form (f_1, f_2) is Hermitian and positive-definite, and so is a scalar product in $L^2(G)$. Thus $L^2(G)$ is an N-dimensional Euclidean space with the linear operations and scalar product just defined.

We now define the operators $T(h)$ for all $h \in G$:

$$T(h)f = f_h, \quad \text{that is,} \quad T(h)f(g) = f(gh). \tag{1.3.2}$$

Plainly $T(h)$ is linear. Furthermore, the correspondence $h \to T(h)$ is a representation of the group G. It is denoted by T. For $h_1, h_2 \in G$ we have

$$\begin{aligned}
T(h_1)T(h_2)f(g) &= T(h_1)(T(h_2)f(g)) = T(h_1)(f(gh_2)) \\
&= f((gh_1)h_2) = f(g(h_1h_2)) = T(h_1h_2)f(g),
\end{aligned}$$

and clearly $T(e)f(g) = f(g)$. The representation T is unitary. Setting $f = f_1\bar{f}_2$, we apply (6), I in 1.1 to show that

$$(T(h)f_1, T(h)f_2) = M(f_{1h}\bar{f}_{2h}) = M(f_h) = M(f) = M(f_1\bar{f}_2) = (f_1, f_2).$$

The unitary representation T in $L^2(G)$ is called *the right regular representation of the group G. The left regular representation* $\tilde{T}$ of G in $L^2(G)$ is defined similarly by the formula

$$\tilde{T}_h f(g) = f(h^{-1}g). \tag{1.3.3}$$

The representation $\tilde{T}$ is also unitary, since the mean $M(f)$ is left invariant. (See (1.1.1b) and (6), I in 1.1.)

I. *The left and right regular representations are unitarily equivalent.*

Proof. For $f \in L^2(G)$, let f' be the function defined by

$$f(g) = f(g^{-1}). \tag{1.3.4}$$

The operator $W: f \to f'$ is clearly linear. It follows from (1.3.4) that

$$W^2 = 1$$

and therefore W maps $L^2(G)$ onto $L^2(G)$. By (7), I in 1.1 for $f, f_1 \in L^2(G)$ we have

$$(Wf, Wf_1) = (f', f_1') = M(f(g^{-1})\overline{f_1(g^{-1})}) = M(f(g)\overline{f_1(g)}) = (f, f_1).$$

Therefore W is unitary. Finally we have

$$(WT(h))(f)(g) = T(h)f(g^{-1}) = f(g^{-1}h), (\tilde{T}(h)W)(f)(g) = Wf(h^{-1}g)$$
$$= f((h^{-1}g)^{-1}) = f(g^{-1}h^{-1-1}) = f(g^{-1}h).$$

That is, we have $WT(h) = \tilde{T}(h)W$, which is to say that W transforms $T(h)$ into $\tilde{T}(h)$. $\square$

1.4. Orthogonality Relations

In accordance with theorem 1 in 1.2, each representation T of a finite group G is equivalent to a unitary representation. Thus we may and shall suppose that X is equipped with a scalar product (x, y) under which T is unitary. We will also write $t_{jk}(g)$ for the matrix elements of the representation T with respect to some fixed orthonormal basis for X. Thus the matrix $t(g)$ of the representation T is unitary (see III in 2.8, chapter I).

Theorem 1. *Let* T^1 *and* T^2 *be irreducible representations of the finite group* G *and let* $t_{jk}^1(g)$, $t_{jk}^2(g)$ *be their matrix elements. Then we have*

$$(t_{jk}^1(g), t_{pq}^2(q)) = 0 \quad \text{if } T^1 \text{ and } T^2 \text{ are inequivalent,} \tag{1.4.1}$$

$$(t_{jk}^1(g), t_{pq}^1(g)) = \begin{cases} 0, & \text{if } j \neq p \text{ or } k \neq q, \\ 1/n_1, & \text{if } j = p \text{ and } k = q. \end{cases} \tag{1.4.2}$$

Here n_1 *is the dimension of the representation* T^1 *and* $(\,,\,)$ *is the scalar product in* $L^2(G)$.

Proof. Let X and Y be the spaces of the representations T^1 and T^2 and let $B \in L(Y, X)$. We define $A(g) = T^1(g)BT^2(g^{-1})$ and

$$C = M(A(g)) = M(T^1(g)BT^2(g^{-1})). \tag{1.4.3}$$

It is clear that C also belongs to $L(Y, X)$. We also have

$$T^1(h)C = CT^2(h). \tag{1.4.4}$$

In fact, properties (d′) and (f′) in III, 1.1 show that

$$
\begin{aligned}
T^1(h)C &= T^1(h)M(A(g)) = M(T^1(h)A(g)T^2(h^{-1})T^2(h)) \\
&= M(T^1(h)T^1(g)BT^2(g^{-1})T^2(h^{-1}))T^2(h) \\
&= M(T^1(hg)BT^2(hg)^{-1})T^2(h) = M((A(hg)))T^2(h) \\
&= M(A(g))T^2(h) = CT^2(h).
\end{aligned}
$$

Suppose that T^1 and T^2 are inequivalent. Applying Schur's lemma to (1.4.4) (see 2.2 in chapter I), we conclude that $C = 0$. That is, we have

$$
M(T^1(g)BT^2(g^{-1})) = 0, \quad \text{for all } B \in L(Y, X). \tag{1.4.5}
$$

Write (1.4.5) in terms of matrix elements. Using property (g′) in III, 1.1, we conclude that

$$
M\left(\sum_{\mu=1}^{n_1}\sum_{\nu=1}^{n_2} t^1_{j\mu}(g)b_{\mu\nu}t^2_{\nu p}(g^{-1})\right) = 0, \tag{1.4.6}
$$

$$
j = 1, \dots, n_1; \qquad p = 1, \dots, n_2
$$

for any $b_{\mu\nu}$. We select

$$
b_{\mu\nu} = \begin{cases} 1, & \text{for } \mu = k \text{ and } \nu = q, \\ 0, & \text{in all other cases.} \end{cases}
$$

Then (1.4.6) becomes the equality

$$
M(t^1_{jk}(g)t^2_{qp}(g^{-1})) = 0. \tag{1.4.7}
$$

But we have $t^2_{qp}(g^{-1}) = \overline{t^2_{pq}(g)}$ (see (2.8.7) chapter I) so that (1.4.7) is the equality

$$
M(t^1_{jk}(g)\overline{t^2_{pq}(g)}) = 0.
$$

This relation just (1.4.1), as the definition of the scalar product in $L^2(G)$ shows. The relation (1.4,1) clearly holds for $T^2 = T^1$, $Y = X$, in the form $T^1(h)C = CT^1(h)$, that is, C commutes with all operators $T^1(g)$. Schur's lemma again shows that $C = \lambda 1$, where λ is a certain number.

Let us find λ. Recall the properties (g′) and (a′) of III in 1.1 and property (5) of I in 2.9, chapter I. These yield

$$
\begin{aligned}
\operatorname{tr} C &= \operatorname{tr} M(T^1(g)BT^1(g^{-1})) = \operatorname{tr} M(T^1(g)B(T^1(g))^{-1} \\
&= M(\operatorname{tr}(T^1(g)B(T^1(g))^{-1})) = M(\operatorname{tr} B) = \operatorname{tr} B;
\end{aligned}
$$

On the other hand, we have $\operatorname{tr} C = \operatorname{tr} \lambda 1 = \lambda n_1$, where $n_1 = \dim X_1$, and therefore $\lambda = 1/n_1 \operatorname{tr} B$. We infer that

$$
C = \left(\frac{1}{n_1}\operatorname{tr} B\right)\cdot 1. \tag{1.4.8}
$$

Combining (1.4.8) with (1.4.3) for $T^2 = T^1$ we conclude

$$M(T^1(g)BT^1(g^{-1})) = \left(\frac{1}{n_1} \operatorname{tr} B\right) \cdot 1 \qquad (1.4.9)$$

for all $B \in L(X)$. Writing (1.4.9) in terms of matrix elements, we obtain

$$M\left(\sum_{j=1}^{n_1} \sum_{v=1}^{n_1} t^1_{j\mu}(g)b_{\mu v}t^1_{vp}(g^{-1})\right) = \begin{cases} \dfrac{1}{n_1} \displaystyle\sum_{\mu=1}^{n_1} b_{\mu\mu}, & \text{for } j = p, \\ 0, & \text{for } j \neq p, \end{cases} \qquad (1.4.10)$$

for all choices of the numbers $b_{\mu v}$. We select

$$b_{\mu v} = \begin{cases} 1, & \text{for } \mu = k, \text{ and } v = q, \\ 0, & \text{in all other cases.} \end{cases}$$

Then (1.4.10) takes the form

$$M(t^1_{jk}(g)t^1_{qp}(g^{-1})) = \begin{cases} 1/n_1, & \text{for } j = p \text{ and } k = q, \\ 0, & \text{for } j \neq p \text{ or } k \neq q. \end{cases} \qquad (1.4.11)$$

Since $t^1_{qp}(g^{-1}) = \overline{t^1_{pq}(g)}$, the relation (1.4.11) is (1.4.2). $\quad\square$

Formulas (1.4.1) and (1.4.2) are called the *orthogonality relations* for irreducible unitary representations of a finite group.

Let $T^1, T^2, \ldots, T^m$ be pairwise inequivalent irreducible representations of a group G of order N and let $n_1, \ldots, n_m$ be their dimensions. By (1.4.1) and (1.4.2) the matrix elements $t^k_{jv}(g), j, v = 1, \ldots, n_k; k = 1, \ldots, m$, of these representations form an orthogonal set in $L^2(G)$ and are therefore linearly independent. Consequently there cannot be more than N of these functions. The inequality $m \leqslant N$ holds a *fortiori*. In other words, the following occurs.

Corollary 1. *The number of pairwise inequivalent irreducible representations of a finite group is finite and does not exceed the order of the group.*

A set $T^1, T^2, \ldots, T^m$ of representations of the group G is called *a complete set of irreducible representations of G* if:

(a) the representations $T^1, T^2, \ldots, T^m$ are irreducible and are pairwise inequivalent;
(b) every irreducible representation of G is equivalent to one of the representations $T^1, T^2, \ldots, T^m$.

Corollary 1 yields the following.

Theorem 2. *If $T^1, T^2, \ldots, T^m$ is a complete set of irreducible representations of the finite group G, then the matrix elements $t^k_{jv}(g)$ of all these representations form a complete orthogonal set of functions in $L^2(G)$.*

Proof. For orthogonality, see theorem 1 *supra*. We will prove completeness. Consider the right regular representation T of G. As in (1.3.2) its operators are the operators of right translation

$$T(h)f(g) = f(gh) \tag{1.4.12}$$

(see (1.3.2)). By theorem 2 in 1.2 this representation is completely reducible. Therefore we have

$$L^2(G) = X_1 + X_2 + \cdots + X_p \tag{1.4.13}$$

where X_k, $k = 1, \ldots, p$ are subspaces invariant under T, for which the restriction $\tilde{T}^k$ of T to each X_k is irreducible. Thus it is equivalent to one of the representations $T^1, T^2, \ldots, T^m$, since by hypothesis these representations are a complete set. Let $\tilde{T}^k$ be equivalent to the representation $T^l, l = l(k)$. We choose an orthonormal basis $f_1(g), \ldots, f_{n_k}(g)$ in X_k such that the matrix elements of $\tilde{T}^k$ in this basis coincide with $t^l_{jv}(g)$ (see VII in 2.8, chapter I). By (1.4.12) and (2.1.3) in chapter I we find that

$$f_v(gh) = T(h)f_v(g) = \tilde{T}^k(h)f_v(g) = \sum_{j=1}^{n_l} t^l_{jv}(h)f_j(g).$$

Setting $g = e$ and $f_j(e) = c_j$ we obtain

$$f_v(h) = \sum_{j=1}^{n_l} c_j t^l_{jv}(h), \quad \text{for all } h \in G.$$

This means that the functions f_j of our basis in X_k, and hence each function f in X_k, is a linear combination of the functions $t^l_{jv}(h)$. From (1.4.13) we conclude that every function in $L^2(G)$ is a linear combination of the functions t^l_{jv}, $v = 1, \ldots, n_l$, $l = 1, \ldots, p$. Thus the functions t^l_{jv} are a basis in $L^2(G)$. □

Remark. The functions

$$e^k_{jv}(g) = \sqrt{n_k} t^k_{jv}(g) \tag{1.4.14}$$

form an ortho*normal* basis in $L^2(G)$: see (1.4.1), (1.4.2) and theorem 2.

Theorem 3 (Burnside's Theorem). *The order N of a group is equal to the sum of the squares of the dimensions of any complete set of irreducible representations of the group*:

$$N = n_1^2 + n_2^2 + \cdots + n_m^2. \tag{1.4.15}$$

Proof. We have already shown that dim $L^2(G) = N$. On the other hand, dim $L^2(G)$ is equal to the number of elements of the basis

$$t^k_{jv}, \quad j, v = 1, \ldots, n_k; \quad k = 1, \ldots, m. \tag{1.4.16}$$

For each fixed k the number of functions t_{jv}^k, j, $v = 1, \ldots, n_k$ is equal to n_k^2. Therefore the number of elements of a basis in $L^2(G)$ is equal to $n_1^2 + n_2^2 + \cdots + n_m^2$. This is (1.4.15). $\square$

1.5. The Factorization of the Regular Representation of a Finite Group into its Irreducible Components

Theorem 1. *The right regular representation of a finite group G decomposes into its irreducible components. Every irreducible representation T^k of G is a component of the regular representation, with multiplicity n_k equal to the dimension of the representation T^k.*

Proof. Let N be the order of G. Let $T^1, \ldots, T^m$ be a complete set of irreducible representations of G and let t_{jv}^k be the matrix elements of T^k. Let M_j^k denote the subspace of $L^2(G)$ spanned by the functions $t_{jv}^k(g)$, $v = 1, \ldots, n_k$, j and k being fixed. Since the t_{jv}^k are orthogonal, they form a basis for M_j^k. The orthogonality relations (1.4.1) and (1.4.2) show that

$$M_j^k \perp M_{j'}^{k'}, \quad \text{for } j \neq j' \text{ or } k \neq k'.$$

Furthermore, the set of all functions $t_{jv}^k(g)$ is a basis in $L^2(G)$ (see Theorem 2 in 1.4). We thus have

$$L^2(G) = \sum_{k=1}^{m} \sum_{j=1}^{n_k} \oplus M_j^k. \tag{1.5.1}$$

Every M_j^k is invariant under the right regular representation T of G. Formula (2.1.6) in chapter I shows that

$$T(h)t_{jv}^k(g) = t_{jv}^k(gh) = \sum_{\mu=1}^{n_k} t_{j\mu}^k(g)t_{\mu v}^k(h) = \sum_{\mu=1}^{n_1} t_{\mu v}^k(h)t_{j\mu}^k(g). \tag{1.5.2}$$

That is, $T(h)t_{jv}^k(g)$ is a linear combination of the functions $t_{j\mu}^k(g)$, $\mu = 1, \ldots, n_k$, and thus $T(h)t_{jv}^k(g) \in M_j^k$.

Let T^{jk} be the restriction of T to M_j^k. Multiply (1.5.2) by $\sqrt{n_k}$ and cite the formulas (1.4.14). We find

$$T^{jk}(h)e_{jv}^k = T(h)e_{jv}^k = \sum_{\mu=1}^{n_k} t_{\mu v}^k(h)e_{j\mu}^k. \tag{1.5.3}$$

The functions $e_{j\mu}^k$, $\mu = 1, \ldots, n_k$, are an orthonormal basis in M_j^k (see the remark in 1.4). Therefore (1.5.3) shows that the matrix elements of T^{jk} in the basis $t_{j\mu}^k$, $\mu = 1, \ldots, n_k$, coincide with $t_{\mu v}^k(h)$. Thus T^{jk} is equivalent to T^k. Hence the restriction of T to each M_j^k, $j = 1, \ldots, n_k$ is equivalent to T^k. This and (1.5.1) imply that T^k is contained in the regular representation T with multiplicity n_k. $\square$

Remark 1. An analogous theorem holds for the left regular representation. The reader may supply the details.

Remark 2. The decomposition (1.5.1) and thus the decomposition of the regular representation into irreducible components depends on the choice of the matrix elements $t_{jv}^k(g)$, that is, on the choice of orthonormal bases in the spaces X_k of the irreducible representation T^k, $k = 1, 2, \ldots, m$. Hence the decomposition (1.5.1) is not unique (barring the case $n_k = 1$). The same is true of decompositions into irreducible components of any representation of a given group, if irreducible components of dimension greater than 1 occur in the decomposition (see *infra*, page 78).

Remark 3. To find a complete set of irreducible representations of a given group G, it is natural to try to decompose the regular representation into irreducible components, since all irreducible representations are contained in this decomposition. Up to the present time, no general methods for finding such decompositions have been found. In order to execute the instructions in Theorem 1 for decomposing the regular representation, we must already be in possession of the matrix elements of the irreducible representations.

1.6. Parseval's Equality and Plancherel's Formula

As noted in 1.4, the functions

$$e_{jv}^k(g) = \sqrt{n_k}\, t_{jv}^k(g) \tag{1.6.1}$$

form an orthonormal basis in $L^2(G)$. For all functions $f(g)$ in $L^2(G)$ we thus have

$$(f, f) = \sum_{k=1}^{m} \sum_{v=1}^{n_k} |(f, e_{jv}^k)|^2. \tag{1.6.2}$$

By (1.6.1) we also have

$$(f, e_{jv}^k) = \sqrt{n_k}(f, t_{jv}^k),$$

and so (1.6.2) can be written as

$$(f, f) = \sum_{k=1}^{m} \sum_{v=1}^{n_k} n_k |(f, t_{jv}^k)|^2, \tag{1.6.3}$$

where the functions $t_{jv}^k(g)$ are matrix elements of a complete set of irreducible representations of the group Gj and n_k are the dimensions of these representations. The numbers (f, t_{jv}^k) are called *Fourier coefficients* of the function f with respect to t_{jv}^k, and the formula (1.6.2) is called *Parseval's equality for the group G*.

Readers familiar with the theory of Fourier series will recognize an analogy with Parseval's equality for Fourier series. As we will see *infra* (for example in Chapter IV) both (1.6.3) and Parseval's equality for Fourier series are special cases of general facts about group representations.

Let $T^{k*}(g)$ be the adjoint operator to $T^k(g)$ with respect to a scalar product under which T^k is unitary. We set

$$T^k(f) = M(f(g)T^{k*}(g)) = \frac{1}{N} \sum_{k=1}^{N} f(g_k)T^{k*}(g). \tag{1.6.4}$$

The operator function $T^k(f)$ defined by formula (1.6.4) is called the *Fourier transform* of the function f. There is one operator for each index k.

Let $t^k_{jv}(g)$ be the matrix elements of the operator $T^k(g)$ under our orthonormal basis $e^k_1, \ldots, e^k_{n_k}$. Thus $\overline{t^k_{vj}(g)}$ are the matrix elements of the operator $T^{k*}(g)$ under the same basis. This fact and (1.6.4) show that

$$\left\| \begin{matrix} (f, t^k_{11}) & \cdots & (f, t^k_{n_k 1}) \\ \cdots\cdots\cdots\cdots \\ (f, t^k_{1 n_k}) & \cdots & (f, t^k_{n_k n_k}) \end{matrix} \right\|$$

is the matrix of the operator $T^k(f)$ in the basis $e^k_1, \ldots, e^k_{n_k}$. Therefore[15] we have

$$\mathrm{tr}(T^{k*}(f)T^k(f)) = \sum_{j,v=1}^{n_k} |(f, t^k_{jv})|^2. \tag{1.6.5}$$

Combining formula (1.6.5) and (1.6.3), we obtain

$$(f, f) = \sum_{k=1}^{m} n_k \, \mathrm{tr}(T^{k*}(f)T^k(f)). \tag{1.6.6}$$

Formula (1.6.6) is called *Plancherel's formula for a finite group*.

1.7. Characters of Representations of a Finite Group

Theorem 1. *Let $T^1, T^2, \ldots, T^m$ be a complete set of irreducible representations of a finite group G. Let $\chi_1, \chi_2, \ldots, \chi_m$ be their characters. Then we have*

$$(\chi_k, \chi_j) = \begin{cases} 1, & \text{for } k = j, \\ 0, & \text{for } k \neq j, \end{cases} \quad k, j = 1, \ldots, m. \tag{1.7.1}$$

Proof. The orthogonality relations (1.4.1) and (1.4.2) yield the identities for $k \neq j$

$$(\chi_k, \chi_j) = \left(\sum_{\mu=1}^{n_k} t^k_{\mu\mu}, \sum_{v=1}^{n_j} t^j_{vv} \right) = \sum_{\mu=1}^{n_k} \sum_{v=1}^{n_j} (t^k_{\mu\mu}, t^j_{vv}) = 0,$$

---

[15] If we replace T^k by an equivalent representation, then $T^k(g)$ will in general change. However, formula (5), I in 2.9, Chapter I for traces shows that the numerical function $\mathrm{tr}(T^{k*}(f)T^k(f))$ does *not* change.

and for $k = j$ the identities

$$(\chi_k, \chi_k) = \left(\sum_{\mu=1}^{n_k} t_{\mu\mu}^k, \sum_{v=1}^{n_k} t_{vv}^k \right) = \sum_{\mu=1}^{n_k} \sum_{v=1}^{n_k} (t_{\mu\mu}^k, t_{vv}^k)$$

$$= \sum_{\mu=1}^{n_k} (t_{\mu\mu}^k, t_{\mu\mu}^k) = \sum_{\mu=1}^{n_k} \frac{1}{n_k} = 1.$$

The formulas (1.6.1) are called the *orthogonality relations for characters of a finite group*. These formulas show in particular that $\chi_1, \chi_2, \ldots, \chi_m$ are linearly independent.

Let T be any representation of the group G. Theorem 2 in 1.2 asserts and T is completely reducible, so that

$$T = r_1 T^1 \dotplus r_2 T^2 \dotplus \cdots \dotplus r_m T^m, \tag{1.7.2}$$

where r_j is the multiplicity with which T^j occurs in T (some of the r_j's may vanish). Applying (e) III in 2.9, Chapter I to (1.6.2), we conclude that

$$\chi_T = r_1 \chi_1 + r_2 \chi_2 + \cdots + r_m \chi_m. \tag{1.7.3}$$

From this and from (1.6.1) we have

$$(\chi_T, \chi_j) = (r_1 \chi_1 + r_2 \chi_2 + \cdots + r_m \chi_m, \chi_j) = r_j. \tag{1.7.4}$$

We summarize.

I. *The Fourier coefficient of the character χ_T of the representation T with respect to the character χ_j of an irreducible representation T^j is equal to the multiplicity with which T^j occurs in T.*

Thus, knowing the characters of a complete set of irreducible representations, we can say (if even these representations are not known) which irreducible representations are contained in a given representation and with what multiplicity. The answer does not depend on the method of factorization of the representation into irreducible representations (cf. Remark 2 in 1.5).

From (1.7.1) and (1.7.3) it follows also that

$$(\chi_T, \chi_T) = r_1^2 + r_2^2 + \cdots + r_m^2. \tag{1.7.5}$$

Indeed, we have $(\chi_T, \chi_T) = 1$ if and only if one of the numbers r_j is unity and all other r_k are zero. By I, this means that T contains only T^j and with multiplicity 1. Again we summarize.

II. *A representation T of a finite group is irreducible if and only if $(\chi_T, \chi_T) = 1$.*

We note one more criterion for irreducibility.

III. *A representation T of a finite group in a space X is irreducible if and only if every linear operator B on X that commutes with all $T(g)$ is a multiple of the identity operator.*

The proof follows directly from V, 2.8 in Chapter I and Theorem 1 in 1.2.

IV. *If the characters χ, χ' of two representations T, T' of a finite group G coincide, then these representations are equivalent.*

If $\chi_T = \chi_{T'}$, then by virtue of (1.6.4) we also have

$$(\chi_{T'}, \chi_j) = (\chi_T, \chi_j) = r_j,$$

so that T' and T contain the same irreducible representations T^j with the same multiplicities r_j. Thus T' is equivalent to T.

Let M denote the set of all functions $f(g)$ on G that are constant on conjugacy classes in G. Let $K_1, K_2, \ldots, K_q$ be all these conjugacy classes. Then we can regard the function $f(g) \in M$ as a function the set whose elements are the classes K_k: we set $f(K_j) = f(g)$ for $g \in K_j$. Thus a function f in M is simply a sequence of q numbers $f(K_1), \ldots, f(K_q)$. Plainly M is a q-dimensional subspace of $L^2(G)$.

Theorem 2. *The characters $\chi_1, \chi_2, \ldots, \chi_m$ of a complete set of pairwise inequivalent irreducible representations of a finite group G form a complete orthonormal set in M.*

Proof. By (b), III in 2.9, Chapter I and Theorem I, all characters χ_i belong to M and are an orthonormal set in M. We will prove that the χ_i are a complete orthonormal set. For $f \in M$, we have $f(h) = f(g^{-1}hg)$ for all $g, h \in G$. We need to prove that f is a linear combination of the characters $\chi_1, \ldots, \chi_m$. Since the set t_{jv}^k is complete in $L^2(G)$ (see Theorem 2 in 1.4) we have

$$f(h) = f(g^{-1}hg) = \sum_{k=1}^{m} \sum_{j,v=1}^{n_k} c_{jv}^k t_{jv}^k(g^{-1}hg),$$

where the c_{jv}^k are certain complex numbers. Thus we have

$$f(h) = M_g(f(h)) = \sum_{k=1}^{m} \sum_{j,v=1}^{n_k} c_{jv}^k M_g(t_{jv}^k(g^{-1}hg)), \qquad (1.7.6)$$

where M_j is the mean value over G with respect to the variable g. The orthogonality relations (1.4.1), (1.4.2) and formulas (2.1.4) in Chapter I show that

$$M_g(t_{jv}^k(g^{-1}hg)) = M_g\left(\sum_{p,q=1}^{n_k} t_{jp}^k(g^{-1})t_{pq}^k(h)t_{qv}^k(g)\right)$$

$$= \sum_{p,q=1}^{n_k} t_{pq}^k(h)M_g(\overline{t_{pj}^k(g)}t_{qv}^k(g))$$

$$= \begin{cases} 0, & \text{for } j \neq v, \\ \dfrac{1}{n_k}\sum_{p=1}^{n_k} t_{pp}^k(h) = \dfrac{1}{n_k}\chi_k(h), & \text{for } j = v. \end{cases}$$

Therefore (1.7.6) can be rewritten as

$$f(h) = \sum_{k=1}^{n} \left(\sum_{j=1}^{n_k} \frac{1}{n_k} c_{jj}^k\right)\chi_k(h). \quad \square$$

Theorem 2 shows that dim $M = m$. On the other hand, as we have seen above, dim $M = q$. Thus we have $m = q$. We state this formally.

Theorem 3. *The number of representations in a complete set of irreducible representations of a finite group is equal to the number of conjugacy classes in the group.*

1.8. Factorization of a Given Representation of a Finite Group into its Irreducible Representations

Let G be a finite group and let $T^1, \ldots, T^m$ be a complete set of irreducible representations of G. Let $n_1, \ldots, n_m$ be their dimensions and let $t_{jv}^k(g)$ be the matrix elements of the representations T^k, $k = 1, \ldots, m$. We will present a method for the actual factorization of a given representation T of G into its irreducible representations. This method is applicable if the functions $t_{jv}^k(g)$ are known.

Let X be the space of the representation T. As usual (see theorem 1 in 1.2) we take T to be unitary. We define operators P_{jv}^k in X, setting

$$P_{jv}^k = n_k(M(\overline{t_{jv}^k(g)}T(g))). \tag{1.8.1}$$

I. *The operators P_{jv}^k have the following properties:*

$$T(h)P_{j\mu}^k = \sum_{v=1}^{n_k} t_{vj}^k(h)P_{v\mu}^k, \tag{1.8.2a}$$

$$P_{j\mu}^k T(h) = \sum_{v=1}^{n_k} t_{\mu v}^k(h)P_{jv}^k \tag{1.8.2b}$$

for all $h \in G$;

$$P_{jl}^{k}P_{\mu v}^{k'} = \begin{cases} 0, & \text{for } k' \neq k \text{ or } l \neq \mu, \\ P_{jv}^{k}, & \text{for } k' = k \text{ and } l = \mu, \end{cases} \qquad (1.8.3)$$

$$(P_{jl}^{k})^{*} = P_{lj}^{k}; \qquad (1.8.4)$$

in particular we have

$$P_{jj}^{k}P_{\mu\mu}^{k'} = \begin{cases} 0, & \text{for } k' \neq k \quad \text{or } \mu \neq j, \\ P_{jj}^{k}, & \text{for } k' = k \quad \text{and } \mu = j, \end{cases} \qquad (1.8.5)$$

$$(P_{jj}^{k})^{*} = P_{jj}^{k}. \qquad (1.8.6)$$

Proof. From (1.8.1), the properties III of an invariant mean (1.4.1), and the identities $t_{jv}^{k}(h^{-1}) = \overline{t_{vj}^{k}(h)}$ (see (2.8.7) chapter I) we see that

$$
\begin{aligned}
T(h)P_{j\mu}^{k} &= n_{k}T(h)M(\overline{t_{j\mu}^{k}(g)}T(g)) = n_{k}M(\overline{t_{j\mu}^{k}(g)}T(h)T(g)) \\
&= n_{k}M(\overline{t_{j\mu}^{k}(g)}T(hg)) = n_{k}M(\overline{t_{j\mu}^{k}(h^{-1}g)}T(g)) \\
&= n_{k}M\left(\sum_{v=1}^{n_{k}} \overline{t_{jv}^{k}(h^{-1})t_{v\mu}^{k}(g)}T(g) \right) \\
&= \sum_{v=1}^{n_{k}} t_{vj}^{k}(h)n_{k}M(\overline{t_{v\mu}^{k}(g)}T(g)) = \sum_{v=1}^{n_{k}} t_{vj}^{k}(h)P_{v\mu}^{k}.
\end{aligned}
$$

This proves (1.8.2a). We prove (1.8.2b) similarly. From (1.8.2a) and the orthogonality relations (1.4.1), (1.4.2) we have

$$
\begin{aligned}
P_{jl}^{k}P_{\mu v}^{k'} &= n_{k}M(\overline{t_{jl}^{k}(g)}T(g))P_{\mu v}^{k'} = n_{k}M(\overline{t_{jl}^{k}(g)}T(g)P_{\mu v}^{k'}) \\
&= n_{k}M\left(\overline{t_{jl}^{k}(g)} \sum_{q=1}^{n_{k'}} t_{q\mu}^{k'}(g)P_{qv}^{k'} \right) = n_{k} \sum_{q=1}^{n_{k'}} (t_{q\mu}^{k'}, t_{jl}^{k})P_{ql}^{k'} \\
&= \begin{cases} 0, & \text{for } k' \neq k \text{ or } l \neq \mu, \\ n_{k}(t_{jl}^{k}, t_{jl}^{k})P_{jv}^{k} = P_{jv}^{k}, & \text{for } k' = k \text{ and } l = \mu. \end{cases}
\end{aligned}
$$

This is (1.8.3). To prove (1.8.4) we note that T is unitary, so that $(T(g))^{*} = T(g^{-1})$ and hence

$$
\begin{aligned}
(P_{jl}^{k})^{*} &= (n_{k}M(\overline{t_{jl}^{k}(g)}T(g)))^{*} = \left(\frac{n_{k}}{N} \sum_{v=1}^{N} \overline{t_{jl}^{k}(g_{v})}T(g_{v}) \right)^{*} \\
&= \frac{n_{k}}{N} \sum_{v=1}^{N} t_{jl}^{k}(g_{v})(T(g_{v}))^{*} = \frac{n_{k}}{N} \sum_{v=1}^{N} \overline{t_{lj}^{k}(g_{v}^{-1})}T(g_{v}^{-1}) \\
&= \frac{n_{k}}{N} \sum_{v=1}^{N} \overline{t_{lj}^{k}(g_{v})}T(g_{v}) = P_{lj}^{k}.
\end{aligned}
$$

The relations (1.8.5) and (1.8.6) follow from (1.8.3) and (1.8.4).

We now define

$$X^k_j = P^k_{jj}X. \tag{1.8.7}$$

Plainly X^k_j is a subspace of X. Some of the X^l_j's may be (0); this means that the corresponding operator P^l_{jj} is 0. ☐

II. *The subspaces X^k_j have the following properties:*

$$X^k_j \perp X^{k'}_{j'} \quad \text{for } k' \neq k \text{ or } j' \neq j; \tag{1.8.8}$$

$$\sum_{k=1}^{m} \sum_{j=1}^{n_k} \oplus X^k_j = X; \tag{1.8.9}$$

$$P^k_{j\mu}X^{k'}_v = \{0\} \quad \text{for } k' \neq k, \text{ or } \mu = v; \tag{1.8.10}$$

$$P^k_{j\mu} \text{ is an isometry of } X^k_\mu \text{ onto } X^k_\mu. \tag{1.8.11}$$

Proof. From (1.8.5), (1.8.6) and (1.8.7) we see that the operators P^k_{jj} are pairwise orthogonal projections mapping the space X onto X^k_j. Property (1.8.8) follows. Let us prove (1.8.9). We write

$$Y = \sum_{k=1}^{m} \sum_{j=1}^{n_k} \oplus X^k_j \tag{1.8.12}$$

and assume that $Y \neq X$. Then there is a nonzero vector x_0 in X that is orthogonal to Y and hence to each X^k_j (see VIII in 2.8, chapter I). From (1.8.7) we infer that $x_0 \perp P^k_{jv}x$ for all $x \in X$, so that $f(g) = (x_0, T(g)x) = 0$. (In fact, one can use (1.8.1) to write

$$0 = (x_0, P^k_{jv}x) = (x_0, n_k M(\overline{t^k_{jv}(g)}T(g)x))$$

$$= n_k\left(x_0, \frac{1}{N} \sum_{\mu=1}^{N} \overline{t^k_{jv}(g_\mu)}T(g_\mu)x\right)$$

$$= \frac{n_k}{N} \sum_{\mu=1}^{N} t^k_{jv}(g_\mu)(x_0, T(g_\mu)x) = n_k(t^k_{jv}, f) \tag{1.8.13}$$

for all $j, v = 1, \ldots, n_k; k = 1, \ldots, m$.) The functions t^k_{jv} are a basis in $L^2(G)$ (Theorem 2 in 1.4). Therefore (1.8.13) shows that

$$0 = f(g) = (x_0, T(g)x), \quad \text{for all } x \in X, \text{ and } g \in G. \tag{1.8.14}$$

Put $g = e$ and $x = x_0$ in (1.8.14). This yields $(x_0, x_0) = 0$, which is impossible for $x_0 \neq 0$. We have proved that $Y = X$, which is (1.8.9).

To prove (1.8.10) we first use (1.8.3) to write

$$P^k_{j\mu}X^{k'}_v = P^k_{j\mu}P^k_{vv}X = (0), \quad \text{for } k' \neq k \text{ or } \mu \neq v.$$

By virtue of (1.8.3) P_v^k is the identity operator on X_j^k. In fact, if x belongs to X_j^k, we may write $x = P_{jj}^k y$ for some $y \in X$; hence $P_{jj}^k x = x$. Finally, let us prove (1.8.11). From (1.8.7), we have $x = P_{jj}^k x_1$, where $x_1 \in X$, and therefore $P_{jj}^k(x) = (P_{jj}^k)^2 x_1 = P_{jj}^k x_1 = x$. Let x belong to X_μ^k, so that $x = P_{\mu\mu}^k x$. By (1.8.3) we have

$$P_{j\mu}^k x = P_{j\mu}^k P_{\mu\mu}^k x = P_{jj}^k P_{j\mu}^k x \in X_j^k.$$

Thus we have

$$P_{j\mu}^k X_\mu^k \subset X_j^k. \tag{1.8.15}$$

Applying the operator $P_{\mu j}^k$ to both sides of (1.8.15), we obtain

$$X_\mu^k = P_{\mu\mu}^k X_\mu^k = P_{\mu j}^k P_{j\mu}^k X_\mu^k \subset P_{\mu j}^k X_j^k \subset X_\mu^k.$$

From this it follows that $P_{\mu j}^k X_j^k = X_\mu^k$. Switching the roles of μ and j, we obtain $P_{j\mu}^k X_\mu^k = X_j^k$, that is, $P_{j\mu}^k$ maps X_μ^k onto X_j^k. Let x, y belong to X_μ^k. Then we have $P_{\mu\mu}^k x = x$ and from (1.8.4) and (1.8.3) we get

$$(P_{j\mu}^k x, P_{j\mu}^k y) = ((P_{j\mu}^k)^* P_{j\mu}^k x, y) = (P_{\mu j}^k P_{j\mu}^k x, y) = (P_{\mu\mu}^k x, y) = (x, y).$$

Thus $P_{j\mu}^k$ preserves inner products, is an isometry and in particular, is one-to-one. □

We conclude from (1.8.11) that

$$\dim X_\mu^k = \dim X_j^k, \quad \text{for all } \mu, j = 1, \ldots, n_k. \tag{1.8.16}$$

We can now decompose the representation T into its irreducible components. First we write

$$Y^k = \sum_{j=1}^{n_k} \oplus X_j^k. \tag{1.8.17}$$

From (1.8.9) we conclude that

$$\sum_{k=1}^{m} Y^k = X. \tag{1.8.18}$$

Choose an orthonormal basis $e_{11}^k, e_{21}^k, \ldots, e_{r_k 1}^k$, (where $r_k = \dim X_1^k = \dim X_j^k, j = 1, \ldots, n_k$) in one of the subspaces X_j^k [16], X_1^k for example. We set

$$e_{jv}^k = P_{v1}^k e_{j1}^k, \quad k = 1, \ldots, m, \quad v, j = 1, \ldots, r_k. \tag{1.8.19}$$

---

[16] We can actually construct X_j^k, noting that X_j^k is the set of all linear combinations

$$\gamma_1 P_{jj}^k f_1 + \cdots + \gamma_n P_{jj}^k f_n,$$

where $f_1, \ldots, f_n$ is any basis in X.

By virtue of (1.8.11),

$$e_{1v}^k, e_{2v}^k, \ldots, e_{r_k v}^k$$

is an orthonormal basis in X_v^k; it follows from (1.8.8) that

$$e_{jv}^k \perp e_{j'v'}^{k'}, \qquad \text{if } k' \neq k, \text{ or } j' \neq j, \text{ or } v' \neq v. \tag{1.8.20}$$

In particular,

$$e_{j1}^k, e_{j2}^k, \ldots, e_{jn_k}^k \tag{1.8.21}$$

is an orthonormal set. Let Y_j^k be the subspace spanned by the set (1.8.21). The system (1.8.21) is obviously a basis in Y_j^k. From (1.8.17) and (1.8.18) we conclude that

$$Y^k = \sum_{j=1}^{r_k} \oplus Y_j^k, \tag{1.8.22}$$

$$X = \sum_{k=1}^{m} \sum_{j=1}^{r_m} \oplus Y_j^k. \tag{1.8.23}$$

III. *Every Y_j^k is invariant under T and the restriction of T to Y_j^k is irreducible and equivalent to T^k. Consequently, formula (1.8.23) describes a factorization of the representation T of G into its irreducible components T^k. The multiplicity with which T^k occurs in T is* dim X_j^m.

Proof. By virtue of (1.8.19) and (1.8.20) we have

$$T(g)e_{j\mu}^k = T(g)P_{\mu 1}^k e_{j1}^k = \sum_{v=1}^{n_k} t_{v\mu}^k(g)P_{v1}^k e_{j1}^k = \sum_{v=1}^{n_k} t_{v\mu}^k(g)e_{jv}^k \in Y_j^k$$

for each j and k. This means that Y_j^k is invariant under T and the matrix elements of the restriction of T to Y_j^k in the basis (1.8.21) in Y_j^k coincide with the functions $t_{jk}^k(g)$. That is, this restriction is equivalent to the representation T^k and is therefore irreducible. From (1.8.23) it is clear that the multiplicity of T^k in T is $r_k = $ dim X_j^k. If we have some $X_j^k = (0)$, then $r_k = 0$ and the corresponding T^k is not contained in T. $\square$

The decomposition described in (1.8.23) depends on the choice of a basis $e_{11}^k, \ldots, e_{r_k 1}^k$ in X_1^k (and also on the choice of matrix elements $t_{jv}^k(g)$, that is, on the choice of a basis in the space X_k of the representation T^k). This decomposition accordingly is not unique (see Remark 2 in 1.5).

On the other hand, the restriction of T to Y^k is a multiple of the representation T^k with multiplicity r_k. Thus (1.8.18) gives a decomposition of T into pairwise inequivalent representations, each a multiple of an irreducible representation.

IV. *The decomposition of a representation of a finite group G into pairwise inequivalent representations that are multiples of irreducible representations is unique.*

Proof. We set

$$P^k = \sum_{j=1}^{n_k} P^k_{jj}. \tag{1.8.24}$$

The operator P^k is a projection (see (1.8.5)) and from (1.8.16) and (1.8.17) we have

$$Y^k = P^k X. \tag{1.8.25}$$

Both P^k and Y^k are independent of the choice of the functions $t^k_{jv}(g)$. Indeed, taking into consideration (1.8.1) and the definition of a character (2.9.3) in chapter I, we conclude that

$$P^k = \sum_{j=1}^{n_k} n_k M(\overline{t^k_{jj}(g)}T(g)) = n_k M\left(\sum_{j=1}^{n_k} \overline{t^k_{jj}(g)}T(g) \right)$$
$$= n_k M(\overline{\chi_T(g)}T(g)). \tag{1.8.26}$$

Since $\chi_T(g)$ is independent of the choice of $t^k_{jv}(g)$ (see 2.9 in chapter I), (1.8.26) shows that P^k is also.

Let Z be a subspace invariant under T on which the restriction S of the representation T is irreducible and equivalent to T^k. Let $f_1, \ldots, f_{n_k}$ be an orthonormal basis in Z in which the matrix elements of the representation S are the functions $t^k_{jv}(g)$, so that

$$S(g)f_v = \sum_{j=1}^{n_k} t^k_{jv}(g)f_j. \tag{1.8.27}$$

To show that

$$f_v \in Y^k, \qquad v = 1, \ldots, n_k \tag{1.8.28}$$

it is sufficient to establish that $P^k f_v = f_v$. This follows from (1.8.26), (1.8.27), and the orthogonality relations (1.4.2). Thus we have the following equalities:

$$P^k f_v = n_k M(\overline{\chi_T(g)}T(g))f_v = n_k M\left(\sum_{\mu=1}^{n_k} \overline{t^k_{\mu\mu}(g)}T(g)f_v \right)$$

$$= n_k M\left(\sum_{j,\mu=1}^{n_k} \overline{t^k_{\mu\mu}(g)}t^k_{jv}(g)f_j \right) = n_k \sum_{j,\mu=1}^{n_k} (t^k_{jv}, t^k_{\mu\mu})f_j$$

$$= n_k(t^k_{vv}, t^k_{vv})f_v = f_v.$$

From (1.8.28) we infer that $Z \subset Y^k$.

Let

$$X = \sum_{k=1}^{m} \oplus Z^k, \qquad Z^k = \sum_{j=1}^{s_k} \oplus Z_j^k, \qquad (1.8.29)$$

where Z_j^k is invariant under T and the restriction of T to Z_j^k is equivalent to T^k. All of the spaces Z_j^k are contained in Y^k, as the foregoing shows. Thus we have

$$Z^k = \sum_{j=1}^{s_k} \oplus Z_j^k \subset Y^k. \qquad (1.8.30)$$

The multiplicity of T^k in T does not depend on the mode of decomposition (see I in 1.7). Therefore we have $s_k = n_k$. From this and (1.8.29), (1.8.17), we infer that

$$\dim Z^k = n_k r_k = \dim Y^k.$$

Thus the inclusion (1.8.30) is actually an equality: $Z^k = Y^k$. $\square$

Examples and Exercises

1. Decompose the regular representation of the group S_3 into irreducible representations.

2. Prove that the dimension of an irreducible representation of a finite-dimensional group is a divisor of the order of the group.

3. Let p be a prime number. Prove that all groups of order p^2 are commutative.

§2. The Group Algebra of a Finite Group

2.1. Facts about Algebras

A set A is called an *algebra* if:

(1) A is a (real or complex) linear space;
(2) the product ab is defined for all pairs of elements $a, b \in A$ and satisfies the following conditions: for all a, b, c in A and all numbers γ in the scalar field of A:

$$ab \in A, \qquad (2.1.1)$$
$$a(b + c) = ab + ac, \qquad (2.1.2)$$
$$(a + b)c = ac + bc, \qquad (2.1.3)$$
$$\gamma(ab) = a(\gamma b) = (\gamma a)b. \qquad (2.1.4)$$

The algebra A is called *real* if A is a real linear space and *complex* if A is a complex linear space. The algebra A is called *finite-dimensional* if A is a finite-dimensional linear space.

A subset B of A is called a *subalgebra* of A if:

(1) B is a linear subspace of A;
(2) if b_1, $b_2 \in B$, then $b_1 b_2 \in B$.

In other words, B is an algebra under the multiplication defined in A. Clearly, *the intersection of any family of subalgebras of A is also a subalgebra of A.* Indeed, *the intersection of all the subalgebras of A containing a given set S is the least subalgebra of A containing S.* It is called the *subalgebra of A generated by the set S.* It is denoted by $A(S)$.

A subset I_1 of an algebra A is called a *left ideal* of A if:

(a) I_1 is a subspace of the linear space A;
(b) $aI_1 \subset I_1$ for all elements a of the entire algebra A.

Right ideals I_r are defined analogously. A set I in A is called a *two-sided ideal* if it is both a left and a right ideal. Plainly, the set $\{0\}$ and the entire algebra A are two-sided ideals of A. These two ideals are called *improper ideals*; all others are called *proper.* A left, right or two-sided ideal in A is plainly a subalgebra of A. An algebra A is called *simple* if it contains no proper two-sided ideals. A left ideal $I_1 \neq \{0\}$ is called *minimal* if it contains no proper left ideals $\tilde{I}_1$ except itself, that is, if it follows from $(0) \neq \tilde{I}_1 \subset I_1$ that $\tilde{I}_1 = I_1$. A left ideal $I_1 \neq A$ is called *maximal* if it is contained in no proper left ideals except itself. Minimal and maximal right and two-sided ideals are defined similarly.

An algebra A is said to be the *direct sum* of algebras $A_1, \ldots, A_n$ and is denoted by $A_1 \dotplus \cdots \dotplus A_k$ if:

(a) the linear space A is the direct sum of the linear spaces $A_1, \ldots, A_n$;
(b) if $a = a_1 + a_2 + \cdots + a_n$, $b = b_1 + b_2 + \cdots + b_n$, where a_j, b_j are in A_j for $j = 1, \ldots, n$, then $ab = a_1 b_1 + a_2 b_2 + \cdots + a_n b_n$.

For each j, let A_j^* be the set of all $0_{(1)} + 0_{(2)} + \cdots + 0_{(j-1)} + a_j + 0_{(j+1)} \cdots + 0_{(n)}$ in $A_1 \dotplus \cdots \dotplus A_n$. It is clear that A_j^* is a two-sided ideal in $A_1 \dotplus \cdots \dotplus A_n$ that is isomorphic with A_j.

Examples and Exercises

1. The set of all real numbers is a one-dimensional real algebra under the usual definition of sums and products. We denote this algebra by A_r^1, to distinguish it from the group $\mathbf{R}^1$. (See example 1 in 1.1, chapter I.) The complex algebra A_c^1 of all complex numbers is defined analogously.

2. Let n be an integer greater than 1. The set of all sequences $\{\gamma_1, \ldots, \gamma_n\}$, $\gamma_j \in A_r^1$, with sums, products, and real multiples defined coordinatewise is a real algebra, denoted by A_r^n. The relation

$$A_r^n = A_r^1 \dotplus \cdots \dotplus A_r^1, \quad (n \text{ summands})$$

is evident. The complex algebra A_c^n is defined similarly, so that $A_c^n = A_c^1 \dotplus \cdots \dotplus A_c^1 (n \text{ summands})$.

3. The set of all real matrices of order n is a real algebra under coordinate-wise addition and multiplication by real numbers and with the product defined as the product of matrices. This algebra is denoted by $A_r(n \times n)$. The algebra $A_c(n \times n)$ of all complex matrices of order n is defined similarly.

4. Let X be a linear space (either complex or real). Let $\mathscr{A}(X)$ be the set of all linear operators in X (complex or real) defined throughout X. In $\mathscr{A}(X)$ we define addition and multiplication by a number as usual. We define the product LM of operators L and M by iteration: $LM(x) = L(M(x))$ for all $x \in X$. Plainly, $\mathscr{A}(X)$ is an algebra, complex if X is complex and real if X is real. Fix a nonzero vector x_0 in X. The set I of all operators $A \in \mathscr{A}(x)$ for which $Ax_0 = 0$ is a left ideal in $\mathscr{A}(X)$. Prove that if X is finite-dimensional, then for every proper left ideal I_1 in $\mathscr{A}(X)$, there exists a vector $x_0 \in X$ such that $Ax_0 = 0$ for all $A \in I_1$.

5. Let X^3 be three-dimensional real vector space. In this space define addition and multiplication by real numbers coordinatewise. As the product of vectors x and y, take their outer or vector product. Then X^3 is an algebra.

2.2. The Quotient Algebra with Respect to a Two-Sided Ideal

Let A be an algebra and I a two-sided ideal in A. Then I is also a linear subspace of the linear space A. Let $\tilde{A}$ be the factor-space A/I: $\tilde{A} = A/I$, so that the elements of $\tilde{A}$ are the cosets $\tilde{a} = I + a$. Here a in A is a *representative* of the set $\tilde{a}$. To define *multiplication in $\tilde{A}$*, we set

$$\tilde{a}\tilde{b} = I + ab, \quad \text{for } a \in \tilde{a}, b \in \tilde{b}. \tag{2.2.1}$$

This definition is independent of the choice of representatives $a \in \tilde{a}, b \in \tilde{b}$. In fact, if $a, a_1 \in \tilde{a}$, then $a_1 - a$ belongs to I and therefore $a_1 b - ab = (a_1 - a)b \in I$, since I is a two-sided ideal. Thus we have $I + a_1 b = I + a_1 b - ab + ab = I + ab$. Similarly, if $b, b_1 \in \tilde{b}$, then $ab_1 + I$ coincides with $ab + I$. Conditions (2.1.1)–(2.1.4) are satisfied for products $\tilde{a}\tilde{b}$. Thus $\tilde{A}$ is an algebra. It is called *the quotient algebra of the algebra A with respect to the ideal I*. It is denoted by A/I. Note that A/I is an algebra only if I is a *two-sided* ideal.

2.3. Homomorphism and Isomorphism of Algebras

A mapping f of the algebra A into the algebra B is called *a homomorphism of A into B* if:

(1) f is a linear mapping of the linear space A into the linear space B;
(2) $f(a_1 a_2) = f(a_1)f(a_2)$, for all $a_1, a_2 \in A$.

The inverse image $f^{-1}(0)$ of the zero element 0 of the algebra B is called the *kernel of the homomorphism f* and is denoted by $\operatorname{Ker} f$.

If the image of the algebra A coincides with the algebra B under the homomorphism f, then f is called *a homomorphism of A onto B*. A one-to-one homomorphism is called an *isomorphism*. Algebras A and B are called *isomorphic* if there is an isomorphism of A onto B.

I. *If f is a homomorphism of the algebra A into B then*:

(1) $f(A)$ *is a subalgebra of B*;
(2) *f maps every subalgebra of A onto a subalgebra of* $f(A)$;
(3) *f maps every left, right, or two-sided ideal in A onto a left, right, or two-sided ideal in* $f(A)$;
(4) *the kernel* Ker f *of the homomorphism f is a two-sided ideal in A.*

A homomorphism f is an isomorphism if and only if Ker $f = \{0\}$.

We leave the proofs to the reader, as they are very simple.

Homomorphisms can be constructed as follows. Let A be an algebra. Let I be a two-sided ideal in A and let $\tilde{A} = A/I$. We use φ to denote the mapping which assigns to every element $a \in A$ the set $\tilde{a}$ that contains a. The definitions of the algebraic operations in $\tilde{A}$ show that φ is *a homomorphism of A onto* $\tilde{A} = A/I$. It is called the *canonical* (or *natural*) *homomorphism of A onto A/I.*

II. *If f is a homomorphism of an algebra A onto an algebra B and I =* Ker f, *then*:

(1) *B is isomorphic to* A/I;
(2) $f = \psi\varphi$, *where φ is the canonical homomorphism of B onto A/I and ψ is an isomorphism of A/I onto B.*

We leave the proof to the reader (compare with that of Proposition II in 1.6, chapter I).

2.4. Associative Algebras

An algebra A is called *associative* if

$$(ab)c = a(bc), \quad \text{for all } a, b, c \in A, \tag{2.4.1}$$

and *nonassociative* in the opposite case. The algebras in Examples 1-4 in 2.1 are associative and the algebra in Example 5 in 2.1 is nonassociative. All the results of 2.1–2.3 are valid whether or not the algebra is associative. However, we will suppose throughout the rest of this section that the algebras under consideration are associative and the term algebra, unless otherwise indicated, will mean an associative algebra.[17]

Products $a_1 a_2 \cdots a_n \in A$ are uniquely defined, since A is associative. If $a_1 = a_2 = \cdots = a_n = a$, this product is called the nth *power of the element* a and is denoted by a^n. From (2.4.1) we infer that

$$a^n a^m = a^{n+m}. \tag{2.4.2}$$

---

[17] The important class of nonassociative algebras will be discussed in chapters IX and X.

An element e of A is called the *unit element* (or simply the *unit*) if $ea = ae = a$ for all $a \in A$. It is easy to see the unit is unique if it exists at all. (See footnote to page 1.) Two elements a, b of A are said to *commute* if $ab = ba$. An algebra A is called *commutative* if all pairs of elements of A commute: $ab = ba$ for all a, b in A. The set of all the elements of A that commute with all elements of A is called the *center of* A. It is denoted by $Z(A)$. Plainly, $Z(A)$ is a commutative subalgebra of A. It coincides with A if and only if A is commutative.

2.5. Algebras with Involution

A mapping $a \to a^*$ of the algebra A in A is called an *involution* if[18]

(1) $a^{**} = a,$
(2) $(\gamma a)^* = \bar{\gamma} a,$
(3) $(a + b)^* = a^* + b^*,$
(4) $(ab)^* = b^* a^*$

for all a, $b \in A$ and all numbers γ.

An involution is called *nondegenerate* if in addition to the above we have:

(5) If $a^* a = 0,$ then $a = 0.$

From (1) it follows that an involution $a \to a^*$ is a mapping of the algebra A onto A. From (2) we infer that $0^* = 0$.

An algebra is called *symmetric* if it admits an involution. From now on all symmetric algebras are *complex* algebras unless the contrary is explicitly stated. Let A be a symmetric algebra. An element $a \in A$ is called *Hermitian* if $a^* = a$.

I. *Let A be a symmetric algebra. Every element $a \in A$ can be represented uniquely in the form*

$$a = a_1 + i a_2, \tag{2.5.1}$$

where a_1, a_2 are Hermitian elements of A.

If (2.5.1) holds, properties (2) and (3) show that $a^* = a_1 - i a_2$ and therefore

$$a_1 = \frac{1}{2}(a + a^*), \qquad a_2 = \frac{1}{2i}(a - a^*). \tag{2.5.2}$$

Thus, if (2.5.1) holds, a_1 and a_2 are defined uniquely. Conversely, for all $a \in A$, the formulas (2.5.2) define Hermitian elements a_1, $a_2 \in A$ for which (2.5.1) holds. □

II. *For every $a \in A$, the element $a^* a$ is Hermitian.*

By (1) and (5) we have $(a^* a)^* = a^* a^{**} = a^* a$. □

---

[18] If A is a real algebra, then γ in property (2) is a real number and therefore $\bar{\gamma} = \gamma$.

III. *If a symmetric algebra A has a unit e, then e is Hermitian.*

We have $e^* = e^*e$, and from (1) and (4)

$$e = e^{**} = (e^*e)^* = e^*e^{**} = e^*e. \quad \square$$

In the following Propositions IV and V we suppose that A is a symmetric algebra with a nondegenerate involution.

IV. *If a is a Hermitian element of A and $a^2 = 0$, then $a = 0$.*

We have $a^*a = a^2 = 0$; now apply property (5). $\quad \square$

V. *If I_l is a left ideal in A and*

$$I_l I_l = \{0\}, \tag{2.5.3}$$

then $I_l = \{0\}$.

Let a belong to I_l. Then we also have $a^*a \in I_l$ and from (2.5.3) we conclude that $(a^*a)^2 = 0$. But then we have $a^*a = 0$ since a^*a is Hermitian (II and IV). Consequently, (5) implies that we have $a = 0$. Thus every element a of I_l is 0; that is, $I_l = \{0\}$. $\quad \square$

2.6. Representations of Algebras

Let A be an algebra. Let X be a complex linear space not equal to (0). Let T be a mapping that assigns to every element $a \in A$ a linear operator $T(a)$ in X such that for all a, a_1, a_2 in A and all numbers γ, we have

(1) $\quad T(\gamma a) = \gamma T(a)$,
(2) $\quad T(a_1 + a_2) = T(a_1) + T(a_2)$
(3) $\quad T(a_1 a_2) = T(a_1)T(a_2)$
(4) $\quad T(e) = 1$ (if A is an algebra with a unit e).

Then T is called a *representation of the algebra A in the space X.*

The space X is called *the space of the representation* and the operators $T(a)$ are called *operators of the representation.* Properties (1)–(4) mean that the mapping $T: a \to T(a)$ is a homomorphism of A into the algebra $A(X)$ (see example 4 in 2.1) which also satisfies condition (4) if A has a unit. All of the fundamental concepts and propositions of the theory of group representations given in 2.1, 2.2, 2.4–2.6 in chapter I, the concepts of irreducibility and equivalence, lemmas 1 and 2 in 2.2, and their corollaries can be transferred *mutatis mutandis* to representations of algebras. We leave the details to the reader. We obtain an important representation T of an algebra A by taking A itself as the representation space X and defining $T(a)$ by the formula

$$T(a)x = ax, \quad \text{for all } a, x \in A, \tag{2.6.1}$$

so that $T(a)$ is simply the operator of multiplication on the left by a. Properties (2.1.1)–(2.1.4) of an algebra A and the associative law show that conditions (1)–(4) hold. That is, T is a bona fide representation. This representation is called the *left regular representation of the algebra A.*

I. *A subspace $M \subset A$ is invariant under the left regular representation of A if and only if M is a left ideal in A.*

This assertion follows directly from the definitions.

II. *The restriction of a left regular representation of the algebra A to a left ideal I_l of A is irreducible if and only if I_l is a minimal left ideal in A.*

See I, the definition of a minimal left ideal, and the definition of an irreducible representation.

III. *The left regular representation T of a finite-dimensional algebra A is completely reducible if and only if A is the linear space direct sum*

$$A = I_{l_1} + \cdots + I_{l_k} \tag{2.6.2}$$

of minimal left ideals. The decomposition of T into irreducible representations is obtained by this decomposition of A.

See II and the definition of a direct sum of representations.

In view of III, the problem of decomposing the left regular representation of a finite-dimensional algebra A into irreducible components is equivalent to the problem of decomposing A into a direct sum of minimal left ideals.

Let A be a symmetric algebra. Let X and Y be linear spaces in duality with respect to the bilinear form (x, y), $x \in X$, $y \in Y$. The representations T and S in X and Y are called *mutually conjugate* with respect to the form (x, y) if

$$(T(a)x, y) = (x, S(a^*)y), \quad \text{for all } a \in A, x \in X, y \in Y. \tag{2.6.3}$$

Reasoning as in 2.3, we conclude the following.

IV. *If A is a symmetric algebra and X and Y are finite-dimensional spaces in duality with respect to (x, y), then for every representation T of A in X there exists a unique representation S of A in Y, conjugate to T with respect to (x, y). With the proper choice of bases $e_1, \ldots, e_n$ and $f_1, \ldots, f_n$ ($n = \dim X = \dim Y$) in X and Y, the matrix elements of T and S under these bases are related by the identities*

$$t_{jk}(a) = \overline{t_{kj}(a^*)}. \tag{2.6.4}$$

V. *If finite-dimensional representation T and S of a symmetric algebra A are conjugate to each other, then T is irreducible if and only if S is irreducible.*

Again let A be a symmetric algebra and let X be a pre-Hilbert space with the scalar product (x, y) for $x, y \in X$. A representation T of A in X is called *symmetric* if

$$(T(a)x, y) = (x, T(a^*)y), \quad \text{for all } a \in A, x, y \in X. \tag{2.6.5}$$

If X is a Euclidean space (that is, finite-dimensional and pre-Hilbert) then (2.6.5) means that

$$T(a^*) = (T(a))^*, \quad \text{for all } a \in A. \tag{2.6.6}$$

The relation (2.6.4) thus means that for the matrix $t(a)$ of the representation T in an orthonormal basis in X we have

$$t(a^*) = (t(A))^*, \tag{2.6.7}$$

that is.

$$t_{jk}(a^*) = \overline{t_{kj}(a)}, \quad j, k = 1, 2, \ldots, \dim X. \tag{2.6.8}$$

VI. *If T is a symmetric representation of a symmetric algebra A in a Euclidean space X and M is a subspace of X invariant under T, then $M^\perp$ is also invariant under T.*

Proof. Let x belong to M and y to $M^\perp$. Then we also have $T(a^*) \in M$, since M is invariant under T. Therefore we have $T(a^*)x \perp y$, that is (see (2.6.5))

$$0 = (T(a^*)x, y) = (x, T(a)y). \tag{2.6.9}$$

The identity (2.6.9) means that $T(a)y \perp x$ for all $x \in M$, or $T(a)y \in M^\perp$. □

VII. *Any symmetric representation of a symmetric algebra in a Euclidean space is completely reducible.*

The proof is like the proof of IX in 2.8, chapter I: replace II in 2.8, chapter I, by Proposition III just proved.

2.7. Definition and Simplest Properties of the Group Algebra of a Finite Group

Unless otherwise stipulated we will suppose in this section that G is a finite group. Let G consist of m distinct elements, say $g_1, g_2, \ldots, g_m$. Consider the set A_G of all formal sums $a = \sum_{k=1}^{m} a(g_k)g_k$ or, more briefly, $a = \sum_g a(g)g$, where $a(g_k)$ are any complex numbers. We define $a = b$ to mean that $a(g) = b(g)$ for all g in G. In particular, $a = 0$ exactly when $a(g) = 0$ for all

g in G. In A_G we define addition, multiplication by a number α and multiplication of sums by the following formulas:

$$\alpha a = \sum_g a(g)g, \tag{2.7.1a}$$

$$a + b = \sum_g [a(g) + b(g)]g, \tag{2.7.1b}$$

$$ab = \sum_{g',g''} a(g')b(g'')g'g'', \tag{2.7.1c}$$

where $a = \sum_g a(g)g$, $b = \sum_g b(g)g$.

It is easy to verify that conditions (2.1.1)–(2.1.4) hold for the definitions in formulas (2.7.1). That is, A_G is actually an algebra. It is called the *group algebra of the group* G. We identify the element $g_0 \in G$ with the sum $\sum_g a(g)g$ for which $a(g) = 0$ for $g \neq g_0$ and $a(g_0) = 1$. The elements of G are thus thought of as elements of the algebra A_G.

I. *The product of g_1 and $g_2 \in G$ as elements of G coincides with their product as elements of A_G.*

This follows from (2.7.1c).

II. *The group algebra A_G is associative.*

We prove this from the identity $(g_1g_2)g_3 = g_1(g_2g_3)$ in G (see (b) of 1.1 in chapter I). The reader may supply the details.

III. *A group algebra contains a unit.*

The unit in A_G is the identity element in G.

Plainly, a group algebra is finite-dimensional and dim $A_G = |G|$. We define an involution in A_G by setting

$$\left(\sum_g a(g)g \right)^* = \sum \overline{a(g)}g^{-1}. \tag{2.7.2}$$

We can easily verify that the conditions (1)–(4) for an involution in 2.5 are satisfied.

IV. *Under the definition of involution in (2.7.2), A_G is a symmetric algebra with a nondegenerate involution.*

Proof. From (2.7.1)–(2.7.2) we infer that

$$a^*a = \sum_{g',g''} \overline{a}(g')a(g'')g'^{-1}g''.$$

The coefficient $(a^*a)(e)$ in this sum is obtained by summing over all g', g'' such that $g' = g''$, so that

$$(a^*a)(e) = \sum_{g'} \bar{a}(g')a(g') = \sum_{g'} |a(g')|^2.$$

If we have $a^*a = 0$, then in particular $(a^*a)(e) = 0$, that is, $\sum_{g'} |a(g')|^2 = 0$. It follows that $a(g') = 0$ for all $g' \in G$ and so $a = 0$. $\square$

V. *For every $a \in A_G$ the element*

$$a' = \sum_g g^{-1}ag$$

belongs to the center $Z(A_G)$ of the algebra A_G.

Proof. For all $g_0 \in G$ we have

$$g_0^{-1}a'g_0 = \sum_g g_0^{-1}g^{-1}agg_0 = \sum_g (gg_0)^{-1}a(gg_0) = \sum_g g^{-1}ag = a'$$

and therefore

$$a'g' = g'a', \qquad \text{for all } g' \in G. \tag{2.7.3}$$

Multiplying both sides of (2.7.3) by $b(g')$ and summing over g', we obtain $a'b = ba'$ for all $b \in A_G$. Thus a' belongs to $Z(A_G)$. $\square$

To specify the sum $\sum_g a(g)g$ is to specify the function $a = \{a(g)\}$ defined on G, where $a(g)$ is the coefficient of g in the sum $\sum_g a(g)g$. Thus A_G can be considered as the set of all functions $a = \{a(g)\}$ on G. From (2.7.1) and (2.7.2) we find that

$$\alpha a = \{\alpha a(g)\}, \tag{2.7.4}$$
$$a + b = \{a(g) + b(g)\}. \tag{2.7.5}$$

To define multiplication of two functions on a group, it suffices to define the coefficient of g in (2.7.1c). To do this, set $g = g'g''$ and $g'' = g'^{-1}g$ in (2.7.1c). We can then rewrite (2.7.1c) as

$$ab = \sum_g \sum_g a(g')b(g'^{-1}g)g.$$

In other words, we have

$$ab = \left\{ \sum_g a(g')b(g'^{-1}g) \right\}. \tag{2.7.6}$$

Replacing g in (2.7.2) by g^{-1} we obtain

$$a^* = \overline{\{a(g^{-1})\}}. \tag{2.7.7}$$

Formulas (2.7.4)–(2.7.7) can be rewritten as

$$(\alpha a)(g) = \alpha a(g), \tag{2.7.4'}$$
$$(a + b)(g) = a(g) + b(g), \tag{2.7.5'}$$
$$(ab)(g) = \sum_{g'} a(g')b(g'^{-1}g), \tag{2.7.6'}$$
$$a^*(g) = \overline{a(g^{-1})}. \tag{2.7.7'}$$

The function $(ab)(g)$ defined in (2.7.6) is called the *convolution of the functions a and b*.

Thus the group algebra A_G of a finite group can be thought of as the set of all complex functions $a(g)$ on G, in which the operations of addition, multiplication by a number, multiplication, and involution are defined in formulas (2.7.4')–(2.7.7'). In the sequel we will use both definitions.

VI. *When we multiply the element $a = \sum_g a(g)g$ on the left (or right) by g_0, the function $a(g)$ becomes $a(g_0^{-1}g)$ (or $a(gg_0^{-1})$).*

Proof. The relations

$$g_0 a = \sum_g a(g)g_0 g = \sum_g a(g_0^{-1}g)g$$

prove the first assertion. The second is proved analogously. $\square$

The functions $a = \{a(g)\}$ of A_G also comprise the space $L^2(G)$. Recall that

$$(a, b) = \frac{1}{n} \sum_g a(g)\overline{b(g)} \tag{2.7.8}$$

is the scalar product of the elements a, b of $L^2(G)$. The expression for (a, b) can be written in terms of the linear functional $f_0(a)$ on A_G defined by

$$f_0(a) = a(e), \quad \text{for } a = \sum_g a(g)g.$$

We then have

$$(a, b) = \frac{1}{n} f_0(b^* a). \tag{2.7.9}$$

Formulas (2.7.1b) and (2.7.2) imply that

$$b^* a = \sum_{g', g''} \overline{b(g')}a(g'')g'^{-1}g'' \tag{2.7.10}$$

and $f_0(b^*a) = (b^*a)(e)$ is the sum of the summands in (2.7.10) for which $g' = g''$. Thus we have

$$f_0(b^*a) = \sum_{g'} \overline{b(g')}a(g') = n(a,b).$$

For a subset E of A_G, let $E^\perp$ denote the orthogonal complement of E in $L^2(G) = A_G$. If E is a subspace of A_G, we have

$$A_G = E \oplus E^\perp. \qquad (2.7.11)$$

VII. *If I_l is a left ideal in A_G, then $I_l^\perp$ is also a left ideal in A_G and*

$$A_G = I_l \oplus I_l^\perp. \qquad (2.7.12)$$

Proof. Let $a \in I_l$, $b \in I_l^\perp$ and $c \in A_G$. Then c^*a is also in I_l and therefore

$$(a, cb) = (c^*a, b) = 0.$$

Consequently, cb is in I_l, i.e., $I_l^\perp$ is a left ideal. The relation (2.7.12) follows from (2.7.11). $\square$

VIII. *Every left ideal I_l in A_G has the form*

$$I_l = A_G\varepsilon, \qquad (2.7.13)$$

where ε is a certain element of I_l. The element ε can be chosen so that

$$\varepsilon^2 = \varepsilon. \qquad (2.7.14)$$

In this case we have

$$I_l = \{a : a \in A_G, a\varepsilon = \varepsilon\}. \qquad (2.7.15)$$

Proof. We apply the decomposition (2.7.12) to the unit element e of A_G. We find

$$e = e' + e'', \qquad e' \in I_l, \qquad e'' \in I_l^\perp. \qquad (2.7.16)$$

Multiplying both sides of (2.7.16) on the right and left by e', we obtain

$$e' = e'^2 + e'e'', \qquad e' = e'^2 + e''e'.$$

Thus we have $e'e'' = e''e'$, and

$$e'e'' = e''e' \in I_l \cap I_l^\perp = \{0\}. \qquad (2.7.17)$$

It follows that $e'e'' = e''e' = 0$ and that $e' = e'^2$.

Since $e' \in I_l$, $e'' \in I_l^{\perp}$, we have

$$A_G e' \subset I_l, \qquad A_G e'' \subset I_l^{\perp}. \tag{2.7.18}$$

We conclude from (2.7.16) that

$$A_G = A_G e = A_G e' \oplus A_G e''.$$

Comparison with (2.7.12) and (2.7.18) gives

$$I_l = A_G e', \qquad I_l^{\perp} = A_G e''.$$

Taking $\varepsilon = e'$, we obtain (2.7.13) and (2.7.14). Suppose that (2.7.13) and (2.7.14) hold and let $a \in I_l$. Then we have $a = b\varepsilon$ for some $b \in A_G$ and thus $a\varepsilon = b\varepsilon^2 = b\varepsilon = a$. Conversely, if $a\varepsilon = a$, then plainly $a = a\varepsilon \in A_G \varepsilon = I_l$. $\square$

Remark. We can easily prove that

$$I_l^{\perp} = \{a : a \in A_G, \, a\varepsilon = 0\},$$

for $\varepsilon = e'$. An element $\varepsilon \in A_G$ is called *idempotent* if $\varepsilon \neq 0$ and $\varepsilon^2 = \varepsilon$. An idempotent ε is called *primitive* if it is not the sum of two nonzero idempotents.

Exercise

1. Let ε be an idempotent. Prove that $I_l = A_G \varepsilon$ is minimal in A_G if and only if ε is primitive.

2.8. Representations of a Group Algebra and Their Connection with Group Representations

Let G be a finite group. Let A_G be its group algebra, let T be a representation of A_G with representation space X. Since we have $G \subset A_G$, we can consider the restriction of the mapping T to G. Applying the relation $T(a_1 a_2) = T(a_1)T(a_2)$ (see (3) in 2.6) to the case $a_1 = g_1$, $a_2 = g_2$ and taking into consideration II in 2.7 and the equality $T(e) = 1$ (see (4) in 2.6), we conclude that this restriction is a representation $g \to T(g)$ of G. Conversely, let the representation $g \to T(g)$ be a representation of G with representation space X. An element $a = \sum_g a(g)g \in A_G$ is mapped into the operator

$$T(a) = \sum_g a(g)T(g). \tag{2.8.1}$$

Let us prove that the mapping $T : a \to T(a)$ is a representation of the algebra A_G. Properties (1), (2) and (4) of a representation of an algebra (see 2.6) are trivial to check. We can also verify that $T(a_1 a_2) = T(a_1)T(a_2)$ (see (3) in 2.6). Let $a_1 = \sum_g a_1(g)g$, $a_2 = \sum_g a_2(g)g$. By the definition of multiplication in A_G (see 2.7.1c), we have

$$a_1 a_2 = \sum_{g',g''} a_1(g') a_2(g'') g' g''.$$

Consequently, by virtue of (2.8.1) we also have

$$T(a_1 a_2) = \sum_{g',g''} a_1(g') a_2(g'') T(g'g'') = \sum_{g',g''} a_1(g') a_2(g'') T(g') T(g'')$$

$$= \sum_{g'} a_1(g') T(g') \sum_{g''} a_2(g'') T(g'') = T(a_1) T(a_2).$$

We have thus proved the following;

Theorem 1. *For every representation $T : a \to T(a)$ of the group algebra A_G of a group G there is a corresponding representation $g \to T(g)$ of this group, which is obtained by the restriction of the representation $a \to T(a)$ to G. Conversely, for every representation $g \to T(g)$ of G there is a corresponding representation $a \to T(a)$ of its group algebra A_G, defined by the formula (2.8.1). The representations of G and its group algebra A_G are said to* correspond to each other if *they are connected by the relation (2.8.1).*

Theorem 2. *A representation $g \to T(g)$ of G in a Euclidean space X is unitary if and only if the corresponding representation $a \to T(a)$ of its group algebra A_G is symmetric.*

Proof. Let the representation $g \to T(g)$ be unitary, so that

$$(T(g))^* = T(g^{-1}). \tag{2.8.2}$$

Then for $a = \sum_g a(g) g$ we have

$$(T(a))^* = \left(\sum_g a(g) T(g) \right)^* = \sum_g a(g) (T(g))^* = \sum_g \overline{a(g)} T(g^{-1}),$$

and on the other hand, $a^* = \sum_g \overline{a(g)} g^{-1}$ (see (2.7.3)) and therefore

$$T(a^*) = \sum_g \overline{a(g)} T(g^{-1}).$$

Thus we have

$$(T(a))^* = T(a^*), \tag{2.8.3}$$

that is, the representation $a \to T(a)$ is symmetric. Conversely, if the representation $a \to T(a)$ is symmetric, we apply relation (2.8.3) to the case $a = g$ and conclude that the corresponding representation $g \to T(g)$ is unitary. $\square$

Using the relation (2.8.1), the reader can easily verify the following statements.

I. *Two representations* $g \to T(g)$, $g \to S(g)$ *of a group G are equivalent (or unitarily equivalent) if and only if the corresponding representations of its group algebra* A_G *are equivalent (or unitarily equivalent).*

II. *A representation* $g \to T(g)$ *of a group G is irreducible if and only if the corresponding representation of the group algebra* A_G *is irreducible.*

III. *The relation* $T = T(1) + \cdots + T(k)$ *for representations of a group is equivalent to the same relation for the corresponding representations of its group algebra.*

Propositions I–III show that the problems of describing all irreducible representations of a group (up to equivalence) and the problem of decomposing a given representation into irreducible components are equivalent to the corresponding problems for its group algebra.

We apply the preceding results to the left regular representation of a group G (see 1.3). Recall that this representation $h \to \tilde{T}(h)$, $h \in G$, is defined in the space $L^2(G)$ by the formula

$$\tilde{T}(h)f(g) = f(h^{-1}g), \quad \text{for } f \in L^2(G). \tag{2.8.4}$$

Let us find the representation of the group algebra A_G that corresponds to the representation $h \to \tilde{T}(h)$. First we note that $L^2(G)$ and A_G consist of exactly the same functions on G, i.e., all functions on G. Thus an f in $L^2(G)$ belongs as well to A_G. Applying (2.8.1) and taking (2.7.6) into consideration, we conclude that

$$\tilde{T}(a)f(g) = \sum_h a(h)\tilde{T}(h)f(g) = \sum_h a(h)f(h^{-1}g) = (af)(g).$$

In other words (see also 2.6) we have the following.

IV. *There is a "left regular representation" of the group algebra that corresponds to the left regular representation of the group G.*

We conclude the following from Proposition III in 2.6 and I–IV above.

V. *A minimal left ideal* I_1 *of the algebra* A_G *is an invariant subspace for the left regular representation of the group G and the restriction of this representation to* I_1 *is irreducible.*

VI. *Corresponding to the decomposition*

$$A_G = I_l^1 + \cdots + I_l^m \tag{2.8.5}$$

*of the group algebra A_G into the direct sum of its minimal left ideals we have
the decomposition*

$$\tilde{T} = \tilde{T}^1 \dotplus \cdots \dotplus \tilde{T}^m \tag{2.8.6}$$

*of the left regular representation of the group G into the direct sum of irreducible
representations $\tilde{T}^1, \ldots, \tilde{T}^m$, where $\tilde{T}^1, \ldots, \tilde{T}^m$ are the restrictions of $\tilde{T}$ to
the ideals $I_l^1, \ldots, I_l^m$.*

According to theorem 1 in 1.5, the decomposition (2.8.6) contains all
(up to equivalence) irreducible representations of the (finite) group G.
Therefore, to find a complete system of irreducible representations it is
sufficient to find a decomposition of the algebra A_G into the direct sum of
its minimal left ideals.

To complete this section, we present a formula for the trace of an irreducible representation, which will be used later on. Remember (see VIII
in 2.7) that every left ideal I_l in A_G has the form $I_l = A_G \varepsilon$, where $\varepsilon \in I_l$.

VII. *Let the irreducible representation T of the group G be the restriction of
its left regular representation $\tilde{T}$ to the minimal left ideal*

$$I_l = A_G \varepsilon, \tag{2.8.7}$$

where ε is an idempotent[19]*; then we have*

$$\chi_T(g) = \sum_{g'} \varepsilon(g'^{-1} g^{-1} g'). \tag{2.8.8}$$

Proof. We define an operator P in A_G by setting $Px = x\varepsilon$ for $x \in A_G$. Since
ε is an idempotent, P is the projection of A_G onto I_l. Along with the operator
$T(a)$ in I_l, consider the operator $T'(a)$ in all of the algebra A_G, defined by

$$T'(a)x = \tilde{T}Px = \tilde{T}(a)x\varepsilon = ax\varepsilon.$$

Thus I_l is invariant under $T'(a)$ and $T(a)$ is the restriction of $T'(a)$ to
I_l. Therefore, (see III in 2.9, chapter I) we have

$$\operatorname{tr}(T(a)) = \operatorname{tr}(T'(a)). \tag{2.8.9}$$

Under the realization of the algebra A_G by functions

$$x(g) = \{x(g_1), \ldots, x(g_n)\}$$

[19] See the Remark after Proposition VIII in 2.7.

on $G = \{g_1, \ldots, g_n\}$, we have

$$T'(a)\{x(g)\} = \sum_{g'g''} a(g')x(g'^{-1}g'')\varepsilon(g''^{-1}g) = \sum_{g',g''} a(g')x(g'')\varepsilon(g''^{-1}g'^{-1}g).$$

Thus $T'(a)$ is the mapping of the n-dimensional space A_G of variables $\{x(g''_1), \ldots, x(g''_n)\}$ whose matrix is

$$t(g, g'') = \sum_{g'} a(g')\varepsilon(g''^{-1}g'^{-1}g).$$

Therefore we have

$$\text{tr}(T'(a)) = \sum_g t(g, g) = \sum_{g,g'} a(g')\varepsilon(g^{-1}g'^{-1}g) = \sum_{g,g'} a(g)\varepsilon(g'^{-1}g^{-1}g),$$

and a comparison with (2.8.9) gives us

$$\sum_{g,g'} a(g)\varepsilon(g'^{-1}g^{-1}g') = \text{tr}(T(a)) = \sum_g a(g)\,\text{tr}(T(g)) = \sum_g a(g)\chi_T(g).$$

This yields (2.8.8).   □

2.9. Further Properties of the Group Algebra

Theorem 1. *Let $T^1, \ldots, T^m$ be a complete system of irreducible representations of the finite group G. Let $n_1, \ldots, n_m$ be their dimensions. Let $t^1(a), \ldots, t^m(a)$ be their matrices in orthonormal bases. Then the mapping*

$$f : a \rightarrow \{t^1(a), \ldots, t^m(a)\}$$

is a symmetric isomorphism of the group algebra onto the direct sum of m complete matrix algebras of dimensions $n_1, \ldots, n_m$.

Proof. Let B be the direct sum of the complete matrix algebras $A(n_1 \times n_1)$, $\ldots$, $A(n_m \times n_m)$ in spaces of dimensions $n_1, n_2, \ldots, n_m$. The definition of a representation of an algebra implies that the mapping

$$f : a \rightarrow \{t^1(a), \ldots, t^m(a)\}$$

is a homomorphism of the algebra A_G into the algebra B; f is symmetric since $T^1, \ldots, T^m$ are unitary (theorem 2 in 2.8). We will prove that f maps A_G onto B. The formula $T(a) = \sum_g a(g)T(g)$ for $a = \sum_g a(g)g$ (see (2.8.1)) shows that

$$t^k(a) = \sum_g a(g)t^k(g), \qquad k = 1, 2, \ldots, m,$$

and therefore

$$t^k_{jv}(a) = \sum_g a(g)t^k_{jv}(g), \qquad j, v = 1, \ldots, n_k. \tag{2.9.1}$$

In (2.9.1) we set $a(g) = a^k_{j'v'}(g) = (n_{k'}/n)t^{k'}_{j'v'}(g)$. By virtue of the orthogonality relations (see (1.4.1), (1.4.2)) we obtain for $a = a^{k'}_{j'v'}$

$$t^k_{jv}(a^{k'}_{j'v'}) = \begin{cases} 1, & \text{for } k = k', j = j', v = v', \\ 0, & \text{in all other cases.} \end{cases}$$

In other words, in the corresponding system

$$\{t^1(a^{k'}_{j'v'}), \ldots, t^m(a^{k'}_{j'v'})\}, \tag{2.9.2}$$

only $t^{k'}(a^{k'}_{j'v'})$ is different from zero and in the matrix $t^{k'}(a^{k'}_{j'v'})$ the j'-th row and v'-th column is 1 and all the other entries are 0. Thus the set of all systems (2.9.2) forms a basis in B and so f maps A_G onto B.

Let us find Ker f. If a belongs to Ker f, then $t^k(a) = 0$ for all k, that is, we have

$$t^k_{jv}(a) = \sum_g a(g)t^k_{jv}(g) = 0 \tag{2.9.3}$$

for all k, j, v. But the functions $t^k_{jv}(g)$, $j, v = 1, \ldots, n_k$, $k = 1, \ldots, m$ are a complete system in $L^2(G) (= A_G)$ (theorem (2.1.4)) and (2.9.3) implies that $a(g) = 0$. Thus we have Ker $f = \{0\}$ and so f is an isomorphism. □

From this theorem we immediately infer the following.

Corollary 1. *The group algebra of a finite group is symmetrically isomorphic to the direct sum of complete matrix algebras.*

Theorem 2. *The center $Z(A_G)$ of the group algebra A_G of a finite group G consists of all functions $a(g)$ of G constant on conjugacy classes. Under the isomorphism $f: a \rightarrow \{t^1(a), \ldots, t^m(a)\}$, every element a of $Z(A_G)$ maps onto an element $\{1/n_1)\chi_1(a)1_{n_1}, \ldots, (1/n_m)\chi_m(a)1_{n_m}\}$, where $\chi_1, \ldots, \chi_m$ are the characters of the representations $T^1, \ldots, T^m$ and $1_{n_1}, \ldots, 1_{n_m}$ are the identity matrices of dimensions $n_1, \ldots, n_m$.*

Proof. If $a = \sum_g a(g)g \in Z(A_G)$, then $g'a = ag'$, which is to say

$$\sum_g a(g)g'g = \sum_g a(g)gg', \quad \text{for all } g' \in G. \tag{2.9.4}$$

Comparing the coefficients of g_1 on both sides of (2.9.4), we conclude that

$$a(g'^{-1}g_1) = a(g_1g'^{-1}), \quad \text{for all } g_1, g' \in G. \tag{2.9.5}$$

Setting $g_1 = gg'$, we obtain

$$a(g'^{-1}gg') = a(g), \quad \text{for all } g, g' \in G. \tag{2.9.6}$$

That is, the functions $a(g)$ are constant on conjugacy classes. Conversely, if (2.9.6) holds, (2.9.5) and (2.9.4) also hold. Multiplying both sides of (2.9.4) by $b(g')$ and summing over g' we conclude that

$$\sum_{g,g'} b(g')a(g)g'g = \sum_{g,g'} a(g)b(g')gg'.$$

That is, we have $ba = ab$ for all $b \in A_G$. This means that $a \in Z(A_G)$ and the first part of our theorem is proved.

Under the isomorphism f the center of the algebra A_G maps onto the center of the algebra B. By Theorem 1 this consists of all $\{t^1(a), \ldots, t^m(a)\}$, where $t^k(a)$ belongs to the center of the algebra $A(n_k \times n_k)$ and therefore $t^k(a) = \lambda_k(a)1$. Taking the trace in both sides of this last equality, we conclude that $\chi_k(a) = n_k\lambda_k(a)$, so that $\lambda_k(a) = (1/n_k)\chi_k(a)$ and $t^k(a) = (1/n_k)\chi_k(a)1_{n_k}$. $\quad\square$

§3. Representations of the Symmetric Group

3.1. Statement of the Problem

Remember that the group of all permutations of n elements $1, 2, \ldots, n$ is called *the symmetric group* S_n. Its order is $n!$ (see Example 5 in 1.7, chapter I). Every permutation $g \in S_n$ is the product of cycles with no common elements. Let $\alpha_1, \alpha_2, \ldots, \alpha_h$ be the lengths of these cycles, so that

$$\alpha_1 + \alpha_2 + \cdots + \alpha_h = n. \tag{3.1.1}$$

Since these cycles are commutative, we can write them as a product such that

$$\alpha_1 \geqslant \alpha_2 \geqslant \cdots \geqslant \alpha_h. \tag{3.1.2}$$

Two permutations $g, g' \in S_n$ are conjugate if and only if for them the number of cycles and the lengths of the corresponding cycles coincide, i.e., $h' = h$, $\alpha'_1 = \alpha_1, \ldots, \alpha'_h = \alpha_h$ (again see Example 5 in 1.7, chapter I). Thus the conjugacy classes of the group S_n are uniquely described by sequences of positive integers $\alpha_1, \alpha_2, \ldots, \alpha_h$ satisfying (3.1.1) and (3.1.2). The number of conjugacy classes in S_n is equal to the number of ways of writing the integer n as the sum of positive integers $\alpha_1, \alpha_2, \ldots, \alpha_h$ with $\alpha_1 \geqslant \alpha_2 \geqslant \cdots \geqslant \alpha_h$. On the other hand, a complete system of irreducible representations of S_n contains the same number of representations as there are conjugacy classes in S_n (theorem 3 in 1.7). Therefore, to obtain a complete system of irreducible representations of S_n, it suffices to construct for every such set an irreducible representation of S_n such that the representations corresponding to the various sets are inequivalent.

The aim of the present section is to carry out this construction explicitly.

3.2. Young Schemes and Diagrams

Consider any sequence $\alpha = (\alpha_1, \alpha_2, \ldots, \alpha_h)$ of positive integers satisfying conditions (3.1.1) and (3.1.2). We construct a scheme as in figure 1 of h rows. The k-th row consists of α_k boxes ($k = 1, \ldots, h$) and the j-th box of the $(k + 1)$-st row is placed directly under the j-th box of the k-th row. This figure is called the *Young scheme* corresponding to the sequence α. We denote it also by α.

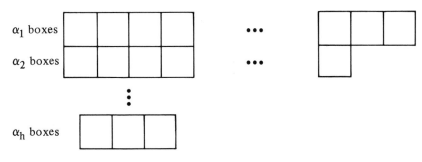

Figure 1

For example, for $n = 3$ we have just three Young schemes. They are shown in figure 2.

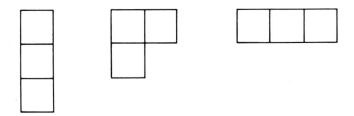

Figure 2

The number of boxes in a Young scheme is equal to $\alpha_1 + \cdots + \alpha_h = n$. Therefore we can place the numbers $1, 2, \ldots, n$ in the boxes. Any placement of the numbers $1, 2, \ldots, n$ in the boxes of the scheme α is called a *Young diagram* corresponding to the scheme α and is denoted by $\sum_\alpha$. Note that there are $n!$ distinct diagrams for each scheme, as the numbers $1, 2, \ldots, n$ can be arranged in $n!$ different orders. Consider a diagram $\sum_\alpha$. For $g = \begin{pmatrix} 1 & 2 & \cdots & n \\ k_1 & k_2 & \cdots & k_n \end{pmatrix} \in S_n$, we denote by $g \sum_\alpha$ the diagram obtained from α by replacing each number j in its box by the number k_j. For a fixed $\sum_\alpha$ and for all possible $g \in G$, $g \sum_\alpha$ runs through all the diagrams that correspond to the scheme α. The rows of the diagram $\sum_\alpha$ can be considered as cycles of lengths $\alpha_1, \alpha_2, \ldots, \alpha_h$. The diagram thus defines a certain permutation, namely, the product of these cycles.

I. *If the diagram* $\sum_\alpha$ *corresponds to a permutation* g_0, *the diagram* $g \sum_\alpha$ *then corresponds to the permutation* gg_0g^{-1}.

The proof is a direct computation, which we leave to the reader. It suffices to verify the proposition for transpositions, since every permutation g is a product of transpositions.

For a given Young diagram $\sum_\alpha$, let P_α be the set of all permutations that permute the numbers in each *row* of $\sum_\alpha$ among themselves. It is clear that P_α is a subgroup of S_n. We denote the permutations in P_α by letters such as p. Let Q_α be the set of all permutations in S_n that permute elements of each *column* of $\sum_\alpha$ among themselves. The set Q_α is also a subgroup of S_n. We denote elements of Q_α by q. Let m_α and m'_α be the orders of the groups P_α and Q_α, respectively. Note that P_α and Q_α depend not only upon the scheme α but also on the actual diagram $\sum_\alpha$.

II. *If we go from* $\sum_\alpha$ *to* $g \sum_\alpha$, *the groups* P_α *and* Q_α *go to* $gP_\alpha g^{-1}$ *and* $gQ_\alpha g^{-1}$.

Like proposition I, this proposition is proved by a direct computation, which we omit. Again it suffices to check II only for transpositions g.

Let A be the group algebra of the group S_n: $A = A_{S_n}$. We consider the following elements of the algebra A:

$$f_\alpha = \sum_p p, \qquad \varphi_\alpha = \sum_q \sigma_q q, \tag{3.2.1}$$

where

$$\sigma_q = \begin{cases} 1, & \text{if } q \text{ is an } even \text{ permutation,} \\ -1, & \text{if } q \text{ is an } odd \text{ permutation.} \end{cases} \tag{3.2.2}$$

III. *The following identities hold:*

$$pf_\alpha = f_\alpha p = f_\alpha; \tag{3.2.3}$$

$$\sigma_q q \varphi_\alpha = \varphi_\alpha \sigma_q q = \varphi_\alpha; \tag{3.2.4}$$

$$f_\alpha^2 = m_\alpha f_\alpha, \qquad \varphi_\alpha^2 = m'_\alpha \varphi_\alpha. \tag{3.2.5}$$

Proof. It is clear that $\sigma_q \sigma_{q'} = \sigma_{qq'}$. Thus (3.2.1) implies that

$$pf_\alpha = p \sum_{p'} p' = \sum_{p'} pp' = \sum_{p'} p' = f_\alpha, \qquad f_\alpha p = \sum_{p'} p'p = \sum_{p'} p' = f_\alpha,$$

$$\sigma_q q \varphi_\alpha = \sigma_q q \sum_{q'} \sigma_{q'} q' = \sum_{q'} \sigma_{qq'} qq' = \sum_{q'} \sigma_{q'} q' = \varphi_\alpha.$$

Analogously, we have $\varphi_\alpha \sigma_q q = \varphi_\alpha$.

It follows that

$$f_\alpha^2 = f_\alpha f_\alpha = f_\alpha \sum_p p = \sum_p f_\alpha = m_\alpha f_\alpha,$$

and similarly $\varphi_\alpha^2 = m_\alpha' \varphi_\alpha$. □

3.3. A Combinatorial Lemma

Let $\alpha = (\alpha_1, \alpha_2, \ldots, \alpha_h)$ and $\beta = (\beta_1, \beta_2, \ldots, \beta_{h'})$. We write $\alpha > \beta$ if the first nonzero difference $\alpha_k - \beta_k$ is positive; we agree that $\alpha_k = 0$ if $k > h$ and $\beta_k = 0$ if $k > h'$. This ordering of schemes is called *lexicographic* since it corresponds to the way we order words in a dictionary.

Lemma. *Suppose that* (a) α *is greater than or equal to* β; *and* (b) *no two elements in the same column of the diagram* $\sum_\beta$ *are in the same row of* $\sum_\alpha$. *We then have:*

(1) $\alpha = \beta$;   (3.3.1)

(2) $\sum_\beta = pq \sum_\alpha$, *where* $p \in P_\alpha$, $q \in Q_\alpha$ *and* P_α, Q_α *are the groups* P *and* Q *for* $\sum_\alpha$.

Proof. The condition $\alpha \geq \beta$ gives in particular

$$\alpha_1 \geq \beta_1.$$ (3.3.2)

The first row of $\sum_\alpha$ consists of α_1 numbers, which by condition (b) are located in α_1 distinct columns of the diagram $\sum_\beta$. As $\sum_\beta$ has β_1 columns, it follows that $\beta_1 \geq \alpha_1$. Combining this with (3.3.2), we see that $\beta_1 = \alpha_1$. Since all the elements of the first row of $\sum_\alpha$ are located in different columns in $\sum_\beta$, there is a permutation $q_1' \in Q_\beta$ (the group Q for $\sum_\beta$), such that the first rows of the diagrams $\sum_\alpha$ and $q_1' \sum_\beta$ consist of the same numbers, but perhaps in different orders.

Since we have $\alpha \geq \beta$ and $\alpha_1 = \beta_1$, we have $\alpha_2 \geq \beta_2$. Discard the first rows of $\sum_\alpha$ and $q_1' \sum_\beta$ and apply the preceding argument to the resulting diagram. We conclude that $\alpha_2 = \beta_2$. Also there is a permutation q_2' that corresponds to the diagram $q_1' \sum_\beta$ and leaves in place the elements of the first line of $q_1' \sum_\beta$ for which the second rows of $\sum_\alpha$ and $q_2' q_1' \sum_\beta$ consist of the same elements, possibly in different orders.

Repeating this reasoning, we conclude that $\alpha = \beta$ and we also obtain the diagram $q_h' q_{h-1}' \cdots q_1' \sum_\beta$, the rows of which consist of the same elements as the corresponding rows of $\sum_\alpha$, but perhaps not in the same orders. Consequently, for some permutation $p \in P_\alpha$ (corresponding to $\sum_\alpha$) we obtain the diagram

$$p \sum_\alpha = q' \sum_\beta$$ (3.3.3)

where $q' = q_h' q_{h-1}' \cdots q_1'$. Clearly, every q_j', and therefore also q', permutes only elements in one and the same column of $\sum_\beta$. It follows that $q' \in Q_\beta$

(where Q_β corresponds to $\sum_\beta$). By applying Proposition II to $\sum_\beta = q'^{-1}p\sum_\alpha$ (see (3.3.3)), we see that $q' = q'^{-1}pq^{-1}(q'^{-1}p)^{-1}$, that is,

$$q' = q'^{-1}pq^{-1}p^{-1}q', \qquad\qquad (3.3.4)$$

where $q \in Q_\alpha$ (it is convenient to write q^{-1} instead of q). From (3.3.4) we find that $q' = pq^{-1}p^{-1}$ and thus (3.3.3) implies that

$$\sum_\beta = q'^{-1}p\sum_\alpha = (pqp^{-1})p\sum_\alpha = pq\sum_\alpha. \quad \square$$

Corollary. *For $\alpha > \beta$ we have*

$$f_\alpha g\varphi_\beta g^{-1} = 0, \quad \text{for all } g \in S_n \qquad\qquad (3.3.5)$$

and

$$f_\alpha A\varphi_\beta = \{0\}. \qquad\qquad (3.3.6)$$

Proof. First we prove that

$$f_\alpha\varphi_\beta = 0, \quad \text{for } \alpha > \beta. \qquad\qquad (3.3.7)$$

Let $\sum_\alpha$ and $\sum_\beta$ be the diagrams from which f_α and f_β are constructed (see 3.2). Since α is greater than β, the preceding lemma shows that there exist two numbers i, k that lie in the same row of $\sum_\alpha$ and in the same column of $\sum_\beta$. Let t be the transposition of these numbers: $t = (i,k)$. Then t belongs to P_α for $\sum_\alpha$ and to Q_β for $\sum_\beta$. The identities (3.2.3) and (3.2.4) show that

$$f_\alpha = f_\alpha t, \qquad \varphi_\beta = \sigma_t t\varphi_\beta = -t\varphi_\beta$$

and therefore[20]

$$f_\alpha\varphi_\beta = f_\alpha t(-t\varphi_\beta) = -f_\alpha t^2\varphi_\beta = -f_\alpha\varphi_\beta.$$

The last equality is possible only for $f_\alpha\varphi_\beta = 0$ and so we have proved (3.3.7). To prove the relations (3.3.5) it suffices to note that $g\varphi_\beta g^{-1}$ is φ_β, constructed for $g\sum_\beta$ (see II in 3.2). Therefore (3.3.5) follows from (3.3.7) for $\varphi'_\beta = g\varphi_\beta g^{-1}$, corresponding to $g\sum_\beta$. By multiplying both sides of (3.3.5) on the right by $a(g)g$ and then summing over g, we obtain $f_\alpha\sum_g a(g)g\varphi_\beta = 0$, that is, $f_\alpha a\varphi_\beta = 0$ for all $a \in A$. This proves (3.3.6). $\square$

3.4. Young Symmetrizers

For a given diagram $\sum_\alpha$, we set

$$h_\alpha = f_\alpha\varphi_\alpha = \sum_{p,q} \sigma_q pq. \qquad\qquad (3.4.1)$$

---

[20] Recall that $t^2 = e$ and $\sigma_t = -1$.

Since the summand $h_\alpha(e)$ (for $p = q = e$) is $\sigma_e = 1$, we have $h_\alpha \neq 0$ and

$$h_\alpha(e) = 1. \tag{3.4.1a}$$

The element h_α was first found by Young. It is called the *Young symmetrizer* corresponding to the diagram $\sum_\alpha$.

I. *For h_α, we have*

$$ph_\alpha\sigma_q q = h_\alpha. \tag{3.4.2}$$

By (3.2.3) and (3.2.4) we have

$$ph_\alpha\sigma_q q = pf_\alpha\varphi_\alpha\sigma_q q = f_\alpha\varphi_\alpha = h_\alpha. \quad \square$$

II. *Suppose that for some $a \in A$ the relations*

$$pa\sigma_q q = a \tag{3.4.3}$$

hold for all $p \in P_\alpha$ and $q \in Q_\alpha$. Then we have

$$a = \lambda h_\alpha, \tag{3.4.4}$$

where λ is a constant.

Proof. Suppose (3.4.3) holds for an element $a = \sum_g a(g)g$, that is,

$$\sum_g \sigma_q a(g)pgq = \sum_g a(g)g, \quad \text{for all } p \in P_\alpha, g \in Q_\alpha. \tag{3.4.5}$$

Comparing the coefficients on both sides of (3.4.5) for $g = pq$ we conclude that

$$\sigma_q a(e) = a(pq). \tag{3.4.6}$$

Suppose that g_0 does not have the form pq. We will prove that $a(g_0) = 0$. Consider $\sum_\alpha$ and $g_0 \sum_\alpha$. According to the lemma in 3.3, there exist two numbers j, k, lying in the same row of $\sum_\alpha$ and in the same column of $g_0 \sum_\alpha$. Again we set $t = (j, k)$. Then we have $t \in P_\alpha$ and $t \in Q'_\alpha$, where Q'_α corresponds to the diagram $g_0 \sum_\alpha$.

From II in 3.2 we see that $g_0^{-1}tg_0 \in Q_\alpha$. We set $p = t, q = g_0^{-1}tg_0$ in (3.4.5) and compare the coefficients of g_0 in both sides of the resulting equality. On the left side g_0 appears for $g = g_0$, since $pg_0q = tg_0g_0^{-1}tg_0 = g_0$ and cannot be obtained for any other g. Therefore we have $\sigma_q a(g_0) = a(g_0)$ and this can happen only if $a(g_0) = 0$, since $\sigma_q = -1$ $(q = (j', k')$, where j', k' are obtained from (j, k) by applying the permutation g_0). Thus we have

$$a(g) = \begin{cases} \sigma_q a(e), & \text{for } g = pq, \\ 0, & \text{if } g \text{ is not of the form } pq. \end{cases}$$

Consequently we have

$$a = \sum_{p,q} \sigma_q a(e)pq = a(e) \sum_{p,q} \sigma_q pq = a(e)h_\alpha,$$

which we wished to prove. □

III. *The element h_α is Hermitian.*

According to the definition of involution in a group algebra (see (2.7.3)) and as a consequence of the lemma in 3.3, we have

$$h_\alpha^* = \left(\sum_{p,q} \sigma_q pq \right)^* = \sum_{p,q} \sigma_q q^{-1}p^{-1} = \sum_{p,q} \sigma_{q^{-1}} pq = \sum_{p,q} \sigma_q pq = h_\alpha$$

($\sigma_q = \sigma_{q^{-1}}$ since q and q^{-1} are both even or both odd.)

IV. *If $\alpha \neq \beta$, we have*

$$h_\alpha h_\beta = 0. \tag{3.4.7}$$

Proof. From the corollary in 3.3 we have

$$h_\alpha h_\alpha = f_\alpha \varphi_\beta f_\beta \varphi_\beta \in f_\alpha A \varphi_\beta = (0). □$$

V. *For every $b \in A$ we have*

$$h_\alpha b h_\alpha = \lambda h_\alpha, \tag{3.4.8}$$

where λ is a number (depending n α). In particular (for $b = e$), we obtain

$$h_\alpha^2 = \mu_\alpha h_\alpha, \tag{3.4.9}$$

where μ_α is a number.

Proof. From (3.4.2) we infer that the element $a = h_\alpha b h_\alpha$ satisfies the condition (3.4.3) so that (3.4.8) follows from Proposition II. □

3.5. Construction of a Complete System of Irreducible Representations of the Group S_n

We set

$$I^\alpha = Ah_\alpha. \tag{3.5.1}$$

Clearly I^α is a left ideal in A and $I^\alpha \neq \{0\}$ since $h_\alpha \neq 0$.

I. *The ideal I^α is a minimal left ideal in A.*

Proof. We first note that

$$h_\alpha I^\alpha \subset Ch_\alpha, \tag{3.5.2}$$

where C is the field of complex numbers. To see this, write

$$h_\alpha I^\alpha = h_\alpha A h_\alpha.$$

Consequently, if $b \in h_\alpha I^\alpha$, we have $b = h_\alpha a h_\alpha$ for some $a \in A$. Therefore for all $p \in P_\alpha, q \in Q_\alpha$ we have

$$p b \sigma_q q = p h_\alpha a h_\alpha \sigma_q q = h_\alpha a h_\alpha = b$$

(see (3.4.1), (3.2.3) and (3.2.4)). By virtue of II in 3.4 we thus conclude that $b = \lambda h_\alpha$ and (3.5.2) is proved.

Let I_l be a left ideal contained in I^α:

$$I_l \subset I^\alpha. \tag{3.5.3}$$

That is, we have

$$h_\alpha I_l \subset h_\alpha I^\alpha \subset Ch_\alpha$$

(see (3.5.2)). But Ch_α is one-dimensional. Therefore only the following two cases are possible:

(1) $h_\alpha I_l = Ch_\alpha$,
(2) $h_\alpha I_l = \{0\}$.

In the first case we have $I^\alpha = Ah_\alpha = ACh_\alpha = AhI_l \subset I_l$ and therefore $I_l = I^\alpha$.

In the second case we have $I_l^2 = I_l I_l \subset I^\alpha I_l = Ah_\alpha I_l = \{0\}$; consequently, $I_l = \{0\}$ by virtue of V in 2.5, since involution in a group algebra is non-degenerate (see IV in 2.7).

Thus, a left ideal contained in I^α either coincides with I_l or is equal to $\{0\}$. This means that I^α is minimal. $\square$

II. *The space I^α is invariant under the left regular representation $\tilde{T}$ of the group S_n and the restriction $\tilde{T}^\alpha$ of the representation $\tilde{T}$ to I^α is irreducible.*

This assertion follows directly from V of 2.8 and from I.

III. *For $\alpha \neq \beta$, the representations $\tilde{T}^\alpha$ and $\tilde{T}^\beta$ are inequivalent.*

Proof. If α does not equal β then $\alpha > \beta$ or $\alpha < \beta$. Suppose that α is greater than β. Then by virtue of (3.3.7) we have

$$f_\alpha A h_\beta = f_\alpha A f_\beta \varphi_\beta \subset f_\alpha A \varphi_\beta = \{0\},$$

so that

$$f_\alpha I^\beta = f_\alpha A h_\beta = \{0\}. \tag{3.5.4}$$

On the other hand, we have

$$f_\alpha I^\alpha = \{0\}. \tag{3.5.5}$$

(Note that $I^\alpha = A h_\alpha$ contains the element h_α, for which $f_\alpha h_\alpha = h_\alpha \neq 0$.)

Assume that $\tilde{T}^\alpha$ and $\tilde{T}^\beta$ are equivalent. The corresponding representations $a \to \tilde{T}^\alpha(a)$ and $a \to \tilde{T}^\beta(a)$ of the algebra A are also equivalent (see I in 2.8). Hence there exists a linear one-to-one mapping U of the space I^α onto I^β for which $\tilde{T}^\alpha(a)$ turns into $\tilde{T}^\beta(a)$:

$$\tilde{T}^\alpha(a) = U^{-1}\tilde{T}^\beta(a)U.$$

In particular, for $a = f_\alpha$ we have

$$\tilde{T}^\alpha(f_\alpha) = U^{-1}\tilde{T}^\beta(f_\alpha)U.$$

But this is impossible, since (3.5.4) and (3.5.5) imply that

$$\tilde{T}^\alpha(f_\alpha)I^\alpha = f_\alpha I^\alpha \neq (0),$$
$$U^{-1}\tilde{T}^\beta(f_\alpha)UI^\alpha = U^{-1}\tilde{T}^\beta(f_\alpha)I^\beta = U^{-1}f_\alpha I^\beta = (0).$$

Consequently, T and T are inequivalent. $\square$

Theorem. *For every Young scheme α choose some Young diagram $\sum_\alpha$. For this diagram construct the element h_α of the group algebra $A = A_{S_n}$:*

$$h_\alpha = \sum_{p,q} \sigma_q pq, \qquad p \in P_\alpha, q \in Q_\alpha,$$

$$\sigma_q = \begin{cases} 1, & \text{if } q \text{ is an even permutation,} \\ -1, & \text{if } q \text{ is an odd permutation.} \end{cases}$$

Then the sets $I^\alpha = A h_\alpha$ are subspaces invariant under the left regular representation $\tilde{T}$ of the group S_n and the restrictions $\tilde{T}^\alpha$ of the representation $\tilde{T}$ to I^α form a complete system of irreducible representations of the group S_n, as α runs through all Young schemes.

Proof. According to Propositions II and III, the representations $\tilde{T}^\alpha$ and $\tilde{T}^\beta$ are inequivalent for $\alpha \neq \beta$. On the other hand, the number of distinct α is equal to the number of conjugacy classes in S_n and consequently coincides with the number of representations in any complete system (see 3.1 and 3.2).

$\square$

From the theorem just proved we obtain the following practical device for constructing a complete system of representations $\tilde{T}$:

(1) enumerate all the elements of the group S_n in some order $g_1, \ldots, g_r (r = n!)$;
(2) choose a Young scheme α and a Young diagram $\sum_\alpha$;
(3) define the subgroups P_α, Q_α and the element h_α from the diagram $\sum_\alpha$;
(4) in the system of elements $g_1 h_\alpha, g_2 h_\alpha, \ldots, g_r h_\alpha$, discard each element that is a linear combination of the preceding elements; the remaining system, which we write as

$$a_1 = g_{k_1} h_\alpha, \qquad a_2 = g_{k_2} h_\alpha, \ldots, a_n = g_{n_\alpha} h_\alpha, \qquad k_1 = 1,$$

forms a basis in $I^\alpha = A h_\alpha$;

(5) therefore we have

$$\tilde{T}(g) a_j = g a_j = \sum_{s=1}^{n_\alpha} t_{sj}(g) a_s, \qquad j = 1, \ldots, n. \tag{3.5.6}$$

This formula defines the matrix elements of the representation $\tilde{T}^\alpha$. Let us apply the process of orthonormalization under the scalar product in $L^2(S_n)$ to $a_1, \ldots, a_n$. We obtain an orthonormal basis in I^α, in which the matrix of $\tilde{T}^\alpha$ is unitary. See Molčanov [1*] for more on the actual computation of matrix elements of the representations $\tilde{T}^\alpha$.

3.6. The Characters of Irreducible Representations of the Symmetric Group

We write

$$\chi_\alpha = \chi_{\tilde{T}^\alpha} \tag{3.6.1}$$

and find expressions for χ_α in terms of the symmetrizers h_α. As a preliminary we prove the following propositions.

I. *The number μ_α in the identity (3.4.9) is given by*

$$\mu_\alpha = n!/n_\alpha, \tag{3.6.2}$$

where n_α is the dimension of the representation $\tilde{T}^\alpha$.

Proof. Let C_α be the linear operator in A defined by

$$C_\alpha a = a h_\alpha. \tag{3.6.3}$$

Clearly we have

$$C_\alpha A = I^\alpha \tag{3.6.4}$$

and in view of (3.4.9) $C_\alpha a h_\alpha = a h_\alpha^2 = \mu a h_\alpha = \mu_\alpha a h_\alpha$, so that

$$C_\alpha = \mu_\alpha 1, \quad \text{on } I^\alpha. \tag{3.6.5}$$

We choose an arbitrary basis $a_1, \ldots, a_p$ in I^α and extend it arbitrarily to a basis $a_1, \ldots, a_{n!}$ in A. By (3.6.4) and (3.6.5), the matrix c_α of the operator C_α in this basis has the form

$$c_\alpha = \left\| \begin{matrix} \mu_\alpha 1_\alpha & * \\ 0 & 0 \end{matrix} \right\|,$$

where 1_α is the identity matrix of order n_α. Hence we have

$$\text{tr } C_\alpha = \mu_\alpha n_\alpha. \tag{3.6.6}$$

On the other hand, (3.6.3) implies that

$$(C_\alpha a)(g) = \sum_{g_1} a(g_1) h_\alpha(g_1^{-1} g).$$

That is, C_α is a linear transformation of the variables $a(g)$, $a \in G$, with matrix $c_\alpha(g_1, g) = h_\alpha(g_1^{-1} g)$. Thus we have

$$\text{tr } C_\alpha = \sum_g c_\alpha(g, g) = \sum_g h_\alpha(g^{-1} g) = \sum_g h_\alpha(e) = \sum_g I = n! \tag{3.6.7}$$

(see (3.4.1a)). A comparison of the right sides of (3.6.6) and (3.6.7) yields the identity (3.6.2).

From (3.6.2) we conclude that $\mu_\alpha > 0$. We set

$$e_\alpha = \frac{1}{\mu_\alpha} h_\alpha = \frac{n_\alpha}{n!} h_\alpha. \quad \square \tag{3.6.8}$$

II. *The element e_α is an idempotent in I^α.*

Proof. It is clear that $e_\alpha \in I^\alpha$ and $e_\alpha \neq 0$; also, from (3.4.9) and (3.6.8) we find

$$e_\alpha^2 = \frac{1}{\mu_\alpha^2} h_\alpha^2 = \frac{1}{\mu_\alpha^2} \mu_\alpha h_\alpha = \frac{-1}{\mu_\alpha} h_\alpha = e_\alpha. \quad \square$$

Theorem 1. *The character χ_α of the irreducible representation T^α of the group S_n is written in terms of the symmetrizer h_α as*

$$\chi_\alpha(g) = \frac{n_\alpha}{n!} \sum_{g_1} h_\alpha(g_1^{-1} g^{-1} g_1), \tag{3.6.9}$$

where n_α is the dimension of the representation T^α.

This theorem follows directly from (2.8.8) and (3.6.8), since $I^\alpha = Ah_\alpha = Ae_\alpha$.

We now find formulas which express n_α and χ_α directly in terms of the numbers $\alpha_1, \alpha_2, \ldots, \alpha_h$, that define the scheme α.

Remember that $\chi_\alpha(g)$ depends only on the conjugacy class containing g. Denote this class (a subset of S_n) by β. We may thus write

$$\chi_\alpha(g) = \chi_\alpha(\beta), \quad \text{for } g \in \beta.$$

Now suppose that g contains β_1 cycles of length 1, β_2 cycles of length $2, \ldots, \beta_q$ cycles of length q, so that

$$1\beta_1 + 2\beta_2 + \cdots + q\beta_q = n. \tag{3.6.10}$$

The set β is uniquely defined by the numbers $\beta_1, \beta_2, \ldots, \beta_q$. We express this fact by writing

$$\beta = (\beta_1, \beta_2, \ldots, \beta_q). \tag{3.6.11}$$

A study of formula (3.6.9),[21] which we omit, leads to the following result.

Theorem 2. *The dimension n_α of the irreducible representation T^α of the group S_n corresponding to the diagram $\alpha = (\alpha_1, \ldots, \alpha_h)$ is given by the formula*

$$n_\alpha = n! \frac{D_\alpha}{\alpha_1! \alpha_2! \cdots \alpha_h!}. \tag{3.6.12}$$

Here we define

$$D_\alpha = \prod_{p < q} (l_p - l_q), \tag{3.6.13a}$$

$$l_1 = \alpha_1 + (h-1), \quad l_2 = \alpha_2 + (h-2), \ldots, l_h = \alpha_h. \tag{3.6.13b}$$

For the characters χ_α of the representations T^α the following identity holds:

$$\sigma_\beta \left| \xi^{h-1}, \ldots, 1 \right| = \sum_\alpha \chi_\alpha(\beta) \left| \xi^{l_1}, \ldots, \xi^{l_h} \right|. \tag{3.6.14}$$

Here we have $\sigma_\beta = (\xi_1 + \cdots + \xi_h)^{\beta_1}(\xi_1^2 + \cdots + \xi_h^2)^{\beta_2} \cdots (\xi_1^q + \cdots + \xi_h^q)^{\beta_q}$,

$$\left| \xi^{p_1}, \ldots, \xi^{p_h} \right| = \begin{vmatrix} \xi_1^{p_1} & \cdots & \xi_1^{p_h} \\ \xi_2^{p_1} & \cdots & \xi_2^{p_h} \\ \cdots & \cdots & \cdots \\ \xi_h^{p_1} & \cdots & \xi_h^{p_h} \end{vmatrix}. \tag{3.6.15}$$

Summation in (3.6.14) is over all α with a fixed h, that is, over all $\alpha_1, \alpha_2, \ldots, \alpha_h$, for which

$$\alpha_1 \geqslant \alpha_2 \geqslant \cdots \geqslant \alpha_h > 0, \quad \alpha_1 + \cdots + \alpha_h = n.$$

---

[21] For a detailed proof of (3.6.12) and of Theorem 3 below, see H. Weyl [1], chapter VII.

We note also the following recurrence rule for computing the characters χ_α.

Theorem 3. *If the set contains a cycle of length v and β^1 is the set obtained from β by deleting this cycle, then we have*

$$\chi_{\alpha_1,\cdots,\alpha_h}(\beta) = \chi_{\alpha_1-v,\alpha_2,\cdots,\alpha_h}(\beta) + \chi_{\alpha_1,\alpha_2-v,\cdots,\alpha_h}(\beta^1) + \cdots. \quad (3.6.16)$$

We agree that if the indices $\alpha'_1, \ldots, \alpha'_h$ for any χ on the right side fail to satisfy the condition

$$\alpha'_1 \geqslant \alpha'_2 \geqslant \cdots \geqslant \alpha'_h \geqslant 0, \quad (3.6.17)$$

then the following steps must be taken.

(1) *If (3.6.17) fails at the last place, that is, $\alpha'_h < 0$, then the corresponding χ is to be deleted;*

(2) *If (3.6.17) fails earlier, i.e., $\alpha'_1 \geqslant \cdots \alpha'_{j-1} \geqslant \alpha'_{j+1} \geqslant \cdots \alpha'_h$, but $\alpha'_j < \alpha'_{j+1}$, then we take the same steps for $\alpha'_{j+1} - \alpha'_{j-1}$ and replace*

$$\chi \cdots {}_{\alpha'_j \alpha'_{j+1}} \cdots \quad (3.6.18)$$

by

$$-\chi \cdots {}_{\alpha'_{j-1} \alpha'_{j+1}} \cdots$$

for $\alpha'_{j+1} - \alpha'_j \geqslant 2$.

In the last case for the new α'_j in (3.6.18) the "irregularity" $\alpha'_j < \alpha'_{j+1}$ is either eliminated or $\alpha'_{j+1} - \alpha'_j$ is reduced by one.[22]

3.7. Decomposition of the Left Regular Representation of the Group S_n into its Irreducible Components

First we decompose the left regular representation $\tilde{T}$ of the group S_n into representations that are multiples of irreducible representations. To do this, we set

$$\varepsilon_\alpha = \frac{1}{\mu^2}\sum_{g_1} g_1 h_\alpha g_1^{-1} = \frac{1}{\mu}\sum_{g_1} g_1 e_\alpha g_1^{-1}, \quad (3.7.1)$$

where h_α is constructed for a fixed diagram $\sum_\alpha$ (see (3.4.1)), and $\mu = \mu_\alpha = n_\alpha/n!$ (see (3.6.2)). Clearly (3.7.1) means that

$$\varepsilon_\alpha(g) = \frac{1}{\mu^2}\sum_{g_1} h_\alpha(g_1^{-1}gg_1) = \frac{1}{\mu}\sum_{g_1} e_\alpha(g_1^{-1}gg_1). \quad (3.7.2)$$

---

[22] This rule was proved in this general form by Murnaghan (see Murnaghan [1]).

Comparison with (3.6.9) leads to

$$\varepsilon_\alpha(g) = \frac{1}{\mu_\alpha} \chi_\alpha(g^{-1}). \qquad (3.7.3)$$

I. *The elements ε_α have the following properties:*

(1) $\varepsilon_\alpha \neq 0$;
(2) ε_α *is a Hermitian idempotent;*
(3) ε_α *belongs to the center $Z(A_{S_n})$ of the algebra A_{S_n};*
(4) $\varepsilon_\alpha \varepsilon_\beta = 0$ *and* $(\varepsilon_\alpha, \varepsilon_\beta) = 0$ *for $\alpha \neq \beta$;*
(5) *the elements ε_α, as α runs through all schemes α, form a basis for $Z(A_{S_n})$.*

Proof.

(1) From (3.4.1) and (3.7.1) we have

$$\varepsilon_\alpha = \frac{1}{\mu^2} \sum_{g_1} \sum_{p,q} \sigma_q g_1^{-1} pqg_1. \qquad (3.7.4)$$

The coefficient $\varepsilon_\alpha(e)$ is obtained in the sum (3.7.4) from the summands for which $g_1^{-1} pqg_1 = e$, that is $pq = e$, and this is possible only for $p = e, q = e$. Thus we have

$$\varepsilon_\alpha(e) = \frac{1}{\mu^2} \sum_{g} \sigma_e g^{-1} g = \frac{1}{\mu^2} \sum_{g} 1 = \frac{n!}{\mu^2} = \frac{n_\alpha^2}{n!}, \qquad (3.7.5)$$

and therefore $\varepsilon_\alpha \neq 0$.

(2) Since h_α is Hermitian (see III in 3.4), the definition of involution (2.7.2) shows that

$$\varepsilon_\alpha^* = \frac{1}{\mu^2} \sum_{g} (g^{-1} h_\alpha g)^* = \frac{1}{\mu^2} \sum_{g} g^{-1} h_\alpha^* g = \frac{1}{\mu^2} \sum_{g} g^{-1} h_\alpha g = \varepsilon_\alpha,$$

that is, ε_α is Hermitian.

(3) Theorem 2 in 2.9 and (3.7.2) imply that ε_α belongs to $Z(A_{S_n})$.

(4) Note that $g_1^{-1} h_\alpha g_1$ and $g_2^{-1} h_\beta g_2$ are the elements h_α and h_β that correspond to the diagrams $g_1^{-1} \sum_\alpha$ and $g_2^{-1} \sum_\beta$. Thus for $\alpha \neq \beta$, IV in 3.4 shows that

$$g_1^{-1} h_\alpha g_1 \cdot g_2^{-1} h_\beta g_2 = 0. \qquad (3.7.6)$$

Summing over g_1 and g_2 in (3.7.6), we obtain

$$\varepsilon_\alpha \varepsilon_\beta = \sum_{g_1} \sum_{g_2} g_1^{-1} h_\alpha g_1 g_2^{-1} h_\beta g_2 = 0, \quad \text{for } \alpha \neq \beta. \qquad (3.7.7)$$

By virtue of (3.7.7) and (3), for $\alpha \neq \beta$ we have

$$(\varepsilon_\alpha, \varepsilon_\beta) = n^{-1} f_0(\varepsilon_\beta^* \varepsilon_\alpha) = n^{-1} f_0(\varepsilon_\beta \varepsilon_\alpha) = n^{-1} f_0(0) = 0. \qquad (3.7.8)$$

(5)   By (3.7.8), all ε_α corresponding to distinct α are mutually orthogonal. In addition (1) states that $\varepsilon_\alpha \neq 0$. Consequently the elements ε_α are linearly independent elements of the center $Z(A_{S_n})$. Their number is equal to the number of conjugacy classes in the group and thus is equal to the dimension of the center $Z(A_{S_n})$. Therefore they form a basis in $Z(A_{S_n})$. $\square$

§4. Induced Representations

4.1. Definition and Simplest Properties of an Induced Representation

Let G be a finite group. Let H be a subgroup of G and let T be a representation of the group H in a space V. We define a representation U of G in the following way.

(1)   The space of the representation U is the set $\mathscr{V}$ of all functions $f(g)$ on G with values in V such that

$$f(hg) = T(h)f(g), \quad \text{for all } h \in H \text{ and } g \in G. \qquad (4.1.1)$$

(2)   The operators of the representation U act in the space $\mathscr{V}$ according to the formula $[U(g_0)f](g) = f(gg_0)$.

We will show that the correspondence $g \to U(g)$ is indeed a representation of the group G.

First we note that $U(g)$ operates in $\mathscr{V}$, that is, for $f \in \mathscr{V}$, $U(g)f$ belongs to $\mathscr{V}$. For $f \in \mathscr{V}$ and $U(g_0)f(g) = f(gg_0) = \varphi(g)$, we have $\varphi(hg) = f((hg)g_0) = f(h(gg_0)) = T(h)f(gg_0) = T(h)\varphi(g)$, so that $\varphi \in \mathscr{V}$. It is clear that $U(g_0)$ is a linear operator in $\mathscr{V}$ for every $g_0 \in G$ and that $U(e) = 1$. Finally, we have $[U(g_0)U(g_1)f](g) = U(g_0)f(gg_1) = f((gg_0)g_1) = f(g(g_0g_1)) = U(g_0g_1)f(g)$, that is, $U(g_0)U(g_1) = U(g_0g_1)$ for all $g_0, g_1 \in G$.

The representation U defined by (1) and (2) is called the representation of the group G, *induced* by the representation T of the subgroup H and is denoted by U^T or $_HU^T$ or $_HU_G^T$ if it is necessary to indicate the subgroup H and the group G.

I. *The dimension of the representation $_HU^T$ is the product of the dimension of the space V and the index of the subgroup H in G.*

First choose an element in each right coset Hg. There are k (the index of H in G) such cosets and so we have "representatives" $g_1, \ldots, g_k$ in the k distinct cosets Hg_i. Consider the mapping p of the space $\mathscr{V}$ into the direct sum of k copies of the space V, defined by the formula $f \to \{f(g_1), \ldots, f(g_k)\}$. The function $f \in \mathscr{V}$ is uniquely defined on each coset Hg_0 by the condition $f(hg) = T(h)f(g)$, provided we know $f(g_0)$. Therefore the mapping p is one-to-one. Clearly it is linear. From this we infer that $\dim \mathscr{V} = k \dim V$. $\square$

II. *If $H = G$, then $_H U^T$ is equivalent to T.*

Proof. If $H = G$, G contains only one coset Hg, and as a representative of this set we take the element $g = e$. The mapping $p: f \to f(e)$ defines an isomorphism of the spaces $\mathscr{V}$ and V. We set $p^{-1} \xi = f_\xi(g)$ for $\xi \in V$, so that $f_\xi(e) = \xi$. For this isomorphism the representation U goes over into the induced representation T. Note that $p^{-1} \xi = f_\xi(g) = T(g) f_\xi(e) = T(g) \xi$ for $\xi \in V$. Therefore $\{[p^{-1} T(g_0) p] f\}(g) = p^{-1} T(g_0) f(e) = p^{-1} f(g_0) = T(g) f(g_0) = T(g) T(g_0) f(e) = f(g g_0)$, which is to say $p^{-1} T(g_0) p = U(g_0)$. $\square$

III. *If $H = \{e\}$ and T is the identity representation of the group H, then $_H U^T$ is equivalent to the right regular representation of the group G.*

Proof. Under our hypotheses the space $\mathscr{V}$ consists of all complex-valued functions on G and the representation U^T, defined by the formula $U(g_0) f(g_0) = f(g g_0)$, is the right regular representation. $\square$

4.2. The Theorem on Induction in Stages

Let K and H be subgroups in G with $K \subset H$. Let T be a representation of the subgroup K in a space V. Let S be the representation of H induced by T. Then the representations $_K U_G^T$ and $_H U_G^S$ are equivalent; that is, $U^{U^T} \sim U^T$.

Proof. The space of the representation $_H U_G^S$ consists of the functions F on the group G with values in the space of the representation S that satisfy the condition $F(hg) = S(h) F(g)$ for all $h \in H$ and $g \in G$. For every $g \in G$ the value $F(g)$ is a function on H with values in V satisfying the identity $\{F(g)\}(kh) = T(k)\{F(g)\}(h)$. Therefore the elements of the space of the representation $_H U_G^S$ can be regarded as those functions f on the Cartesian product $G \times H$ with values in V that satisfy the relations $f(h_0 g, h) = f(g, h h_0)$ for all h_0, $h \in H$, $g \in G$ and $f(g, kh) = T(k) f(g, h)$ for all $k \in K$, $h \in H$, $g \in G$. We define a mapping p of the space $\mathscr{V}^S$ of the representation $_H U_G^S$ into the space $\mathscr{V}^T$ of the representation $_K U_G^T$, by the formula

$$(pf)(g) = f(g, e), \quad \text{for } f \in \mathscr{V}^S.$$

The reader can easily verify that p is an isomorphism of the space $\mathscr{V}^S$ onto the space $\mathscr{V}^T$ and that p yields equivalence of the representations $_H U_G^S$ and $_K U_G^T$. $\square$

4.3. Frobenius' Duality Theorem

Let T and L be irreducible representations of the group G and its subgroup H respectively. The multiplicity of the representation T in U^L is equal to the multiplicity of the representation L in $T|_H$.

Proof. We will show that the linear space $\text{Hom}(T, U^L)$ of the operators that intertwine T with U^L is isomorphic to the linear space $\text{Hom}(T|_H, L)$ of the operators that intertwine $T|_H$ with L:

$$\text{Hom}(T|_H, L) \sim \text{Hom}(T, U^L) \qquad (4.3.1)$$

for all representations T of G and L of the subgroup H. If T and L are irreducible, the dimensions of these spaces are equal to the multiplicities considered in the theorem, so that the formula (4.3.1) proves the theorem.

We proceed to verify (4.3.1). Let K be a linear operator mapping the space V_T of the representation T into the space V_{U^L} of the representation U^L that intertwines T and U^L, that is, $KT(g) = U^L(g)K$ for all $g \in G$. Let $\tilde{K}$ be the operator mapping V_T into V_L defined by $\tilde{K}\xi = (K\xi)(e)$, where $K\xi$ is the function on G with values in V_L that corresponds to $\xi \in V_T$. Since $KT(h) = U^L(h)K$ for $h \in H$, we also have $\tilde{K}T(h)\xi = (K(T(h)\xi))(e) = (U^L(h)K\xi)(e) = (K\xi)(h) = L(h)(K\xi)(e) = L(h)\tilde{K}\xi$ for all $\xi \in H_T$. That is, $\tilde{K}$ intertwines $T|_H$ and L, and so the correspondence $K \to \tilde{K}$ defines a mapping of $\text{Hom}(T, U^L)$ into $\text{Hom}(T|_H, L)$. Since $(K\xi)(g) = (U^L(g)K\xi(e) = (KT(g)\xi)(e) = \tilde{K}T(g)\xi$, $\tilde{K}$ is defined by K and the mapping $K \to \tilde{K}$ is an isomorphism. $\square$

4.4. The Character of an Induced Representation

Theorem. *The character χ of the induced representation $_HU_G^T$ is defined by the formula*

$$\chi(g) = \sum_{\{\delta_i : \delta_i g \in H\delta_i\}} \psi(\delta_i g \delta_i^{-1}) \qquad (4.4.1)$$

where $\{\delta_i\}$ is a set of representatives of all cosets Hg_0, $g_0 \in G$, and ψ is the character of the representation T.

Proof. Let $\{e_j\}$ be a basis in the space V of the representation T. Then the functions

$$f_{ij}(g) = \begin{cases} T(h)e_j, & \text{for } g = h\delta_i, \, h \in H, \\ 0, & \text{for } g \notin H\delta_i, \end{cases} \qquad (4.4.2)$$

form a basis in the space $\mathscr{V}$ of the representation U^T. Indeed, $\mathscr{V}$ is the direct sum of the subspaces $\mathscr{V}_i = \{f : f(g) = 0 \text{ for } g \notin H\delta_i\}$. On the other hand, the functions $f_{ij}(g)$ for a fixed i form a basis in $\mathscr{V}_i$.

If $g = h\delta_k$, then for $\delta_k g_0 = \tilde{h}\delta_q$, $\tilde{h} \in H$, we find that

$$U^T(g_0)f_{ij}(g) = f_{ij}(gg_0) = f_{ij}(h\delta_k g_0) = f_{ij}(h\tilde{h}\delta_q)$$
$$= \begin{cases} T(h)T(\tilde{h})e_j, & \text{for } q = i, \\ 0, & \text{for } q \neq i. \end{cases} \qquad (4.4.3)$$

Thus for $g = h\delta_k$, the coefficient of $f_{ij}(g)$ in the decomposition of $U^T(g_0)f_{ij}(g)$ in the basis $\{f_{ij}\}$ (that is, the corresponding diagonal element of the matrix of the operator $U^T(g_0)$ in the basis (4.4.2)) can be different from zero only in the case when $\delta_k g_0 = \tilde{h}\delta_i$, that is, when $\delta_k g_0 \in H\delta_i$. We set $T(\tilde{h})e_j = \sum_r \alpha_{rj}(\tilde{h})e_r$. By virtue of (4.4.2) and (4.4.3) we have $U^T(g_0)f_{ij}(g) = \sum_k \sum_r \alpha_{rj}(\tilde{h})f_{kr}(g)$ for $g = h\delta_k$, $\delta_k g_0 \in H\delta_i$. Consequently, the coefficient of $f_{ij}(g)$ is equal to $\alpha_{jj}(\tilde{h})$ for $g = h\delta_k$, $\delta_i g_0 = \tilde{h}\delta_i$, $\tilde{h} \in H$. Thus we have

$$\chi(g_0) = \sum_{\{\delta_i : \delta_i g_0 \in H\delta_i\}} \sum_j \alpha_{jj}(\tilde{h}). \qquad (4.4.4)$$

Since $\sum_j \alpha_{jj}(\tilde{h}) = \psi(\tilde{h})$ and $\tilde{h} = \delta_i g_0 \delta_i^{-1}$, the formula (4.4.1) follows directly from (4.4.4). $\quad\square$

4.5 The Restriction of an Induced Representation to a Subgroup

Theorem. *Let G be a finite group with subgroups H and K, and let L be a representation of the group H. We set $G_g = K \cap g^{-1}Hg$, for $g \in G$. Let T^g be the representation of the group K induced by the representation L^g: $x \to L(gxg^{-1})$ of the subgroup G_g. The representation T^g is uniquely defined (up to equivalence) by the double coset $HgK = D(g)$ that contains g. We write T^g as T^D. The restriction of the representation U^L to K is a direct sum of representations T^D (where the sum is taken over the set of all double cosets HgK, $g \in G$).*

Proof. Let $g_1 = h_0 g k_0$. We have

$$G_{g_1} = K \cap g_1^{-1}Hg_1 = K \cap k_0^{-1}g^{-1}h_0^{-1}Hh_0 g k_0 = k_0^{-1}G_g k_0.$$

That is, G_{g_1} and G_g are conjugate in K under the inner automorphism defined by the element k_0. If $\delta_1, \ldots, \delta_m$ is a set of representatives of the (pairwise disjoint) sets $G_g g_0$, $g_0 \in G$, then the elements $k_0^{-1}\delta_i k_0$, $i = 1, \ldots, m$, are a set of representatives of the pairwise disjoint sets $G_{g_1} g_0$, $g_0 \in G$. Since every character is constant on conjugacy classes, (4.4.1) shows that the characters of the representations T_g and T_{g_1} are equal. Hence the representations T^g and T^{g_1} are equivalent.

Let $\mathscr{V}_j$ be the subspace of the space $\mathscr{V}$ of the representation U^L formed by all functions that vanish outside of $Hg_j K^n$ where $g_1, \ldots, g_l$ is a complete set of representatives of the distinct double cosets HgK, $g \in G$. Obviously every subspace $\mathscr{V}_j$ is invariant under the operators $U^L(k)$ for $k \in K$, and also the representation $U^L|_K$ is the direct sum of its subrepresentations defined on the subspaces $\mathscr{V}_j$, $j = 1, \ldots, l$. (Note that elements of the same coset HgK occur in the formulas $f(hg) = L(h)f(g)$ and $U^L(k)f(g) = f(gk)$, $h \in H$, $k \in K$.)

We will show that the subrepresentation $U^L|_K$ in the space $\mathscr{V}_j$ is equivalent to the representation of the group K induced by the representation T^D, where $D = Hg_j K$. Let p be the mapping carrying each function f in $\mathscr{V}_j$ into the function $p(f)$ on K with values in the space V of the representation L according to the formula $p(f)(k) = f(g_jk)$. A given function φ on K is the image of some function $f \in \mathscr{V}_i$ under p if and only if there exists a function $\bar{\varphi}$ on the set D such that $\bar{\varphi}(g_jk) = \varphi(k)$ and the relation $\bar{\varphi}(hh_0g_ik) \equiv L(h)\bar{\varphi}(h_0g_ik)$ holds for the function $\bar{\varphi}$. The last relation holds for all $h, h_0 \in H$ and $k \in K$ if and only if it holds for $h_0 = e$ and $hg_ik \in g_iK$. Suppose that $hg_ik = g_i\tilde{k}$. The relation $\bar{\varphi}(hg_ik) = L(h)\varphi(k) = \varphi(\tilde{k})$ must be satisfied, but we have $g_i^{-1}hg_i = \tilde{k}k^{-1}$, that is, $\tilde{k}k^{-1} \in G_{g_i}$, and furthermore we have $L_{g_i}(\tilde{k}k^{-1}) = L(g_i\tilde{k}k^{-1}g_i^{-1}) = L(h)$. Hence the function φ on k is the image of a function $f \in \mathscr{V}$ under the mapping p if and only if φ lies in the space of the induced representation $T^{g_i} = G_{g_i}U_K^L$. It is immediate that p is an isomorphism yielding the equivalence of the representation T^{g_i} and the subrepresentation of the representation $U^{La}|_K$ in the space $\mathscr{V}_i$. $\square$

4.6. Induced Representations of a Direct Product of Groups

Theorem. *Let G_1, G_2 be finite groups. Let H_1, H_2 be subgroups of G_1 and G_2 respectively, and let L_1, L_2 be representations of the groups H_1, H_2 in spaces V_1, V_2. Then $_{H_1 \times H_2}U_{G_1 \times G_2}^{L_1 \otimes L_2}$ is equivalent to $_{H_1}U_{G_1}^{L_1} \otimes {}_{H_2}U_{G_2}^{L_2}$.*

Proof. It is easy to verify that the canonical operator producing an isomorphism between the space of functions on $G_1 \times G_2$ with values in $V_1 \otimes V_2$ and the tensor product of the spaces of functions on G_1 with values in V_1 and on G_2 with values in V_2 also yields equivalence of the representations $_{H_1 \times H_2}U_{G_1 \times G_2}^{L_1 \otimes L_2}$ and $_{H_1}U_{G_1}^{L_1} \otimes {}_{H_2}U_{G_2}^{L_2}$. $\square$

4.7. Tensor Products and Induced Representations

The tensor product of two representations T_1, T_2 of a group G can be considered as the restriction of the representation $T_1 \otimes T_2$ of the group $G \times G$ to the "diagonal" subgroup $\bar{G} = \{(g, g) : g \in G\} \subset G \times G$. The theorem in 4.5 thus gives us information about the decomposition of tensor products.

Theorem. *Let G be a finite group with subgroups H and K. Let T and S be representations of the groups H and K respectively. Let G_{g_1,g_2} be a subgroup in G of the form $g_1^{-1}Hg_1 \cup g_2^{-1}Kg_2$, where $g_1, g_2 \in G$. Let $T^{g_1}:x \to T(g_1xg_1^{-1})$, $S^{g_2}:x \to S(g_2xg_2^{-1})$ be representations of the group G_{g_1,g_2}. Let L^{g_1,g_2} be the tensor product of the representations T^{g_1} and S^{g_2}. Let $U^{L_{g_1,g_2}}$ be the corresponding induced representation of the group G. Then $U^{L_{g_1,g_2}}$ is defined uniquely (up to equivalence) by the double coset $Hg_1g_2^{-1}K$ containing $g_1g_2^{-1}$ and the direct sum of the representations $U^{L_{g_1b_{g_2}}}$ (over all double cosets of the form HgK) is equivalent to $U^T \otimes U^S$.*

Proof. Apply the theorem in 4.5 to the representation $U_{G \times G}^{T \otimes S}$ (see 4.6), induced from the subgroup $H \times K$, and then restrict to $\bar{G}$. We leave the details of the proof to the reader. $\square$

4.8. The Imprimitivity Theorem

Let T be a representation of the group G, and let H be a subgroup of G. We say that T *admits a system of imprimitivity with basis* G/H if there exists a mapping P of the set of subsets of G/H into the set of projections in the space V of the representation T with the following properties: $P(\varnothing) = 0$; $P(G/H) = I$; if we have E, $F \subset G/H$ and $E \cap F = \varnothing$,[23] then $P(E \cup F) = P(E) + P(F)$; and $P(E \cap F) = P(E)P(F)$ for all E, $F \subset G/H$; $P(Eg) = T^{-1}(g)P(E)T(g)$ for all $E \subset G/H$ and $g \in G$.

Theorem (Criterion for inducibility of a representation). *A representation T of a group G is equivalent to a representation of the form $_H U_G^L$ for a given subgroup H and some representation L of H if and only if T admits a system of imprimitivity with basis G/H.*

Proof. See for example the treatise Serre [2].

Examples and Exercises

   1. Using the theorem of 4.3, show that the order of a group is equal to the sum of the squares of the dimensions of its inequivalent irreducible representations.

   2. *Criterion for irreducibility of an induced representation.* Let G be a finite group and H a subgroup of G. Let T be an irreducible representation of H. Show that the representation U^T is irreducible if and only if, for all $g \notin H$, the representations of the group $H^g = g^{-1}Hg \cap H$ defined by the formulas $h \to T(ghg^{-1})$ and $h \to T(h)$ have no common irreducible components.

   3. Let G be a finite group with H a subgroup of G. Show that the representation of G induced by the identity representation of H contains the identity representation of G.

   4. Let G be a finite group with subgroups H and K. Let T and S be representations of H and K respectively. Suppose that $_H U^T$ and $_K U^S$ are irreducible. We have $_H U^T \sim {}_K U^S$ if and only if there exists an element $g \in G$ such that the representations $h \to T(ghg^{-1})$ and $h \to S(h)$ of the subgroup $g^{-1}Hg \cap K$ have a common irreducible component.

   5. Let G be the group of nonsingular matrices $\left\| \begin{matrix} a & b \\ c & d \end{matrix} \right\|$ with elements in a finite field. Let H be the subgroup of matrices $\left\| \begin{matrix} a & b \\ 0 & d \end{matrix} \right\|$ with $ad \neq 0$. Find the representations of G that are induced by one-dimensional representations of H and identify those that are irreducible. (For this exercise, see §5 below.)

---

[23] Recall that the symbol $\varnothing$ denotes the void set.

§5. Representations of the Group $SL(2,\mathbf{F}_q)$

5.1. Fields

A set K is called a *field* if operations of *addition* and *multiplication* (with a number of conditions) are defined in K. That is, for each ordered pair of elements (x, y) of K there is a corresponding element $x + y \in K$, called the *sum* of x and y and an element $xy \in K$ called the *product* of x and y. The following conditions must hold.

(1) K is a commutative group under addition. (This group is called *the additive group of the field K*.)

(2) If 0 is the identity element under addition, then the set $K^* = K\backslash\{0\}$ is a commutative group under multiplication. (This group is called *the multiplicative group of the field K*.)

(3) The *distributive law* holds:

$$x(y + z) = (xy) + (xz)$$

for all x, y, z in K.

Some simple examples of fields are: the set $\mathbf{R}$ of real numbers and the set $\mathbf{C}$ of complex numbers with the usual operations of addition and multiplication.

Let us consider another example of a field, a field with p elements, where p is a prime number. In the set of numbers $0, 1, \ldots, p - 1$, we define the "sum" of the numbers m and n as the remainder after dividing the usual sum $m + n$ by p and the "product" as the remainder after dividing the usual product mn by p. The reader can easily verify that these operations satisfy conditions (1)–(3). The resulting field is called the *field of residues* modulo p and is denoted by $\mathbf{F}_p$. This is an example of a *finite field*.

One can show (see for example van der Waerden [1]) that if K is a finite field and K contains q elements, then $q = p^n$ where p is prime and n is a natural number. The number p is called the *characteristic of the field K*. A field with $q = p^n$ elements is uniquely defined up to an *isomorphism*, that is, a one-to-one mapping preserving addition and multiplication. This field is denoted by $\mathbf{F}_q$.

The multiplicative group of the field $\mathbf{F}_q$ is cyclic. If q is odd, then $\mathbf{F}_q$ contains $(q + 1)/2$ distinct squares, $u = v^2$, where $v \in \mathbf{F}_q$. No generator of the multiplicative group of $\mathbf{F}_q$ can be a square.

5.2. "Circles" in a Finite Field

Choose an element of the field $\mathbf{F}_q$ that is not a square and denote it by the symbol ε. The set of pairs (x, y), $x, y \in \mathbf{F}_q$ that satisfy the condition $x^2 - \varepsilon y^2 = c$ for a given $c \in \mathbf{F}_q$ is called a *circle in the field* $\mathbf{F}_q$. We note that the "circle of radius zero", that is, the set of pairs (x, y) such that $x^2 - \varepsilon y^2 = 0$, consists of the identity element $(0,0)$ alone, since for $y \neq 0$ the element εy^2 is not a square if ε is not a square.

I. *The circle $x^2 - \varepsilon y^2 = 1$ consists of $q + 1$ elements.*

Proof. It is clear that the point $(-1, 0)$ lies on this circle, and if the point (x, y) belonging to the circle satisfies the condition $x = -1$, then $y = 0$. Let (x, y) lie on the circle and let $x \neq -1$. We set $t = y/(x + 1)$. Then we have $x^2 - 1 = \varepsilon y^2$, $x + 1 = yt^{-1}$, $x - 1 = (x^2 - 1)/(x + 1) = \varepsilon y t$, $(x - 1)/(x + 1) = \varepsilon t^2$, $x = (1 + \varepsilon t^2)/(1 - \varepsilon t^2)$, $y = t(x + 1) = 2t/(1 - \varepsilon t^2)$. Since t is arbitrary, the number of distinct elements of the circle such that $x \neq -1$ is equal to q, that is, the circle consists of $q + 1$ elements. $\square$

II. *For any $c \in \mathbf{F}_q^*$ the circle $x^2 - \varepsilon y^2 = c$ consists of $q + 1$ elements.*

To prove this we consider the set $\mathcal{O}$ of matrices of the form $a = \begin{Vmatrix} \sigma & v \\ \varepsilon v & \sigma \end{Vmatrix}$ such that $\det a = \sigma^2 - \varepsilon v^2 \neq 0$. Since $\sigma^2 - \varepsilon v^2 = 0$ only for $\sigma = v = 0$, the set $\mathcal{O}$ consists of $q^2 - 1$ elements. The reader can easily verify that $\mathcal{O}$ is a group under matrix multiplication and the mapping $\det : a \to \det a$ is a homomorphism of the group $\mathcal{O}$ into the group $\mathbf{F}_q^*$. Proposition I implies that the kernel of this homomorphism consists of $q + 1$ elements and thus the image of the homomorphism $\det$ consists of $(q^2 - 1)/(q + 1) = q - 1$ elements. Since $\mathbf{F}_q^*$ consists of $q - 1$ elements, the image of the homomorphism $\det$ coincides with $\mathbf{F}_q^*$, that is, for every $c \neq 0$ there exists a coset of the kernel of the homomorphism $\det$ that maps into c under the homomorphism $\det$. Since the number of elements in the coset is equal to the number of elements in the kernel of the homomorphism, we have proved our assertion. $\square$

5.3. Definition of the Group $G = SL(2, \mathbf{F}_q)$. The Order of the Group G and of Conjugacy Classes in G

Again, let $\mathbf{F}_q$ be a finite field with $q = p^n$ elements, where p is a prime. We use $G = SL(2, \mathbf{F}_q)$ to denote the group of all unimodular matrices g of the second order with elements in the field $\mathbf{F}_q$:

$$g = \left\{ \begin{Vmatrix} a & b \\ c & d \end{Vmatrix}; \quad a, b, c, d \in \mathbf{F}_q; ad - bc = 1 \in \mathbf{F}_q \right\}.$$

In the sequel we will suppose that $p \neq 2$.

We will compute the order of the group G and describe the conjugacy classes in G.

If $g = \begin{Vmatrix} a & b \\ c & d \end{Vmatrix}$ and $a \in \mathbf{F}_q^*$, then for any b and c we determine d from

the relation $ad - bc = 1$ by the formula $d = a^{-1}(1 + bc)$. That is, if $a \in \mathbf{F}_q^*$ then b and c run through all $\mathbf{F}_q$. The number of such elements in G is equal to $(q - 1)q^2$. If $a = 0$, then $bc = -1$, that is, $b \in \mathbf{F}_q^*$ and $c = -b^{-1}$ and d is arbitrary. The number of these elements in G is $q(q - 1)$. Thus, the order of the group G is $q^2(q - 1) + q(q - 1) = q(q^2 - 1)$.

Let us find the conjugacy classes in G. It is clear that $\{e\}$ and $\{-e\}$ are conjugacy classes, and they also comprise the center of G. To describe the other conjugacy classes in G, we note that if M is a conjugacy class in G containing an element $g_0 \in G$, then M is a homogeneous space under the action of G on M according to the formula $h \to ghg^{-1}$, $g \in G$, $h \in M$.

Let Q be the stationary subgroup corresponding to the element g_0, that is, the set of elements $g \in G$ such that $gg_0g^{-1} = g_0$. The number of elements in the set M is equal to the index of Q in G (see III in 1.8, chapter I).

Consider the conjugacy classes defined by the element $g = \begin{Vmatrix} \lambda^{-1} & 0 \\ 0 & \lambda \end{Vmatrix}$

where $\lambda^2 \neq 1$, and denote it by $A\lambda$.

I. *The number of classes A_λ is equal to $(q - 3)/2$. Each of the classes A_λ consists of $q(q + 1)$ elements.*

Proof. The stationary subgroup Q corresponding to the class A_λ consists of the elements $g \in G$ such that $gg_\lambda g^{-1} = g$, that is, $gg_\lambda = g_\lambda g$. As is easy to verify, this means that the subgroup Q consists of diagonal matrices. Therefore Q is a subgroup of order $q - 1$ and the index of the subgroup Q in G is equal to $q(q^2 - 1)/(q - 1) = q(q + 1)$. Consequently, every class A_λ consists of $q(q + 1)$ elements. The number of these sets is equal to $(q - 3)/2$, since g_λ and g_μ ($\lambda, \mu \neq 0$, $\lambda^2 \neq 1$, $\mu^2 \neq 1$) belong to the same class if and only if $\lambda = \mu^{-1}$. $\square$

Consider the conjugacy classes defined by the elements

$$e_1^+ = \begin{Vmatrix} 1 & 1 \\ 0 & 1 \end{Vmatrix}, \quad e_\varepsilon^+ = \begin{Vmatrix} 1 & \varepsilon \\ 0 & 1 \end{Vmatrix}, \quad e_1^- = \begin{Vmatrix} -1 & 1 \\ 0 & -1 \end{Vmatrix}, \quad e_\varepsilon^- = \begin{Vmatrix} -1 & \varepsilon \\ 0 & -1 \end{Vmatrix};$$

ε is as in 5.2. We denote these four classes by $B^+(1)$, $B^+(\varepsilon)$, $B^-(1)$ and $B^-(\varepsilon)$, respectively.

II. *Each of the classes $B^+(1)$, $B^+(\varepsilon)$, $B^-(1)$, $B^-(\varepsilon)$ contains $(q^2 - 1)/2$ elements.*

Proof. The stationary subgroups Q are the elements $g \in G$ such that g commutes with e_1^+ (and e_ε^+, e_1^-, e_ε^-, respectively). We easily verify that the subgroup Q consists of all the matrices of the form $\begin{Vmatrix} \alpha & \mu \\ 0 & \alpha \end{Vmatrix}$ where $\alpha = \pm 1$. Thus Q has order $2q$ and the index of Q in G is $q(q^2 - 1)/2q = (q^2 - 1)/2$, that is, each of the classes $B^+(1)$, $B^+(\varepsilon)$, $B^-(1)$, $B^-(\varepsilon)$ consists of $(q^2 - 1)/2$ elements. $\square$

Consider the conjugacy classes defined by the element $g_\sigma = \begin{Vmatrix} \sigma & v \\ \varepsilon v & \sigma \end{Vmatrix}$,

where the pair (σ, v) belongs to a circle of radius 1, that is, $\sigma^2 - \varepsilon v^2 = 1$, where $\sigma^2 \neq 1$. This class is denoted by C_σ.

III. *The class C_σ is uniquely defined by the element σ. Each class C_σ contains $q(q - 1)$ elements. The number of classes C_σ is equal to $(q - 1)/2$.*

Proof. Let $\sigma^2 - \varepsilon v^2 = 1$. For a given σ, $\sigma^2 \neq 1$, the element v is uniquely defined up to its sign. We will show that the matrices $\left\|\begin{matrix} \sigma & v \\ \varepsilon v & \sigma \end{matrix}\right\|$ and $\left\|\begin{matrix} \sigma & -v \\ -\varepsilon v & \sigma \end{matrix}\right\|$ belong to the same conjugacy class in G. Let (α, β) be a point on the circle $\alpha^2 - \varepsilon \beta^2 = -1$ (according to II in 5.2 this circle is nonvoid). An easy calculation shows that

$$\left\|\begin{matrix} \alpha & \beta \\ -\varepsilon\beta & -\alpha \end{matrix}\right\| \left\|\begin{matrix} \sigma & v \\ \varepsilon v & \sigma \end{matrix}\right\| = \left\|\begin{matrix} \sigma & -v \\ -\varepsilon v & \sigma \end{matrix}\right\| \times \left\|\begin{matrix} \alpha & \beta \\ -\varepsilon\beta & -\alpha \end{matrix}\right\|$$

and det $\left\|\begin{matrix} \alpha & \beta \\ -\varepsilon\beta & -\alpha \end{matrix}\right\| = -\alpha^2 + \varepsilon\beta^2 = 1$, that is $\left\|\begin{matrix} \alpha & \beta \\ -\varepsilon\beta & -\alpha \end{matrix}\right\| \in G$. Thus the class C_σ is uniquely defined by the number σ. If we also have $g_\tau \in C_\sigma$, then the traces of the matrices g_τ and g_σ are equal. Therefore we have $\sigma = \tau$, that is, all the classes C_σ are distinct and the number of classes C_σ is equal to the number of $\sigma \in \mathbf{F}_q$ for which $\sigma^2 - \varepsilon v^2 = 1$ for some $v \neq 0$. But the circle of radius 1 contains $q + 1$ elements, including the pairs $(1,0)$ and $(-1,0)$. Thus the number of pairs (σ, v), $\sigma^2 \neq 1$, such that $\sigma^2 - \varepsilon v^2 = 1$ is equal to $q - 1$. On the other hand, for a given σ, an element $v \neq 0$ is defined by the equality $\sigma^2 - \varepsilon v^2 = 1$ up to its sign, and therefore the number of classes C_σ is $(q - 1)/2$.

The subgroup Q consists of the elements $g \in G$ commuting with g_σ. As we easily verify, these are exactly the matrices $\left\|\begin{matrix} \alpha & \beta \\ \varepsilon\beta & \alpha \end{matrix}\right\|$ where $\alpha^2 - \varepsilon\beta^2 = 1$. According to I in 5.2 the subgroup Q has order $q + 1$ and therefore the index of Q in G is equal to $q(q - 1)$. That is, C_σ consists of $q(q - 1)$ elements. $\square$

Thus, the number of conjugacy classes in G is $2 + (q - 3)/2 + 4 + (q - 1)/2 = q + 4$.[24] According to theorem 3 in 1.7, chapter II, the group G has precisely $q + 4$ pairwise inequivalent irreducible representations. Let us find these representations.

5.4. The Representations T_π

Let π be a one-dimensional representation (that is, a character) of the group $\mathbf{F}_q^*$. (Sometimes π is called a *multiplicative character of the field $\mathbf{F}_q$*.) Consider the space of all complex-valued functions $f(x, y)$, defined for (x, y) in $\mathbf{F}_q \times \mathbf{F}_q$ different from $(0, 0)$ and satisfying the condition

$$f(\lambda x, \lambda y) = \pi(\lambda) f(x, y), \qquad (\lambda \in \mathbf{F}_q^*, x, y \in \mathbf{F}_q, (x, y) \neq (0,0)). \quad (5.4.1)$$

---

[24] Adding up the numbers of elements in the classes $\{e\}, \{-e\} A_\lambda, B^+(1), B^+(\varepsilon), B^-(1), B^-(\varepsilon),$ C_σ, we get $2 + \frac{1}{2}(q - 3)q(q + 1) + 4((q^2 - 1)/2) + \frac{1}{2}(q - 1)q(q - 1) = q(q^2 - 1)$, the order of G, just as we should.

We denote this space by H_π. We define the representation T_π of G in the space H_π by the formula

$$(T_\pi(g)f)(x, y) = f(ax + cy, bx + dy), \qquad g = \left(\left\|\begin{matrix} a & b \\ c & d \end{matrix}\right\|\right). \qquad (5.4.2)$$

We easily verify that $g \to T_\pi(g)$ is a representation of the group G.

Condition (5.4.1) implies that a function on H_π is uniquely defined by its values at the $q + 1$ points $(0, 1)$ and $(1, a)$ ($a \in \mathbf{F}_q$). The dimension of the representation T_π is thus $q + 1$.

Let us find the character φ_π of T_π. Consider the basis in H_π, consisting of the vectors f_∞ and f_a ($a \in \mathbf{F}_q$), where f_∞ and f_a are defined by

$$f_\infty(x, y) = \begin{cases} \pi(x), & \text{if } y = 0, \\ 0, & \text{if } y \neq 0; \end{cases}$$

$$f_a(x, y) = \begin{cases} \pi(y), & \text{if } x = ay, \\ 0, & \text{if } x \neq ay. \end{cases} \qquad (5.4.3)$$

Let χ be a fixed character of the additive group of the field $\mathbf{F}_q$ (called an *additive character of the field* $\mathbf{F}_q$) not identically 1. We set $e_\infty = f_\infty$, $e_u = \sum_{a \in \mathbf{F}_q} \chi(ua)f_a$ ($u \in \mathbf{F}_q$). It is easy to check that

$$T_\pi(g_\lambda)e_u = \pi(\lambda)e_{\lambda^{-2u}},$$
$$T_\pi(g_\lambda)e_\infty = \pi(\lambda^{-1})e_\infty, \qquad (5.4.4a)$$

$$T_\pi\left(\left\|\begin{matrix} 1 & 0 \\ b & 1 \end{matrix}\right\|\right)e_u = \chi(ub)e_u,$$

$$T_\pi\left(\left\|\begin{matrix} 1 & 0 \\ b & 1 \end{matrix}\right\|\right)e_\infty = e_\infty \qquad (5.4.4b)$$

for all $g = \left\|\begin{matrix} \lambda^{-1} & 0 \\ 0 & \lambda \end{matrix}\right\|$, $\lambda \neq 0$ and all $b \in \mathbf{F}_q$. Since the matrices $\left\|\begin{matrix} 1 & b \\ 0 & 1 \end{matrix}\right\|$ and $\left\|\begin{matrix} 1 & 0 \\ -b & 1 \end{matrix}\right\|$ are conjugate in G $\left(\text{use the element } s = \left\|\begin{matrix} 0 & 1 \\ -1 & 0 \end{matrix}\right\|\right)$, formulas (5.4.4) allow us to find the value of the character φ_π of T_π on the conjugacy classes $\{e\}, \{-e\}, A_\lambda, B^\pm(1), B^\pm(\varepsilon)$. It is clear that $\varphi_\pi(e) = \dim T_\pi = q + 1$, $\varphi_\pi(-e) = \pi(-1)(q + 1)$. Let us find $\varphi_\pi(e_1^+)$. By (5.4.4b) we have

$$\varphi_\pi(e_1^+) = \varphi_\pi\left(\left\|\begin{matrix} 1 & 0 \\ -1 & 1 \end{matrix}\right\|\right) = \sum_{u \in \mathbf{F}_q} \chi(-u) + 1. \qquad (5.4.5)$$

But χ is not the identity character and the relations of orthogonality imply that $\sum_{u \in \mathbf{F}_q} \chi(-u) = 0$. Thus we have $\varphi_\pi(e_1^+) = 1$. Similarly we have $\varphi_\pi(e_\varepsilon^+) = 1$; $\varphi_\pi(e_1^-) = \varphi_\pi(e_\varepsilon^-) = \pi(-1)$. Let us find $\varphi_\pi(g_\lambda)$, $\lambda \neq \pm 1$. Since $\lambda^2 \neq 1$, we have

$u \neq \lambda^2 u$ for $u \neq 0$. Therefore the diagonal matrix element in the decomposition of the vector $T_\pi(g_\lambda)e_u$, $u \in F_q$ (see (5.4.4a)) is nonzero only for $u = 0$. Thus we have $\varphi_\pi(g_\lambda) = \pi(\lambda^{-1}) + \pi(\lambda)$.

It remains to determine φ_π on the set C_σ. Let ρ be the one-dimensional representation of the subgroup $K = \left\{ \left\| \begin{matrix} \lambda & 0 \\ \mu & \lambda^{-1} \end{matrix} \right\|, \lambda \in F_q^*, \mu \in F_q \right\}$ defined by $\rho(k) = \pi(\lambda)$ for $k = \left\| \begin{matrix} \lambda & 0 \\ \mu & \lambda^{-1} \end{matrix} \right\|$. The representation T_π is equivalent to the representation of G induced by the representation ρ. To see this let ψ be a function on G such that $\psi(kg) = \rho(k)\psi(g)$ for all $k \in K$, $g \in G$; that is,

$$\psi\left(\left\| \begin{matrix} \lambda & 0 \\ \mu & \lambda^{-1} \end{matrix} \right\| \left\| \begin{matrix} a & b \\ c & d \end{matrix} \right\| \right) = \psi\left(\left\| \begin{matrix} \lambda a & \lambda b \\ \mu a + \lambda^{-1}c & \mu b + \lambda^{-1}d \end{matrix} \right\| \right)$$

$$= \pi(\lambda)\psi\left(\left\| \begin{matrix} a & b \\ c & d \end{matrix} \right\| \right). \tag{5.4.6}$$

We set $F(a, b) = \psi\left(\left\| \begin{matrix} a & b \\ c & d \end{matrix} \right\| \right)$ and denote by X the set of all pairs (x, y), $x, y \in F_q$, distinct from $(0, 0)$. For a fixed $(a, b) \in X$ the general solution of the equation $\begin{vmatrix} a & b \\ c & d \end{vmatrix} = 1$ has the form $c = c_0 + \mu a$, $d = d_0 + \mu b$, where $\mu \in F_q$, $\begin{vmatrix} a & b \\ c_0 & d_0 \end{vmatrix} = 1$. Therefore condition (5.4.6) implies that the function F is well defined, that is, it does not depend on the choice of c and d for which $ad - bc = 1$. Thus we establish the correspondence $\psi \to F$ between the space of the induced representation U^ρ and the space H_π. The reader can easily verify that this correspondence is an isomorphism that carries U^ρ onto T_π. It follows from formula (4.4.1) for the character of an induced representation that the character of the representation U^ρ is zero on all elements $g \in G$, not conjugate to elements of the subgroup K. Thus we have $\varphi_\pi(g_\sigma) = 0$. Combining the relations we have found for φ_π, we obtain the following.

I. *Let* φ_π *be the character of the representation* T_π. *We have*

$$\varphi_\pi(e) = q + 1, \qquad \varphi_\pi(-e) = (q + 1)\pi(-1);$$
$$\varphi_\pi(g_\lambda) = \pi(\lambda) + \pi(\lambda^{-1}), \qquad (\lambda^2 \neq 1);$$
$$\varphi_\pi(e_1^+) = \varphi_\pi(e_\varepsilon^+) = 1; \qquad \varphi_\pi(e_1^-) = \varphi_\pi(e_\varepsilon^-) = \pi(-1);$$
$$\varphi_\pi(g_\sigma) = 0 \qquad (\sigma^2 - \varepsilon v^2 = 1, \sigma^2 \neq 1).$$

A direct calculation shows that for $\pi^2 \neq 1$ we have $\sum_{g \in G} |\varphi_\pi(g)|^2 = q(q^2 - 1)$. From this and from II in 1.8 we infer the following.

II. *All representations* T_π, *corresponding to characters* π *such that* $\pi^2 \neq 1$, *are irreducible.*

We can infer more from the formulas for φ_π.

III. *Two representations* T_π, T_{π_1} *are equivalent if and only if* $\pi = \pi_1$ *or* $\pi = \pi_1^{-1}$ ($\pi^2 \neq 1$, $\pi_1^2 \neq 1$).

Consider the representation T_1, where 1 is the identity character of the group $\mathbf{F}_q^*$, that is, $1(\lambda) \equiv 1$ for all $\lambda \in \mathbf{F}_q^*$. Plainly the constant functions ($f(x, y) = c$ for all $(x, y) \neq (0, 0)$) belong to H_1 and form a one-dimensional invariant subspace of H_1, in which the identity representation 1_G of the group G operates. The character φ_{1_G} of this representation is identically equal to 1. The space of functions $f \in H_1$ such that $\sum_{(x,y) \neq (0,0)} f(x, y) = 0$ is also invariant under T_1, that is, a certain subrepresentation $\tilde{T}_1$ of T_1 operates in this subspace. The dimension of $\tilde{T}_1$ is q. The character $\varphi_{\tilde{T}_1}$ of $\tilde{T}_1$ is equal to $\varphi_1 - \varphi_{1_G} = \varphi_1 - 1$, where φ_1 is the character of T_1. We thus have the following.

IV. *Let* $\tilde{\varphi}$ *be the character of the representation* $\tilde{T}_1$. *We then have*

$$\tilde{\varphi}(e) = \tilde{\varphi}(-e) = q;$$
$$\tilde{\varphi}(e_1^+) = \tilde{\varphi}(e_1^-) = \tilde{\varphi}(e_\varepsilon^+) = \tilde{\varphi}(e_\varepsilon^-) = 0;$$
$$\varphi(g_\lambda) = 1; \qquad \tilde{\varphi}(g_\sigma) = -1.$$

A direct calculation shows that $\sum_{g \in G} |\tilde{\varphi}(g)|^2 = q(q^2 - 1) = |G|$ and Proposition II in 1.8 gives the following.

V. *The representation* $\tilde{T}_1$ *is irreducible.*

We will consider the representation T_{π_0}, where $\pi_0^2 \equiv 1$, $\pi_0 \neq 1$, in 5.6.

5.5. A Trigonometric Sum in $\mathbf{F}_q$

Let $\pi_0 \neq 1$ be a character of the group $\mathbf{F}_q^*$, assuming only the values 1 and -1. Then π_0 assumes the value 1 at all squares in $\mathbf{F}_q^*$ ($\pi_0(a^2) = (\pi_0(a))^2 = +1$). The set of squares in $\mathbf{F}_q^*$ forms a subgroup of index 2 in $\mathbf{F}_q^*$. Thus it follows from $\pi \neq 1$ that $\pi_0(x) = -1$ for all elements $x \in \mathbf{F}_q^*$ that are *not* squares. The character π_0 is called a *quadratic residue* in the field $\mathbf{F}_q^*$. Note that $\pi_0(\varepsilon) = -1$.

We will examine the function $f(a) = \sum_{u \in \mathbf{F}_q^{*2}} \chi(au)$, where $\mathbf{F}_q^{*2}$ is the subgroup of $\mathbf{F}_q^*$ formed by the nonzero squares in $\mathbf{F}_q$. If $a = 0$, then $f(a) = (q - 1)/2$, since the order of the subgroup $\mathbf{F}_q^{*2}$ is equal to $(q - 1)/2$. Let $a \neq 0$. We then have

$$f(a) = \sum_{u = v^2 \neq 0} \chi(au) = \sum_{u \neq 0} \chi(au) \frac{\pi_0(u) + 1}{2}. \tag{5.5.1}$$

It follows from the orthogonality relations (1.4.2) that the character χ ($\neq 1$) is orthogonal (as a function of the group) to the character identically 1.

That is, $\sum_{u \in F_q} \chi(u) = 0$ for $\chi \not\equiv 1$. Therefore we have (see (5.5.1))

$$f(a) = \frac{1}{2} \sum_{u \neq 0} \chi(au)\pi_0(u) + \frac{1}{2} \sum_{u \neq 0} \chi(au) = \frac{1}{2} \sum_{u \neq 0} \chi(au)\pi_0(u) - \frac{1}{2}. \quad (5.5.2)$$

We set

$$\Phi(a) = \sum_{u \neq 0} \chi(au)\pi_0(u),^{25} \quad (5.5.3)$$

and thus we have

$$\Phi(a) = \sum_{v \neq 0} \chi(v)\pi_0(va^{-1}) = \pi_0(a^{-1}) \sum_{v \neq 0} \chi(v)\pi_0(v). \quad (5.5.4)$$

For $\sum_{v \neq 0} \chi(v)\pi_0(v)$ we write $\Gamma(\pi_0)$. Since we have $\pi_0(a^{-1}) = \pi_0(a)$ we also have $\Phi(a) = \Gamma(\pi_0)\pi_0(a)$ and therefore

$$\sum_{a \neq 0} \Phi(a)\chi(av) = \Gamma(\pi_0) \sum_{a \neq 0} \pi_0(a)\chi(av)$$
$$= \Gamma(\pi_0)\Phi(v) = (\Gamma(\pi_0))^2\pi_0(v). \quad (5.5.5)$$

On the other hand we have

$$\sum_{a \neq 0} \Phi(a)\chi(av) = \sum_{u \neq 0} \sum_{a \neq 0} \chi(au + av)\pi_0(u)$$
$$= \sum_{w \neq 0} \pi_0(-w) \sum_{a \neq 0} \chi(a(v-w))$$
$$= \sum_{w \neq 0} \pi_0(-w)(-1) + q\pi_0(-v) = q\pi_0(-v). \quad (5.5.6)$$

Comparing (5.5.2), (5.5.4), (5.5.5) and (5.5.6) we conclude that

$$\Gamma(\pi_0)^2 = \left(\sum_{v \neq 0} \chi(v)\pi_0(v) \right)^2 = q\pi_0(-1), \quad (5.5.7)$$

$$\sum_{u \in F_q^{*2}} \chi(au) = \frac{1}{2} \Gamma(\pi_0)\pi_0(a) - \frac{1}{2}, \quad (a \neq 0). \quad (5.5.8)$$

5.6. The Representation T_{π_0}

Consider the representation T_{π_0}. Let χ_0 be the character of the representation T_{π_0}. We infer from I in 5.4 and from the equality $\pi_0(x)^2 \equiv 1$ that $\sum_{g \in G} |\chi_0(g)|^2 = 2|G|$. Consequently T_{π_0} is reducible and according to (1.7.5) it can be decomposed into two irreducible subrepresentations.

---

[25] Note that if $\mathbf{F}$ is the field of residues modulo p and χ is the character for which $\chi(k) = e^{2\pi(k/p)i}$, then $\Phi(a)$ is the classical *Gaussian sum*. (See J.-P. Serre [3].)

Consider the restriction of T_{π_0} to the subgroup K. Clearly, any subspace invariant under T_{π_0} is also invariant under $T_{\pi_0}|_K$. On the other hand, (5.4.4) shows that the subspace $H_{\pi_0}^+$ spanned by $\Gamma(\pi_0)e_\infty + e_0$ and e_{v^2}, $v \in \mathbf{F}_q^*$ and the subspace $H_{\pi_0}^-$ spanned by $\Gamma(\pi_0)e_\infty - e_0$ and e_u, $u \in \mathbf{F}_q^* \backslash \mathbf{F}_q^{*2}$ are invariant under $T_{\pi_0}|_K$. Let θ^+, θ^- be the characters of the corresponding representations of the group K. Formula (5.5.4) implies that $\theta^+ \neq \theta^-$ (for example, $\theta^+(e_1^+) = (1 + \Gamma(\pi_0))/2 \neq (1 - \Gamma(\pi_0))/2 = \theta^-(e_1^+)$). Consequently, the restrictions of $T_{\pi_0}|_K$ to $H_{\pi_0}^+$ and $H_{\pi_0}^-$ are inequivalent. On the other hand, using Schur's lemma or computing $\sum_{k \in K} |\theta^+(k)|^2$, one can easily verify that the restrictions of $T_{\pi_0}|_K$ to $H_{\pi_0}^+$ and $H_{\pi_0}^-$ are irreducible. Therefore the subspaces $H_{\pi_0}^+$ and $H_{\pi_0}^-$ are uniquely defined as the only eigenspaces contained in H_{π_0} that are invariant under $T_{\pi_0}|_K$. Since T_{π_0} is reducible, there exists an eigenspace $\tilde{H} \subset H_{\pi_0}$ invariant under T_{π_0}. Then $\tilde{H}$ is invariant under $T_{\pi_0}|_K$ and $\tilde{H} = H_{\pi_0}^+$ or $\tilde{H} = H_{\pi_0}^-$. Hence the subspaces $H_{\pi_0}^+$ and $H_{\pi_0}^-$ are invariant under T_{π_0}. Let $T_{\pi_0}^+$ and $T_{\pi_0}^-$ be the restrictions of T_{π_0} to the subspaces $H_{\pi_0}^+$ and $H_{\pi_0}^-$ respectively. It is clear that the dimensions of these representations are equal to $(q + 1)/2$.

Let us find the characters of the representations $T_{\pi_0}^+$ and $T_{\pi_0}^-$. Let φ^+ be the character of $T_{\pi_0}^+$. We infer from formulas (5.4.4), (5.4.5), and (5.5.8) that:

$$\varphi^+(-e) = \pi_0(-1)(q+1)/2;$$
$$\varphi^+(e_1^+) = 1 + \sum_{u=v^2 \neq 0} \chi(u) = (1 + \Gamma(\pi_0))/2;$$
$$\varphi^+(e_\varepsilon^+) = 1 + \sum_{u=v^2 \neq 0} \chi(u\varepsilon) = (1 - \Gamma(\pi_0))/2;$$
$$\varphi^+(e_1^-) = (1 + \Gamma(\pi_0))\pi_0(-1)/2;$$
$$\varphi^+(e_\varepsilon^-) = \pi_0(-1)(1 - \Gamma(\pi_0))/2;$$
$$\varphi^+(g_\lambda) = \pi_0(\lambda).$$

In addition, we have $\varphi^+(e) = (q+1)/2 \, (= \dim T_{\pi_0}^+)$. Consequently, we have

$$\sum_{g \in G} |\varphi^+(g)|^2 \geq ((q+1)/2)^2 + ((q+1)/2)^2$$
$$+ \{|1 + \Gamma(\pi_0)|^2/2 + |1 - \Gamma(\pi_0)|^2/2\}(q^2 - 1)/2$$
$$+ (q-3)q(q+1)/2 = q(q^2 - 1);$$

and therefore $\sum_{g_\sigma} |\varphi^+(g)|^2 \leq 0$; that is, $\varphi^+(g_\sigma) = 0$ for all g_σ. Since the character φ^- of $T_{\pi_0}^-$ satisfies the relation $\varphi^- = \varphi - \varphi^+$, we obtain the following.

I. *Let φ^+ and φ^- be the characters of the representation $T_{\pi_0}^+$ and $T_{\pi_0}^-$ respectively. We then have*

$$\varphi^+(e) = (q+1)/2; \qquad \varphi^+(-e) = (q+1)\pi_0(-1)/2;$$
$$\varphi^+(e_1^+) = (\Gamma(\pi_0) + 1)/2; \qquad \varphi^+(e_\varepsilon^+) = (1 - \Gamma(\pi_0))/2;$$

$$\varphi^+(e_1^-) = (\Gamma(\pi_0) + 1)\pi_0(-1)/2; \qquad \varphi^+(e_\varepsilon^-) = (1 - \Gamma(\pi_0))\pi_0(-1)/2;$$
$$\varphi^+(g_\lambda) = \pi_0(\lambda); \qquad \varphi^+(g_\sigma) = 0; \qquad \varphi^-(e) = (q+1)/2;$$
$$\varphi^-(-e) = (q+1)\pi_0(-1)/2; \qquad \varphi^-(e_1^+) = (1 - \Gamma(\pi_0))/2;$$
$$\varphi^-(e_\varepsilon^+) = (1 + \Gamma(\pi_0))/2; \qquad \varphi^-(e_1^-) = (1 - \Gamma(\pi_0))\pi_0(-1)/2;$$
$$\varphi^-(e_\varepsilon^-) = (1 + \Gamma(\pi_0))\pi_0(-1)/2; \qquad \varphi^-(g_\lambda) = \pi_0(\lambda); \qquad \varphi^-(g_\sigma) = 0.$$

Comparing the formulas for the characters of the representations T_π ($\pi^2 \neq 1$), 1_G, $\tilde{T}_1$, $T_{\pi_0}^+$, T_{π_0}, we see that we have constructed a set of $2 + 2 + (q-3)/2 = (q+5)/2$ pairwise inequivalent irreducible representations of the group G.

5.7. The Representations S_π

Let us construct yet another set of representations of G. Consider the *quadratic extension* $\mathbf{F}_q(\sqrt{\varepsilon})$ of the field $\mathbf{F}_q$, that is, the set of all elements of the form $x + y\sqrt{\varepsilon}$, x, $y \in \mathbf{F}_q$ with the usual operations of addition and multiplications of polynomials in $\sqrt{\varepsilon}$ with coefficients from $\mathbf{F}_q$, where $\sqrt{\varepsilon}\sqrt{\varepsilon} = \varepsilon$. Note that $\mathbf{F}_q(\sqrt{\varepsilon})$ is a field containing $\mathbf{F}_q$ as a subfield. We introduce into $\mathbf{F}_q(\sqrt{\varepsilon})$ an operation of conjugation, setting $\bar{t} = x - y\sqrt{\varepsilon}$ for $t = x + y\sqrt{\varepsilon} \in \mathbf{F}_q(\sqrt{\varepsilon})$. The set $U = \{t : t \in \mathbf{F}_q(\sqrt{\varepsilon}), t\bar{t} = 1\}$ is a subgroup of the multiplicative group $\mathbf{F}_q(\sqrt{\varepsilon})^*$ of the field $\mathbf{F}_q(\sqrt{\varepsilon})$.

Let ρ be a character of the group $\mathbf{F}_q(\sqrt{\varepsilon})$ such that $\rho(t) \not\equiv 1$ for $t \in U$. Consider the vector space H of all functions f on $\mathbf{F}_q^*$ and define the representation S_ρ of G in H by

$$(S_\rho(g)f)(u) = \sum_{v \in \mathbf{F}_q^*} K_\rho(u, v; g)f(v),$$

where $g = \begin{Vmatrix} a & b \\ c & d \end{Vmatrix}$, and

$$K_\rho(u, v; g)$$

$$= \begin{cases} -\dfrac{1}{q}\chi\left(\dfrac{du + av}{b}\right) \displaystyle\sum_{t\bar{t} = vu^{-1}} \chi\left(-\dfrac{ut + vt^{-1}}{b}\right)\rho(t), & \text{if } b = 0; \quad (5.7.1a) \\ \rho(d)\chi(dcu)\delta(d^2 u - v), & \text{if } b = 0. \quad (5.7.1b) \end{cases}$$

Here δ is the *Kronecker symbol* ($\delta(x) = 0$ for $x \neq 0$, $\delta(0) = 1$) and χ is a fixed additive character of the field $\mathbf{F}_q$, not identically equal to 1.

The fact that S_ρ is a representation (that is, $S(e) = 1$, $S_\rho(g_1 g_2) = S_\rho(g_1)S_\rho(g_2)$), is easy to verify. The dimension of S_ρ is $q - 1$.

Let π be the restriction of the character ρ to the subgroup U. Then π is a character of U, not identically 1.

Let us find the character φ_ρ of S_ρ. It is evident that $\varphi_\rho(e) = q - 1$, $\varphi_\rho(-e) = (q-1)\pi(-1)$ (since $-1 \in U$, we have $\rho(-1) = \pi(-1)$). We also infer from formula (5.7.1b) that $\varphi_\rho(g_\lambda) = 0$ (since $\lambda^2 \neq 1$ for all g_λ);

$$\varphi_\rho(e_1^+) = \sum_{u \neq 0} \chi(u) = \sum_u \chi(u) - 1 = -1.$$

Analogously, we have $\varphi_\rho(e_\varepsilon^+) = -1$, $\varphi_\rho(e_1^-) = \varphi_\rho(\varepsilon_\varepsilon^-) = -\pi(-1)$. The orthogonality relations (1.4.2) for characters χ_p defined by the formula $\chi_p(u) = \chi(pu)$ show that $\sum_{u \neq 0} \chi(pu) = q - 1$ for $p = 0$ and $\sum_{u \neq 0} \chi(pu) = -1$ for $p = 0$ ($p \in \mathbf{F}_q$). Therefore for $g = g_\sigma$ we set $t_0 + t_0^- = 2\sigma = a + d$, $t_0 \in U$, to obtain

$$\varphi_\rho(g_\sigma) = \sum_{u \neq 0} K_\rho(u, u; g)$$

$$= -\frac{1}{q} \sum_{u \neq 0} \chi\left(\frac{2\sigma u}{v}\right) \sum_{t\bar{t}=1} \chi\left(-\frac{ut + ut^{-1}}{v}\right) \rho(t)$$

$$= -\frac{1}{q} \sum_{t\bar{t}=1} \rho(t) \sum_{u \neq 0} \chi\left(\frac{2\sigma - (t + t^{-1})}{v} u\right)$$

$$= -\frac{1}{q} \sum_{t\bar{t}=1} \rho(t)(-1) - \frac{1}{q}(\rho(t_0)q + \rho(t_0^{-1})q)$$

$$= -(\rho(t_0) + \rho(t_0^{-1})).$$

(It follows from the condition $\rho \neq 1$ on U that $\sum_{t\bar{t}=1} \rho(t) = 0$.) Since $t_0 \in U$ and $t_0 + t_0^{-1} = 2\sigma$, we find $\varphi_\rho(g_\sigma) = -\pi(t_0) - \pi(t_0^{-1})$, where $g_\sigma = \begin{Vmatrix} \sigma & v \\ \varepsilon v & \sigma \end{Vmatrix}$, $t_0\bar{t}_0 = 1$, $t_0 + t_0^{-1} = 2\sigma$. This implies that $t_0 = \sigma \pm v\sqrt{\varepsilon}$. Assembling these facts, we obtain the following.

I. Let φ_ρ be the character of the representation S_ρ. Let π be the restriction of the character ρ to the subgroup U. We then have:

$$\varphi_\rho(e) = q - 1; \qquad \varphi_\rho(-e) = (q-1)\pi(-1);$$
$$\varphi_\rho(g_\lambda) = 0; \qquad \varphi_\rho(e_1^+) = \varphi_\rho(e_\varepsilon^+) = -1;$$
$$\varphi_\rho(e_1^-) = \varphi_\rho(e_\varepsilon^{-1}) = -\pi(-1);$$
$$\varphi_\rho(g_\sigma) = -\pi(t_0) - \pi(t_0^{-1}), \quad \text{where } t_0\bar{t}_0 = 1, t_0 + \bar{t}_0 = 2\sigma.$$

A corollary follows.

II. If $\rho_1 = \rho_2$ on U or $\rho_1 = \rho_2^{-1}$ on U, it follows that $\varphi_{\rho_1} \approx \varphi_{\rho_2}$.

The symbol S_π denotes the representation S_ρ, where π is the restriction of ρ to U. The formula for $\varphi_\rho(g_\sigma)$ uniquely defines $\pi(t) + \pi(t^{-1})$ for all t such that $t\bar{t} = 1$. This has a useful consequence.

III. The representation S_{π_1} and S_{π_2} are equivalent if and only if $\pi_1 = \pi_2$ or $\pi_1 = (\pi_2)^{-1}$.

We easily verify that $\sum_{g \in G} |\varphi_\rho(g)|^2 = q(q^2 - 1)$, if $\rho^2 \not\equiv 1$ on U. That is, if $\pi^2 \not\equiv 1$ on U then S_π is irreducible. If the character π_1 of U is not identically 1 and if $\pi_1^2 = 1$, then π_1 assumes exactly the values ± 1. Therefore π_1 is identically 1 on the subgroup $U^2 \subset U$, formed by squares of the elements of U. Consequently, we have $\pi_1 = -1$ on $U \backslash U^2$, that is, π_1 is uniquely defined. Let φ_1 be the character of the representation S_{π_1}. We immediately verify that $\sum_{g \in G} |\varphi_1(g)|^2 = 2q(q^2 - 1)$, so that S_{π_1} is the direct sum of two irreducible representations of G (see (3.7.5)). The formula (5.7.1b) shows that the subspace $H^+ \subset H$, consisting of the functions on $\mathbf{F}_q^*$ that vanish on $\mathbf{F}_q^* \backslash \mathbf{F}_q^{*2}$ and the space $H^- \subset H$, consisting of the functions vanishing on $\mathbf{F}_q^{*2}$, are invariant under the operators $S_{\pi_1}(k)$, $k \in K$. Furthermore a matrix of the form $\begin{Vmatrix} 1 & 0 \\ c & 1 \end{Vmatrix}$ operates in H by multiplication multiplying by the character $\chi(cu)$. Therefore any operator A in H commuting with the operators of S_{π_1} commutes with the operators of multiplication by $\chi(cu)$ and also with the operators of multiplication by any function $\alpha(u)$ (since $\alpha(u)$ is a linear combination of characters). Hence A is an operator of multiplication by a function. Indeed, if we have $f_0(u) \equiv 1$ and $A f_0 = \psi_A$, then

$$(A\varphi)(u) = (A(\varphi f_0))(u) = \varphi(u)(A f_0)(u) = \psi_A(u)\varphi(u).$$

That is, A is the operator of multiplication by $\psi_A(u)$. Since $(S_{\pi_1}(g)f)(u) = \rho(d)f(d^2 u)$ for $g = \begin{Vmatrix} d^{-1} & 0 \\ 0 & d \end{Vmatrix}$, we see that $\psi_A(d^2 u) = \psi_A(u)$. That is, the function ψ_A is constant on the cosets of $\mathbf{F}_q^{*2}$ in $\mathbf{F}_q^*$. Hence the restriction of S_{π_1} to K is irreducible on H^+ and H^-. Since S_{π_1} is reducible, H^+ and H^- are invariant under S_{π_1}.[25a] Let $S_{\pi_1}^+$ and $S_{\pi_1}^-$ be the subrepresentations of S_{π_1} on the spaces H^+ and H^- respectively. The dimension of each of the representations $S_{\pi_1}^+$ and $S_{\pi_1}^-$ is equal to $(q - 1)/2$. Let us find their characters φ_1^+, φ_1^-.

Let φ_1^+ be the character of the representation $S_{\pi_1}^+$. We have

$$\varphi_1^+(e) = (q - 1)/2; \qquad \varphi_1^+(-e) = \pi_1(-1)(q - 1)/2;$$
$$\varphi_1^+(e_1^+) = \sum_{u = v^2 \neq 0} \chi(u) = (\Gamma(\pi_0) - 1)/2;$$
$$\varphi_1^+(e_\varepsilon^+) = \sum_{u = v^2 \neq 0} \chi(\varepsilon u) = -(\Gamma(\pi_0) + 1)/2;$$
$$\varphi_1^+(e_1^-) = \pi_1(-1)(\Gamma(\pi_0) - 1)/2;$$
$$\varphi_1^+(e_\varepsilon^-) = -\pi_1(-1)(\Gamma(\pi_0) + 1)/2; \qquad \varphi_1^+(g_\lambda) = 0.$$

[25a] It follows in particular that $K_{\pi_1}(u, v; g) = 0$ if $vu^{-1} \in \mathbf{F}_q^* \backslash \mathbf{F}_q^{*2}$, i.e.,

$$\sum_{ti = uv^{-1}} \chi(-(1/b)(ut + vt^{-1}))\rho(t) = 0,$$

for all multiplicative characters ρ of $\mathbf{F}_q^*(\sqrt{\varepsilon})$ that coincide with π_1 on U.

(Since $\lambda^2 \neq 1$, (5.7.1b) shows that $K(u, u; g_\lambda) = 0$ for all $u \neq 0$.) We also have

$$\varphi_1^+(g_\sigma) = \sum_{u = w^2 \neq 0} K_{\pi_0}(u, u; g)$$

$$= -(1/q) \sum_{t\bar{t} = 1} \pi_1(t) \times \sum_{u = w^2 \neq 0} \chi((2\sigma - t - t^{-1})/v)u)$$

and from formula (5.4.2) we obtain

$$\varphi_1^+(g_\sigma) = -\frac{1}{q} \sum_{\substack{t\bar{t} = 1 \\ t \neq t_0, \bar{t}_0}} \pi_1(t) \left\{ \frac{1}{2} \Gamma(\pi_0)\pi_0 \left(\frac{2\sigma - (t + t^{-1})}{v} \right) - \frac{1}{2} \right\}$$

$$- \frac{q - 1}{2q} \{\pi_1(t_0) + \pi_1(t_0^{-1})\}$$

$$= -\frac{1}{q} \frac{\Gamma(\pi_0)}{2} \sum_{\substack{t\bar{t} = 1 \\ t \neq t_0, \bar{t}_0}} \pi_1(t)\pi_0 \left(\frac{2\sigma - (t + t^{-1})}{v} \right) - \frac{1}{2} (\pi_1(t_0) + \pi_1(t_0^{-1})),$$

$$(5.7.2)$$

where $t_0 + \bar{t}_0 = 2\sigma$, $t_0 \bar{t}_0 = 1$. Since π_1 and π_0 assume only real values and $\Gamma(\pi_0) = (q\pi_0(-1))^{1/2}$ is either real or purely imaginary, it follows from the relations $\sum_{g \in G} |\varphi_1^+(g)|^2 = q(q^2 - 1)$, $\sum_{g \in G} \varphi_1^+(g)\varphi_1^-(g) = 0$ (which are orthogonality relations) that

$$\sum_{\substack{t\bar{t} = 1 \\ t \neq t_0, \bar{t}_0}} \pi_1(t)\pi_0 \left(\frac{2\sigma - (t + t^{-1})}{v} \right) = 0$$

for $g = \begin{Vmatrix} \sigma & v \\ \varepsilon v & \sigma \end{Vmatrix}$, $t_0 + t_0^{-1} = 2\sigma$, $t_0 \bar{t}_0 = 1$. From this and (5.7.2) we infer that $\varphi_1^+(g_\sigma) = -\frac{1}{2}(\pi_1(t_0) + \pi_1(t_0^{-1}))$. Finally, since $\varphi_1^- = \varphi_1 - \varphi_1^+$, we obtain the following.

IV. Let φ_1^+ and φ_1^- be the characters of the representations $S_{\pi_1}^+$ and $S_{\pi_1}^-$, respectively. We then have

$$\varphi_1^+(e_1^+) = (\Gamma(\pi_0) - 1)/2, \qquad \varphi_1^+(e_\varepsilon^+) = -(\Gamma(\pi_0) + 1)/2;$$
$$\varphi_1^+(e_1^-) = (\Gamma(\pi_0) - 1)\pi_1(-1)/2;$$
$$\varphi_1^+(e_\varepsilon^-) = -(\Gamma(\pi_0) + 1)\pi_1(-1)/2;$$
$$\varphi_1^+(g_\lambda) = 0; \qquad \varphi_1^+(g_\lambda) = -(\pi_1(t_0) + \pi_1(t_0^{-1}))/2 = -\pi_1(t_0),$$

where $t_0 + t_0^{-1} = 2\sigma$, $t_0 \bar{t}_0 = 1$;

$$\varphi_1^-(e) = (q - 1)/2; \qquad \varphi_1^-(-e) = (q - 1)\pi_1(-1)/2;$$
$$\varphi_1^-(e_1^+) = -(\Gamma(\pi_0) + 1)/2;$$

$$\varphi_1^-(e_\varepsilon^+) = (\Gamma(\pi_0) - 1)/2; \qquad \varphi_1^-(e_1^-) = -(\Gamma(\pi_0) + 1)\pi_1(-1)/2;$$
$$\varphi_1^-(e_\varepsilon^-) = (\Gamma(\pi_0) - 1)\pi_1(-1)/2; \qquad \varphi_1^-(g_\lambda) = 0;$$
$$\varphi_1^-(g_\sigma) = -(\pi_1(t_0) + \pi_1(t_0^{-1}))/2 = -\pi_1(t_0),$$

where $t_0 + t_0^{-1} = 2\sigma$, $t_0 \bar{t}_0 = 1$.

The representations of the group G of the form S_π ($\pi^2 \neq 1$), $S_{\pi_1}^+$, $S_{\pi_1}^-$ form a set of $(q-1)/2 + 2 = (q+3)/2$ pairwise inequivalent representations. Comparing the characters of these representations with the characters of the representations T_π and their subrepresentations, we see that we have constructed $q + 4$ pairwise inequivalent irreducible representations. Since the number of conjugacy classes in G is $q + 4$, we have completely classified the irreducible representations of G.

V. *The set of $q + 4$ representations* 1_G; T_π ($\pi^2 \neq 1$); $\tilde{T}_1$; $T_{\pi_0}^+$; $T_{\pi_0}^-$; S_π ($\pi^2 \neq 1$); $S_{\pi_1}^+$; $S_{\pi_1}^-$ *forms a complete system of irreducible representations of* G.

The characters of the irreducible unitary representations of the group $SL(n, \mathbf{F}_q)$ for $n > 3$ have been found by Green [1*]. The work [1*] of S. I. Gel'fand was devoted to the study of various properties of representations of the group $SL(n, \mathbf{F}_q)$.

Chapter III

Basic Concepts of the Theory of Representations of Topological Groups

§1. Topological Spaces

1.1. Definition of a Topological Space

A set X is called a *topological space* if there is given a family $\mathcal{U} = \{U\}$ of subsets U of X having the following properties:

(1) $\varnothing \in \mathcal{U}$ and $X \in \mathcal{U}$;

(2) the union of any family of sets in $\mathcal{U}$ also belongs to $\mathcal{U}$;

(3) the intersection of any finite number of sets belonging to $\mathcal{U}$ also belongs to $\mathcal{U}$.

The sets $U \in \mathcal{U}$ are called *open* sets of the topological space X and the elements $x \in X$ are called *points* of the space. We say that the system of sets $\mathcal{U}$ defines a *topology* T in X and that $\mathcal{U}$ *is* a topology on X.

A given set X can admit different systems $\mathcal{U}$. They then define different topologies in X. A fixed set X with different topologies can be viewed as different topological spaces. If we consider all subsets of X to be open, we obtain a particular topology. It is clear that conditions (1)–(3) are then satisfied. This topology on X is called *discrete*.

A system $\mathcal{V} = \{V\}$ of sets $V \subset X$ is called a *basis* for the topology $\mathcal{U}$ if every open set $\mathcal{U}$ in X is the union of certain sets $V \in \mathcal{V}$. We can identify a topology $\mathcal{U}$ in X by giving a basis for $\mathcal{U}$. Then all the possible unions of sets from $\mathcal{V}$ will be all of the open sets. Clearly a system $\mathcal{V}$ is a basis of a topology if and only if: (1) $\varnothing \in \mathcal{V}$; (2) the intersection of any finite number of sets from $\mathcal{V}$ is the union of sets in $\mathcal{V}$; (3) the union of all sets in $\mathcal{V}$ is X.

Examples

1. Let $\mathbf{R}^1$ be the set of all real numbers. In $\mathbf{R}^1$ we define a topology by taking all intervals $a < x < b$ and the empty set $\varnothing$ as a basis. Conditions (1)–(3) hold and the usual open sets in $\mathbf{R}^1$ are the open sets. The topology in $\mathbf{R}^1$ is called the *natural topology* of $\mathbf{R}^1$. This topology is obviously different from the discrete topology in $\mathbf{R}^1$. Unless otherwise noted, $\mathbf{R}^1$ will denote the topological space of all real numbers with its natural topology.

2. Let $X = \mathbf{R}^n$ be the set of all sequences $x = \{x_1, \ldots, x_n\}$, $x_k \in \mathbf{R}^1$. We define a topology in $\mathbf{R}^n$, defining a basis to be the system that consists of the void set and of all open parallelepipeds:

$$a_k < x_k < b_k, \qquad k = 1, 2, \ldots, n. \tag{1.1.1}$$

The void set and all unions of parallelepipeds (1.1.1) will be open sets. These are the usual open sets in n-dimensional space (for $n = 2$ and $n = 3$ the usual open sets in the plane and in 3-dimensional space respectively). Conditions (1)–(3) are obviously satisfied. This topology described is called *the natural topology in* $\mathbf{R}^n$.

3. Let $X = \mathbf{C}^n$ be the set of all sequences $z = \{z_1, \ldots, z_n\}$ of complex numbers $z_k = x_k + iy_k$, $x_k, y_k \in \mathbf{R}^1$. The sequence $z = \{x_1 + iy_1, \ldots, x_n + iy_n\}$ may be associated with the sequence $x = \{x_1, y_1, \ldots, x_n, y_n\} \in \mathbf{R}^{2n}$. We then have a one-to-one mapping of $\mathbf{C}^n$ onto $\mathbf{R}^{2n}$. The inverse images of open sets in $\mathbf{R}^{2n}$ are the open sets in $\mathbf{C}^n$. Clearly the void set and all sets

$$a_k < x_k < b_k, \qquad c_k < y_k < d_k, \qquad k = 1, 2, \ldots, n$$

are a basis in $\mathbf{C}^n$. This topology is called *the natural topology in* $\mathbf{C}^n$.

Unless otherwise stated, $\mathbf{R}^n$ and $\mathbf{C}^n$ will denote the sets $\mathbf{R}^n$ and $\mathbf{C}^n$ provided with their natural topologies.

4. Let X consist of two distinct points: $X = \{a, b\}$. Let X, the void set $\varnothing$ and the set $\{a\}$, consisting of the point a, be the open sets. Conditions (1)–(3) are satisfied.

Exercise. Prove that the void set and all open balls

$$U(a, r) = \{x : (x_1 - a_1)^2 + \cdots + (x_n - a_n)^2 < r^2\},$$

$r > 0$, $a = \{a_1, \ldots, a_n\} \in \mathbf{R}^n$ form a basis in $\mathbf{R}^n$.

1.2. Neighborhoods

Any open set containing x is called *a neighborhood of the point* x. We frequently write $U(x)$ for neighborhoods of x.

I. *A subset M of X is open if and only if every point $x \in M$ admits a neighborhood $U(x)$ such that $U(x) \subset M$.*

If M is open, then it is itself a neighborhood of each of its points. Conversely, if every point $x \in M$ admits a neighborhood $U(x) \subset M$, then M is the union of such $U(x)$ and thus by (2) in 1.1, M is open.

A system $\mathscr{W}$ of neighborhoods of a point x is called a *basis at* x if every neighborhood $U(x)$ of x contains some $W(x) \in \mathscr{W}$. Plainly a basis at x

must have the following properties: (1) $x \in W(x)$; (2) every intersection $W_1(x) \cap W_2(x)$ of neighborhoods in the basis contains a neighborhood that belongs to the basis; (3) if $y \in W(x)$, then there is a neighborhood $W(y) \subset W(x)$ in the basis of neighborhoods of the point y.

Conversely, suppose that for every x in X there is a family of sets $W(x)$ satisfying conditions (1)–(3). We can then define a topology in X by taking the void set and all unions of sets $W(x)$ as open sets. Clearly conditions (1)–(3) of 1.1 are satisfied. The sets $W(x)$, $x \in X$, form a basis for the topology. Thus we can also describe a topology in X by specifying a neighborhood basis for every point $x \in X$. Thus in example 1 of 1.1, we can take all intervals $(x_0 - \varepsilon, x_0 + \varepsilon)$, $\varepsilon > 0$, as a neighborhood basis at the point x_0.

1.3. Closed Sets; the Closure of Sets

The complement of an open set is called a *closed set*. Properties (1)–(3) of open sets (see 1.1) show that:

(1)   the void set and the entire space X are closed;
(2)   the intersection of any family of closed sets is closed;
(3)   the union of a finite number of closed sets is closed.

Closed sets are frequently denoted by the letter F and the family of all closed sets by the script letter $\mathscr{F}$. The intersection of all closed sets containing a given subset of X is called the *closure of M* and is denoted by $\bar{M}$. By (2), $\bar{M}$ is closed and it clearly is the smallest closed set containing M. It is also clear that $\bar{M} = M$ if and only if M is closed. The following facts are obvious:

(1′) $M \subset \bar{M}$;
(2′) $\bar{\bar{M}} = \bar{M}$ (since $\bar{M}$ is closed);
(3′) $\overline{M_1 \cup M_2 \cup \cdots \cup M_n} = \bar{M}_1 \cup \bar{M}_2 \cup \cdots \cup \bar{M}_n$ for any finite number of sets;
(4′) $\bar{\varnothing} = \varnothing$.

Every point x of $\bar{M}$ is called a *limit point of M*. Note that $x \in \bar{M}$ if and only if every neighborhood $U(x)$ contains at least one point of M. The concept of the limit of a sequence is related to the concept of a limit point. A point x is called *a limit of the sequence* $x_1, x_2, x_3, \ldots$, written $x = \lim_{n \to \infty} x_n$, if for every neighborhood $U(x)$ of x, there is a positive integer n_0 (depending on $U(x)$) such that all elements $x_{n_0}, x_{n_0+1}, \ldots$ of the sequence belong to $U(x)$. A set $M \subset X$ is called *dense in X* if $\bar{M} = X$. The space X is called *separable* if X contains a countable dense subset.

Example

In the spaces $\mathbf{C}^n$ and $\mathbf{R}^n$ closed sets are usual closed sets, the closure coincides with the usual closure, and the limit with the usual limit. These spaces are separable since the points with rational coordinates are countable dense subsets.

1.4. Comparison of Topologies

Let two topologies T_1 and T_2 be defined on a set X with systems $\mathscr{U}_1$ and $\mathscr{U}_2$ of open sets. The topology T_2 is said to *majorize* the topology T_1, written $T_1 \leqslant T_2$, if $\mathscr{U}_1 \subset \mathscr{U}_2$. If in addition $\mathscr{U}_1 \neq \mathscr{U}_2$, then we say that T_2 is *stronger than* T_1 (and also that T_1 is *weaker than* T_2). In this case, we write $T_1 < T_2$ and $T_2 > T_1$. It is clear that the discrete topology on X majorizes every topology on X.

   Let $\mathscr{F}_1$ and $\mathscr{F}_2$ be the families of closed sets in T_1 and T_2 respectively. Denote the closures of a set $M \subset X$ in the topologies T_1 and T_2 by $\bar{M}^1$ and $\bar{M}^2$ respectively.

I. *If $T_1 \leqslant T_2$, then we have $\mathscr{F}_1 \subset \mathscr{F}_2$ and hence*

$$\bar{M}^1 \supset \bar{M}^2$$

for all subsets M of X.

1.5. The Interior and Boundary of a Set

The union of all open sets contained in a given set $M \subset X$ is the largest open subset of M. It is called the *interior of M* and is denoted by int M. The set $\bar{M} \setminus \text{int } M$ is called the *boundary of M* and is denoted by ∂M.

Examples

Let $X = \mathbf{R}^n$ and

$$M_1 = \{x : x \in \mathbf{R}^n, (x_1 - a_1)^2 + \cdots + (x_n - a_n)^2 \leqslant r^2\},$$
$$M_2 = \{x : x \in \mathbf{R}^n, (x_1 - a_1)^2 + \cdots + (x_n - a_n)^2 < r^2\},$$
$$M_3 = \{x : x \in \mathbf{R}^n, 0 < (x_1 - a_1)^2 + \cdots + (x_n - a_n)^2 < r^2\},$$
$$M_4 = \{x : x \in \mathbf{R}^n, (x_1 - a_1)^2 + \cdots + (x_n - a_n)^2 = r^2\},$$

where $a = \{a_1, \ldots, a_n\}$ and $r > 0$ are fixed. Clearly M_1 is a closed ball, M_2 is an open ball, M_3 is an open ball with its center removed, and M_4 is the sphere in $\mathbf{R}^n$ with center a and radius r. It is easy to verify that

$$
\begin{array}{lll}
\bar{M}_1 = M_1, & \text{int } M_1 = M_2, & \partial M_1 = M_4, \\
\bar{M}_2 = M_1, & \text{int } M_2 = M_2, & \partial M_2 = M_4, \\
\bar{M}_3 = M_1, & \text{int } M_3 = M_3, & \partial M_3 = \{a, M_4\}, \\
\bar{M}_4 = M_4, & \text{int } M_4 = \varnothing, & \partial M_4 = M_4.
\end{array}
$$

1.6. Subspaces

Every subset Y of a topological space X can be made a topological space by taking the intersections $Y \cap U$ (U open in X) as open sets in Y. The space Y with this topology is called *a subspace of the space X* and the topology in Y is said to be *induced by the topology in* X. The definition implies that:

(1) if $\mathscr{V} = \{V\}$ is a basis in X, then $\{V \cap Y\}$ is a basis in Y;
(2) a closed set in Y is the intersection with Y of a closed set in X;
(3) the closure in Y of any set $M \subset Y$ is the intersection with Y of the closure of M in X.

From (2) we conclude the following.

I. *If Y is closed in X, then every subset of Y that is closed in Y is also closed in X.*

II. *If we have $M_1 \subset M_2 \subset X$, then the closure in M_1 of every set $A \subset M_1$ is the intersection with M_1 of the closure of the set A in M_2.*

Examples

1. Sets $[a,b] = \{x : x \in \mathbf{R}^1, a \leqslant x \leqslant b\}$ and $(a,b) = \{x : x \in \mathbf{R}^1, a < x < b\}$, taken as subspaces of the space $\mathbf{R}^1$, are called *closed intervals* and *open intervals*, respectively.

2. Analogously, the sets $[a_1, b_1; \ a_2, b_2] = \{x : x \in \mathbf{R}^2, \ a_1 \leqslant x_1 \leqslant b_1, a_2 \leqslant x_2 \leqslant b_2\}$ and $(a_1, b_1 : a_2, b_2) = \{x : x \in \mathbf{R}^2, a_1 < x_1 < b_1, a_2 < x_2 < b_2\}$, taken as subspaces of $\mathbf{R}^2$, are called *closed* and *open rectangles* respectively.

Remark. An open (or closed) set in a subspace Y of X may fail to be open (or closed) in X. Thus for $X = \mathbf{R}^1$ and $Y = [a, b]$, every set $[a, \alpha) = \{x : x \in \mathbf{R}^1, a \leqslant x < \alpha\}$, $a < \alpha < b$ is open in Y but not open in X.

1.7. Mappings of Topological Spaces

Let f be a mapping of a topological space X into a topological space Y. In analogy with the usual definition of a continuous function, the mapping f is called *continuous at the point* $x_0 \in X$ if the inverse image of each neighborhood $V(y_0)$ of the point $y_0 = f(x_0)$ contains some neighborhood $U(x_0)$ of the point x_0. A mapping f of X into Y is called *continuous* if f is continuous at every point $x \in X$.

Continuity can be characterized as follows.

I. *A mapping f of the space X into the space Y is continuous if and only if the inverse image of every open set in Y is an open set in X (or if the inverse image of every closed set in Y is a closed set in X).*

A second useful fact is the following.

II. *If a continuous mapping f of X into Y maps a set $M \subset X$ into a set $N \subset Y$, then it also maps $\bar{M}$ into $\bar{N}$.*

Indeed, $f^{-1}(\bar{N})$ is closed by virtue of I and it contains M and therefore also $\bar{M}$.

If f is a continuous mapping of X into Y then f is also called a *continuous function* on X with values in Y, and $f(X)$ is called a *continuous image* of the space X (in the space Y). If X is a closed interval $[a, b]$, a continuous image $f(X) = f([a, b])$ in Y is called a *continuous curve in* Y; $f(a)$ is called the *initial point of the curve* and $f(b)$ the *terminal point*. We also say that this curve *joins* the points $f(a)$ and $f(b)$. A curve is *closed* if its initial point and terminal point coincide.

A mapping f of X onto Y is called a *homeomorphism* if:

(1) f is a one-to-one mapping;
(2) the mappings f and f^{-1} are continuous.

A mapping f of the space X into the space Y is called a *homeomorphism of X into Y* if f is a homeomorphism of X onto $f(X)$, where $f(X)$ has its subspace topology. Two topological spaces X and Y are called *homeomorphic* or *homeomorphs* if there is a homeomorphism of X onto Y. Proposition I gives the following.

III. *Under a homeomorphism of X onto Y, open sets are mapped onto open sets and closed sets onto closed sets. The closure of every set in X is mapped onto the closure of the image of that set in Y.*

Properties of a topological space that are preserved under homeomorphisms are said to be *topological*. The branch of mathematics concerned with topological properties is called *topology*.

1.8. Separated Spaces

A topological space X is called *separated* (or *a Hausdorff space*) if it satisfies the following separation axiom: every two distinct points in X admit disjoint neighborhoods.

The spaces $\mathbf{R}^n$ and $\mathbf{C}^n$ are separated. The space $X = \{a, b\}$ of example 4 in 1.1 is not. The only neighborhood of the point b is all of X, which intersects every neighborhood of the point a.

I. *If f and φ are two continuous mappings of the space X into the separated space Y, then the set $M_1 = \{y: f(x) \neq \varphi(x)\}$ is open and the set $M_2 = \{x: f(x) = \varphi(x)\}$ is closed.*

Proof. It suffices to prove that M_1 is open since M_2 is the complement of M_1 in X. Let x belong to M, that is, $f(x_0) \neq \varphi(x_0)$. There exist disjoint neighborhoods $U = U(f(x_0))$ and $V = V(\varphi(x_0))$. Since f and φ are continuous, there is a neighborhood $W(x_0)$ whose images under the mappings f and φ are contained in U and V respectively. Since U and V do not intersect, we have $W(x_0) \subset M_1$. Thus for every point $x_0 \in M_1$ there exists a neighborhood $W(x_0) \subset M_1$ and by virtue of I in 1.2, M_1 is open. $\square$

Corollary. *If f_j, φ_j, $j = 1, \ldots, m$ are continuous mappings of the space X into the separated space Y, then the set $M = \{x : f_j(x) = \varphi_j(x), j = 1, \ldots, m\}$ is closed.*

Note that M is the intersection of the sets $M_j = \{x : f_j(x) = \varphi_j(x)\}$, $j = 1, \ldots, m$, which are closed, as shown in I. $\square$

II. *If two continuous mappings f and φ of X into the separated space Y coincide on a set $N \subset X$, then they also coincide on $\bar{N}$.*

Proof. Let M_2 be as in I. We then have $N \subset M_2$ and therefore $\bar{N} \subset \bar{M}_2 = M_2$. Consequently we obtain $f(x) = \varphi(x)$ on $\bar{N}$. $\square$

III. *If f and φ are continuous real-valued functions on a separated space X, then the set $M_1 = \{x : f(x) < \varphi(x)\}$ is open and the set $M_2 = \{x : f(x) \geqslant \varphi(x)\}$ is closed.*

The proof is like the proof of Proposition I (note that $Y = \mathbf{R}^1$).

IV. *All finite subsets of a separated space are closed.*

Proof. We may consider only a set $\{x_0\}$ consisting of one point x_0, since the union of a finite number of closed sets is closed (property 3) in 1.3. If $x_1 \neq x_0$, then there is a neighborhood $U(x_1)$ that does not contain x_0, since X is separated. Therefore $X \backslash \{x_0\}$ is open and $\{x_0\}$ is closed. $\square$

1.9. Products of Topological Spaces

Let $X_1, \ldots, X_n$ be topological spaces. We write $X_1 \times X_2 \times \cdots \times X_n$ for the set of all sequences $x = \{x_1, x_2, \ldots, x_n\}$, where $x_j \in X_j$ for $j = 1, \ldots, n$. If $M_1, \ldots, M_n$ are arbitrary sets in $X_1, \ldots, x_n$ respectively, then $M_1 \times M_2 \times \cdots \times M_n$ denotes the set of all sequences $x = \{x_1, \ldots, x_n\}$ for $x_j \in M_j$ $(j = 1, 2, \ldots, n)$.

We introduce a topology into $X_1 \times \cdots \times X_n$ by taking the family of all sets $U_1(x_1^0) \times \cdots \times U_n(x_n^0)$, as a neighborhood basis at the point $x^0 = \{x_1^0, \ldots, x_n^0\}$. Here $U_i(x_i^0)$ are selected from neighborhood bases at x_i^0 $(i = 1, 2, \ldots, n)$. Clearly, axioms (1)–(3) for a neighborhood basis (see 1.2) are satisfied, so that $X_1 \times \cdots \times X_n$ becomes a topological space. The space is called the *product*[26] of the spaces $X_1, \ldots, X_n$.

If $x = \{x_1, \ldots, x_n\}$, then x_j is called the *j-th coordinate of the point x* and the mapping $x \to x_j$ is called *the projection of the point x onto X_j.*

---

[26] This definition can be extended also to the case of an infinite' number of spaces (see Bourbaki [3] or Naĭmark [1]). However this extension is not needed here.

I. *The projection* $\{x_1, \ldots, x_n\} \to x_j$ *of the space* $X_1 \times \cdots \times X_n$ *onto* X_j *is continuous.*

The proof is obvious.

Examples and Exercises

1. The space $\mathbf{R}^n$ is the product of n replicas of the space $\mathbf{R}^1$ (see examples 1 and 2 in 1.1). Indeed, the open intervals (a_j, b_j) that contain a real number x_j^0 are a neighborhood basis at x_j and the parallelepipeds $(a_1, b_1) \times \cdots \times (a_n, b_n)$ that contain $(x_1^0, \ldots, x_n^0)$ are a neighborhood basis at the point $(x_1^0, \ldots, x_n^0)$. Analogously, $\mathbf{C}^n$ is the product of n replicas of the space $\mathbf{C}^1$ (see example 3 in 7.1).

2. Prove that a product of separated spaces is separated.

§2. Topological Groups

2.1. Definition of a Topological Group

A set G is called a *topological group* if:

(a) G is a group;
(b) G is a separated[27] topological space;
(c)   the function $f_1(g) = g^{-1}$ is a continuous map of G onto G and the function $f_2(g, h) = gh$ is a continuous map of $G \times G$ onto G.

Any group G is a topological group under the discrete topology since all mappings, in particular f_1 and f_2 in condition (c), are continuous on a discrete topological space (see I. in 1.7). Such a topological group is called *discrete*.

I. *The mapping* $g \to g^{-1}$ *is a homeomorphism of the topological space* G *onto itself.*

The mapping $f_1 : g \to g^{-1}$ is continuous by virtue of condition (c). The inverse mapping to $g \to g^{-1}$ coincides with f_1 (since $f_1 : g^{-1} \to (g^{-1})^{-1} = g$) and is therefore also continuous. $\square$

We conclude from I that if U is a neighborhood of the identity then U^{-1} is also a neighborhood of the identity.

II. *Every neighborhood* U *of the identity contains a symmetric neighborhood of the identity, that is, a neighborhood* V *such that* $V^{-1} = V$.

---

[27] Actually the separation property of the space G (and even stronger properties follow from conditions (a) and (c) and a condition weaker than (b): for every two distinct elements $g_1, g_2 \in G$ there exists a neighborhood of each element that does not contain the other (see Pontryagin [1], §17).

The neighborhood $V = U \cap U^{-1}$ is such a neighborhood. $\square$

III. *Any neighborhood U of the identity contains a neighborhood V such that $VV \subset U$.*

Proof. Since the mapping $f_2(g_1 g_2) = g_1 g_2$ is continuous for $g_1 = e$, $g_2 = e$ there exist neighborhoods V_1, V_2 of the identity, for which $V_1 V_2 \subset U$. Thus the neighborhood $V = V_1 \cap V_2$ satisfies the condition $VV \subset U$. $\square$

Examples

1. $\mathbf{R}^1$ is a group (see example 1 in 1.1) and $\mathbf{R}^1$ with its natural topology is a separated topological space. Condition (c) is obviously satisfied. That is, $f_1(x) = -x$ and $f(x_1, x_2) = x_1 + x_2$ are continuous functions. Consequently, $\mathbf{R}^1$ with its natural topology is a topological group. Likewise, $\mathbf{R}^n$ and $\mathbf{C}^n$ with their natural topologies are topological groups.

2. $\mathbf{R}_0^1$ is a group and $\mathbf{R}_0^1$ is a topological space in its natural topology (the topology of a subspace of $\mathbf{R}^1$). Condition (c) is satisfied, that is, $f_1(x) = 1/x$ and $f_2(x_1, x_2) = x_1 x_2$ are continuous functions on $\mathbf{R}_0^1$ and $\mathbf{R}_1^0 \times \mathbf{R}_1^0$ respectively. That is, $\mathbf{R}_0^1$ with its natural topology is a topological group. Analogously, $\mathbf{R}_0^+$ and $\mathbf{C}_0^1$ with their natural topologies are topological groups.

3. Consider the full linear group $GL(n, \mathbf{C})$. Each element

$$g = \left\| \begin{matrix} g_{11} & \cdots & g_{1n} \\ \cdots & \cdots & \cdots \\ g_{n1} & \cdots & g_{nn} \end{matrix} \right\| \in GL(n, \mathbf{C})$$

can be considered as a point $(g_{11}, \ldots, g_{1n}, g_{21}, \ldots, g_{2n}, g_{n1}, \ldots, g_{nn})$ of the space $\mathbf{C}^{n^2}$ and consequently the group $GL(n\ \mathbf{C})$ can be regarded as a subset of $\mathbf{C}^{n^2}$. Provide $GL(n, \mathbf{C})$ with the subspace topology induced by the natural topology of $\mathbf{C}^{n^2}$ (see example 3 in 1.1). We call this topology *the natural topology of the group $GL(n, \mathbf{C})$*. We will now prove the continuity of the functions f_1 and f_2 in condition (c). Let G_{jl} be the determinant of the $(n-1) \times (n-1)$ matrix obtained by deleting the j-th row and l-th column from the matrix g. Since

$$(g^{-1})_{jl} = \frac{G_{jl}}{\det g}, \tag{2.1.1}$$

the function $f_1(g) = g^{-1}$ is continuous. Similarly we have

$$(gh)_{jl} = \sum_{s=1}^{n} g_{js} h_{sl}, \tag{2.1.2}$$

and so f_2 is continuous. (Note that G_{jl} and $\det g$ are polynomials in the variables $g_{11}, \ldots, g_{nn}$). Hence $GL(n, \mathbf{C})$ with its natural topology is a topological group.

The natural topology in $GL(n, \mathbf{R})$ is defined similarly. In this topology, $GL(n, \mathbf{R})$ is also a topological group. We need only to substitute $\mathbf{R}^{n^2}$ for $\mathbf{C}^{n^2}$.

From the definition of the topology in $GL(n, \mathbf{C})$ it follows that the sets

$$W(g^0, \varepsilon) = \{g : g \in GL(n, \mathbf{C}), \qquad |g_{jl} - g_{jl}^0| < \varepsilon, j, l = 1, 2, \ldots, n\}$$

are a neighborhood basis at the element $g^0 \in GL(n, \mathbf{C})$. We also note that $GL(n, \mathbf{C})$ and $GL(n, \mathbf{R})$ are open sets in $\mathbf{C}^{n^2}$ and $\mathbf{R}^{n^2}$ respectively. We have

$$GL(n, \mathbf{C}) = \{g : g \in \mathbf{C}^{n^2}, \det g \neq 0\},$$
$$GL(n, \mathbf{R}) = \{g : g \in \mathbf{R}^{n^2}, \det g \neq 0\},$$

and our assertion follows from I in 1.8 and the continuity of the function $\det g$. In the sequel, unless otherwise noted, all groups in examples (1)–(3) will be regarded as topological groups with their natural topologies.

2.2. Translations of a Topological Group

I. *The right translation $g \to gg_0$ and the left translation $g \to g_0 g$ of a topological group G are homeomorphisms of the space G onto itself. (Here g_0 is any fixed element of G.)*

Proof. The right translation $g \to gg_0$ is a one-to-one mapping of G onto itself and is continuous by condition (c) in 2.1. The inverse mapping to $g \to gg_0$ is the translation $g \to gg_0^{-1}$; it too is continuous. That is, the right translation is a homeomorphism. Left translations are handled similarly. $\square$

II. *If $\{W(g)\}$ is a neighborhood basis at an element g of the topological group G, then $\{W(g)g_0\}$ and $\{g_0 W(g)\}$ are neighborhood bases at the elements gg_0 and $g_0 g$ respectively. If $\{W(e)\}$ is a neighborhood basis at the identity element e of the group G, then $\{W(e)g_0\}$ and $\{g_0 W(e)\}$ are neighborhood bases at the element $g_0 \in G$.*

The proof follows directly from I and III in 1.7. $\square$

We infer from Proposition II that a topology on a topological group is fully defined by describing a neighborhood basis $\{W(e)\}$ at e. Neighborhood bases at the other elements are obtained from them by right or left translations.

2.3. Subgroups of a Topological Group

Let G be a topological group and let H be a subgroup of G. We provide H with its topology as a topological subspace (see 1.6) of G. Condition (c) is satisfied in H since it is satisfied in all of G. With this definition of topology, H is thus a topological group. Unless otherwise noted, *a subgroup of the topological group G will be given its subspace topology.*

A subgroup H in the group G is called *closed* if it is a closed subset of the space G. A topological group G is called a *linear group* if it is isomorphic and homeomorphic to some subgroup of a group $GL(n, \mathbf{C})$ or $GL(n, \mathbf{R})$.

Examples

1. $GL(n, \mathbf{R})$ is a subgroup of $GL(n, \mathbf{C})$. By the corollary in 1.8, it is a closed subgroup. In fact, we have

$$GL(n, \mathbf{R}) = \{g : g \in GL(n, \mathbf{C}), \operatorname{Im} g_{jl} = 0, j, l = 1, \ldots, n\}$$

and the functions $\operatorname{Im} g_{jl}$ are continuous functions on $GL(n, \mathbf{C})$.

2. $SL(n, \mathbf{C})$ is a subgroup of the group $GL(n, \mathbf{C})$. It is closed since

$$SL(n, \mathbf{C}) = \{g : g \in GL(n, \mathbf{C}), \det g = 1\}$$

and $\det g$ is a continuous function on $GL(n, \mathbf{C})$. Similarly, $SL(n, \mathbf{R})$ is a closed subgroup of $GL(n, \mathbf{R})$ and also of $GL(n, \mathbf{C})$.

3. $U(n)$ and $SU(n)$ are subgroups of the groups $GL(n, \mathbf{C})$ and $SL(n, \mathbf{C})$ respectively. They are closed since we have

$$U(n) = \left\{ g : g \in GL(n, \mathbf{C}), \sum_{s=1}^{n} \bar{g}_{sj} g_{sl} = \delta_{jl}, j, l = 1, \ldots, n \right\},$$

$$SU(n) = \left\{ g : g \in SL(n, \mathbf{C}), \sum_{s=1}^{n} \bar{g}_{sj} g_{sl} = \delta_{jl}, j, l = 1, \ldots, n \right\},$$

where

$$\delta_{jl} = \begin{cases} 1, & \text{for } j = l, \\ 0, & \text{for } j \neq l, \end{cases}$$

and the functions $\sum_{s=1}^{n} \bar{g}_{sj} g_{sl}$, $j, l = 1, \ldots, n$ are continuous on $GL(n, \mathbf{C})$ and $SL(n, \mathbf{C})$.

2.4. Factor Spaces and Factor Groups

Let G be a topological group. Let H be a subgroup of G. We set $\tilde{G} = G/H$ where G/H is the space of right cosets of H in G. Let φ denote the canonical mapping of G onto $\tilde{G}$. We define a topology in G by considering as open sets in G the images $\varphi(U)$ of open sets U in G under the mapping φ. Conditions (1)–(3) in 1.1 are clearly satisfied. That is, the set of all $\varphi(U)$ is a topology in $\tilde{G}$. We will always give $\tilde{G}$ this topology.

A mapping of a topological space X into a topological space Y is called *open* if the image of every open set in X is open in Y.

I. *The canonical mapping φ of the space G onto $\tilde{G} = G/H$ is open and continuous.*

Proof. We infer from the definition of our topology in $\tilde{G}$ that φ is open. Let $\tilde{U}$ be an open set in $\tilde{G}$. By definition $\tilde{U}$ has the form $U = \varphi(U)$, where U is open in G. Therefore the inverse image of the set $\tilde{U}$,

$$\varphi^{-1}(\tilde{U}) = UH = \bigcup_{h \in H} Uh$$

is open since it is the union of open sets Uh (see I in 2.2). Thus ψ is continuous. □

II. *If H is a closed subgroup of the topological group G, then the factor space $G = G/H$ is separated.*

Proof. Let $\tilde{g}_1, \tilde{g}_2 \in \tilde{G}$ and $\tilde{g}_1 \neq \tilde{g}_2$. Let $g_1 \in \tilde{g}_1, g_2 \in \tilde{g}_2$. Since $\tilde{g}_1 \neq \tilde{g}_2, g_1^{-1}g_2$ does not belong to H. Since H is closed, there is a neighborhood U of the element $g_1^{-1}g_2$, disjoint from H:

$$U \cap H = \varnothing. \tag{2.4.1}$$

On the other hand, since the functions $f_1(g) = g^{-1}$ and $f_2(g, h) = gh$ are continuous, there are neighborhoods U_1 and U_2 of the elements g_1 and g_2 such that

$$U_1^{-1}U_2 \subset U. \tag{2.4.2}$$

We define $\tilde{U}_1 = \varphi(U_1)$ and $\tilde{U}_2 = \varphi(U_2)$. By definition, the sets $\tilde{U}_1$ and $\tilde{U}_2$ are neighborhoods of the elements $\tilde{g}_1$ and $\tilde{g}_2$ respectively. We will prove that $\tilde{U}_1 \cap \tilde{U}_2 = \varnothing$. Assume the contrary:

$$\tilde{g}_0 \in \tilde{U}_1 \cap \tilde{U}_2 = \varphi(U_1) \cap \varphi(U_2) \tag{2.4.3}$$

We conclude from (2.4.3) that there are elements

$$g_0, g_0' \in \tilde{g}_0 \tag{2.4.4}$$

such that $g_0 \in U_1, g_0' \in U_2$. Thus we have

$$g_0^{-1}g_0' \in U_1^{-1}U_2. \tag{2.4.5}$$

On the other hand, (2.4.4) implies that

$$g_0^{-1}g_0' \in H. \tag{2.4.6}$$

But (2.4.5) and (2.4.6) contradict (2.4.1). Thus we have $\tilde{U}_1 \cap \tilde{U}_2 = \varnothing$ and so G is separated. □

III. *Let H be a subgroup of the topological group G. For every $g_0 \in G$ the mapping $\bar{g}_0 : \tilde{g} \to \tilde{g}g_0$ is a homeomorphism of the space $\tilde{G}$ onto itself.*

Proof. The mapping $g \to \tilde{g}g_0$ of $\tilde{G}$ onto $\tilde{G}$ is one-to-one (see 1.7, chapter I). We will prove that it is continuous.

Let $\tilde{U}$ be an open set in $\tilde{G}$. We need to prove that its inverse image $\tilde{U}\overline{g_0^{-1}}$ is open in $\tilde{G}$. Since $\tilde{U} = \varphi(U)$, where U is open in G, the definition of the mapping $\overline{g}_0$ shows that

$$\tilde{U}\overline{g_0^{-1}} = \varphi(Ug_0^{-1}).$$

The set Ug_0^{-1} is open by virtue of I in 2.2. Thus $\tilde{U}\overline{g_0^{-1}}$ is open in G and $\overline{g}_0$ is continuous. Apply this fact to g_0^{-1} instead of g_0 to conclude that $(\overline{g}_0)^{-1} = \overline{g_0^{-1}}$ is continuous. That is, the mapping $\overline{g}_0$ is a homeomorphism. $\square$

We can topologize the space of *left* cosets of H in the same way and obtain results analogous to II and III.

IV. *If H is a closed normal divisor of a topological group G, then G/H is a topological group.*

Proof. G/H is a group and G/H is a separated topological space. We must prove that condition (c) in 2.1 is satisfied for G/H. We set $\tilde{G} = G/H$ and denote by φ the canonical mapping $G \to \tilde{G}$. Let $\tilde{U}$ be a neighborhood of the element $\tilde{g}_1^{-1} \in \tilde{G}$. There is an element $g_1^{-1} \in \tilde{g}_1^{-1}$ and a neighborhood U of g_1^{-1} such that $\tilde{U} = \varphi(U)$. Since the function $f(g) = g^{-1}$ is continuous, there is a neighborhood U_1 of g_1 such that

$$U_1^{-1} \subset U. \tag{2.4.7}$$

We set $\tilde{U}_1 = \varphi(U_1)$. Thus $\tilde{U}_1$ is a neighborhood of the element $\tilde{g}_1$ and (2.4.7) shows that

$$\tilde{U}_1^{-1} = \varphi(U_1)^{-1} \subset \varphi(U) = \tilde{U}.$$

Consequently the function $f_1(\tilde{g}) = \tilde{g}^{-1}$ is continuous on $\tilde{G}$. Let $\tilde{U}$ be a neighborhood of the element $\tilde{g}\tilde{h}$, where $\tilde{g}, \tilde{h} \in G$. We then have elements $g \in \tilde{g}$, $h \in \tilde{h}$ and a neighborhood U of gh such that $\tilde{U} = \varphi(U)$. Since the function $f_2(g, h) = gh$ is continuous, there are neighborhoods U_1, U_2 of g and h respectively such that

$$U_1 U_2 \subset U. \tag{2.4.8}$$

We set $\tilde{U}_1 = \varphi(U_1)$, $\tilde{U}_2 = \varphi(U_2)$. Thus $\tilde{U}_1$, $\tilde{U}_2$ are neighborhoods of the elements g, h respectively and (2.4.8) shows that

$$\tilde{U}_1\tilde{U}_2 = \varphi(U_1)\varphi(U_2) = \varphi(U_1 U_2) \subset \varphi(U) = \tilde{U}.$$

That is, the function $f_2(\tilde{g}, \tilde{h}) = \tilde{g}\tilde{h}$ is continuous on G. Condition (c) in 2.1 is satisfied and so $\tilde{G}$ is a topological group. $\square$

2.5. Homomorphisms and Isomorphisms of Topological Groups

Let G and G' be topological groups. A mapping f of the group G into the group G' is called *a continuous homomorphism of G into G'* if:

(a) f is a continuous mapping of the space G into the space G';
(b) f is a homomorphism of G into G'.

If $f(G) = G'$, f is called *a continuous homomorphism of G onto G'*.

I. *If H is a closed normal divisor of a topological group G, then the canonical mapping φ of the group G onto the factor group G/H is an open continuous homomorphism of G onto G/H.*

The mapping φ is a homomorphism (see 1.6, chapter I) and φ is open and continuous by I in 2.4.   □

A mapping f of the group G into the group G' is called *a continuous isomorphism of G into G'* if:

(a') f is a continuous mapping of the space G into the space G';
(b') f is an isomorphism of the group G into the group G'.

If $f(G) = G'$, the mapping f is called *a continuous isomorphism of the group G onto the group G'*.
A mapping f of the group G into the group G' is called *a topological isomorphism of G into G'* if:

(a'') f is a homeomorphism of the space G into the space G';
(b'') f is an isomorphism of the group G into the group G'.

If $f(G) = G'$, then f is called *a topological isomorphism of the group G onto the group G'*. Two topological groups G and G' are called *topologically isomorphic* if there exists a topological isomorphism of G onto G'.

II.

(1)   *If f is a continuous homomorphism of the group G onto the group G', and $H = \ker f$, then:*
(a)   *H is a closed normal subgroup of G;*
(b)   *$f = \psi\varphi$, where φ is the canonical homomorphism of the group G onto the group G/H and ψ is a continuous isomorphism of the group $\tilde{G}$ onto the group G'.*
(2)   *If in addition to this f is open, then ψ is a topological isomorphism of the group $\tilde{G} = G/H$ onto the group G'. In this case $\tilde{G}$ and G' are topologically isomorphic.*

Proof.

(1)   The subgroup H is closed, as the inverse image of a single point (the identity element e' of the group G). The relation $f = \psi\varphi$, where ψ is an isomorphism of $\tilde{G}$ onto G', is proved in III in 1.5, chapter I.

To prove (b), it remains only to show that ψ is continuous. Let U' be an open set in G'. We need to prove that $\tilde{U} = \psi^{-1}(U')$ is open. The set $f^{-1}(U') = U$ is open since f is continuous and $U' = f(U) = (\psi\varphi)(U) = \psi(\varphi(U))$. Since the isomorphism ψ is one-to-one, we have

$$\tilde{U} = \psi^{-1}(U') = \varphi(U).$$

Hence $\tilde{U} = \varphi(U)$ is open by the definition of the topology of $\tilde{G}$.

(2)   Suppose that f is continuous and open. We will show that ψ is a topological isomorphism. By the preceding paragraph, we need only show that ψ^{-1} is continuous.

Let $\tilde{U}$ be an open set in $\tilde{G}$. We set

$$U' = (\psi^{-1})^{-1}(\tilde{U}) = \psi(\tilde{U}). \tag{2.5.1}$$

We need to prove that U' is open in G'. The definition of the topology of $\tilde{G}$ shows that

$$\tilde{U} = \varphi(U), \tag{2.5.2}$$

U being open in G. From (2.5.1) we have

$$U' = \psi(\varphi(U)) = f(U),$$

so that U' is open in G', since f is an open mapping. $\square$

Remark. For many topological groups, (for example, $GL(n, \mathbf{C})$ and its closed subgroups) all continuous homomorphisms are open. (See Pontryagin [1], §20, theorem 12.)

A topological isomorphism of the topological group G onto itself is called a *topological automorphism* (compare 1.6, chapter I). For example, every mapping

$$g \to g_1^{-1} g g_1 \tag{2.5.3}$$

is a topological isomorphism (see I in 2.2) for fixed $g_1 \in G$. Topological automorphisms of the form (2.5.3) are called *inner*; all others are called *outer*.

In the sequel the terms *homomorphism, isomorphism* and *automorphism* will mean *continuous homomorphism, topological isomorphism* and *topological automorphism*, respectively. If homomorphism, isomorphism or automorphism are used in the sense of chapter I, the adjective "algebraic" will be added. The same convention is to hold for the corresponding adjectives.

Examples and Exercises

   1. Prove that: (a) the mapping $f : g \to \det g$ (example 1 in 1.6, chapter I) is an open continuous homomorphism of the group $GL(n, \mathbf{C})$ onto the group $\mathbf{C}_0^1$; (b) the factor group $\tilde{G} = GL(n, \mathbf{C})/SL(n, \mathbf{C})$ is isomorphic to the group $\mathbf{C}_0^1$.

2. Prove that the groups G_X for dim $X = n$ and $GL(n, \mathbf{C})$ are isomorphic (see example 3 in 1.6, chapter I).

3. The one-dimensional torus $\mathcal{T}^1$ is isomorphic to the factor group $\mathbf{R}^1/\mathbf{N}$ (see example 4 in 1.6, chapter I). We topologize $\mathcal{T}^1$ by considering the images of the open sets in $\mathbf{R}^1/\mathbf{N}$ as open sets in $\mathcal{T}^1$. Thus $\mathcal{T}^1$ becomes a topological group isomorphic to the group $\mathbf{R}^1/\mathbf{N}$. Using the algebraic isomorphism of the groups $\mathcal{T}^1$ and Γ^1 (the rotations of a circle, as described in example 1 in 1.7, chapter I) define a topology in Γ^1 such that $\mathcal{T}^1$ and Γ^1 are topologically isomorphic. Unless otherwise noted, $\mathcal{T}^1$ and Γ^1 will be considered topological groups with these topologies, which we call the *natural topologies*. Prove that both of the spaces $\mathcal{T}^1$ and Γ^1 are homeomorphic to a circle with its natural topology.

4. Prove that *if an algebraic homomorphism f of a topological group G into a topological group G_1 is continuous at the identity element $e \in G$, then f is continuous.*

Hint. Use Proposition I in 2.2.

2.6. Groups of Transformations of a Topological Space

Any homeomorphism of a space X onto itself we will call a *transformation of the topological space X*. The product of two transformations of X and the inverse of a transformation of X are again transformations. The identity transformation is obviously a transformation. That is, *the set of all transformations of the topological space X is a group*. Its identity element is the identity transformation. Any subgroup G of this group is called a *group of transformations* of the space X and the pair (X, G) is called a *topological space X with the transformation group G.*

Any transformation of the topological space X is also a transformation of the set X (see 1.5, chapter I). We can thus apply to transformations of X the notation, terminology and results found in 1.5 and 1.7, chapter I. We will use both the right notation: $x \to xg$ and left notation $x \to gx$ for the transformation g. If we need to emphasize this, we will write G_r for the group of right transformations and G_l for the group of left transformations. In the sequel we will for definiteness consider the group of right transformations and write G instead of G_r. Our results, with appropriate changes, remain valid for G_l.

The transformation group G of the space X is often a topological group. This, along with the fact that transformations of X are homeomorphisms, leads to a number of complements of a topological character for the results of 1.5, 1.7, chapter I.

We can represent every topological group G as a transformation group, as follows.

We set $X = G$ and with every element $g_0 \in G$ we associate the right translation $\hat{g}_0 : g \to g g_0$. Thus $\hat{g}_0$ is a homeomorphism of the space G (see I in 2.2) and the mapping $g_0 \to \hat{g}_0$ is an algebraic isomorphism of the group G

onto the group $\hat{G}$ of all right translations (see II in 1.6, chapter I). We introduce a topology into $\hat{G}$ by taking the images of all open sets in G as the open sets for $\hat{G}$. It is easy to see that $\hat{G}$ is then a topological group and that the mapping $g_0 \to \hat{g}_0$ is an isomorphism of the topological group G onto the topological group $\hat{G}$. We therefore have the following.

I. *Any topological group G is isomorphic to the topological group $\hat{G}$ of right translations of G.*

An analogous result is true for the group of left translations.

II. *Let G be a topological group. Let H be a closed subgroup of G and let $\hat{G} = G/H$ be the topological space of right cosets of H in G. The following assertions hold.*

    (1)   *For every $g_0 \in G$, the mapping g_0 defined by the formula*

$$\{g\}\bar{g}_0 = \{gg_0\}, \tag{2.6.1}$$

        where $\{g\} \in \tilde{G}$, is a transformation of the topological space $\tilde{G}$.
    (2)   *The space $\tilde{G}$ is a homogeneous space under the group $\tilde{G}$ of all transformations $\bar{g}_0$, $g_0 \in G$.*
    (3)   *The mapping $f:g \to \bar{g}$ is a homomorphism of the group G onto the group $\bar{G}$ and the kernel N of this homomorphism is a normal subgroup of the group G that is contained in H.*

Proof. Statement (1) is merely III in 2.4 and (2) and (3) are proved in 1.7, chapter I. $\square$

We now *assume* that the kernel N of the homomorphism $f:g \to \bar{g}$ is closed in G, and we write $\dot{G} = G/N$. Then $\dot{G}$ is a topological group (see IV in 2.4) and $f = \psi\dot{\phi}$ where $\dot{\phi}$ is the canonical mapping of G onto $\dot{G}$ and ψ is an algebraic isomorphism of $\dot{G}$ onto $\bar{G}$ (see 1.7, chapter I). We define a topology in $\bar{G}$, taking all images under ψ of open sets in $\dot{G}$ as open sets in $\bar{G}$. Plainly $\bar{G}$ is a topological group isomorphic to the group $\dot{G}$. We will take $\bar{G}$ to be topologized as above (provided of course that N be closed). If $N = \{e\}$ then $\dot{G}$ coincides with G and $\bar{G}$ is isomorphic with the group G.

III. *Let (X, G) be a topological space with a topological group G of right transformations, homogeneous under G. Let H be the stationary group of a fixed point $x_0 \in X$ and let f be the mapping that maps every point $x \in X$ onto the right coset $\tilde{g} = \{g\} \in G/H$ consisting of all elements g of G for which $x_0 g = x$. Suppose that the following condition is satisfied.*

    (α) The correspondence $\omega:g \to x_0 g$ is an open continuous mapping of the group G onto X.

    Then f is a homeomorphism of the space X onto $\tilde{G} = G/H$ for which every transformation $g_0 \in G$ maps onto the transformation $\bar{g}_0$ of the space $\tilde{G}$, defined *by the formula*

$$\{g\}\bar{g}_0 = \{gg_0\} \quad \text{for } \{g\} \in \tilde{G}.$$

In other words, we have $f(xg) = f(x)\bar{g}$. *The resulting correspondence* $\psi : g_0 \to \bar{g}_0$ *is a topological isomorphism of the group* G *onto the group* $\bar{G}$.

Proof. In III, 1.7, chapter I, we proved that f is one-to-one and that ψ is an algebraic isomorphism of the group G onto $\bar{G}$. We now need only to prove that f and ψ are homeomorphisms. First we note that H is closed, as the inverse image of the single point x_0 under the continuous mapping ω (see condition (α)). Therefore $\tilde{G} = G/H$ is a separated topological space (see II in 2.4). Let $\tilde{U}$ be an open set in $\tilde{G}$. We have $\tilde{U} = \varphi(U)$ where U is an open set in G and φ is the canonical mapping of G onto $\tilde{G}$ (see 2.4). The definitions of the mappings f and ω show that

$$f\omega = \varphi, \tag{2.6.2}$$

so that $f\omega(U) = \varphi(U) = \tilde{U}$. Therefore $f^{-1}(\tilde{U}) = \omega(U)$ is open since ω is an open mapping (again see condition (α)). Hence f is continuous.

Let V be an open set in X. We write $U = \omega^{-1}(V)$; U is open in G since the mapping ω is continuous (again condition (α)). On the other hand, (2.6.2) shows that $f(V) = f\omega(U) = \varphi(U)$ is open in G by the definition of the topology of G. That is, f^{-1} is also continuous and f is a homeomorphism. The definition of the topology of $\bar{G}$ (see above) shows that ψ is also a homeomorphism. $\square$

Proposition III leads to the following definition.

Two homogeneous spaces X and X' with topological transformation groups G and G', respectively, are called *topologically equivalent* if there exist: (a) a topological isomorphism $\varphi : g \to g'$ of G onto G' and (b) a homeomorphism $f : x \to x'$ of X onto X' such that if $x \to x'$ then $xg \to x'g'$, that is,

$$f(xg) = f(x)\varphi(g).$$

Proposition III means that *any homogeneous space* X *with a topological transformation group* G *satisfying condition* (α) *for some* $x_0 \in X$ *is topologically isomorphic to the space* $\tilde{G} = G/H$ *with the transformation group* $\bar{G}$, *where* H *is the stationary group of the point* $x_0 \in X$. Usually X is identified with $\tilde{G}$ and G with $\bar{G}$. This identification is called *the canonical realization of the pair* (X, G). The pair $(\tilde{G}, \bar{G})$ is called *the canonical model* of the homogeneous space.

Thus, *any homogenous space satisfying condition* (α) *for some* $x_0 \in X$ *is topologically isomorphic to its canonical model.* In the sequel, when referring to topological homogeneous spaces with a topological transformation group the term "isomorphism" will mean a topological isomorphism. When referring to an isomorphism as defined in 1.7, chapter I we will add the term *algebraic*.

Remark. *Condition* (α) *in Proposition* III *shows that for every* $x \in X$ *the correspondence* $\omega_x : g \to xg$ *is an open mapping of the group* G *into* X. Let condition (α) be satisfied for a point x_0. Since X is homogeneous, there exists an element $g_0 \in G$ such that $x_0 g_0 = x$. The relation $xg = (x_0 g_0)g = x_0(g_0 g)$ implies that the mapping ω_x is open as the product of the homeomorphism $g \to g_0 g$ (I in 2.1) and the open mapping ω.

§3. Definition of a Finite-Dimensional Representation of a Topological Group; Examples

3.1. Continuous Functions on a Topological Group

Let $f = f(g)$ be a complex-valued function defined on a topological group G. The function f is called *continuous* if f is continuous on the topological space G.

Next, let X be a finite-dimensional complex vector space of dimension n and let $e_1, e_2, \ldots, e_n$ be a fixed basis in X. Let f be a vector-valued function with values in X defined on the topological group G and for $g \in G$, let $(f_1(g), \ldots, f_n(g))$ be the components of the point $f(g)$ in the basis $e_1, \ldots, e_n$. The function f is called *continuous on the group* if the complex-valued functions $f_1(g), \ldots, f_n(g)$ are continuous on G. This definition is independent of the choice of a basis $e_1, \ldots, e_n$ in X. For each $g \in G$, the components $f'_j(g)$ of the element $f(g)$ in another basis $e'_1, \ldots, e'_n$ are linear combinations with constant coefficients of the components $f_j(g)$. If the functions f_j are continuous, so are the functions f'_j.

Finally, let $A = A(g)$ be an operator-valued function on G whose values are linear operators on X. The function A is said to be *continuous on* G if $g \to A(g)x$ is a continuous vector function on G for every $x \in X$. The following is clear from this definition.

I. *The function* $A(g)$ *is continuous if and only if its matrix elements* $a_{jl}(g)$, *in any basis* $e_1, \ldots, e_n$ *in* X, *are continuous complex-valued functions on* G.

3.2. Definition of a Finite-Dimensional Representation of a Topological Group

Let G be a topological group and let X be a finite-dimensional complex linear space different from $\{0\}$. *A representation of* G *in* X is any mapping that carries every $g \in G$ into a linear operator $T(g)$ on X in such a way that:

(1) $T(e) = 1$, where 1 is the identity operator in X;
(2) $T(g_1 g_2) = T(g_1)T(g_2)$;
(3) $T(g)$ is a continuous operator-valued function on G.

Thus we add to the definition in 2.1, chapter I the condition (3) of continuity[28] of the operator-valued function $T(g)$. This requirement will play a significant role in the sequel. Representations as defined in 2.1, chapter I (that is, not necessarily satisfying condition (3)) will now be called *algebraic representations*.

I. *If T is a representation of a topological group G in a space X, then its restriction to every subgroup of G and also its restrictriction to invariant subspaces of X also satisfy condition (3). That is, these representations are also representations of the topological group G.*

Thus all definitions and propositions found in §2, chapter I are applicable to representations of a topological group.

II. *The matrix elements $t_{jl}(g)$ and the character $\chi_T(g)$ of a finite-dimensional representation of a topological group G are continuous complex-valued functions on G.*

The continuity of the matrix elements $t_{jl}(g)$ follows from condition (3) and Proposition I in 3.1. The continuity of the character $\chi_T(g)$ follows from the formula (see (2.9.3), chapter I)

$$T(g) = \sum_{j=1}^{n} t_{jj}(g). \quad \square$$

We conclude from Proposition I and from the definition of the topology of $GL(n, \mathbf{C})$ that a representation of a topological group G in an n-dimensional space is a continuous homomorphism of G into $GL(n, \mathbf{C})$. This description can be viewed as defining a finite-dimensional representation of a topological group.

III. *All the one-dimensional representations*

$$g_1 \times g_2 \times \cdots \times g_n \to f(g_1, g_2, \ldots, g_n),\, g_1 \in G_1, \ldots, g_n \in G_n \quad (3.2.1)$$

of a direct product $G_1 \times G_2 \times \cdots \times G_n$ of topological groups $G_1, G_2, \ldots, G_n$ are described by the formula

$$f(g_1, g_2, \ldots, g_n) = f_1(g_1) f_2(g_2) \cdots f_n(g_n), \quad (3.2.2)$$

where

$$g_j \to f_j(g_j), \qquad g_j \in G_j, \qquad j = 1, 2, \ldots, n \quad (3.2.3)$$

[28] To emphasize this, representations of a topological group are often called *continuous*.

are one-dimensional representations of the groups $G_1, \ldots, G_n$. The representation (3.2.1) is unitary if and only if each of the representations (3.2.3) is unitary.

Proof. The algebraic part of the statements in Proposition III is proved in Proposition II in 2.7, chapter I. It is easy to see that the function $f(g_1, \ldots, g_n)$ in (3.2.2) is continuous if and only if all of the functions $f_1(g_1), \ldots, f_n(g_n)$ on $G_1, \ldots, G_n$, respectively, are continuous. $\square$

3.3 Description of all One-Dimensional Representations of the Simplest Commutative Topological Groups

Every irreducible finite-dimensional representation of a commutative group is one-dimensional (see the corollary to Lemma 2 in 2.2, chapter I).

(a) One-dimensional representations of the group $\mathbf{R}^1$. Every one-dimensional representation of the group $\mathbf{R}^1$ can be described by a (complex-valued function $\alpha \to f(\alpha)$, $\alpha \in \mathbf{R}^1$, that satisfies the conditions

$$f(0) = 1, \tag{3.3.1}$$
$$f(\alpha_1 + \alpha_2) = f(\alpha_1)f(\alpha_2), \tag{3.3.2}$$

and is continuous (see condition (3) in 3.2 and (2.1.4), chapter I). We will find all functions that satisfy these conditions and hence all one-dimensional representations of $\mathbf{R}^1$. From the equalities

$$f(\alpha)f(-\alpha) = f(\alpha - \alpha) = f(0) = 1$$

it follows that

$$f(\alpha) \neq 0, \quad \text{for all } \alpha \in \mathbf{R}^1. \tag{3.3.3}$$

We now prove more.

I. *The function $f(\alpha)$ is differentiable and its derivative is continuous[29] throughout $\mathbf{R}^1$.*

Proof. Let $\omega(\alpha)$ be a continuously differentiable function on $\mathbf{R}^1$ equal to zero outside of some neighborhood of the point $\alpha_0 \in \mathbf{R}^1$ for which

$$c = \int_{-\infty}^{\infty} f(\alpha)\omega(\alpha)\,d\alpha \neq 0. \tag{3.3.4}$$

By virtue of (3.3.3) such a function $\omega(\alpha)$ exists. Multiply both sides of (3.3.2) by $\omega(\alpha_2)$ and integrate both sides of the resulting equality from $-\infty$ to $+\infty$

---

[29] Briefly, $f(\alpha)$ is *continuously differentiable on* $\mathbf{R}^1$.

with respect to α_2. We obtain[30]

$$\int_{-\infty}^{\infty} f(\alpha_1 + \alpha_2)\omega(\alpha_2)\,d\alpha_2 = f(\alpha_1)\int_{-\infty}^{\infty} f(\alpha_2)\omega(\alpha_2)\,d\alpha_2. \qquad (3.3.5)$$

and so

$$f(\alpha_1) = \frac{1}{c}\int_{-\infty}^{\infty} f(\alpha_1 + \alpha_2)\omega(\alpha_2)\,d\alpha_2 = \frac{1}{c}\int_{-\infty}^{\infty} f(\alpha)\omega(\alpha - \alpha_1)\,d\alpha. \qquad (3.3.6)$$

But the right side of (3.3.6) is a continuously differentiable function of α_1, since $\omega(\alpha - \alpha_1)$ has this property and the integral is over a finite interval. That is, $f(\alpha_1)$ is continuously differentiable. □

Remark 1. Since ω can be taken as infinitely differentiable, f is infinitely differentiable.

II. *The function $f(\alpha)$ satisfies the differential equation*

$$\frac{df}{d\alpha} = kf, \qquad (3.3.7)$$

where k is a certain constant.

Proof. Differentiating both sides of (3.3.2) by α_1 and setting $\alpha_1 = 0$, $\alpha_2 = \alpha$, we obtain

$$f'(\alpha) = f'(0)f(\alpha) = kf(\alpha), \qquad (3.3.8)$$

where $k = f'(0)$. □

Any solution of the equation (3.3.7) that satisfies (3.3.1) has the form

$$f(\alpha) = e^{k\alpha}.$$

Thus we have proved the following.

III. *All one-dimensional representations $\alpha \to f(\alpha)$ of the group $\mathbf{R}^1$ are described by the formula*

$$f(\alpha) = e^{k\alpha}, \qquad (3.3.9)$$

where k is a (complex) constant.

The unitary one-dimensional representations f of the group $\mathbf{R}^1$ are now easily identified. The property of being unitary in the one-dimensional case means that $|f(\alpha)| = 1$. For $f(\alpha) = e^{k\alpha}$ this holds if and only if k is purely imaginary, $k = i\tau$, for some $\tau \in \mathbf{R}^1$.

---

[30] This device (in a more general situation) is due to I. M. Gel'fand (Gel'fand, I. M., [2*]).

IV. *All unitary one-dimensional representations* $\alpha \to f(\alpha)$ *of the group* $\mathbf{R}^1$ *have the form*

$$f(\alpha) = e^{i\tau\alpha}, \tag{3.3.10}$$

where τ *is a real constant. Conversely, for every* $\tau \in \mathbf{R}^1$ *the formula (3.3.10) defines a one-dimensional unitary representation of the group* $\mathbf{R}^1$.

Remark 2. The reader has already noted that continuity of a representation $\alpha \to f(\alpha)$ implies differentiability (and by virtue of (3.3.9) even analyticity!) of the function $f(\alpha)$. As we will see below, analogous facts hold for large classes of groups.

(b) One-dimensional representations of the group $\mathbf{R}^n$. The group $\mathbf{R}^n$ is the direct product of n copies of the group $\mathbf{R}^1$. Combining Proposition III of 3.2 and Propositions III and IV of example (a) we conclude the following.
   All one-dimensional representations $(\alpha_1, \ldots, \alpha_n) \to f(\alpha_1, \ldots, \alpha_n)$ *of the group* $\mathbf{R}^n$ *are described by*

$$f(\alpha_1, \ldots, \alpha_n) = e^{k_1\alpha_1 + \cdots + k_n\alpha_n}, \tag{3.3.11}$$

where $k_1, \ldots, k_n$ *are arbitrary complex constants. Such a representation is unitary if and only if*

$$f(\alpha_1, \ldots, \alpha_n) = e^{i(\tau_1\alpha_1 + \cdots + \tau_n\alpha_n)} \qquad \tau_1, \ldots, \tau_n \in \mathbf{R}^1. \tag{3.3.12}$$

(c) One-dimensional representations of the group $\mathbf{C}^n$. The group $\mathbf{C}^n$ is isomorphic to the group $\mathbf{R}^{2n}$ and the mapping $(z_1, \ldots, z_n) \to (\alpha_1, \beta_1, \ldots, \alpha_n, \beta_n)$ is an isomorphism of the group $\mathbf{C}^n$ onto the group $\mathbf{R}^{2n}$ where

$$z_j = \alpha_j + i\beta_j. \tag{3.3.13}$$

Therefore a one-dimensional representation

$$(z_1, \ldots, z_n) \to \varphi(z_1, \ldots, z_n)$$

of the group $\mathbf{C}^n$ defines a one-dimensional representation of the group $\mathbf{R}^{2n}$

$$f(\alpha_1, \beta_1, \ldots, \alpha_n, \beta_n) \to f(\alpha_1, \beta_1, \ldots, \alpha_n, \beta_n)$$

where

$$f(\alpha_1, \beta_1, \ldots, \alpha_n, \beta_n) = \varphi(\alpha_1 + i\beta_1, \ldots, \alpha_n + i\beta_n) \tag{3.3.14}$$

Thus (3.3.11) give us

$$f(\alpha_1, \beta_1, \ldots, \alpha_n, \beta_n) = e^{k_1\alpha_1 + i_1\beta_1 + \cdots + k_n\alpha_n + i_n\beta_n}. \tag{3.3.15}$$

But (3.3.13) implies that

$$\alpha_j = \frac{z_j + \bar{z}_j}{2}, \qquad \beta_j = \frac{z_j - \bar{z}_j}{2i}. \tag{3.3.16}$$

Let us write

$$p_j = \tfrac{1}{2}(k_j - il_j), \qquad q_j = \tfrac{1}{2}(k_j + il_j) \tag{3.3.17}$$

and conclude from (3.3.14) and (3.3.17) that

$$\varphi(z_1, \ldots, z_n) = e^{p_1 z_1 + q_1 \bar{z}_1 + \cdots + p_n z_n + q_n \bar{z}_n}. \tag{3.3.18}$$

Thus *all one-dimensional representations* $(z_1, \ldots, z_n) \to \varphi(z_1, \ldots, z_n)$ *of the group* $\mathbf{C}^n$ *are as in* (3.3.18), *where* $p_1, q_1, \ldots, p_n, q_n$ *are arbitrary complex constants.*

A representation (3.3.18) is unitary if and only if k_j, l_j are pure imaginaries. This and (3.3.17) give the following.

All one-dimensional unitary representations $(z_1, \ldots, z_n) \to \varphi(z_1, \ldots, z_n)$ *of the group* $\mathbf{C}^n$ *have the form*

$$\varphi(z_1, \ldots, z_n) = e^{p_1 z_1 - \bar{p}_1 \bar{z}_1 + \cdots + p_n z_n - \bar{p}_n \bar{z}_n}, \tag{3.3.19}$$

where $p_1, \ldots, p_n$ *are arbitrary complex constants*

(d) One-dimensional representations of the group Γ^1 (the group of rotations of a circle; see example 1 in 1.7, chapter I). Every rotation of a circle is described by an angle α of the rotation, where rotations by α and by $a + 2\pi$ are the same rotation. Therefore, in a one-dimensional representation

$$\alpha \to f(\alpha) \tag{3.3.20}$$

of the group Γ^1, the function $f(\alpha)$ must, besides the conditions in example (a), also satisfy the condition

$$f(\alpha + 2\pi) = f(\alpha). \tag{3.3.21}$$

From example (a) we see that $f(\alpha) = e^{k\alpha}$ and from (3.3.21) that $k = im$, where m is an integer.

Thus, *all one-dimensional representations* $\alpha \to f(\alpha)$ *of the group* Γ^1 *have the form*

$$f(\alpha) = e^{im\alpha} \tag{3.3.22}$$

where m *is an arbitrary integer.*

Plainly, *all of these representations are unitary.*

(e) One-dimensional representations of the torus $\mathcal{T}^1$. The torus $\mathcal{T}^1$ is isomorphic to the group Γ^1 and the mapping $\alpha \to \beta = (1/2\pi)\alpha$ is an isomorphism of Γ^1 onto $\mathcal{T}^1$ (see example 1 in 1.6, chapter I). Therefore a one-dimensional representation

$$\beta \to \varphi(\beta) \tag{3.3.23}$$

of the group $\mathcal{T}^1$ defines a one-dimensional representation

$$\alpha \to f(\alpha)$$

of the group Γ^1 where

$$f(\alpha) = \varphi\left(\frac{1}{2\pi}\alpha\right). \tag{3.3.24}$$

As in example (d), we have

$$f(\alpha) = e^{im\alpha}$$

where m is an integer. Combining this with (3.3.24), we find the following.

All one-dimensional representations $\beta \to \varphi(\beta)$ of the torus $\mathcal{T}^1$ have the form

$$\varphi(\beta) = e^{2\pi im\beta}, \tag{3.3.25}$$

where m is an integer.

All of these representations are obviously unitary.

(f) One-dimensional representations of the torus $\mathcal{T}^n$. The torus $\mathcal{T}^n$ is the direct product of n copies of the group $\mathcal{T}^1$. Combine (e) with III in 2.3 to obtain the following.

All one-dimensional representations $(\beta_1, \ldots, \beta_n) \to \varphi(\beta_1, \ldots, \beta_n)$ of the torus $\mathcal{T}^n$ have the form

$$(\beta_1, \ldots, \beta_n) = e^{2\pi i(m_1\beta_1 + \cdots m_n\beta_n)}, \tag{3.3.26}$$

where $m_1, \ldots, m_n$ are integers. All of these representations are unitary.

(g) One-dimensional representations of the group $\mathbf{R}_0^+$. Let $\mathbf{R}_0^+$ denote the set of all positive real numbers. It is a topological subgroup of the topological group $\mathbf{R}_0$. The mapping $\beta \to \alpha = \ln \beta$ is an isomorphism of the group $\mathbf{R}_0^+$ onto $\mathbf{R}^1$. Every one-dimensional representation $\beta \to \varphi(\beta)$ of the group $\mathbf{R}_0^+$ defines a one-dimensional representation $\alpha \to f(\alpha)$ of the group $\mathbf{R}^1$ if we set

$$f(\alpha) = \varphi(e^\alpha), \quad \text{which is to say } \varphi(\beta) = f(\ln \beta). \tag{3.3.27}$$

By example (a) we have $f(\alpha) = e^{k\alpha} = e^{\alpha \ln \beta} = \beta^k$. From this and (3.3.27) we conclude the following.

All one-dimensional representations $\beta \to \varphi(\beta)$ *of the group* $\mathbf{R}_0^+$ *have the form*

$$\varphi(\beta) = \beta^k = e^{k \ln \beta}, \qquad (3.3.28)$$

where k is any complex constant. Such a representation is unitary if and only if $k = i\tau, \tau \in \mathbf{R}^1$.

(h) One-dimensional representations of the group $\mathbf{R}_0$. The set of all negative numbers is denoted by $\mathbf{R}_0^-$. We have $\mathbf{R}_0 = \mathbf{R}_0^+ \cup \mathbf{R}_0^-$ and every number $\beta \in \mathbf{R}_0^-$ has the form $\beta = (-1)|\beta|$. Let $\beta \to \varphi(\beta)$ be a one-dimensional representation of the group $\mathbf{R}_0$. Its restriction to $\mathbf{R}_0^+$ is a one-dimensional representation of the group $\mathbf{R}_0^+$; (3.3.28) shows that

$$\varphi(\beta) = \beta^k, \quad \text{for } \beta > 0, \qquad (3.3.29)$$

where k is some complex number. For $\beta < 0$ (3.3.29) implies that

$$\varphi(\beta) = \varphi((-1)|\beta|) = \varphi(-1)\varphi(|\beta|) = \varphi(-1)|\beta|^k. \qquad (3.3.30)$$

Since

$$\varphi(-1)\varphi(-1) = \varphi((-1)^k) = \varphi(1) = 1$$

we infer that $\varphi(-1) = \pm 1 = (-1)^\varepsilon$ where $\varepsilon = 0$ or $\varepsilon = 1$. We may rewrite the formula (3.3.30) as

$$\varphi(\beta) = (-1)^\varepsilon |\beta|^k \quad \text{for } \beta < 0. \qquad (3.3.31)$$

Formulas (3.3.29) and (3.3.31) can be combined into one formula

$$\varphi(\beta) = (\text{sign } \beta)^\varepsilon |\beta|^k. \qquad (3.3.32)$$

Thus *all one-dimensional representations of the group* $\mathbf{R}_0$ *are as in* (3.3.32) *where k is a complex constant and $\varepsilon = 0$ or $\varepsilon = 1$. These representations are unitary if and only if k is a pure imaginary.*

(i) One-dimensional representations of the group $\mathbf{C}_0^1$. The group $\mathbf{C}_0^1$ is isomorphic to the direct product $\mathbf{R}_0^1 \times \Gamma^1$ and the mapping $z \to (r, \alpha)$ for $z = |r|, \alpha = |z|$, is an isomorphism of the group $\mathbf{C}_0^1$ onto $\mathbf{R}_0^+ \times \Gamma^1$. Combining Proposition II of 3.2 with (3.3.22) and (3.3.28) we conclude the following.

All one-dimensional representations $z \to \varphi(z)$ *of the group* $\mathbf{C}_0^1$ *have the form*

$$\varphi(z) = |z|^k e^{im \arg z}, \qquad (3.3.33)$$

where k is a complex constant and m is an integer. Such a representation is unitary if and only if k is a pure imaginary.

(j) One-dimensional representations of the group $\mathbf{C}_0^n$. The group $\mathbf{C}_0^n$ is the direct product of n copies of the group $\mathbf{C}_0^1$. We may thus combine Proposition III of 3.2 with (3.3.33), as follows.

All one-dimensional representations $(z_1, \ldots, z_n) \to \varphi(z_1, \ldots, z_n)$ *of the group* $\mathbf{C}_0^n$ *have the form*

$$\varphi(z_1, \ldots, z_n) = |z_1|^{k_1} \cdots |z_n|^{k_n} e^{i(m_1 \arg z_1 + \cdots + m_n \arg z_n)}, \qquad (3.3.34)$$

where $k_1, \ldots, k_n$ *are complex constants and* $m_1, \ldots, m_n$ *are integers. Such a representation is unitary if and only if all of the numbers* $k_1, \ldots, k_n$ *are pure imaginaries.*

3.4. Tensor Representations of Linear Groups

The matrix elements of every tensor representation of the group $GL(n, \mathbf{C})$ (the tensor product of m "copies" of the identity representation $g \to g$, $g \in GL(n, \mathbf{C})$) are monomials in the matrix elements of the group $GL(n, \mathbf{C})$ and consequently are continuous functions on $GL(n, \mathbf{C})$. That is, the operators of tensor representations of the group $GL(n, \mathbf{C})$ satisfy the condition of continuity (3) in 2.2.

I. *Tensor representations of the group* $GL(n, \mathbf{C})$ *are representations of the topological group* $GL(n, \mathbf{C})$.

This fact and Proposition I of 3.2 yield the following.

II. *The tensor representations of a linear group* G *(see 2.3), (considered a subgroup of the topological group* $GL(n, \mathbf{C})$), *as well as the restrictions of these representations to an invariant subspace, are representations of the topological group* G.

§4. General Definition of a Representation of a Topological Group

4.1. Topological Linear Spaces

A set X is called a *topological linear space* if:

(1) X is a linear space;
(2) X is a separated topological space;
(3) the mapping $\{x_1, x_2\} \to x_1 + x_2, x_1, x_2 \in X$, is a continuous mapping of the topological space $X \times X$ onto itself;
(4) the mapping $\{\alpha, x\} \to \alpha x, \alpha \in \mathbf{C}^1, x \in X$, is a continuous mapping of the topological space $\mathbf{C}^1 \times X$ onto the topological space X.

A very simple example of a topological linear space is the linear space $\mathbf{C}^n$, provided with its own natural topology. Other examples are given below.

Conditions (3) and (4) imply that X is a topological group with the operation of addition in X. Throughout the rest of this section X, Y, and Z will denote topological linear spaces.

A subset M of X is called a *closed subspace in* X if:

(a) M is a linear subspace of the linear space X;
(b) M is a closed subset of the topological space X.

I. *The closure $\bar{M}$ of a linear subspace M of X is also a linear subspace of X.*

Proof.

(1)   For each fixed x_1 in M, (3) implies that the correspondence $x \to x + x_1$, $x \in M$, is a continuous mapping of M onto itself that also carries $\bar{M}$ onto itself. We thus have $x + x_1 \subset \bar{M}$ for $x \in \bar{M}$, $x_1 \in M$. The correspondence $x_1 \to x + x_1$, $x_1 \in \bar{M}$ is a continuous mapping of M into itself, carrying M into $\bar{M}$ and $\bar{M}$ into $\bar{\bar{M}} = \bar{M}$. Thus we have $x + x_1 \in \bar{M}$ for $x, x_1 \in \bar{M}$.

(2)   By virtue of (4), the correspondence $x \to \alpha x$, $\alpha \in \mathbf{C}$, $x \in X$, is a continuous mapping of X into itself, carrying M into itself and hence $\bar{M}$ into itself. That is, we have $\alpha x \in \bar{M}$ for $\alpha \in \mathbf{C}$, $x \in \bar{M}$.

Assertions 1 and 2 show that M is a linear subspace of the space X. Plainly $\bar{M}$ is closed and thus it is a closed linear subspace of X. $\square$

A (linear) operator A mapping X into Y is called *continuous* if the correspondence $A : x \to Ax$ is a continuous mapping of X into Y. A continuous operator A carrying X into itself is called a *continuous (linear) operator on X*.

II. *If A is a continuous linear operator on X and M is a linear subspace of X invariant under A, then $\bar{M}$ is also invariant under A.*

Proof. The correspondence $A : x \to Ax$, $x \in X$, is a continuous mapping of X into itself, carrying M into itself and thus $\bar{M}$ into itself. $\square$

Examples

1. Normed spaces. Let X be a normed space (see Šilov [1]). We topologize X by taking the set of all open balls

$$W(x_0, \varepsilon) = \{x : \|x - x_0\| < \varepsilon\}, \qquad \varepsilon > 0 \qquad (4.1.1)$$

as a neighborhood basis at each point x_0 of X. All the axioms for a neighborhood basis (see 1.1) and the separation axiom (see 1.8) are satisfied for the sets (4.1.1). This is easy to prove. Thus the sets (4.1.1) define a separated topology in X. This topology for a normed space X is called *the strong topology in X*.

A normed space with its strong topology is a topological linear space.

We omit the proof.

Exercise. Prove that *a linear operator from a normed space X into a normed space Y is continuous if and only if it is bounded.*

2. Euclidean spaces (in particular Hilbert spaces). A *Euclidean space* is a normed linear space with norm $\|x\| = \sqrt{(x, x)}$, where (x, y) is a scalar product in X.

A Euclidean (Hilbert) space X with the strong topology defined by the norm $\|x\| = \sqrt{(x, x)}$ is a topological linear space.

3. Locally convex spaces. Let X be a linear space. For x_1, x_2 in X, the set $tx_1 + (1 - t)x_2$, $0 \leqslant t \leqslant 1$, is called the *line segment joining x_1 and x_2*; we write $[x_1, x_2]$. The points x_1 and x_2 are called *ends* of the segment. (Think geometrically of the cases $X = \mathbf{R}^2$ and $X = \mathbf{R}^3$.) A subset Q of X is called *convex* if it contains the entire line segment $[x_1, x_2]$ for all x_1, x_2 in Q. A subset Q of X is called *symmetric* if $\alpha x \in Q$ for all x in Q and complex numbers such that $|\alpha| = 1$.

A real-valued function $p(x)$ on X is called a *seminorm* if:

(1) $p(x) \geqslant 0$, for every $x \in X$;
(2) $p(\alpha x) = |\alpha| p(x)$, for all $x \in X$ and $\alpha \in \mathbf{C}$;
(3) $p(x_1 + x_2) \leqslant p(x_1) + p(x_2)$, for all $x_1, x_2 \in X$.

Condition (2) implies that $p(0) = 0$.

III. *If p is a seminorm in X, then for every $c > 0$ the set*

$$Q = \{x : x \in X, p(x) < c\}$$

is convex and symmetric.

Proof. If x_1, x_2 belong to Q, then $p(x_1) < c$, $p(x_2) < c$. For $0 \leqslant t \leqslant 1$, this and properties (2) and (3) of a seminorm yield

$$p(tx_1 + (1-t)x_2) \leqslant p(tx_1) + p((1-t)x_2) = tp(x_1) + (1-t)p(x_2) < ct + c(1-t) = c.$$

Thus we have $[x_1, x_2] \subset Q$, i.e., Q is convex. Also if x belongs to Q and $|\alpha| = 1$, we get

$$p(\alpha x) = |\alpha| \cdot p(x) < |\alpha| \cdot c = c;$$

that is, $\alpha x \in Q$ and Q is symmetric. □

Consider a family P of seminorms p on a linear space X. The set P is called *sufficient* if for every $x \in X$ there exists a seminorm $p \in P$ such that $p(x) > 0$. Let P be a sufficient set of seminorms on X. We define a topology in X by defining a neighborhood basis at every point $x_0 \in X$ to be all sets

$$W(x_0, p_1, \ldots, p_n, \varepsilon) = \{x : x \in X, p_1(x - x_1) < \varepsilon, \ldots, p_n(x - x_0) < \varepsilon\}, \quad (4.1.2)$$

where $p_1, \ldots, p_n$ is any finite number of the seminorms in P and ε is a positive number. For the sets (4.1.2), the axioms of a neighborhood basis and the separation axiom are easily verified. It is also easy to see that X, provided with this topology, is a topological linear space. Such topological linear spaces are called *locally convex*.

Note that $W(x_0, p_1, \ldots, p_n, \varepsilon)$ is the intersection of the convex sets

$$W(x, p_j, \varepsilon) = \{x : x \in X, p_j(x - x_0) < \varepsilon\}, \qquad j = 1, \ldots, n$$

and therefore is also convex. Thus the basis (4.1.2) consists of convex sets, and the term "locally convex spaces" is justified. *Normed spaces* (see example 1) *are locally convex*. In this case, P consists of the single element $p(x) = \|x\|$ and P is sufficient since, according to the definition of a norm, $\|x\|$ is positive if $x \neq 0$.

4. The space $C(-\infty, \infty)$. The set of *all* complex-valued continuous functions $x = x(t)$ on the real line $(-\infty, \infty)$ is denoted by $C(-\infty, \infty)$. The operations of addition and multiplication by a complex number are defined pointwise. For every closed bounded interval $[a, b]$, $a < b$, and $x \in C(-\infty, \infty)$, we define

$$p_{[a,b]}(x) = \sup_{a \leqslant t \leqslant b} |x(t)|.$$

Plainly $p_{[a,b]}(x)$ is a seminorm on $C(-\infty, \infty)$. The set of all $p_{[a,b]}$, for $-\infty < a < b < \infty$, is denoted by P. It is obvious that P is sufficient. If $x \in C(-\infty, \infty)$, we have $p_{[a,b]}(x) > 0$ if x does not vanish identically on the interval $[a, b]$. That is, P defines a locally convex topology on $C(-\infty, \infty)$. We may also consider instead of $C(-\infty, \infty)$ the space of all complex-valued functions on $(-\infty, \infty)$ that are bounded on every interval $[a, b]$ $(-\infty < a < b < \infty)$.

5. The space $D^\infty(a, b)$. Let $D^\infty(a, b)$ denote the set of all complex-valued functions $x = x(t)$, that are continuous and have continuous derivatives of all orders on an interval $[a, b]$. We define addition and multiplication by a number in $D^\infty(a, b)$ as in example 4. For $x \in D^\infty(a, b)$, we set

$$p_k(x) = \sup_{a \leqslant t \leqslant b} |x^{(k)}(t)|,$$

$$P = \{p_k, k = 0, 1, \ldots\}.$$

It is easy to verify that the functions p_k are seminorms and that P is a sufficient set of seminorms on $D^\infty(a, b)$. Thus P defines a locally convex topology on $D^\infty(a, b)$ and $D^\infty(a, b)$, provided with this topology, is a locally convex space.[31]

4.2. Representations of a Topological Group; Basic Concepts

We now give the general definition of a representation of a topological group. Let G be a topological group and let X be a topological linear space. We say that *a representation $T : g \to T(g)$ of G is defined in X* if for every

---

[31] Editor's note. None of the seminorms of example 4 is a genuine norm. The seminorm p_0 in example 5 is a norm, while $p_1, p_2, \ldots$ are not.

element $g \in G$ we have a continuous linear operator $T(g)$ carrying X into itself for which the following conditions are satisfied:

(1) $T(e) = 1$ (e is the identity element of G and 1 is the identity operator in X);
(2) $T(g_1 g_2) = T(g_1)T(g_2)$;
(3)   The correspondence $\{x, g\} \rightarrow T(g)x$ is a continuous mapping of the topological product $X \times G$ onto X.

The space X is called *the space of the representation* T and the operators $T(g)$ are called *operators of the representation*. We add the condition (3) of continuity to the conditions set down in §2 of chapter I. Thus $T(g)x$ is a vector-valued function on G with values in X, continuous in the pair of variables g, x. The continuity of the operator $T(g)$ in X follows from (3). The continuity condition (3) is part of the definition of a representation of a topological group. Occasionally we will emphasize this by calling these representations *continuous*, and representations as defined in §2, chapter I, *algebraic*. In the case of a finite-dimensional X, condition (3) coincides with condition (3) of 3.2. This follows from the linearity of each coordinate function

$$(T(g)x)_j = \sum_{l=1}^{r} t_{jl}(g)x_l, \qquad r = \dim X,$$

of the vector $T(g)x$ with respect to coordinates x_l of a vector x for a fixed basis in X. (See Fihtengol'c [1].) Two representations T^1 and T^2 of G in spaces X^1 and X^2 are called equivalent (in symbols, $T^1 \sim T^2$) if there is a linear, one-to-one mapping S of the space X^1 onto X^2 having the following properties:

(1′) S is a homeomorphism of the topological linear space X^1 onto the topological linear space X^2;
(2′)   for every $g \in G$

$$ST^1(g) = T^2(g)S. \tag{4.2.1}$$

A representation T of a topological group G in a topological linear space X is called *irreducible* if there are no *closed* subspaces in X, distinct from $\{0\}$ and X, that are invariant under all operators $T(g)$, $g \in G$. To the definition of equivalence of representations of topological groups given in §2 of chapter I, we add the condition (1′) and to the definition of irreducibility we add the requirement that invariant subspaces be closed.

To emphasize this, we occasionally write *topologically equivalent* (of representations) and *topologically irreducible* (of a single representation). Equivalence and irreducibility as defined in §2, chapter I, may be called *algebraic equivalence* and *algebraic irreducibility*, respectively. It is clear that *topological equivalence implies algebraic equivalence and algebraic irreducibility implies topological irreducibility*. The converse is not true.

A representation T of G in X is called *unitary* if X is a Hilbert (or Euclidean) space and all operators $T(g)$ of T are unitary.

I. *Let T be a unitary algebraic representation of a topological group G in a space X. If the vector-function $T(g)x$ is continuous in g for every $x \in X$, then T is continuous in the pair of variables g, x.*

Proof. Let ε be greater than 0, $x_0 \in X$, and $g_0 \in G$. Since the function $T(g)$ is continuous at g_0, there exists a neighborhood $U(g_0)$ such that $\|T(g)x_0 - T(g_0)x_0\| < \varepsilon$, for $g \in U(g_0)$. For $\|x - x_0\| < \varepsilon$ and $g \in U(g_0)$ we then have

$\|T(g)x_0 - T(g_0)x_0\| < \varepsilon$. For $\|x - x_0\| < \varepsilon$ and $g \in U(g_0)$ we then have

$$\|T(g)x - T(g_0)x_0\| = \|T(g)x - T(g)x_0 + T(g)x_0 - T(g_0)x_0\|$$
$$\leqslant \|T(g)(x - x_0)\| + \|T(g)x_0 - T(g_0)x_0\| < \|x - x_0\| + \varepsilon < 2\varepsilon,$$

since $\|T(g)\| = 1$. $\square$

Some generalizations of Proposition I are found below.

Example

Let $G = \Gamma^1$. Elements of the group Γ^1 are described by angles of the rotation γ, so that functions f on Γ^1 can be considered as functions $f(\gamma)$, $\gamma \in \mathbf{R}^1$. Note that

$$f(\gamma + 2\pi) = f(\gamma) \tag{4.2.2}$$

since $\gamma + 2\pi$ and γ define one and the same element of Γ^1.

In this example we will consider functions f that are Lebesgue measurable and defined almost everywhere on $\mathbf{R}^1$. Naturally, for such functions condition (4.2.2) need be satisfied only for almost every $\gamma \in \mathbf{R}^1$. But every function f of this kind is completely defined (that is, for almost every $\gamma \in \mathbf{R}^1$) by its values on any interval of length 2π, for example on $[-\pi, \pi]$. We denote by $L^2(\Gamma^1)$ or $L^2(-\pi, \pi)$ the set of all Lebesgue measurable functions $f(\gamma)$, $\gamma \in [-\pi, \pi]$, defined for almost all $\gamma \in [-\pi, \pi]$ and satisfying the condition

$$\int_{-\pi}^{\pi} |f(\gamma)|^2 \, d\gamma < \infty. \tag{4.2.3}$$

Under these conditions, two such functions differing only on a set of measure zero yield the same element of the space $L^2(\Gamma^1)$. (For more details, see Šilov [1].) We define addition pointwise and multiplication by a number in $L^2(\Gamma^1)$ also pointwise. We define the scalar product (f_1, f_2) of functions $f_1, f_2 \in L^2(\Gamma^1)$ by the formula

$$(f_1, f_2) = \frac{1}{2\pi} \int_{-\pi}^{\pi} f_1(\gamma)\overline{f_2(\gamma)} \, d\gamma. \tag{4.2.4}$$

Thus $L^2(\Gamma^1)$ becomes a Hilbert space (again see Šilov [1]). We define a representation T of the group Γ^1 in the following manner. Let $X = L^2(\Gamma^1)$ and let $T(\gamma)$ be the operators in $L^2(\Gamma^1)$, defined by the formula

$$T(\gamma_0)f(\gamma) = f(\gamma + \gamma_0). \tag{4.2.5}$$

Let us show that the correspondence $T: \gamma \to T(\gamma)$ is a unitary representation of the group Γ^1 in the space $L^2(\Gamma^1)$. It is clear that for $f \in L^2(\Gamma^1)$ we have
(α) $T(0)f(\gamma) = f(\gamma)$, that is, $T(0) = 1$ (0 is the identity element of Γ^1), and
(β) $T(\gamma_1)T(\gamma_2)f(\gamma) = T(\gamma_1)f(\gamma + \gamma_2) = f(\gamma + \gamma_1 + \gamma_2) = T(\gamma_1 + \gamma_2)f(\gamma)$, that is, $T(\gamma_1 + \gamma_2) = T(\gamma_1)T(\gamma_2)$. Hence T is an algebraic representation of the group Γ^1.

Furthermore, for $f_1, f_2 \in L^2(\Gamma^1)$ we have

$$(T(\gamma_0)f_1, T(\gamma_0)f_2) = \frac{1}{2\pi} \int_{-\pi}^{\pi} f_1(\gamma + \gamma_0)\overline{f_2(\gamma + \gamma_0)}\, d\gamma = \frac{1}{2\pi} \int_{-\pi}^{\pi} f_1(\gamma)\overline{f_2(\gamma)}\, d\gamma,$$

and so T is unitary.

Finally, we show that T is continuous. By I, it suffices to show that the correspondence $\gamma \to T(\gamma)f$ is a continuous mapping of Γ^1 into $L^2(\Gamma^1)$. Let $C(\Gamma^1)$ be the set of all continuous functions on Γ^1. Functions in $C(\Gamma^1)$ can be considered as continuous functions on $\mathbf{R}^1$ that satisfy the condition (4.2.2). Such functions are *uniformly* continuous. Thus, if $\varphi \in C(\Gamma^1)$, for every $\varepsilon > 0$ there exists a neighborhood $U(0)$ of the element 0 in Γ^1 such that

$$|\varphi(\gamma + \gamma_1) - \varphi(\gamma)| < \varepsilon \quad \text{for all } \gamma_1 \in U(0) \text{ and all } \gamma \in \mathbf{R}^1. \tag{4.2.6}$$

It follows that

$$\|T(\gamma_1)\varphi - T(\gamma_0)\varphi\|_2^2 = \frac{1}{2\pi} \int_{-\pi}^{\pi} |\varphi(\gamma + \gamma_1) - \varphi(\gamma + \gamma_0)|^2\, d\gamma < \varepsilon^2, \tag{4.2.7}$$

if $\gamma_1 - \gamma_0 \in U(0)$. Now let f be any function in $L^2(\Gamma^1)$. Since $C(\Gamma^1)$ is dense in $L^2(\Gamma^1)$, for every $\varepsilon > 0$, there exists a function $\varphi \in C(\Gamma^1)$ such that

$$\|f - \varphi\|_2 < \varepsilon. \tag{4.2.8}$$

From (4.2.7) and (4.2.8) we obtain

$$\|T(\gamma_1)f - T(\gamma_0)f\|_2 \leqslant \|T(\gamma_1)(f - \varphi)\|_2 + \|T(\gamma_1)\varphi - T(\gamma_0)\varphi\|_2 + \|T(\gamma_0)(\varphi - f)\|_2$$
$$= \|f - \varphi\|_2 + \|T(\gamma_1)\varphi - T(\gamma_0)\varphi\|_2 + \|f - \varphi\|_2 < 3\varepsilon,$$

which proves that T is continuous.

The representation T is called the *regular representation* of the group Γ^1. For the general definition of regular representations see chapter IV *infra*.

Chapter IV

Representations of Compact Groups

§1. Compact Topological Groups

1.1. Compact Topological Spaces

Let M be a subspace of a topological space X. A system $\{G\}$ of sets is called a *cover* of the set M if the union of all the sets G contains M.

A topological space X is called *compact*[32] if every open cover $\{G\}$ of X admits a finite subsystem $\{G_1, \ldots, G_n\}$ that is also a cover of X.

A subset M of X is called *compact* if it is compact when regarded as a subspace of X[33].

I. *A space X is compact if and only if every system $\{F\}$ of closed subsets of X with void intersection admits a finite subsystem $\{F_1, F_2, \ldots, F_n\}$ with void intersection.*

This is proved simply by going over to complementary sets. $\square$

II. *Every closed subset F_0 of a compact space X is compact.*

Proof. Let $\{F\}$ be a system of closed subsets of F_0 with void intersection. Since F_0 is closed, $\{F\}$ is also a system of closed sets in X (see I in 1.6 of chapter III) with void intersection and by I it admits a finite subsystem $\{F_1, \ldots, F_n\}$ with void intersection. $\square$

The following equivalent definition of a compact space is occasionally useful. A system $\{M\}$ of sets is called *centered* if every finite subsystem of $\{M\}$ has nonvoid intersection. By Proposition I, a space X is compact if and only if every centered system of closed subsets of X has a nonvoid intersection. A third description of compactness follows.

III. *A space X is compact if and only if every centered system of subsets of X has at least one common limit point.*

---

[32] P. S. Aleksandrov, who introduced this concept, called such spaces *bicompact*. We will use the term *compact*, which has been all but universally adopted.

[33] Clearly all finite sets are compact.

Proof. Let X be compact and $\{M\}$ be a centered system of subsets of X. Then $\{\overline{M}\}$ is a centered system of closed subsets of X and consequently, $\{\overline{M}\}$ has nonvoid intersection. Any point of this intersection is a common limit point of the sets M.

To prove the converse, let $\{F\}$ be a centered system of closed subsets of F. This system admits a common limit point. Since the sets F are closed, this point belongs to their intersection. That is, X is compact. $\square$

IV. *If F is a compact subset of a separated space X and $x \notin F$, then there exist disjoint open sets U and V that contain x and F respectively.*

Proof. For every point $y \in F$ we have nonintersecting neighborhoods U_y, V_y of the points x and y. Since F is compact, a finite number of these sets, say $V_{y_1}, \ldots, V_{y_n}$, cover the set F. Thus the sets $U = \bigcap_{k=1}^{n} U_{y_k}$ and $V = \bigcup_{k=1}^{n} V_{y_k}$ are as required. $\square$

V. *A compact subset of a separated space is closed.*

Proof. Let F be compact and $x \notin F$. Proposition IV shows that there is a neighborhood U of the point x disjoint from F and thus (see 1.3, chapter III) we have $x \notin \overline{F}$. We infer that $\overline{F} \subset F$ and so F is closed. $\square$

VI. *A subset M of $\mathbf{R}^n$ is compact in $\mathbf{R}^n$ if and only if it is closed and bounded.*

Proof. The sufficiency coincides with the assertion of the well-known lemma of Borel and Lebesgue (see for example Fihtengol'c [1], vol. I). Conversely, let M be compact in $\mathbf{R}^n$. By virtue of V, M is closed. To prove that M is bounded, we cover every point of M with an open ball of radius 1. Since M is compact, a finite number of these balls covers M. A ball containing this finite family of balls plainly contains M. $\square$

VII. *A continuous image of a compact space is compact.*

Proof. Let f be a continuous mapping of a compact space X onto a space Y and let $\{G'\}$ be any open cover of Y. The sets $G = f^{-1}(G')$ are open in X and plainly cover X. Since the space X is compact, the cover $\{G\}$ admits a finite subcover $\{G_1, G_2, \ldots, G_n\}$. The system $\{G_1', G_2', \ldots, G_n'\}$ is a finite cover of Y, and plainly a subcover of $\{G'\}$. Since $\{G'\}$ is an arbitrary open cover of Y, Y is compact. $\square$

VIII. *For a continuous mapping of a compact space X into a separated space, images of closed sets are closed.*

Proof. Let M be a closed subset of X. By II, M is compact and consequently its continuous image $f(M)$ is also compact (see VII). By V, $f(M)$ is closed in the separated space Y. $\square$

A continuous mapping f of a topological space X into $\mathbf{R}^1$ is called a *real-valued continuous function on X*.

IX. *A real-valued continuous function f on a compact space X is bounded and assumes its infimum and supremum.*

Proof. By VII and VI, $f(X)$ is a closed bounded subset of $\mathbf{R}^1$ and so it contains its infimum and supremum. $\quad\square$

Let f be a complex-valued function on a topological space X. The union of all open sets on which $f(x) = 0$ is denoted by U_f. The complement of U_f is written Q_f. It is clear that U_f is the greatest open set on which $f(x) = 0$ and thus Q_f is the smallest closed set[34] outside of which $f(x) = 0$. The set Q_f is called *the support of the function f*. The function f is said to be *compactly supported* if its support Q_f is a compact set.

X. *A complex-valued function f on a compact topological space X is compactly supported.*

Note that Q_f is compact as a closed subset of the compact space X: see II. $\quad\square$

Let X, Y be topological spaces and let M be a subset of the product space $X \times Y$. The set of all $x \in X$ for which $x \times y \in M$ for some $y \in M$ is called *the projection of the set M onto X*. The projection of M onto Y is defined similarly.

XI. *If Y is compact, then the projection onto X of every closed set $F \subset X \times Y$ is closed.*

Proof. Let M be the projection of the set F onto X and let $x_0 \in \bar{M}$. Thus any neighborhood $U(x_0)$ has a nonvoid intersection with M. Therefore the sets

$$N_U = \{\, y : x \times y \in F, x \in U(x_0)\}$$

form a centered system in the compact space Y and thus have a common limit point y_0. But we then have $x_0 \times y_0 \in \bar{F} = F$, so that x_0 belongs to M. Thus, if x_0 belongs to M, it also belongs to M. This means that M is closed. $\quad\square$

XII. *The topological product of a finite number[35] of compact spaces is compact.*

---

[34] Plainly we can have $U_f = \varnothing$ and thus $Q_f = X$.

[35] Proposition XII holds also for an infinite number of compact spaces with an appropriate topology in their product (see Naĭmark [1], for example). This theorem, which we will not need in the sequel, is due to A. N. Tihonov.

Proof. Let X and Y be compact. Let $\{F_\alpha\}$ be a centered system of closed sets in the space $X \times Y$ and let Φ_α be the projection of the set F_α onto the space X. We may suppose that the system $\{F_\alpha\}$ is closed under the formation of finite intersections. The sets Φ_α are closed, as shown in XI. The system of closed subsets $\{\Phi_\alpha\}$ of the space X is centered and therefore has nonvoid intersection. Let x_0 belong to $\bigcap_\alpha \Phi_\alpha$ and let M_{x_0} be the set $\{x_0\} \times Y \subset X \times Y$. The mapping $(x_0, y) \to y$ is a homeomorphism of M_{x_0} onto Y. Hence M_{x_0} is a compact subspace of $X \times Y$. By our choice of x_0, the system of intersections $F_\alpha \cap M_{x_0}$ is centered and the sets $F_\alpha \cap M_{x_0}$ are closed in M_{x_0}. Since M_{x_0} is compact, the intersection of all sets $F_\alpha \cap M_{x_0}$ contains some point $\{x_0, y_0\}$. Thus, $X \cap Y$ is compact. The proof for a finite number of compact spaces is similar; or one may use induction. $\square$

XIII. *Let G and F be subsets of the product $X \times Y$ of topological spaces X and Y and suppose that G is open and F is closed. Let Q be a compact subset of X. Then the set $\bigcup_{x \in Q} \{y : x \times y \in F\}$ is closed and the set*

$$\bigcap_{x \in Q} \{y : x \times y \in G\}$$

is open.

Proof. The first assertion follows from the fact that $\bigcup_{x \in Q} \{y : x \times y \in F\}$ is the projection onto Y of the closed set $(Q \times Y) \cap F$ (see XI). The second assertion is obtained from the first by considering complementary sets. $\square$

XIV. *Let X, Y, Z be topological spaces and let $f(x, y)$ be a continuous mapping of $X \times Y$ into Z. Let Q be a compact subset of X and let G be an open subset of Z. The set $W = \{y : f(x, y) \in G$ for all $x \in Q\}$ is open in Y.*

Proof. Let $\hat{G}$ be the inverse image of G under the mapping f. Thus $\hat{G}$ is open in $X \times Y$. Now apply Proposition XIII. $\square$

XV. *If Q_1 and Q_2 are compact subsets of a topological group G, then Q_1^{-1} and $Q_1 Q_2$ are also compact.*

Proof. Use VII, noting that Q_1^{-1} and $Q_1 Q_2$ are the images of the compact sets Q_1 and $Q_1 \times Q_2$ (see XII) under the continuous mappings $g \to g^{-1}$ and $\{g_1, g_2\} \to g_1 g_2$ respectively.

1.2. Locally Compact Spaces; Locally Compact Groups

A topological space X is called *locally compact* if every point $x \in X$ has a neighborhood whose closure is compact.

For example, the spaces $\mathbf{R}^n$ are locally compact, since each point $x^0 = \{x_1^0, \ldots, x_n^0\}$ has a neighborhood $U(x_0) = \{x : |x_j - x_j^0| < \varepsilon, j = 1, \ldots, n\}$, whose closure is a closed bounded set in $\mathbf{R}^n$ and so is compact.

Spaces $\mathbf{C}^n$ are homeomorphic to $\mathbf{R}^{2n}$ and so are locally compact.
Prove that a Hilbert space of infinite dimension is *not* locally compact.

I. *A closed subspace F of a locally compact space X is locally compact.*

Proof. Let $x_0 \in F$ and let V be a neighborhood of the point x_0 in X such
that $\bar{V}$ is compact. Then $V \cap F$ is a neighborhood of the point x_0 in F and
$\overline{V \cap F}$ is a closed subset of the compact space $\bar{V}$ and so is compact (II of
1.1). $\square$

A topological group G is called *locally compact* if the topological space
G is locally compact.

II. *A topological group G is locally compact if there exists a neighborhood
V of the identity element e of the group G whose closure is compact.*

The homeomorphism $g \to g_0 g$ maps V onto the neighborhood $g_0 V$ of
the element g_0 and $g_0 V$ is compact since it is a homeomorphic image of the
compact set $\bar{V}$.

For example, the group $GL(n, \mathbf{C})$ is locally compact. We set

$$V = \{g : |g_{jl} - \delta_{jl}| < \varepsilon\}, \quad \text{where } \delta_{jl} = \begin{cases} 1 & \text{for } j = l, \\ 0 & \text{for } j \neq l. \end{cases}$$

For sufficiently small ε, we have $V \subset GL(n, \mathbf{C})$ and so V is a neighborhood of
the identity element $e = \|\delta_{jl}\|_{j,l=1}^n$. Since V is a closed bounded set in $\mathbf{C}^{n^2}$,
it is compact (VI in 1.1). Consequently, $GL(n, \mathbf{C})$ is locally compact by virtue
of II. The groups $SL(n, \mathbf{C})$, $GL(n, \mathbf{R})$ and $SL(n, \mathbf{R})$ are closed in $GL(n, \mathbf{C})$ and
are therefore locally compact by virtue of I.

A complex-valued function f on a topological group G is said to be
uniformly continuous on G if for every $\varepsilon > 0$ there exists a neighborhood
V of the identity element V of G such that

$$|f(g_1) - f(g_2)| < \varepsilon, \quad \text{for } g_2 \in g_1 V. \tag{1.2.1}$$

If G is the additive group $\mathbf{R}^1$ and V has the form $V = \{x : |x| < \delta\}$, where
$\delta > 0$, then $f = f(x)$ is a function on $\mathbf{R}^1$ and the condition $g_1 \in g_2 V$ is rewritten in the form $x_1 - x_2 \in V$, for real numbers x_1 and x_2. Thus the
condition (1.2.1) means that $|f(x_1) - f(x_2)| < \varepsilon$ for $|x_1 - x_2| < \delta$. This is
the common definition of uniform continuity for functions on $\mathbf{R}^1$.

III (Theorem of uniform continuity). *A compactly supported continuous
function f on a locally compact group G is uniformly continuous on G.*

Proof. Let Q be the support of f and let U be a symmetric neighborhood of
the identity element such that $\bar{U}$ is compact. By hypothesis, Q is compact

and therefore $Q\bar{U}$ is also compact (see XV in 1.1). We set

$$W = \{g : |f(g_1 g) - f(g_1)| < \varepsilon \quad \text{for all } g_1 \in Q\bar{U}\}.$$

Plainly e belongs to W. By virtue of XIV in 1.1, W is open and thus W is a neighborhood of e. If $g \in U$, then $f(g_1 g)$ and $f(g_1)$ vanish for $g_1 \notin Q\bar{U}$ and thus $|f(g_1 g) - f(g_1)|$ is less than ε for $g \in W \cap U$ for all $g_1 \in G$. Setting $g_1 g = g_2$ and $V = W \cap U$, we obtain $|f(g_2) - f(g_1)| < \varepsilon$ for $g_2 \in g_1 V$. $\square$

1.3. Uryson's Lemma

A topological space X is said to be *normal* if for any two disjoint closed subsets F_1, F_2 of X there exist disjoint open sets U_1, U_2 containing F_1 and F_2 respectively. This condition is equivalent to the following. For any closed set F and open set $U \supset F$ there exists an open set V such that $F \subset V$ and $\bar{V} \subset U$. (Consider the closed set $X \backslash U$, which is disjoint from F.)

I. *A compact separated space X is normal.*

Proof. Let F_1 and F_2 be disjoint closed (and therefore compact) sets in X. By IV in 1.1, for every point $y \in F_2$ there are disjoint open sets U_y and V_y that contain F_1 and y respectively. The sets V_y, $y \in F_2$ are an open cover of F_2. Since F_2 is compact, we have a finite number of these sets, say V_{y_1}, $\ldots$, V_{y_n} that cover the set F_2. Therefore

$$U = \bigcap_{k=1}^{n} U_{y_k}, \qquad V = \bigcup_{k=1}^{n} V_k$$

are disjoint open sets that contain F_1 and F_2 respectively. $\square$

II (Uryson's lemma). *For any two disjoint closed sets F_0 and F_1 of the normal space X, there exists a real-valued continuous function f on X that satisfies the following conditions:*

  (1) $0 \leqslant f(x) \leqslant 1$;
  (2) $f(x) = 0$ *on* F_0;
  (3) $f(x) = 1$ *on* F_1.

Proof. The proof is trivial if one of the sets F_0, F_1 is void. We may suppose that neither F_0 nor F_1 is void.

Writing $V_1 = X \backslash F_1$, we then have $F_0 \subset V$, and since the space X is normal, we have an open set, which we denote by V_0, such that $F_0 \subset V_0$ and $\bar{V}_0 \subset V_1$.

Likewise, we have an open set, written $V_{1/2}$, such that $\bar{V}_0 \subset V_{1/2}$ and $\bar{V} \subset V_1$. Repeating this construction, we obtain for every number r of the form $m/2^n$, $0 \leqslant m \leqslant 2^n$, an open set V_r such that $\bar{V}_{r_1} \subset V_{r_2}$ for $r_1 < r_2$.

For every real number t in the interval $0 < t \leqslant 1$ we define

$$V_t = \bigcup_{r \leqslant t} V_r,$$

and also $V_t = \varnothing$ for $t < 0$ and $V_t = X$ for $t > 1$. Thus the open sets V_t are defined for all real values t. Note that

$$\bar{V}_{t_1} \subset V_{t_2}, \quad \text{if } t_1 < t_2. \tag{1.3.1}$$

(For $0 \leqslant t_1 < t_2 \leqslant 1$ there are rational numbers r_1, r_2 of the form $m/2^n$ such that $t_1 < r_1 < r_2 < t_2$ and therefore $\bar{V}_{t_1} \subset \bar{V}_{r_1} \subset V_{r_2} \subset V_{t_2}$. If $t_1 < 0$ or $t_2 > 1$, the inclusion (1.3.1) is obvious, since $V_{t_1} = \varnothing$ and $V_{t_2} = X$ for $t_2 > 1$.)
We now define

$$f(x) = \inf\{t : x \in V_t\}. \tag{1.3.2}$$

Let us show that f fulfills all of our requirements.
For $t > 1$, the set V_t is all of X and so (1.3.2) shows that $f(x) \leqslant 1$ for all x in X. For $t < 0$, V_t is void, and so $f(x) < 0$ for no x in X. Thus (1) holds. For $x \in V_0$, (1.3.2) gives us $f(x) \leqslant 0$, and so $f(x) = 0$ for x in V_0 and hence (2) holds (since $F_0 \subset V_0$). If x is in F_1, then x is not in V_1 (by definition) and so (1.3.2) gives us $f(x) \geqslant 1$. In view of (1), we have $f(x) = 1$, so that (3) holds.
It remains to prove that the function f is continuous. Choose any point x_0 in X and any positive number ε. We set $y_0 = f(x_0)$. From (1.3.1) and (1.3.2) we see that $x_0 \in V_t$ for $t > y_0$ and $x_0 \notin \bar{V}_t$ for $t < y_0$. In particular we have $x_0 \in V_{y_0 + \varepsilon} \backslash \bar{V}_{y_0 - \varepsilon}$, so that, being an open set, $V_{y_0 + \varepsilon} \backslash \bar{V}_{y_0 - \varepsilon}$ is a neighborhood of the point x_0. If $x \in V_{y_0 + \varepsilon} \backslash \bar{V}_{y_0 - \varepsilon}$, it follows that $y_0 + \varepsilon \in \{t : x \in V_t\}$ and $y_0 - \varepsilon \notin \{t : x \in V_t\}$. Again (1.3.1) and (1.3.2) show that $y_0 - \varepsilon \leqslant f(x) \leqslant y_0 + \varepsilon$, that is, $f(x_0) - \varepsilon \leqslant f(x) \leqslant f(x_0) + \varepsilon$. This means that f is continuous.
□

Remark. By Proposition I, a compact space is normal and thus Uryson's lemma holds for closed subsets of a compact space.

1.4. Stone's Theorem

Let Q be any set. For real-valued functions $f_1, f_2, \ldots, f_n$ on Q we define functions

$$(f_1 \cup f_2 \cup \cdots \cup f_n)(q) = \max\{f_1(q), f_2(q), \ldots, f_n(q)\},$$
$$(f_1 \cap f_2 \cap \cdots \cap f_n)(q) = \min\{f_1(q), f_2(q), \ldots, f_n(q)\}.$$

A set A of real-valued functions on Q is called a *lattice* if it contains $f_1 \cup f_2$ and $f_1 \cap f_2$ whenever f_1, f_2 are in A. If A is a lattice, A contains $f_1 \cup f_2 \cup \cdots \cup f_n$ and $f_1 \cap f_2 \cap \cdots \cap f_n$ along with $f_1, f_2, \ldots, f_n$. A set A of real-valued functions on X is called a (*real*) *algebra of functions* if along with a function, it also contains its product by any real number and along with any two functions, it contains their sum and their product. A real algebra A of functions on Q is called *uniformly closed* if the limit of every uniformly convergent sequence of functions $f_n \in A$ belongs to A. If an algebra A is not uniformly closed, then by joining in it all such limits, we obtain a new set of functions that contains A, which is a uniformly closed algebra. This algebra is called the *uniform closure* of the algebra A and is denoted by $\bar{A}$.

The set $C'(X)$ of all real-valued continuous functions on a topological space X is an example of a uniformly closed real algebra of functions.

I. *Any uniformly closed real algebra A of bounded functions, containing all real constants, forms a lattice.*

Proof. It is sufficient to prove that if $f \in A$, then we also have $|f| \in A$. For, if this holds and f_1, f_2 are in A, then the functions

$$f_1 \cup f_2 = \tfrac{1}{2}(f_1 + f_2 + |f_1 - f_2|), \qquad f_1 \cap f_2 = \tfrac{1}{2}(f_1 + f_2 - |f_1 - f_2|)$$

also belong to A.

Let $f \in A$ and $|f(x)| \leqslant c$ for all $x \in X$. We then have

$$|f(x)| = \sqrt{c^2 - [c^2 - (f(x))^2]} = c \sqrt{1 - \left(1 - \frac{(f(x))^2}{c^2}\right)}$$

$$= c\left\{1 - \sum_{n=1}^{\infty} \frac{1.1.3 \ldots (2n-3)}{2.4.6 \ldots 2n}\left(1 - \frac{(f(x))^2}{c^2}\right)^n\right\}.$$

The series in the right side converges uniformly on X since $0 \leqslant 1 - (f(x))^2/c^2 \leqslant 1$. Note that all the terms of the series belong to A. Thus $|f|$ belongs to A. $\square$

II. *Let A be a set of continuous real-valued functions f on a compact separated space X that satisfies the following conditions:*

(1) *A is a lattice;*
(2) *for any two distinct points $\xi, \eta \in X$ and any real numbers a, b, there exists a function $f_{\xi\eta} \in A$ such that $f_{\xi\eta}(\xi) = a$, $f_{\xi\eta}(\eta) = b$.*[36]

Then every continuous real-valued function on X is the limit of a uniformly convergent sequence of functions $f_n \in A$.

---

[36] Editor's note. This condition can be satisfied only for topological spaces satisfying Hausdorff's separation axiom.

Proof. Let f be any continuous real-valued function on X. Let ε be any positive number, let ξ and η be any two points of X, and let $f_{\xi\eta}(x)$ be a function in A that satisfies condition (2) for $a = f(\xi)$, $b = f(\eta)$. We set

$$U_{\xi\eta} = \{x : f_{\xi\eta}(x) < f(x) + \varepsilon\}, \qquad V_{\xi\eta} = \{x : f_{\xi\eta}(x) > f(x) - \varepsilon\}.$$

Plainly $U_{\xi\eta}$ and $V_{\xi\eta}$ are open sets in X that contain both ξ and η. For a fixed ξ, the open sets $U_{\xi\eta}$ (as η runs through all of $X \setminus \{\xi\}$) cover the compact space X. Select a finite subcover $\{U_{\xi_1\eta}, \ldots, U_{\xi_n\eta}\}$ from this cover and set $\varphi_\xi = f_{\xi_1\eta} \cap f_{\xi_2\eta} \cap \cdots \cap f_{\xi_n\eta}$ and $V_\xi = \bigcap_{j=1}^{n} V_{\xi_j\eta}$. The function φ_ξ belongs to A and satisfies the following conditions:

$$\varphi_\xi(x) < f(x) + \varepsilon, \quad \text{throughout } X;$$
$$\varphi_\xi(x) > f(x) - \varepsilon, \quad \text{for } x \in V_\xi.$$

The sets V_ξ as ξ runs through all of X form an open cover of X. Select a finite subcover $\{V_{\xi_1}, \ldots, V_{\xi_m}\}$ from $\{V_\xi\}$ and set $\psi = \varphi_{\xi_1} \cup \varphi_{\xi_2} \cup \cdots \cup \varphi_{\xi_m}$. The function ψ is in A and satisfies the conditions $f(x) - \varepsilon < \psi(x) < f(x) + \varepsilon$ for all x in X.

We set $\varepsilon = 1/n$, $n = 1, 2, \ldots$, and write ψ_n for the corresponding function ψ. We have $\psi_n \in A$ and $f(x) - 1/n < \psi_n(x) < f(x) + 1/n$ for all x in X. Thus ψ_n converges uniformly to f. $\square$

A set A of functions on Q is said to be *separate the points of Q* if for any two distinct points q_1, q_2 in Q, there exists a function $f \in A$ such that $f(q_1) \neq f(q_2)$.

Theorem 1 (M. H. Stone [1*]). *Let A be a real algebra of continuous functions on a compact space X that contains all real constants and that separates points. The uniform closure $\bar{A}$ of the algebra A is the algebra $C^r(X)$ of all real-valued continuous functions on X.*

Proof. It is easy to see that $\bar{A}$ is an algebra. By Proposition I, the algebra $\bar{A}$ is a lattice. We will prove that $\bar{A}$ (actually A) satisfies condition (2) of Proposition II. It then follows from I that $\bar{A} = C^r(X)$.

Let ξ and η be distinct points of X. There is a function ψ in A such that

$$\psi(\xi) \neq 0 \quad \text{and} \quad \psi(\xi) \neq \psi(\eta). \tag{1.4.1}$$

To see this, let a and b be any real numbers, and let φ be a function in A such that $\varphi(\xi) \neq \varphi(\eta)$. Define a function ψ on X by

$$\psi(\xi) = \frac{b - a}{\varphi(\eta) - \varphi(\xi)} [\varphi - \varphi(\xi)] + a. \tag{1.4.2}$$

Since A is an algebra containing all constants, ψ belongs to A. It is obvious that $\psi(\xi) = a$ and $\psi(\eta) = b$. Hypothesis (2) of II is thus satisfied, and as already noted, this proves our result.[37] $\quad\square$

Theorem 2 (Theorem of Weierstrass). *Let X be a closed bounded subset of the space $\mathbf{R}^n$. Every real-valued continuous function f on X is the limit of a uniformly convergent sequence of polynomials in the variables $x_1, x_2, \ldots, x_n$ with real coefficients.*

To prove this, apply Stone's theorem. The family of functions $(x_1, \ldots, x_n) \to x_j$ $(j = 1, 2, \ldots, n)$ separates points of X. Polynomials are the smallest algebra containing these functions and also constants. $\quad\square$

A set A of complex-valued functions on Q is called a *complex algebra of functions* if along with every function it contains the product of the function and any complex number and along with every two functions it contains their sum and their product. For a complex algebra of functions, uniform convergence and the uniform closure are defined just as for a real algebra of functions.

The algebra $C(X)$ of all continuous complex-valued functions on a given topological space X is an example of a uniformly closed complex algebra of functions.

Theorem 3. *Let A be a complex algebra of continuous functions on a compact separated space X that satisfies the following conditions:*

(1) *A separates points in X;*
(2) *if $f(x) \in A$, then also $\overline{f(x)} \in A$;*
(3) *A contains all complex constants.*

Then the uniform closure $\bar{A}$ of the algebra A is $C(X)$.

Proof. Let A_1 be the set of all real-valued functions belonging to A. Plainly A_1 is a real algebra of functions. If f belongs to A, $\operatorname{Re} f = \frac{1}{2}(f + \bar{f})$ and $\operatorname{Im} f = (1/2i)(f - \bar{f})$ also belong to A and thus to A_1, as (2) implies. It follows that A_1 separates points on X. Indeed, if x_1 and x_2 belong to X and $x_1 \neq x_2$, then condition (1) implies that there exists a function $f \in A$ such that $f(x_1) \neq f(x_2)$. Thus at least one of the nonequalities $\operatorname{Re} f(x_1) \neq \operatorname{Re} f(x_2)$, $\operatorname{Im} f(x_1) \neq \operatorname{Im} f(x_2)$ holds. By Stone's theorem we have

$$\bar{A}_1 = C^r(X). \tag{1.4.3}$$

Let f be any function in $C(X)$. By (1.4.3), there exist sequences $\varphi_n \in A_1$, $\psi_n \in A_1$ such that $\varphi_n \to \operatorname{Re} f$ and $\psi_n \to \operatorname{Im} f$ uniformly on X. Hence $\varphi_n + i\psi_n$ is in A for all n and $\varphi_n + i\psi_n$ converges to f uniformly on X. $\quad\square$

---

[37] Editor's note. We have simplified Professor Naĭmark's original proof.

1.5. Definition of a Compact Topological Group; Examples

A topological group G is called *compact* if the topological space G is compact (and of course separated).

I. *Any closed subgroup of a compact topological group is compact.*

See II in 1.1.

II. *A continuous complex-valued function on a compact group G is uniformly continuous on G.*

See III in 1.2 and X in 1.1.

Examples

1. Finite groups. Any finite group (provided with the discrete topology) is compact.

2. The group $U(n)$. Recall that a matrix u is called *unitary* if

$$u^*u = 1, \tag{1.5.1}$$

where 1 is the identity matrix. The set of all n by n unitary matrices is denoted by $U(n)$. If u_1 and u_2 belong to $U(n)$, that is, $u_1^*u_1 = 1$, $u_2^*u_2 = 1$, then we also have $(u_1u_2)^*(u_1u_2) = u_2^*u_1^*u_1u_2 = 1 \cdot 1 = 1$, that is, $u_1u_2 \in U(n)$. Furthermore, if u belongs to $U(n)$, (1.5.1) gives us $u^{-1} = u^*$ and $(u^{-1})^*u^{-1} = u^{**}u^{-1} = uu^{-4} = 1$, that is, u^{-1} also belongs to $U(n)$. Thus $U(n)$ is a subgroup of the group $GL(n, \mathbf{C})$. We topologize the group $U(n)$ as a subspace of the topological space $GL(n, \mathbf{C})$. In the sequel we will consider $U(n)$ as a topological group with this topology. The group $U(n)$ is called *the n by n unitary group*.

III. *The group $U(n)$ is compact.*

Proof. As noted in example 3 in 2.1, chapter III, the topology of $U(n)$ is the topology of a subspace of the topological space $\mathbf{C}^{n^2}$. We will prove that $U(n)$ is a bounded closed set in $\mathbf{C}^{n^2}$. This and VI in 1.1 will show that $U(n)$ is compact. Condition (1.5.1) states that $U(n)$ is the set of exactly those points $(u_{11}, \ldots, u_{1n}, \ldots, u_{n1}, \ldots, u_{nn}) \in \mathbf{C}^{n^2}$ for which

$$\sum_{l=1}^{n} \bar{u}_{lj}u_{lk} = \begin{cases} 1, & \text{for } j = k, \\ 0, & \text{for } j \neq k, \end{cases} \qquad j, k = 1, 2, \ldots, n. \tag{1.5.2}$$

Inasmuch as the left and right sides of (1.5.2) are continuous functions on $\mathbf{C}^{n^2}$, $U(n)$ is closed; see I in 1.8, chapter III. Next set $j = k$ in (1.5.2) and sum over k from 1 to n. We conclude that

$$\sum_{k=1}^{n} \sum_{l=1}^{n} |u_{lk}|^2 = n.$$

This means that $U(n)$ is contained in a sphere (or more precisely, on the *surface* of a sphere) of the space $\mathbf{C}^{n^2}$ with center at $(0, \ldots, 0)$ and radius $\sqrt{n}$. Hence the set $U(n)$ is bounded in $\mathbf{C}^{n^2}$. Therefore $U(n)$ is compact. □

In particular the group $U(1)$ is compact. Note that $U(1)$ is the multiplicative group of complex numbers of absolute value 1. Obviously $U(1)$ is (topologically) isomorphic to the group Γ^1 of rotations of a circle and thus to the one-dimensional torus $\mathcal{T}^1$ (see example 3 in 2.5, chapter III). Consequently, *the groups Γ^1 and $\mathcal{T}^1$ are compact.*

Remark 1. It follows from (1.5.1) that $\det u^* \det u = 1$, that is,

$$|\det u| = 1, \tag{1.5.3}$$

for all $u \in U(n)$. The mapping $u \to \det u$ is a continuous homomorphism of the group $U(n)$ onto the group $U(1)$.

3. The group $SU(n)$. The set of all matrices $u \in U(n)$ that satisfy the condition $\det u = 1$ is denoted by $SU(n)$.

The obvious equality $SU(n) = U(n) \cap SL(n, \mathbf{C})$ shows that $SU(n)$ is a closed subgroup of the group $U(n)$. From this and from I we infer the following.

IV. *The group $SU(n)$ is compact.*

For a matrix $u \in U(n)$, consider an element $\{\alpha, v\}$ of the group $U(1) \times SU(n)$, where

$$\alpha = \det u, \qquad v_{1k} = \frac{1}{\alpha} u_{1k}, \qquad v_{jk} = u_{jk}, \quad \text{for } j > 1.$$

The reader can easily see that this correspondence is a homeomorphism of the group $U(n)$ onto the group $U(1) \times SU(n)$.

V. *The groups $U(n)$ and $U(1) \times SU(n)$ are homeomorphic.*

Remark 2. Clearly the homeomorphism defined above is not the only possible one. For example, we may set

$$\alpha = \det u, \qquad v_{2k} = \frac{1}{\alpha} u_{2k}, \qquad v_{jk} = u_{jk}, \quad \text{for } j \neq 2.$$

We consider in more detail the case $n = 2$. The group $SU(2)$ consists of all matrices $u = \begin{Vmatrix} a_1 & b_1 \\ a & b \end{Vmatrix}$ that satisfy the conditions

$$a\bar{a} + b\bar{b} = 1, \qquad a_1\bar{a}_1 + b_1\bar{b}_1 = 1,$$
$$a\bar{a}_1 + b\bar{b}_1 = 0, \qquad a_1 b - b_1 a = 1. \tag{1.5.4}$$

From this we have

$$\bar{b}_1 = \bar{b}_1(a_1 b - b_1 a) = -a_1 a\bar{a}_1 - \bar{b}_1 b_1 a = -a(a_1\bar{a}_1 + b_1\bar{b}_1) = -a,$$
$$\bar{a}_1 = \bar{a}_1(a_1 b - b_1 a) = \bar{a}_1 a_1 b + b_1(\bar{b}_1 b) = b(\bar{a}_1 a_1 + \bar{b}_1 b_1) = b.$$

Conversely, if we have $\bar{b}_1 = -a$ and $\bar{a}_1 = b$ and also $|a|^2 + |b|^2 = 1$, then u belongs to $SU(2)$.

VI. *The group $SU(2)$ consists of all matrices*

$$u = \begin{Vmatrix} \bar{b} & -\bar{a} \\ a & b \end{Vmatrix}, \tag{1.5.5}$$

for which

$$|a|^2 + |b|^2 = 1. \tag{1.5.6}$$

We set

$$a = x_1 + ix_2, \qquad b = x_3 + ix_4; \qquad x_1, x_2, x_3, x_4 \in \mathbf{R}, \tag{1.5.7}$$

and rewrite (1.5.6) in the form

$$|x_1|^2 + |x_2|^2 + |x_3|^2 + |x_4|^2 = 1. \tag{1.5.8}$$

The formulas in (1.5.7) describe a mapping $u \to (x_1, x_2, x_3, x_4)$ of the topological space $SU(2)$ onto the unit sphere in $\mathbf{R}^4$. Also, upon comparing the topologies of $SU(2)$ and $\mathbf{R}^4$, we conclude the following.

VII. *The correspondence $u \to (x_1, x_2, x_3 x_4)$ defined in (1.5.7) is a homeomorphism of $SU(2)$ onto the unit sphere in $\mathbf{R}^4$.*

4. The group $O(n, \mathbf{R})$. Let $O(n, \mathbf{R})$ denote the set of all n by n matrices g with real entries that satisfy the condition

$$g'g = 1, \tag{1.5.9}$$

where g' is the transpose of g. For real matrices, the conditions (1.5.1) and (1.5.9) are equivalent. That is,

$$O(n, \mathbf{R}) = U(n) \cap GL(n, \mathbf{R}). \tag{1.5.10}$$

Since $U(n)$ and $GL(n, \mathbf{R})$ are closed subgroups of the group $GL(n, \mathbf{C})$ (see examples 1 and 3 in 2.3, chapter III), (1.5.10) shows that $O(n, \mathbf{R})$ is a closed subgroup of $U(n)$. We have proved the following fact.

VIII. *The group $O(n, \mathbf{R})$ is compact.*

The group $O(n, \mathbf{R})$ is called a *real orthogonal group* and its elements are called *real orthogonal matrices*.

Remark 3. We infer from (1.5.9) that $\det g' \cdot \det g = 1$, that is, $(\det g)^2 = 1$. Thus we have

$$\det g = \pm 1, \quad \text{for } g \in O(n, \mathbf{R}). \tag{1.5.11}$$

The mapping $g \to \det g$ is a homomorphism of the group $O(n, \mathbf{R})$ onto the 2-element multiplicative group $\{-1, 1\}$.

5. The group $SO(n, \mathbf{R})$. The set of all real orthogonal matrices g with $\det g = 1$ is denoted by $SO(n, \mathbf{R})$. Clearly we have

$$SO(n, \mathbf{R}) = O(n, \mathbf{R}) \cap SL(n, \mathbf{R}). \tag{1.5.12}$$

Arguments like the foregoing give us another useful fact.

IX. *The group $SO(n, \mathbf{R})$ is compact.*

The group $SO(n, \mathbf{R})$ is called a *unimodular real orthogonal group*.

X. *The group $O(n, \mathbf{R})$ is homeomorphic to the group $\{-1, 1\} \times SO(n, \mathbf{R})$.*

The proof is like the proof of Proposition V, except in place of (1.5.2) we apply (1.5.10).

We work out in detail the cases $n = 2$ and $n = 3$. The group $SO(2, \mathbf{R})$ consists of all matrices

$$g = \begin{Vmatrix} g_{11} & g_{12} \\ g_{21} & g_{22} \end{Vmatrix}$$

with real entries such that

$$\left. \begin{array}{ll} g_{11}^2 + g_{12}^2 = 1, & g_{21}^2 + g_{22}^2 = 1, \\ g_{11}g_{21} + g_{12}g_{22} = 0, & g_{11}g_{22} - g_{12}g_{21} = 1, \end{array} \right\} \tag{1.5.13}$$

Thus we may set

$$g_{11} = \cos \theta, \qquad g_{12} = -\sin \theta, \qquad (1.5.14)$$

where θ is defined only up to an integral multiple of 2π. The third condition in (1.5.13) shows that $g_{21} = \lambda \sin \theta$ and $g_{22} = \lambda \cos \theta$. Substituting these expressions in the last condition in (1.5.13), we conclude that $\lambda = 1$, so that $g_{21} = \sin \theta$, $g_{22} = \cos \theta$. The second condition in (1.5.14) is clearly satisfied. $\square$

We have proved the following.

XI. *The group* $SO(2, \mathbf{R})$ *consists of all matrices of the form*

$$g = \begin{Vmatrix} \cos \theta & -\sin \theta \\ \sin \theta & \cos \theta \end{Vmatrix} \qquad (1.5.15)$$

where θ *is a real number.*

We set

$$\left. \begin{aligned} x_1' &= x_1 \cos \theta - x_2 \sin \theta, \\ x_2' &= x_1 \sin \theta + x_2 \cos \theta. \end{aligned} \right\} \qquad (1.5.16)$$

The equalities $(x_1 + ix_2)e^{i\theta} = (x_1 + ix_2)(\cos \theta + i \sin \theta) = x_1' + ix_2'$ show that x_1' and x_2' are the coordinates of the point obtained from (x_1, x_2) by a counterclockwise rotation of the plane $\mathbf{R}^2$ about the origin $(0, 0)$ through the angle θ. The linear transformation (1.5.16) with matrix g is merely a rotation of the plane $\mathbf{R}^2$ through the angle θ. Therefore the group $SO(2, \mathbf{R})$ may be identified with *the group of rotations of two-dimensional space about* $(0, 0)$. The rotation g_θ of $\mathbf{R}^2$ about $(0, 0)$ through the angle θ, corresponds to the rotation γ_θ of a circle with center $(0, 0)$ through the same angle (see example 1 in 1.7, chapter I). The correspondence $g_\theta \to \gamma_\theta$ is obviously an isomorphism. Comparing the topologies of $SO(2, \mathbf{R})$ and Γ^1, we see that this isomorphism is topological.

XII. *The group* $SO(2, \mathbf{R})$ *is isomorphic to the group* Γ^1 *and therefore to the group* $\mathscr{T}^1$.

The group $SO(3, \mathbf{R})$. The set of all rotations of three-dimensional space $\mathbf{R}^3$ about the origin $O = (0, 0, 0)$ is denoted by G_0. We choose a fixed orthogonal system of coordinates with origin at the point O. We write the vectors of length 1 ("unit vectors") along the coordinate axes as e_1, e_2, e_3. The rotation $g \in G_0$ carries e_1, e_2, e_3 into three mutually orthogonal vectors of length 1, g_1, g_2, g_3 respectively, with the same orientation as e_1, e_2, e_3.

Let g_{ik} be the projection of the vector g_k onto the i-th axis. The vectors g_1, g_2, g_3 are completely defined by their projections g_{ik}, that is, by the matrix

$$\left\| \begin{matrix} g_{11} & g_{12} & g_{12} \\ g_{21} & g_{22} & g_{23} \\ g_{31} & g_{32} & g_{33} \end{matrix} \right\|$$

We also denote this matrix by the letter g and we call it *the matrix of the rotation* g. The entries g_{ik} are real as $\mathbf{R}^3$ is a *real vector space*. Since the vectors g_1, g_2, g_3 are orthonormal, the matrix g is orthogonal. Since the orientations e_1, e_2, e_3 and g_1, g_2, g_3 coincide, we have $\det g = 1$. That is, g belongs to $SO(3, \mathbf{R})$.

The rotation g is completely defined by its matrix g.

Indeed, every vector $x \in \mathbf{R}^3$ has the form $x = x_1 e_1 + x_2 e_2 + x_3 e_3$, where the numbers x_1, x_2, x_3 are projections of the vector x onto the corresponding coordinate axes. Because the rotation g is linear and carries e_1, e_2, e_3 into g_1, g_2, g_3 respectively, it carries the vector x into the vector

$$x' = x_1 g_1 + x_2 g_2 + x_3 g_3. \tag{1.5.17}$$

Let x_1', x_2', x_3' be the projections of the vector x' onto the original coordinate axes. Projecting both sides of (1.5.17) onto these axes, we obtain

$$\left. \begin{matrix} x_1' = g_{11}x_1 + g_{12}x_2 + g_{13}x_3, \\ x_2' = g_{21}x_1 + g_{22}x_2 + g_{23}x_3, \\ x_3' = g_{31}x_1 + g_{32}x_2 + g_{33}x_3. \end{matrix} \right\} \tag{1.5.18}$$

That is, the rotation g is a linear transformation of the space $\mathbf{R}^3$ with the matrix $g \in SO(3, \mathbf{R})$. Conversely, for every matrix $g \in SO(3, \mathbf{R})$ we define the rotation $g \in G_0$ according to formula (1.5.18). Therefore the group $SO(3, \mathbf{R})$ can be identified with *the group of rotations of three-dimensional space*. Its irreducible representations play an important role in many applications[38]. (We will take up some of these in §3, chapter IV and in chapter X.)

In many situations it is convenient to represent rotations of three-dimensional space by using the three following independent parameters, which go by the name of *Euler's angles*. Under the rotation g, suppose that the coordinate axes Ox, Oy, Oz go into the axes Ox', Oy' and Oz' respectively (Fig. 3). The line of intersection of the planes xOy and $x'Oy'$ is denoted by Ol. We orient Ol in such a way that an observer at O oriented in the positive direction will see the counterclockwise rotation that carries the axis Oz onto

---

[38] By analogy with the cases $n = 2$ and $n = 3$ the group $SO(n, R)$ is called *the group of rotations of the space R^n*.

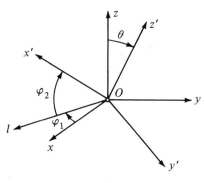

Fig. 3

the axis Oz' as having an angle $\leqslant \pi$. The orientation of the line Ol is defined uniquely by this condition, with the exception of the cases when Oz and Oz' coincide or when they form the angle π. The angle formed by the axis Ox and the line Ol is called φ_1, and the angle formed by the line Ol and the axis Ox' is called φ_2. The angle between Oz and Oz' is called θ. The angles φ_1, φ_2 and θ are *Euler's angles* for the rotation g. The definition itself shows that $0 \leqslant \varphi_1 \leqslant 2\pi$, $0 \leqslant \varphi_2 \leqslant 2\pi$, $0 \leqslant \theta \leqslant \pi$. Distinct rotations correspond to distinct triples of numbers φ_1, φ_2 and θ in the appropriate intervals, with the exception of $\theta = 0$ and $\theta = \pi$. In these cases the planes xOy and $x'Oy'$ coincide and therefore the line Ol is undefined. It is easy to see that triples $(\varphi_1, \varphi_2, 0)$ and $(\varphi_1 + \alpha, \varphi_1 - \alpha, 0)$ define the same rotation and that triples $(\varphi_1, \varphi_2, \pi)$ and $(\varphi_1 + \alpha, \varphi_2 + \alpha, \pi)$ also define the same rotation.

XIII. *A rotation g is completely defined by its Euler's angles φ_1, φ_2 and θ.*

Proof. Let g_φ and g_θ denote rotations about the axes Oz and Ox through the angles φ and θ, respectively. The rotation g can be represented as the product

$$g = \tilde{g}_{\varphi_2} \tilde{g}_\theta g_{\varphi_1}. \tag{1.5.19}$$

of three rotations g_{φ_1}, $\tilde{g}_\theta$, and $\tilde{g}_{\varphi_2}$ around the axes Oz, Ol, and Oz' respectively. Indeed the rotation g_{φ_1} carries the axis Ox onto the axis Ol, after which the rotation $\tilde{g}_\theta$ carries the axis Oz onto Oz'. Finally the rotation $\tilde{g}_{\varphi_2}$ carries the axes Ox and Oy onto Ox' and Oy' respectively. Thus formula (1.5.19) and Proposition XIII are proved. $\square$

A rotation $\tilde{g}_\theta$ is rotation g_θ of the auxiliary system of coordinate axes, obtained from the original rotation g_{φ_1}. Therefore we have $\tilde{g}_\theta = g_{\varphi_1} g_\theta g_{\varphi_1}^{-1}$ and analogously, $\tilde{g}_{\varphi_2} = (\tilde{g}_\theta g_{\varphi_1}) g_{\varphi_2} (\tilde{g}_\theta g_{\varphi_1})^{-1}$. From this and from (1.5.19) we conclude

$$g = \tilde{g}_{\varphi_2}\tilde{g}_\theta g_{\varphi_1} = (\tilde{g}_\theta g_{\varphi_1})g_{\varphi_2}(\tilde{g}_\theta g_{\varphi_1})^{-1}\tilde{g}_\theta g_{\varphi_1} = \tilde{g}_\theta g_{\varphi_1}g_{\varphi_2} = g_{\varphi_1}g_\theta g_{\varphi_2}. \quad (1.5.20)$$

From (1.5.20) we easily obtain expressions for the matrix elements g_{ik} of the rotation g in terms of its Euler's angles. The matrices of the rotations g and g have the form

$$g_\varphi = \begin{Vmatrix} \cos \varphi & -\sin \varphi & 0 \\ \sin \varphi & \cos \varphi & 0 \\ 0 & 0 & 1 \end{Vmatrix}, \qquad g_\theta = \begin{Vmatrix} 1 & 0 & 0 \\ 0 & \cos \theta & -\sin \theta \\ 0 & \sin \theta & \cos \theta \end{Vmatrix}.$$

Substituting these expressions in (1.5.20) and recalling that for multiplication of rotations their matrices are also multiplied, we obtain

$$g = \begin{Vmatrix} \cos \varphi_1 \cos \varphi_2 - \cos \theta \sin \varphi_1 \sin \varphi_2 & -\cos \varphi_1 \sin \varphi_2 - \cos \theta \sin \varphi_1 \cos \varphi_2 & \sin \varphi_1 \sin \theta \\ \sin \varphi_1 \cos \varphi_2 + \cos \theta \cos \varphi_1 \sin \varphi_2 & \sin \varphi_1 \sin \varphi_2 + \cos \theta \cos \varphi_1 \cos \varphi_2 & -\cos \varphi_1 \sin \theta \\ \sin \varphi_2 \sin \theta & \cos \varphi_2 \sin \theta & \cos \theta \end{Vmatrix}. \quad (1.5.21)$$

Exercise

Prove that the mapping

$$f : \begin{Vmatrix} \bar{b} & -\bar{a} \\ a & b \end{Vmatrix} \to \begin{Vmatrix} \frac{1}{2}(b^2 - a^2 + \bar{b}^2 - \bar{a}^2) & \frac{i}{2}(b^2 + a^2 - \bar{b}^2 - \bar{a}^2) & ab + \bar{a}\bar{b} \\ \frac{i}{2}(-\bar{b}^2 + \bar{a}^2 + b^2 - a^2) & \frac{1}{2}(b^2 + a^2 + \bar{b}^2 + \bar{a}^2) & i(ab - \bar{a}\bar{b}) \\ -a\bar{b} - \bar{a}b & i(\bar{a}b - a\bar{b}) & b\bar{b} - a\bar{a} \end{Vmatrix}$$

is an open continuous homomorphism of the group $SU(2)$ onto the group $SO(3, \mathbf{R})$ with kernel $\left\{ \begin{Vmatrix} 1 & 0 \\ 0 & 1 \end{Vmatrix}, \begin{Vmatrix} -1 & 0 \\ 0 & -1 \end{Vmatrix} \right\}$.

Hint. Prove first that f maps the matrices

$$\begin{Vmatrix} e^{-(i\varphi)/2} & 0 \\ 0 & e^{i\varphi/2} \end{Vmatrix} \quad \text{and} \quad \begin{Vmatrix} \cos \dfrac{\theta}{2} & i \sin \dfrac{\theta}{2} \\ i \sin \dfrac{\theta}{2} & \cos \dfrac{\theta}{2} \end{Vmatrix}$$

into g_φ and g_θ respectively.

§2. Representations of Compact Groups

2.1. The Invariant Mean on a Compact Group

Let G be a compact topological group. On G there exists a two-sided in-variant measure μ and this measure can be chosen so that $\mu(G) = 1$ (see Šilov [1]). An invariant mean on G is defined in the following way. Let f belong to $L^1(G)$ (that is, f is integrable on G with respect to the measure μ). The number

$$M(f) = \int_G f(g)\, d\mu(g)$$

is called the *invariant mean* of the function f on G. Sometimes we write $M_g(f(g))$ instead of $M(f)$.

I. *An invariant mean $M(f)$ has the following properties:*

(1) $M(1) = 1$, *where 1 on the left is the function $f \equiv 1$ on G, and 1 on the right is the real number 1;*

(2) $M(\bar{f}) = \overline{M(f)}$;

(3) $M(f) \geqslant 0$ *for $f \geqslant 0$ and $M(f) > 0$ if f is continuous, $f \geqslant 0$ and $f \not\equiv 0$;*

(3') $|M(f_1)| \leqslant M(|f_1|) \leqslant M(|f_2|)$ *for $|f_1| \leqslant |f_2|$;*

(4) $M(f_1 + f_2) = M(f_1) + M(f_2)$;

(5) $M(\alpha f) = \alpha M(f)$, *for all complex numbers α;*

(6) $M(f_h) = M(f)$ *and* $M(f^h) = M(f)$, *where* $f_h(g) = f(gh)$, $f^h(g) = f(hg)$;

(7) $M_g(f(g^{-1})) = M_g(f(g))$.

Properties (1), (2), (4) and (5) follow directly from properties of all inte-grals. Properties (3) and (3') follow from the properties of an integral and the fact that the measure of a nonvoid open set is positive. Properties (6) and (7) are connected with the fact that the measure μ is invariant. $\square$

Exercise

Show that property (3') follows from properties (3)–(5).

The concept of an invariant mean can be extended to vector and operator functions on the group G (see Naĭmark [1]). For such means, propositions analogous to II and III of 1.1, chapter I are valid.

2.2. Complete Reducibility of Representations of a Compact Group

Theorem 1. *Any continuous representation $g \to T(g)$ of a compact group G in a pre-Hilbert space is equivalent to a continuous unitary representation.*

Proof. Let H be a pre-Hilbert space with respect to a scalar product $(x, y)_1$ and let $g \to T(g)$ be a continuous representation of the group G in H. We define a form (x, y) in H by setting

$$f(g) = (T(g)x, T(g)y)_1, \tag{2.2.1}$$

$$(x, y) = M(f) = M_g((T(g)x, T(g)y)_1). \tag{2.2.2}$$

The form (x, y) is a scalar product in H. It is bilinear by virtue of properties (1), (4) and (5) of the invariant mean $M(f)$, the linearity of the operator $T(g)$ and the bilinearity of the form $(x, y)_1$. Since $(x, y)_1$ is Hermitian, property (2) of I gives

$$(y, x) = M((T(g)y, T(g)x)_1) = M(\overline{(T(g)x, T(g)y)_1}) = \overline{(x, y)}.$$

That is, (x, y) is also Hermitian. Finally we have $(T(g)x, T(g)x)_1 \geqslant 0$, since $(x, y)_1$ is a scalar product. From this and from property (3) in I, it follows that

$$(x, y) = M((T(g)x, T(g)x)_1) \geqslant 0.$$

Since $g \to T(g)$ is a continuous representation, the function $f(g)$ defined by formula (2.2.1) is continuous. Therefore (3) in I implies that $(x, x) = 0$ if and only if $(T(g)x, T(g)x)_1 = 0$ for all $g \in G$. Setting $g = e$ we obtain $(x, x)_1 = 0$ and thus $x = 0$. So we have $(x, x) \geqslant 0$ and $(x, x) = 0$ only for $x = 0$. Thus (x, y) is a genuine scalar product in H. By (6) in I, (2.2.1) and (2.2.2), we have

$$
\begin{aligned}
(T(h)x, T(h)y) &= M((T(g)T(h)x, T(g)T(h)y)_1) \\
&= M((T(gh)x, T(gh)y)_1) = M(f_h) = M(f) \\
&= M((T(g)x, T(g)y)_1) = (x, y),
\end{aligned}
$$

for all h in G. That is, T is unitary with respect to the scalar product (x, y). The identity mapping $x \to x$ of the space H with its scalar product $(x, y)_1$ onto itself with the scalar product (x, y) carries $T(g)$ into $T(g)$, which is unitary under (x, y).

It remains to prove that the representation $g \to T(g)$ is continuous with respect to (x, y). For a given $x \in H$, we write

$$\varphi(g, h) = \|T(hg)x - T(h_0 g)x\|_1, \qquad g, h, h_0 \in G, \tag{2.2.3}$$

where $\|x\|_1 = \sqrt{(x, x)_1}$. Since the representation $g \to T(g)$ is continuous, $\varphi(g, h)$ is a continuous function on $G \times G$ and

$$\varphi(g, h_0) = 0. \tag{2.2.4}$$

We also set

$$V_1 = \{\lambda : \lambda \in \mathbf{C}^1, |\lambda| < \varepsilon\}, \tag{2.2.5}$$
$$V_2 = \{h : h \in G, \varphi(g, h) \in V_1 \quad \text{for all } g \in G\}. \tag{2.2.6}$$

Since V_1 is a neighborhood of zero in $\mathbf{C}^1$, V_2 is an open set in G containing h_0 (see XIV in 1.1 and (2.2.4)). From (2.2.6) and (2.2.3) we infer that

$$\|T(hg)x - T(h_0 g)x\|_1 < \varepsilon, \quad \text{for } h \in V_2 \text{ and all } g \in G. \tag{2.2.7}$$

Furthermore, $\|T(g)y\|_1$ is a continuous function of g on the compact group G^{39} and thus we have

$$\|T(g)y\|_1 \leqslant C(y), \quad \text{for all } g \in G, \qquad (2.2.8)$$

where $C(y)$ is a function of y. Combining (2.2.7) and (2.2.8) we conclude that for $g \in G$, $h \in V_2$

$$
\begin{aligned}
&|(T(h)T(g)x, T(g)y)_1 - (T(h_0)T(g)x, T(g)y)_1| \\
&\quad = |(T(hg)x - T(h_0g)x, T(g)y)_1| \\
&\quad \leqslant \|T(hg)x - T(h_0g)x\|_1 \, \|T(g)y\|_1 \leqslant \varepsilon C(y).
\end{aligned}
$$

For $h \in V_2$ we get

$$
\begin{aligned}
|(T(h)x, y) - (T(h_0)x, y)| &= |M_g\{(T(h)T(g)x, T(g)y) - (T(h_0)T(g)x, T(g)y)\}| \\
&\leqslant M(\varepsilon C(y)) = \varepsilon C(y). \qquad (2.2.9)
\end{aligned}
$$

The inequality (2.2.9) proves that the function $(T(h)x, y)$ is continuous on G. $\quad\square$

According to theorem 1, every finite-dimensional continuous representation of the compact group G can be considered unitary.

I. *If H is a Hilbert space, the scalar product (x, y) in H is equivalent to the original scalar product $(x, y)_1$, that is, there exists a number $c > 0$ such that $(x, x) \leqslant c(x, x)_1$, $(x, x)_1 \leqslant c(x, x)$ for all $x \in H$.*

Proof. The inequalities (2.2.8) show that $\|T(g)x\|_1^2 = (T(g)x, T(g)x)_1 \leqslant C(x)^2$. By the Banach-Steinhaus theorem (see, for example, Šilov [1]), there is a constant c independent of x such that $(T(g)x, T(g)x)_1 = \|T(g)x\|_1^2 \leqslant c^2\|x\|_1^2$ for all $x \in H$. Therefore we have

$$\|x\|^2 = (x, x) = M((T(g)x, T(g)x)_1 \leqslant M(c^2(x, x)_1) = c^2\|x\|_1^2$$

so that $\|x\| \leqslant c\|x\|_1$. Conversely, we have $\|x\|_1 = \|T(g^{-1})T(g)x\|_1 \leqslant c\|T(g)x\|_1$ and therefore

$$M(\|x\|_1^2) \leqslant M(c^2\|T(g)x\|_1^2) = c^2\|x\|^2.$$

Hence we have $\|x\|_1^2 \leqslant c^2\|x\|^2$, which is to say that $\|x\|_1 \leqslant c\|x\|$. $\quad\square$

Proposition I shows again that in the case of a Hilbert space H, the representation $g \to T(g)$ is continuous under (x, y).

---

[39] We have $\|T(g)y\|_1 = \sqrt{(T(g)y, T(g)y)_1}$ and $(T(g)y, T(g)y)_1$ is continuous since the function $T(g)$ is continuous (by the definition of a representation) and since the scalar product is continuous in both variables.

Remember that any finite-dimensional representation, equivalent to a unitary representation, is completely reducible. (See II in 2.8, chapter I.) Thus we have the following.

Theorem 2. *Every finite-dimensional continuous representation of a compact group is completely reducible.*

2.3. The Space $L^2(G)$; the Regular Representation

Consider the set of all complex-valued functions f on a compact group G such that:

(1) f is measurable under the invariant measure on G;
(2) $\int_G |f(g)|^2 \, dg < +\infty$.[40] This set of functions is called $L^2(G)$.

We identify two functions that differ only on a set of measure zero. (For more details, see Šilov [1].) The operations of addition and multiplication are defined in $L^2(G)$ in the usual way. Thus $L^2(G)$ is a linear space. We set

$$(f_1, f_2) = M(f_1 \bar{f_2}) = \int_G f_1(g)\overline{f_2(g)} \, dg. \tag{2.3.1}$$

The form (f_1, f_2) is Hermitian and positive definite on $L^2(G)$. Under this scalar product, $L^2(G)$ is a Euclidean space. Also, $L^2(G)$ is complete with respect to this scalar product, which to say that it is a Hilbert space (see Šilov [1]). It is easy to see that the space $L^2(G)$ is finite-dimensional if and only if G is finite.

Editor's note. For $G = \{g_1, g_2, \ldots, g_m\}$, $L^2(G)$ is all complex-valued functions on G, $M(f) = 1/m \sum_{g \in G} f(g)$, and $L^2(G)$ has dimension m. For infinite G, let $\{g_0, g_1, g_2, g_3, \ldots, g_n, \ldots\}$ be a subset of G such that $g_j \neq g_k$ for $j \neq k$. By Uryson's lemma, there is a function f_n in $C(G)$ such that $f_n(g_n) = 1$ and $f_n(g_0) = f_n(g_1) = \cdots = f_n(g_{n-1}) = 0$ (n runs through all positive integers). If a linear combination

$$\sum_{j=1}^{m} \alpha_j f_j = \varphi$$

vanishes identically, we find $\alpha_1 = \varphi(g_1) = 0$, $\alpha_2 = \varphi(g_2) = 0$, and so on. Thus there are an infinite number of linearly independent functions in $C(G) \subset L^2(G)$, and so $L^2(G)$ is infinite-dimensional.

We now define the operators $T(h)$ for $h \in G$. We write

$$T(h)f = f_h, \quad \text{that is, } T(h)f(g) = f(gh). \tag{2.3.2}$$

---

[40] In the sequel let us agree to write dg instead of $d\mu(g)$.

It is clear that the functions f_h are measurable if f is measurable. Since the measure is invariant, it follows that

$$\int_G |f_h(g)|^2 \, dg = \int_G |f(gh)|^2 \, dg = \int_G |f(g)|^2 \, dg < +\infty \qquad (2.3.3)$$

That is, f_h belongs to $L^2(G)$ if $f \in L^2(G)$ and $h \in G$. In addition, the operator $T(h)$ is clearly linear. The mapping $h \to T(h)$ is a representation of the group G. For h_1, h_2 G we have

$$\begin{aligned}
T(h_1)T(h_2)f(g) &= T(h_1)(T(h_2)f(g)) = T(h_1)(f(gh_2)) \\
&= f((gh_1)h_2) = f(g(h_1h_2)) = T(h_1h_2)f(g),
\end{aligned}$$

and $T(e)f(g) = f(ge) = f(g)$. This representation is unitary. Indeed, for $f_1, f_2 \in L^2(G)$ the function $f_1 \bar{f_2}$ is integrable and

$$(T(h)f_1, T(h)f_2) = M((f_1)_h(\overline{f_2})_h) = M(f_h) = M(f) = M(f_1\bar{f_2}) = (f_1, f_2).$$

Finally, the representation $T : h \to T(h)$ is continuous. That is, for every $f \in L^2(G)$, the function $h \to T(h)f$ is a continuous mapping of G into $L^2(G)$. If φ is continuous on G, then φ is left uniformly continuous on G. That is, for every $\varepsilon > 0$, there exists a neighborhood of the identity element e of g such that for $g_1 \in g_2 U$ the relation $|\varphi(g_1) - \varphi(g_2)| < \varepsilon$ is satisfied. If h belongs to $h_0 U$, then gh belongs to $gh_0 U$ and therefore

$$|T(h)\varphi(g) - T(h_0)\varphi(g)| = |\varphi(gh) - \varphi(gh_0)| < \varepsilon$$

for $h \in h_0 U$ and all $g \in G$. Thus we have

$$\begin{aligned}
\|T(h)\varphi - T(h_0)\varphi\|^2 &= (T(h)\varphi - T(h_0)\varphi, T(h)\varphi - T(h_0)\varphi) \\
&= M(|T(h)\varphi - T(h_0)\varphi|^2 \leqslant \varepsilon^2
\end{aligned}$$

for $h \in h_0 U$. Let f belong to $L^2(G)$. Since the set $C(G)$ of all complex-valued continuous functions on G is dense in $L^2(G)$, there exists a $\varphi \in C(G)$ such that $\|f - \varphi\| < \varepsilon/3$. Since the representation T is unitary, it follows that $\|T(h)f - T(h)\varphi\| < \varepsilon/3$ and $\|T(h_0)f - T(h_0)\varphi\| < \varepsilon/3$ for all $h, h_0 \in G$. Using the continuity of φ, we find a neighborhood U of the element $e \in G$ such that, for $h \in h_0 U$, the inequality $\|T(h)\varphi - T(h_0)\varphi\| \leqslant \varepsilon/3$ is satisfied. Thus we have

$$\begin{aligned}
\|T(h)f &- T(h_0)f\| \\
&\leqslant \|T(h)f - T(h)\varphi\| + \|T(h)\varphi - T(h_0)\varphi\| + \|T(h_0)\varphi - T(h_0)f\| < \varepsilon
\end{aligned}$$

for $h \in h_0 U$, which proves that the representation T is continuous.

The continuous unitary representation T of G in $L^2(G)$ is called the *right regular representation of the group G*. The *left regular representation S* of

G in $L^2(G)$ is defined analogously by the formula

$$S(h)f(g) = f(h^{-1}g). \tag{2.3.4}$$

The representation S is unitary. This follows from the *left* invariance of the mean $M(f)$ on G (see (6) of Proposition I in 2.1).

I. *The left and right regular representations of G are unitarily equivalent.*

Proof. For every function $f \in L^2(G)$ we consider the function f' defined by

$$f'(g) = f(g^{-1}). \tag{2.3.5}$$

The operator W in $L^2(G)$ is defined by $W(f) = f'$. Plainly W is linear and maps $L^2(G)$ onto itself. For $f, f_1 \in L^2(G)$, property (7) of I in 2.1 shows that

$$(Wf, Wf_1) = (f', f'_1) = M(f(g^{-1})\overline{f_1(g^{-1})}) = M(f(g)\overline{f_1(g)}) = (f, f_1).$$

That is, W is unitary. Finally, for $h \in G$ and $f \in L^2(G)$, we have

$$\begin{aligned}(W\,T(h)f)(g) &= (T(h)f)(g^{-1}) = f(g^{-1}h) = f((h^{-1}g)^{-1})\\ &= (Wf)(h^{-1}g) = (S(h)Wf)(g).\end{aligned}$$

Thus we have $WT(h) = S(h)W$ for all $h \in G$. That is, W carries T into S. $\quad\square$

2.4. The Orthogonality Relations

Let T be a finite-dimensional continuous representation of the compact group G in a space H. Let $e_1, \ldots, e_n$ be an orthonormal basis in H. The condition that T be unitary is equivalent to the condition that the matrix $t(g)$ of the operator $T(g)$ in the basis $e_1, \ldots, e_n$ be a unitary matrix for all $g \in G$ (see III in 2.8, chapter I). The representation T is continuous if and only if all of the entries $t_{jk}(g)$ are continuous functions of g, where $(t_{jk}(g))$ is the matrix of $T(g)$ in the basis $e_1, e_2, \ldots, e_n$.

Consider a set of pairwise inequivalent finite-dimensional irreducible continuous unitary representations T^α, $\alpha \in A$, of the group G. The set T^α, $\alpha \in A$, is called *complete* if every finite-dimensional irreducible continuous representation of G is equivalent to one of the representations T^α. In the sequel, we will suppose the set T^α, $\alpha \in A$, to be complete.

Let $t^\alpha_{jk}(g)$, $j, k = 1, \ldots, \dim T^\alpha$ be matrix elements of the representation T^α in some orthonormal basis in the space of the representation T^α. Let $n_\alpha = \dim T^\alpha$ be the dimension of the representation T^α.

Theorem 1. *Let T^α and $T^{\alpha'}$ be continuous finite-dimensional irreducible unitary representations of a compact group G. Their matrix elements satisfy the* *following relations*:

$$\int_G t_{ij}^\alpha(g)\,\overline{t_{i'j'}^{\alpha'}(g)}\,dg = \begin{cases} 0 & \text{for } \alpha' \neq \alpha, \quad \text{or for } \alpha' = \alpha, \\ & \text{and } l' \neq l, \quad \text{or } j' \neq j; \\ 1/n_\alpha & \text{if } \alpha' = \alpha, \quad l' = l, \quad j' = j. \end{cases} \tag{2.4.1}$$

Remark. The relations (2.4.1) are called *orthogonality relations for a compact group*. They generalize the orthogonality relations (1.4.1) and (1.4.2), chapter II, for a finite group.

The proof of theorem 1 is a word-for-word repetition of the proof of theorem 1 in 1.4, chapter II.

Theorem 1 implies that the functions

$$e_{ij}^\alpha(g) = \sqrt{n_\alpha}\,t_{ij}^\alpha(g) \qquad (\alpha \in A,\, l,j = 1,\ldots,n_\alpha) \tag{2.4.2}$$

form an orthonormal system in $L^2(G)$.

Recall that an orthonormalized system $\{e_n\}$ in a Hilbert space H is called *complete* if the set of finite linear combinations of vectors of the system is dense in H (see Kolmogorov and Fomin [1] and Šilov [1]).

Theorem 2. *The orthonormal system $\{e_{ij}^\alpha\}$ is complete in $L^2(G)$.*

Proof. It is sufficient to verify that the set of finite linear combinations of the elements of the system $\{e_{ij}^\alpha\}$ is dense in $L^2(G)$ under the L^2-norm in $L^2(G)$. Explicitly, we must prove that for every function $f \in L^2(G)$ and every $\varepsilon > 0$, there exists a finite linear combination $\sum_{\alpha,j,l} \gamma_{j,l}^\alpha t_{jl}^\alpha(g)$ of matrix elements of continuous irreducible unitary representations of the group G for which

$$\left\| f(g) - \sum_{\alpha,j,l} \gamma_{j,l}^\alpha t_{jl}^\alpha(g) \right\|_{L^2(G)} < \varepsilon.$$

We will break up the proof of this into steps.

(a) Let $\chi(g)$ be a real-valued continuous functions on G, not identically equal to zero, for which $\chi(g^{-1}) = \chi(g)$.[41] Consider the function K on $G \times G$ defined by the function $K(g_1, g_2) = \chi(g_1 g_2^{-1})$. Plainly K is a continuous symmetric function on $G \times G$ ($K(g_2, g_1) = \chi(g_2 g_1^{-1}) = \chi((g_1 g_2^{-1})^{-1}) = \chi(g_1 g_2^{-1}) = K(g_1, g_2)$). Consider the integral equation

$$\varphi(g) = \lambda \int_G K(g, g_1)\varphi(g_1)\,dg_1. \tag{2.4.3}$$

---

[41] For example, if $\psi(g) \neq 0$ is a nonnegative continuous function, the function $\chi(g) = \psi(g) + \psi(g^{-1})$ is a χ as required.

(This equation is a generalization of Fredholm's integral equation for functions on a closed interval. The theory of integral equations of the form (2.4.3) is very like the theory of integral equations on a closed interval. See for example Kolmogorov and Fomin [1], chapter IX.)

The function $K(g, g_1)$ is continuous. Hence for all functions $\varphi \in L^2(G)$, the function $\psi(g)$, defined by

$$\psi(g) = \int_G K(g, g_1)\varphi(g_1)\,dg_1,$$

is continuous. In particular, any solution φ in $L^2(G)$ of the integral equation (2.4.3) is continuous. Since K is continuous, the integral

$$\iint\limits_{G \times G} |K(g, g_1)|^2\,dg\,dg_1$$

is finite. Thus we can use the Hilbert-Schmidt theory of integral equations with symmetric quadratically integrable kernels. We recall that a real number λ is called an *eigenvalue* and a function $\varphi \neq 0$ is called an *eigenfunction* of the integral equation (2.4.3) if λ and φ satisfy the relation (2.4.3). According to the Hilbert-Schmidt theory, any integral equation with a symmetric quadratically integrable kernel has at least one eigenvalue λ. Let M_λ be the subspace in $L^2(G)$ consisting of the function 0 and all eigenfunctions of the equation (2.4.3) with corresponding eigenvalue λ. It is known that M_λ is a finite-dimensional subspace. It consists of continuous functions on G.

I. *The subspace M_λ is invariant under right translations. That is, if $\varphi(g) \in M_\lambda$, then $\varphi(gg_0) \in M_\lambda$.*

Let $\varphi(g)$ belong to M_λ; that is,

$$\varphi(g) = \lambda \int_G K(g, g_1)\varphi(g_1)\,dg_1 = \lambda \int_G \chi(gg_1^{-1})\varphi(g_1)\,dg_1.$$

We then have

$$\varphi(gg_0) = \lambda \int_G \chi((gg_0)g_1^{-1})\varphi(g_1)\,dg_1 = \lambda \int_G \chi(gg_0g_1^{-1})\varphi(g_1)\,dg_1.$$

We set $h = g_1 g_0^{-1}$, obtaining $g_1 = hg_0$. Because the measure dg_1 is invariant, we also have

$$\varphi(gg_0) = \lambda \int_G \chi(g^{-1}h)\varphi(hg_0)\,dh,$$

that is, $\varphi(gg_0) \in M_\lambda$.

Since the finite-dimensional space $M_\lambda \subset L^2(G)$ is invariant under right translation on G, M_λ is the representation space of a subrepresentation of the right regular representation of G. The representation T_λ of G in M_λ thus obtained is continuous, finite-dimensional, and unitary. Therefore it is the direct sum of a finite number of continuous irreducible unitary representations $T_\lambda^{(1)}, \ldots, T_\lambda^{(p)}$ of G (see IX in 2.9, chapter I). If $M_\lambda^{(1)}, \ldots, M_\lambda^{(p)}$ are the spaces of these irreducible representations, we have

$$M_\lambda = M_\lambda^{(1)} + \cdots + M_\lambda^{(p)}. \tag{2.4.4}$$

II. *All functions in the space $M_\lambda^{(k)}$ of the representation $T_\lambda^{(k)}$ defined above are linear combinations of the matrix elements of this representation.*

Let $e_1(g), \ldots, e_n(g)$ be a basis in the space $M_\lambda^{(k)}$. Since $M_\lambda^{(k)}$ is contained in M_λ, the functions $e_1(g), \ldots, e_n(g)$ are continuous. The representation $T_\lambda^{(k)}$ is a subrepresentation of the right regular representation. Therefore we have $T_\lambda^{(k)}(g_0)f(g) = f(gg_0)$ for all $f \in M_\lambda^{(k)}$. Let $c_{ij}(g_0)$ be the matrix elements of the operator $T_\lambda^{(k)}(g_0)$ in the basis $e_1(g), \ldots, e_n(g)$. We then have

$$e_1(gg_0) = c_{11}(g_0)e_1(g) + \cdots + c_{n1}(g_0)e_n(g),$$
$$\vdots \qquad \vdots \qquad \qquad \vdots$$
$$e_n(gg_0) = c_{1n}(g_0)e_1(g) + \cdots + c_{nn}(g_0)e_n(g)$$

for all $g, g_0 \in G$. Setting $g = e$, $g_0 = h$, we find that

$$e_1(h) = e_1(e)c_{11}(h) + \cdots + e_n(e)c_{n1}(h),$$
$$\vdots \qquad \vdots \qquad \qquad \vdots$$
$$e_n(h) = e_1(e)c_{1n}(h) + \cdots + e_n(e)c_{nn}(h),$$

That is, all elements of a basis in $M_\lambda^{(k)}$ and therefore all elements of $M_\lambda^{(k)}$ are linear combinations of matrix elements of the irreducible continuous unitary representation $T_\lambda^{(k)}$. This representation is equivalent to a representation T^α for some α in A. Therefore we can choose a basis $e_1(g), \ldots, e_n(g)$ such that the functions $c_{jk}(g)$ coincide with $t_{jk}^\alpha(g)$. □

We summarize.

III. *All functions in the space M_λ are finite linear combinations of matrix elements of irreducible unitary representations of the group G.*

Consider any function on G of the form

$$\psi(g) = \int_G K(g, g_1)\varphi(g_1)\,dg_1, \qquad (\varphi \in L^2(G)). \tag{2.4.5}$$

The Hilbert-Schmidt theory shows that ψ is the sum of an absolutely and uniformly convergent series of eigenfunctions of the integral equation (2.4.3). Combining this fact with II and III we have the following.

IV. *Let ψ be a function on G of the form* (2.4.5). *Then ψ can be arbitrarily and uniformly approximated by finite linear combinations of matrix elements of irreducible continuous finite-dimensional unitary representations of G.*

Theorem 3. *Every continuous function on G is arbitrarily uniformly approximable by functions of the form* (2.4.5).

Proof. Let f be a continuous function on G. Since f is uniformly continuous on G, there exists a neighborhood U of the identity element e of G such that $|f(g) - f(g')| < \varepsilon$ for all g, g' in G for which $g' \in Ug$. Replacing U by $U \cap U^{-1}$, we may suppose that $U = U^{-1}$. We select a neighborhood V of the element e such that $\bar{V} \subset U$. Let ψ be a nonnegative continuous function on G such that $\psi = 1$ on V and $\psi = 0$ outside of U. (Use Uryson's lemma; see II in 1.3.) Writing $\chi(g) = c(\psi(g) + \psi(g^{-1}))$, we have $\chi(g^{-1}) = \chi(g)$. Choose the number c so that $\int \chi(g)\, dg = 1$. Since U is symmetric and $\psi(g) = 0$ outside of U, we also have $\psi(g^{-1}) = 0$ outside of U. Define the function φ by

$$\varphi(g) = \int_G f(g_1)\chi(gg_1^{-1})\, dg_1;$$

Setting $h = g_1 g^{-1}$ and using the invariance of Haar measure, we obtain

$$\varphi(g) = \int_G f(hg)\chi(h^{-1})\, dh.$$

Because $\int_G \chi(g)\, dg = 1$, we also have $\int_G \chi(h^{-1})\, dh = 1$, so that $f(g) = \int_G f(g)\chi(h^{-1})\, dh$ and

$$|f(g) - \varphi(g)| = \left| \int_G f(g)\chi(h^{-1})\, dh - \int_G f(hg)\chi(h^{-1})\, dh \right|$$

$$= \left| \int_G (f(g) - f(hg))\chi(h^{-1})\, dh \right| \leqslant \int_G |f(g) - f(hg)|\chi(h^{-1})\, dh$$

$$= \int_U |f(g) - f(hg)|\chi(h^{-1})\, dh \tag{2.4.6}$$

(recall that $\chi(h^{-1}) = \chi(h) = 0$ outside of U). By the construction of the neighborhood U, we have $|f(g) - f(hg)| < \varepsilon$ for all $g \in G$ and $h \in U$, since $hg \in Ug$ for $h \in U$. It follows that $\int_U |f(g) - f(hg)|\chi(h^{-1})\, dh \leqslant \varepsilon \int_U \chi(h^{-1})\, dh = \varepsilon$ and (2.4.6) implies that $|f(g) - \varphi(g)| \leqslant \varepsilon$ for all $g \in G$. $\square$

Theorem 4. *Every continuous function on a compact topological group G is arbitrarily uniformly approximable on G by finite linear combinations of* *matrix elements of continuous irreducible finite-dimensional unitary representations of the group G.*

For the proof, combine theorem 3 and Proposition IV. □

(b) We now complete the proof of theorem 2. Let f belong to $L^2(G)$. The set of continuous functions on G is dense in $L^2(G)$. That is, for every $\varepsilon > 0$ there exists a continuous function $\tilde{f}$ on G such that $\|f - \tilde{f}\|_{L^2(G)} < \varepsilon/2$. By theorem 3 we can find a finite linear combination φ of matrix elements of continuous irreducible finite-dimensional unitary representations of the group G such that $|\tilde{f}(g) - \varphi(g)| < \varepsilon/2$ for all $g \in G$. Therefore we have $\|f - \varphi\|_{L^2(G)} = (\int_G |\tilde{f}(g) - \varphi(g)|^2 \, dg)^{1/2} < (\frac{1}{4}\varepsilon^2)^{1/2} = \varepsilon/2$ and hence

$$\|f - \varphi\|_{L^2(G)} \leqslant \|f - \tilde{f}\|_{L^2} + \|\tilde{f} - \varphi\|_{L^2} < \varepsilon/2 + \varepsilon/2 = \varepsilon.$$

That is, finite linear combinations of functions from the system $\{e^\alpha_{lj}\}$ are dense in $L^2(G)$. Therefore $\{e^\alpha_{lj}\}$ is a complete orthonormal system and theorem 2 is proved. □

Theorem 5. *Every function $f \in L^2(G)$ is representable in the form*

$$f = \sum_{\alpha \in A} \sum_{l,j=1}^{n_\alpha} (f, e^\alpha_{lj})e^\alpha_{lj} = \sum_{\alpha \in A} \sum_{l,j=1}^{n_\alpha} n_\alpha(f, t^\alpha_{lj})t^\alpha_{lj}, \tag{2.4.7}$$

where the series converges in $L^2(G)$. Also the following identity holds:

$$(f, f) = \sum_{\alpha \in A} \sum_{l,j=1}^{n_\alpha} |(f, e^\alpha_{lj})|^2 = \sum_{\alpha \in A} \sum_{l,j=1}^{n_\alpha} n_\alpha |(f, t^\alpha_{lj})|^2. \tag{2.4.8}$$

Theorem 5 follows directly from theorem 2 and from the general theory of orthogonal expansions in a Hilbert space. (See for example Kolmogorov and Fomin [1], §4, chapter III.) □

The equality (2.4.8) is called *Plancherel's formula*.

2.5. Decomposition of the Regular Representation

Theorem. *Let T be the right regular representation of the group G in the Hilbert space $L^2(G)$ and let $\{T^\alpha : \alpha \in A\}$ be the set of all pairwise inequivalent finite-dimensional irreducible unitary representations of G. The set of matrix elements of the representation T^α in a fixed but arbitrary orthonormal basis $\{e_k\}$ is denoted by t^α_{jk} $(j, k = 1, 2, \ldots, n_\alpha; n_\alpha = \dim T^\alpha)$. Let H^α denote the subspace of the space $L^2(G)$ generated by the functions $t^\alpha_{jk}(g)$ $(j, k = 1, 2, \ldots, n_\alpha)$. The Hilbert space $L^2(G)$ is the Hilbert space direct sum of the finite-dimensional subspaces H^α $(\alpha \in A)$. Each subspace H^α is invariant under the right* *regular representation T and the restriction $\dot{T}^\alpha$ of T to the subspace H^α is a multiple of the representation T^α with multiplicity equal to n_α.*

Proof. The definition of representation implies that

$$t_{ij}^\alpha(gg_0) = \sum_{k=1}^{n_\alpha} t_{ik}^\alpha(g)t_{kj}^\alpha(g_0),$$ (2.5.1)

which may also be written as

$$[T(g_0)t_{ij}^\alpha](g) = \sum_{k=1}^{n_\alpha} t_{ik}^\alpha(g)t_{kj}^\alpha(g_0).$$ (2.5.2)

For each fixed α, the functions $t_{jk}^\alpha(g)$ are a basis for the linear subspace H^α. The relations (2.5.2) show that each of them goes into a linear combination of all of them under the operators of the right regular representation. Hence every element of H^α is carried into an element of H^α by all operators of the right regular representation. The restriction of the representation T to the subspace H^α is denoted by $\dot{T}^\alpha$. We can rewrite (2.5.2) as

$$[\dot{T}^\alpha(g_0)t_{ij}^\alpha](g) = \sum_{k=1}^{n_\alpha} t_{kj}^\alpha(g_0)t_{ik}^\alpha(g).$$ (2.5.3)

The subspace generated by the functions of the form $t_{ij}^\alpha(g)$ ($j = 1, \ldots,$ dim T^α) for a fixed i is written as H_i^α, $i = 1, \ldots, n_\alpha$. The subspace H^α is an orthogonal direct sum of the subspaces H_i^α, $i = 1, \ldots, n_\alpha$. It follows from (2.5.3) that each subspace H_i^α is invariant under all the operators of the representation $\dot{T}^\alpha$. Let $\dot{T}_i^\alpha$ be the restriction of $\dot{T}^\alpha$ to the subspace H_i^α. Formula (2.5.3) shows that the matrix elements of $\dot{T}_i^\alpha$ in the basis $\{t_{ij}^\alpha(g), j = 1, \ldots, n_\alpha\}$ coincide with the matrix elements of T^α in the basis $\{e_k\}$. Therefore every representation $\dot{T}_i^\alpha$ is equivalent to the representation T^α and the theorem is proved. □

2.6. Characters of Irreducible Unitary Representations of Compact Groups

Let T be a continuous finite-dimensional representation of a compact group G. Let $e_1, \ldots, e_n$ ($n = $ dim T) be a basis in the space H of T. The matrix elements of T in the basis $e_1, \ldots, e_n$ are denoted by $t_{ij}(g)$, $i, j = 1, \ldots, n$. The function

$$\chi(g) = t_{11}(g) + \cdots + t_{nn}(g)$$ (2.6.1)

is the character of the representation T. From the general properties of characters of finite-dimensional representations (see IV in 2.9, chapter I) it follows that the (continuous) function χ does not depend on the choice of the basis $e_1, \ldots, e_n$. Also, χ does not change when T is replaced by an equivalent representation. The function χ is constant on conjugacy classes in G. The value of χ in the identity element of the group is equal to the dimension of the representation; and so on.

The orthogonality relations for matrix elements (theorem 1 in 4) give the following.

Theorem 1. *The characters χ^α and $\chi^{\alpha'}$ of irreducible unitary representations T^α and $T^{\alpha'}$ of the compact group G satisfy the relations*

$$\int_G \chi^\alpha(g)\overline{\chi^{\alpha'}(g)}\, dg = \begin{cases} 0, & for \ \alpha' \neq \alpha, \\ 1, & for \ \alpha' = \alpha. \end{cases} \tag{2.6.2}$$

Proof. We have

$$\int_G \chi^\alpha(g)\overline{\chi^{\alpha'}(g)}\, dg = \int_G \sum_{i=1}^{n_\alpha} t_{ii}^\alpha(g) \sum_{j=1}^{n_{\alpha'}} \overline{t_{jj}^{\alpha'}(g)}\, dg$$

$$= \sum_{i=1, j=1}^{n_\alpha, n_{\alpha'}} \int_G t_{ii}^\alpha(g)\overline{t_{jj}^{\alpha'}(g)}\, dg. \tag{2.6.3}$$

The right side of (2.6.3) is determined from formulas (2.4.1), and we verify (2.6.2) at once. $\square$

I. *Let S be a finite-dimensional continuous unitary representation of the group G, with character χ. Let*

$$S = n_1 T^{\alpha_1} \oplus \cdots \oplus n_p T^{\alpha_p} \tag{2.6.4}$$

be a decomposition of S as a direct sum of multiples of irreducible unitary representations $T^{\alpha_1}, \ldots, T^{\alpha_p}$ of G. Let $\chi_1, \ldots, \chi_p$ be the characters of $T^{\alpha_1}, \ldots, T^{\alpha_p}$ respectively. We have

$$\chi = n_1 \chi_1 + \cdots + n_p \chi_p, \tag{2.6.5}$$

and the numbers $n_1, \ldots, n_p$ are defined by the relations

$$n_k = (\chi, \chi_k), \qquad k = 1, \ldots, p. \tag{2.6.6}$$

Conversely, if the equality (2.6.5) holds, then the representation S admits the decomposition (2.6.4).

Proof. Plainly (2.6.5) follows from (2.6.4). If (2.6.5) holds, (2.6.2) shows that

$$(\chi, \chi_k) = n_k(\chi_k, \chi_k) = n_k$$

for all $k = 1, \ldots, p$, thus proving (2.6.6). Suppose now that (2.6.5) holds. Suppose that

$$S = m_1 T^{\alpha'_1} \oplus \cdots \oplus m_k T^{\alpha'_k} \tag{2.6.7}$$

is any decomposition of S as the direct sum of multiples of irreducible representations $T^{\alpha'_1}, \ldots, T^{\alpha'_k}$. Let χ'_i be the character of $T^{\alpha'_i}$ ($i = 1, \ldots, k$). From (2.6.7) we find

$$\chi = m_1 \chi'_1 + \cdots + m_k \chi'_k, \tag{2.6.8}$$

Thus (2.6.5) and (2.6.8) define two expansions of the function χ in the orthonormal system of characters of the complete system $\{T^\alpha\}$ of continuous irreducible finite-dimensional unitary representations of the group G. Two such expansions differ only in the order of the sums. Thus, excluding null summands from (2.6.5) and (2.6.8), we have $k = p$ and in some ordering of $\alpha'_1, \ldots, \alpha'_p$, we have $T^{\alpha'_i} = T^{\alpha_i}$, $m_i = n_i$. That is, (2.6.4) holds. $\square$

II. *The characters of two continuous finite-dimensional unitary representations T^1 and T^2 of the group G coincide if and only if the representations T^1 and T^2 are equivalent.*

Proof. Let χ^1 and χ^2 be the characters of the representations T^1 and T^2 respectively. If $\chi^1 = \chi^2$, the decompositions (2.6.5) for χ^1 and χ^2 coincide and by I the decompositions (2.6.4) for T^1 and T^2 also coincide. That is, T^1 is equivalent to T^2. The converse is obvious. $\square$

III. *Let S be a continuous finite-dimensional unitary representation of the group G. The representation S is irreducible if and only if $\int_G |\chi_S(g)|^2 \, dg = 1$.*

Proof. If S is irreducible, we have $\int_G |\chi_S(g)|^2 \, dg = 1$ by theorem 1. Conversely, let $\int_G |\chi_S(g)|^2 \, dg = 1$. Use the decompositions (2.6.4) and (2.6.5) for χ_S. The orthogonality relations show that

$$\int_G |\chi_S(g)|^2 \, dg = (\chi_S, \chi_S) = n_1^2(\chi_1, \chi_1) + \cdots + n_p^2(\chi_p, \chi_p)$$

$$= n_1^2 + \cdots + n_p^2. \tag{2.6.9}$$

Since $n_1, \ldots n_p$ are natural numbers and the left side of (2.6.9) is equal to 1 by hypothesis, we must have $p = 1$ and $n_1 = 1$. That is, S is equivalent to T^{α_1} and is irreducible. $\square$

Let T be an irreducible continuous unitary representation of the group G and let χ be its character. The function $(\dim T)^{-1}\chi$ is called *the normalized character of the representation T.*

Theorem 2. *Let ψ be a continuous complex function on G. The function ψ is the normalized character of some continuous irreducible unitary representation T of G if and only if $\psi \neq 0$ and*

$$\int_G \psi(ghg^{-1}k)\,dg = \psi(h)\psi(k) \qquad (2.6.10)$$

for all $h, k \in G$.

Proof.

(a) Suppose that $\psi = (\dim T)^{-1}\chi$, where χ is the character of a representation T as above. Let A be the linear operator in the representation space H of T, defined by the formula

$$A = \int_G T(ghg^{-1})\,dg. \qquad (2.6.11)$$

Because T is continuous, the integral in (2.6.11) exists. For all $k \in G$ the following relation is satisfied:

$$\begin{aligned}
T(k)A &= T(k)\int_G T(ghg^{-1})\,dg \\
&= \int_G T(k)T(ghg^{-1})\,dg = \int_G T(kghg^{-1})\,dg. \qquad (2.6.12)
\end{aligned}$$

Setting $g_1 = kg$, we rewrite the last expression in (2.6.12) as

$$\begin{aligned}
\int_G T(g_1hg_1^{-1}k)\,dg_1 &= \int_G T(g_1hg_1^{-1})T(k)\,dg_1 \\
&= \left(\int_G T(g_1hg_1^{-1})\,dg_1\right)T(k) = AT(k). \qquad (2.6.13)
\end{aligned}$$

Combining (2.6.12) and (2.6.13), we obtain

$$T(k)A = AT(k)$$

for all $k \in G$. By hypothesis T is irreducible. The identity (2.6.14) and Schur's lemma show that the operator A is a multiple of the identity operator:

$$A = \lambda \cdot 1_H, \qquad (2.6.15)$$

where λ is a number and 1_H is the identity operator in the space H. Thus we have

$$\operatorname{tr} A = \lambda \cdot \dim H = \lambda \cdot \dim T. \qquad (2.6.16)$$

On the other hand, computing the trace of both sides of (2.6.11) we obtain

$$\operatorname{tr} A = \int_G T(ghg^{-1}) \, dg = \int_G \operatorname{tr} T(ghg^{-1}) \, dg$$
$$= \int_G \chi(ghg^{-1}) \, dg = \int_G \chi(h) \, dg = \chi(h). \qquad (2.6.17)$$

From (2.6.11), (2.6.15) and (2.6.17) we conclude that

$$\int_G T(ghg^{-1}) \, dg = (\dim T)^{-1} \chi(h) 1_H, \qquad (2.6.18)$$

and hence

$$\int_G T(ghg^{-1}k) \, dg = \left(\int_G T(ghg^{-1}) \, dg \right) T(k) = (\dim T)^{-1} \chi(h) T(k). \quad (2.6.19)$$

Computing the trace of both sides of (2.6.19), we obtain

$$\int_G \chi(ghg^{-1}k) \, dg = (\dim T)^{-1} \chi(h) \chi(k) \qquad (2.6.20)$$

for all $h, k \in G$. Dividing both sides of (2.6.20) by $\dim T$ and replacing $(\dim T)^{-1} \chi$ by the function ψ we obtain (2.6.10).

(b) Suppose now that ψ is a nonzero continuous function on G that satisfies (2.6.10). Since the matrix elements of the irreducible unitary representations of G form a complete orthogonal system in $L^2(G)$, there exists an irreducible unitary representation T of G in a finite-dimensional Hilbert space H, such that for some matrix element $t_{ij}(g)$ of T the nonequality

$$(t_{ij}, \bar{\psi}) = \int_G t_{ij}(g) \psi(g) \, dg \neq 0 \qquad (2.6.21)$$

holds. It follows from (2.6.21) that the operator $T(\psi)$, defined by the equality

$$T(\psi) = \int_G \psi(g) T(g) \, dg, \qquad (2.6.22)$$

has a nonzero matrix element and is therefore distinct from zero. Multiplying (2.6.22) by $\psi(h)$, $h \in G$, we obtain the equality

$$\psi(h) T(\psi) = \int_G \psi(h) \psi(k) T(k) \, dk. \qquad (2.6.23)$$

By (2.6.10), formula (2.6.23) can be written as

$$\psi(h) T(\psi) = \int_G \int_G \psi(ghg^{-1}k) T(k) \, dg \, dk. \qquad (2.6.24)$$

Make the change of variable $k_1 = ghg^{-1}k$ in (2.6.24) and change the order of integration to find

$$\psi(h)T(\psi) = \int_G \int_G \psi(k_1)T(gh^{-1}g^{-1}k_1)\,dg\,dk_1$$
$$= \int_G \psi(k_1)\left(\int_G T(gh^{-1}g^{-1})\,dg\right)T(k_1)dk_1. \qquad (2.6.25)$$

Let χ be the character of the representation T. Writing the relation (2.6.18) in (2.6.25), we get

$$\psi(h)T(\psi) = \int_G \psi(k_1)(\dim T)^{-1}\chi(h^{-1})T(k_1)\,dk_1$$
$$= (\dim T)^{-1}\chi(h^{-1})T(\psi). \qquad (2.6.26)$$

Since $T(\psi) \neq 0$ (see (2.6.21)), (2.6.26) and IV (d) in 2.9, chapter I show that

$$\psi(h) = (\dim T)^{-1}\chi(h^{-1}) = (\dim T)^{-1}\overline{\chi(h)} \qquad (2.6.27)$$

for all $h \in G$. Let $\bar{T}$ be a representation of the group G in the Hilbert space H such that in some orthonormal basis the matrix elements of the operator $\bar{T}(g)$ for all $g \in G$ are the complex conjugates of the corresponding elements of the matrix of the operator $T(g)$. The reader can easily verify that $\bar{T}$ is an irreducible continuous unitary representation of G in H. The character of $\bar{T}$ is defined by the formula

$$\chi_{\bar{T}}(g) = \overline{\chi(g)} \qquad (2.6.28)$$

for all $g \in G$. The representation $\bar{T}$, defined uniquely to within equivalence, is called the representation *conjugate to* T. Comparing (2.6.27) and (2.6.28), we obtain

$$\psi(g) = (\dim T)^{-1}\chi_{\bar{T}}(g) = (\dim \bar{T})^{-1}\chi_T(g) \qquad (2.6.29)$$

for all $g \in G$. Thus the proof of theorem 2 is completed. $\qquad \square$

2.7. Decomposition of an Arbitrary Continuous Unitary Representation of the Group G into Irreducible Representations

Let S be a continuous unitary representation of the group G in a Hilbert space H. Let T be a finite-dimensional irreducible continuous unitary representation of G and let χ_T be the character of T. Let $e_1, \ldots, e_n$ ($n = \dim T$) be a basis in the space of the representation T and let $t_{ij}^T(g)$, $i, j = 1, \ldots, n$, be the matrix elements of T in the basis $e_1, \ldots, e_n$. We define

$$E_{ij}^T = (\dim T)\int_G \overline{t_{ij}^T(g)}S(g)\,dg;$$
$$E^T = (\dim T)\int_G \overline{\chi_T(g)}S(g)\,dg. \qquad (2.7.1)$$

By (2.6.28) we can write the second formula in (2.7.1) as

$$E^T = (\dim T) \int_G \chi_{\bar{T}}(g) S(g) \, dg, \qquad (2.7.2)$$

where $\chi_{\bar{T}}$ is the character of the conjugate representation $\bar{T}$ of T. Since $\overline{t_{ij}^T(g)} S(g)$ and $\overline{\chi_T(g)} S(g)$ are continuous operator-valued functions on G, the integrals on the right sides of (2.7.1) and (2.7.2) exist, and E_{ij}^T and E^T are continuous linear operators in H. We list the fundamental properties of E_{ij}^T and E^T.

I. *The operators E_{ij}^T and E^T satisfy the following relations:*

$$E^T = \sum_{i=1}^n E_{ii}^T; \qquad (2.7.3a)$$

$$S(g) E_{jl}^T = \sum_{i=1}^n t_{ij}^T(g) E_{il}^T, \quad \text{for all } g \in G; \qquad (2.7.3b)$$

$$E_{jl}^T S(g) = \sum_{i=1}^n t_{li}^T(g) E_{ji}^T, \quad \text{for all } g \in G; \qquad (2.7.3c)$$

$$E_{ij}^T E_{lk}^{T'} = 0, \qquad (2.7.3d)$$

if T' is an irreducible unitary representation of the group G not equivalent to T. We also have

$$E_{ij}^T E_{lk}^T = \begin{cases} 0, & \text{if } j \neq l, \\ E_{ik}^T, & \text{if } j = l; \end{cases} \qquad (2.7.3e)$$

$$E_{ii}^T E_{jj}^T = \begin{cases} 0, & \text{for } i \neq j, \\ E_{ii}^T, & \text{for } i = j; \end{cases} \qquad (2.7.3f)$$

$$(E_{ij}^T)^* = E_{ji}^T; \qquad (2.7.3g)$$

$$E^T E^{T'} = 0, \qquad (2.7.3h)$$

(again T' is an irreducible unitary representation of G not equivalent to T). Finally, we have

$$E^{T*} = E^T = E^{T^2}; \qquad (2.7.3i)$$

$$E^T S(g) = S(g) E^T, \quad \text{for all } g \in G. \qquad (2.7.3j)$$

Proof. Relation (2.7.3a) follows at once from (2.7.1) and (2.6.1). Let us prove (2.7.3b). From the orthogonality relations, the invariance of the integral, and from familiar properties of integrals of operator-valued functions, we find that

$$S(g)E_{jl}^T = S(g)(\dim T) \int_G \overline{t_{jl}^T(h)} S(h)\, dh = (\dim T) \int_G \overline{t_{jl}^T(h)} S(gh)\, dh$$

$$= (\dim T) \int_G \overline{t_{ji}^T(g^{-1}h)} S(h)\, dh$$

$$= (\dim T) \int_G \sum_{i=1}^n \overline{t_{ji}^T(g^{-1}) t_{il}^T(h)} S(h)\, dh$$

$$= \sum_{i=1}^n \overline{t_{ji}^T(g^{-1})}(\dim T) \int_G \overline{t_{il}^T(h)} S(h)\, dh = \sum_{i=1}^n \overline{t_{ji}^T(g^{-1})} E_{il}^T. \qquad (2.7.4)$$

Formula (2.7.3b) follows from (2.7.4) and the identity $\overline{t_{ji}^T(g^{-1})} = t_{ij}^T(g)$. The equality (2.7.3c) is proved analogously. We now prove (2.7.3d) and (2.7.3e). From (2.7.3b) we obtain

$$E_{ij}^T E_{lk}^{T'} = \left((\dim T) \int_G \overline{t_{ij}^T(g)} S(g)\, dg \right) E_{lk}^{T'}$$

$$= (\dim T) \int_G \overline{t_{ij}^T(g)} (S(g) E_{lk}^{T'})\, dg$$

$$= (\dim T) \int_G \overline{t_{ij}^T(g)} \left(\sum_{p=1}^{\dim T'} t_{pl}^{T'}(g) E_{pk}^{T'} \right) dg$$

$$= (\dim T) \sum_{p=1}^{\dim T'} \left(\int_G t_{pl}^{T'}(g) \overline{t_{ij}^T(g)}\, dg \right) E_{pk}^{T'}. \qquad (2.7.5)$$

The orthogonality relations (2.4.1) imply that if T and T' are inequivalent, then the integral in the last term of (2.7.5) is zero. If T' is equal to T, (2.4.1) shows that for $l \neq j$ all summands in the last term of (2.7.5) are zero. For $l = j$ the only nonzero summand in the right side of (2.7.5) corresponds to $p = i$. This term is equal to $(\dim T)(\dim T)^{-1} E_{ik}^T = E_{ik}^T$. Thus we have

$$E_{ij}^T E_{jk}^T = E_{ik}^T \qquad (2.7.6)$$

and the proofs of (2.7.3d) and (2.7.3e) are complete. The relation (2.7.3f) is a special case of (2.7.3e). Since the operators $S(g)$ are unitary we have

$$(E_{ij}^T)^* = \left((\dim T) \int_G \overline{t_{ij}^T(g)} S(g)\, dg \right)^*$$

$$= (\dim T) \int_G t_{ij}^T(g) S^*(g)\, dg = (\dim T) \int_G t_{ij}^T(g) S(g)^{-1}\, dg$$

$$= (\dim T) \int_G t_{ij}^T(g) S(g^{-1})\, dg = (\dim T) \int_G t_{ij}^T(g^{-1}) S(g)\, dg$$

$$= (\dim T) \int_G \overline{t_{ji}^T(g)} S(g)\, dg = E_{ji}^T. \qquad (2.7.7)$$

The above holds because the mapping $T \to \bar{T}$ is continuous and so conjugation may go under the integral sign in (2.7.7). From (2.7.7), (2.7.3g) is evident. We also have

$$(E^T)^* = \left(\sum_{i=1}^{n} E_{ii}^T \right)^* = \sum_{i=1}^{n} (E_{ii}^T)^* = \sum_{i=1}^{n} E_{ii}^T = E^T, \qquad (2.7.8)$$

which proves the first equality in (2.7.3i). The second equality in (2.7.3i), as well as (2.7.3h), are proved similarly. Finally, (2.7.3a) and (2.7.3b) show that

$$S(g)E^T = \sum_{i,j=1}^{n} t_{ij}^T(g)E_{ij}^T, \qquad E^T S(g) = \sum_{i,j=1}^{n} t_{ij}^T(g)E_{ij}^T; \qquad (2.7.9)$$

the equality (2.7.3j) follows directly. $\square$

We now list a number of corollaries of Proposition I.

Let M_i^T be the set of all vectors $x \in H$ such that $E_{ii}^T x = x$, and let M^T be the set of all vectors $x \in H$ such that $E^T x = x$.

II. *The sets M_i^T and M^T are closed linear subspaces of the Hilbert space H. If T' and T are inequivalent, then $M_k^{T'}$ and M_i^T are orthogonal and $M^{T'}$ and M^T are orthogonal. If $i \neq k$, then M_i^T is orthogonal to M_k^T.*

Proof. According to I the operators E_{ii}^T and E^T are projection operators in the Hilbert space H. By definition the sets M_i^T and M^T are the images of H under E_{ii}^T and E^T respectively. Hence M_i^T and M^T are closed linear subspaces of H. Let x belong to M_i^T and y to $M_k^{T'}$, where T is inequivalent to T' or $T = T'$, but $k \neq i$. From (2.7.3d) and (2.7.3e) we infer that

$$(x, y) = (E_{ii}^T x, E_{kk}^{T'} y) = (E_{kk}^{T'} E_{ii}^T x, y) = (0, y) = 0.$$

Therefore M_i^T and $M_k^{T'}$ are orthogonal. In like manner we prove that M^T is orthogonal to $M^{T'}$ if T is inequivalent to T'. $\square$

III. *The space M^T is the orthogonal direct sum of the spaces M_i^T ($i = 1, \ldots,$ dim T).*

Proof. For a vector x_j in M_j^T, the relations (2.7.3) imply that

$$E^T x_j = E^T \cdot E_{jj}^T x_j = \left(\sum_{i=1}^{n} E_{ii}^T \right) E_{jj}^T x_j = \sum_{i=1}^{n} (E_{ii}^T E_{jj}^T) x_j = E_{jj}^T x_j = x_j,$$

that is, $x_j \in M^T$. Thus M^T contains all sums $x_1 + \cdots + x_n$ for $x_j \in M_j^T$. Conversely, if x belongs to M^T, then, setting $x_i = E_{ii}^T x$, we see that $x =$

$E^T x = (\sum E^T_{ii})x = \sum x_i$ and $E^T_{ii}x_i = (E^T_{ii})^2 x = E^T_{ii}x = x_i$, that is, $x_i \in M^T_i$. Finally, the subspaces M^T_i are pairwise orthogonal by II. $\square$

IV. *The Hilbert space H is the orthogonal sum of subspaces M^T where T runs through a complete set of pairwise inequivalent irreducible representations*[42] *of the group G.*

Proof. We know that the spaces M^T are pairwise orthogonal. Let H_1 be the closure of the linear subspace of H formed by finite sums of elements from the subspaces M^T. We will show that $H_1 = H$, which will complete the proof of Proposition IV.

Assume that $H_1 \subsetneqq H$. There then exists a nonzero vector $x \in H$ orthogonal to H_1, and in particular the vector x is orthogonal to all subspaces M^T. By (2.7.3e), we have

$$E^T_{ii}E^T_{ik}x = E^T_{ik}x \qquad (2.7.10)$$

for all T and all $i, k = 1, \ldots, \dim T$, and so $E^T_{ik}x$ belongs to M^T_i. Therefore for all T and all i, k the definition of x implies that

$$(x, E^T_{ik}x) = 0. \qquad (2.7.11)$$

We rewrite (2.7.11) as

$$(\dim T) \int_G \overline{t^T_{ik}(g)}(x, S(g)x) \, dg = 0; \qquad (2.7.12)$$

this equality holds for all T and all $i, k = 1, \ldots, \dim T$. According to theorem 2 in 2.4, the functions $t^T_{ik}(g)$ form a complete system in $L^2(G)$. The equalities (2.7.12) show that the continuous function $g \to (x, S(g)x)$ on G is the zero element of the space $L^2(G)$, so that $(x, S(g)x) = 0$ for all $g \in G$. For $g = e$, we find $(x, x) = 0$ and so x is the zero vector, contrary to construction. Thus IV is proved completely. $\square$

A representation S of G in a Hilbert space H is called a *multiple of the irreducible representation T* if the space H is the orthogonal direct sum of closed linear subspaces H^i, $i = 1, \ldots, \dim T$, having the following property: for every H^i and every $x \in H^i$ there exist vectors $x_j \in H^j$, $j = 1, \ldots, \dim T$ such that $x_i = x$ and

$$S(g)x_j = \sum_{k=1}^{\dim T} t^T_{kj}(g)x_k \qquad (2.7.13)$$

for all $g \in G$.

---

[42] We obviously discard all T for which $M^T = \{0\}$.

V. *Let* $x_1, \ldots, x_{\dim T}$ *be a set of vectors in the space H. The relation (2.7.13) holds if and only if*

$$E_{kj}^T x_j = x_k \quad \text{for all } j, k = 1, \ldots, \dim T. \tag{2.7.14}$$

Indeed, if S is a multiple of T, then $H^i = M_i^T$.

Proof. If (2.7.13) is satisfied, then applying (2.4.1) we obtain

$$E_{kj}^T x_j = (\dim T) \int_G \overline{t_{kj}^T(g)} S(g) x_j \, dg$$

$$= (\dim T) \int_G \overline{t_{kj}^T(g)} \sum_{p=1}^{\dim T} t_{pj}^T(g) x_p \, dg$$

$$= (\dim T) \sum_{p=1}^{\dim T} \left(\int_G t_{pj}^T(g) \overline{t_{kj}^T(g)} \, dg \right) x_p = x_k.$$

That is, (2.7.14) holds. Conversely, if (2.7.14) holds, we apply (2.7.3b) to obtain

$$S(g) x_j = S(g) E_{jj}^T x_j = \sum_{k=1}^{\dim T} t_{kj}^T(g) E_{kj}^T x_j = \sum_{k=1}^{\dim T} t_{kj}^T(g) x_k. \quad \square$$

VI. *If the representation S of G in a Hilbert space H is a multiple of T, then H is the orthogonal direct sum of a certain family of subspaces* M_λ *such that: (1) every* M_λ *is invariant under S; (2) the restriction of S to every subspace* M_λ *is equivalent to T.*

Proof. Let $\{e_\lambda^1, \lambda \in \Lambda\}$ be an orthonormal basis in the space H^1. It is clear that $e_\lambda^1 = E_{11}^T e_\lambda^1$ for all $\lambda \in \Lambda$. We set $e_\lambda^i = E_{i1}^T e_\lambda^1$ for all $i = 2, \ldots, \dim T$. As shown in the proof of Proposition IV, the vectors e_λ^i belong to the subspace $M_i^T = H^i$. The reader can easily verify that for a fixed i, the vectors e_λ^i form an orthonormal basis in the space M_i^T. Since H is the direct sum of the subspaces H^i, the set of vectors $\{e_\lambda^i, i = 1, \ldots, \dim T, \lambda \in \Lambda\}$ forms an orthonormal basis in H. Let $M_\lambda, \lambda \in \Lambda$, be the subspace of H generated by the vectors $e_\lambda^i, i = 1, \ldots, \dim T$. From the construction of the vectors e_λ^i and from (2.7.3) it follows that

$$E_{kj}^T e_\lambda^j = E_{kj}^T E_{j1}^T e_\lambda^1 = E_{k1}^T e_\lambda^1 = e_\lambda^k. \tag{2.7.15}$$

That is, the vectors e_λ^k satisfy the relations (2.7.14). We infer from V that

$$S(g) e_\lambda^j = \sum_{k=1}^{\dim T} t_{kj}^T(g) e_\lambda^k. \tag{2.7.16}$$

The equality (2.7.16) means that M_λ is invariant under $S(g)$ and that the restriction of S to M_λ is equivalent to T. Since the set $\{e_\lambda^i, i = 1, \ldots, \dim T, \lambda \in \Lambda\}$ is an orthonormal basis in H, the orthogonal direct sum of the subspaces M_λ, $\lambda \in \Lambda$, is the entire space H. $\square$

Proposition VI shows that the above definition of a representation that is a multiple of an irreducible representation is equivalent to our earlier definition (see chapter I, 2.4).

VII. *Let S be a continuous unitary representation of the group G in a Hilbert space H. Every subspace M^T is invariant under all operators $S(g)$ and if $M^T \neq \{0\}$, then the restriction of S to the subspace M^T is a multiple of the irreducible representation T. Thus, any continuous unitary representation of a compact group G in a Hilbert space is the orthogonal direct sum of unitary representations of this group that are multiples of finite-dimensional irreducible unitary representations of G.*

Proof. In view of IV, it suffices to prove that every subspace M^T is invariant under all operators $S(g)$ and that the restriction of S to M^T is a multiple of the irreducible representation T. Let x belong to M^T. We then have $x = E^T x$ and by (2.7.3j)

$$S(g)x = S(g)E^T x = E^T S(g)x \qquad (2.7.17)$$

for all $g \in G$. That is, $S(g)x$ belongs to M^T for $x \in M^T$: M^T is invariant under $S(g)$, $g \in G$. Furthermore, the space M^T is the orthogonal direct sum of the subspaces $H^i = M_i^T$, $i = 1, \ldots, \dim T$. Let x belong to M_i^T. We set $x_j = E_{ji}^T x$ for all $j = 1, \ldots, \dim T$. We then have: $x_j \in M_j^T$ for all j; $x_i = x$; and therefore $x_k = E_{kj}^T x_j$ for all $j, k = 1, \ldots, \dim T$. According to V, this means that the restriction of S to M^T is a multiple of the irreducible representation T. $\square$

Propositions I-VII together not only guarantee that decompositions into irreducible representations exist, but also provide a rule for such decompositions. By VII the projection E^T projects the entire space H onto the subspace M^T, on which S is a multiple of T. Proposition VI provides a way to decompose multiples of irreducible representations into irreducible representations. The decomposition of a multiple of an irreducible representation as a direct sum of irreducible representations is not in general unique. Nevertheless, the intrinsic description of the operator E^T shows that the decomposition of a given continuous unitary representation S as a direct sum of representations that are multiples of pairwise inequivalent irreducible representations is unique.

2.8. Sufficient Conditions for Completeness of a Given Set of Irreducible Unitary Representations of the Group G

We will produce a result that is in a sense the converse of theorems 2 and 4 of 2.4.

Theorem. *Let $\{T^\alpha, \alpha \in A\}$ be a set of irreducible unitary representations of a compact group G. Let $\{t_{ij}^\alpha(g), i, j = 1, \ldots, \dim T^\alpha\}$ be the set of matrix elements of the representation T^α in some orthonormal basis. If finite linear combinations of the set of functions $\{t_{ij}^\alpha(g), \alpha \in A, i, j = 1, \ldots, \dim T^\alpha\}$ form a dense subset of the space $L^2(G)$ or of the space $C(G)$, then every irreducible unitary representation of G is unitarily equivalent to one of the representations of the set $\{T^\alpha\}$.*

Proof. If linear combinations of the functions $\{t_{ij}^\alpha\}$ are dense in $C(G)$, then they are *a fortiori* dense in $L^2(G)$. We may thus suppose that linear combinations of the functions $\{t_{ij}^\alpha\}$ are dense in $L^2(G)$. Let S be an irreducible (finite-dimensional) unitary representation of G in a Hilbert space H. Let χ be the character of S. The orthogonal system $\{t_{ij}^\alpha\}$ is complete in the Hilbert space $L^2(G)$.[43] Therefore there exists a function $t_{ij}^\alpha(g)$ such that

$$(\chi, t_{ij}^\alpha) \neq 0. \tag{2.8.1}$$

We construct an operator $E_{ij}^{T^\alpha}$ for S, writing

$$E_{ij}^{T^\alpha} = (\dim T^\alpha) \int_G \overline{t_{ij}^\alpha(g)} S(g) \, dg. \tag{2.8.2}$$

A continuity argument shows that the trace of $E_{ij}^{T^\alpha}$ can be found by taking the trace under the integral sign in (2.8.2). We find

$$\text{tr}(E_{ij}^{T^\alpha}) = (\dim T^\alpha) \int_G \overline{t_{ij}^\alpha(g)} \, \text{tr} \, S(g) \, dg$$

$$= (\dim T^\alpha) \int_G \overline{t_{ij}^\alpha(g)} \chi(g) \, dg = (\chi, t_{ij}^\alpha)(\dim T^\alpha). \tag{2.8.3}$$

By (2.8.1) the last term in (2.8.3) is not zero, and so $E_{ij}^{T^\alpha}$ is not zero. Using (2.7.3e) we obtain

$$E_{ii}^{T^\alpha} \neq 0, \tag{2.8.4}$$

since $E_{ii}^{T^\alpha} E_{ij}^{T^\alpha} = E_{ij}^{T^\alpha} \neq 0$. *A fortiori*, we find

$$E^{T^\alpha} \neq 0, \quad \text{and thus } M^{T^\alpha} \neq \{0\}. \tag{2.8.5}$$

---

[43] Editor's note. This fact requires proof. It follows from elementary Hilbert space theory. An incomplete orthogonal system cannot span a dense subspace.

The nonequalities (2.8.5) and VII in 2.7 show that S admits a nonzero subrepresentation equivalent to T^α. The representation S is irreducible and so is equivalent to T^α. □

2.9. Examples

1. Let G be the group of rotations of a circle ($G = \Gamma^1$: see example 1 in 1.7, chapter I). A function f on G can be regarded as a function on the real line $\mathbf{R}$ for which

$$f(\varphi + 2\pi) = f(\varphi) \tag{2.9.1}$$

for all $\varphi \in \mathbf{R}$. One-dimensional representations of G can be identified with complex-valued functions on G (see 2.1, chapter I). Consider the set χ_n, $n = 0, \pm 1, \pm 2, \ldots$ of one-dimensional representations of G that are defined by

$$\chi_n(\varphi) = e^{in\varphi}, \qquad \varphi \in \mathbf{R}, \qquad n \in \mathbf{Z}. \tag{2.9.2}$$

The functions $\chi_n(\varphi)$ satisfy condition (2.9.1) and therefore the definition of χ_n as functions on G is correct. The functions $\chi_n(\varphi)$ are matrix elements of the corresponding irreducible representations. In example (d) in 3.3, chapter III, we proved that the functions $\chi_n(\varphi)$ form a complete set of irreducible representations of G. We offer here a second proof. Consider the set A of finite linear combinations of the functions $\chi_n(\varphi)$ on G. The set A forms an algebra of functions on G since the product of the functions χ_n and χ_m is χ_{n+m}. The algebra A separates points in G since the single function $\chi_1(\varphi) = e^{i\varphi}$ separates points in G. (It is a faithful representation of the group G.) In addition, the algebra A contains the constant functions (since $\chi_0(g) \equiv 1$). Finally we have $\bar{\chi}_n = \chi_{-n}$ and therefore A contains along with any function its complex conjugate. By the Stone-Weierstrass Theorem, A is dense in $C(G)$. The theorem of 2.8 shows that the set χ_n, $n \in \mathbf{Z}$, defines a complete set of irreducible unitary representations of G.

Let us see what becomes of theorem 5, 2.4, in the present case. It is easy to verify that the measure $d\varphi/2\pi$ is an invariant measure on G with total measure 1. We can therefore identify the space $L^2(G)$ with the set of all measurable 2π-periodic functions f on $\mathbf{R}$ such that $(1/2\pi) \int_0^{2\pi} |f(\varphi)|^2 \, d\varphi < +\infty$. According to theorem 5 in 2.4, the set of functions χ_n, $n \in \mathbf{Z}$, is an orthonormal basis in $L^2(G)$ and formula (2.4.7) assumes the form

$$\frac{1}{2\pi} \int_0^{2\pi} |f(\varphi)|^2 \, d\varphi = \sum_{n=-\infty}^{+\infty} |c_n|^2, \quad \text{for all } f \in L^2(G), \tag{2.9.3}$$

where

$$c_n = \frac{1}{2\pi} \int_0^{2\pi} f(\varphi) \overline{\chi_n(\varphi)} \, d\varphi = \frac{1}{2\pi} \int_0^{2\pi} f(\varphi) e^{-in\varphi} \, d\varphi. \tag{2.9.4}$$

Formula (2.9.3) is the well-known *Parseval's equality* in the theory of (trigonometric) Fourier series.

The completeness of the trigonometric system in $L^2[-\pi, \pi]$ follows at once.

2. Let G be the group of 2×2 unitary matrices with determinant 1, that is, $G = SU(2)$ (see example 3 in 1.2, §1). We will construct a complete set of irreducible unitary representations of G.

Let m be a nonnegative integer. Let H_m be the linear space of homogeneous polynomials of degree m in two complex variables z_1 and z_2, provided with the scalar product

$$(p(z_1, z_2), q(z_1, z_2)) = \int_{|z_1|^2 + |z_2|^2 \leqslant 1} p(z_1, z_2)\overline{q(z_1, z_2)} \, dz_1 \, dz_2, \quad (2.9.5)$$

where $dz_1 = dx_1 \, dy_1$, $dz_2 = dx_2 \, dy_2$ for $z_1 = x_1 + iy_1$, $z_2 = x_2 + iy_2$. The reader can easily verify that the linear space H_m, provided with the scalar product (2.9.5), is a finite-dimensional Hilbert space. The formula

$$(T_m(g)p)(z_1, z_2) = p(g_{11}z_1 + g_{21}z_2, g_{12}z_1 + g_{22}z_2), \quad (2.9.6)$$

where

$$g = \begin{Vmatrix} g_{11} & g_{12} \\ g_{21} & g_{22} \end{Vmatrix} \in SU(2),$$

defines a continuous unitary representation $T_m(g)$ of G in H_m. (Note in particular that the operators $T_m(g)$ carry homogeneous polynomials of degree m into homogeneous polynomials of degree m.)

I. *The representations T_m are irreducible.*

Proof. Let Γ be the subgroup of G consisting of all diagonal matrices, that is,

$$\Gamma = \left\{ \gamma_\varphi = \begin{Vmatrix} e^{i\varphi} & 0 \\ 0 & e^{-i\varphi} \end{Vmatrix}, \varphi \in \mathbf{R} \right\}. \quad (2.9.7)$$

The restriction of T_m to Γ can be decomposed into the direct sum of pairwise inequivalent one-dimensional representations. For $P_k(z_1, z_2) = z_1^k z_2^{m-k}$ ($k = 0, 1, \ldots, m$), we get

$$[T_m(\gamma_\varphi)p_k](z_1, z_2) = (e^{i\varphi}z_1)^k(e^{-i\varphi}z_2)^{m-k}$$
$$= e^{i(2k-m)\varphi}p_k(z_1, z_2). \quad (2.9.8)$$

Thus, any nonzero subspace L of the space H_m invariant under T must be the direct sum of subspaces invariant under the restriction of T to Γ. That is to say, L is the linear span of a subset of the polynomials $p_0, p_1, \ldots, p_m$.

Consider the subgroup $\varDelta$ of the group G defined by

$$\varDelta = \left\{ \delta_\theta = \left\| \begin{matrix} \cos\theta & -\sin\theta \\ \sin\theta & \cos\theta \end{matrix} \right\|, \theta \in \mathbf{R} \right\}. \tag{2.9.9}$$

It is clear that

$$(T_m(\delta_\theta)p_k)(z_1, z_2) = (z_1\cos\theta + z_2\sin\theta)^k(-z_1\sin\theta + z_2\cos\theta)^{m-k}. \tag{2.9.10}$$

Use the formula

$$T_m(\delta_\theta)p_k = \sum_{j=0}^{m} t_{kj}(\theta)p_j \tag{2.9.11}$$

to write the right side of (2.9.10) in the basis $p_0, p_1, \ldots, p_m$. Clearly the functions $t_{kj}(\theta)$ are polynomials in $\cos\theta$ and $\sin\theta$ and so are continuously differentiable. We will show that $t_{k,\,k+1}(\theta)$ and $t_{k,\,k-1}(\theta)$ are not identically equal to zero if $k + 1 \leqslant m$ and $k - 1 \geqslant 0$, respectively. It suffices to show that the derivatives of these functions with respect to θ do not vanish for $\theta = 0$. Differentiating (2.9.10) and setting $\theta = 0$ we obtain

$$\frac{d}{d\theta}(T_m(\delta_\theta)p_k)(z_1, z_2)\big|_{\theta=0} = kz_1^{k-1}z_2^{m-k+1} - (m-k)z_1^{k+1}z_2^{m-k+1}$$

$$= kp_{k-1}(z_1, z_2) - (m-k)p_{k+1}(z_1, z_2). \tag{2.9.12}$$

Comparing (2.9.12) and (2.9.11), we obtain

$$\sum_{j=0}^{m} (d/d\theta)(t_{kj}(\theta))\big|_{\theta=0} \cdot p_j = kp_{k-1} - (m-k)p_{k+1}. \tag{2.9.13}$$

That is, we have $t_{k,\,k+1}(0) \not\equiv 0$ for $k < m$ and $t_{k,\,k-1}(0) \not\equiv 0$ for $k > 0$. We infer that if L contains a certain polynomial p_{k_0}, then L contains all polynomials p_k such that $k_0 < k \leqslant m$ or $0 \leqslant k < k_0$. That is, $L = H_m$, and T_m is irreducible. $\square$

In the sequel we shall need the following lemma.

II. *For all elements $g \in G$ there are elements $v \in G$ and $\gamma \in \Gamma$ such that $g = v\gamma v^{-1}$.*

Proof. We know from linear algebra that any unitary matrix g can be reduced by a unitary transformation w to a diagonal form γ:

$$g = w\gamma w^{-1}.$$

If g belongs to $SU(2)$, that is, $\det u = 1$, then $\det \gamma = 1$ and therefore $\gamma \in \Gamma$. If $\det w \neq 1$, we consider the matrix $v = w\tau$, where $\tau = \begin{Vmatrix} (\det \omega)^{-1} & 0 \\ 0 & 1 \end{Vmatrix}$. It is plain that $\det v = 1$, that is, v belongs to $SU(2)$. The matrix τ commutes with all matrices $\gamma \in \Gamma$ since both are diagonal matrices. Therefore we have $\tau \gamma \tau^{-1} = \gamma$ and

$$v\gamma v^{-1} = w\tau\gamma(\omega\tau)^{-1} = w\gamma w^{-1} = g. \quad \square$$

III. *The character χ_m of the irreducible unitary representation T_m is defined by the formula*

$$\chi_m(g) = \frac{e^{i(m+1)\varphi} - e^{-i(m+1)\varphi}}{e^{i\varphi} - e^{-i\varphi}}, \tag{2.9.14}$$

for

$$g = v\gamma_\varphi v^{-1}, \qquad v \in SU(2), \qquad \gamma_\varphi = \begin{Vmatrix} e^{i\varphi} & 0 \\ 0 & e^{-i\varphi} \end{Vmatrix}. \tag{2.9.15}$$

Proof. We use (2.9.15). The character of a finite-dimensional representation is a function constant on conjugacy classes of G, and so $\chi_m(g) = \chi_m(v^{-1}gv) = \chi_m(\gamma_\varphi)$. To prove (2.9.14), we need only to compute $\chi_m(\gamma_\varphi)$. From (2.9.8) we get

$$\chi_m(\gamma_\varphi) = \sum_{k=0}^{m} e^{i(2k-m)\varphi}. \tag{2.9.16}$$

Sum the geometric progression on the right side of (2.9.16) to obtain (2.9.14). $\quad \square$

IV. *The characters χ_m of the representations T_m satisfy the identities*

$$\chi_m(g)\chi_n(g) = \chi_{m+n}(g) + \chi_{m+n-2}(g) + \cdots + \chi_{|m-n|}(g) \tag{2.9.17}$$

for all $g \in G$ and all nonnegative integers m and n.

Proof. We need only to verify (2.9.17) for $g = \gamma_\varphi$. From (2.9.14) and (2.9.16) we obtain

$$\chi_m(\gamma_\varphi)\chi_n(\gamma_\varphi) = \frac{e^{i(m+1)\varphi} - e^{-i(m+1)\varphi}}{e^{i\varphi} - e^{-i\varphi}} (e^{in\varphi} + e^{i(n-2)\varphi} + \cdots + e^{-in\varphi}). \tag{2.9.18}$$

Substitute (2.9.14) in the right side of (2.9.17) to get the right side of (2.9.18). $\quad \square$

V. *The tensor product of representations T_m and T_n can be decomposed into the direct sum of $T_{m+n}, T_{m+n-2}, \ldots, T_{m-n}$.*

This assertion follows at once from IV in 2.9, chapter I and I in 2.6.

VI. *Finite linear combinations of matrix elements* $t_{ij}^m(g)$, $m \geqslant 0$, $i, j = 1, \ldots, \dim T_m$ *of the irreducible unitary representations* T_m *are dense in* $C(G)$.

Proof. It follows from V that finite linear combinations of matrix elements of the representations T_m form an algebra A of continuous functions on G. This algebra contains all constants, since it contains the matrix element of the one-dimensional representation T_0 of G, $T_0(g) = 1$ for all $g \in G$. Furthermore, the algebra A separates points in G since the matrix of T_1 in the basis $p_1 = z_1$, $p_2 = z_2$ has elements g_{ij}, $i, j = 1, 2$. That is, T_1 is equivalent to the self-representation of G. Finally, if $\bar{T}_k$ is the representation conjugate to T_k, then $\chi_{\bar{T}_k} = \bar{\chi}_k$ and (2.9.14) shows that $\bar{\chi}_k = \chi_k$. Thus $\bar{T}_k$ is equivalent to T_k. Consequently, along with a function, the algebra A also contains its conjugate function. By the Stone-Weierstrass Theorem, the algebra A is dense in $C(G)$. $\square$

VII. *The set of representations* T_m *is a complete system*[44] *of irreducible unitary representations of the group* G.

This follows from VI and from the theorem of 2.8.
Applying theorem 5 in 2.4 we find the following.

Theorem. *The functions* $\varphi_{ij}^m(g) = \sqrt{m+1}\, t_{ij}^m(g)$, *where* $m \geqslant 0$ *and* $t_{ij}^m(g)$ *are the matrix elements of the irreducible representation* T_m *in some orthonormal basis, form an orthonormal system in* $L^2(G)$. *For all functions* $f \in L^2(G)$ *we have the equality*

$$\int_G |f(g)|^2 \, dg = \sum_{m=0}^{\infty} (m+1) \sum_{i,j=0}^{\infty} |(f, t_{ij}^m)|^2. \qquad (2.9.19)$$

§3. The Group Algebra of a Compact Group

3.1. Definition and the Simplest Properties of a Group Algebra

Let G be a compact topological group. We denote the set of all integrable complex-valued functions on G by the symbol $L^1(G)$. These are the functions f such that: (1) f is measurable under the invariant measure on G, (2) $\int_G |f(g)| \, dg < +\infty$. We identify those functions that differ only on a set of measure zero. In $L^1(G)$ we define the operations of addition and multiplication by a number in the usual way. Thus $L^1(G)$ is a linear space. To

---

[44] Below (see chapter XII) the completeness of the system T_m is obtained by different means, as a corollary of a general fact about representations of the groups $SU(n)$.

define the operation of multiplication in $L^1(G)$ we set

$$(f_1 * f_2)(g) = \int_G f_1(h) f_2(h^{-1}g)\, dh \qquad (3.1.1)$$

for all $f_1, f_2 \in L^1(G)$. The function $f_1 * f_2$ is called *the convolution of the functions f_1 and f_2.*

We will show that the function $f_1 * f_2$ belongs to the space $L^1(G)$. Consider the mapping φ of the compact group $G \times G$ onto itself defined by the formula $(g, h) \to (h^{-1}g, h)$. This mapping is clearly a homeomorphism. It is easy to verify that φ carries open sets of the form $M_1 \times M_2$ into sets with the same measure. Therefore the mapping φ carries measurable functions on $G \times G$ into measurable functions. Since functions of the form $f_1(h)f_2(g)$ are measurable on $G \times G$, functions of the form $f_1(h)f_2(h^{-1}g)$ are also measurable on $G \times G$. On the other hand, the iterated integral of the function $|f_1(h)f_2(h^{-1}g)|$ is equal to

$$\int_G dh \int_G |f_1(h)|\,|f_2(h^{-1}g)|\, dg = \int_G |f_1(h)|\, dh \int_G |f_2(h^{-1}g)|\, dg$$

$$= \int_G |f_1(h)|\, dh \int_G |f_2(g)|\, dg < +\infty. \qquad (3.1.2)$$

That is, functions $f_1(h)f_2(h^{-1}g)$ are integrable on $G \times G$. Applying Fubini's Theorem to such a function, we see that the integral $\int_G f_1(h)f_2(h^{-1}g)\, dh$ exists for almost all $g \in G$ and defines an integrable function on G. Thus, formula (3.1.1) defines a legitimate multiplication in $L^1(G)$. The reader can easily verify that for any complex number λ and any $f_1, f_2, f_3 \in L^1(G)$ the following identities hold:

$$\begin{aligned}
\lambda(f_1 * f_2) = \lambda f_1 * f_2, \qquad & f_1 * (\lambda f_2) = \lambda(f_1 * f_2), \\
f_1 * (f_2 * f_3) = (f_1 * f_2) * f_3, & \\
(f_1 + f_2) * f_3 = f_1 * f_3 + f_2 * f_3, & \\
f_1 * (f_2 + f_3) = f_1 * f_2 + f_1 * f_3. &
\end{aligned} \qquad (3.1.3)$$

We infer from (3.1.3) that $L^1(G)$ is an associative algebra over the complex numbers.

We set

$$f^*(g) = \overline{f(g^{-1})}, \quad \text{for all } g \in G \quad \text{and} \quad f \in L^1(G). \qquad (3.1.4)$$

Since

$$\int_G |f^*(g)|\, dg = \int_G |\overline{f(g^{-1})}|\, dg = \int_G |f(g)|\, dg \qquad (3.1.5)$$

see (7) in Proposition I in (2.1), f^* belongs to $L^1(G)$. Plainly the mapping $f \to f^*$ satisfies conditions (1)–(4) in 2.5, chapter II. Thus $L^1(G)$ is a symmetric algebra.

We introduce a *norm* into $L^1(G)$:

$$\|f\|_{L^1(G)} = \int_G |f(g)|\, dg \qquad (3.1.6)$$

for all $f \in L^1(G)$. From (3.1.5) we have

$$\|f^*\|_{L^1(G)} = \|f\|_{L^1(G)}. \qquad (3.1.7)$$

Furthermore, from Fubini's Theorem and the invariance of the measure dg we conclude that

$$\|f_1 * f_2\|_{L^1(G)} = \int_G \left| \int_G f_1(h) f_2(h^{-1}g)\, dh \right| dg \leqslant \int_G \left\{ \int_G |f_1(h)|\, |f_2(h^{-1}g)|\, dh \right\} dg$$

$$= \int_G \left\{ \int_G |f_2(h^{-1}g)|\, dg \right\} |f_1(h)|\, dh = \int_G |f_2(g)|\, dg \int_G |f_1(h)|\, dh$$

$$= \|f_1\|_{L^1(G)} \|f_2\|_{L^1(G)}$$

and thus

$$\|f_1 * f_2\|_{L^1(G)} \leqslant \|f_1\|_{L^1(G)} \|f_2\|_{L^1(G)}. \qquad (3.1.8)$$

Compare the definition of the algebra $L^1(G)$ (in particular formula (3.1.1)) with the definition of the group algebra of a finite group G (specifically formula (2.7.6), chapter II). We see that for a finite group G, the algebra $L^1(G)$ coincides with the algebra A_G.

The algebra $L^1(G)$ is called the *group algebra of the group G*.

I. *The group algebra* $L^1(G)$ *contains a multiplicative unit if and only if the group* G *is finite.*

Proof. If the group G is finite, $L^1(G)$ contains a unit, as shown in Proposition II in 2.7, chapter II. Conversely, suppose that $L^1(G)$ has a unit function, say $e(g)$. We will show that the measure of nonvoid open sets has a positive lower bound. Assume that the group identity has neighborhoods of arbitrarily small measure. Then for any $\varepsilon > 0$ there exists a neighborhood U of the identity element of G such that $\int_U |e(g)|\, dg < \varepsilon$. Let V be a symmetric neighborhood of the identity element of G such that $V^2 \subset U$. Let ξ_V be the characteristic function of V. For $g \in V$, we have

$$1 = \xi_V(g) = (e * \xi_V)(g) = \int_G e(h) \xi_V(h^{-1}g)\, dh$$

$$= \int_{gV} e(h)\, dh \leqslant \int_U |e(h)|\, dh < \varepsilon,$$

which is impossible for $\varepsilon < 1$.

Thus the measure of any nonvoid open subset M of G cannot be smaller than some $a > 0$. Assume that the group G is infinite. For every positive integer n, there are n distinct points $g_1, \ldots, g_n \in G$. These points have pairwise disjoint neighborhoods $M_1, \ldots, M_n$. Therefore the measure of the entire group G is at least na. Since $\mu(G)$ is equal to 1, we have a contradiction. $\square$

II. *If f belongs to $L^1(G)$, the functions*

$$f^{g_0}(g) = f(g_0 g), \qquad f_{g_0}(g) = f(g g_0) \tag{3.1.9}$$

are continuous functions of g_0 in the sense of the norm in $L^1(G)$.

Proof. If f is a continuous function on G, then f is uniformly continuous. (See III in 1.2.) For any $\varepsilon > 0$ there exists a neighborhood U of the identity element of G such that $|f(h_1) - f(h_0)| < \varepsilon$ for all h_1, h_0 such that $h_1 \in h_0 U$. Thus for $g_1 \in g_0 U$, $g g_1$ belongs to $g g_0 U$ and therefore

$$\|f_{g_1} - f_{g_0}\|_{L^1(G)} = \int_G |f_{g_1}(g) - f_{g_0}(g)| \, dg$$
$$= \int_G |f(g g_1) - f(g g_0)| \, dg \leqslant \varepsilon \int_G dg = \varepsilon.$$

That is, Proposition II holds for all continuous functions f on G. If f is any function in $L^1(G)$, then there exists a continuous function f_1 on G such that $\|f - f_1\| < \varepsilon/3$. We choose a neighborhood V of the identity element in G such that for $g \in hU$, we have $\|(f_1)_g - (f_1)_h\|_{L^1} < \varepsilon/3$. Therefore we have

$$\|f_g - f_h\|_{L^1} \leqslant \|f_g - (f_1)_g\|_{L^1} + \|(f_1)_g - (f_1)_h\|_{L^1}$$
$$+ \|(f_1)_h - f_h\|_{L^1} < \varepsilon/3 + \varepsilon/3 + \varepsilon/3 = \varepsilon.$$

The continuity of the function f^{g_0} is proved analogously. $\square$

Let f belong to $L^2(G)$, i.e., f is a measurable function on G and $\int_G |f(g)|^2 \, dg$ is finite. From the inequality $|f| \leqslant (1 + |f|^2)/2$ it follows that $\int_G |f(g)| \, dg$ is finite. We conclude that

$$L^2(G) \subset L^1(G). \tag{3.1.10}$$

From the Cauchy-Bunjakovskiĭ inequality we infer that

$$\left(\int_G |f(g)| \, dg \right)^2 \leqslant \left(\int_G 1 \, dg \right) \int_G |f(g)|^2 \, dg = \int_G |f(g)|^2 \, dg,$$

which is to say that

$$\|f\|^2_{L^1(G)} \leqslant \int_G |f(g)|^2 \, dg = \|f\|^2_{L^2(G)}. \tag{3.1.11}$$

III. *If $f \in L^2(G)$, then the functions f^{g_0}, f_{g_0} defined by the formulas (3.1.9) are continuous functions of g_0 in the sense of the norm in $L^2(G)$.*

The proof is analogous to the proof of Proposition II.

IV. *If f_1, f_2 belong to $L^2(G)$, then $f_1 * f_2$ is a continuous function on G.*

Proof. It is clear that the function f_2^* belongs to $L^2(G)$. According to III, the function $(f_2^*)^{g^{-1}}$ is a continuous function in the sense of the norm in $L^2(G)$. On the other hand we have

$$(f_1 * f_2)(g) = \int_G f_1(h) f_2(h^{-1}g)\, dh$$

$$= \int_G f_1(h)\overline{(f_2^*)^{g^{-1}}(h)}\, dh = (f_1, (f_2^*)^{g^{-1}}). \tag{3.1.12}$$

Since $(f_2^*)^{g^{-1}}$ is a continuous function in $L^2(G)$, $(f_1, (f_2^*)^{g^{-1}})$ is a continuous complex-valued function on G and (3.1.12) shows that $f_1 * f_2$ is continuous.   □

V. *For any function $f \in L^1(G)$ and any $\varepsilon > 0$ there exists a function $\varphi \in L^1(G)$ such that $\|f * \varphi - f\|_{L^1(G)} < \varepsilon$, $\|\varphi * f - f\|_{L^1(G)} < \varepsilon$.*

Proof. Let U be a neighborhood of the identity in G. Let φ_U be a nonnegative function from $L^1(G)$ vanishing outside of U and satisfying the condition

$$\int_G \varphi_U(g)\, dg = 1. \tag{3.1.13}$$

(For example, we can take $\varphi_U = (1/\mu(U))\xi_U$, where $\mu(U)$ is the Haar measure of the set U and ξ_U is the characteristic function of U.) We then have

$$(\varphi_U * f)(g) = \int_G \varphi_U(h) f(h^{-1}g)\, dh,$$

and therefore

$$\varphi_U * f = \int_G \varphi_U(h) f^{h^{-1}}\, dh. \tag{3.1.14}$$

From (3.1.13) and (3.1.14) we infer that

$$\|\varphi_U * f - f\|_{L^1(G)} = \left\| \int_G \varphi_U(h) f^{h^{-1}}\, dh - f \right\|_{L^1(G)}$$

$$= \left\| \int_U \varphi_U(h)(f^{h^{-1}} - f)\, dh \right\|_{L^1(G)}$$

$$\leqslant \int_U \varphi_U(h) \| f^{h^{-1}} - f \|_{L^1(G)}\, dh. \tag{3.1.15}$$

By II, we can choose a neighborhood U such that $\|f^{h-1} - f\|_{L^1(G)} < \varepsilon$ for $h \in U$. From (3.1.15) we obtain

$$\|\varphi_U * f - f\|_{L^1(G)} \leqslant \varepsilon \int_U \varphi_U(h)\, dh = \varepsilon.$$

The proof that $\|f * \varphi_U - f\|_{L^1(G)} \to 0$ is similar. □

VI. *A closed subspace I in $L^1(G)$ is a left (or right) ideal in $L^1(G)$ if and only if it is invariant under all left (or right) translations.*

Proof. Suppose that I is a left ideal in $L^1(G)$ and that f belongs to I. Choose functions φ_U as in V for the function f. The function $(\varphi_U)^{g_0} * f$ also belongs to I. On the other hand we have

$$((\varphi_U)^{g_0} * f)(g) = \int_G \varphi_U(g_0 h) f(h^{-1}g)\, dh$$

$$= \int_G \varphi_U(h) f(h^{-1}g_0 g)\, dh = (\varphi_U * f)^{g_0}(g). \qquad (3.1.16)$$

From V we have $\varphi_U * f \to f$. It follows this and II that $(\varphi_U * f)^{g_0} \to f^{g_0}$. Since $(\varphi_U * f)^{g_0}$ is equal to $(\varphi_U)^{g_0} * f \in I$ and the ideal I is closed, f^{g_0} belongs to I. Thus a closed left ideal is invariant under left translations.

Conversely, let I be a closed subspace in $L^1(G)$ that is invariant under left translations. If the function ψ is continuous on G, the function $\psi(h)f^{h-1}$ is continuous on G under the norm in $L^1(G)$ for all functions $f \in L^1(G)$. The usual arguments show that the integral

$$\psi * f = \int_G \psi(h) f^{h-1}\, dh$$

is the limit in the L^1 norm of finite sums of the type.

$$\sum_k \psi(h_k) f^{h_k^{-1}} \mu(\Delta_k),$$

where the sets $\{\Delta_k\}$ form a decomposition of the group G into pairwise disjoint measurable subsets. Let f belong to I. Since I is invariant under left translations, the function $\sum_k \psi(h_k) f^{h_k^{-1}} \mu(\Delta_k)$ belongs to I. Because $\psi * f$ is the limit of these sums and I is closed, $\psi * f$ belongs to I for all continuous functions ψ. If ψ is any function in $L^1(G)$, there exists a continuous function φ on G such that $\|\psi - \varphi\|_{L^1(G)} < \varepsilon$. From (3.1.8) we have

$$\|\psi * f - \varphi * f\|_{L^1(G)} \leqslant \|\psi - \varphi\|_{L^1(G)} \times \|f\|_{L^1(G)} < \varepsilon \|f\|_{L^1(G)}.$$

Since $\varphi * f$ belongs to I for all continuous functions φ, and since the subspace I is closed, it follows that $\psi * f \in I$ for all $\psi \in L^1(G)$. That is, I is a left ideal in $L^1(G)$. □

3.2. Representations of a Group Algebra and Their Connection with Representations of the Group

Let T be a representation of the group algebra $L^1(G)$ in a Hilbert space H. The representation T is called *nonsingular* if for every nonzero $\xi \in H$, there is an f in $L^1(G)$ such that $T(f)\xi \neq 0$.

I. *Let $f \to T(f)$ be a nonsingular symmetric continuous representation of the group algebra $L^1(G)$ in a Hilbert space H. Let φ_U be the set of elements of $L^1(G)$ constructed in Proposition V in 3.1. We then have $\|T(\varphi_U)\xi - \xi\| \to 0$ for any vector $\xi \in H$. The set H' of finite linear combinations of vectors of the form $T(f)\eta$, $f \in L^1(G)$, $\eta \in H$, is dense in H.*

Proof. Let H_1 be the closure of the linear span of vectors of the form $T(f)\eta$, $f \in L^1(G)$, $\eta \in H$. If H_1 is not equal to H, then there exists a nonzero vector $\xi \in H$ orthogonal to H_1. That is, $(\xi, T(f)\eta)$ is equal to zero for all $f \in L^1(G)$, $\eta \in H$. Thus we have

$$(T(f)\xi, \eta) = (\xi, T(f)^*\eta) = (\xi, T(f^*)\eta) = 0 \qquad (3.2.1)$$

for all $\eta \in H$, $f \in L^1(G)$. We infer from this that $T(f)\xi = 0$ for all $f \in L^1(G)$. Since the representation $f \to T(f)$ is nonsingular, ξ is equal to zero. The resulting contradiction shows that $H_1 = H$.

If $\xi = T(f)\eta$ for some $\eta \in H$, $f \in L^1(G)$, we have

$$\begin{aligned}
\|T(\varphi_U)\xi - \xi\| &= \|T(\varphi_U)T(f)\eta - T(f)\eta\| \\
&= \|T(\varphi_U * f - f)\| \|\eta\| \to 0, \qquad (3.2.2)
\end{aligned}$$

since $\|\varphi_U * f - f\|_{L^1(G)} \to 0$ and the representation $f \to T(f)$ is continuous. From (3.2.2) it follows that $\|T(\varphi_U)\xi - \xi\| \to 0$ for all vectors $\xi \in H'$ (that is, all vectors that can be represented as finite linear combinations of vectors of the form $T(f)\eta$). The set H' is dense in $H_1 = H$. If ξ is any vector in H, there exists a vector $\xi' \in H'$ such that $\|\xi - \xi'\| < \varepsilon$. Therefore we have

$$\begin{aligned}
\|T(\varphi_U)\xi - \xi\| &\leqslant \|T(\varphi_U)\xi - T(\varphi_U)\xi'\| + \|T(\varphi_U)\xi' - \xi'\| + \|\xi' - \xi\| \\
&< (1 + \|T(\varphi_U)\|)\|\xi - \xi'\| + \|T(\varphi_U)\xi' - \xi'\|.
\end{aligned}$$

Since the mapping $f \to T(f)$ is continuous and $\|\varphi_U\|_{L^1(G)} = 1$, there is a positive constant C such that $\|T(\varphi_U)\| \leqslant C$ for all U. Choosing a function φ_U such that $\|T(\varphi_U)\xi' - \xi'\| < \varepsilon$, we find that $\|T(\varphi_U)\xi - \xi\| < (C + 2)\varepsilon$. $\square$

II. *There is a one-to-one correspondence between the symmetric nonsingular representations $f \to T(f)$ of the group algebra $L^1(G)$ of the group G and the* *continuous unitary representations* $g \to T(g)$ *of G. This correspondence is defined by the formula*

$$T(f) = \int_G f(g)T(g)\,dg. \tag{3.2.3}$$

Proof. Let $g \to T(g)$ be a continuous unitary representation of G. We define operators $T(f)$ by (3.2.3). The correspondence $f \to T(f)$ is plainly linear. We also have

$$
\begin{aligned}
T(f^*) &= \int_G f^*(g)T(g)\,dg = \int_G \overline{f(g^{-1})}T(g)\,dg \\
&= \int_G \overline{f(g)}T(g^{-1})\,dg = \int_G \overline{f(g)}T^*(g)\,dg \\
&= \int_G (f(g)T(g))^*\,dg = T(f)^* \tag{3.2.4}
\end{aligned}
$$

and

$$
\begin{aligned}
T(f_1 * f_2) &= \int_G \left(\int_G f_1(h)f_2(h^{-1}g)\,dh \right) T(g)\,dg \\
&= \int_G f_1(h) \left(\int_G f_2(h^{-1}g)T(g)\,dg \right) dh \\
&= \int_G f_1(h) \left(\int_G f_2(k)T(hk)\,dk \right) dh \\
&= \int_G f_1(h) \left(\int_G f_2(k)T(h)T(k)\,dk \right) dh \\
&= \int_G f_1(h)T(h)\,dh \int_G f_2(k)T(k)\,dk = T(f_1)T(f_2). \tag{3.2.5}
\end{aligned}
$$

These relations show that the correspondence $f \to T(f)$ defined by (3.2.3) is a symmetric representation of the group algebra $L^1(G)$. We will show that it is continuous. Since $\|T(g)\| = 1$, we have

$$
\begin{aligned}
\|T(f)\| &= \left\| \int_G f(g)T(g)\,dg \right\| \leqslant \int_G \|f(g)T(g)\|\,dg \\
&= \int_G |f(g)|\,dg = \|f\|_{L^1(G)} \tag{3.2.6}
\end{aligned}
$$

for all $f \in L^1(G)$. The mapping $f \to T(f)$ is therefore continuous.

Let g be an element of G. Proposition I implies that $\|T(\varphi_U)\xi - \xi\| \to 0$ and therefore

$$\|T(g)T(\varphi_U)\xi - T(g)\xi\| \to 0. \tag{3.2.7}$$

But we also have

$$T(g)T(\varphi_U) = T(g)\left(\int_G \varphi_U(h)T(h)\,dh \right)$$
$$= \int_G \varphi_U(h)T(gh)\,dh = \int_G \varphi_U(g^{-1}h)T(h)\,dh$$
$$= \int_G (\varphi_U)^{g^{-1}}(h)T(h)\,dh = T((\varphi_U)^{g^{-1}}).$$

Thus (3.2.7) and (3.2.8) give us

$$T((\varphi_U)^{g^{-1}})\xi \to T(g)\xi \qquad (3.2.9)$$

for all $g \in G$.

We will show that the representation $f \to T(f)$ is nonsingular. For all $\xi \in H$ we have

$$T(\varphi_U)\xi \to \xi$$

from which the nonsingularity of $f \to T(f)$ follows.

Conversely, suppose that we have a nonsingular symmetric continuous representation T of the group algebra $L^1(G)$ in a Hilbert space H. We will show that there exists a unique continuous unitary representation $g \to T(g)$ of G for which (3.2.3) holds.

The uniqueness of the desired representation of G follows from (3.2.9), which shows that if the representation $f \to T(f)$ is defined by (3.2.3), then the operators $T(g)$ are uniquely defined by the representation $f \to T(f)$. We will prove the existence of some representation $g \to T(g)$ of G that satisfies (3.2.3). From (3.1.16) we have $(\varphi_U)^{g_0} * f = (\varphi_U * f)^{g_0}$. On the other hand, since $\varphi_U * f$ converges to f in $L^1(G)$, we also have $(\varphi_U * f)^{g_0} \to f^{g_0}$. It follows that

$$(\varphi_U)^{g_0} * f - f^{g_0} \underset{U}{\to} 0 \qquad (3.2.10)$$

in $L^1(G)$. Because the representation $f \to T(f)$ is continuous, (3.2.10) implies that

$$T((\varphi_U)^{g_0})T(f)\eta - T(f^{g_0})\eta \to 0 \qquad (3.2.11)$$

for all $\eta \in H$. Let H' be the set of finite linear combinations of vectors $T(f)\zeta$, $f \in L^1(G)$, $\zeta \in H$. According to (3.2.11), the limit of the vectors $T((\varphi_U)^{g_0})\xi$ exists for all $\xi \in H'$ and belongs to H'. We write $T'(g_0)$ for the operator in H' defined by the formula

$$T'(g_0)\xi = \lim_U T((\varphi_U)^{g_0^{-1}})\xi \qquad (3.2.12)$$

for all $\xi \in H'$, $g_0 \in G$. The equality (3.2.11) also yields

$$T'(g_0)T(f) = T(f^{g_0^{-1}}) \tag{3.2.13}$$

for all $g_0 \in G$, $f \in L^1(G)$. Since $\|(\varphi_U)^{g_0^{-1}}\|_{L^1(G)} = \|\varphi_U\|_{L^1(G)} = 1$, we have $\|T((\varphi_U)^{g_0^{-1}})\| \leqslant C$ for some $C > 0$, all $g_0 \in G$ and all φ_U. Therefore the operator $T'(g_0)$ defined by (3.2.12) can be uniquely extended over all H to be a continuous linear operator in all of the space H. This operator is denoted by $T(g_0)$. Thus we have

$$\|T(g_0)\| \leqslant C, \quad \text{for all } g_0 \in G. \tag{3.2.14}$$

From (3.2.13) we infer that

$$T'(gh)T(f) = T(f^{(gh)^{-1}}) = T((f^{h^{-1}})^{g^{-1}})$$
$$= T'(g)T(f^{h^{-1}}) = T'(g)T'(h)T(f) \tag{3.2.15}$$

for all $g, h \in G$, $f \in L^1(G)$. The last equality shows that $T'(gh) = T'(g)T'(h)$ on H', and so

$$T(gh) = T(g)T(h) \tag{3.2.16}$$

for all $g, h \in G$. Furthermore (3.2.13) implies that

$$T(e) = 1. \tag{3.2.17}$$

If ξ is equal to $T(f)\eta$, then

$$T(g_0)\xi = T'(g_0)T(f)\eta = T(f^{g_0^{-1}})\eta; \tag{3.2.18}$$

also $f^{g_0^{-1}}$ depends continuously on g_0, according to II in 3.1. Therefore the right side of (3.2.18) depends continuously on g_0. Hence $T(g_0)\xi$ is a continuous function of g_0 for all $\xi \in H'$. For all $\xi \in H$, there exists a $\xi' \in H'$ such that $\|\xi - \xi'\| < \varepsilon$. Thus (3.2.14) implies that

$$\|T(g_1)\xi - T(g_0)\xi\| \leqslant \|T(g_1)(\xi - \xi')\| + \|T(g_1)\xi' - T(g_0)\xi'\| + \|T(g_0)(\xi' - \xi)\|$$
$$\leqslant 2C\varepsilon + \|T(g_1)\xi' - T(g_0)\xi'\|.$$

Observe that $\|T(g_1)\xi' - T(g_0)\xi'\| \to 0$ for $g_1 \to g_0$. Therefore the function $T(g)\xi$ is a continuous function of $g \in G$ for all $\xi \in H$. This fact, together with (3.2.16) and (3.2.17), shows that the mapping $g \to T(g)$ is a continous representation of the group G in the space H.

We now show that this representation is unitary. Set $V = U \cap U^{-1}$. Then V is a symmetric neighborhood of the identity element in G. We set

$\varphi_V(g) = (\mu(V))^{-1}\xi_V$, where ξ_V is the characteristic function of the set V and $\mu(V)$ is the Haar measure of the set V. We then have $(\varphi_V)^*(g) = \overline{\varphi_V(g^{-1})} = \overline{\varphi_V(g)} = \varphi_V(g)$ and $((\varphi_V)^{g_0^{-1}})^*(g) = \overline{\varphi_V(g_0^{-1}g^{-1})} = \varphi_V(gg_0) = (\varphi_V)_{g_0}(g)$. Computations like (3.2.16) show that $f * (\varphi_V)^{g_0} = f_{g_0} * \varphi_V$. Applying the involution, $*$ we obtain

$$(\varphi_V)_{g_0} * f^* = \varphi_V * (f^*)^{g_0}, \qquad (3.2.19)$$

which shows that $(\varphi_V)_{g_0} * f \to f^{g_0}$ as V shrinks to the identity, for all f in $L^1(G)$. If follows that $T((\varphi_V)_{g_0^{-1}})T(f) \to T(f^{g_0^{-1}}) = T'(g_0)T(f)$ for all $g \in G$, $f \in L^1(G)$ and so

$$T'(g_0)\xi = \lim T((\varphi_V)_{g_0^{-1}})\xi \qquad (3.2.20)$$

for all $\xi \in H'$. Comparing (3.2.19) with (3.2.12) and using the relation $(\varphi_V)_{g_0^{-1}} = ((\varphi_V)_{g_0})^*$, we see that $T'(g_0) = (T'(g_0^{-1}))^*$ on H' and therefore

$$T(g_0) = (T(g_0^{-1}))^*. \qquad (3.2.21)$$

Hence the representation $g \to T(g)$ is unitary.

Finally, we will prove that (3.2.3) holds. From the equality

$$f_1 * f_2 = \int_G f_1(h)(f_2)^{h-1}\, dh$$

(see (3.1.13)) and from (3.2.13) we infer that

$$T(f_1)T(f_2) = T(f_1 * f_2) = \int_G f_1(h)T((f_2)^{h-1})\, dh$$

$$= \int_G f_1(h)T(h)T(f_2)\, dh = \left(\int_G f_1(h)T(h)\, dh\right)T(f_2).$$

Therefore the bounded linear operators $T(f_1)$ and $\int f_1(h)T(h)\, dh$ coincide on H' and thus also on H. $\square$

3.3. The Center of a Group Algebra

Let φ be a function in $L^1(G)$ such that

$$\varphi * f = f * \varphi \qquad (3.3.1)$$

for all $f \in L^1(G)$, i.e., φ belongs to the center $Z_1(G)$ of the group algebra $L^1(G)$. For continuous f, (3.3.1) is

$$\int_G \varphi(gh)f(h^{-1})\, dh = \int_G f(h)\varphi(h^{-1}g)\, dh \quad \text{for almost all } g \text{ in } G, \quad (3.3.2)$$

and both sides are continuous functions of $g \in G$. Thus (3.3.2) holds for *all* $g \in G$. Substituting h^{-1} for h in the right side, we find that

$$\varphi(gh) = \varphi(hg) \tag{3.3.3}$$

for all $g \in G$ and almost all $h \in G$. Since the function $\varphi(gh)$ is measurable as a function on $G \times G$, (3.3.3) holds almost everywhere on $G \times G$. The mapping $(g, h) \rightarrow (h^{-1}g, h)$ of the group $G \times G$ into itself preserves measurability and measure (see (3.1.2)). In (3.3.3) we can replace g by $h^{-1}g$ and obtain

$$\varphi(g) = \varphi(h^{-1}gh) \tag{3.3.4}$$

almost everywhere on $G \times G$. Thus for almost all $g \in G$ (and *a fortiori* as an equality in $L^1(G)$) we have

$$\varphi(g) = \int_G \varphi(h^{-1}gh)\, dh. \tag{3.3.5}$$

Conversely, if a function $\varphi \in L^1(G)$ satisfies the identity (3.3.4), then (3.3.3) holds for almost all $g \in G$. That is, (3.3.1) holds for all $f \in L^1(G)$, and therefore φ belongs to $Z_1(G)$. We have proved the following proposition.

I. *A function $\varphi \in L^1(G)$ belongs to $Z_1(G)$ if and only if (3.3.4) holds for φ almost everywhere on $G \times G$.*

II. *The sum and convolution of two functions in $Z_1(G)$ also belong to $Z_1(G)$.*

We omit the (obvious) proof.

III. *The character of any continuous unitary finite-dimensional representation of the group G belongs to $Z_1(G)$.*

Proof. The character of a continuous unitary finite-dimensional representation of G is a continuous function on G constant on conjugacy classes and therefore satisfies (3.3.4). $\square$

IV. *The characters of continuous irreducible unitary representations of G form a complete orthonormal system in the Hilbert space $L^2(G) \cap Z_1(G)$.*

Proof. The orthogonality relations for characters (2.6.2) show that the characters χ^α of the irreducible continuous unitary representations of G form an orthonormal system in $L^2(G)$. According to III, the characters χ^α belong to $Z_1(G)$, which is to say that $\chi^\alpha \in L^2(G) \cap Z_1(G)$. From (3.3.3) and the continuity of multiplication in $L^1(G)$ (see (3.1.8)) we infer that the center $Z_1(G)$ is closed in $L^1(G)$. Then (3.1.10) and (3.1.11) imply that the subspace $L^2(G) \cap Z_1(G)$ is closed in $L^2(G)$. Therefore $L^2(G) \cap Z_1(G)$ is a Hilbert space.

It remains to prove that the orthonormal system $\{\chi^\alpha\}$ is complete in $L^2(G) \cap Z_1(G)$. Let φ belong to $L^2(G) \cap Z_1(G)$. By applying (2.4.6) we find that the equality

$$\varphi(g) = \sum_\alpha n_\alpha \sum_{i,j} (\varphi, t_{ij}^\alpha) t_{ij}^\alpha(g) \tag{3.3.6}$$

holds in $L^2(G)$ and hence almost everywhere. On the other hand, (3.3.4) holds for almost all $(g, h) \in G \times G$. If we substitute (3.3.6) in (3.3.4) we obtain

$$\sum_\alpha n_\alpha \sum_{i,j} (\varphi, t_{ij}^\alpha) t_{ij}^\alpha(g) = \sum_\alpha n_\alpha \sum_{i,j} (\varphi, t_{ij}^\alpha) t_{ij}^\alpha(h^{-1}gh)$$

$$= \sum_\alpha n_\alpha \sum_{k,l} \left(\sum_{i,j} (\varphi, t_{ij}^\alpha) t_{ik}^\alpha(h^{-1}) t_{lj}^\alpha(h) \right) t_{kl}^\alpha(g) \tag{3.3.7}$$

for almost all $(g, h) \in G \times G$. In particular, (3.3.7), for almost all $h \in G$, holds for almost all $g \in G$. On the other hand, for a fixed h the right and left sides of (3.3.7) belong to $L^2(G)$ (as functions of g). Thus, for almost all $h \in G$, the coefficients of $t_{kl}^\alpha(g)$ in the right and left sides of (3.3.7) must be equal. Hence we have

$$(\varphi, t_{kl}^\alpha) = \sum_{i,j} (\varphi, t_{ij}^\alpha) t_{ik}^\alpha(h^{-1}) t_{lj}^\alpha(h) = \sum_{i,j} (\varphi, t_{ij}^\alpha) \overline{t_{ki}^\alpha(h)} t_{lj}^\alpha(h) \tag{3.3.8}$$

for almost all $h \in G$. But the right and left sides of (3.3.8) depend continuously on h and therefore (3.3.8) holds everywhere on G. Integrating (3.3.8) with respect to h and applying the orthogonality relations, (3.4.1), we obtain

$$(\varphi, t_{kl}^\alpha) = \begin{cases} 0 & \text{for } i \neq j, \\ \dfrac{1}{n_\alpha} \displaystyle\sum_{i=1}^{n_\alpha} (\varphi, t_{ii}^\alpha) = \dfrac{1}{n_\alpha} \left(\varphi, \sum_i t_{ii}^\alpha \right) = \dfrac{1}{n_\alpha}(\varphi, \chi^\alpha), & \text{for } i = j. \end{cases} \tag{3.3.9}$$

Substituting (3.3.9) in (3.3.6), we obtain

$$\varphi = \sum_\alpha n_\alpha \sum_{i=1}^{n_\alpha} (\varphi, t_{ii}^\alpha) t_{ii}^\alpha = \sum_\alpha n_\alpha \sum_{i=1}^{n_\alpha} \frac{1}{n_\alpha}(\varphi, \chi^\alpha) t_{ii}^\alpha$$

$$= \sum_\alpha \sum_{i=1}^{n_\alpha} (\varphi, \chi^\alpha) t_{ii}^\alpha = \sum_\alpha (\varphi, \chi^\alpha) \chi^\alpha. \tag{3.3.10}$$

That is, every function $\varphi \in L^2(G) \cap Z_1(G)$ can be expressed as a series of characters. $\quad\square$

Remark. For φ in $Z_1(G)$, the series $\sum_\alpha (\varphi, \chi^\alpha)\chi^\alpha$ need not converge pointwise to φ, but the function φ is determined by the coefficients (φ, χ^α). Furthermore, every function $\varphi \in L^1(G)$ is uniquely determined by its coefficients (φ, t_{ij}^α). We now prove this.

Let U be a neighborhood of the identity element in G. Let U_1 be a neighborhood of the identity element such that $U_1^3 \subset U$ and $U_1^{-1} = U_1$. Let $g_1, \ldots, g_n$ be elements of G such that the sets $g_i U_1$ $(i = 1, \ldots, n)$ cover G. Let U_2 be a neighborhood of the identity element such that $g_i^{-1} U_2 g_i \subset U_1$ for all $i = 1, \ldots, n$, and let U_3 be a neighborhood of the identity element such that $U_3^2 \subset U_2$. Denote the function $(\mu(U_3))^{-1}\xi_{U_3}$ by θ_U and set $\psi_U = \theta_U * \theta_U$. According to IV in 3.1, the function ψ_U is continuous. Plainly ψ_U is nonnegative and vanishes outside the neighborhood U_2. We also have

$$\int_G \psi_U(g)\, dg = \int_G \left(\int_G \theta_U(h)\theta(h^{-1}g)\, dh \right) dg = \left(\int_G \theta_U(g)\, dg \right)^2 = 1.$$

We set $\varphi_U(g) = \int_G \psi_U(h^{-1}gh)\, dh$. The function φ_U is nonnegative, continuous, and satisfies (3.3.4). For each $h \in G$ there exists an element g_k such that h belongs to $g_k U_1$ and therefore $h^{-1}U_2 h \subset U_1^{-1}g_k^{-1}U_2 g_k U_1 \subset U_1^3 \subset U$. That is, the function φ_U vanishes outside the neighborhood U. Note as well that $\int_G \psi_U(g)\, dg = 1$.

Consider the convolution $\varphi * \varphi_U$. Since φ_U is continuous, we infer from the identity $(\varphi * \varphi_U)(g) = \int_G \varphi(h)\varphi_U(h^{-1}g)\, dh$ that the function $\varphi * \varphi_U$ is also continuous. *A fortiori* $\varphi * \varphi_U$ belongs to $L^2(G)$. Consider the expansion of $\varphi * \varphi_U$ in the series (2.4.6). We compute the coefficients in this series expansion:

$$(\varphi * \varphi_U, t_{ij}^\alpha) = \int_G \left(\int_G \varphi(h)\varphi_U(h^{-1}g)\, dh \right) \overline{t_{ij}^\alpha(g)}\, dg$$

$$= \iint_{G \times G} \varphi(h)\overline{t_{ij}^\alpha(g)}\varphi_U(h^{-1}g)\, dh\, dg$$

$$= \iint_{G \times G} \varphi(h)\varphi_U(k)\overline{t_{ij}^\alpha(hk)}\, dh\, dk$$

$$= \iint_{G \times G} \varphi(h)\varphi_U(k) \sum_{l=1}^{n_\alpha} \overline{t_{il}^\alpha(h)t_{lj}^\alpha(k)}\, dh\, dk$$

$$= \sum_{l=1}^{n_\alpha} \left(\int_G \varphi(h)\overline{t_{il}^\alpha(h)}\, dh \right)\left(\int_G \varphi_U(k)\overline{t_{lj}^\alpha(k)}\, dk \right)$$

$$= \sum_{l=1}^{n_\alpha} (\varphi, t_{il}^\alpha)(\varphi_U, t_{lj}^\alpha). \qquad (3.3.11)$$

Putting (3.3.9) for $(\varphi, t^{\alpha}_{ij})$ in (3.3.11), we obtain

$$(\varphi * \varphi_U, t^{\alpha}_{ij}) = \frac{1}{n_{\alpha}} (\varphi, t^{\alpha}_{ij})(\varphi_U, \chi^{\alpha}). \qquad (3.3.12)$$

Therefore (2.4.6) and (3.3.12) yield

$$\varphi * \varphi_U = \sum_{\alpha} \sum_{i,\,j=1}^{n_{\alpha}} (\varphi, t^{\alpha}_{ij})(\varphi_U, \chi^{\alpha}) t^{\alpha}_{ij}. \qquad (3.3.13)$$

Thus $\varphi * \varphi_U$ is uniquely defined by the coefficients $(\varphi, t^{\alpha}_{ij})$. Since $\varphi * \varphi_U$ converges to φ in $L^1(G)$ as U shrinks to e, the function φ is determined by the coefficients $(\varphi, t^{\alpha}_{ij})$.

3.4. The Structure of a Group Algebra

I. *The characters of irreducible continuous unitary representations of the group G enjoy the following properties:*

$$\chi^{\alpha} * \chi^{\alpha'} = \begin{cases} 0, & \text{if } T^{\alpha} \text{ and } T^{\alpha'} \text{ are inequivalent;} \\ n_{\alpha}^{-1}\chi^{\alpha}, & \text{if } T^{\alpha} = T^{\alpha'}. \end{cases} \qquad (3.4.1)$$

Proof. The orthogonality relations in (2.4.1) give us

$$(t^{\alpha}_{ij} * t^{\alpha'}_{kl})(g) = \int_G t^{\alpha}_{ij}(h) t^{\alpha'}_{kl}(h^{-1}g)\, dh$$

$$= \sum_{m=1}^{n_{\alpha'}} \left(\int_G t^{\alpha}_{ij}(h) t^{\alpha'}_{km}(h^{-1})\, dh \right) t^{\alpha'}_{ml}(g)$$

$$= \sum_{m=1}^{n_{\alpha}} \left(\int_G t^{\alpha}_{ij}(h) \overline{t^{\alpha'}_{mk}(h)}\, dh \right) t^{\alpha'}_{ml}(g)$$

$$= \begin{cases} 0 & \text{for } \alpha \neq \alpha', \text{ or } j \neq k, \\ n_{\alpha}^{-1} t^{\alpha}_{il}(g), & \text{for } \alpha = \alpha', \text{ and } j = k. \end{cases} \qquad (3.4.2)$$

The equalities (3.4.1) follow at once. $\square$

II. *The following equalities hold:*

$$t^{\alpha}_{ij} * \chi^{\alpha'} = \begin{cases} 0, & \text{if } T^{\alpha} \text{ and } T^{\alpha'} \text{ are inequivalent} \\ n_{\alpha}^{-1} t^{\alpha}_{ij}, & \text{if } T^{\alpha} = T^{\alpha'}. \end{cases} \qquad (3.4.3)$$

See (3.4.2). $\square$

III. *Let T^α be an irreducible continuous unitary representation of the group G. Let χ^α be its character and let I_α be the set of functions $f \in L^1(G)$ satisfying the relation $f * n_\alpha \chi^\alpha = f$. Then I_α is a finite-dimensional (hence closed) minimal symmetric two-sided ideal in $L^1(G)$. It consists of all linear combinations of the matrix elements of the representation T^α. The ideal I_α is isomorphic (as a symmetric algebra) to the algebra of all linear operators in the space of the representation T^α.*

Proof. Plainly I_α is a linear subspace of $L^1(G)$. Let f belong to I_α and let φ belong to $L^1(G)$. Then we have $(\varphi * f) * n_\alpha \chi^\alpha = \varphi * (f * n_\alpha \chi^\alpha) = \varphi * f$. Therefore $\varphi * f$ belongs to I_α and so I_α is a left ideal in $L^1(G)$. Since χ^α belongs to $Z_1(G)$ (see III in 3.3), we also have $(f * \varphi) * n_\alpha \chi^\alpha = n_\alpha \chi^\alpha * (f * \varphi) = (n_\alpha \chi^\alpha * f) * \varphi = f * \varphi$. That is, I_α is a two-sided ideal in $L^1(G)$. The relation $f = f * n_\alpha \chi^\alpha$ means that

$$f(g) = n_\alpha \int_G f(h)\chi^\alpha(h^{-1}g)\,dh = \sum_{i=1}^{n_\alpha} n_\alpha \int_G f(h)t_{ii}^\alpha(h^{-1}g)\,dh$$

$$= \sum_{i=1}^{n_\alpha} \sum_{j=1}^{n_\alpha} n_\alpha \left(\int_G f(h)t_{ij}^\alpha(h^{-1})\,dh \right) t_{ji}^\alpha(g), \qquad (3.4.4)$$

Thus f is a linear combination of matrix elements of the representation T^α. Conversely, if f is a linear combination of matrix elements of T^α, then f belongs to I_α by Proposition II. Therefore I_α coincides with the set of linear combinations of matrix elements of T^α. Since $(t_{ij}^\alpha)^*(g) = \overline{t_{ij}^\alpha(g^{-1})} = t_{ji}^\alpha(g)$, I_α is symmetric.

To prove that I_α is minimal (that is, that there are no nonzero ideals $I'_\alpha \subsetneqq I_\alpha$ of the algebra $L^1(G)$), we need to show that I_α is a simple algebra. Let f belong to I_α. We then have $f = \sum c_{ij}t_{ij}^\alpha$, where c_{ij} are certain complex numbers. We map f into the square $n_\alpha \times n_\alpha$ matrix having elements $(n_\alpha^{-1}c_{ij})$. This correspondence is clearly one-to-one and is a linear mapping of I_α onto the algebra of all square matrices of order n_α. Suppose that $f_1 = \sum c_{ij}^1 t_{ij}^\alpha$ and that $f_2 = \sum c_{ij}^2 t_{ij}^\alpha$. The relations (3.4.2) give

$$f_1 * f_2 = \sum_{i,j,k,l} c_{ij}^1 c_{kl}^2 (t_{ij}^\alpha * t_{kl}^\alpha) = \sum_{i,j,l} c_{ij}^1 c_{jl}^2 n_\alpha^{-1} t_{il}^\alpha$$

$$= \sum_{i,l} \left(n_\alpha^{-1} \sum_{j=1}^{n_\alpha} c_{ij}^1 c_{jl}^2 \right) t_{il}^\alpha. \qquad (3.4.5)$$

The equalities (3.4.5) show that the image of a convolution is the matrix product of the image matrices. Therefore I_α is algebra-isomorphic to the algebra of all $n_\alpha \times n_\alpha$ matrices and so is isomorphic to the algebra of all linear operators in the space of the representation T^α. Since the latter algebra of matrices is simple, we have proved that I_α is a simple algebra. $\square$

Let B be a Banach space. Let $\{B_\alpha\}$ be a family of closed subspaces of B. The space B is called the *closed direct sum* of the subspaces $\{B_\alpha\}$ if: (1) B is the closure of the set of all finite sums $x_{\alpha_1} + \cdots + x_{\alpha_n}$, $x_{\alpha_k} \in B_{\alpha_k}$, $k = 1, \ldots, n$, $n = 1, 2, \ldots$; (2) if a sequence $x^{(m)} = x^{(m)}_{\alpha_1} + x^{(m)}_{\alpha_2} + \cdots + x^{(m)}_m$ converges to zero in B (norm metric) and for a fixed α, the sequence $x^{(m)}_\alpha$, $m = 1, 2, \ldots$ converges in norm, then $\|x^{(m)}_\alpha\| \to 0$.

IV.

(a) Any nonzero minimal closed two-sided ideal in $L^1(G)$ coincides with one of the ideals I_α.

(b) The algebra $L^1(G)$ is the closed direct sum of its minimal closed two-sided ideals.

Proof.

(a) Let I be a nonzero minimal closed two-sided ideal in $L^1(G)$. Let φ be a nonzero element of I. If (φ, t^α_{ij}) vanishes for all α, i, j, then $\varphi * \varphi_U = 0$ for all U, by (3.3.13). Since $\|\varphi * \varphi_U - \varphi\|_{L^1(G)} \to 0$ it follows that $\varphi = 0$. Therefore there exist an α and i, j such that $(\varphi, t^\alpha_{ij}) \neq 0$. We next write

$$(\varphi * t^\alpha_{pq})(g) = \int_G \varphi(h) t^\alpha_{pq}(h^{-1}g)\, dh = \sum_{k=1}^{n_\alpha} \left(\int_G \varphi(h) t^\alpha_{pk}(h^{-1})\, dh \right) t^\alpha_{kq}(g)$$

$$= \sum_{k=1}^{n_\alpha} \left(\int_G \varphi(h) \overline{t^\alpha_{kp}(h)}\, dh \right) t^\alpha_{kq}(g) = \sum_{k=1}^{n_\alpha} (\varphi, t^\alpha_{kp}) t^\alpha_{kq}(g). \tag{3.4.6}$$

This identity shows that for $p = j$, at least one of the coefficients (φ, t^α_{kp}) in (3.4.6) is nonzero. Therefore we have $\varphi * t^\alpha_{jq} \neq 0$ for all q. But III shows that t^α_{jq} belongs to I_α and thus the convolution $\varphi * t^\alpha_{jq}$ belongs to both I and I_α. Since $\varphi * t^\alpha_{jq}$ is not zero, we have $I \cap I_\alpha \neq (0)$. Since $I \cap I_\alpha$ is a nonzero closed two-sided ideal that is contained in both of the minimal ideals I and I_α, we must have $I \cap I_\alpha = I = I_\alpha$.

(b) Let φ belong to $L^1(G)$. By (3.3.13) all functions $\varphi * \varphi_U$ are the limits in $L^2(G)$ of finite linear combinations of matrix elements of the representations T^α. By III, these matrix elements belong to I_α and therefore there exists a sequence of finite sums $\psi_{\alpha_1} + \cdots + \psi_{\alpha_n}$, $\psi_{\alpha_k} \in I_{\alpha_k}$, $k = 1, \ldots, n$, $n = 1, 2, \ldots$, converging to $\varphi * \varphi_U$ in $L^2(G)$. By (3.1.10) and (3.1.11), this sequence also converges to $\varphi * \varphi_U$ in $L^1(G)$. But V in 3.1 implies that $\varphi * \varphi_U - \varphi \to 0$ in $L^1(G)$ and thus $L^1(G)$ is the closure of the set of all finite sums $\psi_{\alpha_1} + \cdots + \psi_{\alpha_n}$, where $\psi_{\alpha_k} \in I_{\alpha_k}$.

Let the sequence $\varphi^{(m)} = \varphi^{(m)}_{\alpha_1} + \cdots + \varphi^{(m)}_{\alpha_m}$ be norm-convergent to zero in $L^1(G)$, where $\varphi^{(m)}_{\alpha_k} \in I_{\alpha_k}$. We then have $\varphi^{(m)} = \varphi^{(m)}_{\alpha_1} * n_{\alpha_1}\chi^{\alpha_1} + \cdots + \varphi^{(m)}_{\alpha_m} * n_{\alpha_m}\chi^{\alpha_m}$ and hence $\varphi^{(m)} * n_\alpha\chi^\alpha = \varphi^{(m)}_{\alpha_1} * n_{\alpha_1}n_\alpha(\chi^{\alpha_1} * \chi^\alpha) + \cdots + \varphi^{(m)}_{\alpha_m} * n_{\alpha_m}n_\alpha(\chi^{\alpha_m} * \chi^\alpha)$ for all α. We infer from (3.4.1) that $\chi^{\alpha_k} * \chi^\alpha = 0$ for $\alpha \neq \alpha_k$ and $\chi^{\alpha_k} * \chi^\alpha = n_\alpha\chi^{\alpha_k}$

for $\alpha = \alpha_k$. Thus we have

$$\varphi^{(m)} * n_\alpha \chi^\alpha = \varphi_\alpha^{(m)} * n_\alpha \chi^\alpha = \varphi_\alpha^{(m)} \tag{3.4.7}$$

for all α. We must also have

$$\left\| \varphi_\alpha^{(m)} \right\| \leqslant n_\alpha \left\| \varphi^{(m)} \right\| \left\| \chi^{\alpha_k} \right\|. \tag{3.4.8}$$

From (3.4.8) and from the hypothesis $\varphi^{(m)} \to 0$ it follows that $\left\| \varphi_\alpha^{(m)} \right\| \to 0$ for all α. This completes the proof of IV. $\square$

The basic results of this chapter remain valid (with appropriate changes) for locally compact groups that are not compact (see for example, Naĭmark [1], chapter VI).

Chapter V

Finite-Dimensional Representations of Connected Solvable Groups; the Theorem of Lie

§1. Connected Topological Groups

1.1. Connected Topological Spaces

Let X be a set. We say that two sets, $M \subset X$ and $N \subset X$, form a *dissection* of the set X, if $M \cup N = X$, $M \cap N = \emptyset$, $M \neq \emptyset$ and $N \neq \emptyset$.

A topological space X is called *connected* if it admits no decomposition into closed sets. Otherwise X is called *disconnected*.

A set M in the topological space X is called *connected* if it is connected when considered as a subspace in X. Otherwise M is called *disconnected*. A connected open set in a topological space is called a *region*.

Remark. In these definitions we can substitute the word "open" for the word "closed". For, if $X = U_1 \cup U_2$, where U_1, U_2 are open and $U_1 \cap U_2 = \emptyset$, then U_1, U_2 are also closed, being the complements of the open sets U_2, U_1 respectively.

I. *Closed intervals $[a, b]$ in* **R** *are connected.*

Proof. Assume that $[a, b] = F_1 \cup F_2$, where F_1, F_2 are closed, $F_1 \cap F_2 = \emptyset$ and $F_1 \neq \emptyset$, $F_2 \neq \emptyset$. The right end point b belongs to F_1 or to F_2; let us say that $b \in F_2$. We set $x_0 = \sup F_1$. Since F_1 is closed, x_0 belongs to F_1 and hence does not belong to F_2. Furthermore, x_0 is less than b since $b \notin F_1$. The definition of x_0 shows that $(x_0, b) \subset F_2$. Thus x_0 belongs to $\bar{F}_2 = F_2$ and we arrive at a contradiction, since $x_0 \notin F_2$. $\square$

II. *A continuous image of a connected set is connected.*

Proof. Let X be connected and let f be a continuous mapping of X onto Y: $Y = f(X)$. Assume that Y has the form $F_1 \cup F_2$, where $F_1 \neq \emptyset$, $F_2 \neq \emptyset$, $F_1 \cap F_2 = \emptyset$ and F_1, F_2 are closed. We set $\tilde{F}_1 = f^{-1}(F_1)$, $\tilde{F}_2 = f^{-1}(F_2)$. Since f is continuous, $\tilde{F}_1$ and $\tilde{F}_2$ are closed. We also have $X = \tilde{F}_1 \cup \tilde{F}_2$, $\tilde{F}_1 \neq \emptyset$, $\tilde{F}_2 \neq \emptyset$, and $\tilde{F}_1 \cap \tilde{F}_2 = \emptyset$. That is, X is not connected. $\square$

III. *A continuous curve is connected.*

The proof follows from I and II since a continuous curve is a continuous image of a closed interval (see 1.2, chapter III). $\square$

IV. *If a subset M of X is dense in X and M is connected, then X is also connected.*

Proof. Assume that X is disconnected: $X = U_1 \cup U_2$, where U_1, U_2 are open in X (see the remark on page 229), $U_1 \neq \emptyset$ and $U_1 \cap U_2 = \emptyset$. We set $M \cap U_1 = \tilde{U}_1$, $M \cap U_2 = \tilde{U}_2$. Thus M is the union $\tilde{U}_1 \cup \tilde{U}^2$ of the open subsets $\tilde{U}_1$ and $\tilde{U}_2$, and $\tilde{U}_1 \cap \tilde{U}_2 = \emptyset$.

Both $\tilde{U}_1$ and $\tilde{U}_2$ are nonvoid. This follows from the hypothesis that M is dense in X: a dense subset of a topological space has nonvoid intersection with every nonvoid open subset. Thus M is disconnected if X is. $\square$

V. *If M is connected, then any set M_1 such that $M \subset M_1 \subset \bar{M}$ is connected. In particular, if M is connected, $\bar{M}$ is also connected.*

The proof follows from IV, since M_1 and M are dense in $\bar{M}$. $\square$

VI. *Let X be a topological space. The union of any family of connected subsets of X that have a common point is connected.*

Proof. Let M equal $\bigcup_\alpha M_\alpha$, where M_α is connected, α runs through some index set and every M_α contains a fixed x. Assume that $M = U_1 \cup U_2$, where U_1, U_2 are open sets in M, $U_1 \neq \emptyset$, $U_2 \neq \emptyset$ and $U_1 \cap U_2 = \emptyset$. There are open subsets V_1 and V_2 of X such that $U_1 = V_1 \cap M$, $U_2 = V_2 \cap M$. Thus we have

$$V_1 \cap M \neq \emptyset, \quad V_2 \cap M \neq \emptyset, \quad \text{and } V_1 \cap V_2 \cap M = \emptyset. \quad (1.1.1)$$

The point x belongs to one of the sets V_1, V_2; suppose that

$$x \in V_1. \quad (1.1.2)$$

On the other hand, (1.1.1) implies that

$$V_2 \cap M_\alpha \neq \emptyset, \quad \text{for some } \alpha. \quad (1.1.3)$$

Write $V_1 \cap M_\alpha = W_{1\alpha}$, $V_2 \cap M_\alpha = W_{2\alpha}$. Then $W_{1\alpha}$ and $W_{2\alpha}$ are open in M_α, $W_{1\alpha} \neq \emptyset$ by (1.1.2) and $W_{2\alpha} \neq \emptyset$ by (1.1.3) (since $x \in M_\alpha$, and

$$M_\alpha = M \cap M_\alpha = (U_1 \cap M_\alpha) \cup (U_2 \cap M_\alpha)$$
$$= (V_1 \cap M_\alpha) \cup (V_2 \cap M_\alpha) = W_{1\alpha} \cup W_{2\alpha}.$$

Hence M_α is disconnected. $\square$

Let X be a topological space and let x be a point of X. The union of all connected subspaces of X that contain x is connected by virtue of VI and is clearly the largest connected set containing x. It is called the *connected component of* x and is denoted by $K(x)$. The connected component of x in a subspace M of X is called the *connected component of* x *in the set* $M \subset X$.

VII. *A connected component is closed.*

For, $\overline{K(x)}$ is connected by V and contains x. Since $K(x)$ is the largest connected set containing x, we have $\overline{K(x)} \subset K(x)$. Since we also have $K(x) \subset \overline{K(x)}$, it follows that $K(x) = \overline{K(x)}$.

VIII. *Two connected components are disjoint or identical.*

If $K(x) \cap K(y) \neq \varnothing$, then $K(x) \cup K(y)$ is connected and contains x. From this and from the definition of a connected component we conclude that $K(x) \supset K(x) \cup K(y)$, so that $K(y) \subset K(x)$. Interchange x and y to see that $K(y) = K(x)$. $\square$

By VIII the entire space X can be decomposed into a family of pairwise disjoint connected components. The space X is connected if and only if it coincides with the connected component of any of its points, or in other words, X consists of one connected component.

IX. *If every two points* $x, y \in X$ *are contained in a connected subset* $M \equiv X$. X *is connected.*

In this case X coincides with the connected component of any point. $\square$

X. *If every two points* $x, y \in X$ *can be joined by a curve in* X, X *is connected.*

In this case X is connected by IX since a continuous curve is connected (see III). $\square$

XI. *The space* X *is disconnected if and only if* X *can be continuously mapped onto a discrete space containing more than one point.*

Proof. Sufficiency follows directly from II, since a discrete space that contains more than one point is disconnected. Conversely, let X be disconnected and let $X = F_1 \cup F_2$, where F_1 and F_2 are closed, $F_1 \neq \varnothing$, $F_2 \neq \varnothing$ and $F_1 \cap F_2 = \varnothing$. Let $Y = \{a, b\}$ be a two-point discrete space, so that all four subsets of Y are closed. Setting $f(F_1) = \{a\}$, $f(F_2) = \{b\}$, we obtain a continuous mapping f of X onto Y. $\square$

XII. *The topological product* $X_1 \times \cdots \times X_n$ *of spaces* $X_1, \ldots, X_n$ *is connected if and only if each of the component spaces* $X_1, \ldots, X_n$ *is connected.*

Proof. Assume that $X_1, \ldots, X_n$ are connected and that $X = X_1 \times \cdots \times X_n$ is disconnected. By XI there is a continuous mapping f of the space X onto a discrete space Y containing more than one point. Consider and fix a point $a = \{a_1, \ldots, a_n\} \in X$. For a fixed $j(1 \leqslant j \leqslant n)$ define a mapping $f_j \colon X_j \to Y$ of the space X_j into Y, defined by the formula

$$f_j(x_j) = f(x'_1 \cdots x'_n),$$

where

$$x'_j = x_j \quad \text{and} \quad x'_k = a_k, \quad \text{for } k \neq j.$$

The mapping is continuous and therefore is a constant by virtue of XI. Thus we have

$$f(a_1 \times \cdots \times a_{j-1} \times x_j \times a_{j+1} \times \cdots \times a_n)$$
$$= f(a_1 \times \cdots \times a_j \times \cdots \times a_n). \tag{1.1.4}$$

In particular we have

$$f(x_1 \times a_2 \times \cdots \times a_n) = f(a_1 \times a_2 \times \cdots \times a_n). \tag{1.1.5}$$

Applying (1.1.4) to $j = 2$ and $a = x_1 \times a_2 \times \cdots \times a_n$, we conclude that $f(x_1 \times x_2 \times a_3 \times \cdots \times a_n) = f(x_1 \times a_2 \times \cdots \times a_n)$ and therefore (see (1.1.5))

$$f(x_1 \times x_2 \times a_3 \times \cdots \times a_n) = f(a_1 \times a_2 \times \cdots \times a_n).$$

Repeating this process, we conclude that $f(x_1 \times \cdots \times x_n) = f(a_1 \times \cdots \times a_n)$, that is, f maps X onto one point alone. This contradiction proves that the product space is connected.

Conversely, if X is connected, then every X_j is connected as the continuous image of the space X under the continuous mapping $x_1 \times \cdots \times x_n \to x_j$ (see II). □

Examples

1. The space $\mathbf{R}^1$ is connected. Indeed, any two points a, b $(a < b)$ are contained in the closed interval $[a, b]$, which is connected by virtue of I. Now apply IX.

2. The space $\mathbf{R}^n$ is connected as the topological product of n copies of the connected space $\mathbf{R}^1$. Note too that each pair of points $a = (a_1, \ldots, a_n)$ and $b = (b_1, \ldots, b_n)$ of $\mathbf{R}^n$ lie in a "closed interval"

$$x_j = (1 - t)a_j + tb_j, \quad 0 \leqslant t \leqslant 1.$$

Then apply IX.

3. The space $\mathbf{C}^1$ is connected as the topological product $\mathbf{R}^1 \times \mathbf{R}^1$.

4. The space $\mathbf{C}^n$ is connected as the product of n copies of the space $\mathbf{C}^1$.

5. The set $z = e^{i\varphi}$, $0 \leqslant \varphi \leqslant \varphi_0$, where $\varphi \leqslant 2\pi$ (an arc of the unit circle or the entire unit circle) is connected as the image of the closed interval $[0, \varphi_0]$ under the continuous mapping $\varphi \to e^{i\varphi}$ (see I, II and III).

6. The sets $M_1 = \{(x_1, x_2): x_1^2 + x_2^2 < 1\}$ (the interior of the unit circle) and $M_2 = \{(x_1, x_2): x_1^2 + x_2^2 > 1\}$ (the exterior of the unit circle) in $\mathbf{R}^2$ are connected. For, every pair of points in M_1 lies in a line segment contained in M_1. Every pair of points a, b in M_2 lies on an arc of a circle with center at $(a + b)/2$, which lies entirely in M_2. The set $M_0 = \{(x_1, x_2): x_1^2 + x_2^2 \neq 0\}$ (the plane with the point $(0, 0)$ deleted) is connected. (Imitate the proof that M_2 is connected.)

7. The set $M = \{(x_1, x_2): x_1^2 + x_2^2 \neq 1\}$ in $\mathbf{R}^2$ is *disconnected*. It is clear that $M = M_1 \cup M_2$ (see example 6) and M_1 and M_2 are connected components of the set M.

8. The set M of all rational numbers, considered as a subspace of $\mathbf{R}^1$, is disconnected. Prove this.

1.2. Connected Subsets of a Topological Group, Connected Topological Groups

A subset S of a topological group G is called *connected* if S is a connected subset of the topological space G.

I. *If S is a connected subset of the topological group G, then the set S^{-1} is connected and the sets $g_0 S$, $S g_0$, $g_0^{-1} S g_0$ are connected for every $g_0 \in G$.*

Note that S^{-1}, $g_0 S$, $S g_0$, $g_0^{-1} S g_0$ are images of the connected set S under the continuous mappings $g \to g^{-1}$, $g \to g_0 g$, $g \to g g_0$, $g \to g_0^{-1} g g_0$ respectively. Then apply I of 1.1. $\square$

II. *If $S_1, S_2, \ldots, S_n$ are connected subsets of a topological group G, then the set $S_1 S_2 \cdots S_n$ is connected. In particular, if S is connected, then S^n is connected for every $n = 1, 2, 3, \ldots$.*

Proof. The correspondence $\{g_1, g_2, \ldots, g_n\} \to g_1 g_2 \cdots g_n$ is a continuous mapping of the topological product $S_1 \times S_2 \times \cdots \times S_n$ onto $S_1 S_2 \cdots S_n$. Since $S_1 \times S_2 \times \cdots \times S_n$ is connected (XII of 1.1), its continuous image $S_1 S_2 \cdots S_n$ is also connected (II of 1.1). For $S_1 = S_2 = \cdots = S_n$ we find that S^n is connected if S is connected. $\square$

A topological group G is called *connected* if the topological space G is connected. Otherwise G is called *disconnected*. By VIII of 1.1, G decomposes into pairwise disjoint connected components, which can be described as follows.

III. *The connected component K of the identity element e of G is a closed normal subgroup of G.*

Proof. If g belongs to K, the set $g^{-1}K$ is connected and contains $g^{-1}g = e$. Thus, $g^{-1}K$ is contained in K for every $g \in K$. Hence K is a subgroup of G. By III of 1.1, K is closed. Furthermore, the mapping $g \to g_0^{-1}gg_0$ is a homeomorphism of G onto itself that carries e into $g_0^{-1}gg_0 = e$. Therefore $g_0^{-1}Kg_0$ is a connected set containing e and thus $g_0^{-1}Kg_0 \subset K$ for every $g \in G$. Thus K is a normal subgroup of G. $\square$

IV. *The connected component of an element $g_0 \in G$ has the form $g_0K = Kg_0$.*

To prove this, use III and the fact that the mapping $g \to g_0g$ is a homeomorphism of G onto itself that carries e into g_0. $\square$

V. *If G is a connected topological group and V is a neighborhood of the identity in G, then $G = \bigcup_{n=1}^{\infty} V^n$.*

Proof. We set $M = \bigcup_{n=1}^{\infty} V^n$. It is clear that $e \in M$ and that M is the union of open sets V^n and so is open. On the other hand, we have $\bar{M} \subset MV = \bigcup_{n=1}^{\infty} V^nV \subset M$. Hence $\bar{M}$ is equal to M, that is, M is closed. If M is not equal to G, then M and $G\backslash M$ form a decomposition of the space G into nonvoid closed disjoint sets, so that G is disconnected. The equality $G = M$ follows. $\square$

VI. *Let G be a connected topological group. Let N be a discrete normal subgroup of G. Then N is contained in the center of the group G.*

Proof. Let n be any element of the group N. Consider the mapping φ of the group G into N, defined by the formula $\varphi(g) = gng^{-1}$. Since φ is continuous and G is connected, $\varphi(G)$ is a connected set. On the other hand, $\varphi(G) \subset N$ and thus $\varphi(G)$ is discrete. Being connected, $\varphi(G)$ must consist of one element. Since $\varphi(e) = n$, we have $\varphi(g) = n$ for all $g \in G$, which is to say that $gng^{-1} = n$, $gn = ng$ for all $n \in N$ and $g \in G$. Thus N is contained in the center of G. $\square$

Examples

1. The groups $\mathbf{R}^1$, $\mathbf{C}^1$, $\mathbf{R}^n$, $\mathbf{C}^n$ are connected, since the topological spaces $\mathbf{R}^1$, $\mathbf{C}^1$, $\mathbf{R}^n$, $\mathbf{C}^n$ are connected (see examples 1–4 in 1.1).

2. The multiplicative group $\mathbf{R}_0$ is *disconnected*. Its connected components are $\mathbf{R}_0^+$ (the component of 1) and $\mathbf{R}_0^-$ the set of all negative real numbers (the component of -1).

3. The multiplicative group $\mathbf{C}_0$ is connected (see example 5 in 1.1).

4. The group K_n. We denote the set of all complex matrices

$$k = \begin{Vmatrix} k_{11} & 0 & 0 & \cdots & 0 \\ k_{21} & k_{22} & 0 & \cdots & 0 \\ \multicolumn{5}{c}{\cdots\cdots\cdots\cdots\cdots} \\ k_{n1} & k_{n2} & k_{n3} & \cdots & k_{nn} \end{Vmatrix},$$

where all $k_{jj} \neq 0$ as K_n. On easily verifies that K_n is a group under matrix multiplication, so that K_n is a subgroup of the group $GL(n, \mathbf{C})$. Topologize K_n with the subspace topology inherited from $GL(n, \mathbf{C})$.

VII. *K_n is a closed subgroup of $GL(n, \mathbf{C})$.*

For, K_n is just the set of matrices $g \in GL(n, \mathbf{C})$ such that

$$g_{jl} = 0, \quad \text{for } l > j.$$

Since g_{jk} are continuous functions on $GL(n, \mathbf{C})$, our claim follows from III in 1.8, chapter III. $\square$

VIII. *The group K_n is connected.*

The parameters k_{jl}, $l < j$, of the elements of K run independently through $\mathbf{C}^1$ and the diagonal elements k_{jj} run independently through $\mathbf{C}_0^1$. A moment's thought shows that K is homeomorphic to the topological product of $n(n-1)/2$ copies of the space $\mathbf{C}^1$ and n copies of the space $\mathbf{C}_0^1$. Inasmuch as $\mathbf{C}^1$ and $\mathbf{C}_0^1$ are connected, K is also connected by XII of 1.1. $\square$

5. The group D_n. The set of all diagonal matrices

$$\delta = \begin{Vmatrix} \lambda_1 & 0 & \cdots & 0 \\ 0 & \lambda_2 & \cdots & 0 \\ \multicolumn{4}{c}{\cdots\cdots\cdots\cdots} \\ 0 & 0 & \cdots & \lambda_n \end{Vmatrix},$$

for which $\det \delta = \lambda_1 \lambda_2 \cdots \lambda_n \neq 0$ is denoted by D_n. Plainly D_n is a closed subgroup of the group $GL(n, \mathbf{C})$. Provide D_n with the subspace topology inherited from $GL(n, \mathbf{C})$.

IX. *The group D_n is connected.*

We need only note that D_n is homeomorphic to the topological product of n copies of the space $\mathbf{C}_0^1$ and apply XII of 1.1. $\square$

6. The group $U(n)$. We will make use of the following simple lemma.

Lemma 1. *Every matrix $u \in U(n)$ can be represented in the form $u = e^{ih}$, where h is Hermitian. Conversely, every matrix e^{ih}, where h is Hermitian, is a unitary matrix.*

Proof. Any matrix $u \in U(n)$ can be reduced to diagonal form by a unitary transformation, that is,

$$
u = v \left\| \begin{matrix} e^{i\varphi_1} & \cdots & 0 \\ \multicolumn{1}{c}{\cdots\cdots\cdots\cdots\cdots} \\ 0 & \cdots & e^{i\varphi_n} \end{matrix} \right\| v^{-1},
$$

where $v \in U(n)$ and $\varphi_1, \ldots, \varphi_n$ are real numbers. We set

$$
h = v \left\| \begin{matrix} \varphi_1 & 0 & \cdots & 0 \\ 0 & \varphi_2 & \cdots & 0 \\ \multicolumn{1}{c}{\cdots\cdots\cdots\cdots\cdots} \\ 0 & 0 & \cdots & \varphi_n \end{matrix} \right\| v^{-1}.
$$

The matrix h is Hermitian and $u = e^{ih}$. Conversely, if $u = e^{ih}$, where h is Hermitian, then $u^* = e^{-ih}$ and $u^*u = uu^* = e$. $\square$

X. *The group $U(n)$ is connected.*

Proof. Let $u_1 = e^{ih_1}, u_2 = e^{ih_2}$, where h_1, h_2 are Hermitian. We set $h(t) = th_1 + (1-t)h_2$, where $0 \leqslant t \leqslant 1$, and we define $u(t) = e^{ih(t)} = e^{ith_1 + i(1-t)h_2}$. Plainly $u(t)$, $0 \leqslant t \leqslant 1$, is a continuous curve in $U(n)$ that joins u_1 and u_2. Hence, every two points in $U(n)$ can be joined by a continuous curve in $U(n)$, so that $U(n)$ is connected. In particular, $U(1)$ is connected. (This is also clear since the group $U(1)$ is homeomorphic to the unit circle.) The groups Γ^1, $\mathcal{T}^1$ and $SO(2, \mathbf{R})$ are topologically isomorphic to $U(1)$ and so are connected. $\square$

7. The group $SU(n)$.

XI. *The group $SU(n)$ is connected.*

Proof. The topological product $SU(n) \times U(1)$ is homeomorphic to the connected group $U(n)$: see V of 1.2, chapter IV, and X. Thus it is connected. From this and XII of 1.1 we infer that $SU(n)$ is connected. $\square$

8. The group $SO(3, \mathbf{R})$. Let I denote the topological product $[0, 2\pi] \times [0, 2\pi] \times [0, \pi]$. As the topological product of closed intervals (see I and XII in 1.1) I is connected. On the other hand, $SO(3, \mathbf{R})$ is a continuous image of I. This is implied by formulas (1.2.20) in chapter IV. The following is now clear.

XII. *The group $SO(3, \mathbf{R})$ is connected.*

9. The group $O(3, \mathbf{R})$. The function $f(g) = \det g$ is continuous on $GL(n, \mathbf{C})$ and so its restriction to $O(3, \mathbf{R})$ is also continuous. But in $O(3, \mathbf{R})$, $\det g$ assumes the values 1 and -1 and no others. The following is now obvious.

XIII. *The group $O(3, \mathbf{R})$ is disconnected.*

The group $SO(3, \mathbf{R})$ is connected and is contained in $O(3, R)$. If $\det g = -1$ and $\det g_0 = -1$, then $\det g_0^{-1} g = 1$ and thus g_0^{-1} belongs to $SO(3, \mathbf{R})$. Therefore we have

$$\{g : \det g = -1\} = g_0 SO(3, \mathbf{R}).$$

We have proved the following.

XIV. *The group $O(3, \mathbf{R})$ is the union of two connected components $SO(3, \mathbf{R})$ and $g_0 SO(3, \mathbf{R})$, where g_0 is any fixed element for which $\det g_0 = -1$. Thus $SO(3, \mathbf{R})$ is the connected component of the identity element of the group $O(3, \mathbf{R})$.*

Exercise

Determine the connectedness or disconnectedness of the groups $O(n, \mathbf{R})$, $SO(n, \mathbf{R})$, $O(n, \mathbf{C})$ and $SO(n, \mathbf{C})$.

10. The group $GL(n, \mathbf{C})$. We first prove a lemma.

Lemma 2. *Every matrix $g \in GL(n, \mathbf{C})$ can be represented in the form*

$$g = ku, \quad \text{where } k \in K_n \text{ and } u \in U(n). \tag{1.2.1}$$

Proof. Consider the rows of the matrix g as vectors in $\mathbf{C}^n$. Since $\det g \neq 0$, these vectors are linearly independent. First we apply the Gram-Schmidt orthonormalization process to these vectors. Multiply $\{g_{11}, \ldots, g_{1n}\}$ be a number k_{11} such that the resulting vector, which we write as $\{u_{11}, \ldots, u_{1n}\}$, has norm 1. Next choose a linear combination of $\{u_{11}, \ldots, u_{1n}\}$ and $\{g_{21}, \ldots, g_{2n}\}$ with coefficients k_{21}, k_{22}, such that the resulting vector, which we write as $\{u_{21}, \ldots, u_{2n}\}$, has norm 1 and is orthogonal to $\{u_{11}, \ldots, u_{1n}\}$. Repeating this process, we obtain

$$kg = u, \tag{1.2.2}$$

where $k \in K$ and the rows of the matrix u are an orthonormal set. Thus u belongs to $U(n)$. From (1.2.2) we conclude that $g = k^{-1}u$ where $k^{-1} \in K_n$ and $u \in U(n)$. This is a factorization as called for in (1.2.1). $\square$

The factorization (1.2.1) is called a *Cramer factorization*.

XV. *The group $GL(n, \mathbf{C})$ is connected.*

By lemma 2, $GL(n, \mathbf{C})$ is equal to $K_n U(n)$ and our assertion follows from II, since K_n and $U(n)$ are connected.

11. The group $SL(n, \mathbf{C})$.

XVI. *The group $SL(n, \mathbf{C})$ is connected.*

Proof. The group $GL(n, \mathbf{C})$ is homeomorphic to the direct product $GL(1, \mathbf{C}) \times SL(n, \mathbf{C}) = \mathbf{C}_0^1 \times SL(n, \mathbf{C})$ (the proof is similar to the proof of V in 1.2, chapter IV). Since $GL(n, \mathbf{C})$ is connected, our assertion follows directly from XII of 1.1. □

§2. Solvable and Nilpotent Groups

2.1. The Algebraic Commutator

Let G be a group. The product $g_1 g_2 g_1^{-1} g_2^{-1}$ is called the *commutator of the elements $g_1, g \in G$.* If $g_1 g_2 g_1^{-1} g_2^{-1}$ is equal to e, then clearly $g_1 g_2 = g_2 g_1$.

Let S_1 and S_2 be nonvoid subsets of G. The smallest algebraic subgroup of G containing all commutators $g_1 g_2 g_1^{-1} g_2^{-1}$ for $g_1 \in S_1$ and $g_2 \in S_2$ is called the *algebraic commutator $K_a(S_1, S_2)$ of S_1 and S_2.* We call $K_a(G, G)$ the *algebraic commutator of the group G.* This subgroup is denoted by G_a'.

I. *The algebraic commutator G_a' of the group G is a normal subgroup of G and the factor group G/G_a' is commutative.*

Proof. Each element $h \in G_a'$ has the form

$$h = g_1 g_2 g_1^{-1} g_2^{-1} \cdots g_{2n-1} g_{2n} g_{2n-1}^{-1} g_{2n}^{-1}.$$

For every $g \in G$, we thus have

$$g^{-1}hg = (g^{-1}g_1 g)(g^{-1}g_2 g)(g^{-1}g_1 g)^{-1}(g^{-1}g_2 g)^{-1}$$
$$\cdots (g^{-1}g_{2n-1} g)(g^{-1}g_{2n} g)(g^{-1}g_{2n-1} g)^{-1}(g^{-1}g_{2n} g)^{-1} \in G_a';$$

Thus G_a' is a normal subgroup of G. Let $\tilde{g}_1, \tilde{g}_2$ belong to G/G_a' and let g_1, g_2 be elements of the cosets $\tilde{g}_1, \tilde{g}_2$, respectively. The coset $\tilde{g}_1 \tilde{g}_2 \tilde{g}_1^{-1} \tilde{g}_2^{-1}$ contains $g_1 g_2 g_1^{-1} g_2^{-1} \in G_a'$. Hence this coset is $G_a' = \tilde{e}$, the identity element in G/G_a'. Accordingly we have $\tilde{g}_1 \tilde{g}_2 = \tilde{g}_2 \tilde{g}_1$. □

2.2. Algebraically Solvable and Nilpotent Groups

For a group G we construct the following two sequences:

$$G_a^{(1)} = G_a', \; G_a^{(2)} = (G_a^{(1)})_a', \ldots, G_a^{(n)} = (G_a^{(n-1)})_a', \ldots; \qquad (2.2.1)$$

$$G_a^{[1]} = K_a(G, G) = G_a', \; G_a^{[2]} = K_a(G, G_a^{[1]}), \ldots,$$
$$G_a^{[n]} = K_a(G, G_0^{[n-1]}), \ldots. \qquad (2.2.2)$$

The group $G_a^{(n)}$ is called the *n-th algebraic derivative of the group* G. Clearly we have

$$G_a^{[n]} \supset G_a^{(n)}. \qquad (2.2.3)$$

The group G is called *algebraically solvable* if there is an n for which

$$G_a^{(n)} = \{e\}. \qquad (2.2.4)$$

The smallest positive integer for which (2.2.4) holds is called the *rank of the solvable group* G. For $n = 1$, G is commutative, so that solvable groups form a natural generalization of commutative groups. The rank of a solvable group measures the degree to which it is noncommutative. A group G is called *algebraically nilpotent* if there is an n for which

$$G_a^{[n]} = \{e\}. \qquad (2.2.5)$$

I. *An algebraically nilpotent group is algebraically solvable.*

This assertion follows at once from (2.2.3).

If G is algebraically solvable, then plainly all of the groups $G_a^{(k)}$ are algebraically solvable. If G is algebraically nilpotent, then all of the groups $G_a^{[k]}$ are algebraically nilpotent.

II. *Every subgroup H of an algebraically solvable (nilpotent) group G is algebraically solvable (nilpotent).*

We infer this from the obvious inclusions

$$H^{(n)} \subset G^{(n)}, \qquad H^{[n]} \subset G^{[n]}.$$

2.3. The Commutator of a Topological Group

Let G be a *topological* group and let S_1 and S_2 be nonvoid subsets of G. The smallest *closed* subgroup of G containing all the commutators

$$g_1 g_2 g_1^{-1} g_2^{-1}, \qquad g_1 \in S_1, \qquad g_2 \in S_2$$

is called the *commutator* $K(S_1, S_2)$ *of the sets* S_1, S_2. It is plain that

$$K(S_1, S_2) = \overline{K_a(S_1, S_2)} \supset K_a(S_1, S_2). \tag{2.3.1}$$

The commutator $K(G, G)$ is called the *commutator subgroup of the group G*. It is denoted by G'. For this case (2.3.1) becomes

$$G' = \overline{G'_a} \supset G'_a. \tag{2.3.2}$$

I. *The commutator subgroup G' of G is a closed normal subgroup of G and the factor group G/G' is commutative.*

Proof. The set G' is the closure of a normal subgroup and so is a closed normal subgroup of G. The proof that G/G' is commutative is the same as the proof that G/G'_a is commutative. $\square$

II. *If S_1, S_2 are connected subsets of the topological group G that have a common element, then:*

(1) $K_a(S_1, S_2)$ *is connected*

and

(2) $K(S_1, S_2)$ *is connected.*

Proof. We set

$$S_{12} = \{g_1 g_2 g_1^{-1} g_2^{-1}: g_2 \in S_1, g_2 \in S_2\}, \tag{2.3.3}$$
$$S_{21} = \{g_2 g_1 g_2^{-1} g_1^{-1}: g_1 \in S_1, g_2 \in S_2\}, \tag{2.3.4}$$
$$S = S_{12} \cup S_{21}. \tag{2.3.5}$$

It is clear that

$$S_{21} = S_{12}^{-1} \quad \text{and that } S_{12} = S_{21}^{-1}. \tag{2.3.6}$$

The mapping $\{g_1, g_2\} \to g_1 g_2 g_1^{-1} g_2^{-1}$ is a continuous mapping of the space $S_1 \times S_2$ onto S_{12}. Since $S_1 \times S_2$ is connected (see XII of 1.1), S_{12} is connected; likewise S_{21} is connected.

Let g_0 be a common element of S_1 and S_2. In this case, S_{12} and S_{21} contain e (consider (2.3.3) and (2.3.4) for $g_1 = g_2 = g_0$). Therefore $S = S_{12} \cup S_{21}$ is connected (VI in 1.1). Using (2.3.6) we check that

$$K_a(S_1, S_2) = \bigcup_{n=1}^{\infty} S^n. \tag{2.3.7}$$

The sets S^n are connected (II in 1.2) and $S = S^1 \subset S^2 \subset \cdots$. Therefore we have

$$S^n \cap S^m \supset S \neq \varnothing \tag{2.3.8}$$

and the union $K_a(S_1, S_2)$ of all S^n is connected by VI of 1.1. Then $K(S_1, S_2)$ is also connected as the closure $\overline{K_a(S_1, S_2)}$ of the connected set $K_a(S_1, S_2)$ (see (2.3.1) and V in 1.1). $\square$

2.4. Solvable and Nilpotent Topological Groups

For a topological group G we construct the following two sequences:

$$G^{(1)} = G', G^{(2)} = (G^{(1)})', \ldots, G^{(n)} = (G^{(n-1)})', \ldots; \qquad (2.4.1)$$
$$G^{[1]} = K(G, G) = G', G^{[2]} = K(G, G^{[1]}), \ldots, G^{[n]} = K(G, G^{[n-1]}), \ldots . \quad (2.4.2)$$

The group $G^{(n)}$ is called the *n-th derivative of* G. Plainly we have

$$G^{[n]} \supset G^{(n)} \qquad (2.4.3)$$

and (see (2.3.1))

$$G^{[n]} \supset G_a^{[n]}, \qquad G^{(n)} \supset G_a^{(n)}. \qquad (2.4.4)$$

The topological group G is called *solvable* if there is an n such that

$$G^{(n)} = \{e\}. \qquad (2.4.5)$$

The smallest positive integer n for which (2.4.5) holds is called the *rank of the solvable group* G. A topological group G is called *semisimple* if it admits no closed solvable normal subgroups except for $\{e\}$. A topological group G is called *simple* if it admits no closed normal subgroups except for G and $\{e\}$. A topological group G is called *nilpotent* if there is an n such that

$$G^{[n]} = \{e\}. \qquad (2.4.6)$$

I. *A nilpotent topological group is solvable.*

II. *A solvable (nilpotent) topological group is also algebraically solvable (algebraically nilpotent).*

III. *A subgroup of a solvable (nilpotent) topological group is solvable (nilpotent).*

The proofs of these propositions are analogous to the proofs of I (see also II) in 2.2 (cf (2.4.4)).

Exercises

Prove the following.

1. *A factor group of a solvable group by any closed normal subgroup is solvable.*

Hint. Prove that under the canonical homomorphism $\varphi: G \to \bar{G} = G/H$, every derivative $G_a^{(k)}$ (or $G^{(k)}$) maps into $G_a^{(k)}$ (or $G^{(k)}$).

2. *If H is a closed (algebraically) solvable normal subgroup of a group G such that the factor group G/H is solvable, then G is also solvable.*

Hint. Let k be the rank of the group G/H. Prove that $G^{(k)} \subset H$.

3. *A group G is solvable if and only if the following exist:*

(a)   *a closed commutative normal subgroup H_0 of G;*
(b)   *a closed commutative normal subgroup H_1 of $G_1 = G/H_0$,*
(c)   *a commutative normal subgroup H_2 of $G_2 = G_1/H_1$, and so on;*

where at some step, say the m-th, $G_m = \{e\}$.

Hint. Let k be the rank of the group G. Set $H_0 = G^{(k-1)}$ and use induction on k, making use also of exercises 1 and 2.

4. *The group K_n (described in example 4 of 1.2) is solvable.*

Hint. Set

$$
H_0 = \left\{ h_0 : h_0 =
\begin{Vmatrix}
1 & 0 & \cdots & 0 & 0 \\
0 & 1 & \cdots & 0 & 0 \\
  &   & \vdots &   &   \\
0 & 0 & \cdots & 1 & 0 \\
k_{n1} & k_{n2} & \cdots & k_{n,n-1} & k_{1,n}
\end{Vmatrix}
\right\},
$$

$$
H_1 = \left\{ h_1 : h_1 =
\begin{Vmatrix}
1 & 0 & \cdots & 0 & 0 \\
0 & 1 & \cdots & 0 & 0 \\
  &   & \vdots &   &   \\
0 & 0 & \cdots & 1 & 0 \\
k_{n-1,1} & k_{n-1,2} & \cdots & k_{n-1,n-2} & k_{n-1,n-1}
\end{Vmatrix}
\right\},
$$

and so on. Then apply exercise 3.

5. *A factor group of a nilpotent group by any closed normal subgroup is nilpotent.*

6. *If H is the center of a group G and G/H is nilpotent, then G is nilpotent.*

7. *A group G is nilpotent if and only if: G has a nontrivial center H_0; $G_1 = G/H$ has nontrivial center H_1; $G_2 = G_1/H_1$ has nontrivial center H_2, and so on, where at some step, say the m-th, $G_m = \{e\}$.*

8. *The set of all matrices z of the form*

$$
z =
\begin{Vmatrix}
1 & z_{12} & \cdots & z_{1n} \\
0 & 1 & \cdots & z_{2n} \\
\multicolumn{4}{c}{\dotfill} \\
0 & 0 & \cdots & 1
\end{Vmatrix}
$$

is a nilpotent group.

9. *The group K_n is not nilpotent (but is solvable: see exercise 4).*

10. *The derivative of a solvable group is nilpotent.*

11. *Prove statements 1–9 for algebraic groups.*

Hint. Consider these groups as topological groups with the discrete topology.

§3. Lie's Theorem

3.1. Basic Results on Representations of Connected Solvable Groups

Theorem 1 (Lie's Theorem). *A finite-dimensional irreducible representation* $T: g \to T(g)$ *of a connected algebraically solvable group G is one-dimensional.*

Proof. We will prove the theorem by induction on the rank of G. If G has rank 1, G is commutative (see I of 2.1) and the theorem follows from Schur's lemma (see the corollary to 2.2 of chapter 1). Suppose that the theorem holds for algebraically solvable groups of rank $\leqslant n - 1$. We will prove the theorem for groups G of rank n. We set $H = G'_a$. The group H is connected (see II in 2.3) and is algebraically solvable with rank $< n$. Let $T: g \to T(g)$ be an irreducible finite-dimensional representation of the group G in the space X and let $h \to T(h)$ be the restriction of the representation T to the group H. By I in 2.1, chapter I, there is a subspace X_1 of X on which $h \to T(h)$ is irreducible. By our inductive hypothesis, X_1 is one-dimensional. Let x_1 be a nonzero vector in X_1. We have $T(h)x_1 \in X_1$, and

$$T(h)x_1 = \lambda_1(h)x_1, \qquad h \in H, \tag{3.1.1}$$

where $\lambda_1(h)$ is a complex-valued function on H, continuous because T is continuous. We write $\lambda_1, \lambda_2, \ldots, \lambda_r$ for all of the distinct functions on H for which the identity

$$T(h)x_j = \lambda_j(h)x_j, \qquad h \in H, \qquad j = 1, \ldots, r \tag{3.1.2}$$

holds for some nonzero x_j in X. The number of such x_j, and so the number of functions λ_j, is finite, since dim $X < \infty$. We set $\Lambda = \{\lambda_1, \lambda_2, \ldots, \lambda_r\}$ and give Λ the discrete topology. Let Y_1 denote the set of all $y \in X$ for which

$$T(h)y = \lambda_1(h)y, \quad h \in H. \tag{3.1.3}$$

By (3.1.1), Y_1 is a nonzero subspace of X. Note that $g^{-1}hg$ belongs to H for all $g \in G$, $h \in H$ (H is a normal subgroup of G by definition). Thus we can apply (3.1.3) to $g^{-1}hg$:

$$T(g^{-1}hg)y = \lambda_1(g^{-1}hg)y, \qquad h \in H, \qquad y \in Y_1. \tag{3.1.4}$$

That is, we have

$$T(g)^{-1}T(h)T(g)y = \lambda_1(g^{-1}hg)y,$$
$$T(h)T(g)y = \lambda_{1g}(h)T(g)y, \qquad y \in Y_1, \tag{3.1.5}$$

where we have defined

$$\lambda_{1g}(h) = \lambda_1(g^{-1}hg). \tag{3.1.6}$$

From (3.1.5) we conclude that $\lambda_{1g}(h) \in \Lambda$. The function $g \to \lambda_{1g}(h) = \lambda_1(g^{-1}hg)$ is continuous on G. Therefore $h \to \lambda_{1g}(h)$ is a continuous mapping of the group G into Λ. The inverse image of a point, and thus of any subset of Λ, is closed in G. Since G is connected and Λ is discrete, the image of G under the mapping $g \to \lambda_{1g}(h)$ is a single point (XI in 1.1); that is, the function $\lambda_{1g}(h)$ is independent of g. This means that $\lambda_{1g}(h) = \lambda_{1e}(h) = \lambda_1(h)$, and so the second equality in (3.1.5) has the form

$$T(h)T(g)y = \lambda_1(h)T(g)y, \qquad y \in Y_1, \qquad g \in G, \qquad h \in H. \tag{3.1.7}$$

The equality (3.1.7) means that Y is invariant under all operators $T(g)$, $g \in G$. Since Y_1 is not $\{0\}$ alone and T is irreducible, Y_1 must be X. Using (3.1.3) we find that

$$T(h)x = \lambda_1(h)x, \quad \text{for all } x \in X. \tag{3.1.8}$$

Hence we have

$$\det T(h) = (\lambda_1(h))^m, \quad \text{where } m = \dim X. \tag{3.1.9}$$

On the other hand, for $h = g_1g_2g_1^{-1}g_2^{-1}$, we have

$$\det T(h) = \det(T(g_1)T(g_2)T(g_1^{-1})T(g_2^{-1})) = 1$$

and by (2.3.7)

$$\det T(h) = 1, \quad \text{for all } h \in H. \tag{3.1.10}$$

Comparing (3.1.9) and (3.1.10) we infer that $(\lambda_1(h))^m = 1$, so that the set of all m-th roots of unity $\varepsilon = \{\varepsilon_0 = 1, \varepsilon_1, \varepsilon_2, \ldots, \varepsilon_{m-1}\}$ are the only possible values of the function $\lambda_1(h)$. Since the function $h \to \lambda_1(h)$ is continuous on H, the group H is connected, and the set ε is discrete, the function $\lambda_1(h)$ is constant on H, i.e., $\lambda_1(h) = \lambda_1(e) = 1$ (see (3.1.1)). From this and (3.1.8) we infer that $T(h) = 1$. For $h = g_1g_2g_1^{-1}g_2^{-1}$ we obtain

$$T(g_1g_2g_1^{-1}g_2^{-1}) = 1, \qquad T(g_1)T(g_2)T(g_1)^{-1}T(g_2)^{-1} = 1,$$

which is to say that

$$T(g_1)T(g_2) = T(g_2)T(g_1).$$

and all operators $T(g)$, $g \in G$, commute with each other. From the irreducibility of the representation $g \to T(g)$, $g \in G$, and Schur's lemma, we infer that X is one-dimensional (see the corollary to 2.2, chapter I). □

Theorem 2. *Let* $T: g \to T(g)$ *be a finite-dimensional representation of a connected algebraically solvable group G in a space X. There is a nonzero vector x_0 in X that is an eigenvector for all of the operators $T(g)$.*

Proof. The space X admits a subspace $X_1 \neq \{0\}$ invariant under all operators $T(g)$, $g \in G$, on which $T|_{X_1}$ is irreducible (see I in 2.1, chapter II). According to theorem 1, X_1 is one-dimensional. Let x_0 be a nonzero vector in X_1. We then have $T(g)x_0 \in X_1$, that is,

$$T(g)x_0 = \lambda(g)x_0, \quad \text{for all } g \in G,$$

$\lambda(g)$ being a complex-valued function on G. This means that x_0 is an eigenvector for all $T(g)$, $g \in G$. □

Remark. Theorems 1 and 2 are in general false for infinite-dimensional representations.

3.2. Corollaries of the Basic Theorems

Corollary 1. *Any finite-dimensional irreducible representation $T: g \to T(g)$ of a connected solvable group G is one-dimensional.*

Corollary 2. *If $T: g \to T(g)$ is a finite-dimensional representation of a connected solvable group G, then the space X of this representation admits a nonzero vector x_0 that is an eigenvector for all of the operators $T(g)$.*

Any solvable topological group is also algebraically solvable (II in 2.4) and so the corollaries follow from theorems 1 and 2 respectively.

Exercise

Let $g \to T(g)$ be a finite-dimensional representation of a connected algebraically solvable group in a space X. Prove that there is a basis in X for which the matrices of all the operators $T(g)$ are triangular.

Chapter VI

Finite-Dimensional Representations of the Full Linear Group

The finite-dimensional irreducible representations of a large class of groups are constructed analogously to those of the full linear group $GL(n, \mathbf{C})$. We will first present this construction for the simplest case, that of the group $GL(n, \mathbf{C})$ itself. Throughout this chapter, G will denote the group $GL(n, \mathbf{C})$.

§1. Some Subgroups of the Group G

1.1. The Group K

The set of all matrices

$$k = \begin{Vmatrix} k_{11} & 0 & \cdots & 0 \\ k_{21} & k_{22} & \cdots & 0 \\ \multicolumn{4}{c}{\dotfill} \\ k_{n1} & k_{n2} & \cdots & k_{nn} \end{Vmatrix}, \qquad k_{11} \neq 0, \ldots, k_{nn} \neq 0,$$

is denoted by K.[45] Recall that K is a closed connected solvable subgroup of G (see example 4 in 1.2 of chapter V).

1.2. The Group H

The set of all matrices h of the form

$$h = \begin{Vmatrix} h_{11} & h_{12} & \cdots & h_{1n} \\ 0 & h_{22} & \cdots & h_{2n} \\ \multicolumn{4}{c}{\dotfill} \\ 0 & 0 & \cdots & h_{nn} \end{Vmatrix}, \qquad h_{11} \neq 0, \ldots, h_{nn} \neq 0,$$

is denoted by H.

I. *H is a closed connected solvable subgroup of G.*

---

[45] Throughout this chapter we will write K instead of K_n (see example 4 in 1.2, chapter V).

Clearly the matrices in H are just the transposes of the matrices of K, and so H, like K, is a closed connected solvable subgroup of G. $\square$

1.3. The Group D

The set of all diagonal matrices

$$\delta = \begin{Vmatrix} \lambda_1 & 0 & \cdots & 0 \\ 0 & \lambda_2 & \cdots & 0 \\ \multicolumn{4}{c}{\cdots\cdots\cdots\cdots} \\ 0 & 0 & \cdots & \lambda_n \end{Vmatrix}, \quad \lambda_1\lambda_2 \ldots \lambda_n \neq 0,$$

is denoted by D. Plainly D is a closed connected subgroup of K and also of H. Note that D is commutative.

I. *The group D is isomorphic to the direct product of n copies of the group $\mathbf{C}_0^1$.*

Observe also that

$$K \cap H = D. \tag{1.3.1}$$

1.4. The Group Z_-

Let Z_- denote the set of all matrices of the form

$$\zeta = \begin{Vmatrix} 1 & 0 & 0 & \cdots & 0 \\ \zeta_{21} & 1 & 0 & \cdots & 0 \\ \multicolumn{5}{c}{\cdots\cdots\cdots\cdots\cdots} \\ \zeta_{n1} & \zeta_{n2} & \zeta_{n3} & \cdots & 1 \end{Vmatrix},$$

where the complex numbers $\zeta_{jl}, j > l$, are arbitrary. It is easy to see that Z_- is a closed connected subgroup of K and thus of G.

1.5. The Group Z_+

Let Z_+ denote the set of all matrices

$$z = \begin{Vmatrix} 1 & z_{12} & z_{13} & \cdots & z_{1n} \\ 0 & 1 & z_{23} & \cdots & z_{2n} \\ \multicolumn{5}{c}{\cdots\cdots\cdots\cdots\cdots} \\ 0 & 0 & 0 & \cdots & 1 \end{Vmatrix},$$

where the complex numbers $z_{jl}, j < l$, are arbitrary. It is easy to see that Z_+ is a closed connected subgroup of G and that

$$K \cap Z_+ = \{e\}, \tag{1.5.1}$$

where e denotes the identity matrix.

I. *The mappings* $\zeta \to \delta^{-1}\zeta\delta$, $z \to \delta^{-1}z\delta$ *are automorphisms of the groups* Z_- *and* Z_+ *respectively.*

Proof. Matrix multiplication shows that the elements g_{pq} of the matrices $\delta^{-1}\zeta\delta$ and $\delta^{-1}z\delta$ are obtained from the corresponding elements of the matrices ζ and z by multiplication by $\lambda_p^{-1}\lambda_q$. $\square$

1.6. Decomposition of Elements of the Group K

I. *Every element k of the group K can be uniquely represented in the form*

$$k = \delta\zeta, \quad where \; \delta \in D \quad and \; \zeta \in Z_-, \tag{1.6.1}$$

and also in the form

$$k = \zeta\delta, \quad where \; \delta \in D \quad and \; \zeta \in Z_-. \tag{1.6.2}$$

Proof. The equality (1.6.1) is equivalent to the system of equalities

$$k_{pq} = \lambda_p\zeta_{pq}, \quad p \geqslant q. \tag{1.6.3}$$

For $q = p$, (1.6.3) yields

$$\lambda_p = k_{pp}. \tag{1.6.4}$$

For $q < p$, (1.6.3) becomes

$$\zeta_{pq} = k_{pq}/k_{pp}. \tag{1.6.5}$$

Thus the system (1.6.3) has a unique solution. This proves (1.6.1). We prove (1.6.2) analogously, replacing (1.6.2) by

$$\lambda_p = k_{pp}, \quad \zeta_{pq} = k_{pq}/k_{qq}. \quad \square \tag{1.6.6}$$

1.7. Decomposition of Elements of the Group H

I. *Every element h of the group H can be uniquely represented in the form*

$$h = \delta z, \quad where \; \delta \in D \quad and \; z \in Z_+, \tag{1.7.1}$$

and also in the form

$$h = z\delta, \quad where \; \delta \in D \quad and \; z \in Z_+. \tag{1.7.2}$$

Proof. Follow the proof of I in 1.6 (formulas (1.7.1) and (1.7.2) can be obtained from (1.6.2), (1.6.1) by transposition). In (1.7.1) we have

$$\lambda_p = h_{pp}, \quad z_{pq} = \frac{h_{pq}}{h_{pp}} \tag{1.7.3}$$

and in (1.7.2)

$$\lambda_p = h_{pp}, \qquad z_{pq} = \frac{h_{pq}}{h_{qq}}. \quad \square \qquad (1.7.4)$$

Exercise

Prove that $H^{(1)} = Z_+$ and that $K^{(1)} = Z_-$.
  Hint. Use Propositions I in 1.5 and 1.7.

1.8. Gauss's Decomposition

For $g \in G$ we set

$$\Delta_l(g) = \begin{vmatrix} g_{11} & \cdots & g_{1\ell} \\ \cdots & \cdots & \cdots \\ g_{l1} & \cdots & g_{ll} \end{vmatrix}. \qquad (1.8.1)$$

The matrix $g \in G$ is called *regular* if

$$\Delta_l(g) \neq 0, \quad \text{for all } l = 1, 2, \ldots, n; \qquad (1.8.2)$$

otherwise, g is called *irregular*. The set G_{reg} of all regular matrices g is an open subset of G and the irregular matrices are a subset of G of lower dimension.

I. *Every regular matrix $g \in G$ can be represented uniquely in the form*

$$g = kz, \quad \text{where } k \in K \quad \text{and } z \in Z_+. \qquad (1.8.3)$$

Proof. First we find an element $\dot{z} \in Z_+$ such that

$$g\dot{z} \in K. \qquad (1.8.4)$$

Since $\dot{z}_{pq}$ is equal to zero for $p > q$ and $k_{pq} = 0$ for $p < q$, (1.8.4) is equivalent to the system of equalities

$$\sum_{s=1}^{q-1} g_{ps}\dot{z}_{sq} + g_{pq} = 0, \quad \text{for } p < q. \qquad (1.8.5)$$

  For a fixed $q > 1$ and $p = 1, 2, \ldots, q-1$, the equalities (1.8.5) are a system of equations in the unknowns $\dot{z}_{1q}, \ldots, \dot{z}_{q-1, q}$. Its determinant coincides with $\Delta_q(g)$ and is nonzero since g is regular. Thus the system (1.8.5) admits a unique solution, *i.e.*, there is an element $\dot{z} \in Z_+$ such that $g\dot{z} = k \in K$. Hence we have $g = k\dot{z}^{-1} = kz$, where $z = \dot{z}^{-1}$. Thus a decomposition (1.8.3) exists. Now suppose that

$$g = kz = k_1z_1, \quad \text{where } k, k_1 \in K \quad \text{and } z, z_1 \in Z_+.$$

From this we find that $kz = k_1 z_1$, $k^{-1}k_1 = zz_1^{-1} \in K \cap Z_+ = \{e\}$ (see (1.5.1)). Thus we have $k^{-1}k_1 = e$, $zz_1^{-1} = e$ and consequently $k = k_1$, $z = z_1$. That is, the decomposition (1.8.3) is unique. $\square$

Formula (1.8.3) is called *Gauss's decomposition*.[46]

Combining I in 1.6 and I above, we conclude the following.

II. *Every regular matrix $g \in G$ can be uniquely represented in the form*

$$g = \delta \zeta z, \quad \text{where } \delta \in D, \quad \zeta \in Z_-, \quad \text{and } z \in Z_+, \tag{1.8.6}$$

and also in the form

$$g = \zeta \delta z, \quad \text{where } \delta \in D, \quad \zeta \in Z_-, \quad \text{and } z \in Z_+. \tag{1.8.7}$$

These decompositions are also called *Gauss's decompositions*.

We will find explicit formulas in terms of the entries of g for the elements of the matrices k and z in (1.8.3) and also for the elements of the matrices δ, ζ and z in (1.8.6) and (1.8.7).

The symbol $g \begin{pmatrix} p_1 & \cdots & p_m \\ q_1 & \cdots & q_m \end{pmatrix}$, $p_1 < \cdots < p_m$, $q_1 < \cdots < q_m$, denotes the minor determinant of the submatrix of the matrix g consisting of the entries in the p_j-th row and q_j-th column of g ($j = 1, 2, \ldots, m$).

When we multiply the matrix g by z, we add to its first row a linear combination of the following rows; to the second row, we add a linear combination of the rows following it; and so on. Thus, the minor $g \begin{pmatrix} p_1 & \cdots & p_m \\ q_1 & \cdots & q_m \end{pmatrix}$ does not change in such a multiplication. That is, this minor is equal to the analogous minor of the matrix k in (1.8.3). In particular, we have

$$\Delta_m(g) = \begin{vmatrix} k_{11} & 0 & \cdots & 0 \\ k_{21} & k_{22} & \cdots & 0 \\ \vdots & \vdots & & \vdots \\ k_{m1} & k_{m2} & \cdots & k_{mm} \end{vmatrix} = k_{11}k_{22}\cdots k_{mm}, \tag{1.8.8}$$

and for $p > q$

$$g \begin{pmatrix} p & p+1 & \cdots & n \\ q & p+1 & \cdots & n \end{pmatrix} = \begin{vmatrix} k_{pq} & 0 & 0 & \cdots & 0 \\ k_{p+1,q} & k_{p+1,p+1} & 0 & \cdots & 0 \\ \vdots & \vdots & \vdots & & \vdots \\ k_{nq} & k_{n,p+1} & k_{n,p+2} & \cdots & k_{nn} \end{vmatrix}$$

$$= k_{pq}k_{p+1,p+1}\cdots k_{nn}. \tag{1.8.9}$$

---

[46] Gauss used this decomposition to obtain a solution by recursion of a system of linear equations.

From (1.8.8) we infer that

$$k_{mm} = \frac{\varDelta_m(g)}{\varDelta_{m-1}(g)}, \qquad m = 2, \ldots, n, \tag{1.8.10}$$

$$k_{11} = \varDelta_1(g) \tag{1.8.11}$$

and from (1.8.9) that

$$k_{pq} = \frac{g\begin{pmatrix} p & p+1 & \cdots & n \\ q & p+1 & \cdots & n \end{pmatrix}}{k_{p+1,\,p+1} \cdots k_{nn}} = \varDelta_n(g)\frac{g\begin{pmatrix} p & p+1 & \cdots & n \\ q & p+1 & \cdots & n \end{pmatrix}}{\varDelta_p(g)}. \tag{1.8.12}$$

Combining (1.8.10)–(1.8.12) with (1.6.4)–(1.6.6), we find that ζ and δ in (1.8.6) are given by

$$\lambda_m = \frac{\varDelta_m(g)}{\varDelta_{m-1}(g)}, \qquad m = 2, \ldots, n; \qquad \lambda_1 = \varDelta_1(g); \tag{1.8.13}$$

and

$$\zeta_{pq} = \frac{g\begin{pmatrix} p & p+1 & \cdots & n \\ q & p+1 & \cdots & n \end{pmatrix}}{\varDelta_p(g)^2}\frac{\varDelta_{p-1}(g)}{\varDelta_n(g)}, \qquad p < q. \tag{1.8.14}$$

Finally, applying the rule of multiplication of minors to the right side of (1.8.3), we obtain for $p < q$:

$$g\begin{pmatrix} 1 & 2 & \cdots & p-1 & p \\ 1 & 2 & \cdots & p-1 & q \end{pmatrix}$$

$$= \begin{vmatrix} k_{11} & 0 & \cdots & 0 \\ k_{21} & k_{22} & \cdots & 0 \\ \cdots & \cdots & \cdots & \cdots \\ k_{p1} & k_{p2} & \cdots & k_{pp} \end{vmatrix} \begin{vmatrix} 1 & z_{12} & \cdots & z_{1,q-1} & z_{1q} \\ 0 & 1 & \cdots & z_{2,q-1} & z_{2q} \\ \cdots \cdots \cdots \cdots \cdots \cdots \cdots \cdots \\ 0 & 0 & \cdots & 1 & z_{p-1,q} \\ 0 & 0 & \cdots & 0 & z_{pq} \end{vmatrix} = \varDelta_p(g)z_{pq}.$$

It follows that

$$z_{pq} = \frac{g\begin{pmatrix} 1 & 2 & \cdots & p-1 & p \\ 1 & 2 & \cdots & p-1 & q \end{pmatrix}}{\varDelta_p(g)}. \tag{1.8.15}$$

We therefore have the following.

III. *The elements of the matrices* k, z, δ, ζ *in the decompositions (1.8.3) and (1.8.7) are expressed in terms of elements of the matrix* g *by the formulas*

(1.8.10)–(1.8.12), (1.8.13)–(1.8.15). *Thus the elements of the matrices k,
z, δ, ζ in these decompositions are rational functions of elements of g.*

Remark. In some cases it is convenient to take K as the group of all matrices

$$
k = \left\|
\begin{matrix}
k_{11} & k_{12} & \cdots & k_{1n} \\
0 & k_{22} & \cdots & k_{2n} \\
\cdots & \cdots & \cdots & \cdots \\
0 & 0 & \cdots & k_{nn}
\end{matrix}
\right\|,
\tag{1.8.16}
$$

Z_+ as the group of all matrices

$$
z = \left\|
\begin{matrix}
1 & 0 & \cdots & 0 \\
z_{21} & 1 & \cdots & 0 \\
\cdots & \cdots & \cdots & \cdots \\
z_{n1} & z_{n2} & \cdots & 1
\end{matrix}
\right\|,
\tag{1.8.17}
$$

and Z_- as the group of all matrices

$$
\zeta = \left\|
\begin{matrix}
1 & \zeta_{12} & \cdots & \zeta_{1n} \\
0 & 1 & \cdots & \zeta_{2n} \\
 & & \vdots & \\
0 & 0 & \cdots & 1
\end{matrix}
\right\|.
\tag{1.8.18}
$$

For this K, Z_+, Z_- (and the original D) Gauss's decompositions also hold.

To see this, it suffices to apply the original decompositions to the transposed matrix g' and then take the transpose of both sides of the resulting identities.

Let us consider more closely the case $n = 2$. Gauss's decomposition takes the form

$$
\left\|
\begin{matrix}
g_{11} & g_{12} \\
g_{21} & g_{22}
\end{matrix}
\right\| =
\left\|
\begin{matrix}
k_{11} & k_{12} \\
0 & k_{22}
\end{matrix}
\right\|
\left\|
\begin{matrix}
1 & 0 \\
z_{21} & 1
\end{matrix}
\right\| =
\left\|
\begin{matrix}
k_{11} + k_{12}z_{21} & k_{12} \\
k_{22}z_{21} & k_{22}
\end{matrix}
\right\|,
$$

which is to say

$$
\begin{aligned}
k_{12} &= g_{12}, & k_{22} &= g_{22}, \\
k_{22}z_{21} &= g_{21}, & k_{11} + k_{12}z_{21} &= g_{11}.
\end{aligned}
\tag{1.8.19}
$$

Rewriting, we get

$$
z_{21} = \frac{g_{21}}{g_{22}}, \qquad k_{11} = g_{11} - k_{12}z_{21} = g_{11} - g_{12}\frac{g_{21}}{g_{22}} = \frac{\det g}{g_{22}}.
\tag{1.8.20}
$$

1.9. Gram's Decomposition

Let Γ denote the set of all diagonal matrices

$$\gamma = \begin{Vmatrix} e^{i\varphi_1} & 0 & \cdots & 0 \\ 0 & e^{i\varphi_2} & \cdots & 0 \\ \cdots\cdots\cdots\cdots\cdots\cdots \\ 0 & 0 & \cdots & e^{i\varphi_n} \end{Vmatrix}, \quad \text{where } \varphi_1, \ldots, \varphi_n \in \mathbf{R}^1. \quad (1.9.1)$$

Plainly Γ is a subgroup of the group D, isomorphic to the direct product of n copies of the group Γ^1. For the sake of brevity, we write $U(n)$ as U. It is clear that

$$\Gamma \subset U \quad (1.9.2)$$

and

$$U \cap K = U \cap D = \Gamma. \quad (1.9.3)$$

Let E denote the set of all diagonal matrices

$$\varepsilon = \begin{Vmatrix} \varepsilon_1 & 0 & \cdots & 0 \\ 0 & \varepsilon_2 & \cdots & 0 \\ \cdots\cdots\cdots\cdots\cdots \\ 0 & 0 & \cdots & \varepsilon_n \end{Vmatrix}, \quad \text{with } \varepsilon_1, \ldots, \varepsilon_n > 0. \quad (1.9.4)$$

Clearly, E is a subgroup of the group D, isomorphic to the direct product of n copies of the group $\mathbf{R}_0^+$.

I. *Every matrix $\delta \in D$ admits a unique representation in the form*

$$\delta = \varepsilon\gamma = \gamma\varepsilon, \quad \text{where } \varepsilon \in E \quad \text{and } \gamma \in \Gamma. \quad (1.9.5)$$

To see this, write $\lambda_j = \varepsilon_j e^{i\varphi_j}$, $\varepsilon_j > 0$, $\varphi_j \in \mathbf{R}^1$, where λ_j are the diagonal entries of δ. $\square$

II. *Every matrix $g \in G$ can be represented in the form*

$$g = ku, \quad \text{where } k \in K \quad \text{and } u \in U. \quad (1.9.6)$$

If we also have $g = k_1 u_1$, $k_1 \in K$, $u_1 \in U$, then $k_1 = k\gamma$ and $u_1 = \gamma^{-1}u$.

Proof. The first statement was proved in lemma 2 in example 10 of 1.2, chapter IV. If we also have $g = k_1 u_1$, $k_1 \in K$, $u_1 \in U$, then $ku = k_1 u_1$ and thus by (1.9.4) we have

$$k^{-1}k_1 = uu_1^{-1} \in U \cap K = \Gamma.$$

Therefore we have $k^{-1}k_1 = \gamma$, $uu_1^{-1} = \gamma$, where $\gamma \in \Gamma$, and $k_1 = k\gamma$, $u_1 = \gamma^{-1}u$. $\square$

III. *Every matrix $g \in G$ admits a representation in the form*

$$g = \zeta\varepsilon u, \quad \text{where } \zeta \in Z_-, \quad \varepsilon \in E, \quad \text{and } u \in U, \tag{1.9.7}$$

and also in the form

$$g = \varepsilon\zeta u, \quad \text{where } \varepsilon \in E, \quad \zeta \in Z_-, \quad \text{and } u \in u. \tag{1.9.8}$$

Proof. By (1.9.6) we have

$$g = ku_1, \quad k \in K, \quad u_1 \in U. \tag{1.9.9}$$

On the other hand (1.6.2) implies that

$$k = \zeta\delta, \quad \zeta \in Z_-, \quad \delta \in D, \tag{1.9.10}$$

and (1.9.5) shows that

$$\delta = \varepsilon\gamma, \quad \varepsilon \in E, \quad \gamma \in \Gamma. \tag{1.9.11}$$

If we substitute these expressions in (1.9.9), we obtain $g = \zeta\varepsilon\gamma u_1 = \zeta\varepsilon u$, where $u = \gamma u_1 \in U$. Hence the decomposition (1.9.7) exists.

If we also have $g = \zeta_1\varepsilon_1 u_1$, then $\zeta_1\varepsilon_1 u_1 = \zeta\varepsilon u$. From this and II, we infer that $u_1 = \gamma u$, $\zeta_1\varepsilon_1 = \zeta\varepsilon\gamma^{-1}$. The uniqueness of (1.6.2) and (1.9.3) imply that $\zeta_1 = \zeta$, $\varepsilon_1 = \varepsilon\gamma^{-1}$. These last equalities are possible only if $\gamma = e$, $\varepsilon_1 = \varepsilon$. $\square$

The decompositions (1.9.6) and (1.9.7)–(1.9.8) are called *Gram's decompositions*.

§2. Description of the Irreducible Finite-Dimensional Representations of the Group $GL(n, \mathbf{C})$

2.1. Weights and Weight Vectors

Let $T: g \to T(g)$ be a representation of the group G in the finite-dimensional space X. A nonzero vector $x \in X$ is called a *weight vector of the representation T* (with respect to the group D) if

$$T(\delta)x = v(\delta)x, \quad \text{for all } \delta \in D, \tag{2.1.1}$$

where $v(\delta)$ is a continuous character of the group D. The character $v(\delta)$ is called a *weight* of the representation T. A nonzero vector $x_0 \in X$ is called a *vector of highest weight of the representation T* if

$$T(\delta)x_0 = \alpha(\delta)x_0, \quad \text{for all } \delta \in D \text{ and}$$
$$T(z)x_0 = x_0, \quad \text{for all } z \in Z_+, \tag{2.1.2}$$

where $\alpha(\delta)$ is a continuous character of the group D. The character $\alpha(\delta)$ in this case is called a *highest weight of the representation T*. Finally, a nonzero vector $x_0' \in X$ is called a *vector of lowest weight of the representation T* if

$$T(\delta)x_0' = \mu(\delta)x_0', \quad \text{for all } \delta \in D$$

and

$$T(\zeta)x_0' = x_0', \quad \text{for all } \zeta \in Z_-. \tag{2.1.3}$$

The continuous character $\mu(\delta)$ is called in this case a *lowest weight of the representation*.

Theorem 1. *The space X of every finite-dimensional representation*

$$T : g \to T(g)$$

of the group $G = GL(n, \mathbf{C})$ admits a vector of highest weight and a vector of lowest weight. If T is irreducible, then X admits only one vector of highest weight and only one vector of lowest weight (both up to nonzero scalar multiples).

Proof. The restriction of the representation T to H is a representation of the connected solvable group H. By Lie's Theorem (3.1, chapter V), X contains a nonzero x_0 that is a common eigenvector for all operators $T(h)$:

$$T(h)x_0 = \alpha(h)x_0, \quad \text{for all } h \in H, \tag{2.1.4}$$

where $h \to \alpha(h)$ is a continuous one-dimensional representation of the group H. From (2.1.4) we infer that $T(\delta)x_0 = \alpha(\delta)x_0$ for all $\delta \in D$ and that $T(z)x_0 = x_0$ for all $z \in Z_+$, since $Z_+ = H^{(1)}$. Therefore x_0 is a vector of highest weight for the representation T. Replacing the group H by the group K, we obtain a vector of lowest weight for the representation T. Suppose now that T is irreducible. Let Y be a linear space dual to X under some nonsingular bilinear form (x, y). Let $\hat{T}$ be the representation of G contragredient to T. Let y_0' be a vector of lowest weight and let $\hat{\mu}$ be a lowest weight of the representation $\hat{T}$. Let $g = \zeta\delta z$ ($\zeta \in Z_-, \delta \in D, z \in Z_+$) be a regular element of G. From (2.1.4) we obtain

$$(T(g)x_0, y_0') = (T(\zeta \, \delta z)x_0, y_0') = (T(\zeta)T(\delta)T(z)x_0, y_0')$$
$$= (T(\delta)x_0, \hat{T}(\zeta^{-1})y_0') = (\alpha(\delta)x_0, y_0') = \alpha(\delta)(x_0, y_0'). \tag{2.1.5}$$

Similarly we find

$$(T(g)x_0, y_0') = (T(\delta)x_0, y_0') = (x_0, \hat{T}(\delta^{-1})y_0') = \hat{\mu}(\delta^{-1})(x_0, y_0'). \quad (2.1.6)$$

It follows that

$$(\alpha(\delta) - \hat{\mu}(\delta^{-1}))(x_0, y_0') = 0. \quad (2.1.7)$$

If (x_0, y_0) is zero, (2.1.5) shows that

$$(T(g)x_0, y_0') = 0 \quad (2.1.8)$$

for all $g \in G_{\mathrm{reg}}$ and thus for all g, since $\overline{G_{\mathrm{reg}}} = G$. The linear span of all $T(g)x_0$, $g \in G$, coincides with X, T being irreducible. Thus (x, y_0') is zero for all $x \in X$, which violates the nonsingularity of the form (x, y). We thus have

$$(x_0, y_0') \neq 0 \quad (2.1.9)$$

and (2.1.7) gives

$$\alpha(\delta) = \hat{\mu}(\delta^{-1}). \quad (2.1.10)$$

Now suppose that x_1 is another nonzero vector of highest weight, where $\alpha_1(\delta)$ is the corresponding weight of the representation T. From (2.1.10) we infer that $\alpha_1(\delta) = \hat{\mu}(\delta^{-1})$. Therefore, we have $\alpha_1(\delta) = \alpha(\delta)$. Multiplying x_0 by $c = (x_1, y_0')/(x_0, y_0')$, we obtain

$$(cx_0 - x_1, y_0') = c(x_0, y_0') - (x_1, y_0') = 0. \quad (2.1.11)$$

If $cx_0 - x_1 \neq 0$, then $cx_0 = x_1$ is also a vector of highest weight with corresponding weight $\alpha(\delta)$. Then (2.1.11) contradicts the equality (2.1.9) (in which x_0 is replaced by the vector $cx_0 - x_1$). Thus we have $cx_0 - x_1 = 0$, $x_1 = cx_0$ and x_0 is defined uniquely up to a scalar multiple. Our assertions concerning vectors of lowest weight are proved similarly. $\square$

Corollary 1. *For every finite-dimensional irreducible representation of the group* $G = GL(n, \mathbf{C})$*, there exist a single highest weight and a single lowest weight.*

Corollary 2. *Let* $T: g \to T(g)$ *be a finite-dimensional representation of the group G in the space X. Let* x_0 *be a nonzero vector of highest weight and let* $\alpha(\delta)$ *be the corresponding highest weight of the representation T. The representation T is irreducible if the following hold:*

(1) *X is the linear span of all* $T(g)x_0$, $g \in G$;
(2) x_0 *is the unique vector of highest weight of the representation T in X (up to scalar multiples).*

Proof. Let M be a nonzero subspace of X, invariant under T. By theorem 1, M admits a vector of highest weight for the restriction of T to M, which is also a vector of highest weight for T in X. By hypothesis (2), this vector is a scalar multiple of x_0. Therefore x_0 belongs to M, as does the linear span of all vectors $T(g)x_0$, $g \in G$. By hypothesis (1), this linear span is equal to X. Thus we have $M = X$, and so the representation T is irreducible. $\square$

2.2. The Canonical Realization of Finite-Dimensional Irreducible Representations of the Group G

Let $T: g \to T(g)$ be an irreducible representation of the group $G = GL(n, \mathbf{C})$ in a finite-dimensional space X. Let Y, $\hat{T}$, x_0, y_0', α, $\hat{\mu}$ be as in the proof of theorem 1 of 2.1.

For every vector $x \in X$ we define a complex-valued function $f(g) = f_x(g)$ by the formula

$$f(g) = f_x(g) = (T(g)x, y_0'). \tag{2.2.1}$$

I. *The correspondence $x \to f_x(g)$ is linear.*

To see this, compute as follows:

$$f_{\alpha_1 x_1 + \alpha_2 x_2}(g) = (T(g)(\alpha_1 x_1 + \alpha_2 x_2), y_0')$$
$$= \alpha_1(T(g)x_1, y_0') + \alpha_2(T(g)x_2, y_0') = \alpha_1 f_{x_1}(g) + \alpha_2 f_{x_2}(g).$$

Let Φ be the image of the space X under the mapping $x \to f_x(g)$. By I, Φ is a linear space of functions on G, and the continuity of the representation T implies that the functions in Φ are continuous.

II. *The correspondence $x \to f_x$ is one-to-one.*

Proof. Let M be the kernel of the mapping $x \to f_x(g)$. Plainly M is a subspace of X. If x belongs to M, then $0 = f_x(g) = (T(g)x, y_0')$ for all $g \in G$. For every $g_1 \in G$ we then find that

$$(T(g)T(g_1)x, y_0') = (T(gg_1)x, y_0') = 0.$$

Therefore $T(g_1)x$ also belongs to M; that is, M is invariant under all $T(g_1)$, $g_1 \in G$. Since the representation T is irreducible, we conclude that either $M = X$ or $M = (0)$. In the first case, $(T(g)x, y_0')$ is equal to 0 for all $x \in X$. This means that $(x, y_0') = 0$ for all $x \in X$, which is impossible, since (x, y) is nonsingular. Therefore M is equal to $\{0\}$, that is, the correspondence $x \to f_x$ is an isomorphism. $\square$

Combining I and II, we obtain the following.

III. *The correspondence $x \to f_x(g)$ defined by the formula $f_x(g) = (T(g)x, y_0')$ is a linear isomorphism of the space X onto the space Φ.*

Let us examine the space Φ in more detail.

IV. *Every function $f(g) \in \Phi$ satisfies the condition*

$$f(kg) = \alpha(k)f(g). \tag{2.2.2}$$

Proof. By the definition of Φ, we have

$$
\begin{aligned}
f_x(kg) &= (T(kg)x, y_0') = (T(k)T(g)x, y_0') \\
&= (T(g)x, T'(k)y_0') = (T(g)x, \hat{T}^{-1}(k)y_0') \\
&= (T(g)x, \hat{\mu}(k^{-1})y_0') = \hat{\mu}(k^{-1})(T(g)x, y_0') \\
&= \alpha(k)f_x(g). \quad \square
\end{aligned}
$$

V. *If a function $f(g)$ belongs to Φ, then all right translates $f(gg_0)$ of $f(g)$ also belong to Φ.*

Proof. If $f(g) = f_x(g)$ belongs to Φ, we have

$$
\begin{aligned}
f(gg_0) &= f_x(gg_0) = (T(gg_0)x, y_0') \\
&= (T(g)T(g_0)x, y_0') = f_{T(g_0)x}(g) \in \Phi. \quad \square
\end{aligned}
\tag{2.2.3}
$$

We have also proved the following useful fact.

VI. *Under the isomorphism $x \to f_x(g)$, the operator $T(g_0)$ goes into a right translation operator in Φ:*

$$T(g_0)f(g) = f(gg_0). \tag{2.2.4}$$

Proof. Replacing x by $T(g_0)x$, we change the function $f_x(g) = (T(g)x, y_0')$ into $(T(g)T(g_0)x, y_0') = (T(gg_0)x, y_0') = f_x(gg_0)$. $\quad \square$

We summarize Propositions III through VI.

VII. *The isomorphism $x \to f_x(g)$ defines an equivalence of the representation T in X and of the representation T in Φ, by the formula*

$$\tilde{T}(g_0)f(g) = f(gg_0), \qquad f \in \Phi. \tag{2.2.5}$$

Thus every irreducible representation T of the group G is equivalent to a representation of G in the space Φ, given by (2.2.5).

VIII. *Let Φ be a finite-dimensional linear space of continuous functions $f(g)$ on G satisfying the following conditions:*

(1) $f(kg) = \alpha(k)f(g)$, *where $k \to \alpha(k)$ is a one-dimensional representation of the subgroup K;*

(2)   *If $f(g)$ belongs to Φ, then $f(gg_0)$ belongs to Φ.*
Let T be the representation of G in Φ defined by the formula

$$T(g_0)f(g) = f(gg_0). \tag{2.2.6}$$

Then Φ admits only one vector (up to scalar multiples) $f_0(g)$ of highest weight, defined by the formula

$$f_0(\delta\zeta z) = c\alpha(\delta), \quad \text{for } g = \delta\zeta z \in G_{\text{reg}}. \tag{2.2.7}$$

Proof. Let $f_0(g)$ be a vector of highest weight for the representation T. Hypothesis (1) gives

$$f_0(\delta\zeta z) = \alpha(\delta\zeta)f_0(z) = \alpha(\delta)\alpha(\zeta)f_0(z) = \alpha(\delta)f_0(z). \tag{2.2.8}$$

On the other hand, (2.1.2) and (2.2.9) show that $f_0(gz_0) = T(z_0)f_0(g) = f_0(g)$, and then (2.2.8) yields

$$f_0(\delta\zeta z) = f_0(\delta\zeta) = \alpha(\delta)f_0(e). \tag{2.2.9}$$

That is, $f_0(g)$ is equal to $f_0(\delta\zeta z) = C\alpha(\delta)$ for $g = \delta\zeta z \in G_{\text{reg}}$ and $C = f(e)$ is not zero. (If $C = 0$, then f_0 is the zero function, a contradiction to the definition of vector of highest weight.)
    If $f_1(g)$ is another vector of weight α, then $f_1(g) = C_1\alpha(\delta)$. Therefore, $f_1(g)$ is equal to $(C_1/C)f_0(g)$ for $g = \delta\zeta z \in G_{\text{reg}}$ and so for all $g \in G$. $\square$

    From Propositions VII and VIII we obtain a second proof of the part of theorem 1 in 2.1 dealing with vectors of highest weight. Interchanging K and H, we obtain a second proof of the statement about vectors of lowest weight.
    A character $\alpha(\delta)$ of the group D is called *inductive for $G = GL(n, \mathbf{C})$* if:

(1)   the formula (2.2.9) defines a continuous function $f_0(g)$ on the entire group G;
(2)   the linear span Φ_α of all right translates $f_0(gg_0)$, $g_0 \in G$, is finite-dimensional.

IX. *All functions in Φ_α satisfy the identity $f(kg) = \alpha(k)f(g)$.*

Proof. Write $f(g)$ for the function $f_0(gg_0)$. We find $f(kg) = f_0(kgg_0)$, and setting $k = \delta\zeta$, $gg_0 = \delta_1\zeta_1 z_1$, we find

$$\begin{aligned}
f(kg) &= f_0(\delta\zeta\,\delta_1\zeta_1 z_1) = C\alpha(\delta\zeta\,\delta_1\zeta_1) \\
&= C\alpha(\delta\zeta)\alpha(\delta_1\zeta_1) = \alpha(k)f_0(gg_0)
\end{aligned}$$

for $gg_0 \in G_{\text{reg}}$ and hence for all $g, g_0 \in G$.

Let α be an inductive character of the group D. Let $T_\alpha : g \to T_\alpha(g)$ denote the representation of G in the space Φ_α defined by

$$T_\alpha(g_0) f(g) = f(gg_0) \quad \text{for } f \in \Phi_\alpha.$$

Theorem 2.

(1) *If α is an inductive character of the group D, then T_α is an irreducible representation of $GL(n, \mathbf{C})$ with highest weight α.*

(2) *The highest weight α of a finite-dimensional irreducible representation of $GL(n, \mathbf{C})$ is inductive, and any finite-dimensional irreducible representation of $GL(n, \mathbf{C})$ with highest weight α is equivalent to the representation T_α.*

(3) *Two finite-dimensional irreducible representation of $GL(n, \mathbf{C})$ are equivalent if and only if their highest weights coincide.*

Proof.

(1) Let α be inductive and let $f_0(g)$ be a vector of highest weight for the representation T_α. Condition (2) in the definition of inductivity states that the linear span of the vectors $T_\alpha(g) f_0$ coincides with Φ_α and VIII implies that $f_0(g)$ is the unique vector (to within a scalar multiple) of highest weight in Φ_α. Therefore T_α is irreducible by corollary 2 in 2.1.

(2) Let T be a finite-dimensional irreducible representation of $GL(n, \mathbf{C})$ with highest weight α. Proposition VII states that T is equivalent to the representation $\tilde{T}$ in the space Φ. Proposition VIII states that Φ contains a vector (unique up to a scalar multiple) having the form $f_0(g) = C\alpha(\delta)$ for $g = \delta \zeta z$ of weight α. Since T is irreducible, $\tilde{T}$ is also irreducible and therefore the linear span of all $\tilde{T}(g_0) f(g) = f(gg_0)$ coincides with Φ. That is, $\Phi = \Phi_\alpha$ and $\tilde{T} = T_\alpha$. This proves that α is inductive and that $T \sim T_\alpha$.

(3) Two finite-dimensional irreducible representations with the same highest weight α are equivalent to T_α and hence to each other. The converse in (3) is obvious. $\square$

Guided by (2) of the preceding theorem, we may call T_α the *canonical realization* of an irreducible finite-dimensional representation of $GL(n, \mathbf{C})$ with highest weight α.

To obtain a complete list (up to equivalence) of all finite-dimensional irreducible representations of $GL(n, \mathbf{C})$, it remains only to identify the inductive characters among all continuous characters $\alpha(\delta)$ of the group D. This task will be carried out in 2.5 *infra*.

We also note that T_α acts in a space of functions on the homogeneous space G under the group of right translations.

In the sequel, T_α will always denote the irreducible finite-dimensional representation of G with highest weight α. Below we present some different realizations of the representation T_α.

2.3. A Realization of Finite-Dimensional Irreducible Representations of the Group G in a Space of Functions on Z_+

According to theorem 2 in 2.2, a finite-dimensional irreducible representation of G is equivalent to some representation T_α, operating in Φ_α. All functions $f(g)$ of the space Φ_α are continuous on G and satisfy the condition $f(kg) = \alpha(k)f(g)$. In particular, we have

$$f(g) = f(kz) = \alpha(k)f(z), \quad \text{for } k \in K, \, z \in Z_+, \, g \in G_{\text{reg}}. \tag{2.3.1}$$

The function $f(z)$ is continuous on Z_+, being the restriction to Z_+ of the continuous function $f(g)$ on G. Formula (2.3.1) defines $f(g)$ uniquely on G_{reg} and, by continuity, on all of G, since G_{reg} is dense in G. The correspondence $f(g) \to f(z)$ in (2.3.1) is obviously linear and furthermore it is one-to-one. For, if $f(z) \equiv 0$, then $f(g)$ vanishes on G_{reg} and so vanishes on all of G. Let F_α denote the image of the space Φ_α under the mapping $f(g) \to f(z)$ of (2.3.1). Under this mapping the operators $T_\alpha(g)$ go into operators $\dot{T}_\alpha(g)$ in F_α.

We now find an explicit expression for $\dot{T}_\alpha(g)$. By definition, we have

$$T_\alpha(g_0)f(g) = f(gg_0) = \alpha(k)\dot{T}_\alpha(g_0)f(z), \quad \text{for } g = kz, \, k \in K, \, z \in Z_+. \tag{2.3.2}$$

We write

$$\alpha(g) = f_0(g), \quad \text{for } g \in G_{\text{reg}} \quad \text{and} \quad zg_0 = k_1 z_1. \tag{2.3.3}$$

Therefore we have

$$\begin{aligned}
f(gg_0) = f(kzg_0) &= f(kk_1 z_1) \\
&= \alpha(kk_1)f(z_1) = \alpha(k)\alpha(k_1)f(z_1).
\end{aligned} \tag{2.3.4}$$

From (2.3.4) and (2.3.2) we conclude that

$$T_\alpha(g_0)f(z) = \alpha(k_1)f(z_1), \quad \text{for } zg_0 = k_1 z_1. \tag{2.3.5}$$

The correspondence $z \to z_1$ described by (2.3.3) is a mapping of Z_+ into itself. We denote this mapping by $\bar{g}_0$[47] and write $z_1 = z\bar{g}_0$. Also, (2.3.3) shows that $\alpha(zg_0) = \alpha(k_1)\alpha(z_1) = \alpha(k_1)$. Thus (2.3.5) takes the form

$$\dot{T}_\alpha(g_0)f(z) = \alpha(zg_0)f(z\bar{g}_0). \tag{2.3.6}$$

---

[47] Strictly speaking, the mapping $z \to z\bar{g}_0$ is defined only for those z and g_0 for which $zg_0 \in G_{\text{reg}}$. This is immaterial, since the functions $f(z) \in F_\alpha$ are extensible by continuity to irregular matrices.

We now identify the space F_α explicitly. By definition Φ_α is the linear span of all functions $T(g_0)f_0(g) = f_\alpha(gg_0)$, $g_0 \in G$, and so its image F_α is the linear span of all $T(g_0)f_0(z)$, $g_0 \in G$. By (2.2.11) and (2.3.1), $f_0(g)$ goes into the constant function $f_0(z) \equiv C$ under the mapping $f(g) \to f(z)$. Without loss of generality, we may suppose that $C = 1$, i.e., $f_0(z) \equiv 1$. Applying (2.3.6), we conclude that F_α is the linear span of all $\alpha(zg)$, $g \in G$.

We combine these results with theorem 2 in 2.2.

Theorem 3. *A finite-dimensional irreducible representation T of the group $G = GL(n, \mathbf{C})$ is equivalent to the representation $\dot{T}_\alpha$, where α is the highest weight of T. The representation $\dot{T}_\alpha$ is defined as follows. The space F_α of the representation $\dot{T}_\alpha$ is the linear span of all functions $\alpha(zg)$, $g \in G$, and the operators of the representation are defined by*

$$\dot{T}_\alpha(g)f(z) = \alpha(zg)f(z\bar{g}), \tag{2.3.7}$$

where $z_1 = z\bar{g}$ is defined by the condition $zg = k_1 z_1$.

2.4. The Case $n = 2$

We look at the case $G = GL(2, \mathbf{C})$ in more detail. It is convenient here to redefine K and Z_+, as the groups of matrices

$$k = \left\| \begin{matrix} k_{11} & k_{12} \\ 0 & k_{22} \end{matrix} \right\|, \qquad z = \left\| \begin{matrix} 1 & 0 \\ z_{21} & 1 \end{matrix} \right\|, \tag{2.4.1}$$

respectively. (See the remark in 1.8.) For $\delta \in D$, we write

$$\delta = \left\| \begin{matrix} \lambda_1 & 0 \\ 0 & \lambda_2 \end{matrix} \right\|.$$

as usual. The matrix z is identified by the single complex parameter $x = z_{21}$ and we may write

$$f(z) = f(x), \quad \text{for } z = \left\| \begin{matrix} 1 & 0 \\ x & 1 \end{matrix} \right\|. \tag{2.4.2}$$

Let us find the corresponding parameter x_1 in the matrix $z_g = z\bar{g}$ for

$$zg = kz_g. \tag{2.4.3}$$

Multiplying the matrices z and g, we obtain

$$zg = \left\| \begin{matrix} 1 & 0 \\ x & 1 \end{matrix} \right\| \left\| \begin{matrix} g_{11} & g_{12} \\ g_{21} & g_{22} \end{matrix} \right\| = \left\| \begin{matrix} g_{11} & g_{12} \\ g_{11}x + g_{21} & g_{12}x + g_{22} \end{matrix} \right\|.$$

From this and (1.8.20) we obtain

$$x_1 = \frac{g_{11}x + g_{21}}{g_{12}x + g_{22}}; \qquad (2.4.4)$$

and therefore[48]

$$f(z\bar{g}) = f\left(\frac{g_{11}x + g_{21}}{g_{12}x + g_{22}}\right). \qquad (2.4.5)$$

We conclude further from (1.8.20) that for $zg = kz_g$, $k = \zeta\delta$,

$$\lambda_2 = k_{22} = g_{12}x + g_{22}, \qquad \lambda_1 = k_{11} = \frac{\det(zg)}{g_{12}x + g_{22}} = \frac{\det g}{g_{12}x + g_{22}}. \qquad (2.4.6)$$

The character $\alpha(\delta)$ is a continuous character of the group D, which is isomorphic to the product of two copies of the group $\mathbf{C}_0^1$. Therefore we have

$$\alpha(\delta) = \lambda_1^{p_1}\bar{\lambda}_1^{q_1}\lambda_2^{p_2}\bar{\lambda}_2^{q_2}, \qquad (2.4.7)$$

where p_1, q_1, p_2, q_2 are any complex numbers for which $p_1 - q_1$ and $p_2 - q_2$ are integers. In (2.4.7), write the expressions for λ_1 and λ_2 from (2.4.6):

$$\alpha(zg) = \left(\frac{\det g}{g_{12}x + g_{22}}\right)^{p_1}\left(\frac{\overline{\det g}}{g_{11}x + g_{22}}\right)^{q_1}(g_{11}x + g_{22})^{p_2}(\bar{g}_{12}\bar{x} + \bar{g}_{22})^{q_2},$$

which is to say

$$\alpha(zg) = \varDelta^{p_1}\bar{\varDelta}^{q_1}(g_{12}x + g_{22})^{p_2 - p_1}(\bar{g}_{12}\bar{x} + \bar{g}_{22})^{q_2 - q_1}. \qquad (2.4.8)$$

Here we have written $\varDelta$ for $\det g$. Formula (2.3.7) takes the form

$$\dot{T}_\alpha(g)f(z) = \varDelta^{p_1}\bar{\varDelta}^{q_1}(g_{12}x + g_{22})^{p_2 - p_1}(\bar{g}_{12}\bar{x} + \bar{g}_{22})^{q_2 - q_1}f\left(\frac{g_{11}x + g_{21}}{g_{11}x + g_{22}}\right). \qquad (2.4.9)$$

It remains to find the inductive characters $\alpha(\delta)$. Suppose that $\alpha(\delta)$ is inductive, so that the linear span of all $\alpha(z)$ is finite-dimensional, let us say of dimension $k - 1$. Then for every $g_1, \ldots, g_k \in G$ there exist constants $C_1, \ldots, C_k$ not

---

[48] Thus, for $n = 2$, the transformation $z \to z\bar{g}$ reduces to a fractional linear transformation of the variable x. For $n > 2$, the elements of the matrix $z\bar{g}$ are rational functions of the entries of the matrices z and g. This follows at once from (1.18.15). We can thus consider the transformation $z \to z\bar{g}$ for $n > 2$ as a generalization of a fractional linear transformation.

all zero (depending on $g_1, \ldots, g_k$, but not on z), such that

$$C_1\alpha(zg_1) + C_2\alpha(zg_2) + \cdots + C_k\alpha(zg_k) = 0 \qquad (2.4.10)$$

for all $g_1, g_2, \ldots, g_k \in G$ and all $z \in Z_+$.

We make a special choice of $g_1, \ldots, g_k$ in (2.4.10):

$$g_j = \begin{pmatrix} x_j^{-1} & 1 \\ 0 & x_j \end{pmatrix}, \qquad j = 1, \ldots, k,$$

where x_j are any nonzero complex numbers. We then have $\Delta_j = 1$ and by (2.4.8)

$$\alpha(zg_j) = (x + x_j)^r(\bar{x} + \bar{x}_j)^s, \qquad j = 1, 2, \ldots, k, \qquad (2.4.11)$$

where

$$r = p_2 - p_1, \qquad s = q_2 - q_1. \qquad (2.4.12)$$

Going back to (2.4.10), we obtain

$$C_1(x + x_1)^r(\bar{x} + \bar{x}_1)^s + \cdots + C_k(x + x_k)^r(\bar{x} + \bar{x}_k)^s = 0 \qquad (2.4.13)$$

for arbitrary complex numbers $x, x_1, \ldots, x_k$ (x_j's nonzero). Differentiating the equalities (2.4.13) $k - 1$ times with respect to[49] x and then setting $x = 0$, we obtain a system of k identities:

$$
\begin{aligned}
&C_1 x_1^r \bar{x}_1^s + \cdots + C_k x_k^r \bar{x}_k^s = 0, \\
&r C_1 x_1^{r-1} \bar{x}_1^s + \cdots + r C_k x_k^{r-1} \bar{x}_k^s = 0, \\
&\qquad \vdots \qquad\qquad \vdots \qquad \vdots \\
&r(r-1)\cdots(r-k+1)C_1 \bar{x}_1^s + \cdots + r(r-1)\cdots(r-k+1)C_k \bar{x}_k^s = 0.
\end{aligned}
\qquad (2.4.14)
$$

As we already know, (2.4.14) holds for certain C_j's that are not all zero. Therefore the determinant of this system must vanish, the determinant being equal to $r^k(r-1)^{k-1}\cdots(r-k+1)\bar{x}_1^s \cdots \bar{x}_k^s w(x_1, \ldots, x_k)$, where $w(x_1, \ldots, x_k)$ is the van der Monde determinant of the numbers $x_1, \ldots, x_k$. But these numbers can be chosen arbitrarily. They can be chosen distinct from zero and from each other, so that one of the numbers $r, r-1, \ldots, r-k+1$ must vanish. Therefore r has to be a nonnegative integer. Similarly, differentiating (2.4.13) with respect to $\bar{x}$ $k - 1$ times and setting $x = 0$, we find that s has to be a nonnegative integer. Thus we have the following.

---

[49] Here we set $\partial/\partial x = \frac{1}{2}(\partial/\partial\xi - i(\partial/\partial\eta))$ and $\partial/\partial\bar{x} = \frac{1}{2}(\partial/\partial\xi + i(\partial/\partial\eta))$, where $\xi = \operatorname{Re} x$, $\eta = \operatorname{Im} x$. The reader may verify that we then have $\partial\bar{x}/\partial x = 0$, $\partial x/\partial\bar{x} = 0$, so that for this differentiation x and $\bar{x}$ can be construed as independent variables.

I. *If the character* $\alpha(\delta)$ *is inductive, then* $r = p_2 - p_1$ *and* $s = q_2 - q_1$ *are nonnegative integers.*

Conversely, let r and s be nonnegative integers. From (2.4.8) we infer that

$$\alpha(zg) = \Delta^{p_1} \bar{\Delta}^{q_1}(g_{12}x + g_{22})^r(\bar{g}_{12}\bar{x} + \bar{g}_{22})^s \qquad (2.4.15)$$

is a polynomial in x of degree not higher than r and in $\bar{x}$ of degree not higher than s. This also holds for any linear combination of the functions $\alpha(zg)$. We conclude the following.

II. *If* $r = p_2 - p_1$ *and* $s = q_2 - q_1$ *are nonnegative integers, then the character* $\alpha(\delta)$ *is inductive.*

In this case F_α is a space of polynomials of degree $\leqslant r$ in x and of degree $\leqslant s$ in $\bar{x}$. We will show that F_α consists of *all* such polynomials. First, it contains all functions of the form (2.4.15). Setting $g_{12} = 1$, $g_{22} = 0$, $g_{11} = 0$, $g_{21} = -1$, we see that F_α contains $x^r\bar{x}^s$. Next, If a polynomial $f(x) = f(\xi + i\eta)$ ($\xi = \operatorname{Re} x$, $\eta = \operatorname{Im} x$) belongs to F_α, then $f(x + h)$ also belongs to F_α. To see this, set $g = \begin{Vmatrix} 1 & h \\ 0 & 1 \end{Vmatrix}$, where $h \in \mathbf{R}^1$. If $f(x)$ belongs to F_α, then by (2.4.9) $T_\alpha(g_0)f(x) = f(x + h)$ belongs to F_α. Therefore $(1/h)(f(x + h) - f(x))$ and $\partial f/\partial \xi = \lim_{h \to 0} (1/h)(f(x + h) - f(x))$ also belong to F_α. Similarly, taking ih instead of h, $h \in \mathbf{R}^1$, we find that $\partial f/\partial \eta \in F_\alpha$. We thus infer that $\partial f/\partial x = \frac{1}{2}(\partial f/\partial \xi - i(\partial f/\partial \eta)) \in F_\alpha$ and $\partial f/\partial \bar{x} = \frac{1}{2}(\partial f/\partial \xi + i(\partial f/\partial \eta)) \in F_\alpha$. Therefore F_α contains the derivatives of all orders $\partial^k/\partial x^k$, $\partial^l/\partial \bar{x}^l$ of the function $x^r\bar{x}^s$. Thus F_α contains all monomials $x^{r_1}\bar{x}^{s_1}$, $0 \leqslant r_1 \leqslant r$, $0 \leqslant s_1 \leqslant s$ and so F_α is the set of all polynomials $f(z, \bar{z})$ of degree $\leqslant r$ in x and degree $\leqslant s$ in $\bar{x}$. We have proved the following fact.

Theorem 4. *A finite-dimensional irreducible representation of the group* $GL(2, \mathbf{C})$ *is described by two nonnegative integers* r, s *and two complex numbers* p, q, *the difference* $p - q$ *being an integer. The representation described by these numbers is equivalent to the representation* $\dot{T}_\alpha$ *operating in the space* F_α *of polynomials* $f(x, \bar{x})$ *in* x *and* $\bar{x}$ *of degree* $\leqslant r$ *in* x *and degree* $\leqslant s$ *in* $\bar{x}$. *The operators* $\dot{T}_\alpha(g)$ *are defined by*

$$\dot{T}_\alpha(g)f(x, \bar{x}) = (\det g)^p(\overline{\det g})^q(g_{12}x + g_{22})^r(\bar{g}_{12}\bar{x} + \bar{g}_{22})^s$$

$$\times f\left(\frac{g_{11}x + g_{21}}{g_{11}x + g_{22}}, \frac{\bar{g}_{11}\bar{x} + \bar{g}_{21}}{\bar{g}_{12}\bar{x} + \bar{g}_{22}}\right). \qquad (2.4.16)$$

The dimension of this representation is $(r + 1)(s + 1)$.

If $s = 0$ and $q = 0$, then p is an integer and F_α consists of polynomials in x alone, which are of course analytic functions in x. In this case the representation is called *analytic*. If $r = 0$ and $p = 0$, then q is an integer and F_α consists of polynomials in $\bar{x}$ alone. In this case the representation is called *antianalytic*.

Remark 1. The pair of numbers r, s that define a representation is often called its *highest weight*. If $f(x)$ is a weight vector of the representation $\dot{T}_\alpha$ with a weight of the form $\lambda_1^{m_1}\lambda_2^{m_2}$, the pair m_1, m_2 is also called the *weight of the vector* $f(x)$.

III. *Every monomial* $f = x^{r_1}\bar{x}^{s_1}$, $0 \leqslant r_1 \leqslant r$, $0 \leqslant s_1 \leqslant s$ *is a weight vector of the representation* $\dot{T}_\alpha$ *with weight* $r - 2r_1$, $s - 2s_1$.

Proof. Put $f = x^{r_1}\bar{x}^{s_1}$ in (2.4.16) and take $g = \delta$. We obtain

$$\dot{T}_\alpha(\delta)x^{r_1}\bar{x}^{s_1} = (\det g)^p(\overline{\det g})^q\lambda_2^r\bar{\lambda}_2^s\left(\frac{\lambda_1}{\lambda_2}x\right)^{r_1}\left(\frac{\bar{\lambda}_1}{\bar{\lambda}_2}\bar{x}\right)^{s_1}$$

$$= (\det g)^{p-2r_1}(\overline{\det g})^{q-2s_1}\lambda_2^{r-2r_1}\bar{\lambda}_2^{s-2s_1}x^{r_1}\bar{x}^{s_1};$$

III follows at once. $\quad\square$

The largest components of the weight are obtained for $r_1 = s_1 = 0$, that is, for $f \equiv 1$. The corresponding weight in III is equal to r, s. In all other cases the components are smaller. Hence the name "highest weight". An analogous situation occurs for $n > 2$. (See for example, Želobenko [1].)

Remark 2. Suppose that we start from the usual definition of the groups K and Z_+, as the groups

$$k = \begin{Vmatrix} k_{11} & 0 \\ k_{21} & k_{22} \end{Vmatrix}, \qquad z = \begin{Vmatrix} 1 & x \\ 0 & 1 \end{Vmatrix},$$

respectively. Carrying out analogous computations, we obtain

$$\dot{T}_\alpha(g)f(x) = (\det g)^p(\overline{\det g})^q(g_{11}+g_{21}x)^r(\bar{g}_{11}+\bar{g}_{21}\bar{x})^s f\left(\frac{g_{12}+g_{22}x}{g_{11}+g_{21}x}, \frac{\bar{g}_{12}+\bar{g}_{22}\bar{x}}{\bar{g}_{11}+\bar{g}_{21}\bar{x}}\right)$$

in place of (2.4.16). Here p, q are complex numbers for which $p - q$ is an integer, r, s are nonnegative integers and F_α is as before. Usually in applications for $n = 2$, the formula (2.4.16) is preferred. In the present case, the character $\alpha(\delta) = \lambda_1^{p_1}\bar{\lambda}_1^{q_1}\lambda_2^{p_2}\bar{\lambda}_2^{q_2}$ is inductive if and only if the numbers

$$r_1 = p_1 - p_2, \qquad s_1 = q_1 - q_2$$

are nonnegative integers and p_2, q_2 are complex numbers whose difference is an integer.

2.5. Inductive Characters of the Group D in the General Case

Let G^0 be a subgroup of the group $G = GL(n, \mathbf{C})$. We write

$$Z_+^0 = Z_+ \cap G^0, \qquad Z_-^0 = Z_- \cap G^0, \qquad D^0 = D \cap G^0. \qquad (2.5.1)$$

We say that Gauss's decomposition of G *induces* Gauss's decomposition of G^0 if when we write

$$g^0 = \delta\zeta z \qquad (2.5.2)$$

for $g^0 \in G_{\mathrm{reg}} \cap G^0$, the relations

$$\delta = \delta^0 \in D^0, \qquad \zeta = \zeta^0 \in Z_+^0, \qquad z = z^0 \quad Z_+^0$$

hold and also $G^0 \cap G_{\mathrm{reg}}$ is dense in G^0.

Lemma. *Suppose that Gauss's decomposition of G induces Gauss's decomposition of G^0. If the character $\alpha(\delta)$ of the group D is inductive with respect to G, then its restriction $\alpha_0(\delta^0)$ to D^0 is inductive with respect to G^0.*

Proof. Suppose that $\alpha(\delta)$ is inductive with respect to G, and that the linear span of all $\alpha(zg)$, $g \in G$, is $(k-1)$-dimensional. Then for every $g_1, g_2, \ldots, g_k$ there exist constants $C_1, \ldots, C_k$ such that

$$C_1\alpha(zg_1) + \cdots + C_k\alpha(zg_k) = 0. \qquad (2.5.3)$$

The equality (2.5.3) holds in particular when $g_1 = g_0^1 \in G^0, \ldots, g_k = g_k^0 \in G^0$ for all $z \in Z_+$ and for $z = z^0 \in Z_+^0$. Since Gauss's decomposition of G induces Gauss's decomposition of G^0, $\alpha(z^0 g_j^0)$ is equal to $\alpha_0(z^0 g_j)$ and (2.5.3) becomes

$$C_1\alpha_0(z^0 g_1^0) + \cdots + C_k\alpha_0(z^0 g_k^0) = 0.$$

Therefore α_0 is inductive. $\quad\square$

Recall again that D is the direct product of n copies of the group $\mathbf{C}_0^1$. Hence (see (j) in 3.3 of chapter III) all characters $\alpha(\delta)$ of D have the form

$$\alpha(\delta) = \lambda_1^{p_1}\overline{\lambda}_1^{q_1}\lambda_2^{p_2}\overline{\lambda}_2^{q_2} \cdots \lambda_n^{p_n}\overline{\lambda}_n^{q_n}. \qquad (2.5.4)$$

Suppose that $\alpha(\delta)$ is inductive under G and apply the preceding lemma to the group G^0 of matrices g^0 having the form

$$g^0 = \begin{Vmatrix} \begin{matrix} a & b \\ c & d \end{matrix} & & \\ & 1 & \\ & & \ddots \\ & & & 1 \end{Vmatrix}, \qquad (2.5.5)$$

where all unmarked entries are zero and $\begin{vmatrix} a & b \\ c & d \end{vmatrix} \neq 0$. Then Z_+^0, Z_-^0 and D^0 consist of all matrices

$$z^0 = \begin{Vmatrix} \begin{matrix} 1 & z_{12} \\ 0 & 1 \end{matrix} & & & \\ & 1 & & \\ & & \ddots & \\ & & & 1 \end{Vmatrix}, \quad \zeta^0 = \begin{Vmatrix} \begin{matrix} 1 & 0 \\ \zeta_{21} & 1 \end{matrix} & & & \\ & 1 & & \\ & & \ddots & \\ & & & 1 \end{Vmatrix},$$

$$\delta^0 = \begin{Vmatrix} \begin{matrix} \lambda_1 & 0 \\ 0 & \lambda_2 \end{matrix} & & & \\ & 1 & & \\ & & \ddots & \\ & & & 1 \end{Vmatrix},$$

respectively. Plainly G^0, Z_+^0, Z_-^0, D^0 are isomorphic to the groups $GL(2, \mathbf{C})$ and its corresponding subgroups Z_+, Z_-, D, respectively. Thus Gauss's decomposition of $GL(n, \mathbf{C})$ induces Gauss's decomposition of $GL(2, \mathbf{C})$. The last lemma shows that if $\alpha(\delta)$ is inductive, then

$$\alpha_0(\delta^0) = \lambda_1^{p_1} \bar{\lambda}_1^{q_1} \lambda_2^{p_2} \bar{\lambda}_2^{q_2}$$

is inductive (with respect to G^0). By remark 2 in 2.4 this holds if and only if $r_1 = p_1 - p_2$ and $s_1 = q_1 - q_2$ are nonnegative integers. Moving the matrix $\begin{pmatrix} a & b \\ c & d \end{pmatrix}$ down along the main diagonal and repeating the argument, we see that $r_2 = p_2 - p_3$, $s_2 = q_2 - q_3, \ldots, r_{n-1} = p_{n-1} - p_n$, $s_{n-1} = q_{n-1} - q_n$ are also nonnegative integers. We arrive at the following result.

I. *If the character*

$$\alpha(\delta) = \lambda_1^{p_1} \bar{\lambda}_1^{q_1} \lambda_2^{p_2} \bar{\lambda}_2^{q_2} \cdots \lambda_n^{p_n} \bar{\lambda}_n^{q_n}$$

of the group D is inductive with respect to $GL(n, \mathbf{C})$, then the numbers

$$r_1 = p_1 - p_2, \qquad s_1 = q_1 - q_2, \ldots, r_{n-1} = p_{n-1} - p_n,$$

$$s_{n-1} = q_{n-1} - q_n \tag{2.5.6}$$

are nonnegative integers.

We now prove the converse.

II. *If the numbers (2.5.6) are nonnegative integers, then the character $\alpha(\delta)$ is inductive with respect to $GL(n, \mathbf{C})$.*

Proof. Let $\Delta_1(g), \Delta_2(g), \ldots, \Delta_n(g)$ denote the consecutive principal minors of the matrix $g \in GL(n, \mathbf{C})$. From (1.8.13) and (2.5.4) we obtain

$$\alpha(zg) = \Delta_1(zg)^{p_1} \left(\frac{\Delta_2(zg)}{\Delta_1(zg)}\right)^{p_2} \overline{\frac{\Delta_2(zg)}{\Delta_1(zg)}}^{q_1} \left(\overline{\frac{\Delta_2(zg)}{\Delta_1(zg)}}\right)^{q_2} \cdots \left(\frac{\Delta_n(zg)}{\Delta_{n-1}(zg)}\right)^{p_n} \left(\overline{\frac{\Delta_n(zg)}{\Delta_{n-1}(zg)}}\right)^{q_n}$$

$$= \Delta_1(zg)^{r_1} \overline{\Delta_1(zg)}^{s_1} \cdots \Delta_{n-1}(zg)^{r_{n-1}} \overline{\Delta_{n-1}(zg)}^{s_{n-1}} \Delta_n(g)^{p_n} \overline{\Delta_n(g)}^{q_n},$$

since $\Delta_n(zg) = \det zg = \det z \det g = \Delta(g)$. $\square$

We obtain the next theorem from I, II and theorem 3 of 2.3.

Theorem 5. *A finite-dimensional irreducible representation T of the group $G = GL(n, \mathbf{C})$ is described by nonnegative integers $r_1, r_2, \ldots, r_{n-1}, s_1, s_2, \ldots, s_{n-1}$ and by complex numbers p, q whose difference is an integer. The corresponding representation T is equivalent to a representation $\dot{T}_\alpha$ of $GL(n, \mathbf{C})$ in a space F_α of polynomials $f(z, \bar{z})$ in z_{jl} of degree not greater than $r_1 + \cdots + r_{n-1}$ and in $\bar{z}_{jl}$ of degree not greater than $s_1 + \cdots + s_{n-1}$. The operators $T_\alpha(g)$ are defined by*

$$\dot{T}_\alpha(g)f(z, \bar{z}) = \Delta_1(zg)^{r_1} \Delta_1\overline{(zg)}^{s_1} \cdots \Delta_{n-1}(zg)^{r_{n-1}} \Delta_{n-1}\overline{(zg)}^{s_{n-1}} \Delta_n^p \Delta_n^q f(zg, \overline{zg}),$$

$$(2.5.7)$$

where $\Delta_j(g)$ is the j-th principal minor of the matrix g. The space F_α is the linear span of the functions

$$\Delta_1(zg)^{r_1} \overline{\Delta_1(zg)}^{s_1} \cdots \Delta_{n-1}(zg)^{r_{n-1}} \overline{\Delta_{n-1}(zg)}^{s_{n-1}}.$$

Thus any finite-dimensional irreducible representation T of the group $GL(n, \mathbf{C})$ is described by a sequence of numbers $\{p_1, p_2, \ldots, p_n; q_1, q_2, \ldots, q_n\}$, where $p_n - q_n$ is an integer and $r_1 = p_1 - p_2, \ldots, r_{n-1} = p_{n-1} - p_n$, $s_1 = q_1 - q_2, \ldots, s_{n-1} = q_{n-1} - q_n$ are nonnegative integers. This sequence is called the *signature* of the representation T and is denoted by α:

$$\alpha = \{p_1, p_2, \ldots, p_n; q_1, q_2, \ldots, q_n\}.$$

The sequence $\{r_1, \ldots, r_{n-1}, p_n; s_1, \ldots, s_{n-1}, q_n\}$ is also often called the *signature*. One writes

$$\alpha = \{r_1, \ldots, r_{n-1}, p_n; s_1, \ldots, s_{n-1}, q_n\}.$$

The irreducible representation T of $GL(n, \mathbf{C})$ with signature α is denoted by T_α.

If $s_1 = s_2 = \cdots = s_{n-1} = q = 0$, then p is an integer and F_α consists of polynomials in $\bar{z}_{jl}$ only. In this case the representation is called *antianalytic*. If $r_1 = r_2 = \cdots = r_{n-1} = p = 0$, then q is an integer and F_α consists of polynomials in $\bar{z}_{jl}$ only. In this case the representation is called *antianalytic*. Clearly, $\dot{T}_\alpha$ can be simultaneously analytic and antianalytic if and only if it is the trivial one-dimensional representation.

From (2.5.7) we infer the following.

III. *Every finite-dimensional irreducible representation T of the group $GL(n, \mathbf{C})$ is equivalent to the tensor product of analytic and antianalytic irreducible representations. Conversely, any such tensor product is an irreducible representation of the group $GL(n, \mathbf{C})$.*

For $n > 2$ the space F_α does *not* contain *all* polynomials in z_{jl} of degree not greater than $r_1 + \cdots + r_{n-1}$ and in $\bar{z}_{jl}$ of degree not greater than $s_1 + \cdots + s_{n-1}$. (Note the difference from the case $n = 2$.) The definition of F_α for $n > 2$ is somewhat complicated. We state without proof [50] one way to construct F_α.

We define differential operators in the space of all polynomials in z_{jl} and $\bar{z}_{jl}$ by the formulas

$$D_1 = \frac{\partial}{\partial z_{12}} + z_{23}\frac{\partial}{\partial z_{13}} + \cdots + z_{2n}\frac{\partial}{\partial z_{1n}},$$

$$D_2 = \frac{\partial}{\partial z_{23}} + \cdots + z_{3n}\frac{\partial}{\partial z_{2n}},$$

$$\vdots$$

$$D_{n-1} = \frac{\partial}{\partial z_{n-1,\,n}},$$

$$\bar{D}_1 = \frac{\partial}{\partial \bar{z}_{12}} + \bar{z}_{23}\frac{\partial}{\partial \bar{z}_{13}} + \cdots + \bar{z}_{2n}\frac{\partial}{\partial \bar{z}_{1n}},$$ (2.5.8)

$$\bar{D}_2 = \frac{\partial}{\partial \bar{z}_{23}} + \cdots + \bar{z}_{3n}\frac{\partial}{\partial \bar{z}_{2n}},$$

$$\vdots$$

$$\bar{D}_{n-2} = \frac{\partial}{\partial \bar{z}_{n-1,\,n}}.$$

IV. *The space F_α is the set of all polynomials in z_{jl} and $\bar{z}_{jl}$ that satisfy the following conditions:*

$$D_1^{r_1+1}f = 0, \qquad D_2^{r_2+1}f = 0, \ldots, D_{n-1}^{r_{n-1}+1}f = 0,$$
$$\bar{D}_1^{s_1+1}f = 0, \qquad \bar{D}_2^{s_2+1}f = 0, \ldots, \bar{D}_{n-1}^{s_{n-1}+1}f = 0.$$ (2.5.9)

---

[50] For the proof, see for example Želobenko [1], chapter X or Želobenko's article [1*].

Indeed, f can be considered as an arbitrary function having a sufficient number of partial derivatives. The set of all such functions that satisfy (2.5.9) is exactly the space F_α.

We will offer another description of the space F_α in §3 *infra*.

Remark. The mapping $z \to zg$ is defined only for $zg \in G_{\text{reg}}$. We can eliminate this awkwardness if instead of the functions $f(z)$, $z \in Z_+$, we consider the functions $f(\tilde{z})$, where $\tilde{z}$ is the right coset $\tilde{z} \in \tilde{Z} = G/K$. The coset $\tilde{z}$ is called *regular* if it contains at least one regular matrix g (in this case all matrices in $\tilde{z}$ are regular). Other cosets are called *irregular*. From Gauss's decomposition (1.8.3), we infer the following.

V. *Every regular coset $\tilde{z}$ contains exactly one element $z \in Z_+$. Conversely, every element $z \in Z_+$ is contained in exactly one regular coset $\tilde{z}$.*

By identifying a matrix $z \in Z_+$ with the regular coset that contains it, we imbed Z_+ into $\tilde{Z}$. Plainly $\tilde{Z}_+$ is a homogeneous space with respect to G, which acts on $\tilde{Z}_+$ by the formula

$$\tilde{z}\bar{g}_0 = \{g\}\bar{g}_0 = \{gg_0\} \tag{2.5.10}$$

(see 2.6 of chapter III). The transformation (2.5.10) is defined for all $\tilde{z} \in \tilde{Z}$. Irregular elements in $\tilde{Z}$ can be regarded as points at infinity that are added to Z_+.

Exercise

Find all of the irregular elements in $\tilde{Z}$ for $G = GL(2, \mathbf{C})$.

2.6. The Realization of Representations in a Space of Functions on U

Consider Gram's decomposition (1.9.7) instead of Gauss's. If f belongs to Φ_α, then IV in 2.2 and (1.9.7) yield

$$f(g) = f(\zeta\varepsilon u) = \alpha(\zeta\varepsilon)f(u) = \alpha(\varepsilon)f(u). \tag{2.6.1}$$

Clearly $f(u)$, as the restriction of f to U, is continuous on U and the correspondence

$$f(g) \to f(u) \tag{2.6.2}$$

established by (2.6.1) is linear and one-to-one. Let F_α denote the image of Φ_α under the mapping (2.6.2). For this mapping, the representation $\dot{T}_\alpha$ goes into an equivalent representation. This representation is again denoted by

$\dot{T}_\alpha$ and the operators of $\dot{T}_\alpha$ by $\dot{T}_\alpha(g)$. Repeat the reasoning that led to (2.3.5) to see the following.

I. *The operators $\dot{T}_\alpha(g)$ of the representation $\dot{T}_\alpha$ are defined by*

$$T_\alpha(g_0)f(u) = \alpha(\varepsilon')f(u_{g_0}), \quad \text{for } ug_0 = \zeta_1\varepsilon'u_{g_0}. \tag{2.6.3}$$

Since $\Gamma = K \cap U$, IV of 2.2 gives the following.

II. *The functions $f \in \dot{F}_\alpha$ satisfy the identity*

$$f(\gamma u) = \alpha(\gamma)f(u). \tag{2.6.4}$$

Formula (2.6.3) is called the *realization of an irreducible representation of the group $GL(n, \mathbf{C})$ in a space of functions on U*. The compactness of U makes this relation useful in many cases for a general study of properties of the representations $\dot{T}_\alpha$. In specific problems the realization on Z_+ is more convenient, since the entries of the matrix u are very complicated to express in terms of independent parameters.

§3. Decomposition of a Finite-Dimensional Representation of the Group $GL(n, \mathbf{C})$ into Irreducible Representations

3.1. The Method of Z-Invariants

Let $T: g \to T(g)$ be a representation of $G = GL(n, \mathbf{C})$ in a finite-dimensional space X. The vector $x \in X$ is called a *Z-invariant* for the representation T if

$$T(z)x = x, \quad \text{for all } z \in Z. \tag{3.1.1}$$

Plainly the set of Z-invariant vectors is a linear subspace of X; we denote it by Ω_T. According to the definition in 2.1 (see 2.1.2), a nonzero Z-invariant x is a *vector of highest weight* if

$$T(\delta)x = \alpha(\delta)x, \quad \text{for all } \delta \in D, \tag{3.1.2}$$

where $\alpha(\delta)$ is the highest weight. The set of all vectors x in Ω_T that satisfy (3.1.2) is denoted by $M(T)$. Note that $M_\alpha(T)$ is a linear subspace of Ω_T. Its dimension dim $M_\alpha(T)$ is called the *multiplicity* of the highest weight α in the representation T. If X admits no vector of highest weight α, then (and only then) we write dim $M_\alpha(T) = 0$.

I. *Let T be a representation of the group $G = GL(n, \mathbf{C})$ in a finite-dimensional space X. If T is completely reducible, then the multiplicity of the appearance* in T of the irreducible representation T_α coincides with the multiplicity of the highest weight α in T.

Proof. Since T is completely reducible, X is a direct sum

$$X = X_1 + X_2 + \cdots + X_m, \tag{3.1.3}$$

in which every X_j is invariant under T and the restriction $T_j = T|_{X_j}$ is irreducible. Let α_j be the highest weight of the representation T_j, so that $T_j \sim T_{\alpha_j}$. Let α_j be equal to α for $j = 1, \ldots, k$ and $\alpha_j \neq \alpha$ for $j > k$. That is, the multiplicity of the appearance of T_α in T is equal to k.

According to theorem 1 in 2.1, every $X_j, j = 1, \ldots, k$, admits exactly one vector x_j of weight α (up to scalar multiples). Plainly all of the vectors $x_j \in M_\alpha(T)$ are linearly independent, since $x_j \in X_j$ and the spaces $X_1, \ldots, X_k$ are linearly independent. We will prove that $x_1, x_2, \ldots, x_k$ form a basis in $M_\alpha(T)$. This will show that dim $M_\alpha(T) = k$ and Proposition I will be proved.

Let x' belong to $M_\alpha(T)$. Then by (3.1.2) and (3.1.1) we have

$$T(\delta)x' = \alpha(\delta)x', \qquad T(z)x' = x'. \tag{3.1.4}$$

On the other hand, (3.1.3) implies that

$$x' = \sum_{j=1}^m x'_j, \tag{3.1.5}$$

where $x'_j \in X_j$. Substituting in (3.1.4), we obtain

$$\sum_{j=1}^m \alpha(\delta)x'_j = \alpha(\delta)x' = \sum_{j=1}^m T(\delta)x'_j,$$

$$\sum_{j=1}^m x'_j = x' = T(z)x' = \sum_{j=1}^m T(z)x'_j,$$

and hence

$$T(\delta)x'_j = \alpha(\delta)x'_j, \qquad T(z)x'_j = x'_j. \tag{3.1.6}$$

Therefore, if $x'_j \neq 0$, x'_j is the vector of highest weight $\alpha(\delta)$ for the representation T_j in X_j. This vector is unique up to scalar multiples. Therefore x'_j is equal to 0 for $\alpha_j \neq \alpha$, that is, for $j > k$. Also, x'_j is equal to $c_j x_j$ for $j = 1, 2, \ldots, k$. From (3.1.5) we obtain $x' = \sum_{j=1}^k c_j x_j$. We have proved that dim $M_\alpha(T) = k$. $\square$

From I we infer the following.

II. *A completely reducible representation T of the group $G = GL(n, \mathbf{C})$ is irreducible if and only if the space X of this representation admits only one vector of highest weight (up to scalar multiples).*

Proof. Necessity follows from theorem 1 in 2.1. To prove sufficiency, observe that if X admits only one vector of highest weight α, then dim $M_\alpha(T) = 1$. Therefore T contains only a single T_α, which is to say that $T \sim T_\alpha$. $\square$

III. *The set of irreducible representations T_α that are contained in a represen-tation T of the group $G = GL(n, \mathbf{C})$, and the multiplicity of their appearance, are independent of the method of decomposing T into irreducible re-presentations.*

A representation T_α appears in T if and only if $M_\alpha(T) \neq \{0\}$. The multi-plicity of its appearance in T coincides with dim $M_\alpha(T)$ (Proposition I). On the other hand, $M_\alpha(T)$ is independent of the method of decomposing T into irreducible representations.

Proposition I thus offers a practical means of determining the set of T_α that are contained in the decomposition of T and their multiplicities in T. It is called the *method of Z-invariants*. This method consists of first finding the space Ω_T of Z-invariants of T and then finding all subspaces $M_\alpha(T) \neq \{0\}$ in Ω_T.

It is far more difficult actually to carry out the decomposition of the representation T of G. As always, the decomposition is not unique if irre-ducible representations contained in T occur with multiplicity exceeding 1. We will later illustrate the solution of this problem for some specific examples.

3.2. Analytic Representations of the Group $GL(n, \mathbf{C})$

Lemma. *Let V be the neighborhood of the identity element of the group $G = GL(n, \mathbf{C})$, defined by the conditions*

$$\sum_{j,l=1}^{n} |g_{jl} - \delta_{jl}|^2 < \varepsilon < 1, \tag{3.2.1}$$

where

$$\delta_{jl} = \begin{cases} 1, & for\ j = l, \\ 0, & for\ j \neq l. \end{cases}$$

Let A be the set of all complex matrices $a = (a_{jl})_{j,l=1}^{n}$. The image of the set A under the mapping $a \to e^a$ contains V.

Proof. Consider the series

$$a = (g - e) - \tfrac{1}{2}(g - e)^2 + \tfrac{1}{3}(g - e)^3 - \cdots. \tag{3.2.2}$$

By (3.2.1), $|g - e|$ is less than $\sqrt{\varepsilon}$. Therefore this series converges absolutely in norm for $g \in V$: we denote its sum by $\ln g$. For absolutely convergent series of matrices, we can carry out operations just as for numerical series. Therefore we have $e^a = e^{\ln g} = g$ for $a = \ln g$.

In view of (3.2.1), we have

$$|a| \leqslant |g - e| + \tfrac{1}{2}|g - e|^2 + \tfrac{1}{3}|g - e|^3 + \cdots$$

$$\leqslant \sqrt{\varepsilon} + \varepsilon^{2/2} + \varepsilon^{3/2} + \cdots = \frac{\sqrt{\varepsilon}}{1 - \sqrt{\varepsilon}}. \tag{3.2.3}$$

A representation $T : g \to T(g)$ of $G = GL(n, \mathbf{C})$ in a finite-dimensional space X is called *analytic* if the matrix elements of T in some basis (hence in all bases) are analytic functions of the matrix elements g_{jl}. It is easy to see that this definition agrees with the definition of an analytic irreducible representation given in 2.5 *supra*.

Theorem 1.

(1) *Any finite-dimensional analytic representation T of $G = GL(n, \mathbf{C})$ is completely reducible.*

(2) *The restriction of a finite-dimensional analytic irreducible representation T of $GL(n, \mathbf{C})$ to the group $U = U(n)$ is irreducible.*

Proof

(a) Let T be an analytic representation of G in a finite-dimensional space X. Let Y be a finite-dimensional space in duality with X under the non-singular bilinear form (x, y) for $x \in X$, $y \in Y$.

Let M be a subspace in X invariant under all operators $T(u)$, $u \in U$. We will show that M is also invariant under all $T(g)$, $g \in G$.

Let $N = M^\perp$ be the orthogonal complement to M in Y under the form (x, y). We have $M = N^\perp$ and

$$(T(u)x, y) = 0, \quad \text{for all } x \in M, \, y \in N. \tag{3.2.4}$$

This holds because M is invariant under all operators $T(u)$, $u \in U$. Note that a matrix $u \in U$ can be written in the form $u = e^{ih}$, where h is a Hermitian matrix.

We can parametrize matrices $u \in U$ by the real numbers $h_{jj}, j = 1, 2, \ldots, n$ and $b_{jl} = \operatorname{Re} h_{jl}$, $c_{jl} = \operatorname{Im} h_{jl}$ for $j < l$, and $l = 2, \ldots, n$. These parameters run independently through $\mathbf{R}$. The matrix h, and thus also u, are completely defined by these parameters, since

$$\begin{aligned} h_{jl} &= b_{jl} + ic_{jl}, & \text{for } j < l, \\ h_{jl} &= \bar{h}_{lj} = b_{lj} - ic_{lj}, & \text{for } j > l. \end{aligned} \tag{3.2.5}$$

For *complex* values h_{jj}, b_{jl} and c_{jl} the matrix $a = ih$ defined by (3.2.5) is an arbitrary matrix in A. Therefore matrices of the form $y = e^{ih}$ fill out the entire neighborhood V (see the lemma). The function $(T(g)x, y)$ is a linear combination with constant coefficients of the matrix entries of the operator $T(g)$. This function is therefore an analytic function of the entries g_{jl} and thus of the complex parameters h_{jj}, b_{jl}, c_{jl}. For *real* values of these parameters we have $g = u$ and by (3.2.4)

$$(T(g)x, y) = (T(u)x, y) = 0, \quad \text{for } x \in M, y \in N.$$

From this and the uniqueness principle for analytic functions we infer that

$$(T(g)x, y) = 0, \quad \text{for } x \in M, y \in N \tag{3.2.6}$$

throughout the domain of the analytic function. Thus (3.2.6) holds at least for all $g \in V$. We conclude that

$$T(g)M \subset M, \quad \text{for all } g \in V.$$

This yields

$$T(g_1 g_2)M = T(g_1)T(g_2)M \subset T(g_1)M \quad \text{for all } g_1, g_2 \in V;$$

which is to say that

$$T(g)M \subset M, \quad \text{for all } g \in V^2.$$

Repeating this process, we conclude that $T(g)M$ is contained in M for all $g \in V^n, n = 1, 2, 3, \ldots$, i.e., for all $g \in \bigcup_{n=1}^{\infty} V^n$. Since G is connected, $\bigcup_{n=1}^{\infty} V^n$ is equal to G. Therefore, M is also invariant under all $g \in G$.

(b) Let T be an analytic representation of G in a finite-dimensional space X. Its restriction $u \to T(u)$ is a representation of the compact group U and is therefore completely reducible. Thus we have a decomposition $X = X_1 + \cdots + X_m$ such that every X_j is invariant under all $T(u)$ and the restriction of $T(u)$ to X_j is irreducible. By the proof of (a), X_j is also invariant under all $T(g), g \in G$, and the restriction of $T(g)$ to X_j is irreducible since the restriction of $T(u)$ to X_j is irreducible. We have thus proved statement (1).

(c) Suppose now that T is irreducible and assume that its restriction $u \to T(u), u \in U$ is reducible. Then X admits a subspace M such that $(0) \subsetneq M \subsetneq X$ and M is invariant under all $T(u), u \in U$. By part (a), M is also invariant under all $T(g), g \in G$, so that T is not irreducible. This proves statement (2). □

The method in the preceding proof is a specific application of a general method. (This method will be described below in chapter XI.) It is called "the unitary method of H. Weyl".

Theorem 1 implies that for an analytic representation of the group $GL(n, \mathbf{C})$ all of the Propositions I–III in 3.1 hold. We have the following in particular.

I. *An analytic representation T of the group $GL(n, \mathbf{C})$ in a finite-dimensional space X is irreducible if and only if X admits only one vector of highest weight (to within a scalar multiple).*

II. *Any tensor representation of the group $GL(n, \mathbf{C})$ is analytic and hence completely reducible.*

This statement follows directly from the formulas for matrix entries of a tensor representation (see 2.6, chapter I) and from theorem 1.

From II we conclude that all the Propositions I–III in 3.1 hold for tensor representations of $GL(n, \mathbf{C})$.

Remark 1. The condition of analyticity is essential. For example, the representation

$$g \to \left\| \begin{matrix} 1 & \ln|\det g| \\ 0 & 1 \end{matrix} \right\|$$

fails to be completely reducible. On the other hand, the representation

$$g \to \left\| \begin{matrix} 1 & \ln \det g \\ 0 & 1 \end{matrix} \right\|$$

is analytic but not single-valued, because the function $\ln \det g$ is not single-valued. The range of the function $\det g$ is the set of all nonzero complex numbers (that is, the non simply connected set $\mathbf{C}_0^1$). Therefore there is no single-valued continuous branch of the function $\ln \det g$ on G.

Remark 2. The relation $GL(n, \mathbf{C}) \sim \mathbf{C}_0^1 SL(n, \mathbf{C})$ shows that every finite-dimensional representation $T: g \to T(g)$ of the group $GL(n, \mathbf{C})$ can be represented in the form $T(g) = T_0(\tau)T_1(g)$ for $g = \tau g_1$, $\tau \in \mathbf{C}_0^1$ and $g_1 \in SL(n, \mathbf{C})$. As we will show below (see §3, chapter X and 7.2, chapter XI) any finite-dimensional representation of the group $SL(n, \mathbf{C})$ is completely reducible. Therefore failure of complete reducibility for a representation of $GL(n, \mathbf{C})$ can occur only if the factor $\mathbf{C}_0^1$ is present.

Remark 3. A function $f(\xi_1, \xi_2, \ldots, \xi_n)$ defined in an open subset V of $\mathbf{C}^n$ is called *antianalytic in V* if for every point $(\xi_1^0, \ldots, \xi_n^0)$ of this domain there exists a neighborhood in which we have

$$f(\xi_1, \xi_2, \ldots, \xi_n) = \sum_{k_1, \ldots, k_n = 1} a_{k_1, \ldots, k_n} (\bar{\xi}_1 - \bar{\xi}_1^0)^{k_1} \cdots (\bar{\xi}_n - \bar{\xi}_0^0)^{k_n},$$

the series on the right side being absolutely convergent. A representation $T:g \to T(g)$ of $GL(n, \mathbf{C})$ in the finite-dimensional space X is called *anti-analytic* if the matrix entires of the operator $T(g)$ in some basis (and hence in all bases) of X are antianalytic functions on $GL(n, \mathbf{C})$ of the entries g_{ji} of the matrix g. This definition plainly agrees with the definition of an anti-analytic irreducible representation given in 2.5 *supra*. It is also clear that the uniqueness theorem holds as well for antianalytic functions and therefore theorem 1 and Propositions I and II are valid for antianalytic representations.

3.3. Young's Product

Let us apply the results of 3.2 to irreducible representations of the group $GL(n, \mathbf{C})$. From I and II in 2.5 we infer the following.

I. *The product $\alpha_1 \alpha_2$ of two inductive characters α_1, α_2 of the group $G = GL(n, \mathbf{C})$ is also an inductive character of this group.*

Therefore, $\alpha_1 \alpha_2$ defines a finite-dimensional irreducible representation $T_{\alpha_1 \alpha_2}$ of the group G. It is called the *Young product* of the representations T_{α_1} and T_{α_2} and is denoted by $T_{\alpha_1} T_{\alpha_2}$. In many realizations, it is easy to construct the space of $T_{\alpha_1} T_{\alpha_2}$ from the spaces of the representations T_{α_1} and T_{α_2}. Consider the realization in the space Φ_α constructed in 2.3.

II. *The space $\Phi_{\alpha_1 \alpha_2}$ is*

$$\Phi_{\alpha_1 \alpha_2} = \Phi_{\alpha_1} \Phi_{\alpha_2} \tag{3.3.1}$$

where $\Phi_{\alpha_1} \Phi_{\alpha_2}$ denotes the linear span of all products $f_1(g) f_2(g)$, $f_1(g) \in \Phi_{\alpha_1}$, $f_2(g) \in \Phi_{\alpha_2}$.

Proof. We first suppose that T_{α_1} and T_{α_2} are analytic. The space $\Phi_{\alpha_1 \alpha_2}$ is the linear span of all functions $\alpha_1(gg_0) \alpha_2(gg_0)$, $g_0 \in G$ and $\Phi_{\alpha_1} \Phi_{\alpha_2}$ is the linear span of all functions $\alpha_1(gg_1) \alpha_2(gg_2)$ for $g_1, g_2 \in G$. Therefore we have

$$\Phi_{\alpha_1 \alpha_2} \subset \Phi_{\alpha_1} \Phi_{\alpha_2} \tag{3.3.2}$$

Writing $\alpha_1 \alpha_2 = \alpha$, we see that every function $f(g) = \alpha_1(gg_1) \alpha_2(gg_2)$ satisfies the condition

$$\begin{aligned} f(kg) &= \alpha_1(kgg_1) \alpha_2(kgg_2) \\ &= \alpha_1(k) \alpha_2(k) \alpha_1(gg_1) \alpha_2(gg_2) = \alpha(k) f(g). \end{aligned} \tag{3.3.3}$$

Thus every function $f(g) \in \Phi_{\alpha_1} \Phi_{\alpha_2}$ also satisfies this condition.

It is also clear that $\Phi_{\alpha_1} \Phi_{\alpha_2}$ is invariant under right translations $g \to gg_0$ and is finite-dimensional. We define a representation $T:g \to T(g)$ of

$G = GL(n, \mathbb{C})$ in $\Phi_{\alpha_1}\Phi_{\alpha_2}$ by the formula

$$T(g_0)f(g) = f(gg_0), \quad \text{for } f(g) \in \Phi_{\alpha_1}\Phi_{\alpha_2}. \tag{3.3.4}$$

Clearly T is an analytic representation in $\Phi_{\alpha_1}\Phi_{\alpha_2}$. By VIII in 2.2, $\Phi_{\alpha_1}\Phi_{\alpha_2}$ admits only one vector of highest weight. By I of 3.2, T is irreducible. By (3.3.2) $\Phi_{\alpha_1\alpha_2}$ is a nonzero subspace of $\Phi_{\alpha_1}\Phi_{\alpha_2}$. Thus $\Phi_{\alpha_1\alpha_2}$ is equal to $\Phi_{\alpha_1}\Phi_{\alpha_2}$ and we have proved (3.3.1) for analytic representations. The proof for anti-analytic representations is analogous. Suppose next that T_{α_1} is analytic and T_{α_2} is antianalytic. Then (3.3.1) holds again, since $T_{\alpha_1}T_{\alpha_2}$ is the tensor product $T_{\alpha_1} \otimes T_{\alpha_2}$ (see II and III in 2.5). It is also clear that the general case reduces to those already considered. $\square$

Exercise

Prove that relations like (3.3.1) hold for the Z-realization and the U-realization of irreducible representations of the group $GL(n, \mathbb{C})$.

3.4. Basic Representations and Their Realization in Spaces of Tensors

An irreducible representation T_α is called *basic* if one of the numbers r_i or s_i in its signature is one and all other numbers are zero. The basic representation such that $r_i = 1$ is denoted by d_i and the basic representation such that $s_i = 1$ is denoted by $\bar{d}_i$. The Young product of a representation T with itself k times is denoted by T^k. Multiplication of characters corresponds to addition of the signatures. Hence for T_α with a signature

$$\{r_1, r_2, \ldots, r_{n-1}, 0, s_1, s_2, \ldots, s_{n-1}, 0\},$$

we obtain the formula

$$T_\alpha = d_1^{r_1} d_2^{r_2} \cdots d_{n-1}^{r_{n-1}} \bar{d}_1^{s_1} \bar{d}_2^{s_2} \cdots \bar{d}_{n-1}^{s_{n-1}}. \tag{3.4.1}$$

In the general case, we must multiply by the number $(\det g)^{p_n}(\overline{\det g})^{q_n}$ (see theorem 5 in 2.5), obtaining

$$T_\alpha = (\det g)^{p_n}(\overline{\det g})^{q_n} d_1^{r_1} d_2^{r_2} \cdots d_{n-1}^{r_{n-1}} \bar{d}_1^{s_1} \bar{d}_2^{s_2} \cdots \bar{d}_{n-1}^{s_{n-1}}. \tag{3.4.2}$$

We will now show how to realize basic representations in a space of tensors. Consider the covariant tensors of rank p. Every such tensor can be considered a polylinear form $f(\xi_1, \ldots, \xi_p)$ of p independent variables $\xi_j \in C^n$, $j = 1, \ldots, p$. The corresponding tensor representation (which we denote by T_p) is defined by the formula

$$T_p(g)f(\xi_1, \ldots, \xi_p) = f(\xi_1 g, \ldots, \xi_p g), \tag{3.4.3}$$

where $\xi_j g$ is the product of the row $\xi_j = (\xi_{j1}, \ldots, \xi_{jn})$ by the matrix g. A tensor f is called *symmetric* if it is invariant under permutation of variables and *antisymmetric* if it changes sign under every transposition of two distinct variables. Antisymmetric tensors are called *polyvectors*. The space of all polyvectors of rank p is denoted by X_p. Clearly, X_p is a finite-dimensional linear space. If one of the coefficients $c_{j_1 j_2 \ldots j_p}$ of the polyvector f is known, the property of antisymmetry yields all coefficients obtained from $c_{j_1 j_2 \ldots j_p}$ by permuting indices. Therefore, in order to define a polyvector, it is sufficient to specify its coefficients $c_{j_1 \ldots j_p}$ for $j_1 < j_2 < \cdots < j_p$. We conclude that polyvectors $e_{j_1 j_2 \ldots j_p}, j_1 < j_2 < \cdots < j_p$ for which $c_{j_1 \ldots j_p} = 1$ and $c_{j_1 \ldots j_p} = 0$ for $\{j_1', \ldots, j_p'\}$, not obtained from $\{j_1, \ldots, j_p\}$ by permutation of indices form a basis in X_p. Clearly we have

$$e_{j_1 j_2 \cdots j_p}(\xi_1, \ldots, \xi_p) = \begin{vmatrix} \xi_{1j_1} & \cdots & \xi_{1j_p} \\ \cdots\cdots\cdots\cdots \\ \xi_{pj_1} & \cdots & \xi_{pj_p} \end{vmatrix}. \tag{3.4.4}$$

By (3.4.3) every basis vector is a weight vector, that is,

$$T_n(\delta)e_{j_1 j_2 \cdots j_p} = \lambda_{j_1} \cdots \lambda_{j_p} e_{j_1 j_2 \cdots j_p}. \tag{3.4.5}$$

Also the weights $\lambda_{j_1} \cdots \lambda_{j_p}$ are distinct for distinct basis vectors. Therefore X_p admits no other weight vectors (up to scalar multiples). Among these weight vectors only the vector

$$e_{12\cdots p} = \begin{vmatrix} \xi_{11} & \cdots & \xi_{1p} \\ \cdots\cdots\cdots\cdots \\ \xi_{p1} & \cdots & \xi_{pp} \end{vmatrix}$$

is Z-invariant. (This is easy to show.) Thus X admits only one vector of highest weight (up to scalar multiples). Therefore T_p is irreducible (II in 3.2, II in 3.1). The corresponding highest weight is $\alpha_p(\delta) = \lambda_1 \lambda_2 \cdots \lambda_p = \Delta_p(g)$. This means that in the signature of the representation T_p, r_p is one and all other r_j and all s_j, p_n, q_n are zero. Thus, the highest weights of the representations d_p and T_p coincide, and so these representations are equivalent. We have proved the following.

I. *The basic representation d_p of the group $G = GL(n, C)$ is equivalent to the representation T_p in the space X_p of polyvectors of rank p, operating according to the formula*

$$T_p(g)f(\xi_1, \ldots, \xi_p) = f(\xi_1 g, \ldots, \xi_p g), \qquad f \in X_p. \tag{3.4.6}$$

The representation $\bar{d}_p$ is the product of the mappings $g \to \bar{g}$ and $\bar{g} \to d_p(g)$. From this we infer the following.

II. *The basic representation $\bar{d}_p$ of the group $G = GL(n, \mathbf{C})$ is equivalent to the representation $\bar{T}_p$ in the space X_p of polyvectors of rank p, operating according to the formula*

$$\bar{T}_p(g)f(\xi_1, \ldots, \xi_p) = f(\xi_1\bar{g}_1, \ldots, \xi_p\bar{g}), \qquad f \in X_p. \tag{3.4.7}$$

3.5. Decomposition of a Tensor Representation of the Group G into Irreducible Representations

Let T be a tensor representation of the group $G = GL(n, \mathbf{C})$. That is, T is the tensor product of the identity representation $g \to g$ of the group G with itself m times. Here m is any positive integer. As we noted in II of 3.2, the representation T is analytic and completely reducible, that is, it is the direct sum of irreducible analytic representations. The signature of any irreducible analytic representation S has the form $\{r_1, \ldots, r_{n-1}, p_n, 0, \ldots, 0\}$, which we abbreviate as $[r_1, \ldots, r_{n-1}, p_n]$. Here $r_1, \ldots, r_n$ are nonnegative integers and p_n is an integer. We set $m_n = p_n$, $m_k = r_k + m_{k+1}$ for $k = n - 1, n - 2, \ldots, 1$. Thus the sequence of numbers $\beta = (m_1, \ldots, m_n)$ satisfies the condition $m_1 \geqslant \cdots \geqslant m_n$. In this case the irreducible analytic representation S of the group G is denoted by $d(\beta)$. Let H be the space of the representation T. The space H can be naturally identified with the tensor product of m copies of the space $\mathbf{C}^n$. Let S_m be the symmetric group on m letters. Clearly the operators of the representation T are commutative with all linear operators $\tilde{\sigma}$ in H, defined for elements of the form $x_1 \otimes \cdots \otimes x_n$ by the formula $\tilde{\sigma}(x_1 \otimes \cdots \otimes x_n) = x_{\sigma(1)} \otimes \cdots \otimes x_{\sigma(n)}$. Here σ is an arbitrary element of S_m. We extend $\tilde{\sigma}$ by linearity over the entire space H. Extend the representation $\sigma \to \tilde{\sigma}$ to a representation of the group algebra $A = A_{S_m}$ of the group S_m in H. This representation of the algebra A is denoted by π.

Let $m = m_1 + \cdots + m_h$ be a partition of the number m into integers that satisfy the condition $m_1 \geqslant \cdots \geqslant m_h$. Let α be the sequence $(m_1, \ldots, m_h)$ and let $\sum_\alpha$ be the corresponding Young diagram (see 3.2, chapter II). Let ε_α be the element of the group algebra that corresponds to the diagram $\sum_\alpha$ (see 3.7), chapter II). The operator in the space H that is the image of the element $\varepsilon_\alpha \in A$ is denoted by $\varepsilon(\alpha)$. Note that $\varepsilon(\alpha) = \pi(\varepsilon_\alpha)$.

Theorem. *If the signature $\alpha = (m_1, \ldots, m_h)$ has more than n nonzero coordinates $(h > n)$, then $\varepsilon(\alpha)$ is equal to zero. If $h \leqslant n$, then the space $H_\alpha = \varepsilon(\alpha)H$ is a maximal subspace of H in which the representation is a multiple of the irreducible representation $d(\tilde{\alpha})$. Here we have $\tilde{\alpha} = (m_1, \ldots, m_h, 0, \ldots, 0)$.[51] The corresponding multiplicity $k(\tilde{\alpha})$ satisfies the condition $\mu(\tilde{\alpha})k(\tilde{\alpha}) = m!$, where the number $\mu(\tilde{\alpha})$ is defined in (3.6.2), chapter II.*

---

[51] Observe that $\tilde{\alpha}$ is a sequence of n numbers. For $h < n$, we fill out the sequence $(m_1, \ldots, m_h)$ with zeros.

For a proof of this theorem, and additional facts, the reader may consult the books of H. Weyl [1] and Želobenko [1].

Remark. According to theorem 1 in 3.2, the representation S_T of the group $U(n)$ (the restriction to $U = U(n)$ of the finite-dimensional analytic irreducible representation T of $G = GL(n, \mathbf{C})$) is irreducible. We obtain all finite-dimensional irreducible continuous representations of the group U in this way.[52] Consider the set A of finite linear combinations of the matrix entries of the restrictions of the tensor representations of G to the subgroup U. Since the identity representation of G is a particular tensor representation and the restriction of the identity representation to G separates elements of the group U, the set A also separates the elements of U. Furthermore, a tensor product of tensor representations of G is itself a tensor representation of G. Hence A is an algebra of continuous functions on U. Finally, the reader can easily verify that the restriction of the basis representation d_{n-1} to the subgroup U is equivalent to the representation $u \to \bar{u}$, where $\bar{u}$ is the matrix whose entries are the complex conjugates of the corresponding entries of u. Stone's theorem shows that the algebra A is dense in $C(U)$. The theorem in 2.8, chapter IV, implies that the set of irreducible representations S_T, where T runs through the set of all tensor representations of G, is a complete set of irreducible unitary representations of the group U.

An analogous argument shows that any compact linear group admits a complete set of irreducible continuous unitary representations that are subrepresentations of tensor products of the group's identity representation.

---

[52] In the sequel we obtain an analogous result for all complex semisimple Lie groups (see IV in 8.2, chapter IX).

Chapter VII

Finite-Dimensional Representations of the Complex Classical Groups

§1. The Complex Classical Groups

1.1. Definition of Complex Classical Groups

The following four sets of groups are called *complex classical groups*.

(1) The group of all complex unimodular linear transformations (that is, with determinant one) of the $n + 1$-dimensional complex linear space X_{n+1}, $n = 1, 2, 3, \ldots$. It is denoted by A_n. Clearly A_n is isomorphic to $SL(n + 1, \mathbf{C})$. We will identify[53] A_n with $SL(n + 1, \mathbf{C})$. The group A_n is called the *complex unimodular orthogonal group of order* $n + 1$.

(2) Consider the group of all complex unimodular linear transformations of an odd-dimensional complex linear space X_{2n+1}, $n = 2, 3, \ldots$, that leave invariant some nonsingular symmetric bilinear form φ on X_{n+1}. This group is denoted by B_n and also by $SO(2n + 1, \mathbf{C})$. It is called the *complex unimodular orthogonal group of order* $2n + 1$.

(3) Consider the group of all complex unimodular linear transformations of an even-dimensional complex linear space X_{2n}, $n = 3, 4, \ldots$, that leave invariant some nonsingular symmetric bilinear form φ on X_{2n}. This group is denoted by D_n and also by $SO(2n, \mathbf{C})$. It is called the *complex unimodular orthogonal group of order* $2n$.

(4) Consider the group of all complex unimodular linear transformations of the even-dimensional complex linear space X_{2n}, $n = 2, 3, 4, \ldots$, that leave invariant some skew-symmetric bilinear form φ on X_{2n}. This group is denoted by C_n and also by $Sp(2n, \mathbf{C})$. This group is called the *complex symplectic group of order* $2n$.

As we will see in the sequel (see §10 in chapter 10), these groups and their representations play an important role in many applications and also in the study of a wide class of groups.

In any fixed basis the matrices of the groups B_n, D_n and C_n are unimodular and we may therefore suppose that

$$B_n \subset SL(2n + 1, \mathbf{C}), \qquad D_n \subset SL(2n, \mathbf{C}), \qquad C_n \subset SL(2n, \mathbf{C}). \quad (1.1.1)$$

---

[53] As above (see the examples in 1.2, chapter V) we denote the matrix $g \in GL(n, \mathbf{C})$, and the linear transformation in X_n with matrix g in a fixed basis, by the same letter g.

1.2. Choice of a Basis

For brevity we set $m = 2n$ in the case of the groups C_n, D_n and $m = 2n + 1$ in the case of the group B_n. We choose a basis in X_m so that the matrices of the groups B_n, C_n, D_n have the simplest possible form. The matrix of the form φ in a fixed but arbitrary basis is denoted by s_0. In the case of a symmetric form φ we choose a basis $c_1, \ldots, c_m$ so that $\varphi(x, y)$ is equal to $x_1 y_m + x_2 y_{m-1} + \cdots + x_m y_1$ and thus for $m = 2n$ the matrix s_0 has the form

$$s_0 = \begin{Vmatrix} 0 & s_1 \\ s_1 & 0 \end{Vmatrix},$$

where s_1 is the matrix of the n-th order given by

$$s_1 = \begin{Vmatrix} 0 & \cdots & 0 & 1 \\ 0 & \cdots & 1 & 0 \\ \cdots\cdots\cdots\cdots \\ 1 & \cdots & 0 & 0 \end{Vmatrix}. \tag{1.2.1}$$

In the case of a skew-symmetric form φ, we choose a basis so that

$$\varphi(x, y) = x_1 y_m + x_2 y_{m-1} + \cdots + x_n y_{n+1} - x_{n+1} y_n - \cdots - x_m y_1.$$

We find that

$$s_0 = \begin{Vmatrix} 0 & -s_1 \\ s_1 & 0 \end{Vmatrix}. \tag{1.2.2}$$

(See Bourbaki [1].) The matrix in (1.2.1) has the entry one exactly once in each row and each column.

Thus, if we write $(x, y) = \sum_{k=1}^{m} x_k y_k$, we obtain

$$\varphi(\xi, \eta) = (s_0 x, y). \tag{1.2.3}$$

The form φ must be invariant under the transformations g of the given group. This means that

$$(s_0 g x, g y) = (s_0 x, y), \tag{1.2.4}$$

which implies that $g' s_0 g = s_0$; that is,

$$g'^{-1} = s_0 g s_0^{-1}, \tag{1.2.5}$$

where g' is the transpose of g.

Plainly, (1.2.5) is equivalent to each of the following conditions:

$$g' s_0 g_0 s_0^{-1} = e, \tag{1.2.6}$$

$$g = s_0^{-1} g'^{-1} s_0. \tag{1.2.7}$$

1.3. Basic Subgroups

Throughout this chapter we use the symbol G to denote one of the complex classical groups. For the dimension of the space in which the transformations of G operate, we write m. We set

$$G_m = GL(m, \mathbf{C}).$$

The subgroups H, Z^+, D, E, Z^-, K, U, constructed in §1 of chapter VI for the group G_m, will be denoted by $H_m, Z_m^+, D_m, E_m, Z_m^-, K_m, U_m, \Gamma_m$. We set

$$H = G \cap H_m, \quad Z^+ = G \cap Z_m^+, \quad D = G \cap D_m, \quad Z^- = G \cap Z_m^-,$$
$$K = G \cap K_m, \quad U = G \cap U_m, \quad E = G \cap E_m, \quad \Gamma = G \cap \Gamma_m.$$

We will establish for these subgroups of G relations analogous to those proved in §1, chapter VI.

1.4. Decomposition of Elements of the Group H

I. *Every element $h \in H$ is represented uniquely in the form*

$$h = \delta z, \quad where \ \delta \in D \quad and \ z \in Z^+, \tag{1.4.1}$$

and also in the form

$$h = z \delta, \quad where \ \delta \in D \quad and \ z \in Z^+. \tag{1.4.2}$$

Proof. We suppose first that $G = SL(m, \mathbf{C})$. The group G consists exactly of those elements $g \in G_m$ for which $\det g = 1$. Since h belongs to H_m, we cite I in 1.7, chapter VI to write $h = \delta z$, $\delta \in D_m$, $z \in Z_m^+$. This decomposition is unique: we need only to prove that δ, z belong to G. It is clear that $\det z = 1$ and thus z belongs to G. Furthermore, h belongs to G by hypothesis. Therefore we have $1 = \det h = \det \delta \det z = \det \delta$ and so δ belongs to G. This proves (1.4.1). The relation (1.4.2) is proved similarly.

We now suppose that G is an orthogonal or symplectic group. Then G is contained in $SL(m, \mathbf{C})$. In (1.4.1), h belongs to $SL(m, \mathbf{C})$ and as we just proved, δ, z also belong to $SL(m, \mathbf{C})$. Furthermore, h is equal to $s_0^{-1} h'^{-1} s_0$. Substituting δz for h, we obtain

$$\delta z = s_0^{-1} \delta'^{-1} s_0 s_0^{-1} z'^{-1} s_0. \tag{1.4.3}$$

We can easily verify that $s_0^{-1}\delta'^{-1}s_0 \in D_m$ and $s_0^{-1}z'^{-1}s_0 \in Z_m^+$. By the uniqueness of the decomposition, we have $h = \delta z$, $\delta \in D_m$, $z \in Z_m^+$. We conclude that $\delta = s_0^{-1}\delta'^{-1}s_0$, $z = s_0^{-1}z'^{-1}\delta_0$. Therefore δ belongs to D and z belongs to Z^+, and we have proved (1.4.1). The proof of (1.4.2) is similar. $\square$

1.5. Decomposition of Elements of the Group K

I. *Every element k of the group K is represented uniquely in the form*

$$k = \delta\zeta, \quad where \ \delta \in D \quad and \ \zeta \in Z^-, \tag{1.5.1}$$

and also in the form

$$k = \zeta\delta, \quad where \ \delta \in D \quad and \ \zeta \in Z^-. \tag{1.5.2}$$

The proof is analogous to that of I in 1.4.

1.6. Gauss's Decomposition

The minor of the matrix g formed from its first p rows and first p columns is denoted by $\Delta_p(g)$. The set of all $g \in G$ for which

$$\Delta_p(g) \neq 0, \quad p = 1, 2, \ldots, m, \tag{1.6.1}$$

is denoted by G_{reg}. It is clear that

$$G_{\mathrm{reg}} = G \cap G_{m\,\mathrm{reg}}. \tag{1.6.2}$$

The matrices $g \in G_{\mathrm{reg}}$ are called *regular*.

I. *The set G_{reg} is open in G and we have*

$$\overline{G}_{\mathrm{reg}} = G. \tag{1.6.3}$$

Proof. The first statement is obvious. To prove the second, we first note that it is obvious for $G = A_n$.

Now let G be one of the groups B_n, C_n, D_n. For $g \in G$, we write

$$\Delta(g) = \Delta_1(g)\Delta_2(g) \cdots \Delta_n(g),$$
$$G_\Delta = \{g : g \in G, \Delta(g) = 0\}.$$

Plainly G_{reg} contains $G\backslash G_\Delta$, and so it suffices to prove that $\overline{G\backslash G_\Delta} = G$. To do this, note that e belongs to $G\backslash G_\Delta$. On the other hand, if $\overline{G\backslash G_\Delta}$ is not G, there exists a nonvoid open set $U \subset G_\Delta$. Since Δ is a polynomial, $\Delta = 0$ on G, that is, G is equal to G_Δ, which contradicts the relation $e \in G\backslash G_\Delta$. $\square$

II. *Every matrix $g \in G_{\text{reg}}$ is represented uniquely in the form*

$$g = kz, \quad \text{where } k \in K \quad \text{and } z \in Z^+. \tag{1.6.4}$$

Proof. According to I in 1.8, chapter VI, every matrix $g \in G_{\text{reg}}$ is represented uniquely in the form

$$g = kz, \quad \text{where } k \in K_m \quad \text{and } z \in Z_m^+. \tag{1.6.5}$$

We need to prove only that k and z belong to G. In the case $G = A_n$, this is obvious. For, if g belongs to A_n, then $1 = \det g = \det k \cdot \det z = \det k$.

Let G be a symplectic or orthogonal group and let g belong to G. By (1.2.7) we have $g = s_0^{-1} g'^{-1} s_0$. Substituting in (1.6.5), we obtain

$$g = kz = s_0^{-1} k'^{-1} s_0 \cdot s_0^{-1} z'^{-1} s_0.$$

We easily verify that $s_0^{-1} k'^{-1} s_0 \in K_m$ and $s_0^{-1} z'^{-1} s_0 \in Z_m^+$. From this and the uniqueness of the decomposition (1.6.5) we infer that $k = s_0^{-1} k'^{-1} s_0$, $z = s_0^{-1} z'^{-1} s_0$, that is, $k \in K$ and $z \in Z^+$. □

Combining II in 1.6 and I in 1.5, we arrive at the following.

III. *Every matrix $g \in G_{\text{reg}}$ is represented uniquely in the form*

$$g = \delta \zeta z, \quad \text{where } \delta \in D, \quad \zeta \in Z^-, \quad \text{and } z \in Z^+, \tag{1.6.6}$$

and also in the form

$$g = \zeta \delta z, \quad \text{where } \zeta \in Z^-, \quad \delta \in D, \quad \text{and } z \in Z^+. \tag{1.6.7}$$

Each of the decompositions (1.6.4), (1.6.6) and (1.6.7) is called a *Gauss decomposition in G.*

Every Gauss decomposition in G is also a Gauss decomposition in G_m and therefore the entries of the matrices k, z, δ, ζ in these decompositions can be computed by the formulas (1.8.10)–(1.8.15) of chapter VI.

1.7. Gram's Decomposition

I. *Every element $g \in G$ can be represented in the form*

$$g = ku, \quad \text{where } k \in K \quad \text{and } u \in U. \tag{1.7.1}$$

If $g = k_1 u_1$ is another such decomposition, we have

$$k_1 = k\gamma \quad \text{and } u_1 = \gamma^{-1} u, \quad \gamma \in \Gamma. \tag{1.7.2}$$

Proof. By II in 1.9, chapter VI, we have

$$g = ku, \quad \text{where } k \in K_m \quad \text{and } u \in U_m; \tag{1.7.3}$$

and if $g = k_1 u_1$ is another such decomposition, we obtain

$$k_1 = k\gamma, \quad \text{where } u_1 = \gamma^{-1} u \quad \text{and } \gamma \in \Gamma_m.$$

Thus we need to prove only that we can choose k and u in G and thus also $\gamma \in G$.

First, a proper choice of γ in (1.7.2) yields a k such that

$$k_{pp} > 0. \tag{1.7.4}$$

Plainly k is defined uniquely by this condition. Setting $r = \det g$, we get

$$r = \prod_p k_{pp} > 0.$$

Furthermore, $|\det u|$ is equal to 1, so that $\det u = e^{i\varphi}$, $\varphi \in \mathbf{R}^1$.

Suppose first that $G = A_n$. For $g \in G$, (1.7.3) gives

$$1 = \det g = \det k \cdot \det u = re^{i\varphi},$$

which is possible only for $r = 1$, $\varphi = 0$, that is, $\det k = 1$, $\det u = 1$, and so k and u belong to A_n.

Next suppose that G is an orthogonal or symplectic group. For $g \in G$, we have $g = s_0^{-1} g'^{-1} s_0$ (see (1.2.7)). Substituting in (1.7.3), we obtain

$$ku = s_0^{-1} k'^{-1} s_0 s_0^{-1} u'^{-1} s_0. \tag{1.7.5}$$

Suppose that k satisfies (1.7.4). We easily verify that $s_0^{-1} k'^{-1} s_0$ belongs to K and also satisfies (1.7.4). This condition uniquely defines k in (1.7.3). Therefore k is equal to $s_0^{-1} k'^{-1} s_0$ and thus $u = s_0^{-1} u'^{-1} s_0$, and so k and u belong to G. If we have $g = k_1 u_1$, $k_1 \in K$, $u_1 \in U$, (1.7.2) is satisfied. In this case $\gamma = k_1^{-1} k$ belongs to G and thus $\gamma \in \Gamma_m \cap G = \Gamma$. $\square$

II. *Every element $g \in G$ can be represented uniquely in the form*

$$g = \varepsilon \zeta u, \quad \text{where } \varepsilon \in E, \quad \zeta \in Z^-, \quad \text{and } u \in U. \tag{1.7.6}$$

Proof. Proposition I implies (1.7.1) and that there exists exactly one matrix k that satisfies (1.7.1) and (1.7.4). By I in 1.5, we have

$$k = \delta\zeta, \quad \text{where } \delta \in D \quad \text{and } \zeta \in Z^-, \tag{1.7.7}$$

where $\delta_{pp} = k_{pp} > 0$. Therefore δ belongs to E_m and we may set $\delta = \varepsilon$. In this case $\varepsilon \in E_m \cap D \subset E_m \cap G = E$. $\square$

1.8. Independent Parameters in the Group Z^+

(a) The unimodular group $(G = SL(m, \mathbf{C}))$. In this case we have $Z^+ = Z_m^+$, so that z_{pq}, $p < q$, are independent parameters. Therefore Z^+ is homeomorphic to C^N, where $N = \sum_{1 \leqslant p < q < m} 1$ and thus *the group* $Z^+ = Z_m^+$ *is connected*.

(b) The orthogonal group of even order $(m = 2n, G = D_n)$. Every matrix $z \in Z^+$ is represented in the form

$$z = \left\| \begin{matrix} \eta & a \\ 0 & \eta_{-1} \end{matrix} \right\| \tag{1.8.1}$$

where η_{-1}, η and a are square matrices of the n-th order and η and η_{-1} belong to $Z_{A_{n-1}}^+$.

Consider next matrices of the form

$$x = \left\| \begin{matrix} 1_n & \xi \\ 0 & 1_n \end{matrix} \right\|, \qquad y = \left\| \begin{matrix} \eta & 0 \\ 0 & \eta_{-1} \end{matrix} \right\|, \tag{1.8.2}$$

where ξ is a square matrix of the n-th order and 1_n is the identity matrix of the n-th order.

I. *Every matrix* $z \in Z^+$ *is represented uniquely in the form*

$$z = xy, \tag{1.8.3}$$

where x, y are as in (1.8.2) and also belong to Z^+.

Proof. Multiplying the matrices in (1.8.2), we obtain

$$xy = \left\| \begin{matrix} \eta & \xi\eta_{-1} \\ 0 & \eta_{-1} \end{matrix} \right\|. \tag{1.8.4}$$

This matrix coincides with z in (1.8.1) for $\xi\eta_{-1} = a$, that is, for $\xi = a\eta_{-1}^{-1}$. We infer the existence and uniqueness of the decomposition (1.8.3). It remains to prove that x and y belong to G.

Since $z = xy$ belongs to G, (1.2.7) gives

$$z = xy = s_0^{-1}x'^{-1}s_0 s_0^{-1}y'^{-1}s_0. \tag{1.8.5}$$

We easily verify that $s_0^{-1}x'^{-1}s_0$ and $s_0^{-1}y'^{-1}s_0$ are also matrices of the same form as x and y. Therefore (1.8.5) and the uniqueness of the decomposition (1.8.3) imply that

$$s_0^{-1}x'^{-1}s_0 = x, \qquad s_0^{-1}y'^{-1}s_0 = y; \tag{1.8.6}$$

that is, x and y belong to G. $\square$

We will now determine conditions under which matrices of the form x and y belong to G. Recall that

$$s_0 = \begin{Vmatrix} 0 & s_1 \\ s_1 & 0 \end{Vmatrix}, \tag{1.8.7}$$

where s_1 is the matrix of the n-th order

$$s_1 = \begin{Vmatrix} 0 & 0 & \cdots & 0 & 1 \\ 0 & 0 & \cdots & 1 & 0 \\ & & \vdots & & \\ 1 & 0 & \cdots & 0 & 0 \end{Vmatrix}. \tag{1.8.8}$$

Replacing x, y, and s_0 in (1.8.6) by their expressions from (1.8.2) and (1.8.7) and multiplying the matrices on the left sides of the resulting equalities, we conclude that the conditions (1.8.6) are equivalent to the following equalities:

$$\xi = -s_1^{-1}\xi's_1; \tag{1.8.9}$$
$$\eta = s_1^{-1}\eta'^{-1}_{-1}s_1; \tag{1.8.10}$$
$$\eta_{-1} = s_1^{-1}\eta'^{-1}s_1. \tag{1.8.11}$$

It is easy to see that (1.8.11) is equivalent to (1.8.10).

From (1.8.11) we infer that the matrices η and η_{-1} are not independent. That is, (1.8.11) defines η_{-1} uniquely from a given matrix $\eta \in Z^+_{A_{n-1}}$. The condition (1.8.9) can be written more conveniently if we set $\hat{\xi} = s_1\xi$. Substituting in (1.8.9), we see that this condition is equivalent to $\hat{\xi}' = -\hat{\xi}$. That is, $\hat{\xi}$ is a skew-symmetric matrix. Hence we have the following.

II. *The matrix* $\eta \in Z^+_{A_{n-1}}$ *and the skew-symmetric matrix* $\hat{\xi}$ *of order n are independent "parameters" in* Z^+.

The matrix $z \in Z^+$ *is expressed through parameters by the formula*

$$z = \begin{Vmatrix} \eta_1 & s_1^{-1}\hat{\xi}\eta \\ 0 & \eta \end{Vmatrix}, \quad \text{where } \eta_1 = s_1^{-1}\eta'^{-1}s_1. \tag{1.8.12}$$

Let $\hat{X}$ be the set of all skew-symmetric matrices of the n-th order. Clearly, the formulas (1.8.12) define a homeomorphism of the space $Z^+_{A_{n-1}} \times \hat{X}$ onto the space Z^+.

III. *The space Z^+ is homeomorphic to the space $Z^+_{A_{n-1}} \times \hat{X}$.*

The skew symmetry of the matrix $\hat{\xi}$ means that

$$\hat{\xi}_{pq} = -\hat{\xi}_{qp}. \tag{1.8.13}$$

In particular $\hat{\xi}_{pp}$ is equal to 0, while the entries $\hat{\xi}_{pq}$, $p < q$, can be arbitrary, and $\hat{\xi}_{pq}$ for $p > q$ is given by (1.8.13). Therefore $\hat{X}$ is homeomorphic to $\mathbf{C}^N$, where $N = \sum_{1 \leqslant p < q < n} 1$. Thus X is connected. Since $Z^+_{A_{n-1}}$ is also connected, we infer the following from III.

IV. *In the case $G = D_n$, the group Z^+ is connected.*

(c) The symplectic group $(G = C_n)$. In this case, all of the preceding arguments are repeated almost *verbatim*. The sole difference is that now

$$s_0 = \begin{Vmatrix} 0 & -s_1 \\ s_1 & 0 \end{Vmatrix},$$

where s_1 is as in (1.8.8). Therefore, $\hat{\xi}$ is now symmetric. We thus arrive at the following.

V. *For $G = C_n$ the group Z^+ is connected.*

(d) The orthogonal group of odd order $(G = B_n, m = 2n + 1)$. In this case we write the matrices $z \in Z$ in the form

$$z = \begin{Vmatrix} \eta & \lambda & a \\ 0 & 1 & \mu \\ 0 & 0 & \eta_{-1} \end{Vmatrix},$$

where η, η_{-1} are as in (b) and μ and λ are rows and columns of n numbers, respectively. Arguing as in (b), we conclude that $z \in Z^+$ can be written in the form

$$z = xy, \tag{1.8.14}$$

where x and y are matrices in Z^+ of the form

$$x = \begin{Vmatrix} 1_n & \eta_0 & \xi \\ 0 & 1 & \xi_0 \\ 0 & 0 & 1_n \end{Vmatrix}, \qquad y = \begin{Vmatrix} \eta & 0 & 0 \\ 0 & 1 & 0 \\ 0 & 0 & \eta_{-1} \end{Vmatrix}. \tag{1.8.15}$$

The matrix s_0 has the form

$$s_0 = \begin{Vmatrix} 0 & 0 & s_1 \\ 0 & 1 & 0 \\ s_1 & 0 & 0 \end{Vmatrix}, \tag{1.8.16}$$

where s_1 is as in (1.8.8). Multiplying matrices and noting that $x, y \in Z^+$, we find

$$\eta = s_1^{-1}\eta'_{-1}s_1; \qquad \eta_0 = -s_1\xi'_0, \qquad s_1\xi' + \xi s_1 + \eta_0\eta'_0 = 0. \tag{1.8.17}$$

The last equality in (1.8.17) means that the matrix $\hat{\xi} = \xi s_1 + (1/2)\eta_0\eta'_0$ is skew-symmetric. Therefore we have the following.

VI. *For $G = B_n$, every matrix $z \in Z^+$ is described by the following indepen-dent "parameters": a matrix $\eta \in Z_{A_{n-1}}$, a skew-symmetric matrix $\hat{\xi}$ of order n, and a row ξ_0 of n numbers.*

VII. *For $G = B_n$, the group Z^+ is connected.*

Combine 1.8 (a), IV, V, and VII to see the following.

VIII. *For every complex classical group G, the group Z^+ is connected.*

1.9. The Group Z^-

This group is the image of the group Z^+ under the homeomorphism $g \to g'$. Proposition VIII of 1.8 gives the following.

I. *For every complex classical group G, the group Z^- is connected.*

1.10. Independent Parameters in the Group D

(a) The unimodular group $(G = SL(n, \mathbf{C}))$. In this case the matrix

$$\delta = \begin{Vmatrix} \lambda_1 & 0 & \cdots & 0 \\ 0 & \lambda_2 & \cdots & 0 \\ \multicolumn{4}{c}{\dotfill} \\ 0 & 0 & \cdots & \lambda_n \end{Vmatrix}$$

is restricted only by the condition $\lambda_1\lambda_2\cdots\lambda_n = 1$, so that $\lambda_1, \lambda_2, \ldots, \lambda_{n-1}$ (for example) are independent parameters, where every $\lambda_j \in \mathbf{C}_0^1$, $j = 1, \ldots, n-1$. When multiplying two matrices δ, the corresponding param-eters also multiply.

I. *For $G = SL(n, \mathbf{C})$, the group D is topologically isomorphic to the group $\mathbf{C}_0^{n-1}$ and is therefore connected.*

(b) For the other complex classical groups, the condition $\delta'^{-1} = s_0 \delta s_0^{-1}$ means that

$$\lambda_{m-v} = \lambda_{v+1}^{-1}, \qquad v = 0, 1, \ldots, m-1. \tag{1.10.1}$$

For the orthogonal group of odd order, (1.10.1) implies that $\lambda_{n+1} = \lambda_{n+1}^{-1}$, so that $\lambda_{n+1} = \pm 1$. From $\det \delta = 1$ and the remaining conditions in (1.10.1) we see that $\lambda_{n+1} = 1$.

II. *For the groups B_n, C_n, and D_n, the diagonal elements $\lambda_1, \lambda_2, \ldots, \lambda_n \in \mathbf{C}_0^1$ of a matrix in D are independent parameters.*

As before, II implies the following.

III. *For the groups B_n, C_n, and D_n, the group D is topologically isomorphic to the group $\mathbf{C}_0^n$, and so is connected.*

We summarize.

IV. *For every complex classical group G, the group D is connected.*

1.11. Connectedness of the Group K

I. *For every complex classical group G, the group K is connected.*

This proposition follows at once from the relation $K = DZ^-$ (see (1.5.1)) and the connectedness of the groups D and Z^- (see II in 1.2, chapter V).

1.12. Connectedness of the Group H

The group H is the image of the group K under the mapping $g \to g'$. This mapping is a homeomorphism, which gives the following.

I. *The group H is connected.*

1.13. Connectedness of the Complex Classical Groups

I. *Every complex classical group G is connected.*

Proof. Gauss's decomposition gives $G_{\text{reg}} = KZ^+$. The groups K and Z^+ are connected, and therefore G_{reg} is connected (see II in 1.2, chapter V). Thus, $G = \overline{G_{\text{reg}}}$ is connected (IV in 1.1, chapter V). □

1.14. Connectedness of the Group U

I. *For every complex classical group G, the group U is connected.*

Proof. By Cramer's decomposition (1.7.6), the correspondence $\varepsilon \times \zeta \times u \to \varepsilon \zeta u = g$ is a one-to-one continuous mapping of the space $E \times Z^- \times U$ onto

G. The inverse mapping is also continuous. This follows from the formulas for ε, ζ, u, obtained by the process of orthonormalization (see the derivation of Cramer's decomposition in example 10 in 1.2, chapter V). Thus the mapping $\varepsilon \times \zeta \times u \to \varepsilon\zeta u$ is a homeomorphism of the space $E \times Z^- \times U$ onto *G*. The group *G* is connected (I in 1.13) and therefore $E \times Z^- \times U$ and hence *U* are connected (XII of 1.1, chapter V).   □

§2. Finite-Dimensional Continuous Representations of the Complex Classical Groups

2.1. Weight Vectors and Weights of a Representation

When studying finite-dimensional representations of the full linear group (see chapter VI), we used first and foremost Gauss's decomposition for this group. Gauss's decomposition exists for every complex classical group *G* (see II and III in 1.6). Hence our description of the finite-dimensional irreducible representations of the group $GL(n, \mathbf{C})$ (with the aid of inductive characters) can be carried over *mutatis mutandis* to the complex classical groups. We will accordingly omit proofs and will concentrate on finding the inductive characters for the groups in question, since these characters contain all that is peculiar to the given group.

Let *T* be a representation of the group *G* in a finite-dimensional space *X*. A nonzero vector $x \in X$ is called a *weight vector of the representation T*, and the function $v(\delta)$ is the *weight of the vector x*, if

$$T(\delta)x = v(\delta)x, \quad \text{for all } \delta \in D. \tag{2.1.1}$$

Plainly $v(\delta)$ is a character of the group *D*. A weight vector *x* of *T* is called a *vector of highest weight* if

$$T(z)x = x, \quad \text{for all } z \in Z^+, \tag{2.1.2}$$

and a *vector of lowest weight* if

$$T(\zeta)x = x, \quad \text{for all } \zeta \in Z^-. \tag{2.1.3}$$

Theorem 1.

    (1)   *The space X of every finite-dimensional irreducible representation T of the group G admits a vector of highest weight and a vector of lowest weight.*

    (2)   *If T is irreducible, then X admits exactly one vector of highest weight and exactly one vector of lowest weight (up to scalar multiples).*

Statement (1) follows directly from Lie's theorem (3.1, chapter V), when applied to the representations $T|_K$, $T|_H$. As subgroups of the solvable groups

K_m and H_m, K and H are solvable, and they are connected (I in 1.11 and I in 1.12).

The proof of Statement (2) is the same as the proof of the analogous assertion in theorem 1 of 2.1, chapter VI. The weight of the vector of highest weight of an irreducible representation is called the *highest weight* of this representation.

A character $\alpha(\delta)$ of the group D is called *inductive under G* if the following conditions hold:

(1)   the function $\alpha(g) = \alpha(\delta\zeta z) = \alpha(\delta)$, defined for $g = \delta\zeta z \in G_{\mathrm{reg}}$, extends to a continuous function $\alpha(g)$ on the entire group G;

(2)   the linear span of the set of functions $\alpha(gg_0)$, $g_0 \in G$, is finite-dimensional.

Theorem 2.

(1)   *A character $\alpha(\delta)$ of the group D is the highest weight of some irreducible representation of the group G if and only if $\alpha(\delta)$ is inductive under G.*

(2)   *Two finite-dimensional irreducible representations of G are equivalent if and only if their highest weights coincide.*

(3)   *A finite-dimensional irreducible representation T of G with highest weight α is equivalent to the representation T_α constructed as follows.*
  (a)   *The space X of the representation T_α is the linear span of all functions $\alpha(gg_0)$, $g_0 \in G$.*
  (b)   *For $f(g) \in X$ we have*

$$T_\alpha(g_0)f(g) = f(gg_0). \qquad (2.1.4)$$

The representation T_α is called the *canonical realization of the representation with highest weight α.* The function $\alpha(g)$ constructed from the inductive character $\alpha(\delta)$ is called the *generating function of the representation T_α.*

Applying Gauss's decomposition (see 1.6) and repeating the argument in 2.3, chapter VI, we arrive at the following realization of the representation T_α in a space of functions on the group Z^+.

Theorem 3. *Every finite-dimensional irreducible representation T of the group G is equivalent to a representation T_α defined as follows.*

(1)   *The space X_α of the representation T_α is the linear span of the functions $\alpha(zg)$, $g \in G$, where $\alpha(g)$ is the generating function defined by the highest weight α of the representation T.*

(2)   *The operators of T_α are defined by the formula*

$$(T_\alpha(g)f)(z) = \alpha(zg)f(z\bar{g})$$

Finally, using Gram's decomposition, we obtain the following realization of the representation T_α in a space of functions on the group U.

The space X_α of the representation T_α is the linear span of all functions $\alpha(ug)$, $g \in G$. The operators of the representation are defined by the formula

$$(T_\alpha(g)f)(u) = \alpha(\varepsilon')f(u_g), \quad \text{for } ug = \varepsilon'\zeta u_g,$$
$$\varepsilon' \in E, \qquad \zeta \in Z^-, \qquad u_g \in U.$$

To complete our description of the representations T_α, it remains to identify the inductive characters of the group G under consideration.

The matrices δ of the group D are defined by their diagonal elements $\lambda_1, \lambda_2, \ldots, \lambda_n$. The mapping $\delta \to (\lambda_1, \ldots, \lambda_n)$ is a topological isomorphism of the group D onto $\mathbf{C}_0^n$ (I in 1.10). Therefore every character $\alpha(\delta)$ of D has the form

$$\alpha(\delta) = \lambda_1^{p_1}\bar\lambda_1^{q_1}\lambda_2^{p_2}\bar\lambda_2^{q_2} \cdots \lambda_n^{p_n}\bar\lambda_n^{q_n}, \tag{2.1.5}$$

where $p_1 - q_1, p_2 - q_2, \ldots, p_n - q_n$ are integers. The sequence $(p_1, \ldots, p_n, q_1, \ldots, q_n)$ is called the *signature of the character* α. If $\alpha(\delta)$ is inductive, then the signature of $\alpha(\delta)$ is also called the signature of the representation T defined by the character $\alpha(\delta)$.

We will find the signatures that yield inductive characters, for each of the complex classical groups.

2.2. The Unimodular Group ($G = A_n$)

Repetition of the argument in 2.5, chapter VI, yields the following.

I. *The character $\alpha(\delta)$ of the group $D \in A_n$ is inductive if and only if its signature* $(p_1, \ldots, p_n, q_1, \ldots, q_n)$ *satisfies the condition that*

$$p_1 - p_2, p_2 - p_3, \ldots, p_n, \qquad q_1 - q_2, q_2 - q_3, \ldots, q_n \tag{2.2.1}$$

be nonnegative integers.

According to formulas (1.6.4), (1.8.10), chapter VI and (2.1.5) for $g = \delta\zeta z$, we have

$$\alpha(g) = \alpha(\delta)$$

$$= \Delta_1(g)^{p_1}\,\overline{\Delta_1(g)}^{q_1}\left(\frac{\Delta_2(g)}{\Delta_1(g)}\right)^{p_2}\left(\frac{\overline{\Delta_2(g)}}{\overline{\Delta_1(g)}}\right)^{q_2}\cdots \Delta_n(g)^{p_n}(\overline{\Delta_n(g)})^{q_n}$$

$$= \Delta_1(g)^{r_1}\,\overline{\Delta_1(g)}^{s_1}\,\Delta_2(g)^{r_2}(\overline{\Delta_2(g)})^{s_2}\cdots(\Delta_n(g))^{r_n}(\overline{\Delta_n(g)})^{s_n}, \tag{2.2.2}$$

where we write

$$r_1 = p_1 - p_2, \qquad s_1 = q_1 - q_2, \ldots, r_n = p_n, \qquad s_n = q_n. \tag{2.2.3}$$

By I, $r_1, s_1, \ldots, r_n, s_n$ are nonnegative integers. Thus from (2.2.2) we conclude the following.

II. *An inductive character $\alpha(g)$ of the group A_n is a polynomial in the entries g_{jl} of degree not greater than $r_1 + \cdots + r_n = p_1$, and in the entries $\bar{g}_{jl}$ of degree not greater than $s_1 + \cdots + s_n = q_1$.*

Combine theorem 2 in 2.1 with Proposition I to find the following result.

Theorem 1. *Every finite-dimensional irreducible representation of the group $A_n = SL(n + 1, \mathbf{C})$ is defined by a system of integers*

$$p_1, \ldots, p_n, \qquad q_1, \ldots, q_n, \tag{2.2.4}$$

that satisfy the conditions

$$\begin{aligned} p_1 - p_2 \geqslant 0, &\qquad p_2 - p_3 \geqslant 0, \ldots, p_n \geqslant 0, \\ q_1 - q_2 \geqslant 0, &\qquad q_2 - q_3 \geqslant 0, \ldots, q_n \geqslant 0. \end{aligned} \tag{2.2.5}$$

The representation T_α with the given system (2.2.4) can be realized in the following manner. The space X of the representation T_α is the linear span of all functions $\alpha(gg_1)$, $g_1 \in A_n$, where

$$\alpha(g) = \Delta_1(g)^{p_1 - p_2} \overline{\Delta_1(g)}^{q_1 - q_2} \Delta_2(g)^{p_2 - p_3} \overline{\Delta_2(g)}^{q_2 - q_3} \cdots \Delta_n(g)^{p_n} \overline{\Delta_n(g)}^{q_n} \tag{2.2.6}$$

is the generating function of the representation T_α. The operators $T_\alpha(g)$ are described by the formula

$$T_\alpha(g_0)f(g) = f(gg_0). \tag{2.2.7}$$

Using Gauss's decomposition, we obtain the following realization of T_α in a space of the functions of the group Z^+.

III. *The space X_α of the representation T_α is the linear span of all functions $\alpha(zg)$, $g \in G$, and the operators of the representation are defined by the formula*

$$T_\alpha(g)f(z) = \Delta_1(zg)^{p_1 - p_2} \overline{\Delta_1(zg)}^{q_1 - q_2} \Delta_2(zg)^{p_2 - p_3} \overline{\Delta_2(zg)}^{q_2 - q_3}$$
$$\cdots \Delta_n(zg)^{p_n} \overline{\Delta_n(zg)}^{q_n} f(z\bar{g}), \tag{2.2.8}$$

Here the element $z\bar{g} \in Z^+$ is defined from Gauss's decomposition $zg = kz\bar{g}$.

Using Gram's decomposition, we obtain an analogous realization of T_α in a space of functions on U.

Comparing theorem 1 in 2.1 with theorem 5 in 2.5, chapter VI, we conclude the following.

IV. *Every irreducible representation of the group* $SL(n + 1, \mathbf{C})$ *is the restriction to* $SL(n + 1, \mathbf{C})$ *of an irreducible representation of the group* $GL(n + 1, \mathbf{C})$.

2.3. The Symplectic Group $(G = C_n)$

I. *The character* $\alpha(\delta)$ *of the group* $D \subset C$ *is inductive under* C_n *if and only if its signature satisfies the condition that*

$$p_1 - p_2, p_2 - p_3, \ldots, p_n, \qquad q_1 - q_2, q_2 - q_3, \ldots, q_n \qquad (2.3.1)$$

be nonnegative integers.

Proof. We set $G = C_n$ and denote by G_0 the set of matrices g_0 of the form

$$g_0 = \begin{Vmatrix} a & 0 \\ 0 & s_1^{-1}a'^{-1}s_1 \end{Vmatrix}, \qquad (2.3.2)$$

where $a \in A_{n-1} = GL(n, \mathbf{C})$ and s_1 is as in (1.2.1). We also set $D_0 = G_0 \cap D$. Multiplying matrices, we see that G_0 is contained in C_n and that the correspondence $g_0 \to a$ is a topological isomorphism of the group G_0 onto $GL(n, \mathbf{C})$, mapping D_0 onto the group D_n for $GL(n, \mathbf{C})$. Suppose that a character $\alpha(\delta)$ of the group D is inductive under G. According to the lemma in 2.5, chapter VI, its restriction $\alpha_0(\delta_0)$ to D_0 is inductive under G_0. By the isomorphism given above, $\alpha_0(\delta_0)$ is an inductive character of the group D_n for $GL(n, \mathbf{C})$. This restriction has the form

$$\alpha_0(\delta_0) = \lambda_1^{p_1}\bar{\lambda}_1^{q_1}\lambda_2^{p_2}\bar{\lambda}_2^{q_2} \cdots \lambda_n^{p_n}\bar{\lambda}_n^{q_n}$$

where

$$\delta = \begin{Vmatrix} \delta_0 & 0 \\ 0 & s_1^{-1}\delta_0'^{-1}s_1 \end{Vmatrix}, \qquad \delta = \begin{Vmatrix} \lambda_1 & & & & & & 0 \\ & \ddots & & & & & \\ & & \lambda_n & & & & \\ & & & \lambda_n^{-1} & & & \\ 0 & & & & \ddots & & \\ & & & & & \lambda_1^{-1} \end{Vmatrix}.$$

From I in 2.5, we conclude that

$$p_1 - p_2, p_2 - p_3, \ldots, p_{n-1} - p_n, \qquad q_1 - q_2, q_2 - q_3, \ldots, q_{n-1} - q_n$$

are nonnegative integers. $\quad\square$

Let G_0 denote the set of all matrices

$$g_1 = \left\| \begin{matrix} 1_{n-1} & & \\ & a & \\ & & 1_{n-1} \end{matrix} \right\|,$$

where $a \in SL(2, \mathbf{C})$ and zeros occupy all unmarked places. We write s_0 in the form

$$s_0 = \left\| \begin{matrix} & & -s \\ & \sigma & \\ s & & \end{matrix} \right\|. \tag{2.3.3}$$

where

$$\sigma = \left\| \begin{matrix} 0 & -1 \\ 1 & 0 \end{matrix} \right\|, \qquad s = \left\| \begin{matrix} 0 & -1_{n-1} \\ 1_{n-1} & 0 \end{matrix} \right\| \tag{2.3.4}$$

and zeros occupy all unmarked places in (2.3.3). Multiplying matrices, we see that:

(1) $g_1'^{-1}$ is equal to $s_0 g_1 s_0^{-1}$, that is, g_1 belongs to C_n;
(2) G_0 is a subgroup of the group C_n;
(3)   the mapping $g_1 \to a$ is a topological isomorphism of the group G_0 onto $SL(2, \mathbf{C})$ carrying D_0 onto the group D for $SL(2, \mathbf{C})$.

According to the lemma in 2.5, chapter VI, the restriction $\alpha_0(\delta_0)$ to $SL(2, \mathbf{C})$ of the inductive character $\alpha(\delta)$ of D is an inductive character of $SL(2, \mathbf{C})$. This restriction has the form

$$\alpha_0(\delta_0) = \lambda_n^{p_n} \bar{\lambda}_n^{-q_n}$$

Thus, by I in 2.2, we conclude that p_n, q_n are also nonnegative integers. Thus

$$p_1 - p_2, \qquad p_2 - p_3, \ldots, p_n, \qquad q_1 - q_2, \qquad q_2 - q_3, \ldots, q_n$$

are nonnegative integers. This proves the necessity of (2.3.1). Conversely, if (2.3.1) is satisfied, then

$$\alpha(g) = \varDelta_1(g)^{p_1 - p_2} \overline{\varDelta_1(g)}^{q_1 - q_2} \varDelta_2(g)^{p_2 - p_3} \overline{\varDelta_2(g)}^{q_2 - q_3} \cdots \varDelta_n(g)^{p_n} \overline{\varDelta_n(g)}^{q_n}$$

is a polynomial of degree not higher than p_1 in the g_{jl} and of degree not higher than q_1 in the $\bar{g}_{jl}$. This implies the inductivity of the character $\alpha(\delta)$

under C_n. From Proposition I and theorem 2 in 2.1, we conclude the following.

Theorem 2. *Every finite-dimensional irreducible representation of the group $C_n = Sp(2n, \mathbf{C})$ is defined by a system of integers*

$$p_1, p_2, \ldots, p_n, \qquad q_1, q_2, \ldots, q_n \qquad (2.3.5)$$

that satisfy the conditions

$$p_1 \geqslant p_2 \geqslant \cdots \geqslant p_n \geqslant 0, \qquad q_1 \geqslant q_2 \geqslant \cdots \geqslant q_n \geqslant 0.$$

The representation T_α with a given system (2.3.5) can be realized as follows. The space X of the representation is the linear span of the functions $\alpha(gg_1)$, $g_1 \in C_n$, where

$$\alpha(g) = \Delta_1(g)^{p_1 - p_2} \overline{\Delta_1(g)}^{q_1 - q_2} \Delta_2(g)^{p_2 - p_3} \overline{\Delta_2(g)}^{q_2 - q_3} \cdots \Delta_n(g)^{p_n} \overline{\Delta_n(g)}^{q_n} \quad (2.3.6)$$

is the generating function of the representation. The operators $T_\alpha(g)$ of the representation are defined by the formula

$$T_\alpha(g_0)f(g) = f(gg_0). \qquad (2.3.7)$$

Applying Gauss's decomposition, we obtain the following realization of the representation T_α in a space of the functions on Z^+.

II. *The space X of the representation is the linear span of all functions $\alpha(zg)$, $g \in C_n$, and the operators of the representation are defined by the formula*

$$T_\alpha(g)f(z) = \alpha(zg)f(z\bar{g}), \qquad (2.3.8)$$

where $z\bar{g} \in Z^+$ is defined from Gauss's decomposition $zg = kz\bar{g}$.

Gram's decomposition produces in like manner a realization of the representation T_α in a space of functions on the group U.

2.4. The Orthogonal Group

(a) Consider first the case $G = D_n$, that is, $G = SO(2n, \mathbf{C})$. Following the same argument as in 2.3, we obtain the following.

I. *The character $\alpha(\delta)$ of the group $D \subset D_n$ is inductive under D_n if and only if all numbers $p_1 - p_2, \ldots, p_{n-1} - |p_n|, q_1 - q_2, \ldots, q_{n-2} - q_{n-1}, q_{n-1} - |q_n|$ are nonnegative integers.*

Proposition I and theorem 2 in 2.1 imply the following.

Theorem 3. *Every finite-dimensional irreducible representation of the group* $G = SO(2n, \mathbf{C})$ *is defined by a system of integers that satisfy the conditions* $p_1 \geqslant p_2 \geqslant \cdots \geqslant p_{n-1} \geqslant |p_n|, q_1 \geqslant q_2 \geqslant \cdots \geqslant q_{n-1} \geqslant |q_n|.$

Applying Gauss's decomposition, we can describe the corresponding representation as in II, 2.3.

(b) Suppose next that $G = B_n$, that is, $G = SO(2n + 1, \mathbf{C})$. By using the same argument as in 2.3 or by applying (a) and restricting to a subgroup, we arrive at the following.

II. *A character* $\alpha(\delta)$ *of the group* $D \subset B_n$ *is inductive under* B_n *if and only if its signature satisfies the condition that* $p_1 - p_2, \ldots, p_{n-1} - p_n, p_n,$ $q_1 - q_2, \ldots, q_{n-1} - q_n, q_n$ *be nonnegative integers.*

From II and theorem 2 in 2.1 we infer the following theorem.

Theorem 4. *Every finite-dimensional irreducible representation of the group* $G = SO(2n + 1, \mathbf{C})$ *is defined by a system of integers* $p_1, \ldots, p_n, q_1, \ldots, q_n,$ *for which* $p_1 \geqslant p_2 \geqslant \cdots \geqslant p_n \geqslant 0, q_1 \geqslant q_2 \geqslant \cdots \geqslant q_n \geqslant 0.$

For more details, see Želobenko's book [1].

Let G be the group $SO(n, \mathbf{C})$. There exists a group $\tilde{G}$ such that G is isomorphic to the factor group $\tilde{G}$ by a two-element normal subgroup N of $\tilde{G}$. The group $\tilde{G}$ is called a *spinor group*. The irreducible representations of $\tilde{G}$ nontrivial on N are called *two-valued representations of the group* G. (See Želobenko [1] and Chevalley [1].)

Chapter VIII

Covering Spaces and Simply Connected Groups

§1. Covering Spaces

A separated topological space X is called *locally connected* if for every point $x \in X$, every neighborhood of x contains a connected neighborhood of x.

I. *Let X be a locally connected space. Let U be an open subset of X and x a point of U. The union $K(x)$ of all connected subsets of U containing x is an open set.*

Proof. If y belongs to $K(x)$ and $V \subset U$ is a connected neighborhood of y, the union $K(x) \cup V$ is connected and contains x. Hence V is contained in $K(x)$. Thus all points of $K(x)$ are interior points. $\square$

We recall from §1, chapter V, that the union of all the connected subsets of a given set M containing a point x of M is called the *(connected) component* of x in M, or simply a *component* of the set M. Proposition I tells us that in a locally connected space every component of an open set is an open set.

Let X, Y be topological spaces and f a continuous mapping of Y onto X. The space Y is called a *covering space for X* (under the mapping f) if Y is connected and locally connected and every point $x \in X$ has a neighborhood $U \subset X$ such that every component of the open set $f^{-1}(U)$ is mapped homeomorphically by f onto all of the set U. Note that the neighborhood U must be connected in view of I.

If X has a covering space, X is obviously connected and locally connected. Conversely, if X is connected and locally connected, it is a covering space for itself under the identity mapping. This covering space is called *trivial*.

II. *If Y is a covering space for X under a mapping f, and G is an open subset of Y, then $f(G)$ is open in X.*

Proof. Let y belong to G and write $x = f(y)$. Let $U \subset X$ be a neighborhood of the point x such that every component of the set $f^{-1}(U)$ is mapped homeomorphically by f onto U. Let V be the component of y in $f^{-1}(U)$. The set $V \cap G$ is open in Y and hence open in V. Since f is a homeomorphic mapping of V onto U, the set $f(V \cap G)$ is open in U. The set $f(V \cap G)$ contains x, i.e., it is a neighborhood of x. On the other hand, $f(V \cap G)$ is contained in $f(G)$ and therefore $f(G)$ contains a neighborhood of each of its points. $\square$

III. *Let Y be locally connected and let f be a continuous mapping of the space Y into a space X. Let y belong to Y and let U be a neighborhood of the point $x = f(y)$ in X. The component V of the point y in $f^{-1}(U)$ is a neighborhood of y.*

Proof. The set $f^{-1}(U)$ is open and by I, V is therefore open. $\square$

IV. *If Y is a covering space for X under a mapping f, then f is a local homeomorphism. That is, every point of Y has a neighborhood V such that $f|V$ is a homeomorphism.*

The proof follows from III and from the definition of a covering space.

V. *Let f be a continuous mapping of a space Y into a space X. Let G be a subspace of X such that every component of the set $f^{-1}(G)$ is mapped homeomorphically onto G by f. Let F be any connected subset of G. Then every component of the set $f^{-1}(F)$ is mapped homeomorphically onto F in X by f. The components of $f^{-1}(F)$ are the intersections of the components of $f^{-1}(G)$ with the set $f^{-1}(F)$.*

Proof. Let $\{H_\alpha\}$ be the family of components of $f^{-1}(G)$. Write Φ_α for

$$H_\alpha \cap f^{-1}(F).$$

Since Φ_α is contained in H_α and f maps H_α homeomorphically onto G, f also maps Φ_α homeomorphically onto F. Therefore the sets Φ_α are connected. On the other hand, every connected subset of the set $f^{-1}(F)$ is contained in some component H_α of the set $f^{-1}(G)$. Thus the sets Φ_α are the components of the set $f^{-1}(F)$. $\square$

VI. *Let φ be a continuous mapping of a locally connected space Z into a connected space X. Let every point $x \in X$ have a neighborhood U such that φ maps every component of the set $\varphi^{-1}(U)$ homeomorphically onto U. Let Y be any component of the space Z, and let f be the restriction of φ to Y. Then Y is open in Z and Y is a covering space for X under the mapping f.*

Proof. First we show that $f(Y)$ is all of X. For x in X, let U be a neighborhood of x such that φ maps the components V_α of the set $\varphi^{-1}(U)$ homeomorphically onto U. If a component V_α of the set $\varphi^{-1}(U)$ intersects Y, then V_α is contained in Y, since Y is a component of the space Z. Thus $Y \cap \varphi^{-1}(U)$ is the union of the components V_α of the set $\varphi^{-1}(U)$ over the indices α for which $V_\alpha \cap Y$ is nonvoid. Therefore, if U intersects $f(Y)$, then $Y \cap \varphi^{-1}(U)$ is the nonvoid union of certain components V_α. On the other hand, $\varphi(V_\alpha)$ is equal to U and so U is contained in $f(Y)$. In particular, if x belongs to $\overline{f(Y)}$, then every neighborhood of x intersects $f(Y)$. This implies that $U \cap f(Y) \neq \varnothing$. Therefore U is contained in $f(Y)$ and every point $x \in \overline{f(Y)}$ is an interior point of $f(Y)$. Thus $f(Y)$ is both open and closed in X. Since X is connected, $f(Y)$ is X.

The sets V_α are maximal connected subsets of the set $\varphi^{-1}(U)$ and $f^{-1}(U) = Y \cap \varphi^{-1}(U)$ is the union of some of the sets V_α. Therefore these sets V_α are the components of $f^{-1}(U)$. Hence Y is a covering space for X under f. By virtue of I, the set Y is open in Z. $\square$

VII. *Let Y be a covering space for X under a mapping f. Let Z be a connected and locally connected subspace of X. Every component W of the set $f^{-1}(Z)$ is relatively open in $f^{-1}(Z)$ and W is a covering space for Z under the restriction of f to W.*

Proof. Let x belong to Z and let U be a neighborhood in X of the point x such that f on each component of the set $f^{-1}(U)$ is a homeomorphism onto U. Since the space Z is locally connected, there is a connected neighborhood V of x in Z that is contained in U. Let W_α be the components of the set $f^{-1}(V)$. By V, every set W_α is the intersection of $f^{-1}(V)$ with a certain component H_α of the set $f^{-1}(U)$. If y_α is a point in W_α such that $f(y_\alpha) = x$, then W_α is a neighborhood of the point y_α in the relative topology of $f^{-1}(Z)$. Therefore $f^{-1}(Z)$ is locally connected and every point x of Z has a neighborhood V such that f on every component of $f^{-1}(V)$ is a homeomorphism onto V. Now VII follows from VI. $\square$

Example

Let X be the unit circle Γ_1; that is, $X = \{e^{i\varphi}, \varphi \in \mathbf{R}\}$. Let $Y = \mathbf{R}$ be the real line. In this case Y is connected (see example 1 of §1, chapter V) and locally connected (since every neighborhood of the point $y \in Y$ contains a connected interval $(y - \delta, y + \delta)$ for some $\delta > 0$). Let f be the mapping of Y into X defined by the formula $f(y) = e^{i\varphi}$, $y \in Y = \mathbf{R}$. Then f maps Y onto X and any point $x = e^{i\varphi} \in X$ has the neighborhood $U = \{e^{iy}, e^{iy} \neq -e^{i\varphi}\}$, for which $f^{-1}(U)$ has the form $\{(\varphi + (2k - 1)\pi, \varphi + (2k + 1)\pi)$, $k = 0, \pm 1, \pm 2, \ldots\}$. This set is the union of a countable number of components $Y_k = (\varphi + (2k - 1)\pi, \varphi + (2k + 1)\pi)$, $k \in \mathbf{Z}$. The mapping f of each component onto U is a homeomorphism. The mapping f is continuous and the continuous mapping $\psi_k : U \to Y_k$ that is the inverse to $f|_{Y_k}$ is defined by the formula

$$\psi_k(e^{iy}) = \varphi + 2k\pi + 2\arctan\frac{e^{iy}e^{-i\varphi} - e^{-iy}e^{i\varphi}}{i(2 + e^{iy}e^{-i\varphi} + e^{-iy}e^{i\varphi})}, \quad e^{iy} \neq -e^{i\varphi}.$$

Thus **R** is a covering space for Γ_1 under the mapping $y \to e^{iy}$, $y \in \mathbf{R}$.

§2. Simply Connected Spaces and the Principle of Monodromy

2.1. Simply Connected Spaces

Let X be a topological space and let Y and Z be covering spaces for X under mappings f and g respectively. The covering spaces Y and Z are called *isomorphic* if there is a homeomorphism φ of the space Y onto Z such that $f = g \circ \varphi$. The space X is called *simply connected* if X is a connected and locally connected space with the property that every covering space of X is isomorphic to the trivial covering space. Thus, if X is simply connected and Y is a covering space for X under the mapping f, then f is a homeomorphism of Y onto X. Clearly, a space homeomorphic to a simply connected space is also simply connected.

The space Γ_1 is an example of a space that is *not* simply connected. The example in §1 shows that there exists a covering space for the space Γ_1, namely, the space **R**, that is not homeomorphic to Γ_1. Below we will give examples of simply connected spaces (see §4) and in particular we will show that **R** is simply connected.

We now study some general properties of simply connected spaces.

I. *Let Y be a covering space for X under the mapping f. If G is an open subset of Y and f is a one-to-one mapping of the set G onto X, then f is a homeomorphism of Y and X. That is, Y is isomorphic to the trivial covering space.*

Proof. The mapping f is continuous by hypothesis. By II in §1, f carries open sets into open sets. Thus the mapping f of G onto X is a homeomorphism. We will prove that $G = Y$. Since Y is connected, it is sufficient to prove that G is closed in Y.

Let y belong to $\bar{G}$ and let U be a connected neighborhood of the point $x = f(y)$ such that f on every component of the set $f^{-1}(U)$ is a homeomorphism onto U. Let V be the component of the set $f^{-1}(U)$ that contains the point y. The mapping f is a homeomorphism of both the set V and the set $V_1 = G \cap f^{-1}(U)$ onto U. On the other hand, V is a neighborhood of the point y, according to I in §1. Therefore V intersects both G and V_1.

Since V_1 is homeomorphic to U, V_1 is connected and thus $V_1 \subset V$. The restriction of f to V is a one-to-one mapping and $f(V_1) = f(V) = U$. Therefore we have $V_1 = V$. Thus y belongs to G. $\square$

II. *If X_1 and X_2 are simply connected spaces, their product $X_1 \times X_2$ is also simply connected.*

Proof. Clearly the space $X_1 \times X_2$ is connected and locally connected. Let X be a covering space for $X_1 \times X_2$ under a mapping f. For all $x_2 \in X_2$, any component of the set $f^{-1}(X_1 \times \{x_2\})$ is a covering space for X_1 (under the restriction of f to this component). This is shown by VII in §1. Since the space X_1 is simply connected, the mapping f is a homeomorphism of each component of the set $f^{-1}(X_1 \times \{x_2\})$ onto X_1. Let $Z_2^0(x_1^0)$ be some fixed component of the set $f^{-1}(\{x_1^0\} \times X_2)$, where $x_1^0 \in X_1$. Consider all components, say $Z_1(x_2)$, of all sets of the form $f^{-1}(X_1 \times \{x_2\})$ as x_2 runs through X_2. Let G be the union of all of these components that intersect $Z_2^0(x_1^0)$. Plainly, f is one-to-one on G and maps G onto $X_1 \times X_2$. We will show that the set G is open. Then II will follow from I.

Let M be the set of interior points of the set G that belong to a given subset $Z_1(x_2)$ of G. The set M is relatively open in $Z_1(x_2)$. We will show that M is closed and nonvoid.

Let z belong to $Z_1(x_2)$ and write $f(z) = (x_1, x_2)$. Let U_1, U_2 be connected neighborhoods of the points x_1 and x_2 in X_1 and X_2 such that f is a homeomorphism of each component of the set $f^{-1}(U_1 \times U_2)$ onto $U_1 \times U_2$. Let V be the component of the point z in $f^{-1}(U_1 \times U_2)$. Let (y_1, y_2) belong to $U_1 \times U_2$. We set

$$V_1(y_2) = V \cap f^{-1}(U_1 \times \{y_2\}), \qquad V_2(y_1) = V \cap f^{-1}(\{y_1\} \times U_2).$$

The mapping f from $V_1(y_2)$ onto $U_1 \times \{y_2\}$ and from $V_2(y_1)$ onto $\{y_1\} \times U_2$ are homeomorphisms. Hence $V_1(y_2)$ is contained in some component $Z_1(y_2)$ and $V_2(y_1)$ is contained in some component $Z_2(y_1)$. These components have at least one common point w in V, for which $f(w) = (y_1, y_2)$. If we choose as z a point in which $Z_1(x_2)$ intersects $Z_2^0(x_1^0)$, then $Z_2(x_1)$ coincides with $Z_2^0(x_1^0)$ and $Z_1(y_2)$ is contained in G for all $y_2 \in U_2$. The equality $V = \bigcup_{y_2 \in U_2} V_1(y_2)$ shows that $V \subset G$ and thus $z \in M$. Therefore M is nonvoid.

If z belongs to $\overline{M}$, then $V_1(x_2)$ has a point w^* in common with M. Write $f(w^*) = (x_1^*, x_2)$, and let U_2^* be a neighborhood of the point x_2 in X_2 such that $U_2^* \subset U_2$ and $V \cap f^{-1}(\{x_1^*\} \times U_2^*)$ is contained in G. For $x_2^* \in U_2^*$, some component $Z_1(x_2^*)$ has nonvoid intersection with G and therefore $Z_1(x_2^*)$ is contained in G. Furthermore, the set $V \cap f^{-1}(U_1 \times U_2^*)$ is the union of the sets $Z_1(x_2^*)$ for all $x_2^* \in U_2^*$. Thus $V \cap f^{-1}(U_1 \times U_2^*)$ is contained in G. The set $V \cap f^{-1}(U_1 \times U_2^*)$ is open, *i.e.*, it is a neighborhood of the point z. Thus z belongs to M and M is closed.

Since $M \subset Z_1(x_2)$ is nonvoid, closed, and open in $Z_1(x_2)$ and $Z_1(x_2)$ is homeomorphic to the connected space X_1, we have $M = Z_1(x_2)$. Therefore all points of the set G are interior points and we have proved II. $\square$

2.2. The Principle of Monodromy

The following theorem presents the fundamental property of simply connected spaces. It is called the *principle of monodromy*.

Theorem 1. *Let X be a simply connected space. Let there be a nonvoid set M_x associated with every $x \in X$. For every point (x, y) of a certain subset D of $X \times X$, let there be a mapping φ_{xy} of the set M_x onto the set M_y, for which the following conditions hold:*

(1) D *is a connected open subset of $X \times X$ containing the diagonal (all points of the form (x, x), $x \in X$);*

(2) φ_{xy} *is a one-to-one mapping for all $(x, y) \in D$; φ_{xx} is the identity mapping for all $x \in X$;*

(3) *if (x, y), (y, z), (x, z) belong to D, then $\varphi_{xz} = \varphi_{yz} \circ \varphi_{xy}$.*

Then there is a mapping ψ that associates an element $\psi(x)$ of M_x with every $x \in X$, in such a way that $\psi(y) = \varphi_{xy}(\psi(x))$, for all $(x, y) \in D$. Furthermore, we can choose the mapping ψ so that at a given point $x_0 \in X$, we have $\psi(x_0) = w^0_{x_0}$, where $w^0_{x_0}$ is a fixed element of the set M_{x_0}. With this additional condition, the mapping ψ is defined uniquely.

Proof. Define Y as $\bigcup_{x \in X} \{x\} \times M_x$. Let O be the family of all subsets G of Y such that for every point $(x, w_x) \in G$ there exists a neighborhood U of x in X for which $U \times U \subset D$ and $(y, \varphi_{xy}(w_x)) \in G$ for all $y \in U$. Clearly Y and $\varnothing$ belong to 0. The reader can easily verify that 0 is closed under the formation of arbitrary unions and finite intersections. If we take O as the family of all open sets, Y becomes a topological space.

We define a mapping π of Y onto X by setting $\pi(x, w_x) = x$. From the definition of O, we see that $\pi^{-1}(U)$ belongs to O for every open set $U \subset X$. That is, π is a continuous mapping of Y onto X and for all $G \in O$, the set $\pi(G)$ is open in X.

Let U be an open set in X such that $U \times U \subset D$. Let x belong to U and let w_x belong to M_x. The set of all elements of the form $(y, \varphi_{xy}(w_x))$ as y runs through U is denoted by $G(x, U, w_x)$. We will prove that $G(x, U, w_x)$ belongs to O. Let $(y, \varphi_{xy}(w_x))$ be any element of the set $G(x, U, w_x)$. Since $U \times U \subset D$, we see that for all $z \in U$, the mappings φ_{xy}, φ_{yz} and φ_{xz} are defined. Condition (3) shows that $(z, \varphi_{yz}(\varphi_{xy}(w_x))) = (z, \varphi_{xz}(w_x)) \in G(x, U, w_x)$. That is, $G(x, U, w_x)$ belongs to O.

Let (x, w_x) and (y, w_y) be distinct points in Y. If $x \neq y$, let U and V be disjoint neighborhoods of x and y in X, respectively. We then have $(x, w_x) \in \pi^{-1}(U)$ and $(y, w_y) \in \pi^{-1}(V)$. The sets $\pi^{-1}(U)$, $\pi^{-1}(V)$ belong to the family

O and are disjoint. If $x = y$, then w_x and w_y are distinct. Let U be a neighborhood of x in X such that $U \times U \subset D$. Then the sets $G(x, U, w_x)$ and $G(x, U, w_y)$ belong to O and are disjoint, since the mappings φ_{xy} are one-to-one. Thus the topological space Y is separated.

Every point $x \in X$ has a connected neighborhood U in X such that $U \times U \subset D$. Since every mapping φ_{xy}, $x, y \in U$, of M_x onto M_y is one-to-one, the set $\pi^{-1}(U)$ is the union of the sets $G(x, U, w_x)$ as w_x runs through M_x. All of these sets $G(x, U, w_x)$ are open in Y. Also the mapping π of these sets onto U is one-to-one. Since π is continuous and carries open sets onto open sets, the mapping π of every set $G(x, U, w_x)$, $w_x \in M_x$, onto U is a homeomorphism. The set U being connected, the pairwise disjoint sets $G(x, U, w_x)$ are components of the set $\pi^{-1}(U)$. Thus, every point $x \in X$ has a neighborhood U for which the mapping π of every component of $\pi^{-1}(U)$ onto U is homeomorphism. Since U is connected, the sets $G(x, U, w_x)$ are connected open sets in Y. Hence Y is locally connected.

Let Y_0 be the component of the point $(x_0, w_{x_0}^0)$ in the space Y and let ω be the restriction of the mapping π to Y_0. Proposition VI in §1 shows that Y_0 is a covering space for the space X under the mapping ω. Since X is simply connected, ω is a homeomorphism of the space Y_0 onto X. We define a mapping ψ by the formula

$$\omega^{-1}(x) = (x, \psi(x)). \tag{2.2.1}$$

Let D^* denote the set of all points $(x, y) \in D$ that satisfy the condition $\psi(y) = \varphi_{xy}(\psi(x))$. Let (x_1, y_1) be a point of the set D. Let U_1, V_1 be connected neighborhoods of the points x_1 and y_1 in the space X such that $U_1 \times U_1 \subset D$, $V_1 \times V_1 \subset D$ and $U_1 \times V_1 \subset D$. Suppose that $U_1 \times V_1$ intersects D^*, and let (x_2, y_2) belong to $D^* \cap (U_1 \times V_1)$. We then have

$$\psi(y_2) = \varphi_{x_2 y_2}(\psi(x_2)), \tag{2.2.2}$$

since (x_2, y_2) belongs to D^*. The set $G(x_1, U_1, \psi(x_1))$ is connected and contains the point

$$(x_1, \psi(x_1)) \in Y_0.$$

Since Y_0 is a component in Y,

$$G(x_1, U_1, \psi(x_1))$$

is contained in Y_0. Thus by the definition of $G(x_1, U_1, \psi(x_1))$ we have

$$\psi(x_2) = \varphi_{x_1 x_2}(\psi(x_1)) \tag{2.2.3}$$

for the point $x_2 \in U_1$.

We obtain similarly

$$\psi(y_2) = \varphi_{y_1 y_2}(\psi(y_1)). \tag{2.2.4}$$

On the other hand, all the mappings $\varphi_{x_1x_2}$, $\varphi_{x_2y_1}$, $\varphi_{x_1y_2}$, $\varphi_{x_2y_2}$, $\varphi_{y_1y_2}$ are defined. Hypothesis (3) of the present theorem, (2.2.2), and (2.2.3) imply that

$$\psi(y_2) = \varphi_{x_2y_2}(\psi(x_2)) = \varphi_{x_2y_2}(\varphi_{x_1x_2}(\psi(x_1)))$$
$$= \varphi_{x_1y_2}(\psi(x_1)) = \varphi_{y_1y_2}(\varphi_{x_1y_1}(\psi(x_1))). \tag{2.2.5}$$

By hypothesis, $\varphi_{x_1y_1}$ is a one-to-one mapping of M_{x_1} onto M_{y_1}. Comparing the right sides of (2.2.4) and (2.2.5), we obtain

$$\varphi_{x_1y_1}(\psi(x_1)) = \psi(y_1), \tag{2.2.6}$$

i.e., (x_1, y_1) belongs to D^*. Thus, if (x_1, y_1) belongs to $\bar{D}^*$, then $(x_1, y_1) \in D^*$, which means that D^* is relatively closed in D. Conversely, if (x_1, y_1) belongs to D^*, then (2.2.6) holds. The identities (2.2.3), (2.2.4) and (2.2.6) and hypothesis (3) show that

$$\psi(y_2) = \varphi_{y_1y_2}(\psi_1(y_1)) = \varphi_{y_1y_2}(\varphi_{x_1y_1}(\psi(x_1))) = \varphi_{x_1y_2}(\psi(x_1))$$
$$= \varphi_{x_2y_2} \circ \varphi_{x_1x_2}(\psi(x_1)) = \varphi_{x_2y_2}(\varphi_{x_1x_2}(\psi(x_1))) = \varphi_{x_2y_2}(\psi(x_2))$$

for all $x_2 \in U_1$ and $y_2 \in V_1$. That is, (x_2, y_2) belongs to D^* for $(x_2, y_2) \in U_1 \times V_1$ and so D^* is relatively open in D. Finally, (x, x) belongs to D^* for all $x \in X$ and therefore D^* is nonvoid. Since D is connected, D^* and D coincide. That is, (2.2.6) holds for all $(x_1, y_1) \in D$.

We will now prove the uniqueness of the mapping ψ. Let χ be a mapping that enjoys all the properties ascribed to ψ in the last paragraph of our theorem, including $\chi(x_0) = w_{x_0}^0$. Let U be the set of all points $x \in X$ for which $\chi(x) = \psi(x)$. Since x_0 belongs to U, U is nonvoid. Let y be a point in X and let V be a neighborhood of y for which $V \times V \subset D$. If V and U have a common point y_1, we have

$$\varphi_{y_1y}(\chi(y)) = \chi(y_1) = \psi(y_1) = \varphi_{y_1y}(\psi(y)).$$

Therefore $\psi(y)$ is equal to $\chi(y)$ and thus U is closed in X. Conversely, if $\psi(y) = \chi(y)$, we have $\varphi_{y_1y}(\chi(y)) = \varphi_{y_1y}(\psi(y))$. Therefore we have $\chi(y_1) = \psi(y_1)$ for $y_1 \in V$, and so U is open in X. Since X is connected, $U = X$. This completes the proof. □

2.3. Some Applications of the Principle of Monodromy to the Theory of Topological Groups

Let G be a topological group. A mapping f of a neighborhood U of the identity element e of G into a group H is called a *local homomorphism*, if for all $g, h \in U$, such that $gh \in U$, we have $f(gh) = f(g)f(h)$.

I. *Let G be a simply connected topological group. Let f be a local homomorphism of the group G into a group H, and suppose that the domain of f is a connected* neighborhood U of $e \in G$. Then there exists a unique homomorphism ψ of the entire group G into H that coincides with f on U.

Proof. Let $D \subset G \times G$ be the set of pairs (g, h) such that $hg^{-1} \in U$. The set D is open in $G \times G$, and contains all pairs (g, g), $g \in G$. The set D is the union of all sets $\{g\} \times Ug$, $g \in G$. Since U is connected, every set $\{g\} \times Ug$ is connected. All these sets intersect the connected set $\{(g, g) : g \in G\}$. Therefore the set D is connected.

Suppose that (g, h) belongs to D. We write the mapping $x \to f(hg^{-1})x$ of the group H onto itself as φ_{gh}. If (g, h), (h, k), and (g, k) are elements of D, then kh^{-1}, hg^{-1} and $kg^{-1} = kh^{-1} \cdot hg^{-1}$ are in U. We then have

$$\varphi_{gk}(x) = f(kg^{-1})x = f(kh^{-1})f(hg^{-1})x = \varphi_{hk}(\varphi_{gh}(x))$$

for all $x \in H$. We apply the principle of monodromy with $M_g = H$ for all $g \in G$. There is thus a mapping ψ of the group G into H such that $\psi(e)$ is the identity element of H and $\psi(h) = f(hg^{-1})\psi(g)$ for all $g, h \in G$, such that $hg^{-1} \in V$. Setting $g = e$, we see that the mapping ψ coincides with f on U. If k belongs to U, we have $\psi(kg) = f(kg \cdot g^{-1})\psi(g) = f(k)\psi(g) = \psi(k)\psi(g)$. For $V = U \cap U^{-1}$, we have $G = \bigcup_{n=1}^{\infty} V^n$, as G is connected. That is, every element $g \in G$ has the form $k_1 \cdots k_n$, where $k_i \in U \cap U^{-1}$ for all $i = 1, \ldots, n$. By induction on n, we obtain $\psi(k_1 \cdots k_n h) = \psi(k_1) \cdots \psi(k_n)\psi(h)$. For $h = e$, we obtain $\psi(k_1 \cdots k_n) = \psi(k_1) \cdots \psi(k_n)$ and therefore $\psi(gh) = \psi(g)\psi(h)$ for all $g, h \in G$. That is, ψ is a homomorphism of G into H. The uniqueness of the homomorphism follows from the fact that

$$\psi(g) = \psi(k_1) \cdots \psi(k_n) \quad \text{for } g = k_1 \cdots k_n, k_i \in V. \quad \square$$

Let G and H be topological groups. Let U and V be neighborhoods of the identity elements of the groups G and H respectively. A homomorphism f that maps the neighborhood U onto the neighborhood V is called a *local isomorphism* of the groups G and H if the following conditions are met.

(1)   If $g \in U$, $g_1 \in U$ and $gg_1 \in U$, then $f(gg_1) = f(g)f(g_1)$.
(2)   If $g \in U$, $g_1 \in U$ and $f(g)f(g_1) \in U$, then $gg_1 \in U$.

II. *Let G be a simply connected topological group. Let H be a connected topological group that is locally isomorphic to G. The group H is isomorphic to a factor group of the group G by a discrete subgroup of the center of G.*

Proof. Let U and V be neighborhoods of the identity elements of the groups G and H respectively. Let f be a homomorphism of U onto V, satisfying conditions (1) and (2) of the preceding definition. According to Proposition I, the mapping f can be extended to a homomorphism ψ of G into H that coincides with f on H. The homomorphism ψ is continuous at the identity element of G and hence is continuous everywhere. The set $\psi(G)$ is a subgroup of the group H, but $\psi(G)$ contains $\psi(U) = f(U) = V$. Therefore $\psi(G)$ contains $\bigcup_{n=1}^{\infty} V^n$. Since H is connected, H is equal to $\bigcup_{n=1}^{\infty} V^n$ and thus $\psi(G) = H$. Because the image of the neighborhood U is the open set V in H, the image of any open set in G is open in H. Thus a subset of H is open if and only if its inverse image in G is open. Therefore the group H is isomorphic as a topological group, with the factor group of G by the kernel N of the homomorphism ψ. Since the mapping ψ of U onto V is a homeomorphism, $U \cap N$ consists only of the identity element. That is, N is a discrete group. Finally, N belongs to the center of the group G by Proposition VI in 1.2, chapter V. □

§3. Covering Groups

3.1. Some Properties of Covering Spaces

I. *Let Y be a covering space for the space X under a mapping f. Let φ, φ' be continuous mappings of a connected space Z into Y, such that $f \circ \varphi = f \circ \varphi'$. Suppose that the mappings φ and φ' coincide at a single point of the space Z. Then we have $\varphi = \varphi'$.*

Proof. Let F be the set of all points $z \in Z$ for which $\varphi(z) = \varphi'(z)$. The set F is clearly closed and as stipulated, it is nonvoid. We will show that F is open. Let z belong to F. The point $x = f(\varphi(z))$ has a neighborhood U such that the mapping f is a homeomorphism of every component of the set $f^{-1}(U)$ onto U. Let V be the component of the point $\varphi(z) = \varphi'(z)$ in the set $f^{-1}(U)$. By I in 1.1, the set V is a neighborhood of the point $\varphi(z) = \varphi'(z)$. Thus there is a neighborhood W of the point z in Z such that $\varphi(W) \subset V$, $\varphi'(W) \subset V$. Since the mapping f is a homeomorphism of V onto U, we see that $\varphi(w) = \varphi'(w)$ for all $w \in W$. Therefore W is contained in F and F is open. Since Z is connected, F is all of Z; i.e., $\varphi(z) = \varphi'(z)$ for all $z \in Z$. □

II. *Let Z be a simply connected space. Let Y be a covering space for the space X under the mapping f. If φ is a continuous mapping of Z into X, there is a continuous mapping ψ of Z into Y such that $f \circ \psi = \varphi$. If $z_0 \in Z$ and $y_0 \in Y$ have the property that $f(y_0) = \varphi(z_0)$, we can choose the mapping ψ so that $\psi(z_0) = y_0$. The mapping ψ is defined uniquely by this restriction.*

Proof. Let W be the set of all pairs (z, y) in $Z \times Y$ such that $\varphi(z) = f(y)$. We define $\chi(z, y) = z$ for all $(z, y) \in W$. For every point $z \in Z$ there is a connected neighborhood U of the point $\varphi(z)$ in X, such that the mapping

f is a homeomorphism of each component V_α of the set $f^{-1}(U)$ onto U. Let G be a connected neighborhood of the point z_0 in Z such that $\varphi(G) \subset U$. Let z belong to G and let y_α be a point in V_α such that $f(y_\alpha) = \varphi(z)$. Then the correspondence $z \to (z, y_\alpha)$ is a continuous mapping of the neighborhood G onto some subset G_α of W. Note that $\chi(z, y_\alpha) = z$. Therefore the mapping χ of the open set $G_\alpha \subset W$ onto $G \subset Z$ is a homeomorphism. The set $\chi^{-1}(G)$ is the union of the sets G_α. Under the mapping $(z, y) \to y$ of W into Y, every connected subset $F \subset \chi^{-1}(G)$ is mapped onto a connected subset of $f^{-1}(U)$. Therefore the image of F is contained in some component V_α of the set $f^{-1}(U)$. Thus the sets G_α are components of the set $\chi^{-1}(G)$. The mapping χ of each of these components onto G is a homeomorphism.

Let W_0 be the component of the point (z_0, y_0) in the space W. By Proposition VII in 1.1, the space W_0 is a covering space for Z under the restriction χ_0 of the mapping χ to W_0. The space Z is simply connected and therefore the mapping χ_0 is a homeomorphism. We define the mapping ψ by the formula $\chi_0^{-1}(z) = (z, \psi(z))$, $z \in Z$. The mapping ψ plainly satisfies the requirements of the present proposition. Uniqueness of the mapping ψ is implied by I. $\square$

III. *Let Y_1, Y_2 be covering spaces for the space X under mappings f_1 and f_2, respectively. If the spaces Y_1 and Y_2 are simply connected, they must be homeomorphic.*

Proof. Let $y_1 \in Y_1$, $y_2 \in Y_2$ be points such that $f_1(y_1) = f_2(y_2)$. By Proposition II, there are continuous mappings $\varphi: Y_1 \to Y_2$ and $\psi: Y_2 \to Y_1$ such that $\varphi(y_1) = y_2$, $\psi(y_2) = y_1$, $f_2 \circ \varphi = f_1$, $f_1 \circ \psi = f_2$. Then the mapping $\theta = \psi \circ \varphi$ carries the space Y_1 continuously into itself and also has the properties $f_1 \circ \theta = f_1$ and $\theta(y_1) = y_2$. By Proposition II, the mapping θ must be the identity mapping of Y_1 onto itself. In like manner we see that $\varphi \circ \psi$ is the identity mapping of Y_2 onto itself. That is, φ and ψ are homeomorphisms and $\varphi = \psi^{-1}$. $\square$

Proposition III shows that if a connected, locally connected topological space X has a simply connected covering space Y, that space Y is defined uniquely to within a homeomorphism. Proposition III is simply a uniqueness theorem for simply connected covering spaces.

A space X is called *locally simply connected* if every point of the space has a simply connected neighborhood. The following proposition provides sufficient conditions for the existence of a simply connected covering space.

IV. *Every connected and locally simply connected space admits a simply connected covering space.*

Proofs can be found in Pontrjagin [1] and Chevalley [1].

3.2. Covering Groups

Let G be a topological group. A topological group $\tilde{G}$ is called a *covering group* for G (under a mapping f) if: (1) $\tilde{G}$ is a covering space for G under f; (2) f is a homomorphism of the group $\tilde{G}$ into G.

I. *Let G be a topological group and let $\tilde{G}$ be a simply connected covering space for G under a mapping f. We can define multiplication in $\tilde{G}$ so that $\tilde{G}$ becomes a topological group and the mapping f becomes a homomorphism of the group $\tilde{G}$ into G.*

Proof. Let e be the identity element in G. Let $\tilde{e}$ be any element of the space $\tilde{G}$ such that $f(\tilde{e}) = e$.

By Proposition II of 2.1, the space $\tilde{G} \times \tilde{G}$ is simply connected. By Proposition II of 3.1, there is a continuous mapping φ of the space $\tilde{G} \times \tilde{G}$ into G such that

$$f(\varphi(\tilde{g}, \tilde{h})) = f(\tilde{g})(f(\tilde{h}))^{-1} \tag{3.2.1}$$

for all $\tilde{g}, \tilde{h} \in \tilde{G}$. We also have

$$\varphi(\tilde{e}, \tilde{e}) = \tilde{e}. \tag{3.2.2}$$

Setting $\tilde{h} = \tilde{e}$ in (3.2.1), we get $f(\tilde{h}) = e$ and thus

$$f(\varphi(\tilde{g}, \tilde{e})) = f(\tilde{g}) \tag{3.2.3}$$

for all $\tilde{g} \in \tilde{G}$. By (3.2.3) the mapping ψ of the space $\tilde{G}$ into itself, defined by $\psi(\tilde{g}) = \varphi(\tilde{g}, \tilde{e})$, satisfies the condition $f \circ \psi = f$. From (3.2.2) we see that $\psi(\tilde{e}) = \tilde{e}$. By the uniqueness assertion in Proposition II of 3.1, the mapping ψ is the identity mapping. It follows that

$$\varphi(\tilde{g}, \tilde{e}) = \tilde{g} \tag{3.2.4}$$

for all $\tilde{g} \in \tilde{G}$. We define

$$\tilde{h}^{-1} = \varphi(\tilde{e}, \tilde{h}), \qquad \tilde{g}\tilde{h} = \varphi(\tilde{g}, \tilde{h}^{-1}). \tag{3.2.5}$$

From (3.2.1) and (3.2.5) we infer that

$$
\begin{aligned}
f(\tilde{h}^{-1}) &= f(\varphi(\tilde{e}, \tilde{h})) = (f(\tilde{h}))^{-1}; \\
f(\tilde{g}\tilde{h}) &= f(\varphi(\tilde{g}, \tilde{h}^{-1})) = f(\tilde{g})(f(\tilde{h}^{-1}))^{-1} = f(\tilde{g})f(\tilde{h}).
\end{aligned}
\tag{3.2.6}
$$

Again applying the uniqueness asserted in Proposition II of 3.1 to the mappings of $\tilde{G} \times \tilde{G} \times \tilde{G}$ into $\tilde{G}$ defined by the formulas $(\tilde{g}, \tilde{h}, \tilde{k}) \to (\tilde{g}\tilde{h})\tilde{k}$ and $(\tilde{g}, \tilde{h}, \tilde{k}) \to \tilde{g}(\tilde{h}\tilde{k})$, we obtain from (3.2.6) the associative law

$$\tilde{g}(\tilde{h}\tilde{k}) = (\tilde{g}\tilde{h})\tilde{k} \tag{3.2.7}$$

for all $\tilde{g}, \tilde{h}, \tilde{k} \in \tilde{G}$. Similarly we find that

$$\tilde{g}\tilde{e} = \tilde{e}\tilde{g} = \tilde{g}. \tag{3.2.8}$$

Consider the mapping $\tilde{g} \to \tilde{g}\tilde{g}^{-1}$. This mapping carries the connected space $\tilde{G}$ into the discrete space $f^{-1}(e)$. Since $\tilde{e}$ goes into $\tilde{e}$, we have

$$\tilde{g}\tilde{g}^{-1} = \tilde{e} \tag{3.2.9}$$

for all $\tilde{g} \in \tilde{G}$. Similarly we have

$$\tilde{g}^{-1}\tilde{g} = \tilde{e} \tag{3.2.10}$$

for all $\tilde{g} \in \tilde{G}$.

The identities (3.2.7)–(3.2.10) state that $\tilde{G}$ is a group under the operations defined in (3.2.5) and that $\tilde{e}$ is the identity element of this group. Since the mapping φ is continuous, the definitions (3.2.5) show that $\tilde{G}$ is a topological group. Finally, from (3.2.5) and (3.2.1), we infer that $f(\tilde{g}, \tilde{h}^{-1}) = f(\tilde{g})(f(\tilde{h}))^{-1}$ for all $\tilde{g}, \tilde{h} \in \tilde{G}$. That is, f is a homomorphism of the group $\tilde{G}$ into the group G. $\square$

The preceding proposition shows that *every topological group with a simply connected covering space has a simply connected covering group*. We now show that this group is unique.

II. *Let G be a topological group. Let G_1 and G_2 be simply connected covering groups for G under mappings f_1 and f_2, respectively. There exists an isomorphism θ_1 of the topological group G_1 onto G_2 such that $f_1 = f_2 \circ \theta_1$.*

Proof. Let e, e_1, e_2 be the identity elements of the groups G, G_1, G_2, respectively. There are neighborhoods U, U_1, U_2 of the points e, e_1, e_2, respectively, such that the mapping f_i defines a local isomorphism φ_i of the group G_i onto $G(i = 1, 2)$ for which the neighborhoods U_i and U satisfy conditions (1) and (2) of the definition of a local isomorphism. Also, φ_i is the restriction of f_i to U_i. Thus there exist mappings $\psi_1 : U_1 \to U_2$ and $\psi_2 : U_2 \to U_1$, defined by the formulas $\psi_1(g_1) = \varphi_2^{-1}(\varphi_1(g_1))$, $\psi_2(g_2) = \varphi_1^{-1}(\varphi_2(g_2))$ for $g_1 \in U_1$, $g_2 \in U_2$. These mappings define local isomorphisms of G_1 into G_2 and G_2 into G_1, respectively. Furthermore, $\psi_2 \circ \psi_1$ and $\psi_1 \circ \psi_2$ are the identity mappings on U_1 and U_2 respectively and $f_2 \circ \psi_1$ coincides with f_1 on U_1. By Proposition I in 2.3, there exist homomorphisms θ_1 and θ_2 of G_1 into G_2 and of G_2 into G_1, respectively. These homomorphisms extend the mappings ψ_1 and ψ_2. Since $\psi_2 \circ \psi_1$ and $\psi_1 \circ \psi_2$ are identity mappings, the statement of uniqueness in Proposition I of 2.3, shows that $\theta_1 \circ \theta_2$ and $\theta_2 \circ \theta_1$ are the identity mappings on the groups G_2 and G_1 respectively. Therefore, the mappings θ_1 and θ_2 are topological isomorphisms. Since f_1 and $f_2 \circ \theta_1$ are homomorphisms of the connected group G_1 into G_2 that coincide on the neighborhood U, and f_1 and $f_2 \circ \theta_1$ coincide everywhere. $\square$

§4. Simple Connectedness of Certain Groups

4.1. The Group **R**

I. *The group* **R** *is simply connected.*

Proof. Clearly **R** is connected and locally connected. Let Y be a covering space for **R** under a mapping f. By Proposition III in §1, every point $x \in$ **R** has a neighborhood U_x in **R** such that: U_x is an open interval; every component of the open set $f^{-1}(U_x)$ is open in Y; and the mapping f is a homeomorphism of every such component onto U_x. Let n be an integer. From the covering of the closed interval $[n, n+1]$ by neighborhoods U_x we select a finite subcovering $V_1^{(n)}, \ldots, V_m^{(n)}$. Combining these finite coverings, we obtain a countable open covering $\mathscr{V}$ of the entire line **R**, finite on every closed interval. We shrink the open sets $V_k^{(n)}$ as needed to obtain the result that every point $x \in$ **R** belongs to no more than two intervals of the covering $\mathscr{V}$. That is, every interval in the covering $\mathscr{V}$ intersects exactly two disjoint intervals of $\mathscr{V}$. The intervals V of the covering $\mathscr{V}$ are then naturally enumerated by integers p, where the following conditions are satisfied: (1) V_0 contains zero; (2) if $V_p = (a_p, b_p)$, then $a_p < b_{p-1} < a_{p+1} < b_p$ for all integers p.

Let W_0 be a fixed component of the open set $f^{-1}(V_0)$. Suppose that we have constructed open set W_r, $r = p, p+1, \ldots, q$, where $p \leqslant 0 \leqslant q$, such that: (a) W_r is a component of the open set $f^{-1}(V_r)$ for all $r = p, p+1, \ldots, q$; (b) the union $\bigcup_{r=p}^{q} W_r$ is connected. Since we have

$$V_{p-1} \cap V_p \neq \varnothing, \qquad V_{q+1} \cap V_q \neq \varnothing,$$

there exist uniquely defined components of the open sets $f^{-1}(V_{p-1})$ and $f^{-1}(V_{q+1})$ that intersect W_p and W_q, respectively. Denote these components by W_{p-1} and W_{q+1} respectively. Plainly $\bigcup_{r=p-1}^{q+1} W_r$ is connected. Thus the sets W_r are defined inductively for all integers r. Let G be equal to $\bigcup_{r=-\infty}^{+\infty} W_r$. The construction of G shows that the restriction φ of the mapping f to G is a one-to-one mapping of G onto **R**. On the other hand, the mapping φ is a local homeomorphism by its very definition. Thus φ is a homeomorphism of G onto **R**. We will show that $G = Y$. Plainly G is nonvoid and open. Let us show that it is closed. Let y belong to $\bar{G}$. Every neighborhood $W(y)$ of the point y intersects G; that is, we have $W(y) \cap W_p \neq \varnothing$ for some p. We write $x = f(y)$ and choose r such that $x \in V_r$. Choose a neighborhood $W(y)$ such that $f(W(y)) \subset V_r$. Then we have $\varnothing \neq f(W(y) \cap W_p) \subset f(W(y)) \cap f(W_p) \subset V_r \cap V_p$. Furthermore, the restriction of f to W_p is a homeomorphism. Since $f(W(y) \cap W_p) \subset V_r \cap V_p$, we see that $W(y) \cap W_p \subset f^{-1}(V_r) \cap W_p$. From the construction of the sets W_r, we see that $f^{-1}(V_r) \cap W_p = W_r \cap W_p$. Thus we have $W(y) \cap W_p \subset W_r \cap W_p$ and $W(y) \cap W_r \supset W(y) \cap W_p \cap W_r = W(y) \cap W_p \neq \varnothing$. That is, the intersection of $W(y)$ and W_r is nonvoid. If $W(y)$ is connected, we infer from the relations $f(W(y)) \subset V_r$, $W(y) \cap W_r \neq \varnothing$

that $W(y) \subset W_r$ (see V in §1). In particular y belongs to $W_r \subset G$; that is, $\bar{G} = G$ and G is closed. Since Y is connected, $G = Y$ and f is a homeomorphism of Y onto $\mathbf{R}$. That is, every covering space for $\mathbf{R}$ is isomorphic to the trivial covering space. $\square$

Exercise

Let X be a simply connected space and let Y be a connected, locally connected, locally compact space. Let f be a one-to-one continuous mapping of Y onto X. Prove that f is a homeomorphism.

4.2. Simple Connectedness of Closed Intervals and Spheres

I. *Let Y be a covering space for X under a mapping f. Let M, N be closed and locally connected subsets of X. Suppose that the mapping f is a homeomorphism of every component of the sets $f^{-1}(M), f^{-1}(N)$ onto M and N respectively. If $M \cap N$ is nonvoid and connected, the mapping f is a homeomorphism of every component of $f^{-1}(M \cup N)$ onto $M \cup N$.*

Proof. Let M_α be the components of the set $F^{-1}(M)$ ($M_{\alpha_1} \neq M_{\alpha_2}$ for $\alpha_1 \neq \alpha_2$). We set $F = M \cap N$, $M_\alpha \cap f^{-1}(F) = F_\alpha$. By hypothesis, the mapping f is a homeomorphism of M_α onto M and thus the mapping f of F_α onto F is also a homeomorphism. Since F is connected, F_α is connected and so F_α belongs to a unique component N_α of the set $f^{-1}(N)$. We set $S_\alpha = M_\alpha \cup N_\alpha$, $T_\alpha = \bigcup_{\alpha' \neq \alpha} S_{\alpha'}$. The set S_α is closed, since it is closed in $f^{-1}(M \cup N)$. We have $T_\alpha = (\bigcup_{\alpha' \neq \alpha} M_{\alpha'}) \cup (\bigcup_{\alpha' \neq \alpha} N_{\alpha'})$. Note too that M_α is relatively open in $f^{-1}(M)$ and N_α is relatively open in $f^{-1}(N)$ (see Proposition VII in §1). Consequently the sets $\bigcup_{\alpha' \neq \alpha} M_{\alpha'}$, $\bigcup_{\alpha' \neq \alpha} N_{\alpha'}$ are closed in $f^{-1}(M)$ and $f^{-1}(N)$, respectively. That is, they are closed in Y. Therefore the set T_α is closed. Since $S_\alpha \cup T_\alpha = f^{-1}(M \cup N)$ is a closed set, the set S_α is not only closed, but also open in $f^{-1}(M \cup N)$. The sets M_α and N_α are connected and have nonvoid intersection ($M_\alpha \cap N_\alpha \supset F_\alpha$). Therefore S_α are components of the set $f^{-1}(M \cup N)$. Let f_α be the restriction of f to S_α. By Proposition VII in §1, the space S_α is a covering space for $M \cup N$ under the mapping f_α. We will show that f_α is one-to-one. If $f(y_1) = f(y_2)$ for $y_1, y_2 \in S_\alpha$, two results are possible. Either y_1, y_2 both belong to one of the sets M_α or N_α, in which case $y_1 = y_2$, or $f(y_1) = f(y_2) \in f(M_\alpha) \cap f(N_\alpha) = M \cap N = F$, and again we have $y_1 = y_2$, since the mapping f_α on both M_α and N_α is a homeomorphism. Thus f_α is one-to-one. Since f_α is a local homeomorphism, it maps S_α onto $M \cup N$ homeomorphically. $\square$

II. *All intervals in $\mathbf{R}$ are simply connected.*

Proof. An open interval (a, b) is homeomorphic to $\mathbf{R}$ and is therefore simply connected. Consider a half-open interval $(a, b]$. Let Y be a covering space for $X = (a, b]$ under the mapping f. By Proposition VII in §1, every component of the set $f^{-1}((a,b))$ is a covering space under the restriction of the mapping f. The interval (a, b) is simply connected and therefore the mapping f is a homeomorphism of every component of the set $f^{-1}((a,b))$ onto (a,b). The point b has a neighborhood of the form $(b - \delta, b]$, with the property that f is a homeomorphism of every component of $f^{-1}((b - \delta, b])$ onto $(b - \delta, b]$. Then the sets $M = (a, b - \delta/4]$ and $N = [b - \delta/2, b]$ satisfy the conditions of Proposition I. Therefore the mapping f is a homeomorphism of every component of $f^{-1}(M \cup N) = f^{-1}((a,b]) = Y$ onto $(a,b]$. Since Y is connected, f is a homeomorphism of Y onto $(a,b]$. That is, $(a,b]$ is simply connected. One proves in like manner that the segment $[a, b]$ is simply connected. □

III. *The product of a finite number of intervals is simply connected.*

The proof follows directly from Proposition II above and Proposition II in 2.1.

IV. *The sphere* $S^{n-1} = \{(x_1, \ldots, x_n), x_k \in \mathbf{R}, \sum_{k=1}^{n} x_k^2 = 1\}$ *is simply connected for* $n > 2$.

Proof. Let Y be a covering space for S^{n-1} under a mapping f. Let M be equal to $\{(x_1, \ldots, x_n):x_n \geqslant 0, (x_1, \ldots, x_n) \in S^{n-1}\}$ and $N = \{(x_1, \ldots, x_n):x_n \leqslant 0, (x_1, \ldots, x_n) \in S^{n-1}\}$. The mapping $(x_1, \ldots, x_n) \to (x_1, \ldots, x_{n-1})$ maps M and N homeomorphically onto the ball $\{(x_1, \ldots, x_{n-1}:\sum_{k=1}^{n-1} x_k^2 \leqslant 1\}$ in $\mathbf{R}^{n-1}$. Since this ball is homeomorphic to an $(n - 1)$-dimensional cube, Proposition III shows that M and N are simply connected. The set $M \cap N = \{(x_1, \ldots, x_n):x_n = 0, \sum_{k=1}^{n-1} x_k^2 = 1\}$ is homeomorphic to the sphere S^{n-2}. It is easy to verify that S^{n-2} is connected for $n > 2$. Thus the sets M and N satisfy the hypotheses of Proposition I. Therefore the mapping f is a homeomorphism of every component of the set $Y = f^{-1}(M \cup N)$ onto $M \cup N = S^{n-1}$. Since Y is connected, f is a homeomorphism of Y onto S^n. That is, S^{n-1} is simply connected. □

4.3. Auxiliary Propositions

I. *Let G be a connected and locally connected topological group. Let H be a closed locally connected subgroup of G and H_0 the component of the identity element in H. There exists a mapping f of the space G/H_0 onto G/H such that G/H_0 is a covering space for G/H under f. If H is a normal subgroup, G/H_0 is a covering group for G/H under a homomorphism f.*

Proof. Since H is locally connected, H_0 is relatively open in H (see Proposition I of §1). Therefore H admits a neighborhood V of the identity element such that $V^{-1}V \cap H \subset H_0$. Choosing if need be a connected neighborhood of e contained in V, we may suppose that V is connected. Let π and π_0 be the canonical mappings of the group G onto the spaces of cosets G/H and G/H_0 respectively. Every element ξ of the space G/H_0 is a coset of the subgroup H_0 and so ξ is contained in some coset of H. If g belongs to G and $\xi = gH_0$, then ξ is contained in η, where $\eta = gH$. We write $\eta = f(\xi)$. This defines a mapping f of the space G/H_0 onto G/H. Since the mappings π and π_0 are continuous and open, the mapping f is continuous and open.

Let g belong to G. We set $W(g) = \pi(gV)$. The set $W(g)$ is a neighborhood of the element $\pi(g)$ in G/H. In each coset hH_0 ($h \in H$) we choose a representative element. Let A be the set of these representatives. We clearly have $H = \bigcup_{a \in A} aH_0$. Consider the set $f^{-1}(W(g))$. This set coincides with $gVH = \bigcup_{a \in A} gVaH_0$. We write $W_a(g) = \pi_0(gVa)$. It is clear that $f^{-1}(W(g)) = \bigcup_{a \in A} W_a(g)$. We will show that the mapping f is one-to-one on every set $W_a(g)$ and maps $W_a(g)$ onto $W(g)$, and that the sets $W_a(g)$ are pairwise disjoint. We have $f(W_a(g)) = f(\pi_0(gVa)) = gVaH = gVH = \pi(gV) = W(g)$. If we have $f(y_1) = f(y_2)$ for $y_1 \in W_a(g)$, $y_2 \in W_{a'}(g)$, then $y_1 = \pi(gx_1a)$, $y_2 = \pi_0(gx_2a')$ for certain $x_1, x_2 \in V$. We also have $\pi(gx_1a) = f(\pi_0(gx_1a)) = f(\pi_0(gx_2a')) = \pi(gx_2a')$; hence $gx_1a = gx_2a'h$ for some $h \in H$. We then obtain

$$x_2^{-1}x_1 = a'ha^{-1}. \tag{4.3.1}$$

Because a', h belong to H, $a'ha^{-1}$ belongs to H and $x_2^{-1}x_1 \in V^{-1}V \cap H \subset H_0$. If we suppose that $a' = a$ (that is, $y_1, y_2 \in W_a(g)$), we can infer from (4.3.1) that

$$h = a^{-1}(x_2^{-1}x_1)a \in a^{-1}H_0a. \tag{4.3.2}$$

The subgroup H_0 is a normal subgroup of H (see Proposition III in 1.2, chapter V), and so (4.3.2) implies that $h \in H_0$. This means that $\pi_0(gx_1a) = \pi_0(gx_2a)$ and $y_1 = y_2$. Thus f is a one-to-one mapping of $W_a(g)$ onto $W(g)$. Finally, if $W_a(g)$ and $W_{a'}(g)$ intersect, that is, $\pi_0(gx_1a) = \pi_0(gx_2a')$ for some $x_1, x_2 \in V$, then $gx_1a = gx_2a'h$ for some $h \in H_0$. We then have

$$a'ha^{-1} \in a'H_0a^{-1} = (a'H_0a'^{-1})a'a^{-1} \subset H_0a'a^{-1},$$

and at the same time $a'ha^{-1} = x_2^{-1}x_1 \in H_0$. That is, the intersection $H_0 \cap H_0a'a^{-1}$ is nonvoid, which means that $H_0a \cap H_0a'$ is nonvoid. Because a and a' are representatives of pairwise disjoint cosets, we obtain $a = a'$. Hence the sets $W_a(g)$ are pairwise disjoint.

Every set $W_a(g)$ is open in G/H_0. Since f is continuous and open and the mapping f of $W_a(g)$ onto $W(g)$ is one-to-one, f is a homeomorphism of $W_a(g)$ onto $W(g)$. Every set $W_a(g)$ is connected, as the continuous image of the connected set gVa. Therefore, the pairwise disjoint sets $W_a(g)$ are the components of the set $f^{-1}(W(g))$. We have proved that G/H_0 is a covering space for G/H under the mapping f.

If H is a normal subgroup of G, then H_0 is also a normal subgroup of G. For, if g is an element of G, the set gH_0g^{-1} is connected, contains the identity element and is contained in $gHg^{-1} = H$. That is, gH_0g^{-1} is contained in H_0 for all $g \in G$. That is, H_0 is a normal subgroup of G. In this case the mapping

f constructed above is a homomorphism of the group G/H_0 onto the group G/H with kernel H/H_0. Thus G/H_0 is a covering group for G/H under the homomorphism f.   □

II. *Retain the hypotheses of* I. *If the space G/H is simply connected, the group H is connected.*

Proof. If G/H is simply connected, the mapping $f: G/H_0 \to G/H$ is a homeomorphism. Therefore we have $H = H_0$ and so H is connected.   □

III. *Let G be a connected and locally connected topological group. Let H be a discrete normal subgroup of G and let π be the canonical homomorphism of G onto G/H. Then G is a covering group for G/H under the mapping π.*

The proof follows from I, since the discreteness of the group H implies that $H_0 = \{e\}$ and $G/H_0 = G$.

IV. *Let G be a connected and locally connected topological group. Let H be a closed, locally connected subgroup of G. Suppose that G is locally simply connected and that H and G/H are simply connected. Then G is simply connected.*

Proof. Since G is connected and locally connected, it has a universal covering group. Let $\tilde{G}$ be a covering group for G under a mapping π. We set $\tilde{H} = \pi^{-1}(H)$. If $\tilde{g} \in \tilde{G}$ and $\pi(\tilde{g}) = g$, then we have $\pi(\tilde{g}\tilde{H}) = \pi(\tilde{g})\pi(\tilde{H}) = gH$. That is, π maps the cosets in $\tilde{G}$ of the subgroup $\tilde{H}$ onto the cosets in G of H. If we have $\pi(\tilde{g}_1\tilde{H}) = \pi(\tilde{g}_2H)$, then $\pi(\tilde{g}_1)H = \pi(\tilde{g}_2)H$, that is, $\pi(\tilde{g}_2^{-1}\tilde{g}_1) \in H$, $\tilde{g}_2^{-1}\tilde{g}_1 \in \tilde{H}$, $\tilde{g}_1 \in \tilde{g}_2H$ and $\tilde{g}_1\tilde{H} = \tilde{g}_2H$. Thus the mapping $\pi^*: \tilde{G}/\tilde{H} \to G/H$, defined by $\pi^*(\tilde{g}\tilde{H}) = \pi(\tilde{g})H$, is one-to-one. Since the canonical mappings $\rho: G \to G/H$ and $\tilde{\rho}: \tilde{G} \to \tilde{G}/\tilde{H}$ are continuous and open, and $\pi^* \circ \tilde{\rho} = \rho \circ \pi$, the mapping π^* is also continuous and open. Therefore the mapping π^* is a homeomorphism of $\tilde{G}/\tilde{H}$ onto G/H. Since G/H is simply connected, $\tilde{G}/\tilde{H}$ is also simply connected. Note that $\tilde{H}$ is closed in G (it is the inverse image of the closed group H). Since H is connected and locally connected, Proposition VII of §1 shows that the group $\tilde{H}$ is locally connected. From the simple connectedness of $\tilde{G}/\tilde{H}$ and from Proposition II, we infer that the group $\tilde{H}$ is connected. Applying Proposition VII of §1, we see that $\tilde{H}$ is a covering space for H under the mapping π. The group H is simply connected and therefore the mapping π of $\tilde{H}$ onto H is an isomorphism. In particular, the kernel of the mapping π is the identity element alone. Hence π is an isomorphism of $\tilde{G}$ onto G and so G is simply connected.   □

4.4. Simple Connectedness of Certain Classical Groups

I. *Let $G = SU(n)$. Let H be the subgroup of the group G consisting of matrices of the form $h = \begin{Vmatrix} 1 & 0 \\ 0 & g \end{Vmatrix}$, where g is a matrix of order $n - 1$. The subgroup H*

is isomorphic with the group $SU(n - 1)$ and the factor space G/H is homeomorphic to the sphere S^{2n-1}.

Proof. We write $h = \begin{Vmatrix} 1 & 0 \\ 0 & u \end{Vmatrix}$, where u is a square matrix of order $n - 1$. The relation $h \in G$ is equivalent to the condition $h^*h = 1_n$, that is, to the condition $u^*u = 1_{n-1}$, where 1_k is the identity matrix of order k. Thus u belongs to $SU(n - 1)$, which means that $H \approx SU(n - 1)$. We will show that $G/H \approx S^{2n-1}$. Note that the unit sphere S in the complex space C^n is homeomorphic to the unit sphere S^{2n-1}. We will show that S is a homogeneous space under the action of the group G. If $x = (x_1, \ldots, x_n) \in S$, that is, $(x, x) = 1$, then we also have $(xg, xg) = 1$ for all $g \in G$, where $xg = (x_1, \ldots, x_n) \begin{Vmatrix} g_{11} & \cdots & g_{1n} \\ \cdots\cdots\cdots\cdots \\ g_{n1} & \cdots & g_{nn} \end{Vmatrix}$.

Next look at $x_0 = (1, \ldots, 0) \in S$. For any point x of S, there is an orthonormal basis $e_1, \ldots, e_n$ in C^n whose first vector is $x : e_1 = x$. The matrix g that carries the basis $(1, 0, \ldots, 0), \ldots, (0, \ldots, 0, 1)$ to the basis $e_1, \ldots, e_n$ belongs to the group G, and of course we have $x = x_0 g$. Hence S is homogeneous under G. The reader may verify that the mapping $g \to x_0 g$ is continuous and open. By Proposition III in 2.6, chapter III, the space S is homeomorphic to the space of cosets $G/\tilde{H}$, where $\tilde{H}$ is the stationary subgroup of the point x_0. We now determine the subgroup $\tilde{H}$. The condition $x_0 g = x_0$ is equivalent to $g_{11} = 1$, $g_{12} = \cdots = g_{1n} = 0$. Since the rows of the matrix g are orthogonal, we have $g_{21} = \cdots = g_{n1} = 0$ for $g \in \tilde{H}$ and thus $\tilde{H} \doteq H$. We therefore have $G/H \approx S \approx S^{2n-1}$. $\square$

The group of all unitary matrices g of order $2n$ for which

$$g' J_n g = J_n, \tag{4.4.1}$$

$\left(g' \text{ is the transpose of } g \text{ and } J_n = \begin{Vmatrix} 0 & 1_n \\ -1_n & 0 \end{Vmatrix} \right)$ is denoted by $Sp(2n)$. The group $Sp(2n)$ is a closed subgroup of the group $U(2n)$ so it is a compact group.

II. *Let $G = Sp(2n)$. Let σ be the square matrix of order $2n$ defined by*

$$\sigma = \begin{Vmatrix} 1 & & & & & & & \\ & \ddots & & & & & 0 & \\ & & 1 & & & & & \\ & & & 0 & & 1 & & \\ & 0 & & 1 & \ddots & & & \\ & & & 0 & & \ddots & & \\ & & & & 1 & 0 & & \end{Vmatrix},$$

and let H be the subgroup of G consisting of all matrices of the form

$$
h = \sigma
\begin{Vmatrix}
1 & 0 & \cdots & 0 & 0 \\
0 & h_{22} & \cdots & h_{2,n-1} & 0 \\
\cdots\cdots\cdots\cdots\cdots\cdots\cdots\cdots\cdots\cdots \\
0 & h_{n-1,2} & \cdots & h_{n-1,n-1} & 0 \\
0 & 0 & \cdots & 0 & 1
\end{Vmatrix}
\sigma^{-1}.
\tag{4.4.2}
$$

The subgroup H is isomorphic to the group $Sp(2n-2)$ for $n > 2$ and $H = \{e\}$ for $n = 2$. The factor space G/H is homeomorphic to the sphere S^{4n-1}.

Proof. If $n = 2$, then plainly $H = \{e\}$. Suppose that $n > 2$ and that

$$
h = \sigma
\begin{Vmatrix}
1 & 0 & 0 \\
0 & u & 0 \\
0 & 0 & 1
\end{Vmatrix}
\sigma^{-1},
\tag{4.4.3}
$$

where u is a square matrix of order $2n - 2$. The condition $h \in G$ is equivalent to $h^*h = 1_{2n}$ and $h'J_n h = J_n$. Since $\sigma = \sigma^{*-1} = \sigma'^{-1}$, the condition $h^*h = 1_{2n}$ is equivalent to $u^*u = 1_{2n-2}$, that is, to u being unitary. On the other hand, a short computation shows that

$$
\sigma^{-1}J_n\sigma =
\begin{Vmatrix}
0 & 0 & 1 \\
0 & J_{n-1} & 0 \\
1 & 0 & 0
\end{Vmatrix}.
\tag{4.4.4}
$$

From (4.4.4) we infer that for $g = \sigma \begin{Vmatrix} 1 & 0 & 0 \\ 0 & u & 0 \\ 0 & 0 & 1 \end{Vmatrix} \sigma^{-1}$, (4.4.1) is equivalent to $u'J_{n-1}u = J_{n-1}$. Therefore for an element g of the form (4.4.3), the relations $g \in Sp(2n)$ and $u \in Sp(2n-2)$ are equivalent. Hence H is isomorphic to $Sp(2n-2)$.

We will show that $G/H \approx S^{4n-1}$. Note that the unit sphere S in the complex space $\mathbf{C}^{2n}$ is homeomorphic to the unit sphere S^{4n-1}. We will show that S is a homogeneous space under the action of G. Since G is contained in $SU(2n)$, xg belongs to S for all $x \in S$ and $g \in G$. We set $x_0 = (1,0,\ldots,0)$. We will show that for every $x \in S$, there is an element $g \in G$ such that $x_0g = x$. We set $e_1 = x$ and $e_{n+1} = \bar{x}J_n$, where $\bar{x} = (\bar{x}_1,\ldots,\bar{x}_{2n})$ for $x = (x_1,\ldots,x_{2n})$. Plainly e_{n+1} belongs to S and $(e_1, e_{n+1}) = 0$. Suppose that we have already constructed a sequence of orthonormal vectors $e_1,\ldots,e_k, e_{n+1},\ldots,e_{n+k}$ such that $e_{n+j} = \bar{e}_j J_n$ for all $j = 1,\ldots,k$. For e_{k+1}, we choose any vector of length one that is orthogonal to all of the vectors $e_1,\ldots,e_k, e_{n+1},\ldots,$

e_{n+k}. Let $e_{n+k+1} = \bar{e}_{k+1}J_n$. We now have

$$(e_{n+k+1}, e_j) = (\bar{e}_{k+1}J_n, -\bar{e}_{n+j}J_n) = -(\bar{e}_{k+1}, \bar{e}_{n+j}) = -\overline{(e_{k+1}, e_{n+j})} = 0$$

and

$$(e_{n+k+1}, e_{n+j}) = (\bar{e}_{k+1}J_n, \bar{e}_jJ_n) = (\bar{e}_{k+1}, \bar{e}_j) = \overline{(e_{k+1}, e_j)} = 0.$$

By induction on k we can construct an orthonormal basis $e_1, \ldots, e_{2n}$ in $\mathbf{C}^{2n}$ such that $x = e_1$ and $e_{n+k} = \bar{e}_kJ_n$ for $k = 1, \ldots, n$. Let g be the matrix that carries the basis $(1, 0, \ldots, 0), \ldots, (0, \ldots, 0, 1)$ onto the basis $e_1, \ldots, e_{2n}$. Thus g is a unitary matrix and the identities $e_{n+k} = \bar{e}_kJ_n$ mean that $J_ng = \bar{g}J_n$, where $\bar{g}$ is the matrix whose entries are the complex conjugates of the corresponding entries of the matrix g. Since g is unitary, the relation

$$J_ng = \bar{g}J_n \tag{4.4.5}$$

is equivalent to (4.4.1) and thus $g \in G$ and $x_0g = x$.

We will now show that H is the stationary subgroup of the element $x_0 = (1, 0, \ldots, 0)$. The condition $x_0g = x_0$ means that $g = \left\| \begin{matrix} 1 & w \\ 0 & v \end{matrix} \right\|$ for some row w and matrix v. Since g is unitary, we have $g = \left\| \begin{matrix} 1 & 0 \\ 0 & v \end{matrix} \right\|$. Condition (4.4.5), as we easily verify, is equivalent to the condition that the $(n + 1)$-st row of the matrix g be equal to $(0, \ldots, 0, 1, 0, \ldots, 0)$, where 1 is in the $(n + 1)$-st place. This condition in turn is equivalent to the condition that the last line of the matrix $\sigma^{-1}g\sigma$ be equal to $(0, \ldots, 0, 1)$. Because g and σ are unitary, we conclude that h satisfies (4.4.2) if and only if h belongs to the stationary subgroup of the element x_0. The mapping $g \to x_0g$ is easily shown to be continuous and open. Applying Proposition III in 2.6, chapter III, we conclude that G/H is homeomorphic to S^{4n-1}. $\square$

III. *The groups $SU(n)$ and $Sp(2n)$ are simply connected for all $n \geqslant 1$.*

Proof. The group $SU(1)$ is equal to $\{e\}$ and so is simply connected. The group $Sp(2)$ is homeomorphic to S^3 (see II) and so is simply connected (see IV in 4.2). We will prove below (in §3, chapter IX and §2, chapter XI) that the groups $SU(n)$ and $Sp(2n)$ are locally simply connected. (Specificially, we will show that $SU(n)$ and $SO(2n)$ are Lie groups and that any element of a Lie group has a neighborhood that is homeomorphic to a ball in a Euclidean space and so is simply connected.) The present proposition is proved by induction on n with the aid of IV in 4.2, IV in 4.3 and I and II above. $\square$

IV. *The group $SO(3, \mathbf{R})$ is not simply connected.*

Proof. The homomorphism $\pi: SU(2) \to SO(3)$, constructed in the exercise in 1.2, chapter IV, has a discrete kernel equal to $\{e, -e\}$. Therefore $SU(2)$ is a covering group for $SO(3)$ (see III in 4.3). Because π is not a homeomorphism, $SO(3, \mathbf{R})$ is not simply connected. $\quad\square$

Below we will show (see III in 7.2, chapter XI) that the groups $SO(n, \mathbf{R})$ and $SO(n, \mathbf{C})$ for $n \geqslant 3$ are not simply connected. The universal covering group for the group $SO(n, \mathbf{C})$ or for the group $SO(n, \mathbf{R})$ is called a *spinor group.* Its construction is found in Žhelobenko [1] and Chevalley [1].

Chapter IX

Basic Concepts of Lie Groups and Lie Algebras

§1. Analytic Manifolds

1.1 Definition of an Analytic Manifold

Let M be a separated topological space with a countable basis of open sets. The space M is called a real (or complex) *analytic manifold* if for every open subset $U \subset M$ there is an algebra $D(U)$ of complex-valued functions on U that contains the function identically 1 for which the following conditions hold.

  (a) If V, U are open subsets of M, $V \subset U$, the restriction of every function $f \in D(U)$ to the set V belongs to $D(V)$.
  (b) If V and V_i $(i \in I)$ are open subsets of M, $V = \bigcup_i V_i$, and f is a complex-valued function on V such that $f | V_i \in D(V_i)$ for all $i \in I$, then f belongs to $D(V)$.
  (c) There is a positive integer m such that for any $x \in M$, there exists an open set $U \subset M$ containing x and m real- (or complex-) valued functions $x_1, \ldots, x_m \in D(U)$ for which the following conditions hold.
  (1) The mapping $\xi : y \to (x_1(y), \ldots, x_m(y))$ is a homeomorphism of the set U onto an open subset of the space $\mathbf{R}^m$ (or $\mathbf{C}^m$).
  (2) For every open subset W of U, the algebra $D(W)$ consists precisely of those functions on W having the form $F \circ \xi$, where F is a real (or complex) analytic function $\xi(W)$.

  Functions in the algebra $D(U)$ are called *analytic functions on U*. Any open set U that satisfies (c) is called a *coordinate neighborhood in M* and the functions $x_1, \ldots, x_m$ are called *analytic coordinates on U*. The number m is called the *dimension of the manifold M*.

I. Let $x_1, \ldots, x_m$ be analytic coordinates on U. Let $y_1, \ldots, y_n$ be a finite number of functions in $D(U)$. The set $y_1, \ldots, y_n$ satisfies (1) and (2) of (c) in some open subset $V \subset U$ if and only if: (1) $m = n$; and (2) if $y^i = F_i \circ \xi$, where F_i is an analytic function of the variables $x_1, \ldots, x_m$, the functional determinant $D(F_1, \ldots, F_m)/D(x_1, \ldots, x_m)$ is distinct from zero at some point of the set $\xi(U)$.

Proof. Let the set $y_1, \ldots, y_n$ satisfy (1) and (2) in (c) in an open subset $V \subset U$. We will prove that $y_1, \ldots, y_n$ satisfies (1) and (2) of our present proposition. We set $y_i = F_i \circ \xi$, where F_i $(i = 1, \ldots, n)$ is an analytic function of the variables $x_1, \ldots, x_m$ on the set $\xi(V)$. By hypothesis, the mapping $\eta: z \to (y_1(z), \ldots, y_n(z))$ is a homeomorphism of the set V onto the open set $\eta(V)$ of the corresponding coordinate space. Furthermore, we have $x_k = G_k \circ \eta$, where G_k $(k = 1, \ldots, m)$ is an analytic function of the variables $y_1, \ldots, y_n$. Therefore we have

$$F_i(G_1(y_1, \ldots, y_n), \ldots, G_m(y_1, \ldots, y_n)) = y_i, \qquad i = 1, \ldots, n, \quad (1.1.1)$$
$$G_k(F_1(x_1, \ldots, x_m), \ldots, F_k(x_1, \ldots, x_m)) = x_k, \qquad k = 1, \ldots, m, \quad (1.1.2)$$

where $(x_1, \ldots, x_m) \in \xi(V)$ and $(y_1, \ldots, y_n) \in \eta(V)$. Differentiating (1.1.1) and (1.1.2), we obtain

$$\sum_{k=1}^{m} \frac{\partial F_i}{\partial x_k} \frac{\partial G_k}{\partial y_j} = \delta_{ij}, \qquad \sum_{i=1}^{n} \frac{\partial G_k}{\partial y_i} \frac{\partial F_i}{\partial x_l} = \delta_{kl}. \quad (1.1.3)$$

The first identity in (1.1.3) can be considered as a family of systems of linear equations with matrix $(\partial F_i / \partial x_k)$ in the unknowns $\partial G_k / \partial y_j$ (where $k = 1, \ldots, m$, and j is fixed). Since these systems of equations are solvable with right sides δ_{ij}, they are solvable with the matrix $(\partial F_i / \partial x_k)$ for *any* right sides. Therefore we have $m \geq n$ and the rank of the matrix $\partial F_i / \partial x_k$ is n. Similarly, if we consider the second identity in (1.1.3) as a family of systems of linear equations with matrix $(\partial F_i / \partial x_1)$ in the unknowns $\partial G_k / \partial y_i$ $(i = 1, \ldots, n$ and k is fixed), we obtain $n \geq m$. Thus $(\partial F_i / \partial x_k)$ is a square matrix with nonzero determinant. That is, the set $y_1, \ldots, y_n$ satisfies (1) and (2) in I.

Conversely, let $y_1, \ldots, y_n \in D(U)$ satisfy (1) and (2) of the present proposition. Suppose that $D(F_1, \ldots, F_m)/D(x_1, \ldots, x_m) \neq 0$ at the point $(x_1^0, \ldots, x_m^0)$. By the implicit function theorem, there is a neighborhood W of the point $(x_1^0, \ldots, x_m^0)$ in the set $\xi(U)$ such that for any point $(y_1, \ldots, y_m) \in W$, the system of equations

$$F_i(x_1, \ldots, x_m) = y_i, \qquad i = 1, \ldots, m \quad (1.1.4)$$

has one and only one solution $(x_1, \ldots, x_m)$. This solution is defined by equalities of the form

$$x_k = G_k(y_1, \ldots, y_m), \qquad k = 1, \ldots, m. \quad (1.1.5)$$

The functions G_k, $k = 1, \ldots, m$, are analytic on the set W. Let V be the inverse image of the set W under the mapping ξ. By hypothesis, V and W are homeomorphic. We set $\eta(z) = (y_1(z), \ldots, y_m(z))$. The identities (1.1.4) and (1.1.5) show that the mapping $\Phi: (x_1, \ldots, x_m) \to (F_1(x_1, \ldots, x_m), \ldots, F_m(x_1, \ldots, x_m))$ is a homeomorphism of the set W onto the set $\eta(V)$. From the relation $\eta = \Phi \circ \xi$ we infer that η is a homeomorphism of the open set $V \subset M$ onto the open set $\eta(V)$ of the coordinate space. Thus (1) in (c) is satisfied. If z belongs to $D(V)$, we have $z = H \circ \xi$, where H is an analytic function. Thus we have $z = H \circ \Phi^{-1} \circ \Phi \circ \xi = (H \circ \Phi^{-1}) \circ \eta$, where Φ^{-1} is the mapping $(y_1, \ldots, y_m) \to (G(y_1, \ldots, y_m), \ldots, G_m(y_1, \ldots, y_m))$. Since the functions G_k are analytic, we have $z = H_1 \circ \eta$, where $H_1 = H \circ \Phi^{-1}$ is an analytic function on $\eta(V)$. Conversely, if we have $z = H_1 \circ \eta$, where H_1 is analytic, then $z = (H_1 \circ \Phi) \circ \xi$, where $H_1 \circ \Phi$ is an analytic function. Thus (2) in (c) is satisfied. $\square$

1.2. Examples of Manifolds

1. Let M be the space $\mathbf{R}^m$. For every open subset $U \subset M$ we take the algebra $D(U)$ to be *all* complex analytic functions on U. It is easy to verify that conditions (a)–(c) are satisfied. The coordinate functions $x_1, \ldots, x_m$, defined by the formulas $x_k(y_1, \ldots, y_m) = y_k, k = 1, \ldots, m, (y_1, \ldots, y_m) \in \mathbf{R}^m$, can be taken as the functions $x_1, \ldots, x_m$. Thus the space $\mathbf{R}^m$ is a (real) analytic manifold. Similarly, the space $\mathbf{C}^m$ is a complex analytic manifold.

2. Let T^1 be the unit circle in the complex plane: $T^1 = \{e^{i\varphi}, \varphi \in \mathbf{R}\}$. If f is a function on T^1, $f(e^{i\varphi})$ is a function of the real variable φ. For every open subset $U \subset T^1$ we take $D(U)$ to be all of the complex-valued functions f on U such that $f(e^{i\varphi})$ is an analytic function of φ on the set where it is defined. The validity of (a) and (b) is obvious. To show that (c) also holds, it suffices to note that for every point $x \in T^1$ different from ± 1, we can set $x_1(\varphi) = \cos \varphi = \frac{1}{2}(e^{i\varphi} + 1/e^{i\varphi})$ and $U = \{\varphi : |x - \varphi| < \min(|1 - x|/2, |1 + x|/2)\}$. For every point different from $\pm i$ we set $x_1(\varphi) = \sin \varphi = (1/2i)(e^{i\varphi} - 1/e^{i\varphi})$ and $U = \{\varphi : |x - \varphi| < \min|x - i|/2, |x + i|/2\}$. Thus T^1 is a (real) analytic manifold of dimension 1.

3. Let M be an analytic manifold and let U be an open subset of M. It is clear that the correspondence $V \to D(V)$, where V is an open subset of U, satisfies (a)–(c). Therefore V is an analytic manifold. It is called an *open submanifold* of the manifold M.

4. Let M be a complex analytic manifold. For every coordinate neighborhood $U \subset M$ we replace the set of functions $x_1, \ldots, x_m \in D(U)$ by the set of $2m$ functions $(y_1, \ldots, y_{2m}) = (\operatorname{Re} x_1, \operatorname{Im} x_1, \ldots, \operatorname{Re} x_m, \operatorname{Im} x_m)$. We replace each algebra $D(U)$ by the algebra $D_r(U)$, defined as follows. A real-valued function f on U belongs to $D_r(U)$ if and only if for every coordinate neighborhood $V \subset U$, the restriction of f to V is a real analytic function of the variables $y_1, \ldots, y_{2m}$. One easily shows that M can be considered as a *real* analytic manifold.

1.3. Mappings of Manifolds and Products of Manifolds

Let M and N be manifolds. Let φ be a mapping of M into N. The mapping φ is called *analytic*, if for every open subset W of N intersecting $\varphi(M)$ and every function $f \in D(W)$, the function $f \circ \varphi$ belongs to $D(\varphi^{-1}(W))$.

If the mapping φ of M onto N is a homeomorphism and if φ and φ^{-1} are analytic on M and N respectively, the mapping φ is called an *analytic isomorphism* of the manifolds M and N.

Let M_1 and M_2 be manifolds of dimensions m_1 and m_2 respectively. Let $M = M_1 \times M_2$ be the product of the topological spaces M_1 and M_2. In this case M is a separated space with a countable basis. Let U be an open subset of M and let f be a complex-valued function on U. We will say that $f \in D(U)$ if and only if for every point $(y_1, y_2) \in U$, there are coordinate neighborhoods V_1, V_2 of the points y_1 and y_2 in M_1 and M_2 respectively and analytic coordinates $x_1^{(1)}, \ldots, x_{m_1}^{(1)}$ and $x_1^{(2)}, \ldots, x_{m_2}^{(2)}$ on V_1 and V_2, respectively, which satisfy the following conditions. (1) The product $V_1 \times V_2$ is contained in U. (2) If $\tilde{V}_i$ is the image of V_i for the mapping $y \to (x_1^{(i)}(y), \ldots, x_{m_i}^{(i)}(y))$, $i = 1, 2$, there is an analytic function g on $\tilde{V}_1 \times \tilde{V}_2$ such that $f(z_1, z_2) = (x_1^{(1)}(z_1), \ldots, x_{m_1}^{(1)}(z_1), x_1^{(2)}(z_2), \ldots, x_{m_2}^{(2)}(z_2))$ for all $(z_1, z_2) \in V_1 \times V_2$. It is easy to verify that the mapping $U \to D(U)$ satisfies (a)–(c). Thus $M_1 \times M_2$ is an analytic manifold. It is called the *product of the manifolds M_1 and M_2*. Clearly the mappings $\varphi_i : M \to M_i$, $i = 1, 2$, defined by $\varphi_i(x_1, x_2) = x_i$ are analytic mappings. The mapping φ_i is called the *projection of the manifold M onto M_i, $i = 1, 2$*.

1.4. Tangent Vectors and Tangent Spaces

Let M be a real (or complex) analytic manifold of dimension m. Let x be a point in M. The union of the algebras $D(U)$ for all open sets U containing x is denoted by $A(x)$. Suppose that $f, g \in A(x)$, where $f \in D(U_1)$, and $g \in D(U_2)$. The function $\lambda f + \mu g$ (λ and μ constants) and the function fg belong to $D(U_1 \cap U_2)$. Thus sums, scalar multiples, and products are defined in the set of functions $A(x)$.

Consider any mapping v of the set $A(x)$ into the field of real (or complex) numbers for which the following conditions hold:

(1)
$$v(\lambda f + \mu g) = \lambda v(f) + \mu v(g) \tag{1.4.1}$$

for all $f, g \in A(x)$ and all real (or complex) numbers λ and μ;

(2)
$$v(fg) = v(f)g(x) + f(x)v(g) \tag{1.4.2}$$

for all $f, g \in A(x)$. We call the mapping v a *tangent vector to M at the point x*.

If v is a tangent vector and f is a function in $A(x)$, the number $v(f)$ is called the *derivative of the function f in the direction v*.

Let v, v' be tangent vectors to the manifold M in the point x. Clearly the mapping $\lambda v + \mu v'$, defined by $(\lambda v + \mu v')(f) = \lambda v(f) + \mu v'(f)$, satisfies (1) and (2) for all real (or complex) numbers λ and μ. In this way the tangent vectors at x form a linear space. This space is called the *tangent space to the manifold M at the point x* and is denoted by $T_x(M)$.

Let U be a coordinate neighborhood of the point $x \in M$. Let $x_1, \ldots, x_m$ be a system of analytic coordinates in U. Let f belong to $A(x)$, so that $f \in D(V)$, where V is some open set in M containing x. Since $U \cap V$ is open and contains x, we may suppose that $V \subset U$. By condition (c), there is an analytic function F on the set $\xi(V)$ such that $f = F \circ \xi$. It is clear that the formula

$$v(f) = \sum_{i=1}^{m} c_i \frac{\partial F}{\partial x_i}\bigg|_{x_k = x_k(x)} \tag{1.4.3}$$

defines a tangent vector to the manifold M at x for all choices of real (or complex) coefficients c_i. We will show that (1.4.3) is the general form of a tangent vector to the manifold M at x.

I. *Let $x_1, \ldots, x_m$ be a system of analytic coordinates in a certain coordinate neighborhood U of the point $x \in M$. Let v be a tangent vector to the manifold M in the point x. For any function $f \in A(x)$ we have the following equality:*

$$v(f) = \sum_{i=1}^{m} v(x_i) \frac{\partial F}{\partial x_i}\bigg|_{x_k = x_k(x)}. \tag{1.4.4}$$

In particular, the tangent vector v is uniquely defined by its values on the functions in a system of analytic coordinates.

Proof. In (1.4.3) suppose that $f = g \equiv 1$ in some neighborhood V of the point x. We find that $v(f) = 2v(f)$, so that $v(f) = 0$. Therefore we have $v(c) = 0$ for any constant function c. Let f be any function in $A(x)$ and let F be the corresponding analytic function of $x_1, \ldots, x_m$ such that $f = F \circ \xi$. Using Taylor's formula for the function F in a neighborhood of the point $(y_1^0, \ldots, y_m^0) = (x_1(x), \ldots, x_m(x))$, we obtain

$$F(y_1, \ldots, y_m) = a_0 + a_1(y_1 - y_1^0) + \cdots + a_m(y_m - y_m^0)$$
$$+ \sum_{i,j=1}^{m} (y_i - y_i^0)(y_j - y_j^0) G_{ij}, \tag{1.4.5}$$

where the G_{ij} are analytic functions of the variables $y_1, \ldots, y_m$ in a neighborhood of the point $(y_1^0, \ldots, y_m^0)$. From (1.4.5) we infer that $f = F \circ \xi$ can be represented as

$$f = a_0 + a_1(x - y_1^0) + \cdots + a_m(x_m - y_m^0)$$
$$+ \sum_{i,j=1}^{m} (x_i - y_i)(x_j - y_j^0) g_{ij}, \tag{1.4.6}$$

where the g_{ij} are functions in $A(x)$. Applying the tangent vector to both sides of (1.4.6) and using (1.4.1), (1.4.2), and the relation $v(c) = 0$, we obtain

$$v(f) = a_1 v(x_1 - y_1^0) + \cdots + a_m v(x_m - y_m^0) + v\left(\sum_{i,j=1}^{m} (x_i - y_i^0)(x_j - y_j^0)g_{ij}\right)$$

$$= a_1 v(x_1) + \cdots + a_m v(x_m) + \sum_{i,j=1}^{m} \{[(x_i(x) - y_i^0)v(x_j - y_j^0)$$

$$+ (x_j(x) - y_j^0)v(x_i - y_i^0)]g_{ij}(x) + v(g_{ij})(x_i(x) - y_i^0)(x_j(x) - y_j^0)\}$$

$$= a_1 v(x_1) + \cdots + a_m v(x_m), \tag{1.4.7}$$

since $x_i(x) = y_i^0$ for all $i = 1, \ldots, m$. Since $a_i = \partial F/\partial x_i(y_1^0, \ldots, y_m^0)$, (1.4.7) implies (1.4.4). $\square$

II. *The tangent space $T_x(M)$ has a basis of m vectors v_i, defined by $v_i(f) = (\partial F/\partial x_i)|_{x_k = x_k(x)}$.*

The proof follows directly from (1.4.4). It suffices to show that the vectors v_i are linearly independent, which is clear from the identities $v_i(x_j) = \delta_{ij}$.

1.5. The Differential of an Analytic Mapping

Let M and N be analytic manifolds and let φ be an analytic mapping of M into N. Let v be a tangent vector to the manifold M at the point x. We write $y = \varphi(x)$. For every function $g \in A(y)$ we set

$$w(g) = v(g \circ \varphi). \tag{1.5.1}$$

This formula defines a tangent vector w to the manifold N at the point y. We write $w = \varphi_*(v)$. The mapping φ_* of the tangent space $T_x(M)$ into the tangent space $T_y(N)$, defined by (1.5.1), is clearly linear. It is called the *differential of the mapping φ at the point x* and is denoted by $d\varphi$ or $d\varphi_x$.

I. *Let M and N be manifolds, let φ be an analytic mapping of M into N, and let x be a point in M. Suppose that the mapping $d\varphi_x$ satisfies the following condition: if $v \in T_x(M)$ and $d\varphi_x(v) = 0$, then $v = 0$. Then for every coordinate neighborhood W of the point $\varphi(x)$ in N and any system of analytic coordinates $y_1, \ldots, y_n$ in W we can choose m functions from the functions $y_1 \circ \varphi, \ldots, y_n \circ \varphi$, that form a system of coordinates in some neighborhood $U \subset \varphi^{-1}(W)$ of x on M. Conversely, if U is a coordinate neighborhood of the point x in M and $x_1, \ldots, x_m$ is a system of analytic coordinates in U, there exists a coordinate neighborhood W of the point $\varphi(x)$ and a system of coordinates $z_1, \ldots, z_n$ in W, for which the identities $x_j = z_j \circ \varphi$ hold for all $j = 1, \ldots, m$.*

Proof. Let $V \subset \varphi^{-1}(W)$ be a coordinate neighborhood of the point x in M and let $x_1, \ldots, x_m$ be a system of analytic coordinates in V. The functions $y_i \circ \varphi$ can be represented in V in the form $y_i \circ \varphi = F_i \circ \xi$, where the functions F_i are analytic on $\xi(V)$. We will show that the matrix $(\partial F_i / \partial x_j)_{x_k = x_k(x)}$ has rank m; that is, its rank is equal to the number of variables $x_1, \ldots, x_m$. It suffices to show that if

$$\sum_{j=1}^{m} \lambda_j \left(\frac{\partial F_i}{\partial x_j} \right)_{x_k = x_k(x)} = 0, \tag{1.5.2}$$

then all of the numbers $\lambda_1, \ldots, \lambda_m$ are zero. We define a tangent vector v to the manifold M at the point x, using (1.4.3) with $c_i = \lambda_i$, $i = 1, \ldots, m$. The equality (1.5.2) means that $v(y_i \circ \varphi) = \sum_{j=1}^{m} \lambda_j (\partial F_i / \partial x_j)_{x_k = x_k(x)} = 0$ for all $i = 1, \ldots, n$. That is, we have $(d\varphi_x(v))(y_i) = 0$ for all $i = 1, \ldots, n$. Formula (1.4.4) shows that $d\varphi_x(v) = 0$. Proposition I shows that $v = 0$, that is, $\lambda_1 = \cdots = \lambda_m = 0$. Hence the matrix $(\partial F_i / \partial x_j)_{x_k = x_k(x)}$ has rank m. Hence there are m distinct indices $i_1, \ldots, i_m$ among the numbers $1, \ldots, n$, for which the determinant of the matrix $(\partial F_{i_p} / \partial x_j)_{x_k = x_k(x)}$, $p, j = 1, \ldots, m$ is nonzero. By Proposition I in 1.1, the functions $y_{i_1} \circ \varphi, \ldots, y_{i_m} \circ \varphi$ are a system of analytic coordinates in some open subset U of V.

The functions x_j, $j = 1, \ldots, m$, can in turn be represented on the set U in the form $x_j = G_j(y_{i_1} \circ \varphi, \ldots, y_{i_n} \circ \varphi)$, where the G_j are analytic functions on their domain of definition and the determinant $(\partial G_j / \partial y_{i_p})_{y_{i_k} = y_{i_k}(\varphi(x))}$ is nonzero. We set $z_j = G_j(y_{i_1}, \ldots, y_{i_m})$, $j = 1, \ldots, m$. For $z_{m+1}, \ldots, z_n$ we take any functions y_i whose indices are not in the set $i_1, \ldots, i_m$. It is easy to show that $z_1, \ldots, z_n$ are a system of coordinates in some neighborhood W of the point $\varphi(x)$. We also have $z_j \circ \varphi = x_j$, for $j = 1, \ldots, m$. $\square$

II. *We retain the hypotheses of* I. *There is a neighborhood* U *of the point* x *in* M *such that the mapping* φ *of* U *onto* $\varphi(U)$ *is a homeomorphism.*

Proof. This follows from Proposition I. If I holds for W and $z_1, \ldots, z_n$ and U is contained in $\varphi^{-1}(W)$, the mapping φ admits an analytic inverse defined on $\varphi(U)$. This inverse is defined by the formula $\psi(z_1, \ldots, z_n) = (z_1, \ldots, z_m)$, because by I, $\psi \circ \varphi$ is the identity mapping of U onto itself. $\square$

A mapping φ of a manifold M into a manifold N is called *regular at a point* $x \in M$ if φ is analytic and $d\varphi_x$ is a one-to-one mapping of $T_x(M)$ into $T_{\varphi(x)}(N)$. (That is, if $d\varphi_x(v) = 0$ for a given $v \in T_x(M)$, then $v = 0$.)

III. *Let* M *and* N *be manifolds. Let* φ *be an analytic mapping of* M *into* N *and let* x *be a point in* M. *Suppose that* $d\varphi_x(T_x(M)) = T_{\varphi(x)}(N)$. *Suppose that* $y_1, \ldots, y_n$ *are a system of analytic coordinates in the neighborhood* W *of the point* $\varphi(x)$ *in* N. *Then there is a coordinate neighborhood* U *of* x *and a system of analytic coordinates* $z_1, \ldots, z_m$ *in* U *such that* $z_j = y_j \circ \varphi$ *for all* $j = 1, \ldots, n$.

Proof. Let $x_1, \ldots, x_m$ be a system of analytic coordinates in a neighborhood V of x such that $V \subset \varphi^{-1}(W)$. We then have $y_i \circ \varphi = F_i \circ \xi$ for all $i = 1, \ldots, n$. We will show that the matrix $(\partial F_i / \partial x_j)_{x_k = x_k(x)}$ has rank n. Suppose that

$$\sum_{i=1}^{n} \lambda_i \left(\frac{\partial F_i}{\partial x_j} \right)_{x_k = x_k(x)} = 0, \qquad j = 1, \ldots, m, \tag{1.5.3}$$

and let w_i be the tangent vector to the manifold N at the point $\varphi(x)$ defined by $w_i(y_k) = \delta_{ik}$, $i, k = 1, \ldots, n$. By hypothesis there is a tangent vector $v \in T_x(M)$ such that $d\varphi_x(v_i) = w_i$. The chain rule for differentiating composite functions shows that

$$\sum_{j=1}^{m} \left(\frac{\partial F_p}{\partial x_j} \right)_{x_k = x_k(x)} v_i(x_j) = v_i(y_p \circ \varphi). \tag{1.5.4}$$

We infer that

$$\sum_{j=1}^{m} \left(\frac{\partial F_p}{\partial x_j} \right)_{x_k = x_k(x)} v_i(x_j) = v_i(y_p \circ \varphi) = (d\varphi_x(v_i))(y_p) = w_i(y_p) = \delta_{ip}. \tag{1.5.5}$$

Multiplying the p-th equality (1.5.5) by λ_p and summing from 1 to n, we apply (1.5.3) to find that $\lambda_i = 0$. That is, the matrix $(\partial F_i / \partial x_j)_{x_k = x_k(x)}$ has rank n. We may suppose that the determinant of the matrix $(\partial F_i / \partial x_j)_{x_k = x_k(x)}$ $i, j = 1, \ldots, n$, is distinct from zero. From Proposition I in 1.1 we infer that the functions $y_1 \circ \varphi, \ldots, y_n \circ \varphi, x_{n+1}, \ldots, x_m$ are a system of analytic coordinates in some neighborhood $U \subset V$ of the point x in M. $\square$

IV. *If the hypotheses of* **III** *hold, there is a neighborhood U of the point x in M such that $\varphi(U)$ is a neighborhood of $\varphi(x)$ in N.*

The proof follows directly from III.

V. *Suppose that the hypotheses of* **I** *hold and that $d\varphi_x(T_x(M)) = T_{\varphi(x)}(N)$, so that $d\varphi_x$ is a linear isomorphism of the space $T_x(M)$ onto the space $T_{\varphi(x)}(N)$. Then there is a neighborhood U of the point x in M for which φ maps U homeomorphically onto a neighborhood W of $\varphi(x)$ in N. The mapping φ^{-1} of the open set W onto U is analytic.*

This proposition follows from Propositions I–IV.

Let M_1 and M_2 be manifolds of dimensions m_1 and m_2 respectively and $M = M_1 \times M_2$ the product of these manifolds. Choose points $x_1 \in M_1$ and $x_2 \in M_2$, $x = (x_1, x_2) \in M$. Let φ_i be the projection of M onto M_i, $i = 1, 2$.

If v belongs to $T_x(M)$, we can define the tangent vectors $v_i = d\varphi_i(v) \in T_{x_i}(M)$, $i = 1, 2$. Let $x_1^{(i)}, \ldots, x_{m_i}^{(i)}$ be a system of analytic coordinates in the neighborhood U_i of the point x_i in the manifold M_i, $i = 1, 2$. The functions $x_1^{(1)} \circ \varphi_1, \ldots, x_{m_1}^{(1)} \circ \varphi_1, x_1^{(2)} \circ \varphi_2, \ldots, x_{m_2}^{(2)} \circ \varphi_2$ then form a system of analytic coordinates in the neighborhood $U = U_1 \times U_2$ of x in M. If $v_i \in T_{x_i}(M_i)$, $i = 1, 2$, are arbitrary tangent vectors, then the formulas $v(x_k^{(1)} \circ \varphi_1) = v_1(x_k^{(1)})$, $k = 1, \ldots, m_1$, $v(x_k^{(2)} \circ \varphi_2) = v_2(x_k^{(2)})$, $k = 1, \ldots, m_2$, define a vector $v \in T_x(M)$; note that $d\varphi_i(v) = v_i$, $i = 1, 2$. The equalities $d\varphi_i(v) = v_i$, $i = 1, 2$, define the vector $v \in T_x(M)$ uniquely. Thus, *the space $T_x(M)$ can be identified with the direct sum of the spaces $T_{x_1}(M_1)$ and $T_{x_2}(M_2)$.*

Let M be a real (or complex) analytic manifold and let f be a real (or complex) analytic function on M. We can regard f as a mapping of the manifold M into the manifold $\mathbf{R}$ (or $\mathbf{C}$) (see example 1 in 1.2). The differential of the function f at a point $x \in M$ is a linear mapping of the tangent space $T_x(M)$ into the tangent space $L = T_{f(x)}(\mathbf{R})$ (or $L = T_{f(x)}(\mathbf{C})$). The tangent space L is one-dimensional. As a single basis vector in L we may take the vector w_0 defined by $w_0(x) = 1$, where x is the identity mapping of the space $\mathbf{R}$ (or $\mathbf{C}$) onto itself. We identify the space L with the space $\mathbf{R}$ (or $\mathbf{C}$) by identifying the vector λw_0 with the number λ. Thus the mapping df can be considered a real (or complex) linear function on $T_x(M)$. By definition we have $df(v) = v(f)$ for all $v \in T_x(M)$.

1.6. Vector Fields

Let M be an analytic manifold and let U be an open subset of M. A mapping X that carries each point in U into a tangent vector to M at x is called a *vector field on U*.

For $U \subset M$ and $f \in D(U)$ and a vector field X on U we set

$$g(x) = X(x)f \qquad (1.6.1)$$

for all $x \in U$. Thus the function g is defined on U. The function g is denoted by Xf. A vector field X is called *analytic* if for all $V \subset U$ and all $f \in D(V)$, we have $Xf \in D(V)$.

Let U be a coordinate neighborhood in M, with a system $x_1, \ldots, x_m$ of analytic coordinates. For $V \subset U$ and $f \in D(V)$ we have $f = F \circ \xi$, where F is an analytic function on $\xi(V)$. We set

$$X_j(x)(f) = \left(\frac{\partial F}{\partial x_j}\right)_{x_k = x_k(x)}, \qquad \text{for } j = 1, \ldots, m \quad \text{and } x \in U. \quad (1.6.2)$$

Formula (1.6.2) defines a tangent vector to the manifold M at x for all $x \in U$. The mapping $x \to X_j(x)$ is an analytic vector field on U.

Let X be a vector field on U. From Proposition I in 1.4 we infer that X can be represented in the form

$$X(x) = \sum_{j=1}^{m} a_j(x) X_j(x), \quad \text{for } x \in U, \tag{1.6.3}$$

where $X_j(x)$ is defined by (1.6.2) and $a_1, \ldots, a_m$ are certain functions on U. If the vector field X is analytic, then the functions $a_j(x)$ are also analytic, since we have

$$a_j(x) = X x_j, \quad \text{for } j = 1, \ldots, m, \tag{1.6.4}$$

where $x_j \in D(U)$. Therefore $a_j(x)$ belongs to $D(U)$. Conversely, if $a_j(x)$ belongs to $D(U)$ for $j = 1, \ldots, m$, then (1.6.3) defines an analytic vector field on U. From (1.6.2) we infer that (1.6.3) is equivalent to

$$(Xf)(x) = \sum_{j=1}^{m} a_j(x) \left(\frac{\partial F}{\partial x_j} \right)_{x_k = x_k(x)}, \quad \text{for } f \in D(U). \tag{1.6.5}$$

Accordingly, we will sometimes write (1.6.3) as the formal equality

$$X = \sum_{j=1}^{m} a_j \frac{\partial}{\partial x_j}. \tag{1.6.6}$$

Let X and Y be analytic vector fields on the manifold M. We set

$$Z = X \circ Y - Y \circ X. \tag{1.6.7}$$

The mapping Z is an analytic vector field on M. In fact, for $x \in M$, a coordinate neighborhood U of the point x, a system $x_1, \ldots, x_m$ of analytic coordinates on U, and a function $f \in D(V)$, where $x \in V \subset U$, we have

$$
\begin{aligned}
(Xf)(y) &= \sum_{j=1}^{m} a_j(x_1(y), \ldots, x_m(y)) \left(\frac{\partial F}{\partial x_j} \right)_{x_k = x_k(y)}, \\
(Yf)(y) &= \sum_{j=1}^{m} b_j(x_1(y), \ldots, x_m(y)) \left(\frac{\partial F}{\partial x_j} \right)_{x_k = x_k(y)}
\end{aligned}
\tag{1.6.8}
$$

for all $y \in V$. Here the a_j, b_j are analytic functions on $\xi(V)$. A direct calculation shows that

$$(Zf)(y) = \sum_{i,j=1}^{m} \left(a_i \frac{\partial b_j}{\partial x_i} - b_i \frac{\partial a_j}{\partial x_i} \right) \left(\frac{\partial F}{\partial x_j} \right)_{x_k = x_k(y)}, \quad \text{for } y \in V, \tag{1.6.9}$$

which means that Z is an analytic vector field on the manifold M. This vector field is denoted by $[X, Y]$ and is often called the *commutator of the vector fields X and Y*.

The vector field X assumes values in the linear space $T_x(M)$ for all $x \in M$. Hence we can define addition and scalar multiplication of vector fields by the formulas

$$(X + Y)(x) = X(x) + Y(x), \qquad (\alpha X)(x) = \alpha X(x). \tag{1.6.10}$$

It is easy to show that (1.6.10) make the set of vector fields on M a linear space in which the set of analytic vector fields is a linear subspace. For analytic vector fields X, Y, and Z on M and (real or complex) numbers λ, μ, it is also simple to verify the identities

$$[\lambda X + \mu Y, Z] = \lambda[X, Z] + \mu[Y, Z],$$
$$[X, \lambda Y + \mu Z] = \lambda[X, Y] + \mu[X, Z], \tag{1.6.11}$$
$$[X, X] = 0, \tag{1.6.12}$$
$$[[X, Y], Z] + [[Y, Z], X] + [[Z, X], Y] = 0. \tag{1.6.13}$$

The equality (1.6.13) is called *Jacobi's identity*. From (1.6.11) and (1.6.12) we infer that

$$0 = [X + Y, X + Y] = [X, Y] + [Y, X],$$

and therefore

$$[X, Y] = -[Y, X] \tag{1.6.14}$$

for all analytic vector fields X, Y on M.

Let φ be an analytic mapping of a manifold M into a manifold N. Let X be a vector field on M. A vector field Y on N is called the *image of the vector field X* if $d\varphi_x(X(x)) = Y(\varphi(x))$ for all $x \in M$. This vector field Y is denoted by $d\varphi(X)$. If the mapping φ is regular at every point $x \in M$ (*regular everywhere*), then M can admit only one vector field X, the image of which is a given vector field Y on N. That is, for a regular mapping, the vector field $X(x)$ is uniquely defined by its image under the mapping $d\varphi_x$. We will prove that there is a vector field X such that $Y(\varphi(x)) = d\varphi_x(X(x))$ for all $x \in M$.

I. *Let φ be an everywhere regular mapping of the manifold M into the manifold N and let x be a point in M. Let Y be an analytic vector field on N such that $Y(\varphi(x)) \in d\varphi_x(T_x(M))$ for all $x \in N$. Then there is a unique analytic vector field X on M whose image is the vector field Y.*

Proof. Our hypothesis implies that for every $x \in M$ there exists a unique element $X(x) \in T_x(M)$ such that $d\varphi_x(X(x)) = Y(\varphi(x))$. We must prove that the mapping $X: x \to X(x)$ is an analytic vector field. From Proposition I in 1.5 we infer that in some neighborhood W of the point $\varphi(x)$ in N there is a system of analytic coordinates $y_1, \ldots, y_n$, such that $y_1 \circ \varphi, \ldots, y_m \circ \varphi$ is a system of analytic coordinates in a neighborhood $U \subset \varphi^{-1}(W)$ of the point x in M. For $y \in U$, the relation $(d\varphi_y)(X(y)) = Y(\varphi(y))$ means that $X(y)(y_i \circ \varphi) = d\varphi_y(X(y))(y_i) = Y(\varphi(y))(y_i) = (Yy_i)(\varphi(y))$. That is, the function $X(y_i \circ \varphi)$ coincides on the set U with the function $Yy_i \circ \varphi$. Since Y is an analytic vector field on N, Yy_i is an analytic function on W. Therefore $Yy_i \circ \varphi$ is analytic on U and $X(y_i \circ \varphi)$ is also an analytic function on U. This proves that X is an analytic vector field. $\quad\square$

If φ maps M strictly *into* N, the image of the vector field on M is in general not uniquely defined. Nevertheless, the following is true.

II. *Let φ be an analytic mapping of a manifold M into a manifold N. Let X_1 and X_2 be analytic vector fields on M and let Y_1 and Y_2 be the images of X_1 and X_2 respectively. Then the vector field $[Y_1, Y_2]$ is the image of the vector field $[X_1, X_2]$.*

Proof. Suppose that $x_0 \in M$ and $y_0 = \varphi(x_0)$. Let V be a neighborhood of the point y_0 in N and $f \in D(V)$. Let U be a neighborhood of the point x_0 in M for which $\varphi(U) \subset V$. The relation $Y_i = d\varphi(X_i)$ means that $Y_i(\varphi(x))f = d\varphi_x(X_i(x))f$. That is, we have

$$(Y_i f)(\varphi(x)) = (X_i(f \circ \varphi))(x) \tag{1.6.15}$$

for all $x \in U$, $i = 1, 2$. It follows that

$$(Y_i f \circ \varphi)(x) = (X_i(f \circ \varphi))(x) \tag{1.6.16}$$

for all $x \in U$, $i = 1, 2$. From (1.6.15) and (1.6.16) we see that

$$(Y_1 Y_2 f)(\varphi(x)) = (X_1(Y_2 f \circ \varphi))(x) = (X_1 X_2(f \circ \varphi))(x)$$

for all $x \in U$. A similar formula holds for $(Y_2 Y_1 f)(\varphi(x))$. Computing, we obtain

$$([Y_1, Y_2]f)(\varphi(x)) = ([X_1, X_2](f \circ \varphi))(x)$$

for all $x \in U$, which is to say that $([Y_1, Y_2])(\varphi(x)) = d\varphi_x([X_1, X_2](x))$. $\quad\square$

1.7. Submanifolds

A manifold N is called a *submanifold* of the manifold M if N is a subset of M and the identity mapping of N into M is regular at every point of N. The identity mapping of N into M is evidently continuous. Note that an open submanifold N of M is a submanifold in the present sense.

Let N be a submanifold of M. Let φ be the identity mapping of N into M. Choose a point $x \in N$. Let U be an open subset of M with $x \in U$. Let f be an element of the algebra $D(U)$. Then the function $f \circ \varphi$ is analytic on the open subset $U \cap N$ of the manifold N. By Proposition I in 1.5 we can choose our open set $U \subset M$ and a system of analytic coordinates $x_1, \dots, x_m$ in the neighborhood U such that the functions $x_1 \circ \varphi, \dots, x_n \circ \varphi$ (where n is the dimension of N) form a system of coordinates in the open set $U \cap N$ of the manifold N. Let g be a function in $D(U \cap N)$. Then g can be represented in the form of an analytic function $G(x_1 \circ \varphi, \dots, x_n \circ \varphi)$. We set $F(y) = G(x_1(y), \dots, x_n(y))$ for $y \in M$. This is an analytic function in $U \subset M$, and we have $F \circ \varphi = g$ on $U \cap N$. Thus any function g that is analytic in a neighborhood of $x \in N$ is represented in some neighborhood $U \subset M$ of x in the form $g = f \circ \varphi$, where $f \in D(U)$ on M.

The differential $d\varphi_x$ of the mapping φ at the point $x \in N$ is an isomorphic mapping of the tangent space $T_x(N)$ onto a linear subspace $\widetilde{T_x(N)}$ of the space $T_x(M)$. Sometimes the space $T_x(N)$ is called the *tangent space to the manifold N at the point x.*

Let X be an analytic vector field on M such that $X(x) \in \widetilde{T_x(N)}$ for all $x \in N$. Because the mapping φ is everywhere regular, Proposition I in 1.6 shows that there is a unique analytic vector field Y on N such that $X(x) = d\varphi_x(Y(x))$ for all $x \in N$. We say that the vector field Y is *induced on N by the vector field X.* One easily verifies that if X_1, X_2 are analytic vector fields on M and Y_1, Y_2 are the vector fields induced by X_1, X_2 on N respectively, the vector field $[Y_1, Y_2]$ is induced by the vector field $[X_1, X_2]$.

I. *Consider a manifold M with a submanifold N. Let x belong to N. There are a neighborhood U of the point x in N, a neighborhood V of x in M containing U, and a set of functions $f_1, \dots, f_k \in D(V)$ for which the following condition holds. The point $z \in V$ belongs to U if and only if $f_1(z) = \dots = f_k(z) = 0$.*

Proof. Suppose that $\dim M = m$ and $\dim N = n$. Proposition I in 1.5 shows that there is a neighborhood V' of x in M and a system of analytic coordinates $x_1, \dots, x_m$ in V' such that the restrictions of the functions $x_1, \dots, x_n$ to $N \cap V'$ form a system of analytic coordinates in some neighborhood U of x in N. We set $\eta(z) = (x_1(z), \dots, x_n(z))$ for all $z \in V'$. Let V be the subset of V' consisting of all points z such that $\eta(z) \in \eta(U)$. Plainly, V is a neighborhood of x in M and the restrictions of $x_1, \dots, x_m$ to V form a system of analytic coordinates in V. The restrictions of the functions $x_{n+1}, \dots, x_m$ to the set U are analytic functions on U. Hence there are analytic functions $F_1, \dots, F_{n-m}$ on $\eta(U)$ such that $x_{n+j}(z) = F_j(x_1(z), \dots, x_n(z))$ for all $z \in U$ and $j = 1, \dots, m - n$. We define $f_j = x_{n+j} - F_j(x_1, \dots, x_n)$ for all $j = 1, \dots, m - n$. We infer that $f_j(z) = 0$ for all $z \in U$ and $j = 1, \dots, m - n$. Conversely, consider a point $z \in V$ such that $f_j(z) = 0$ for all $j = 1, \dots, m - n$. Let w be the point in U defined by $\eta(w) = \eta(z)$. (We know that such a point

w exists because $\eta(U) = \eta(V)$) We then have

$$x_{n+j}(w) = F_j(x_1(w), \ldots, x_n(w)) = F_j(\eta(w)) = F_j(\eta(z)).$$

We also have $f_j(z) = 0$, which means that $x_{n+j}(z) - F_j(x_1(z), \ldots, x_n(z)) = 0$. Thus we have $F_j(\eta(z)) = x_{n+j}(z)$, and the coordinates of the points w and z in V coincide. This means that $w = z$ and since $w \in U$, z must belong to U. This completes our proof (and also shows that $k = m - n$). $\square$

§2. Lie Algebras

Let K be the field of real or complex numbers. A set L is called a *Lie algebra over the field K* if:

(a) L is a linear space over K;
(b) for every pair x, $y \in L$ there is an element of L, denoted by $[x, y]$, such that the following conditions are satisfied:

(b$_1$) $[x, y]$ is linear in x and in y (this means that $[\alpha x, y] = \alpha[x, y]$, $[x, \alpha y] = \alpha[x, y]$ for $\alpha \in K$ and $[x_1 + x_2, y] = [x_1, y] + [x_2, y]$, $[x, y_1 + y_2] = [x, y_1] + [x, y_2]$ for all x, x_1, x_2, y, y_1, $y_2 \in L$);

(b$_2$) $$[x, x] = 0, \quad \text{for all } x \in L; \tag{2.1.1}$$

(b$_3$) $$[[x, y], z] + [[y, z], x] + [[z, x], y] = 0 \tag{2.1.2}$$

for all x, y, $z \in L$.

The identity (2.1.2) is called *Jacobi's identity*. The element $[x, y]$ is often called the *commutator of the elements x, $y \in L$.*

If K is the real (complex) numbers, the Lie algebra L is called a *real (complex)* Lie algebra. In the sequel, unless otherwise stipulated, all Lie algebras will be considered finite-dimensional (as linear spaces).

I.   *In any Lie algebra L we have*

$$[x, y] = -[y, x] \tag{2.1.3}$$

for all x, $y \in L$.

Proof. By conditions (b$_1$) and (b$_2$) we have

$$0 = [x + y, x + y] = [x, x] + [y, x] + [x, y] + [y, y] = [x, y] + [y, x]. \square$$

Let L be a Lie algebra over K and let $e_1, \ldots, e_n$ be a basis of the vector space L. Expanding the elements $[e_i, e_j]$ of L in the basis $e_1, \ldots, e_n$ we obtain

$$[e_i, e_j] = \sum_{1 \leqslant k \leqslant n} c_{ijk} e_k. \tag{2.1.4}$$

The numbers $c_{ijk} \in K$ are called *the structural constants of the Lie algebra in the basis* $e_1, \ldots, e_n$. It is easy to see that (2.1.2) and (2.1.3) are equivalent to:

$$c_{ijk} = -c_{jik} \tag{2.1.5}$$

for all $i, j, k = 1, \ldots, n$; and

$$\sum_{k=1}^n (c_{ijk} c_{klm} + c_{jlk} c_{kim} + c_{lik} c_{kjm}) = 0 \tag{2.1.6}$$

for all $i, j, l, m = 1, \ldots, n$.

The general definitions given in 2.1–2.3 in chapter II take the following form for Lie algebras.

Let L be a Lie algebra. A subset $M \subset L$ is called an *ideal in* L if M is a linear subspace of L and $[x, y] \in M$ for all $x \in M$ and $y \in L$. A subset L' of L is called a *Lie subalgebra of the Lie algebra* L if L' is a linear subspace of L and $[x, y] \in L'$ for all $x, y \in L'$. Plainly, an ideal is also a Lie subalgebra. Let L, L_1 be Lie algebras over the field K and let π be a linear mapping of L into L_1. The mapping π is called a *homomorphism* if we have

$$[\pi(x), \pi(y)] = \pi([x, y]) \tag{2.1.7}$$

for all $x, y \in L$. If Ker π is equal to zero, π is called *exact*. A one-to-one homomorphism of L onto L_1 is called an *isomorphism* of L and L_1, and L and L_1 are called *isomorphic* if they are connected by an isomorphism.

II. *Let* L, L_1 *be Lie algebras and let* $\pi: L \to L_1$ *be a homomorphism. Then* $\pi(L)$ *is a Lie subalgebra of the Lie algebra* L_1. *The kernel of the mapping* π *is an ideal in* L. *Let* L *be a Lie algebra with an ideal* M. *Let* $L' = L/M$ *be a factor space of* L *by the linear subspace* M *and* π *the canonical mapping of* L *onto* L'. *We set*

$$[x', y'] = \pi([x, y]) \tag{2.1.8}$$

for $x', y' \in L'$ *and* $x' = \pi(x)$, $y' = \pi(y)$, $x, y \in L$. *The element* $[x', y']$ *is independent of the choice of representatives* $x \in x'$, $y \in y'$, *and* (b) *holds for* $[x', y']$. *With this definition of* $[x', y']$, *the set* L' *is a Lie algebra over* K.

The proof is like that of the corresponding statements for associative algebras (2.2, chapter II).

The Lie algebra L' is called the *factor algebra of L by the ideal M*.

III. *Let $L_1, \ldots, L_m$ be Lie algebras over K. The linear space $L = L_1 + \cdots + L_m$, in which we define*

$$[(x_1, \ldots, x_m), (y_1, \ldots, y_m)] = ([x_1, y_1], \ldots, [x_m, y_m]), \qquad (2.1.9)$$

for $x_i, y_i \in L_i$, $i = 1, \ldots, m$, is a Lie algebra over K.

We omit the proof. The Lie algebra L is called the *direct sum of the Lie algebras L_i, $i = 1, \ldots, m$.*

Examples of Lie Algebras

1. Let L be a finite-dimensional vector space over K. We set $[x, y] = 0$ for all $x, y \in L$. It is clear that L is a Lie algebra over K. It is called a *commutative* or *abelian* Lie algebra.

2. Let V be a finite-dimensional vector space over K and let L be the linear space consisting of all linear mappings of V into or onto itself. For $x, y \in L$, we define

$$[x, y] = xy - yx. \qquad (2.1.10)$$

Then L is a Lie algebra over K. It is denoted by the symbol $gl(V)$. If $V = K^n$,[54] the Lie algebra $gl(V)$ is isomorphic to the Lie algebra of $n \times n$ matrices with entries in K, where $[\cdot, \cdot]$ is defined by (2.1.10). This Lie algebra is denoted by $gl(n, K)$. We list some Lie subalgebras of $gl(n, K)$ that will be important in the sequel: the subalgebra $sl(n, K)$ of all matrices with trace zero; the subalgebra $so(n, K)$ of all skew-symmetric matrices (that is, matrices A such that $A^t = -A$, where A^t is the transpose of A); for $n = 2m$, the subalgebra $sp(n, K)$, formed by the matrices A such that $A^t J + JA = 0$, where $J = \left\| \begin{matrix} 0 & I_m \\ -I_m & 0 \end{matrix} \right\|$ and I_m is the identity matrix of order m. We leave it to the reader to verify that these subsets are indeed Lie subalgebras of the Lie algebra $gl(n, K)$

3. Example (2) can be generalized. Let A be any associative algebra over the field K. We set $[a, b] = ab - ba$ for all $a, b \in A$. This operation provides A with the structure of a Lie algebra.

4. Let L be three-dimensional real Euclidean space. For $x, y \in L$ we define x, y as the vector product of the vectors x and y. The properties of a vector product show that L is a real Lie algebra.

5. A complex Lie algebra L can also be considered as a real Lie algebra, since a complex linear space is also a real linear space. If $e_1, \ldots, e_n$ is a

---

[54] K^n is the linear space over K consisting of all sequences of length n of elements of K with the usual componentwise linear operations.

basis in the complex linear space L, then $e_1, \ldots, e_n, ie_1, \ldots, ie_n$ is a basis in the real linear space L. Therefore the dimension of the real Lie algebra L is twice the dimension of the complex Lie algebra L. If (c_{ijk}) are the structural constants of the complex Lie algebra L for the basis $e_1, \ldots, e_n$, we can infer from (2.1.4) that the structural constants of the real Lie algebra for the basis $e_1, \ldots, e_n, ie_1, \ldots, ie_n$ are equal to $\pm \mathrm{Re}\, c_{ijk}, \pm \mathrm{Im}\, c_{ijk}$:

$$[e_k, e_l] = -[ie_k, ie_l] = \sum_{1 \leqslant m \leqslant n} \mathrm{Re}(c_{klm})e_m + \mathrm{Im}(c_{klm})ie_m;$$

$$[ie_k, e_l] = [e_k, ie_l] = \sum_{1 \leqslant m \leqslant n} \mathrm{Re}(c_{klm})ie_m - \mathrm{Im}(c_{klm})e_m. \qquad (2.1.11)$$

In particular, the Lie algebra $gl(n, \mathbf{C})$ is also a real Lie algebra.

Let L be a Lie algebra over K and let V be a finite-dimensional complex vector space. A homomorphism π of L into the algebra $gl(V)$ (as a Lie algebra over the field K) is called a *finite-dimensional representation of the Lie algebra L in the space V*. The dimension of V is called the dimension of the representation π.

Examples of Representations

1. Let π be the mapping of L into $gl(V)$ defined by the formula $\pi(x) = 0$ for all $x \in L$. This π is a representation of L in V, called the *zero representation of dimension n*.

2. Let L be a Lie algebra. For every $x \in L$, the linear transformation of the space L defined by

$$(\mathrm{ad}\ x)(y) = [x, y] \qquad (2.1.12)$$

for all $y \in L$, is denoted by ad x. We can write (2.1.2) in the form

$$[[x, y], z] = [x, [y, z]] - [y, [x, z]]. \qquad (2.1.13)$$

Substituting (2.1.12) in (2.1.13), we obtain

$$(\mathrm{ad}[x, y])z = \mathrm{ad}\ x\ \mathrm{ad}\ y(z) - \mathrm{ad}\ y\ \mathrm{ad}\ x(z) = [\mathrm{ad}\ x, \mathrm{ad}\ y](z) \quad (2.1.14)$$

for all $x, y, z \in L$. Therefore we have

$$\mathrm{ad}[x, y] = [\mathrm{ad}\ x, \mathrm{ad}\ y]. \qquad (2.1.15)$$

Thus (2.1.12) defines a homomorphism of the Lie algebra L into the Lie algebra $gl(L)$. The kernel of this homomorphism is the ideal of all elements $x \in L$ for which $[x, y] = 0$ for all $y \in L$. This ideal is called the *center* of L. If L is a complex Lie algebra, the homomorphism $x \rightarrow \mathrm{ad}\ x$ defines a representation of L in L. This representation is called the *adjoint representation of the Lie algebra L*.

For representations of Lie algebras one can define direct sums, tensor products, equivalent representations, subrepresentations, representations in a factor space, and irreducible representations. All of this goes by analogy with what we have done in chapters I and II for groups and associative algebras. We leave the details to the reader (but see also 1.2 in chapter X).

§3. Lie Groups

3.1. Definition of a Lie Group

A set G is called a *Lie group* if:

    (1) G is a topological group;
    (2) G is an analytic manifold;
    (3)   the mapping $(g, h) \to gh^{-1}$ of the product $G \times G$ onto G is an analytic mapping of manifolds.

If G is a real (complex) analytic manifold, G is called a *real (complex)* Lie group.

Examples

1. Consider the finite-dimensional vector space $\mathbf{R}^n$ (or $\mathbf{C}^n$) as a group under addition, provided with the usual topology, and as a manifold in accordance with example 1 in 1.2. It is a real (or complex) Lie group.

2. Consider the group $GL(n, \mathbf{C})$. Let $g = (x_{ij}(g))$ be a matrix in this group. Map g into the point $\varphi(g)$ in the space $\mathbf{C}^{n^2}$ with coordinates $x_{ij}(g)$ (arranged in some fixed order). The mapping φ thus defined is a homeomorphism of the space $GL(n, \mathbf{C})$ onto the subset M of those points of $\mathbf{C}^{n^2}$ for which $\det(x_{ij}) \neq 0$. The set M is an open subset in $\mathbf{C}^{n^2}$ and so is an open submanifold of $\mathbf{C}^{n^2}$. Thus $GL(n, \mathbf{C})$ is an n^2-dimensional complex analytic manifold; that is, the functions $x_{ij}(g)$ form a system of analytic coordinates on the manifold $GL(n, \mathbf{C})$.

Consider the functions

$$x_{ij}(gh^{-1}) = \sum_{k=1}^{n} x_{ik}(g)x_{kj}(h^{-1}) = \sum_{k=1}^{n} \frac{x_{ik}(g)A_{jk}(h)}{\det(x_{ij}(h))},$$

where $A_{jk}(h)$ is the algebraic complement of the element $x_{jk}(h)$. These functions are rational functions of the coordinate functions $x_{ij}(g)$, $x_{jk}(h)$. The denominator $\det(x_{ij}(h))$ vanishes nowhere on $GL(n, \mathbf{C})$. Hence the mapping $(g, h) \to gh^{-1}$ is an analytic mapping of $GL(n, \mathbf{C}) \times GL(n, \mathbf{C})$ onto $GL(n, \mathbf{C})$. Thus $GL(n, \mathbf{C})$ is a complex Lie group.

3. Consider the group T^1. Example 2 in 1.2 shows that T^1 is a one-dimensional real manifold. The functions $\cos \varphi = (1/2)(e^{i\varphi} + e^{-i\varphi})$ and $\sin \varphi = (1/2i)(e^{i\varphi} - e^{-i\varphi})$ are analytic functions on T^1. On the other hand, every point of the manifold T^1 has a neighborhood in which at least one of the functions $\cos \varphi$, $\sin \varphi$ forms a one-member system of analytic coordinates. The formulas

$$\cos(\varphi - \psi) = \cos \varphi \cos \psi + \sin \varphi \sin \psi,$$
$$\sin(\varphi - \psi) = \sin \varphi \cos \psi - \cos \varphi \sin \psi$$

show that T^1 is a real Lie group.

4. Let G be a complex Lie group. By considering the complex manifold G as a real analytic manifold (see example 4 in 1.2), we obtain a real Lie group. Thus, *any complex Lie group is also a real Lie group.*

5. Let G, H be Lie groups. The product $G \times H$ is a topological group and an analytic manifold. One can easily verify that the mapping φ of the manifold $(G \times H) \times (G \times H)$ onto $G \times H$, defined by the formula $\varphi((g,h),(g_1,h_1)) = (gg_1^{-1}, hh_1^{-1})$, is analytic. Therefore, $G \times H$ is a Lie group. We call it the *product of the Lie groups G and H.*

3.2. The Lie Algebra of a Lie Group

Let G be a Lie group. By definition, the mapping $g \to g^{-1}$ of the manifold G onto G is analytic and thus for any $h \in G$, the mapping $\varphi_h : g \to hg = h(g^{-1})^{-1}$ of G onto G is analytic. Let $d\varphi_h$ be the differential of the mapping φ_h (see 1.5). A vector field X on G is called *left invariant* if

$$(d\varphi_{gh^{-1}})_h X(h) = X(g) \quad \text{for all } g, h \in G. \tag{3.2.1}$$

I. *A vector field X on G is left invariant if and only if $(d\varphi_g)_e X(e) = X(g)$ for all $g \in G$.*

Proof. If X is left invariant, we infer from (3.2.1) for $h = e$, that $(d\varphi_g)_e X(e) = X(g)$ for all $g \in G$. Conversely, suppose that this relation holds. The mappings $\varphi_{h^{-1}}$ and φ_h are inverses of each other and so $d\varphi_{h^{-1}}$ and $d\varphi_h$ are also inverses. Therefore we have $X(e) = (d\varphi_{h^{-1}})_h X(h)$ and thus

$$X(g) = (d\varphi_g)_e (d\varphi_{h^{-1}})_h X(h)$$
$$= (d(\varphi_g \circ \varphi_{h^{-1}}))_h X(h) = (d\varphi_{gh^{-1}})_h X(h). \quad \square$$

II. *For any element $X(e) \in T_e(G)$ (see 1.4) there exists a unique left invariant vector field X on G that has the value $X(e)$ at the point e.*

This follows immediately from I.

III. *Every left invariant vector field on G is analytic.*

Proof. Let g_0 be an element of G. Let U be a coordinate neighborhood of g_0 and let $x_1, \ldots, x_m$ be a system of analytic coordinates in U. There exists an open subset V of U containing g_0 such that $gg_0^{-1}h \in U$ for all $g, h \in V$. Let g be any element of V. The definition of the differential of a mapping shows that

$$X(g)x_i = ((d\varphi_{gg_0^{-1}})_{g_0}X(g_0))x_i = X(g_0)(x_i \circ \varphi_{gg_0^{-1}}).$$

The functions $x_i(gg_0^{-1}h)$ are defined and analytic in g and h on $V \times V$. Thus we have

$$x_i(gg_0^{-1}h) = F_i(x_1(g), \ldots, x_m(g), x_1(h), \ldots, x_m(h)),$$

where the functions $F_i(y_1, \ldots, y_m, z_1, \ldots, z_m)$ are analytic functions of their coordinate variables in $\xi(V) \times \xi(V)$. Therefore we have

$$X(g)x_i = \sum_{j=1}^{m} (X(g_0)x_j) \left(\frac{\partial F_i}{\partial z_j} \right)_{y_k = x_k(g), z_k = x_k(g_0)} \tag{3.2.2}$$

In the right side of (3.2.2), the values of $X(g_0)x_j$ are constant. Since the functions $\partial F_i / \partial z$ are analytic in V, so are the functions $X(g)x_i$. Hence the mapping X is analytic. $\square$

IV. *If X and Y are left invariant vector fields, the fields $X + Y, \lambda X,$ and $[X, Y]$ are also left invariant.*

Proof. The assertion is obvious for $X + Y$ and λX. From Proposition II in 1.6, we infer that

$$(d\varphi_{gh^{-1}})_h([X, Y](h)) = [(d\varphi_{gh^{-1}})_h(X), (d\varphi_{gh^{-1}})_h(Y)](g) = [X, Y](g),$$

and so $[X, Y]$ is left invariant. $\square$

V. *The set of left invariant vector fields on a real (or complex) Lie group G is a real (or complex) Lie algebra under the linear operations of addition and scalar multiplication and commutation of vector fields. The dimension of this Lie algebra is equal to the dimension of the manifold G.*

This assertion follows directly from II–IV and from II in 1.4.

The definition of a Lie algebra using right invariant vector fields is similar. We omit the details.

Examples

1. Let $\mathbf{R}$ be the additive group of real numbers. Let x be the coordinate function on $\mathbf{R}$, defined by the identity mapping of $\mathbf{R}$ onto itself. Let

X be the vector field on $\mathbf{R}$ defined by $X(a) = 1$ for all $a \in \mathbf{R}$. The vector field X is left invariant. If φ_a is translation by $a \in \mathbf{R}$, we have $\varphi_a(b) = b + a$, $((d\varphi_a)_0 X(0))x = X(0)(x \circ \varphi_a) = X(0)(x + a) = 1 = X(a)x$. Thus X is a basis for the Lie algebra of $\mathbf{R}$. This algebra consists of all real multiples of the element X and is a one-dimensional (abelian) real Lie algebra isomorphic to $\mathbf{R}$. Similarly, the Lie algebra of the additive group $\mathbf{C}$, considered as a complex Lie group, is isomorphic to $\mathbf{C}$.

2. Since the group T^1 is one-dimensional, its Lie algebra is also one-dimensional and is therefore isomorphic to the abelian Lie algebra of $\mathbf{R}$.

3. Consider $G = GL(n, \mathbf{C})$ as a complex Lie group. For any analytic vector field X, the result of applying the vector field X to the analytic function x_{ij} on G is denoted by $X x_{ij}$. We set $a_{ij}(X) = X(e)x_{ij}$. The mapping $X \to a_{ij}(X)$ is a linear mapping of the Lie algebra L of G into the complex vector space $M_n(\mathbf{C})$ of complex square matrices of order n. If $a_{ij}(X) = 0$ for all i, j, then $X(e)$ is 0 (since the x_{ij} are a system of analytic coordinates on G). From Proposition I we infer that $X = 0$. Thus the mapping $X \to a_{ij}(X)$ is a linear isomorphism of the space L onto some subspace of $M_n(\mathbf{C})$. Note however that

$$\dim L = \dim G = n^2 = \dim M_n(\mathbf{C}).$$

Therefore the image of the Lie algebra L under the mapping $X \to a_{ij}(X)$ is the entire space $M_n(\mathbf{C})$.

Let X, Y be left invariant vector fields on G. Let us find the matrix $a_{ij}([X,Y])$. The formula

$$X(g)x_{ij} = d\varphi_g X(e)x_{ij} = X(e)(x_{ij} \circ \varphi_g), \qquad g \in G,$$

shows that

$$X(g)x_{ij} = \sum_{k=1}^{n} x_{ik}(g)(X(e)x_{kj}) = \sum_{k} x_{ik}(g)a_{kj}(X). \qquad (3.2.3)$$

Consider $X(g)x_{ij}$ as a function of g. The relation (3.2.3) implies that

$$Y(e)(X(g)x_{ij}) = \sum_{k=1}^{n} a_{ik}(Y)a_{kj}(X). \qquad (3.2.4)$$

Similarly we have

$$X(e)(Y(g)x_{ij}) = \sum_{k=1}^{n} a_{ik}(X)a_{kj}(Y). \qquad (3.2.5)$$

Let $\tilde{X}$, $\tilde{Y}$ be the matrices $(a_{ij}(X))$, $(a_{ij}(Y))$ respectively. From (3.2.4) and (3.2.5) we find that the matrix $(a_{ij}([X, Y]))$ is the matrix $\tilde{X}\tilde{Y} - \tilde{Y}\tilde{X}$. We have proved the following.

VI. *The Lie algebra of the group* $GL(n, \mathbf{C})$ *is isomorphic to the Lie algebra* $gl(n, \mathbf{C})$ *of all complex matrices of order* n, *in which the operation of commutation is defined by the formula* $[\tilde{X}, \tilde{Y}] = \tilde{X}\tilde{Y} - \tilde{Y}\tilde{X}$.

An analogous argument shows that the Lie algebra of the group $GL(n, \mathbf{R})$ is isomorphic to the Lie algebra $gl(n, \mathbf{R})$.

4. Let G, H be Lie groups with Lie algebras L, M. We know that the tangent space $T_{(g,h)}(G \times H)$ is isomorphic to the product $T_g(G) \times T_h(H)$ for all $g \in G$, $h \in H$. Let X, Y be left invariant vector fields on G and H respectively. We set $Z(g, h) = (X(g), Y(h)) \in T_{(g,h)}(G \times H)$. Thus Z is a vector field on $G \times H$. One shows easily that the vector field Z is left invariant. If X_1, Y_1 are also left invariant vector fields on G and H and $Z_1(g, h) = (X_1(g), Y_1(h))$, then we have $[Z, Z_1](g, h) = ([X, X_1](g), [Y, Y_1](h))$. That is, *the Lie algebra of* $G \times H$ *is the direct sum of the Lie algebras of the groups* G *and* H.

The Lie algebra of $\mathbf{C}^n$ is isomorphic to $\mathbf{C}^n$ and the Lie algebras of the groups $\mathbf{R}^n$ and T^n are isomorphic to $\mathbf{R}^n$ (see examples 1 and 2 in 3.2).

3.3. Subgroups, Homomorphisms, and Factor Groups of Lie Groups

Let G be a Lie group and let H be a subgroup in G (in general not closed). We call H a *Lie subgroup* of G if: (1) H is a Lie group; (2) H is a submanifold of the analytic manifold G. A connected Lie subgroup of G is called an *analytic subgroup* of G.

I. *Let G be a Lie group with Lie algebra L. Let H be a Lie subgroup in G and let M be the set of all elements $X \in L$ such that $X(e) \in \widetilde{T_e(H)}$ (the space $\widetilde{T_e(H)}$ is defined in 1.7). Then M is a Lie subalgebra of L and is isomorphic to the Lie algebra of the group H.*

Proof. Consider any $h \in H$. The left translation φ_h is an analytic isomorphism of H onto H. Hence we have $(d\varphi_h)(\widetilde{T_e(H)}) = \widetilde{T_h(H)}$, $(d\varphi_h)(T_e(H)) = T_h(H)$. If X belongs to M, then $X(h) \in \widetilde{T_h(H)}$ for all $h \in H$. Therefore the field X defines an analytic vector field in the submanifold H. This vector field on H is left invariant. If X, Y belong to M, the fields X and Y define analytic vector fields on H and thus the commutator $[X, Y]$ also defines a vector field on H. In particular, $[X, Y]$ belongs to M. Hence M is a Lie subalgebra of L. Map each element X of M into the vector field on H that it defines. This is an isomorphism of the Lie algebra M onto the Lie algebra of the group H. $\square$

II. *Let G be a Lie group with Lie algebra L. Let M be a Lie subalgebra of L. The group G admits exactly one analytic subgroup H for which the following holds: The Lie subalgebra M is the set of all elements $X \in L$ such that $X(e) \in \widetilde{T_e(H)}$.*

Proofs of this theorem can be found, for example, in Pontrjagin [1], Serre [1], Helgason [1], Chevalley [1].

Thus the correspondence between analytic subgroups of G and Lie subalgebras of L described in Proposition I is one-to-one. In the sequel, we will say that the analytic subgroup H and the Lie subalgebra M defined as in I *correspond to each other*.

We will show in §2 of chapter XI *infra* that every closed subgroup of a Lie group is a Lie group. In particular, *the classical groups*

$$SL(n, \mathbf{C}), \ SL(n, \mathbf{R}), \ U(n), \ SU(n), \ Sp(2n), \ O(n, \mathbf{C}), \ O(n, \mathbf{R}), \ SO(n, \mathbf{C}),$$
$$SO(n, \mathbf{R}), \ Sp(2n, \mathbf{R}),$$

which are closed subgroups of the Lie group $GL(n, \mathbf{C})$, *are Lie groups*.

A homomorphism φ of a Lie group G into a Lie group H is called an *analytic homomorphism* if φ is an analytic mapping of the manifold G into the manifold H.

Let φ be an analytic homomorphism of G into H. Let X be a left invariant vector field on G. Then $d\varphi_{e_G} X(e_G)$ is a tangent vector to H at the point e_H, where e_G, e_H are the identity elements of the groups G and H respectively. Let Y be a left invariant vector field on H for which $Y_{e_H} = d\varphi_{e_H} X(e_G)$. We will show that

$$Y(\varphi(g)) = d\varphi_g X(g) \tag{3.3.1}$$

for all $g \in G$. Let ψ_g be left translation by g on G. Let $\chi_{\varphi(g)}$ be left translation by $\varphi(g)$ on H. Since φ is a homomorphism, we have $\varphi \circ \psi_g = \chi_{\varphi(g)} \circ \varphi$, and so

$$d\varphi_g X(g) = d(\varphi \circ \psi_g) X(e_G) = d(\chi_{\varphi(g)} \circ \varphi) X(e_G)$$
$$= d\chi_{\varphi(g)} (d\varphi_{e_G} X(e_G)) = d\chi_{\varphi(g)} Y(e_H) = Y(\varphi(g)).$$

This proves (3.3.1). In turn, (3.3.1) shows that the vector field Y is the image of the vector field X. We denote Y by $d\varphi(X)$. The mapping $d\varphi$ is clearly linear. From Proposition II in 1.6 we infer that for all X_1, X_2 in the Lie algebra of G, we have

$$d\varphi([X_1, X_2]) = [d\varphi(X_1), d\varphi(X_2)].$$

We summarize the foregoing.

III. *Let G, H be Lie groups with Lie algebras L, M respectively. Let φ be an analytic homomorphism of G into H. Let $d\varphi$ be the mapping of L into M such that for any $X \in L$ the element $d\varphi(X) \in M$ is defined by $d\varphi(X)(e_H) = d\varphi_{e_G} X(e_G)$. Thus $d\varphi$ is a homomorphism of L into M.*

IV. *Keep the hypotheses of* **III.** *Let $N_1 \subset L$ be the kernel of the homomorphism $d\varphi$. We set $N_2 = d\varphi(L)$. Let K_1, K_2 be the analytic subgroups of the*

Lie groups G, H corresponding to the Lie subalgebras N_1, N_2 *of the Lie algebras L, M, respectively. We then have the following*:

(1) $\varphi(G)$ *is a Lie subgroup of H and* φ *is an analytic mapping of G onto* $\varphi(G)$;

(2) *the subgroup* K_2 *is the image of the component of the identity element of G; also,* K_2 *coincides with the component of the identity element of the group* $\varphi(G)$;

(3) K_1 *is a closed analytic subgroup of G and is the component of the identity in the kernel of the mapping* φ.

Proofs can be found, for example, in Varadarajan [1], Pontrjagin [1] and Serre [1].

V. *Keep the hypotheses of* III. *The mapping* $d\varphi$ *carries L onto M if and only if* φ *maps the component of the identity element of G onto the component of the identity element of H. The mapping* $d\varphi$ *is one-to-one (that is, has zero kernel) if and only if the kernel of the homomorphism* φ *is discrete.*

This proposition follows at once from IV.

VI. *Let G, H be connected, locally simply connected topological groups. Let* $\varphi: G \to H$ *be a continuous homomorphism of G onto H with discrete kernel. If G (or H) is a Lie group there is exactly one way to give H (or G) the structure of an analytic manifold, so that H (or G) is a Lie group and the homomorphism* φ *is an analytic homomorphism of the Lie group G into the Lie group H.*

This theorem is proved in Varadarajan [1] and Pontrjagin [1].

VII. *If G is a connected Lie group, it admits a universal covering group* $\tilde{G}$. *The group* $\tilde{G}$ *is a Lie group and the Lie algebras of G and* $\tilde{G}$ *are isomorphic.*

Proof. From Propositions IV in 3.1 and I in 3.2, chapter VIII, we see that a universal covering group exists, since a Lie group is locally connected and locally simply connected. The kernel of the homomorphism π of the group $\tilde{G}$ in G is discrete, since $\tilde{G}$ and G are locally isomorphic. By VI, the group $\tilde{G}$ is a Lie group. According to V, $d\pi$ maps the Lie algebra of $\tilde{G}$ onto the Lie algebra of G with zero kernel. That is, $d\pi$ is an isomorphism of the Lie algebras of $\tilde{G}$ and G. □

VIII. *Let G be a Lie group with a closed Lie subgroup H. There is exactly one structure of an analytic manifold on the coset space G/H such that the canonical mapping* $\pi: G \to G/H$ *is an analytic mapping of manifolds. If H is a closed normal subgroup of G, the topological group G/H, provided with the above structure of an analytic manifold, is a Lie group. The mapping* π *is an analytic homomorphism of G onto G/H.*

This theorem is proved in Varadarajan [1], Pontrjagin [1] and Serre [1]. The Lie group G/H, defined in VII is called the *Lie factor group* of G by the closed normal divisor H.

Let G be a real (or complex) Lie group. Let V be a finite-dimensional complex linear space. Write dim V as n. The group G_V (see example 4 in 1.1, chapter I) is isomorphic to $GL(n, \mathbf{C})$ and can be considered both as a real and as a complex Lie group. By VI in 3.2, the Lie algebra of G_V is isomorphic to the Lie algebra $gl(V)$ (see example 3 in 3.2). An analytic homomorphism of G into G_V, considered as a real (or complex) Lie group, is called a *real* (or *complex*) *analytic representation of G in the space V*. From III we infer the following.

IX. *If π is a real (or complex) analytic representation of a Lie group G in the space V, $d\pi$ is a representation of the real (or complex) Lie algebra L of G in V.*

In §2 of chapter XI *infra* we will show that every continuous finite-dimensional representation of a Lie group is analytic. Furthermore, we will prove a fact that is to some extent the converse of Proposition III. Namely, we will construct a representation of a Lie group for every representation of its Lie algebra (see 1.4, chapter XI).

3.4. One-Parameter Subgroups

Let G be a Lie group, let U be a coordinate neighborhood of the point e, and let $x_1, \ldots, x_m$ be a system of analytic coordinates in U. Replacing the functions $x_1, \ldots, x_m$ by the functions $x_1 - x_1(e), \ldots, x_m - x_m(e)$ and if necessary reducing the neighborhood U, we can take $\xi(U)$ to be the set of all $(y_1, \ldots, y_m)$ such that $|y_i| < a$ for all $i = 1, \ldots, m$, for some $a > 0$. Note that

$$(x_1(e), \ldots, x_m(e)) = (0, \ldots, 0).$$

The mapping $G \times G$ onto G defined by $(g, h) \rightarrow g(h^{-1})^{-1} = gh$ is analytic. Hence there is a neighborhood V of e (we may take $\xi(V)$ to be the set of all $(y_1, \ldots, y_m)$ such that $|y_i| < b$ for some positive $b < a$) such that for g, h in V we have

$$x_k(gh) = F_k(x_1(g), \ldots, x_m(g), x_1(h), \ldots, x_m(h)), \qquad k = 1, \ldots, m. \quad (3.4.1)$$

Here the $F_k(y_1, \ldots, y_m, z_1, \ldots, z_m)$ are analytic functions in $\xi(V) \times \xi(V)$. We define

$$u_{ij}(y_1, \ldots, y_m) = \frac{\partial}{\partial z_j} F_i(y_1, \ldots, y_m, 0, \ldots, 0) \qquad (3.4.2)$$

for all $i, j = 1, \ldots, m$.

I. *Let G be a real (or complex) Lie group, and let a be an element of $T_e(G)$. There is a unique analytic homomorphism $\tilde{a}: t \to \tilde{a}(t)$ of the additive group $\mathbf{R}$ (or $\mathbf{C}$) into the group G such that in some neighborhood of the point $t = 0$ we have*

$$(d/dt)x_i(\tilde{a}(t)) = \sum_{j=1}^{m} u_{ij}(x_1(\tilde{a}(t)), \ldots, x_m(\tilde{a}(t)))q_j, \qquad i = 1, \ldots, m. \quad (3.4.3)$$

Here $a_j = a(x_j)$, $j = 1, \ldots, m$, and $x_1, \ldots, x_m$ are a system of analytic coordinates in some neighborhood V of the point e for which $(x_1(e), \ldots, x_m(e)) = (0, \ldots, 0)$ and $\xi(V) = \{(y_1, \ldots, y_m): |y_i| < b\}$.

Proof. Consider the system of ordinary differential equations

$$(dy_i(t)/dt) = \sum_{j=1}^{m} u_{ij}(y_1(t), \ldots, y_m(t))a_j \qquad (3.4.4)$$

with the initial conditions

$$y_i(0) = 0 \qquad (3.4.5)$$

in the region $\xi(V) = \{(y_1, \ldots, y_m): |y_i| < b\}$. We use the theorem of existence and uniqueness of solution of the Cauchy problem for a system of ordinary differential equations. There is a positive number δ such that in the domain $|t| < \delta$ there is a unique solution of the problem $(3.4.4)$–$(3.4.5)$. That is, there is a unique sequence of functions $y_i(t)$, $i = 1, \ldots, m$, for which $(y_1(t), \ldots, y_m(t)) \in \xi(V)$ for $|t| < \delta$ and $(y_1, \ldots, y_m)$ satisfies both $(3.4.4)$ and $(3.4.5)$. Since the functions u_{ij} are analytic on $\xi(V)$ the functions $y_i(t)$ are analytic in the domain $|t| < \delta$.

We set $\tilde{a}(t) = \xi^{-1}(y_1(t), \ldots, y_m(t))$, $|t| < \delta$. It is clear that $\tilde{a}(0) = e$ and that $\tilde{a}(t)$ satisfies $(3.4.4)$. We will show that if $|t| < \delta$, $|s| < \delta$, and $|s + t| < \delta$, we have $\tilde{a}(t)\tilde{a}(s) = \tilde{a}(t + s)$. We set $b(t, s) = \tilde{a}(t)\tilde{a}(s)$ and $\xi(b(t, s)) = (z_1(t, s), \ldots, z_m(t, s))$. We then get $z_i(t, s) = F_i(y_1(t), \ldots, y_m(t), y_1(s), \ldots, y_m(s))$. Expand $z_i(t, s)$ as a Taylor series in s and use $(3.4.2)$:

$$z_i(t, s) = y_i(t) + \sum_{j=1}^{m} u_{ij}(y_1(t), \ldots, y_m(t))a_j s + o(s). \qquad (3.4.6)$$

On the other hand, $(3.4.3)$ implies that

$$y_i(t + s) = y_i(t) + \sum_{j=1}^{m} u_{ij}(y_1(t), \ldots, y_m(t))a_j s + o(s). \qquad (3.4.7)$$

Combine $(3.4.6)$ and $(3.4.7)$ to conclude that

$$z_i(t, s) - y_i(t + s) = o(s) \qquad (3.4.8)$$

as $s \to 0$. We will now show that the functions $z_i(t, s)$ satisfy the system of differential equations

$$\frac{\partial z_i(t, s)}{\partial s} = \sum_{j=1}^{m} u_{ij}(z_1(t, s), \ldots, z_m(t, s))a_j \tag{3.4.9}$$

and the initial conditions

$$z_i(t, 0) = y_i(t). \tag{3.4.10}$$

The definition of the functions $z_i(t, s)$ implies (3.4.10). Let us find $(\partial z_i(t, s))/\partial s$. Since $z_i(t, s + u) = F_i(y_1(t), \ldots, y_m(t), y_1(s + u), \ldots, y_m(s + u))$, (3.4.8) gives us

$$z_i(t, s + u) = F_i(y_1(t), \ldots, y_m(t), z_1(s, u), \ldots, z_m(s, u)) + o(u) \tag{3.4.11}$$

as $u \to 0$. Since multiplication in G is associative, we infer that

$$\begin{aligned} F_i(y_1(t), \ldots, y_m(t), z_1(s, u), \ldots, z_m(s, u)) \\ = x_i(\tilde{a}(t)(\tilde{a}(s)\tilde{a}(u))) = x_i((\tilde{a}(t)\tilde{a}(s))\tilde{a}(u)) \\ = F_i(z_1(t, s), \ldots, z_m(t, s), y_1(u), \ldots, y_m(u)) + o(u). \end{aligned} \tag{3.4.12}$$

Substituting (3.4.12) in the right side of (3.4.11) and applying (3.4.2), we now obtain

$$z_i(t, s + u) = z_i(t, s) + \sum_{j=1}^{m} u_{ij}(z_1(t, s), \ldots, z_m(t, s))a_j u + o(u). \tag{3.4.13}$$

The equalities (3.4.9) are immediate from (3.4.13).

Thus the functions $z_i(t, s)$ satisfy the system of differential equations (3.4.9) with the initial conditions (3.4.10). On the other hand, the functions $\bar{z}_i(t, s) = y_i(s + t)$ also satisfy the system of differential equations

$$\frac{\partial \bar{z}_i(t, s)}{\partial s} = \sum_{j=1}^{m} u_{ij}(\bar{z}_1(t, s), \ldots, \bar{z}_m(t, s))a_j,$$

which is immediately obtained from the system (3.4.4). They also satisfy the initial conditions $\bar{z}_i(t, 0) = y_i(t + 0) = y_i(t)$. Therefore the functions $z_i(t + s)$ and $y_i(t + s)$ of the variable t (for a fixed s) satisfy the same system of equations (3.4.9) with the same initial conditions (3.4.10). By the uniqueness theorem for solutions of this system, we have $z_i(t, s) = y_i(t + s)$ and therefore

$$\tilde{a}(t)\tilde{a}(s) = \tilde{a}(t + s) \tag{3.4.14}$$

for $|t| < \delta, |s| < \delta, |t + s| < \delta$. In particular, we get

$$\tilde{a}(t)\tilde{a}(s) = \tilde{a}(s)\tilde{a}(t) \tag{3.4.15}$$

for $|t| < \delta, |s| < \delta, |t + s| < \delta$.

Consider an arbitrary t and a natural number n such that $|t/n| < \delta$. We define

$$\tilde{a}(t) = (\tilde{a}(t/n))^n. \tag{3.4.16}$$

We will show that (3.4.16) defines the required analytic mapping of the group $\mathbf{R}$ (or $\mathbf{C}$) into G. We first show that (3.4.16) is well defined. If we have $|t/m| < \delta$ for another natural number m, then $|t/(mn)| < \delta$. On the other hand, (3.4.14) implies that

$$(\tilde{a}(t/(mn)))^m = \tilde{a}(t/n), \qquad (\tilde{a}(t/(mn)))^n = \tilde{a}(t/m).$$

Therefore we have

$$(\tilde{a}(t/m))^m = (\tilde{a}(t/(mn)))^{mn} = (\tilde{a}(t/n))^n.$$

Hence the mapping $t \to \tilde{a}(t)$ is unambiguously defined by (3.4.16). For $|t| < \delta$, the function $\tilde{a}(t)$ defined by (3.4.16) coincides with the original function $\tilde{a}$. We will show that

$$\tilde{a}(t + s) = \tilde{a}(t)\tilde{a}(s) \tag{3.4.17}$$

for all t, s. There is a natural number n such that $|t/n| < \delta, |s/n| < \delta, |(t + s)/n| < \delta$. It is clear that $\tilde{a}((t + s)/n) = \tilde{a}(t/n)\tilde{a}(s/n)$. Raising this equality to the n-th power, and using (3.4.15) and (3.4.16), we obtain (3.4.17). Finally, $\tilde{a}(t)$ is analytic for all t, since $\tilde{a}(t)$ is analytic for $|t| < \delta$ and (3.4.16) obviously preserves analyticity.   □

II. *Let G be a real (or complex) Lie group. Let Y be the element of the Lie algebra of the group $\mathbf{R}$ (or $\mathbf{C}$) that carries the function $f(z) = z$ to the function identically 1. For every element X of the Lie algebra L of G, there exists a unique analytic homomorphism $\theta_X : t \to \theta_X(t)$ of the Lie group $\mathbf{R}$ (or $\mathbf{C}$) into G, whose differential $d\theta_X$ carries Y into X.*

Proof. Let $a = X(e)$ be a tangent vector to the manifold G at the point e. Let $t \to \tilde{a}(t)$ be the analytic homomorphism of the group $\mathbf{R}$ (or $\mathbf{C}$) into the Lie group G that we constructed in Proposition I. From (3.4.3) and the definition of Y, we obtain

$$X(e)x_i = ax_i = a_i = (d/dt)x_i(\tilde{a}(t))|_{t=0}$$
$$= Y(0)(x_i \circ \tilde{a}) = (d\tilde{a}_0)Y(0)x_i \tag{3.4.18}$$

for all $i = 1, \ldots, m$. Therefore we have $d\tilde{a}(X) = Y$. We define θ_X as $\tilde{a}$. If $\tilde{\theta}_X$ is another analytic homomorphism of $\mathbf{R}$ (or $\mathbf{C}$) into G such that $(d\tilde{\theta}_X)Y = X$, we have $\tilde{\theta}_X(t + s) = \tilde{\theta}_X(t)\tilde{\theta}_X(s)$ for all t, s. Both sides of this equality are analytic functions of s for fixed t. Setting $x_i(\tilde{\theta}_X(t)) = y_i(t)$, we obtain

$$y_i(t + s) = F_i(y_1(t), \ldots, y_m(t), \quad y_1(s), \ldots, y_m(s)), \quad \text{for } i = 1, \ldots, m, \quad (3.4.19)$$

for all sufficiently small t and s. Differentiating (3.4.19) with respect to s and setting $s = 0$, we see from (3.4.2) that the functions $y_1(t), \ldots, y_m(t)$ satisfy (3.4.4), where $a_j = (dy_j(s)/ds)|_{s=0} = X(e)x_j$. Applying I, we obtain $\tilde{\theta}_X = \tilde{a}$, where $a = X(e)$. That is, the tangent vector $X(e)$ defines the corresponding analytic homomorphism θ_X uniquely.

The element $\theta_X(1) \in G$ is denoted by $\exp(X)$. The mapping $\exp: L \to G$, defined by $X \to \exp(X)$, is called the *exponential mapping of the Lie algebra L into the Lie group G*.

III. *The equality* $\exp(\lambda x) = \theta_X(\lambda)$ *holds for all* $X \in L$ *and all* $\lambda \in \mathbf{R}$ *(or* $\lambda \in \mathbf{C}$*).*

Proof. Consider the mapping $\pi: t \to \theta_X(\lambda t)$ of the group $\mathbf{R}$ (or $\mathbf{C}$) into G. Since θ_X is analytic, so is the mapping π. We also have

$$\pi(t)\pi(s) = \theta_X(\lambda t)\theta_X(\lambda s) = \theta_X(\lambda t + \lambda s) = \pi(t + s)$$

for all t, s. Therefore π is an analytic homomorphism of $\mathbf{R}$ (or $\mathbf{C}$) into G. It is clear that $(d\pi_0)Y(0)x_i = (d/dt)x_i(\theta_X(\lambda t))|_{t=0} = \lambda X(e)x_i$. Proposition II shows that $\pi = \theta_{\lambda X}$ and thus $\theta_{\lambda X}(1) = \theta_X(\lambda)$. $\square$

Proposition III yields at once the identities

$$\exp((t_1 + t_2)X) = \exp(t_1 X) \exp(t_2 X), \qquad (3.4.20a)$$
$$\exp(-tX) = (\exp(t(X))^{-1} \qquad (3.4.20b)$$

for all t_1, t_2 and all $X \in L$.

The linear space L is isomorphic to the space $\mathbf{R}^m$ (or $\mathbf{C}^m$) and therefore can be considered as an analytic manifold. Let $X_1, \ldots, X_m$ be the basis in the Lie algebra L for which $X_i(e)(x_j) = \delta_{ij}, i, j = 1, \ldots, m$. If X belongs to L and $X(e)x_j = a_j$, then $X = a_1 X_1 + \cdots + a_m X_m$. The functions $a_j(X) = a_j$ for $X = a_1 X_1 + \cdots + a_m X_m$ are a system of analytic coordinates on L.

IV. *The mapping* $\exp: L \to G$ *is an analytic mapping of manifolds that is regular at the point* $0 \in L$.

Proof. We denote the solution (y_i) of (3.4.4) with the initial conditions (3.4.5) by $(\Phi_i(t, a_1, \ldots, a_m))$. Since the right sides of (3.4.4) are analytic with respect to all variables $y_1, \ldots, y_m, a_1, \ldots, a_m$, the functions $\Phi_i(t, a_1, \ldots, a_m)$ are analytic in a certain domain of the form $|t| < \delta$, $|a_j| < \varepsilon$, $j = 1, \ldots, m$. We may suppose that $|\Phi_i| \leqslant c$ for $|t| < \delta$, $|a_j| < \varepsilon$, where C is a constant.

By the definition of the mapping θ_X, we have

$$\theta_X(t) = \xi^{-1}(\Phi_1(t, a_1, \ldots, a_m), \ldots, \Phi_m(t, a_1, \ldots, a_m))$$

for $|t| < \delta$, $|a_j| < \varepsilon$, $j = 1, \ldots, m$, where $X = a_1 X_1 + \cdots + a_m X_m$. Using Proposition III, we obtain

$$\theta_X(t) = \xi^{-1}(\Phi_1((\delta t)/2, (2a_1)/\delta, \ldots, (2a_m)/\delta),$$
$$\ldots, \Phi_m((\delta t)/2, (2a_1)/\delta, \ldots, (2a_m)/\delta)),$$

for $|t| < 2$ and $|a_j| < (\varepsilon\delta)/2$. It follows that

$$\exp(X) = \xi^{-1}(\Phi_1(\delta/2, (2a_1)/\delta, \ldots, (2a_m)/\delta),$$
$$\ldots, \Phi_m(\delta/2, (2a_1)/\delta, \ldots, (2a_m)/\delta)). \tag{3.4.21}$$

From (3.4.21) and the analyticity of the functions Φ_i, we see that the mapping exp is analytic in some neighborhood V of the zero element 0 of the manifold L. Furthermore, for any element $X \in L$, there exists a natural number n such that $X/n \in V$. From (3.4.20) we infer that $\exp(X) = (\exp(X/n))^n$. Therefore the exponential mapping is analytic everywhere.

We will now show that the exponential mapping is regular at the point $0 \in L$. We must establish that the differential of the exponential mapping carries a basis for the tangent space $T_0(L)$ into a basis for the tangent space $T_e(G)$. Let $\tilde{X}_i$ be the tangent vector to L at the point 0 defined by

$$\tilde{X}_i a_j = \delta_{ij} \qquad (i, j = 1, \ldots, m).$$

We set $\eta(X) = (a_1, \ldots, a_m)$ for $X = a_1 X_1 + \cdots + a_m X_m$. We will show that $d(\exp)_0 \tilde{X}_i = X_i(e)$. The vector $\tilde{X}_i$ carries a function f, defined and analytic in some neighborhood W of the zero element of the manifold L, into the number $\tilde{X}_i f$ defined by

$$\tilde{X}_i(f) = \sum_{j=1}^{m} (\partial F/\partial a_j)_{0, \ldots, 0} \tilde{X}_i a_j = (\partial F/\partial a_i)_{0, \ldots, 0}. \tag{3.4.22}$$

Here F is an analytic function of $(a_1, \ldots, a_m)$ such that $f = F \circ \eta$. Consider the vector $v_i = d(\exp)_0 \tilde{X}_i$, which is tangent to G at the point e. For any function $\varphi \in D(V)$, where V is some neighborhood of the identity element e in G, we have

$$v_i(\varphi) = \tilde{X}_i(\varphi \circ \exp).$$

If $\varphi = \Phi \circ \xi$, we infer from (3.4.21), (3.4.22) and (3.4.18) that

$$
\begin{aligned}
v_i(\varphi) &= (\partial/\partial a_i)\Phi(\Phi_1(\delta/2,(2a_1)/\delta,\ldots,(2a_m)/\delta), \\
&\qquad \ldots, \Phi_m(\delta/2,(2a_i)/\delta,\ldots,(2a_m/\delta)))|_{0,\ldots,0} \\
&= (\partial/\partial a_i)\Phi(\Phi_1(\delta/2,0,\ldots,(2a_i)/\delta,\ldots,0), \\
&\qquad \ldots, \Phi_m(\delta/2,0,\ldots,(2a_i)/\delta,\ldots,0))|_{a_i=0} \\
&= (d/da_i)(\varphi \circ \exp(a_iX_i))|_{a_i=0} = (d\theta_{X_i})_0 Y(0)\varphi = X_i(e)\varphi. \quad (3.4.23)
\end{aligned}
$$

We have proved that $v_i = X_i(e)$. That is, $d(\exp)_0$ carries the basis Y_i in the space $T_0(L)$ into the basis $X_i(e)$ in $T_e(G)$. Hence the mapping exp is regular at the point $0 \in L$. $\quad\square$

V. *Let $X_1, \ldots, X_m$ be a basis for the Lie algebra L of the Lie group G. There exist a neighborhood U of the identity element e on G, a system of analytic coordinates $x_1, \ldots, x_m$ in U, and a number $\delta > 0$, for which the following holds:*

$$
x_i\left(\exp\left(\sum_{k=1}^{m} u_kX_k\right)\right) = u_i, \quad for \; i = 1, 2, \ldots, m \qquad (3.4.24)
$$

for all $(u_1, \ldots, u_m) \in V$, where V is the set $(u_1, \ldots, u_m)$ such that $|u_k| < \delta$, $k = 1, \ldots, m$.

This proposition follows from Propositions IV, I in 1.1, and V in 1.5.

The system of analytic coordinates that satisfies (3.4.24) is called *the canonical system of coordinates on a group* (or *the canonical system of coordinates of the first kind*). The set $W = \xi^{-1}(V)$ is called *the canonical neighborhood of the element $e \in G$.*

VI. *Let π be an analytic homomorphism of the Lie group G into the Lie group H. For every element X of the Lie algebra L of G, we have*

$$
\pi(\exp(X)) = \exp(d\pi(X)). \qquad (3.4.25)
$$

Proof. Let Y be the vector field on the group $\mathbf{R}$ (or $\mathbf{C}$) defined by $Y\psi = 1$, where ψ is the identity mapping of $\mathbf{R}$ onto $\mathbf{R}$ (or $\mathbf{C}$ onto $\mathbf{C}$). Let X be an element of the Lie algebra L of G and let θ_X be the analytic homomorphism of $\mathbf{R}$ (or $\mathbf{C}$) into G defined in Proposition II. We then have $d\theta_X(Y) = X$ and thus $d(\pi \circ \theta_X)(Y) = d\pi(X)$. Also, $d\theta_{d\pi(X)}(Y)$ is equal to $d\pi(X)$ and therefore, by Proposition II, the homomorphisms $\pi \circ \theta_X$ and $\theta_{d\pi(X)}$ must coincide. Thus we have $(\pi \circ \theta_X)(1) = \theta_{d\pi(X)}(1)$ and we have proved (3.4.25). $\quad\square$

VII. *Let π be an analytic homomorphism of a Lie group G into a Lie group H. If the mapping π is one-to-one, it is regular everywhere.*

Proof. Let L be the Lie algebra of G. If $d\pi(X) = 0$ for some $X \in L$, we then have $\pi(\theta_X(t)) = \pi(\exp(tX)) = \exp(d\pi(tX)) = \exp(td\pi(X)) = e$ for all t. Because π is one-to-one, we have $\exp(tX) = e$ and so $X = 0$. Therefore $d\pi$ is a one-to-one mapping. $\square$

VIII. *Let H be a Lie subgroup of the Lie group G. Let M be the Lie algebra of H. If U is a neighborhood of the zero element in M, the set of elements $\exp(X)$, $X \in U$, contains a neighborhood of the element e in H.*

This proposition follows directly from VI (consider the identity mapping of H into G).

Example

Let G be the Lie group $GL(n, \mathbf{R})$. As we showed in example 3 in 3.2, the Lie algebra L of G can be identified with the Lie algebra $gl(n, \mathbf{R})$. The mapping

$$X \to (X(e)x_{ij}) \tag{3.4.26}$$

is an isomorphism, which we call the *canonical isomorphism*. The matrix $(X(e)x_{ij})$ is denoted by $a(X)$. By Proposition I in 3.2 and the rule for multiplying matrices, we have

$$\begin{aligned} X(g_0)(x_{ij})(g) &= (d\varphi_{g_0})_e X(e)(x_{ij})(g) = X(e)(x_{ij})(g_0 g) \\ &= g_0 X(e)(x_{ij})(g) = (g_0 a(X))_{ij} \end{aligned} \tag{3.4.27}$$

for all $i, j = 1, \ldots, n$ and all $g_0 \in G$, $X \in L$. Let θ_X be the analytic homomorphism of the Lie group $\mathbf{R}$ into G that is described in Proposition II. The relation $d\theta_X(Y) = X$ has to hold. This relation is equivalent to the relation

$$Y(t)(x_{ij} \circ \theta_X) = X(\theta_X(t))x_{ij}, \tag{3.4.28}$$

for all $i, j = 1, \ldots, n$. Combining (3.4.27) and (3.4.28), we see that

$$(d/dt)x_{ij}(\theta_X(t)) = (\theta_X(t)a(X))_{ij}. \tag{3.4.29}$$

From (3.4.29) we infer that the matrix $\theta_X(t)$ has the property that

$$(d/dt)\theta_X(t) = \theta_X(t)a(X), \tag{3.4.30}$$

the derivative of the matrix being the matrix formed by the derivatives of the matrix entries. The matrix function $\theta_X(t)$ satisfies not only (3.4.30), but also the initial condition

$$\theta_X(0) = 1_n, \tag{3.4.31}$$

where 1_n is the identity matrix of order n. The solution of the differential equation (3.4.30) with the initial condition (3.4.31) is the ordinary matrix exponential

$$\theta_X(t) = e^{ta(X)}$$

$$= 1_n + ta(X) + \frac{t^2}{2!} a(X)^2 + \cdots + \frac{t^n}{n!} a(X)^n + \cdots, \qquad t \in \mathbf{R}. \quad (3.4.32)$$

All the matrix entries of the series on the right side of (3.4.32) converge absolutely for all $t \in \mathbf{R}$. Thus, *in the group $GL(n, \mathbf{R})$, the exponential mapping coincides with ordinary matrix exponentiation*:

$$\exp(X) = e^{a(X)}, \quad \text{for all } X \in L. \quad (3.4.33)$$

A similar argument shows that the exponential mapping in the group $GL(n, \mathbf{C})$ also coincides with ordinary matrix exponentiation.

Let G be an analytic subgroup of the Lie group $GL(n, \mathbf{R})$. Let π be the identity mapping of G into $GL(n, \mathbf{R})$. Identifying the Lie algebra of G with a Lie subalgebra of the Lie algebra $gl(n, R)$ as in I, 3.3, we apply Proposition VI to see that the exponential mapping in G coincides with matrix exponentiation (see (3.4.25)).

The example of an exponential mapping that we have examined in the groups $GL(n, \mathbf{C})$ and $GL(n, \mathbf{R})$ allows us to prove an important theorem connecting representations of a Lie group with the corresponding representations of its Lie algebra (see Proposition IX in 3.3).

Theorem. *Let G be a connected Lie group with Lie algebra L. Let T be a finite-dimensional analytic representation of G in a vector space E and let dT be the corresponding representation of L in E. The following assertions hold.*

(a) *A vector subspace V of E is invariant under the representation T if and only if it is invariant under dT.*

(b) *The representation T is irreducible (reducible, completely reducible) if and only if the representation dT is irreducible (reducible, completely reducible).*

Proof. Since (a) implies (b), let us prove (a). Consider the representation T as an analytic homomorphism of the Lie group G into the Lie group G_E (considered as a real Lie group). By (3.4.25) we have

$$T(\exp(X)) = \exp(dT(X)) \quad (3.4.34)$$

for all $X \in L$. Note that

$$\exp(a) = e^a = 1 + a + \frac{1}{2!} a^2 + \cdots + \frac{1}{n!} a^n + \cdots \quad (3.4.35)$$

for all $a \in gl(E)$. From (3.4.34) and (3.4.35) we infer that

$$T(\exp(X)) = 1 + dT(X) + \frac{1}{2!}(dT(X))^2 + \cdots + \frac{1}{n!}(dT(X))^n + \cdots \quad (3.4.36)$$

for all $X \in L$. From (3.4.36) we see that any vector subspace V of E invariant under an operator $dT(X)$, $X \in L$, is also invariant under the operator $T(\exp(X))$. Thus, by VIII and the connectedness of G, every subspace V of E, invariant under the representation dT of L, is also invariant under the representation T of G. On the other hand, (3.4.34) and (3.4.35) imply that

$$T(\exp(tX)) = e^{t\,dT(X)} \quad (3.4.37)$$

for all t. Differentiating both sides of (3.4.37) and then setting $t = 0$, we see that

$$dT(X) = (d/dt)(T(\exp(tX)))|_{t=0}. \quad (3.4.38)$$

The relation (3.4.38) shows that every vector subspace V of E invariant under the operators $T(\exp(tX))$ for all t, is also invariant under the operator $dT(X)$. Thus a subspace invariant under the representation T is also invariant under the representation dT. $\square$

3.5. The Adjoint Representation

Let G be a Lie group with Lie algebra L. If α is an analytic isomorphism of G onto itself, $d\alpha$ is a homomorphism of L into itself. From the relation $\alpha^{-1} \circ \alpha = 1_G$ (1_G is the identity mapping of G onto G), we see that the mapping $d(\alpha^{-1})$ is inverse to $d\alpha$. Therefore $d\alpha$ is an *automorphism* of the Lie algebra L. That is, $d\alpha$ is a linear mapping of L onto L such that

$$d\alpha([X, Y]) = [d\alpha(X), d\alpha(Y)],$$

for all $X, Y \in L$. The mapping $\alpha \to d\alpha$ is a linear representation of the group A of all analytic isomorphisms of G onto itself. We will show that if G is connected, then the representation is exact. Indeed, if $d\alpha$ is the identity automorphism, that is, if $d\alpha(X) = X$ for all $X \in L$, (3.4.25) shows that

$$\alpha(\exp(X)) = \exp(X)$$

for all $X \in L$. Therefore by Proposition VIII in 3.4, the mapping α leaves fixed all elements of some neighborhood V of the identity element in G. Because G is connected, we have $G = \bigcup_{n=1}^{\infty} V^n$ and thus $\alpha(x) = x$ for all $x \in G$. For $g \in G$, let α_g be the analytic isomorphism of G onto itself defined by the formula $\alpha_g(h) = ghg^{-1}$ for all $h \in G$. The mapping $\mathrm{Ad}:g \to d\alpha_g$ is called the *adjoint representation of the Lie group* G. The adjoint representation is a homomorphism of G into G_L. Since L is isomorphic to $\mathbf{R}^m$ or $\mathbf{C}^m$, G_L is isomorphic to $GL(m, \mathbf{R})$ or $GL(m, \mathbf{C})$ and so is a Lie group.

I. *The adjoint representation of the group G is an analytic homomorphism of G into G_L.*

Proof. Let $X_1, \ldots, X_m$ be a basis in L and let $(x_1, \ldots, x_m)$ be the canonical system of coordinates in G defined from this basis, which is to say that

$$x_i\left(\exp\left(\sum_{j=1}^m u_j X_j\right)\right) = u_j, \quad \text{for } i = 1, \ldots, m. \tag{3.5.1}$$

Suppose that

$$da_g(X_i) = \sum_{j=1}^m a_{ji}(g)X_j, \tag{3.5.2}$$

i.e., $(a_{ij}(g))$ is the matrix of the linear transformation da_g. For $h = \exp(tX_i)$, (3.4.25) yields the equality

$$ghg^{-1} = \exp\left(\sum_{j=1}^m ta_{ji}(g)X_j\right). \tag{3.5.3}$$

For sufficiently small t, the values $ta_{ji}(g)$ are canonical coordinates of the element ghg^{-1}. We will find a formula for the canonical coordinates of $x_i(gh)$, $x_i(ghg^{-1})$ and $x_i(h^{-1}g^{-1}hg)$.

Let U be a neighborhood of the element e such that $\xi(U)$ is the set of all $(y_1, \ldots, y_m)$ for which $|y_i| < a$ for some $a > 0$ and all $i = 1, \ldots, m$. Let $\xi(W)$ be the set of all $(y_1, \ldots, y_m)$ such that $|y_i| < b$ for some $b > 0$ and all $i = 1, \ldots, m$, the constant b being chosen so that $WW \subset U$. For $g, h \in W$, we set

$$x_i(gh) = F_i(x_1(g), \ldots, x_m(g), x_1(h), \ldots, x_m(h)),$$
$$i = 1, \ldots, m, \tag{3.5.4}$$

where the functions F_i are analytic in the region $\xi(W) \times \xi(W)$. Expand the function $F_i(y_1, \ldots, y_m, z_1, \ldots, z_m)$ as a power series in the variables $z_1, \ldots, z_m$. We thus represent F_i in the form

$$F_i(y_1, \ldots, y_m, z_1, \ldots, z_m) = \sum_{l=0}^\infty P_{il}(y_1, \ldots, y_m, z_1, \ldots, z_m), \tag{3.5.5}$$

where P_{il} is a homogeneous polynomial of degree l in the variables $z_1, \ldots, z_m$, whose coefficients are analytic functions of $y_1, \ldots, y_m$.

Remember that we have

$$g = \exp\left(\sum_{i=1}^{m} x_i(g)X_i \right) \qquad (3.5.6)$$

for all $g \in W$. We set

$$g(t) = \exp\left(\sum_{i=1}^{m} tx_i(g)X_i \right), \qquad |t| \leqslant 1. \qquad (3.5.7)$$

Then (3.5.5) implies that

$$x_i(gh(t)) = \sum_{l=0}^{\infty} P_{il}(x_1(g), \ldots, x_m(g); x_1(h), \ldots, x_m(h))t^l \qquad (3.5.8)$$

for all $g, h \in W$ and $|t| \leqslant 1$. Let f be a function in $D(U)$. We have

$$(df(g\theta_X(t))/dt) = (Xf)_{g\theta_X(t)}$$

for any one-parameter subgroup θ_X, $X \in L$, if $\theta_X(t) \in U$ (see (3.4.3)). From (3.5.7) and Proposition I in 3.4, we infer that $h(t) = \theta_Y(t)$, where $Y = \sum_{i=1}^{m} x_i(h)X_i$. Therefore we have

$$(df(gh(t))/dt) = \left(\left(\sum_{i=1}^{m} x_i(h)X_i \right)f \right)_{gh(t)} \qquad (3.5.9)$$

for all $|t| \leqslant 1$. We set $Y = \sum_{i=1}^{m} x_i(h)X_i$. Applying (3.5.9) to the functions $Y^n f$ we find by induction that

$$(d^n f(gh(t))/dt^n) = (Y^n f)_{gh(t)} \qquad (3.5.10)$$

for all natural numbers n. Thus the analytic function $f(gh(t))$ can be expanded in a Taylor series in the variable t in a neighborhood of the point $t = 0$, whose coefficients are defined by (3.5.10). It follows that

$$f(gh(t)) = f(g) + \sum_{n=1}^{\infty} \frac{t^n}{n!} (Y^n f)_g \qquad (3.5.11)$$

for all sufficiently small t. If $f = x_i$, (3.5.8) shows that the series (3.5.11) converges for $|t| \leqslant 1$ and thus

$$P_{in}(x_i(g), \ldots, x_m(g), x_1(h), \ldots, x_m(h)) = \frac{1}{n!} (Y^n x_i)_g. \qquad (3.5.12)$$

Similarly, setting $X = \sum_{i=1}^{m} x_i(g)X_i$ and $g(t) = \theta_X(t) = \exp(tX)$, we obtain, for every function $f \in D(U)$,

$$f(g(t)) = f(e) + \sum_{k=1}^{\infty} \frac{t^k}{k!} (X^k f)_e. \qquad (3.5.13)$$

Applying (3.5.13) to the functions $(Y^n x_i)_g$ and substituting the result in (3.5.11), we see that

$$x_i(gh) = \sum_{k,n=0}^{\infty} \frac{1}{k!n!} (X^k Y^n x_i)_e \qquad (3.5.14)$$

for all $g, h \in V$. The expression $(X^k Y^n x_i)_e$ is a homogeneous polynomial of degree $k + n$ in the variables $x_1(g), \ldots, x_m(g), x_1(h), \ldots, x_m(h)$. Therefore the formulas (3.5.14) define Taylor series for the functions $x_i(gh)$.

From (3.5.13) and the analogous relation for $f(h(t))$, we infer that

$$x_i(g(t)) = \sum_{k=1}^{\infty} \frac{1}{k!} (X^k x_i)_e, \qquad x_i(h(t)) = \sum_{n=1}^{\infty} \frac{1}{n!} (Y^n x_i)_e.$$

(Recall that $x_i(e) = 0$ for all $i = 1, \ldots, m$.) Thus (3.5.14) implies that

$$x_i(gh) = x_i(g) + x_i(h) + (XYx_i)_e + \rho(g,h), \qquad (3.5.15)$$

where ρ is a series containing no monomials in $x_1(g), \ldots, x_m(h)$ of degree less than three. On the other hand, the difference $y_i(g, h) = x_i(gh) - x_i(g) - x_i(h)$ can be expanded in a power series containing no summands of degree 0 or 1. We also have $y_i(g, h) = 0$ if g or h is e. Thus we have

$$x_i(gh) - x_i(g) - x_i(h) = \sum_{j,k=1}^{m} x_j(g)x_k(h)f_{ijk}(g,h),$$

where the functions $f_{ijk}(g, h)$ are analytic in a neighborhood of the point (e, e). Replace g by gh and h by $h^{-1}g^{-1}hg$: we obtain

$$x_i(hg) - x_i(gh) - x_i(h^{-1}g^{-1}hg)$$

$$= \sum_{j,k=1}^{m} x_j(gh)x_k(h^{-1}g^{-1}hg)f_{ijk}(gh, g^{-1}h^{-1}gh). \qquad (3.5.16)$$

Since the function $x_k(h^{-1}g^{-1}hg)$ vanishes for $h = e$ or $g = e$, its power series expansion has no terms of degree 0 or 1 in $x_1(g), \ldots, x_m(h)$. Therefore the expansion of the function $x_i(hg) - x_i(gh) - x_i(h^{-1}g^{-1}hg)$ contains no terms of degree less than three. On the other hand, (3.5.15) implies that

$$x_i(hg) - x_i(gh) = ([Y, X]x)_e + \rho(g, h) - \rho(h, g).$$

Thus we have

$$x_i(h^{-1}g^{-1}hg) = ([Y, X]x_i)_e + \rho_1(g, h), \tag{3.5.17}$$

where ρ_1 contains no summand of degree less than three. Replace h by $g^{-1}h$ in (3.5.16) and use (3.5.17) to obtain

$$x_i(g^{-1}hg) = x_i(h) + ([Y, X]x_i)_e + \rho_2(g, h), \tag{3.5.18}$$

where $\rho_2(g, h)$ contains no summands of degree less than three.

We now finish the proof of I. We already know that for sufficiently small t, the values $ta_{ji}(g)$ are canonical coordinates of the element ghg^{-1}. By (3.5.18), these coordinates are analytic functions of g in a neighborhood of the point e. Therefore the adjoint representation is analytic in a neighborhood of e. Since left translations in G are analytic and $d\alpha_{gg_0} = d\alpha_g d\alpha_{g_0}$, analyticity of the representation $Ad: g \to d\alpha_g$ in a neighborhood of e shows that it is analytic everywhere. □

The operator $d\alpha_g$ is denoted by $ad(g)$.

II. *The differential of the adjoint representation of a Lie group G is the adjoint homomorphism $X \to ad\ X$ of the Lie algebra L of G.*

Proof. The mapping $d(Ad)$ is a homomorphism of a Lie algebra into the Lie algebra $gl(L)$. The reader can easily verify that (3.5.18) implies that

$$Ad(\exp(tX))X_i = X_i + t[X, X_i] + \rho(t) \tag{3.5.19}$$

for all $i = 1, \ldots, m$ and all $X \in L$, where $\rho(t)$ is a function analytic in a neighborhood of the point $t = 0$. The Taylor series expansion of ρ contains no summands of degrees 0 or 1. Thus we find

$$(d(Ad))(X)X_i = [X, X_i] \tag{3.5.20}$$

for all $X \in L$ and $i = 1, \ldots, m$, so that

$$((d(Ad))(X))(Y) = [X, Y] = (ad\ X)(Y) \tag{3.5.21}$$

for all $X, Y \in L$. □

III. *For any element $X \in L$, we have*

$$\mathrm{Ad}(\exp(X)) = e^{\mathrm{ad}\, X}. \tag{3.5.22}$$

This proposition follows from Proposition VI in 3.4, II, and (3.4.33).

IV. *Let G be a Lie group with Lie algebra L. Let N be a Lie subalgebra of L and let H be the analytic subgroup of G corresponding to N. The inclusion $[a, N] \subset N$ for some $a \in L$ is equivalent to the inclusion $(\exp ta)H(\exp ta)^{-1} \subset H$, for all $t \in \mathbf{R}$.*

Proof. We write $b = \exp(ta)$. Plainly, $\mathrm{Ad}(b)H = bHb^{-1}$ is an analytic subgroup of G. From (3.5.18) we infer that the image of the Lie algebra N of H under the mapping $d\alpha_b = \mathrm{Ad}(b)$ is the Lie algebra of the subgroup bHb^{-1}. Thus bHb^{-1} is contained in H if and only if $\mathrm{Ad}(b)N \subset N$. The inclusion $\mathrm{Ad}(\exp(ta))N \subset N$ is equivalent to the condition that the subspace N of L be invariant under the restriction of the adjoint representation of G to the subgroup $\exp(ta)$. According to the theorem in 3.4, this condition is satisfied if and only if N is invariant under the representation $d(\mathrm{Ad}) = \mathrm{ad}$ of the Lie subalgebra $\{ta\} \subset L$. $\square$

V. *Let G be a connected Lie group with Lie algebra L. Let H be the analytic subgroup of G with corresponding subalgebra N. The subgroup H is normal in G if and only if N is an ideal in L.*

Proof. The subgroup H is normal if and only if $gHg^{-1} \subset H$ for all g in some neighborhood U of the identity element of G. If gHg^{-1} is contained in H for all $g \in U$, then $gHg^{-1} \subset H$ for all $g \in U^n$. Since G is connected, we have $G = \bigcup_{n=1}^{\infty} U^n$. On the other hand, the image of the Lie algebra L under the exponential mapping contains a neighborhood of the identity element. Thus H is normal in G if and only if $\exp(ta)H \exp(ta)^{-1} \subset H$ for all $a \in L$ and $t \in \mathbf{R}$. Therefore Proposition IV implies the present proposition. $\square$

The image of G under the mapping Ad is called the *adjoint group of* G. Clearly the group $\mathrm{Ad}(G)$ is isomorphic to the factor group of G by the kernel of the mapping Ad.

VI. *The kernel of the adjoint representation of a connected Lie group G is the center of G.*

Proof. If $\mathrm{Ad}(g)$, $g \in G$, is the identity operator on L, then α_g is the identity on G. Since G is connected, a representation π of G is the identity representation if and only if it is the identity on some neighborhood of the identity element in G. From (3.4.25) we infer that π is the identity on a canonical neighborhood of the element e if and only if $d\pi$ is the identity mapping of the Lie algebra L onto itself. Therefore the kernel of the representation $g \to d\alpha_g = \text{Ad}(g)$ is the kernel of the mapping $g \to \alpha_g$, which is the center of G. $\square$

3.6. The Differential of the Exponential Mapping

Let G be a Lie group with Lie algebra L. Since the exponential mapping exp is an analytic mapping of the manifold L into G at every point $x \in L$, the differential $(d \exp)_x$ is a mapping of the tangent space $T_x(L)$ into the tangent space $T_{\exp(x)}(G)$.

The tangent space $T_x(L)$ can be canonically identified with the Lie algebra L as follows. Consider any $y \in L$. We associate with y the tangent vector $v_y \in T_x(L)$ defined by

$$v_y(f) = f(y) \tag{3.6.1}$$

for all linear functions f on the Lie algebra L. By Propositions I and II in 3.2, (3.6.1) uniquely defines a tangent vector $v_y \in T_x(L)$. The mapping $y \to v_y$ is an isomorphism of L onto $T_x(L)$.

We will now identify the tangent space $T_g(G)$, $g \in G$, with the Lie algebra L. For each element z of L we define a tangent vector $w_z(g) \in T_g(G)$ by the rule

$$w_z(g) = z(g). \tag{3.6.2}$$

Propositions I and II in 3.2 show that (3.6.2) is an isomorphism of L onto $T_g(G)$.

In the sequel we will write y instead of v_y and z instead of w_z. We now find an explicit formula for $(d \exp)_x(y)$.

Consider the mapping

$$\to \exp(-x) \exp(x + ty) \tag{3.6.3}$$

for $t \in \mathbf{R}$ or $t \in \mathbf{C}$, depending on whether G is a real or complex Lie group. From (3.6.3) we have $\psi(0) = e$. Since the element $z \in L$ is a left invariant vector field on G, we have $(d\varphi_g)_e z(e) = z(g)$, in agreement with (3.2.1). From the relation $\varphi_{\exp(x)}\psi(t) = \exp(x + ty)$ and (1.5.1) we obtain in particular that

$$(d\varphi)_{\exp x} \circ (d\psi/dt)|_{t=0} = ((d \exp)_x y)(\exp x) = (d\varphi_{\exp x} \circ ((d \exp)_x y))(e),$$

so that

$$(d \exp)_x(y) = ((d/dt)(\exp(-x) \exp(x + ty)))|_{t=0}. \tag{3.6.4}$$

Note too that

$$(t, x, y) \to \exp(-x) \exp(x + ty)$$

is an analytic mapping of the manifold $\mathbf{R} \times L \times L$ (or $\mathbf{C} \times L \times L$) into G. Thus (3.6.4) implies that the correspondence Φ that sends $(x, y) \in L \times L$ to the element $(d \exp)_x(y) \in L$ is an analytic mapping of $L \times L$ into L.

I. *Let A be an associative algebra and let a be an element of A. Let δ_a be the mapping of A into A defined by $\delta_a(x) = ax - xa$, $a \in A$. For all positive integers n, we have*

$$(\delta_a)^n(x) = (-1)^n \sum_{0 \leqslant k \leqslant n} (-1)^k C_n^k a^k x a^{n-k} \qquad (3.6.5)$$

for all $x \in A$.

Use induction on n and argue as in the proof of the binomial theorem. We omit the details.

II. *Let G be a Lie group with Lie algebra L. Let $d(\exp)_x$ be the differential of the exponential mapping at the point $x \in L$. We have*

$$(d \exp)_x = \sum_{n=0}^{\infty} \frac{(-1)^n}{(n+1)!} (\operatorname{ad} x)^n. \qquad (3.6.6)$$

The mapping $d(\exp)_x$ is an isomorphism if and only if no eigenvalue of the linear operator $\operatorname{ad} x$ has the form $2k\pi i$ for a nonzero integer k.

Proof. Consider x, y in L. Let U be a neighborhood of the identity element in G. Choose a number $a > 0$ such that

$$\exp(ux) \exp(vx + wy) \in U$$

for all u, v, w, such that $|u| < a$, $|v| < a$, $|w| < a$. For $f \in D(U)$, the function F defined by

$$F(u, v, w) = f(\exp(ux) \exp(vx + wy)) \qquad (3.6.7)$$

is analytic in the domain I_a defined by $|u| < a$, $|v| < a$, $|w| < a$. We may suppose that the number a is so small that the Taylor series for F at the point $(0, 0, 0)$ converges absolutely and uniformly on I_a. Let p, q, r be non-negative integers. We define

$$F_{p,q,r} = \left(\frac{\partial^{p+q+r} F}{\partial u^p \, \partial v^q \, \partial w^r} \right)(0, 0, 0).$$

Plainly we have

$$F(u, v, w) = \sum_{p,q,r \geqslant 0} \frac{F_{p,q,r}}{p! \, q! \, r!} u^p v^q w^r, \qquad ((u, v, w) \in I_a), \qquad (3.6.8)$$

and so

$$\frac{\partial F}{\partial w}(u, v, 0) = \sum_{p,q \geqslant 0} \frac{F_{p,q,1}}{p!q!} u^p v^q \qquad (3.6.9)$$

for all $|u| < a$, $|v| < a$. For $u = -t$, $v = t$, (3.6.9) yields

$$\frac{\partial F}{\partial w}(-t, t, 0) = \sum_{n=0}^{\infty} \frac{c_n}{n!} t^n \qquad (3.6.10)$$

for $|t| < a$. The coefficients c_n are defined by

$$c_n = \sum_{0 \leqslant k \leqslant n} (-1)^k C_n^k F_{k, n-k, 1}. \qquad (3.6.11)$$

From (3.6.4) and (3.6.7) we infer

$$\frac{\partial F}{\partial w}(-t, t, 0) = ((d \exp)_{tx}(y)f)(e) \qquad (3.6.12)$$

for $|t| < a$. On the other hand, for $(u, v, w) \in I_a$, (3.6.7) and (3.6.8) imply that

$$((vx + wy)^k f)(\exp(ux)) = \left(\frac{d^k}{dt^k} F(u, tv, tw) \right) \bigg|_{t=0}$$

$$= k! \sum_{p \geqslant 0, \, 0 \leqslant q \leqslant k} \frac{F_{p, q, k-q}}{p!q!(k-q)!} u^p v^q w^{k-q}. \qquad (3.6.13)$$

We set $k = q + 1$. Equating the coefficients in the right and left sides of (3.6.13) for $v^q w$, we obtain

$$\left(\left(\sum_{0 \leqslant s \leqslant q} x^s y x^{q-s} \right) f \right)(\exp ux) = (q + 1)! \sum_{p \geqslant 0} \frac{F_{p,q,1}}{p!q!} u^p$$

$$= (q + 1) \sum_{p \geqslant 0} \frac{F_{p,q,1}}{p!} u^p$$

$$= (q + 1) \left(\frac{\partial^{q+1} F}{\partial v^q \, \partial w} \right)(u, 0, 0) \qquad (3.6.14)$$

for all $|u| < a$. Differentiate (3.6.14) p times with respect to u and then set $u = 0$. For all integers $p \geqslant 0$, $q \geqslant 0$, we obtain (from Proposition I in 3.4) that

$$F_{p,q,1} = \frac{1}{q+1} \left(\left(\sum_{0 \leqslant s \leqslant q} x^{p+s} y x^{q-s} \right) f \right)(e). \qquad (3.6.15)$$

Substituting (3.6.15) in (3.6.11) we find that

$$
\begin{aligned}
c_n &= \left(\left(\sum_{k=0}^{n}(-1)^k C_n^k \frac{1}{n-k+1}\sum_{s=0}^{n-k} x^{k+s}yx^{n-k-s}\right)f\right)(e) \\
&= \frac{1}{n+1}\left(\left(\sum_{k=0}^{n}(-1)^k C_{n+1}^k \sum_{k\leqslant m\leqslant n} x^{m}yx^{n-m}\right)f\right)(e) \\
&= \frac{1}{n+1}\sum_{m=0}^{n}(x^{m}yx^{n-m})(e)\left(\sum_{k=0}^{m}(-1)^k C_{n+1}^k\right).
\end{aligned}
\tag{3.6.16}
$$

It is easy to verify by induction on m that

$$
\sum_{k=0}^{m}(-1)^k C_{n+1}^k = (-1)^m C_n^m.
\tag{3.6.17}
$$

Substituting (3.6.17) in (3.6.16), we obtain

$$
c_n = \frac{1}{n+1}\left(\left(\sum_{m=0}^{n}(-1)^m C_n^m x^{m}yx^{n-m}\right)f\right)(e).
\tag{3.6.18}
$$

On the other hand, (3.6.5) implies that in the enveloping algebra of L, the identity

$$
(\mathrm{ad}\, x)^n(y) = (-1)^n \sum_{m=0}^{n}(-1)^m C_n^m x^{m}yx^{n-m}
\tag{3.6.19}
$$

holds. From (3.6.19) and (3.6.18) we infer that

$$
c_n = \frac{(-1)^n}{n+1}(((\mathrm{ad}\, x)^n y)f)(e).
\tag{3.6.20}
$$

By (3.6.20), (3.6.10) and (3.6.12) we now get

$$
((d\,\exp)_{tx}(y)f)(e) = \sum_{n=0}^{\infty}\frac{(-1)^n}{(n+1)!}t^n((\mathrm{ad}\, x)^n(y)f)(e)
\tag{3.6.21}
$$

for all $|t| < a$. The relation (3.6.21) holds in particular if U is a coordinate neighborhood of the element e and f is any of the functions that form a system of analytic coordinates in the neighborhood U. Thus (3.6.21) implies that

$$
(d\,\exp)_{tx}(y) = \sum_{n=0}^{\infty}\frac{(-1)}{(n+1)!}t^n(\mathrm{ad}\, x)^n(y)
\tag{3.6.22}
$$

for $|t| < a$. Both sides of (3.6.22) are entire analytic functions and therefore they coincide everywhere. For $t = 1$, we obtain

$$(d \exp)_x(y) = \sum_{n=0}^{\infty} \frac{(-1)^n}{(n+1)!} (\operatorname{ad} x)^n(y). \tag{3.6.23}$$

Since $y \in L$ is arbitrary, (3.6.23) implies (3.6.6).

Consider the function $F(z)$ of the complex variable z defined by

$$F(z) = \sum_{n=0}^{\infty} \frac{(-1)^n}{(n+1)!} z^n$$

for all $z \in \mathbf{C}$. We then have $zF(z) = 1 - e^{-z}$ and $F(z) = 0$ if and only if $z = 2k\pi i$ for some nonzero integer k. Consider any x in L. The mapping $(d \exp)_x$ is an isomorphism if and only if all eigenvalues of the transformation $(d \exp)_x$ are distinct from zero. The formula (3.6.6) shows that

$$(d \exp)_x = F(\operatorname{ad} x), \tag{3.6.24}$$

so that all eigenvalues of the operator $(d \exp)_x$ are equal to $F(\lambda_1), \ldots, F(\lambda_m)$, where $\lambda_1, \ldots, \lambda_m$ are eigenvalues of the operator $\operatorname{ad} x$. Hence $(d \exp)_x$ is an isomorphism if and only if $\lambda_j \neq 2k\pi i$ for all $j = 1, \ldots, m$, where $k \neq 0$ are integers. $\square$

Chapter X

Lie Algebras

The basic definitions dealing with Lie algebras and their representations are given in §2, chapter IX, along with the most important examples. In this chapter we present the general theory of Lie algebras.

§1. Some Definitions

1.1. Differentiation in a Lie Algebra

Let A be an algebra over the field K (A can be either a Lie algebra or an associative algebra). A linear mapping d of A into itself that satisfies

$$d(xy) = x(dy) + (dx)y \tag{1.1.1}$$

is called a *differentiation in the algebra A*.

I. *The set* Der(A) *of all differentiations of the algebra A is a Lie algebra under the ordinary linear operations and the operation of commutation defined by*

$$[d, d_1] = dd_1 - d_1 d$$

for all d, $d_1 \in$ Der(A). If L is a Lie algebra and $x \in L$, the operator ad $x : L \to L$ *defined by* ad $x(y) = [x, y]$, *is a differentiation of L. The mapping $x \to$ ad x is a homomorphism of L into* Der (L) *and the image of L under this homomorphism is an ideal in* Der(L).

Proof. Let d, d_1 be differentiations of the algebra A. We prove that $[d, d_1]$ is a differentiation by direct computation. We have in fact

$$
\begin{aligned}
[d, d_1](xy) &= dd_1(xy) - d_1 d(xy) \\
&= d((d_1 x)y + x(d_1 y)) - d_1((dx)y + x(dy)) \\
&= ((dd_1)x)y + (d_1 x)(dy) + (dx)(d_1 y) + x((dd_1)y) \\
&\quad - ((d_1 d)x)y - (dx)(d_1 y) - (d_1 x)(dy) - x((d_1 d)y) \\
&= (dd_1 x)y + x(dd_1 y) - (d_1 dx)y - x(dd_1 y) \\
&= ([d, d_1]x)y + x([d, d_1]y). \tag{1.1.2}
\end{aligned}
$$

From (1.1.2) and (1.1.1) we infer that $[d, d_1]$ is a differentiation in A. Suppose that L is a Lie algebra, and that $x, y, z \in L$. Jacobi's identity gives

$$\begin{aligned}
\operatorname{ad} x([y, z]) = [x, [y, z]] &= -[y, [z, x]] - [z, [x, y]] \\
&= [[x, y], z] + [y, [x, z]] \\
&= [(\operatorname{ad} x)(y), z] + [y, (\operatorname{ad} x)(z)].
\end{aligned} \tag{1.1.3}$$

Comparing (1.1.3) and (1.1.1), we see that the mapping ad $x: L \to L$ is a differentiation in L. The mapping $x \to \operatorname{ad} x$ is a homomorphism of L into Der(L) according to (2.1.5) in chapter IX. Finally, for $x \in L$ and $d \in \operatorname{Der}(L)$ we have

$$\begin{aligned}
[d, \operatorname{ad} x](y) &= d(\operatorname{ad} x(y)) - (\operatorname{ad} x)(dy) \\
&= d([x, y]) - [x, dy] \\
&= [dx, y] + [x, dy] - [x, dy] = [dx, y]
\end{aligned} \tag{1.1.4}$$

for all $y \in L$. From (1.1.4) we infer that

$$[d, \operatorname{ad} x] = \operatorname{ad}(dx) \tag{1.1.5}$$

for all $x \in L$ and $d \in \operatorname{Der}(L)$. Thus the linear space of differentiations ad x, $x \in L$, is an ideal in Der(L).

The ideal of differentiations ad x, $x \in L$, in the Lie algebra Der(L) is called the *ideal of inner differentiations of L*.

1.2. The Contragredient Representation and the Tensor Product of Representations of Lie Algebras

Let π be a representation of a Lie algebra L in a space V. Let the spaces V and V^* be in duality under the bilinear form (v, v^*) $(v \in V, v^* \in V^*)$. The representation π^* of L in V^*, defined by

$$(v, \pi^*(x)v^*) = -(\pi(x)v, v^*) \tag{1.2.1}$$

for all $x \in L$, $v \in V$, $v^* \in V^*$, is called the representation *contragredient to the representation* π.

Let $\pi_1, \ldots, \pi_n$ be representations of the Lie algebra L in spaces $V_1, \ldots, V_n$ respectively. The representation π of L in the space $V_1 \otimes \cdots \otimes V_n$ defined by

$$\pi(x)(v_1 \otimes \cdots \otimes v_n) = \sum_{i=1}^{n} v_1 \otimes \cdots \otimes \pi_i(x)v_i \otimes \cdots \otimes v_n \tag{1.2.2}$$

for all $x \in L$, $v_i \in V_i$ $(i = 1, \ldots, n)$ is called the *tensor product of the representations* $\pi_1, \ldots, \pi_n$ and is denoted by $\pi_1 \otimes \cdots \otimes \pi_n$.

It is easy to verify that π^* and $\pi_1 \otimes \cdots \otimes \pi_n$ are representations of L.

1.3. The Canonical Mapping of $V_1^* \otimes V_2$ onto $L(V_1, V_2)$

Let V_1, V_2 be finite-dimensional vector spaces. Let V_1^* be the space conjugate to V_1. Consider the mapping τ of the space $V_1^* \otimes V_2$ into the space $L(V_1, V_2)$ of all linear mappings of V_1 into or onto V_2, defined by

$$\tau\left(\sum_{i=1}^{n} f_i \otimes y_i \right)(z) = \sum_{i=1}^{n} f_i(z)y_i, \qquad z \in V_1, \qquad y_i \in V_2, \qquad f_i \in V_1^*. \quad (1.3.1)$$

The mapping τ is plainly linear. Any linear operator T from V_1 in V_2 belongs to the image of τ, since we have

$$Tz = \sum_{j=1}^{m} \varphi_j(z)e_j, \qquad z \in V_1, \qquad (1.3.2)$$

where (e_j) is a fixed basis in V_2 and φ_j are certain linear functionals on V_1. Since the dimensions of the spaces $V_1^* \otimes V_2$ and $L(V_1, V_2)$ are equal and the image of the mapping τ is the entire space $L(V_1, V_2)$, τ is an isomorphism of $V_1^* \otimes V_2$ onto $L(V_1, V_2)$. This isomorphism τ is called the *canonical isomorphism of $V_1^* \otimes V_2$ onto $L(V_1, V_2)$*.

Let L be a Lie algebra. Let π be a homomorphism of L into the algebra $gl(V)$, where V is a finite-dimensional vector space. We define the homomorphism π^* of L into $gl(V^*)$ by

$$\pi^*(x) = -(\pi(x))^* \qquad (1.3.3)$$

for all $x \in L$, where $(\pi(x))^*$ is the operator conjugate to $\pi(x)$. It is easy to verify that the mapping π^* is indeed a homomorphism. This homomorphism is called *conjugate to π*.

Consider the homomorphism $\pi^* \otimes \pi$ of the Lie algebra L into the Lie algebra $gl(V^* \otimes V)$ defined by

$$(\pi^* \otimes \pi)(x)(v^* \otimes v) = \pi^*(x)(v^* \otimes v) + v^* \otimes \pi(x)v$$
$$= -(\pi(x))^*v^* \otimes v + v^* \otimes \pi(x)v \qquad (1.3.4)$$

for all $x \in L$, $v \in V$, $v^* \in V^*$. Let θ be the homomorphism of L into $gl(L(V, V))$ corresponding to $\pi^* \otimes \pi$ under the isomorphism τ. For an operator T of the form (1.3.2) we have

$$(\theta(x)T)z = \theta(x)\tau\left(\sum_{j=1}^{m} \varphi_j \otimes e_j\right)z$$

$$= \tau\left((\pi^* \otimes \pi)(x)\left(\sum_{j=1}^{m} \varphi_j \otimes e_j\right)\right)(z)$$

$$= \tau\left(\sum_{j=1}^{m}\left((-\pi(x)^*)\varphi_j \otimes e_j + \varphi_j \otimes \pi(x)e_j\right)\right)(z)$$

$$= -\sum_{j=1}^{m}(\pi(x)^*\varphi_j)(z)e_j + \sum_{j=1}^{m}\varphi_j(z)\pi(x)e_j$$

$$= -\sum_{j=1}^{m}\varphi_j(\pi(x)z)e_j + \sum_{j=1}^{m}\varphi_j(z)\pi(x)e_j$$

$$= -T\pi(x)z + \pi(x)Tz = [\pi(x), T]z. \tag{1.3.5}$$

From (1.3.5) we infer that

$$\theta(x)T = [\pi(x), T] \tag{1.3.6}$$

for all $x \in L$, $T \in L(V, V)$.

1.4. The Complex Hull of a Real Lie Algebra

Let L be a real Lie algebra. Let $L_{\mathbf{C}}$ denote the real linear space $L + L$. Let J be the linear operator on $L + L$ defined by $J\{x, y\} = \{-y, x\}$. We define

$$(\alpha + i\beta)\{x, y\} = \alpha\{x, y\} + \beta J\{x, y\} = \{\alpha x - \beta y, \alpha y + \beta x\} \tag{1.4.1}$$

for all α, $\beta \in \mathbf{R}$, x, $y \in L$, as well as

$$[\{x, y\}, \{x_1, y_1\}] = \{[x, x_1] - [y, y_1], [y, x_1] + [x, y_1]\} \tag{1.4.2}$$

for all $x, y, x_1, y_1 \in L$. We provide the set $L_{\mathbf{C}}$ with componentwise addition, multiplication by complex numbers defined by (1.4.1), and the operation of commutation defined by (1.4.2). Then $L_{\mathbf{C}}$ is a complex Lie algebra, also denoted by $L_{\mathbf{C}}$. The complex Lie algebra $L_{\mathbf{C}}$ is called the *complexification* or the *complex hull of the real Lie algebra* L. Plainly L can be identified with the set $\tilde{L}$ of elements of $L_{\mathbf{C}}$ of the form $\tilde{x} = \{x, 0\}$, where $x \in L$.

If $e_1, \ldots, e_n$ is a basis in L, the elements $\tilde{e}_1, \ldots, \tilde{e}_n$ are a basis in the complex linear space $L_{\mathbf{C}}$. The structural constants of the complex Lie algebra in the basis $\tilde{e}_1, \ldots, \tilde{e}_n$ are real, since they coincide with the structural constants of the real Lie algebra L in the basis $e_1, \ldots, e_n$.

Let V be a real linear space. We denote by $L_{\mathbf{C}}$ the complex linear space $V + V$ with ordinary addition and multiplication by complex numbers defined by

$$(\alpha + i\beta)\{x, y\} = \{ax - \beta y, ay + \beta x\}.$$

If π is a homomorphism of L into $gl(V)$, the formula

$$(\pi_{\mathbf{C}}(x,y))(\xi,\eta) = (\pi(x)\xi - \pi(y)\eta, \pi(x)\eta + \pi(y)\xi), \qquad x, y \in L, \xi, \eta \in V,$$

defines a representation $\pi_{\mathbf{C}}$ of $L_{\mathbf{C}}$ in the space $V_{\mathbf{C}}$. We may suppose that V is embedded in $V_{\mathbf{C}}$. The representation $\pi_{\mathbf{C}}$ is called the *complexification the representation* π.

The correspondence $\pi \rightarrow \pi_{\mathbf{C}}$ is plainly a one-to-one correspondence between the homomorphisms of L into $gl(V)$ and the representations $\pi_{\mathbf{C}}$ of $L_{\mathbf{C}}$ in the space $V_{\mathbf{C}}$ for which $\pi_{\mathbf{C}}(L)V \subset V$. This fact often allows us to reduce the study of the homomorphisms $\pi: L \rightarrow gl(V)$ to the study of the representations of the complex Lie algebra $L_{\mathbf{C}}$.

1.5. The Jordan-Hölder Series

The following proposition will be useful in the sequel.

I. *Let π be a homomorphism of the Lie algebra L into the Lie algebra $gl(V)$, where V is a finite-dimensional vector space. There exists a sequence of subspaces $V_0 = \{0\} \subset V_1 \subset V_2 \subset \cdots \subset V_n = V$ of V such that each of the subspaces V_k is invariant under $\pi(L)$ and the set of operators induced by the set $\pi(L)$ in V_k/V_{k-1} is irreducible.*

Proof. If the set $\pi(L)$ is irreducible, our assertion is trivial. Suppose that $\pi(L)$ is reducible. Thus V contains a nonzero proper subspace V^1 invariant under $\pi(L)$. The dimension of V^1 must be less than the dimension of V. If the restriction of $\pi(L)$ to V^1 is reducible, V^1 contains a nontrivial invariant subspace, with dimension less than the dimension of V^1. Since V is finite-dimensional, V contains a nonzero subspace V_1 on which the restriction of $\pi(L)$ is irreducible. The homomorphism π thus defines a homomorphism π_1 of the Lie algebra L into the Lie algebra $gl(V/V_1)$. Let $\tilde{V}_2$ be a nonzero subspace of V/V_1 on which $\pi_1(L)$ is irreducible. (The preceding argument shows that this subspace exists.) The inverse image of $\tilde{V}_2$ in V is denoted by V_2. It is invariant under $\pi(L)$. Continuing this construction, we obtain a finite sequence of subspaces of V:

$$\{0\} = V_0 \subset V_1 \subset V_2 \subset \cdots \subset V_n = V. \tag{1.5.1}$$

Every V_k is invariant under $\pi(L)$ and the set of operators induced by the set of operators $\pi(L)$ in V_{k+1}/V_k $(k = 0, 1, \ldots, n-1)$, is irreducible. $\qquad\square$

The chain of subspaces (1.5.1) is called a *Jordan-Hölder series for the homomorphism* π.

1.6. Definitions of Solvable and Nilpotent Lie Algebras

Let A, B be subsets of a Lie algebra L. The linear subspace of L spanned by all elements of the form $[a, b]$, $a \in A$, $b \in B$, is denoted by $[A, B]$.

I. *If M and N are ideals in L, then $[M, N]$ is an ideal in L.*

Proof. Consider $a \in M$ and $b \in N$. For every $x \in L$ we have

$$[[a, b], x] = [[a, x], b] - [[b, x], a] \tag{1.6.1}$$

by Jacobi's identity. Because M and N are ideals, we have $[a, x] \in M$ and $[b, x] \in N$. Therefore the right side of (1.6.1) belongs to $[M, N]$. Since any element $y \in [M, N]$ is a finite linear combination of elements of the form $[a, b]$ ($a \in M$, $b \in N$), we have $[y, x] \in [M, N]$ for all $y \in [M, N]$. That is, the linear subspace $[M, N]$ is an ideal. $\square$

Let L be a Lie algebra. We define

$$L = L^{(0)} = L_0; \qquad L^{(n)} = [L^{(n-1)}, L^{(n-1)}]; \qquad L_n = [L, L_{n-1}]. \tag{1.6.2}$$

Proposition I shows that $L^{(n)}$ (L_n) is an ideal in $L^{(r)}$ (L_r) for $0 \leqslant r \leqslant n$. The sequence of ideals L_n is called the *descending central series* of the Lie algebra L. The sequence $L^{(n)}$ is called the *derived series of L*. The ideal $L^{(1)}$ is called the *derived ideal of L*. The Lie algebra L is called *solvable* (*nilpotent*) if there is a natural number n such that $L^{(n)} = \{0\}$ ($L_n = \{0\}$). The smallest n is called the *height* (*rank*) *of the solvable* (*nilpotent*) *Lie algebra L*. $\square$

II. *Every nilpotent Lie algebra is solvable.*

The result is implied by the obvious relation $L^{(n)} \subset L_n$.

III. *Any subalgebra M of a solvable* (*nilpotent*) *Lie algebra is solvable* (*nilpotent*).

The result is implied by the inclusions $M^{(n)} \subset L^{(n)}$, $M_n \subset L_n$.

IV. *A homomorphic image* (*factor algebra*) *of a solvable* (*nilpotent*) *Lie algebra L is solvable* (*nilpotent*).

Proof. Let π be a homomorphism of the Lie algebra L onto the Lie algebra L'. It is clear that $L'^{(n)} = \pi(L^{(n)})$, $L'_n = \pi(L_n)$, and the result follows. $\square$

V. *Every factor algebra $L^{(n-1)}/L^{(n)}$ is commutative.*

Proof. Let π be the canonical homomorphism of $L^{(n-1)}$ onto $L^{(n-1)}/L^{(n)}$. It is clear that $[x, y] \in L^{(n)}$ for $x, y \in L^{(n-1)}$ and thus we have $[\pi(x), \pi(y)] = \pi([x, y]) = 0$. $\square$

VI. *Let M be an ideal in $L^{(n-1)}$ such that the factor algebra $L^{(n-1)}/M$ is commutative. Then M contains $L^{(n)}$.*

Proof. Let π be the canonical homomorphism of $L^{(n-1)}$ onto $L^{(n-1)}/M$. For $x, y \in L^{(n-1)}$, we have $\pi([x, y]) = [\pi(x), \pi(y)] = 0$ and thus $\pi(L^{(n)}) = \{0\}$. The relation $L^{(n)} \subset M$ follows. $\square$

§2. Representations of Nilpotent and Solvable Lie Algebras

2.1. Nilpotent Lie Algebras and Engel's Theorem

The principal aim of this section is to prove the following two theorems.

Theorem 1. *A Lie algebra L is nilpotent if and only if for every $x \in L$, the linear transformation $\operatorname{ad} x$ is nilpotent (that is, there exists a natural number p such that $(\operatorname{ad} x)^p = 0$).*

Theorem 2 (Engel's Theorem). *Let L be a Lie algebra over the field K. Let V be a finite-dimensional nonzero vector space over K and let π be a homomorphism of L into the Lie algebra $gl(V)$. If all of the linear transformations $\pi(x) (x \in L)$ are nilpotent, there exists a nonzero vector $v \in V$ such that $\pi(x)v = 0$ for all $x \in L$.*

Proof of theorem 2.

(a) Let N be the kernel of the representation π. We set $L' = L/N$ and define the representation $\tilde{\pi}$ of the Lie algebra L', setting $\tilde{\pi}(\tilde{x}) = \pi(x)$ for $\tilde{x} \in L$ and $x \in \tilde{x}$. It suffices to prove theorem 2 for the representation $\tilde{\pi}: \tilde{x} \to \pi(\tilde{x})$ of L'. From our construction it is clear that $\tilde{\pi}: \tilde{x} \to \tilde{\pi}(\tilde{x})$ is an exact representation of the Lie algebra L' and that this representation satisfies our hypotheses. We may therefore suppose that $N = \{0\}$. If L is one-dimensional, the theorem is simple. In fact, if $\pi(x)$ is a nonzero element in $\pi(L)$, every element in $\pi(L)$ is a scalar multiple of the element $\pi(x)$. Therefore it suffices to show that the kernel of the operator $\pi(x)$ is nonzero. By hypothesis we have $(\pi(x))^p = 0$ for some natural number p. That is, the kernel of the operator $(\pi(x))^p$ is equal to V and so is nonzero. The kernel of the operator $\pi(x)$ must also be distinct from zero. Thus we can apply induction to the dimension of the Lie algebra L. Prove the theorem using the following inductive hypothesis: Theorem 2 is valid for all Lie algebras M with dimension less than $\dim(L)$.

(b) We will prove that the Lie algebra L (of dimension exceeding 1) admits an ideal of codimension 1. Let M be a proper subalgebra of L. (For example every one-dimensional subspace is an abelian Lie subalgebra, since $[x, x] = 0$). Consider the homomorphism $x \to \operatorname{ad} x$ of L into $gl(L)$. The set of operators ad x, $x \in M$, leaves the subspace $M \subset L$ invariant. Going to the operators in the factor space L/M induced by the operators and x $(x \in M)$. we obtain a homomorphism ρ of the Lie algebra M into the Lie algebra $gl(L/M)$. By hypothesis, every operator $\pi(x)$, $x \in L$, is nilpotent. Since the representation π is exact, every operator ad x, $x \in L$, is also nilpotent. Indeed, by identifying the Lie algebra L with the Lie subalgebra $\pi(L) \subset gl(V)$ (with the aid of the exact representation π), we see that $(\operatorname{ad} x)y = xy - yx$ for all $x, y \in \pi(L)$. The identity

$$(\operatorname{ad} x)^n(y) = \sum_{p=0}^{n} (-1)^{n-p} C_n^p x^p y x^{n-p}$$

is easily verified by induction (see also (3.6.19) in chapter IX). From this identity, we infer that if $x^q = 0$, then $(\operatorname{ad} x)^{2q} = 0$. Thus the operators ad x, $x \in L$, are nilpotent. In this case, the operators $\rho(x)$, $x \in M$, being factor operators of nilpotent operators, are also nilpotent. Since dim $M < $ dim L, we may apply our inductive hypothesis. Let S be a one-dimensional subspace of L/M annihilated by all operators $\rho(x)$, $x \in M$. Let T be the inverse image of the subspace S under the canonical mapping of L onto L/M. It is clear that dim $T = $ dim $M + 1$ and that $[M, T] \subset M$. Thus T is a Lie subalgebra of L in which M is an ideal of codimension 1.

Up to this point in the proof, the subalgebra M has been an *arbitrary* proper subalgebra of L. We now suppose that M is a maximal proper subalgebra of L (the existence of such a subalgebra is clear because L is finite-dimensional). The subalgebra T defined in the preceding paragraph contains M as a proper subalgebra. Because M is maximal, T must be all of L. Therefore M is an ideal in L and it has codimension 1 in L.

(c) Let y be any element in $L \backslash M$, where M is an ideal of codimension 1 in L. The algebra L is generated as a linear space by the ideal M and the element y. Let W be the subspace of V consisting of all vectors annihilated by all operators $\pi(x)$, $x \in M$. Since dim $M < $ dim L, our inductive hypothesis shows that $W \neq \{0\}$. We will show that the subspace W is invariant under all operators $\pi(x)$, $x \in L$. For $x \in L$, $h \in M$, $v \in W$, we have

$$\pi(h)\pi(x)v = \pi(x)\pi(h)v + \pi([h, x])v = 0, \qquad (2.1.1)$$

inasmuch as M is an ideal. Since $h \in M$ is arbitrary, (2.1.1) implies that $\pi(x)v$ is contained in W and W is invariant under $\pi(y)$. On the other hand, $\pi(y)$ is nilpotent. Thus there is a nonzero vector $v_0 \in W$ annihilated by the operator $\pi(y)$. Hence the vector v_0 is annihilated by all operators $\pi(x)$, $x \in L$. This completes the proof. $\square$

Corollary 1. *Let L be a Lie algebra such that all operators* ad x, $x \in L$, *are nilpotent. Then the center Z of L is distinct from zero.*

Proof. Apply theorem 2 to the homomorphism $x \to$ ad x of the Lie algebra L into $gl(L)$. A nonzero element $x \in L$ annihilated by all operators ad x is contained in Z. Therefore we have $Z \neq \{0\}$. $\square$

Corollary 2. *Let L be a Lie algebra such that all operators* ad x, $x \in L$, *are nilpotent. There is a chain of ideals*

$$L = A_1 \supset A_2 \supset \cdots \supset A_n = \{0\} \tag{2.1.2}$$

such that A_i/A_{i+1} is contained in the center of the algebra L/A_{i+1}.

Proof. Apply corollary 1 to the Lie algebras L, L/Z (the operators ad y, $y \in L/Z$ are factor operators of nilpotent operators and are therefore nilpotent). Continue this process. In particular, we may take $A_{n-1} = Z$. $\square$

Corollary 3. *Let π be a homomorphism of a Lie algebra L into the Lie algebra $gl(V)$ that satisfies the hypotheses of theorem 2. There is a natural number q such that the product of any q operators $\pi(x)$, $x \in L$, is equal to zero.*

Proof. Consider the subspace W_0 of V formed by the vectors annihilated by all operators $\pi(x)$, $x \in L$. We know that $W_0 \neq \{0\}$. If $W_0 \neq V$, we set $V_1 = V/W_0$. The operators $\pi(x)$, $x \in L$, induce a homomorphism π_1 of L into $gl(V_1)$. The homomorphism π_1 satisfies the hypotheses of theorem 2. Let W_1' be the subspace of V_1 of all vectors annihilated by all operators $\pi_1(x)$, $x \in L$. Let W_1 be the inverse image of W_1' under the canonical mapping of V onto V_1. It is clear that W_1 is a subspace of V invariant under all operators $\pi(x)$, $x \in L$. The nonequality $W_1' \neq \{0\}$ implies that $W_1 \neq W_0$. We set $V_2 = V/W_1$ if $W_1 \neq V$, and so on. Thus the subspaces W_i, $i = 1, 2, \ldots$ are defined by induction. The subspace W_{i+1} consists of those vectors $v \in V$ for which $\pi(x)v \in W_i$ for all $x \in L$. Theorem 2 shows that if $W_i \neq V$, then $W_i \neq W_{i+1}$. Since the space V is finite-dimensional, we have $W_p = V$ for some natural number p. Thus for any $x_0, x_1, \ldots, x_p \in L$ we have

$$\pi(x_0) \cdots \pi(x_p)W_p \subset \pi(x_0) \cdots \pi(x_{p-1})W_{p-1} \subset \cdots \subset \pi(x_0)W_0 = \{0\},$$

and so we have proved corollary 3. $\square$

Proof of theorem 1. Suppose that all of the operators $\mathrm{ad}(x)$, $x \in L$, are nilpotent. By corollary 3 there is a natural number q such that the product of any q operators ad x, $x \in L$, is equal to zero. Consider the descending central series of the Lie algebra L. Since we have $[L, L_n] = L_{n+1}$, we see that

$L_{q+1} = \{0\}$. Conversely, if $L_{q+1} = \{0\}$, then the definitions of $\text{ad}(x)$ and of the ideals L_i show that $(\text{ad}(x))^q = 0$ for all $x \in L$. $\square$

Theorem 1 gives a criterion for nilpotency of a Lie algebra L.

2.2. Some Properties of Solvable Lie Algebras

I. *A Lie algebra L is solvable if and only if there is a nested sequence of Lie subalgebras*

$$L = M_0 \supset M_1 \supset \cdots \supset M_m \supset M_{m+1} = \{0\},$$

where M_{k+1} is an ideal in M_k $(0 \leqslant k \leqslant m)$ and the factor algebra M_k/M_{k+1} is commutative.

Proof. The factor algebras $L^{(k)}/L^{(k+1)}$ are commutative (Proposition V in 1.6) and $L^{(k+1)}$ is an ideal in $L^{(k)}$ (Proposition I in 1.6). Therefore if L is solvable, we may define M_k as $L^{(k)}$. Conversely, let L be a Lie algebra with Lie subalgebras that satisfy our hypotheses. Because the factor algebras M_k/M_{k+1} are commutative, we have $[M_k, M_k] \subset M_{k+1}$. By induction we find that $L^{(k)} \subset M_k$ for all $k = 1, \ldots, m + 1$. Since we have $M_{m+1} = \{0\}$, we see that $L^{(m+1)} = \{0\}$, that is, the Lie algebra L is solvable. $\square$

II. *Let L be a Lie algebra and M an ideal in L. If the algebras M and L/M are solvable, then L is also solvable.*

Proof. Let $M = M_0 \supset M_1 \supset \cdots \supset M_p \supset M_{p+1} = \{0\}$ and $L/M = N_0 \supset N_1 \supset \cdots N_q \supset N_{q+1} = \{0\}$ be sequences of subalgebras as described in Proposition I. Let $\tilde{M}_k$ be the inverse image of N_k in L $(k = 0, \ldots, q + 1)$. We then have $\tilde{M}_{q+1} = M = M_0$, $\tilde{M}_{k+1}$ is an ideal in $\tilde{M}_k$ $(0 \leqslant k \leqslant q)$, and the factor algebra $\tilde{M}_k/\tilde{M}_{k+1}$ is isomorphic to N_k/N_{k+1} and is therefore commutative. Thus the sequence of subalgebras

$$L = \tilde{M}_0 \supset \tilde{M}_1 \supset \cdots \supset \tilde{M}_{q+1} \supset M_1 \supset \cdots \supset M_p \supset M_{p+1} = \{0\}$$

is as in Proposition I. Hence L is solvable. $\square$

2.3. Lie's Theorem

Theorem (S. Lie). *Let π be an irreducible linear representation of the solvable Lie algebra L in a nonzero complex vector space V. We then have $\dim V = 1$.*

Proof. It suffices to prove that the space V admits a nonzero vector that is an eigenvector for all operators of the representation π. If $\dim L = 1$, our theorem follows at once, because a linear operator on a finite-dimensional complex linear space admits an eigenvector. We will prove the theorem by induction on the dimension of L.

Suppose that dim $L > 1$. Let M be linear subspace of L of codimension 1 such that $M \supset L^{(1)}$. (Since L is solvable, we have $L^{(1)} \neq L$ and such a subspace M exists). We then have $[L, M] \subset [L, L] = L^{(1)} \subset M$ and so M is an ideal in the Lie algebra L of codimension 1. Choose a y_0 in $L \backslash M$. Then y_0 and M generate the vector space L. By our inductive hypothesis, there is a nonzero vector v_0 in V such that $\pi(x)v_0 = \lambda(x)v_0$ for all $x \in M$, where the $\lambda(x)$ are numbers. Plainly λ is a linear functional on M. We write v_n for $(\pi(y_0))^n v_0$. Since V is finite-dimensional, we have an integer $p > 0$ such that the vectors $v_0, v_1, \ldots, v_p$ are linearly independent and the vectors v_0, $v_1, \ldots, v_{p+1}$ are linearly dependent. Consider the subspace W of V that is generated by the vectors $v_0, v_1, \ldots, v_p$. By our definition of p, the subspace W is invariant under the operator $\pi(y_0)$. We will prove that W is also invariant under all operators of the form $\pi(h)$, $h \in M$. To do this, we will show by induction on q that the vector $\pi(h)v_q$ differs from $\lambda(h)v_q$ by a linear combination of the vectors $v_0, v_1, \ldots, v_{q-1}$ for all $h \in M$. For $q = 0$ we have $\pi(h)v_0 = \lambda(h)v_0$, and take our linear combination of the (void set of) preceding vectors as 0. Suppose that our claim holds for some integer $q \geqslant 0$. Then we have

$$\pi(h)v_{q+1} = \pi(h)\pi(y_0)v_q = [\pi(h), \pi(y_0)]v_q + \pi(y_0)\pi(h)v_q$$
$$= \pi([h, y_0])v_q + \pi(y_0)\pi(h)v_q.$$

Since M is an ideal, $[h, y_0]$ belongs to M and by the inductive hypothesis, $\pi([h, y_0])v_q$ is a linear combination of $v_0, \ldots, v_q$. This hypothesis ensures also that $\pi(h)v_q$ differs from $\lambda(h)v_q$ by a linear combination of $v_0, v_1, \ldots, v_{q-1}$. Therefore $\pi(y_0)\pi(h)v_q$ differs from $\pi(y_0)(\lambda(h)v_q) = \lambda(h)\pi(y_0)v_q = \lambda(h)v_{q+1}$ by a linear combination of $v_1, v_2, \ldots, v_q$. Hence the subspace W is invariant under $\pi(h)$, $h \in M$. Because W is also invariant under $\pi(y_0)$, it is invariant under all of $\pi(L)$.

Let $\text{tr}_W(x)$ be the trace of the restriction of the operator $\pi(x)$, $x \in L$, to the subspace W. Let h be in M. Because $\pi(h)v_q$ is the sum of $\lambda(h)v_q$ and a linear combination of the vectors $v_0, v_1, \ldots, v_{q-1}$, we see that $\text{tr}_W(h) = \lambda(h) \dim W$. On the other hand, the trace of any commutator is zero, and so $\text{tr}_W(h)$ is 0 for all $h \in L^{(1)}$; that is, $\lambda(h) = 0$ for $h \in L^{(1)}$. We will show that $\pi(h)v_q = \lambda(h)v_q$ for all $h \in M$. For $q = 0$, this holds by our choice of v_0. If the equality holds for a given q, we have

$$\pi(h)v_{q+1} = \pi(h)\pi(y_0)v_q = \pi([h, y_0])v_q + \pi(y_0)\pi(h)v_q$$
$$= \lambda([h, y_0])v_q + \pi(y_0)\lambda(h)v_q = \lambda(h)v_{q+1},$$

since $\lambda([h, y_0]) = 0$ as already noted. Thus the space W consists entirely of eigenvectors for all of the operators $\pi(h)$, $h \in M$. Let $w \in W$ be an eigenvector for the operator $\pi(y_0)$. Plainly w is an eigenvector for all operators in $\pi(L)$. $\square$

Corollary 1. *A Lie algebra is solvable if and only if its derived ideal $L^{(1)}$ is nilpotent.*

Proof. Suppose first that L is a complex Lie algebra. It is easy to show (see (1.6.2)) that $L^{(k)} \subset (L^{(1)})_{k-1}$ for all $k = 1, 2, \ldots$. Therefore if $L^{(1)}$ is nilpotent, L is solvable.

Suppose conversely that L is solvable. The proof of Lie's Theorem shows that the elements $h \in L^{(1)}$ go into nilpotent operators under all linear representations. This applies in particular to the adjoint representation. From this and from theorem 1 in 2, we see that the derived ideal $L^{(1)}$ is nilpotent.

Suppose next that L is a real Lie algebra. We reduce to the complex case by considering the complexification L_C: the details, which are easy enough, are omitted. □

Corollary 2. *Let L be a solvable Lie algebra with a representation π in a complex linear space V. There is a basis in V such that all operators $\pi(x)$, $x \in L$, have upper triangular matrices in this basis.*

Proof. This assertion follows from a construction in Lie's theorem. On the invariant subspace W the operators $\pi(x)$, $x \in L$, have upper triangular matrices in the basis $(v_0, v_1, \ldots, v_p)$. To complete the present proof, go to the factor space V/W, and continue the process. □

2.4. The Structure of Linear Representations of Nilpotent Lie Algebras

Let L be a Lie algebra with a linear representation π in the space V. A function λ on L is called a *weight of the representation* π if there is a nonzero vector $v \in V$ such that $\pi(x)v = \lambda(x)v$ for all $x \in L$. Plainly a weight is a linear functional on L.

I. *If L is a solvable Lie algebra, any representation of L in a nonzero vector space has at least one weight.*

To prove this we need only apply Lie's Theorem.

Let L be a Lie algebra with a linear representation π in the space V. Let λ be a linear functional on V. Let $V(L, \lambda)$ or V^λ denote the subspace of V formed by all vectors $v \in V$ such that for some integer $n \geq 0$ (depending on v), we have

$$(\pi(x) - \lambda(x)1)^n v = 0 \qquad (2.4.1)$$

for all $x \in L$.

II. *Let π be a linear representation of a nilpotent Lie algebra L in a space V. Then:*

(1) *the subspaces $V(L, \lambda)$ are invariant under $\pi(L)$;*
(2) *if $V(L, \lambda) \neq \{0\}$, λ is a weight of the representation π and the restriction of π to $V(L, \lambda)$ has no other weights;*
(3) *the space V is the direct sum of the subspaces V^λ.*

Proof. Let A be a linear operator in V and μ a number. The subspace of V formed by vectors $v \in V$ such that $(A - \mu 1)^n v = 0$ for some $n \geq 0$ (depending on v), is denoted by $V(A, \mu)$. From (2.4.1), we see that $V(L, \lambda) = \bigcap_{x \in L} V(\pi(x), \lambda(x))$. To prove that V^λ is invariant, it thus suffices to show that $V(\pi(x), \lambda(x))$ is invariant under $\pi(L)$ for all $x \in L$. Choose elements $y \in L$ and $v \in V(\pi(x), \lambda(x))$. By hypothesis L is nilpotent, which is to say that $(\operatorname{ad} x)^k y = 0$ for some $k \geq 0$. We will show by induction on k that $\pi(y)v \in V(\pi(x), \lambda(x))$. For $k = 0$, the assertion is obvious. We also have

$$(\pi(x) - \lambda(x))\pi(y) = \pi(y)(\pi(x) - \lambda(x)) + \pi([x, y]),$$

and so

$$(\pi(x) - \lambda(x))^n \pi(y) = \pi(y)(\pi(x) - \lambda(x))^n$$
$$+ \sum_{m=0}^{n-1} (\pi(x) - \lambda(x))^{n-m-1} \pi([x, y])(\pi(x) - \lambda(x))^m. \quad (2.4.2)$$

(This is easy to prove by induction on n.)

We now show that for all $v \in V(\pi(x), \lambda(x))$ we have

$$(\pi(x) - \lambda(x)1)^{\dim V(\pi(x), \lambda(x))} v = 0.$$

By hypothesis the operator $A = \pi(x) - \lambda(x)1$ is nilpotent in $V(\pi(x), \lambda(x))$. Write p for $\dim V(\pi(x), \lambda(x))$. The characteristic polynomial of A in $V(\pi(x), \lambda(x))$ is equal to λ^p. The Hamilton-Cayley Theorem shows that $A^p = 0$. Choose $n > 2p$. Let us apply both sides of (2.4.2) to a vector $v \in V(\pi(x), \lambda(x))$. It is clear that $(\pi(x) - \lambda(x)1)^m v = 0$ for $m \geq p$. On the other hand, we have $\pi([x, y])v \in V(\pi(x), \lambda(x))$, since $(\operatorname{ad} x)^{k-1}[x, y] = (\operatorname{ad} x)^k y = 0$ and one may apply the inductive hypothesis. If $m < p$, then $n - m - 1 \geq p$. Therefore, after applying (2.4.2) to the vector $v \in V(\pi(x), \lambda(x))$, all summands on the right side vanish. Consequently we have $(\pi(x) - \lambda(x)1)^n(y)v = 0$ for all $v \in V(\pi(x), \lambda(x))$ and $n > 2p$. This means that $\pi(y)v \in V(\pi(x), \lambda(x))$. This proves that $V(\pi(x), \lambda(x))$ and $V(L, \lambda)$ are invariant.

We prove (2) from Lie's Theorem. The restriction of π to $V(L, \lambda)$ has a weight according to Proposition I. Write this weight as μ. The operator $\pi(x) - \lambda(x)1$ is nilpotent in $V(L, \lambda)$. At the same time, $\pi(x) - \mu(x)1$ has zero for an eigenvalue in $V(L, \lambda)$. Thus we have $\lambda(x) = \mu(x)$ for all $x \in L$.

We will now show that the subspaces V^λ are linearly independent. Let $\lambda_1, \ldots, \lambda_r$ be distinct weights of the representation π. The relation $\lambda_i(x) = \lambda_j(x)$ holds for $i \neq j$ on a subspace of L of codimension 1. No finite union of such hyperplanes can fill up L. Thus there is an $x \in L$ such that $\lambda_i(x) \neq \lambda_j(x)$ for $i \neq j$. If $v_i \in V^i$, $i = 1, \ldots, r$, then $(\pi(x) - \lambda_i(x))^n v_i = 0$ for some n. Reducing the operator $\pi(x)$ to its Jordan canonical form, we see that the vectors v_i lie in different root subspaces of the operator $\pi(x)$ and are therefore linearly independent. Thus the sum of all of the subspaces $V(L, \lambda)$ is direct.

We will now use induction on the dimension of V to prove that $V = \sum V^\lambda$. For $V = \{0\}$ this is trivial. Suppose that dim $V > 0$. If every operator $\pi(x)$, $x \in L$, has a unique eigenvalue $\lambda(x)$, Lie's Theorem shows that the function λ is a weight of the representation π and $V = V^\lambda$. If there is an $x \in L$ such that the operator $\pi(x)$ has at least two different eigenvalues, the space V is the direct sum of the subspaces $V(\pi(x), \lambda_i)$, which are invariant under $\pi(L)$ and have dimension strictly smaller than dim V. Therefore we may apply our inductive hypothesis to obtain $V = \sum V^\lambda$. $\square$

§3. Radicals of a Lie Algebra

I. *Every Lie algebra L contains a greatest solvable ideal.*

Proof. Let A and B be solvable ideals in L. The sum $A + B$ of these ideals in an ideal in L. We will show that the ideal $A + B$ is solvable. It is clear that B is an ideal in $A + B$. Furthermore, the factor algebra $(A + B)/B$ is isomorphic to the Lie algebra $A/(A \cap B)$. Therefore the Lie algebra $A + B$ contains the solvable ideal B. The factor algebra by this ideal is isomorphic to a factor algebra of a solvable Lie algebra and is therefore solvable. Hence $A + B$ is a solvable ideal in L. Since L is finite-dimensional, the sum of *all* solvable ideals of L coincides with the sum of a finite number of such ideals and so the sum is a solvable ideal. This ideal is the greatest solvable ideal in L. $\square$

II. *Let M be an ideal in a Lie algebra L. Let π be a homomorphism of L into the Lie algebra gl(V) of a finite-dimensional space V. Let E be the algebra of linear operators in V generated by the identity operator and the operators $\pi(x)$, $x \in L$. The following conditions are equivalent.*

(1) *For every $x \in M$, the operator $\pi(x)$ is nilpotent.*
(2) *There is a two-sided ideal I in the associative algebra E that has the following properties:*
   (a) *$\pi(h) \in I$ for $h \in M$;*
   (b) *the ideal I consists of nilpotent operators.*
(3) *If $\{0\} = V_0 \subset V_1 \subset \cdots \subset V_n = V$ is a Jordan-Hölder series for the homomorphism π, then $\pi(M)V_k$ is contained in V_{k-1} for all $k = 1, 2, \ldots, n$.*

Proof. Plainly (2) implies (1). If (3) holds, then for all $x \in M$ we have

$$(\pi(x))^n V = (\pi(x))^n V_n$$
$$= (\pi(x))^{n-1} \pi(x) V_n \subset (\pi(x))^{n-1} V_{n-1} \subset \cdots \subset \pi(x) V_1 \subset V_0 = \{0\},$$

Therefore (1) holds. We will now show that if (1) holds, (2) and (3) also hold. Let $\{0\} = V_0 \subset V_1 \subset \cdots \subset V_n = V$ be a Jordan-Hölder series for the homomorphism π (see Proposition I in 1.5). We set $W_k = V_k/V_{k-1}$. If x belongs to M, the operator $\pi(x)$ is nilpotent in V and so is nilpotent in V_k. Since the subspace V_{k-1} is invariant under $\pi(x)$ for all $x \in L$, each operator $\pi(x)$, $x \in L$, induces an operator $\rho(x)$ in $W_k = V_k/V_{k-1}$. For $x \in M$, the operator $\pi(x)|_{V_k}$ is nilpotent and therefore the operator $\rho(x)$ in the space W_k is also nilpotent for $x \in M$. Consider the subspace H of W_k consisting of the vectors $w \in W_k$ such that $\rho(x)w = 0$ for all $x \in M$. By Engel's Theorem (theorem 2 in 2.1), H is not $\{0\}$. We will show that the subspace H is invariant under the set of operators induced by operators from $\pi(L)$ in the space W_k. We have $\rho(x)w = 0$ for all $x \in M$, $w \in H$. On the other hand, since M is an ideal in L, we have $[x, y] \in M$ for $x \in M$, $y \in L$. We thus have

$$\rho(x)\rho(y)w = \rho([x, y])w + \rho(y)\rho(x)w = 0, \quad \text{for all } y \in L.$$

Therefore $\rho(y)w$ belongs to H for $w \in H$ and $y \in L$. The set of operators induced by operators from $\pi(L)$ in the space W_k is irreducible according to the definition of a Jordan-Hölder series. Thus we have $H = W_k$ and $\pi(M)V_k \subset V_{k-1}$; that is, condition (3) holds.

Let I be the set of elements T of the associative algebra E for which $TV_k \subset V_{k-1}$ for all $k = 1, \ldots, n$. Clearly, I is a linear subspace of E. Note that for every operator $S \in E$, we have $SV_k \subset V_k$ (since the algebra E is generated by operators from $\pi(L)$ and for any $S \in \pi(L)$, we have $SV_k \subset V_k$ by the definition of a Jordan-Hölder series). Therefore, for $S \in E$, $T \in I$, we have $STV_k \subset SV_{k-1} \subset V_{k-1}$ and $TSV_k \subset TV_k \subset V_{k-1}$; that is, $ST \in I$, $TS \in I$. Thus I is a two-sided ideal in the associative algebra E. We have already proved that $\pi(M)V_k \subset V_{k-1}$, that is, $\pi(M)$ is contained in I. Furthermore, we have $T^n V = T^n V_n \subset T^{n-1}(TV_n) \subset T^{n-1}V_{n-1} \subset \cdots \subset TV_1 \subset V_0 = \{0\}$ for all $T \in I$. Thus every element of the ideal I is a nilpotent operator in the space V. $\quad\square$

III. *Let E be the algebra of linear operators on the Lie algebra L generated by the operators ad x, $x \in L$ and the operator 1. Let $\{0\} = V_0 \subset V_1 \subset \cdots \subset V_n = L$ be a Jordan-Hölder series for the homomorphism $x \to \text{ad } x$ of L. The set N of elements $x \in L$ such that $(\text{ad } x)V_k \subset V_{k-1}$ for all $k = 1, \ldots, n$ is the greatest nilpotent ideal in L.*

Proof. Since we have $(\text{ad } y)V_k \subset V_k$ for all $y \in L$ and $(\text{ad } x)V_k \subset V_{k-1}$ for all $x \in N$, we see that $(\text{ad } x \, \text{ad } y)V_k \subset V_{k-1}$ and $(\text{ad } y \, \text{ad } x)V_k \subset V_{k-1}$. It follows that $\text{ad}([x, y])V_k \subset V_{k-1}$ for all $k = 1, \ldots, n$; that is, $[x, y] \in N$ for all $x \in N$ and $y \in L$. Thus N is an ideal in the Lie algebra L. Furthermore, it is obvious that ad x is nilpotent for all $x \in N$. By theorem 1 in 2.1, N is a nilpotent Lie algebra. Let N' be a nilpotent ideal in L. For any $n \in N'$, there is a k such that $(\text{ad } n)^k(N') = \{0\}$. On the other hand, $(\text{ad } n)(L)$ is contained in N', since N' is an ideal in L. Hence the operator ad n in the space L is nilpotent. Proposition II shows that $(\text{ad } n)V_k \subset V_{k-1}$ for all $k = 1, \ldots, n$; that is, N' is contained in N. $\quad\square$

The greatest solvable ideal of L is called the *radical of* L. The greatest nilpotent ideal is called the *nilradical*. The intersection of the kernels of all irreducible representations of L is called the *nilpotent radical of* L.
These three radicals are denoted by R, N and S, respectively. In Propositions I and III, we proved that R and N exist.
    A Lie algebra is called *semisimple* if it contains no nonzero solvable ideals. It is called *simple* if it is not one-dimensional and contains no nontrivial ideals.

IV. *Any simple Lie algebra is semisimple.*

Proof. If the simple Lie algebra L is not semisimple, it contains a nonzero solvable ideal. This ideal coincides with L, since L is simple. Therefore L is solvable. From the proof of Lie's theorem in 2.3, we see that a solvable algebra contains an ideal of codimension 1. This contradicts the simplicity of L, since dim $L > 1$. $\square$

V. *The radical R of a Lie algebra L is the smallest among the ideals A for which the factor algebra L/A is semisimple.*

Proof. Let A be an ideal in L and let L/A be semisimple. The image of R under the canonical homomorphism of L onto L/A is a solvable ideal in L/A. This ideal must be equal to zero, since L/A is semisimple. Thus R is contained in A. Furthermore, the Lie algebra L/R contains no nonzero solvable ideals. For, if R'/R is a nonzero solvable ideal in L/R, the ideal R' in the Lie algebra R is a solvable algebra (since R and R'/R are solvable). Therefore we have $R' \subset R$, that is, $R' = R$ and so L/R is a semisimple Lie algebra. $\square$

    We introduce here a property of differentiations of a Lie algebra that will be important in the sequel.

VI. *A differentiation of the Lie algebra L maps R into N.*

Proof. First we show that a differentiation d of the Lie algebra L maps R into R. Since d is a linear transformation of the space L, we can construct the linear transformation $A_t = e^{td} = 1 + td + (t^2/2!)/d^2 + \cdots$, for $t \in \mathbf{R}$. Since any linear transformation in a finite-dimensional space is continuous, the function $\varphi(t) = [A_t x, A_t y] - A_t[x, y]$ is analytic in t for $t \in \mathbf{R}$. We have $(d/dt)A_t = dA_t$ and therefore

$$\varphi'(t) = [dA_t x, A_t y] + [A_t x, dA_t y] - dA_t[x, y] = d\varphi(t).$$

Hence we have $\varphi^{(n)}(t) = d^n \varphi(t)$. We also have $\varphi(0) = 0$ and therefore $\varphi^{(n)}(0) = 0$ for all $n \geqslant 1$, which is to say that $\varphi(t) \equiv 0$. It follows that $[A_t x, A_t y] = A_t[x, y]$ for all $t \in R$. The identity $e^{td}e^{-td} = 1_L$ shows that the transformation A_t is one-to-one, *i.e.*, A_t is an isomorphism of the Lie algebra L onto itself; A_t carries an ideal onto an ideal and a solvable ideal onto a solvable ideal. Because R is the smallest solvable ideal, we have $A_t R \subset R$, that is, $A_t x \in R$ for $x \in R$, Thus we have $dx = (d/dt)(A_t x)|_{t=0} \in R$ for $x \in R$, that is, $dR \subset R$.

Now let D be the Lie algebra of differentiations of L. Let Q be the direct product of the vector spaces D and L. We define in Q the operation $[(d, x),$ $(d', x')] = ([d, d'], dx' - d'x + [x, x'])$ for $d, d' \in D$ and $x, x' \in L$. Thus Q is a Lie algebra, L is isomorphic to the ideal $\tilde{L} = \{(0, x), x \in L\}$ in Q and D is isomorphic to the Lie subalgebra $\tilde{D} = \{(d, 0), d \in D\}$ in Q. For $d \in D$ and $x \in L$, we have $[(d, 0), (0, x)] = [0, dx]$ in Q. Since dx belongs to R for all $x \in R$, $\tilde{R} = \{(0, x), x \in R\}$ is an ideal in the Lie algebra Q and $\tilde{R}$ is solvable. Consider the ideal $I = [Q, \tilde{R}]$ in Q. We will show that I is nilpotent.

If L is real, we go to its complexification. We may thus suppose that all Lie algebras under study are complex. Let $\tilde{R}_1$ be a solvable Lie subalgebra of Q. Let π be the restriction of the homomorphism $x \to \text{ad } x$, $x \in Q$ to $\tilde{R}_1$. Let $\{0\} = V_0 \subset \cdots \subset V_n = Q$ be a Jordan-Hölder series for the homomorphism π of $\tilde{R}_1$ into $gl(Q)$. By Lie's Theorem (see 2.3), every space V_k/V_{k-1}, $k = 1, \ldots, n$, is one-dimensional. Let $h_k(x)$ be the linear functional on $\tilde{R}_1$ defined by π in the factor space V_k/V_{k-1}. The set $M = \{x \in \tilde{R}_1, h_k(x) = 0$ for all $k = 1, \ldots, n\}$ is an ideal in $\tilde{R}_1$, since the linear subspace M of $\tilde{R}_1$ plainly contains $[\tilde{R}_1, \tilde{R}_1]$. The definition of the ideal M implies that the operator $\pi(x)$ is nilpotent for all $x \in M$.

We now choose the Lie subalgebra $\tilde{R}_1$ in a special way. Let y be an element of the Lie algebra Q and let $\tilde{R}_1$ be the Lie subalgebra of Q generated by $\tilde{R}$ and the element y. Since $\tilde{R}$ is an ideal in $\tilde{R}$, $\tilde{R}$ is an ideal in $\tilde{R}_1$. Since $\tilde{R}$ is a solvable ideal and the factor algebra $\tilde{R}_1/\tilde{R}$ is commutative, $\tilde{R}_1$ is a solvable Lie algebra (Proposition II in 2.2). Apply the preceding construction to $\tilde{R}_1$. Let M be the corresponding ideal in the Lie algebra $\tilde{R}_1$. We see that $[y, \tilde{R}] \subset [\tilde{R}_1, R_1] \subset M$. Thus the set $[y, \tilde{R}]$ is contained in the set of elements $z \in Q$ such that the operator $\pi(z)$ is nilpotent. From Proposition II we infer that the ideal $I = [Q, \tilde{R}]$ is a nilpotent ideal in Q.

We set $\tilde{N} = \{(0, n), n \in N\}$. Since $[Q, \tilde{R}] \subset [Q, \tilde{L}] \subset \tilde{L}$, the ideal $I = [Q, \tilde{R}]$ is a nilpotent ideal in $\tilde{L}$. Therefore we have $[Q, \tilde{R}] \subset \tilde{N}$. Hence a differentiation d of the Lie algebra L maps R into N. $\square$

§4. The Theory of Replicas

Let V be a finite dimensional complex vector space, V^* the space conjugate to V. The tensor product of r copies of V and s copies of V^* is denoted by $V_{r,s}$. The linear representation $x \to x$ of the Lie algebra $gl(V)$ in V defines a representation of $gl(V)$ in the space $V_{r,s}$; specifically, the tensor product of r copies of the representation $x \to x$ in V and s copies of the representation $x \to -x^*$ in the space V^*. The image of the operator $x \in gl(V)$ in this representation is denoted by $x_{r,s}$. Furthermore, we set $V_{0,0} = \mathbf{C}$ and $x_{0,0} = 0$. As we saw in 1.3 (see 1.3.6), the adjoint representation of the Lie algebra $gl(V)$ is equivalent to the representation $x_{1,1}$ under the canonical mapping τ of the space $V \otimes V^*$ onto $L(V)$.

An element $x' \in gl(V)$ is called a *replica of an element* $x \in gl(V)$, if the kernel of the operator $x_{r,s}$ is contained in the kernel of the operator $x'_{r,s}$ for all nonnegative integers r and s.

I. *If* x' *is a replica of* x *and* x'' *is a replica of* x', x'' *is a replica of* x.

The proof is obvious.

II. *If* x' *is a replica of* x, $x'_{r,s}$ *is a replica of* $x_{r,s}$.

Proof. It is easy to see that the space $(V_{r,s})_{r',s'}$ is canonically isomorphic with the space $V_{rr'+ss', rs'+sr'}$. Furthermore, this canonical isomorphism—denoted by ζ—has the property that

$$\zeta \circ (x_{r,s})_{r',s'} \circ \zeta^{-1} = x_{rr'+ss', rs'+r's}$$

for all $x \in gl(V)$. The present proposition follows. $\square$

We now cite a fact from the theory of matrices.

III. *Let* x *be a linear transformation of a finite-dimensional complex linear space* V. *Any linear transformation* y *that commutes with all transformations that commute with* x *can be represented in the form* $y = p(x)$, *where* p *is a polynomial with complex coefficients.*

The proof of this statement is not difficult if we reduce the transformation x to the Jordan canonical form. The details can be found in the book of F. R. Gantmaher [1].

IV. *The operator* x' *is a replica of the operator* x *if and only if for all pairs* (r, s), *the operator* $x'_{r,s}$ *is a polynomial in* $x_{r,s}$ *with no constant term.*

Proof. If $x'_{r,s}$ is a polynomial in $x_{r,s}$ with no constant term, the kernel of $x_{r,s}$ is contained in the kernel of $x'_{r,s}$. That is, x' is a replica of x. Conversely, suppose that x' is a replica of x. Let y commute with x, that is, $(\text{ad } x)y = [x, y] = 0$. Using the isomorphism described earlier between the adjoint representation and the representation $x \to x_{1,1}$, we see that $x_{1,1}(y) = 0$. Since x' is a replica of x, we have $x'_{1,1}(y) = 0$, that is, $(\text{ad } x')y = 0$ and x' commutes with y. Proposition III shows that $x' = p(x)$, where p is a polynomial. We will show that we can choose p so that $p(0) = 0$. This is easy to see if the operator x is invertible. For, if x is invertible we have $\det x \neq 0$ and the constant term $q(0)$ of the characteristic polynomial q is nonzero. The

Hamilton-Cayley Theorem shows that we have $q(x) = 0$; that is, $q(0)1_V = -q(x) + q(0)1_V = q_1(x)$, where $q(0) \neq 0$ and the polynomial q_1 has constant term zero. The constant term of the polynomial $p(x)$ can be replaced by a polynomial divisible by q_1. Now suppose that x is not invertible, so that $xv = 0$ for some nonzero vector $v \in V$. Since $x'v = p(x)v = 0$ and $p(x)v = p(0)v$, we see that $p(0)v = 0$ and thus $p(0) = 0$. Hence if x' is a replica of x, x' has the form $p(x)$, where p has constant term 0. By Proposition II, $x'_{r,s}$ is a replica of $x_{r,s}$ and therefore we have $x'_{r,s} = p_{r,s}(x_{r,s})$, where $p_{r,s}(0) = 0$. This completes the proof. $\square$

We note one more fact from matrix theory.

V. *A linear operator x on the space V can be represented uniquely in the form $x = y + z$, where y and z commute, z is nilpotent, and y has diagonal form in some basis. The operators y and z are polynomials in x and the kernels of the operators y and z contain the kernel of the operator x.*

Proof. Let $e_1, \ldots, e_n$ be a basis in the space V such that the matrix (a_{ij}) of the transformation x in this basis has Jordan canonical form. Let y be the operator on V described in the basis $e_1, \ldots, e_n$ by the matrix (b_{ij}) such that $a_{ii} = b_{ii}$ for all $i = 1, \ldots, n$, and $b_{ij} = 0$ for $i \neq j$. We define z as $x - y$. It is clear that z is nilpotent and that y and z commute. Let $\lambda_1, \ldots, \lambda_k$ be all of the distinct eigenvalues of x. Let V_i $(i = 1, \ldots, k)$ be the root subspace of x, corresponding to the eigenvalue λ_i; that is, V_i is the subspace generated by the eigenvectors and adjoint vectors of x that correspond to λ_i. Observe that V_i is the kernel of the operator $(x - \lambda_i 1)^n$. We set $x_1 = (x - \lambda_2 1)^n \cdots (x - \lambda_k 1)^n$. The subspace $V_2 + \cdots + V_k$ is the kernel of the operator x_1 and the subspace V_1 is invariant under x_1. The restriction of x_1 to V_1 is nonsingular. If we apply the Hamilton-Cayley Theorem to this restriction, we see that the operator E_1, projecting V onto V_1 parallel to $V_2 + \cdots + V_k$, is a polynomial in x_1 and so also a polynomial in x. If we define $E_2, \ldots, E_k$ similarly, and note as well that

$$y = \lambda_1 E_1 + \cdots + \lambda_k E_k,$$

we see that y is a polynomial in x. Therefore z is also a polynomial in x. Our claim concerning the kernels of x, y and z follows from the definition of y and z. We will prove that the decomposition $x = y + z$ is unique. We set $x = y_1 + z_1$, where y_1 and z_1 commute, z_1 is nilpotent and y_1 diagonalizable. Thus y_1 and z_1 commute with x and so commute with all polynomials in x. In particular the polynomial y commutes with y_1 and z commutes with z_1. On the other hand, from $y + z = y_1 + z_1$ we infer that $y - y_1 = z_1 - z$. Since y and y_1 are diagonalizable and commute with each other, $y - y_1$ is diagonalizable. Inasmuch as z_1 and z are nilpotent and commute, $z_1 - z$ is nilpotent. A nilpotent diagonalizable matrix is the zero matrix and thus we have $y - y_1 = z_1 - z = 0$. $\square$

Let $x = y + z$ be the decomposition of x defined in Proposition V. The operators y and z are called respectively the *semisimple* and *nilpotent parts of x*.

VI. *Let x be an operator on V and let y and z be the semisimple and nilpotent parts of x, respectively. Then $y_{r,s}$ and $z_{r,s}$ are the semisimple and nilpotent parts of $x_{r,s}$, respectively, for all pairs (r, s).*

Proof. Since y is diagonalizable, $y_{r,s}$ is diagonalizable. It is also clear that $z_{r,s}$ is nilpotent. Since $x \to x_{r,s}$ is a representation of the Lie algebra $gl(V)$, we have $[y_{r,s}, z_{r,s}] = [y, z]_{r,s} = 0$. Note that $x_{r,s} = y_{r,s} + z_{r,s}$. Thus the present proposition follows from the uniqueness of the decomposition into semisimple and nilpotent components. □

VII. *The semisimple and nilpotent components of an operator are replicas of the operator.*

Proof. Let $x = y + z$ be the decomposition of x into semisimple and nilpotent parts. Proposition VI shows that $x_{r,s} = y_{r,s} + z_{r,s}$ is the decomposition of $x_{r,s}$ into semisimple and nilpotent parts. The last part of Proposition V shows that the kernels of $y_{r,s}$ and $z_{r,s}$ contain the kernel of $x_{r,s}$. That is, y and z are replicas of x. □

Theorem. *The operator x on V is nilpotent if and only if $\mathrm{tr}(xx') = 0$ for all replicas x' of x.*

Proof. If the operator x is nilpotent and x' is a replica of x, x' commutes with x (see IV). Therefore the operator xx' is also nilpotent and so has trace zero.

Suppose conversely that the condition in the theorem holds. Let $x = y + z$ be the decomposition of x into semisimple and nilpotent parts. We must prove that $y = 0$. Proposition VII shows that the operator y is a replica of x. By I, every replica y' of y is also a replica of x. We infer that y' commutes with z, so that the operator $y'z$ is nilpotent and thus $\mathrm{tr}(y'z) = 0$. On the other hand, $\mathrm{tr}(y'x)$ is zero by hypothesis, so that $\mathrm{tr}(y'y) = 0$. To complete the proof, it suffices to prove the following: if y is a diagonalizable operator and $\mathrm{tr}(y'y) = 0$ for all replicas y' of y, then $y = 0$. Assume that this statement is false, and let y be a nonzero diagonalizable operator and let $e_1, \ldots, e_n$ be a basis in V such that $ye_i = \lambda_i e_i$. If y' is a replica of y, then y' is also diagonal in the basis $e_1, \ldots, e_n$. (This is implied by IV.) The operator $y_{r,s}$ is diagonal in the basis of the space $V_{r,s}$ formed by vectors of the form $e_{i_1} \otimes \cdots \otimes e_{i_r} \otimes e_{j_1}^* \otimes \cdots \otimes e_{j_s}^*$, where (e_j^*) is the basis of the space V^* biorthogonal to the basis $\{e_i\}$ of V. The eigenvalue of the operator $y_{r,s}$ corresponding to the vector $e_{i_1} \otimes \cdots \otimes e_{i_r} \otimes e_{j_1}^* \otimes e_{j_s}^*$, is clearly equal to $\lambda_{i_1} + \cdots + \lambda_{i_r} - \lambda_{j_1} - \cdots \lambda_{j_s}$. The kernel of the operator $y_{r,s}$ is the subspace generated by those basis vectors for which

$$\sum_{k=1}^{r} \lambda_{i_k} - \sum_{j=1}^{s} \lambda_{j_s} = 0. \tag{4.1.1}$$

If we have $y'e_i = \mu_i e_i$ and any relation of the form (4.1.1) implies the corresponding relation for the numbers $\mu_1, \ldots, \mu_n$, the kernel of $y'_{r,s}$ contains the kernel of $y_{r,s}$. That is, y' is a replica of y.

Suppose that we have $\operatorname{tr}(yy') = 0$ and $y \neq 0$. The numbers $\lambda_1, \ldots, \lambda_n$ generate a nonzero subspace P of $\mathbf{C}$ over the field $\mathbf{Q}$ of rational numbers. Let h be a nonzero $\mathbf{Q}$-linear mapping of the linear space P over $\mathbf{Q}$ into $\mathbf{Q}$. We set $\mu_i = h(\lambda_i)$. Any integral linear relation among the numbers λ_i, such as that in (4.1.1), implies a corresponding relation among the numbers μ_i. Therefore the operator y', for which $y'e_i = h(\lambda_i)e_i$, is a replica of y. Thus we have $0 = \operatorname{tr}(yy') = \sum \lambda_i h(\lambda_i)$. Because the mapping h is $\mathbf{Q}$-linear and $h(\lambda_i) \in \mathbf{Q}$, we have $\sum h(\lambda_i)h(\lambda_i) = \sum h(\lambda_i)^2 = 0$, that is, $h(\lambda_i) = 0$, which contradicts the selection of h as a nonzero mapping. This contradiction completes the proof. $\square$

VIII. *Let A be a finite-dimensional (not necessarily associative) algebra. If D is a differentiation of A and D' is a replica of the operator D, D' is also a differentiation of A.*

Proof. Multiplication in A defines a bilinear mapping of the space $A \times A$ into A. The space of bilinear mappings $A \times A$ into A is canonically identified with the space $A_{1,2}$. Let μ be the element of $A_{1,2}$ defined by multiplication in A for this canonical isomorphism. Let x be a linear operator on A. We have

$$x_{1,2}\mu = x\mu - \mu(1 \otimes x + x \otimes 1),$$

that is,

$$\begin{aligned} x_{1,2}\mu(a \otimes b) &= x\mu(a \otimes b) - \mu(a \otimes xb + xa \otimes b) \\ &= x(ab) - a(xb) - (xa)b \end{aligned}$$

for all $a, b \in A$. Thus the operator x is a differentiation on A if and only if $x_{1,2}\mu = 0$, that is, if μ is in the kernel of $x_{1,2}$. If x' is a replica of x, the equality $x_{1,2}\mu = 0$ implies that $x'_{1,2}\mu = 0$. Hence x' is also a differentiation on A. $\square$

§5. The Killing Form; Criteria for Solvability and Semisimplicity of a Lie Algebra

5.1. Definition of the Killing Form

Let L be a Lie algebra. Let π_1 and π_2 be homomorphisms of L into the Lie algebras $gl(V_1)$ and $gl(V_2)$, respectively. A bilinear form B on $V_1 \times V_2$ is called *invariant (under π_1 and π_2)* if

$$B(\pi_1(x)v_1, v_2) + B(v_1, \pi_2(x)v_2) = 0 \qquad (5.1.1)$$

for all $x \in L$, $v_1 \in V_1$, $v_2 \in V_2$.

I. *Let π be a homomorphism of the Lie algebra L into the Lie algebra $gl(V)$. We define $B_\pi(x, y) = \mathrm{tr}(\pi(x)\pi(y))$ for $x, y \in L$. The bilinear form B_π on $L \times L$ is symmetric and invariant under the adjoint homomorphism $x \to$ ad x.*

Proof. The relation $\mathrm{tr}(\alpha\beta) = \mathrm{tr}(\beta\alpha)$ implies that the form B_π is symmetric. We will prove that the form B_π is invariant by showing that

$$\mathrm{tr}(\pi([x_1, x_2])\pi(x_3)) + \mathrm{tr}(\pi(x_2)\pi([x_1, x_3]))$$
$$= \mathrm{tr}(\pi(x_1)\pi(x_2)\pi(x_3) - \pi(x_2)\pi(x_3)\pi(x_1))$$

is equal to zero. This too follows from the identity $\mathrm{tr}(\alpha\beta) = \mathrm{tr}(\beta\alpha)$, applied to $\alpha = \pi(x_1)$ and $\beta = \pi(x_2)\pi(x_3)$.

The invariant symmetric bilinear form $B(x, y) = \mathrm{tr}(\mathrm{ad}\ x\ \mathrm{ad}\ y)$ on L that corresponds to the adjoint homomorphism of L into $gl(L)$ is called the *Killing form on L*. We write $B(x, y) = (x, y)$. $\square$

5.2. Cartan's Criterion for Solvability of a Lie Algebra

Theorem 1 (Cartan's criterion). *Let L be a Lie algebra; L is solvable if and only if*

$$(x, [y, z]) = 0, \quad \text{for all } x, y, z \in L. \tag{5.2.1}$$

In particular, if $(x, y) = 0$ for all $x, y \in L$, L is solvable.

Proof.

(a) Suppose that L is solvable. Let $\{0\} = V_0 \subset V_1 \cdots \subset V_n = L$ be a Jordan-Hölder series for the homomorphism $x \to$ ad x. By corollary 1 in 2.3, the derived algebra $L^{(1)}$ of L is nilpotent. That is, $L^{(1)}$ is a nilpotent ideal in L. Proposition III in §3 shows that $L^{(1)}$ is contained in the nilpotent ideal N consisting of all $x \in L$ such that (ad x) $V_k \subset V_{k-1}$ for all $k = 1, \ldots, n$. In particular, $\mathrm{ad}([y, z])V_k$ is contained in V_{k-1} and so we have

$$\mathrm{ad}\ x\ \mathrm{ad}([y, z])V_k \subset V_{k-1}, \quad \text{for } k = 1, \ldots, n.$$

This implies that $\mathrm{tr}(\mathrm{ad}\ x\ \mathrm{ad}([y, z])) = 0$ for all $x, y, z \in L$. This is equivalent to (5.2.1).

(b) Suppose that (5.2.1) holds. To prove that L is solvable, it suffices to show that the Lie algebra $L^{(1)}$ is solvable. The identity (5.2.1) implies that $(x, y) = 0$ for $x, y \in L^{(1)}$, since $y \in L^{(1)}$ is a linear combination of elements of the form $[u, v]$. Thus we may suppose that the Killing form (x, y) vanishes on the original Lie algebra L. We will show that under this hypothesis, the Lie algebra $L^{(1)}$ is nilpotent. That is, (see theorem 1 in 2.1) the image of the adjoint homomorphism of $L^{(1)}$ into $gl(L^{(1)})$ consists entirely of nilpotent operators. Plainly, we need only to prove that the restriction of the adjoint homomorphism $x \to$ ad x of L to the ideal $L^{(1)}$ maps the elements of the Lie algebra $L^{(1)}$ into nilpotent operators (see Proposition II of §3). Let x be an element of $L^{(1)}$, i.e., $x = \sum_{i=1}^{r} [y_i, z_i]$ for certain $y_i, z_i \in L$. We must show that the operator ad x is nilpotent. According to the theorem of §4, it suffices to show that tr(ad xZ) = 0 for all replicas Z of the operator ad x. By Proposition I in 1.1, ad x is a differentiation of the Lie algebra L and by Proposition VIII in §4, a replica of a differentiation is itself a differentiation. Therefore it suffices to show that tr(ad $x \cdot d$) = 0 for every differentiation d of the Lie algebra L.

We have

$$\text{tr}(\text{ad } x \cdot d) = \sum_{i=1}^{r} \text{tr}([\text{ad } y_i, \text{ad } z_i]d)$$

$$= \sum_{i=1}^{r} \text{tr}(\text{ad } y_i \text{ ad } z_i d - \text{ad } z_i \text{ ad } y_i d)$$

$$= \sum_{i=1}^{r} \text{tr}(\text{ad } z_i d \text{ ad } y_i - \text{ad } z_i \text{ ad } y_i d), \tag{5.2.2}$$

since tr($\alpha\beta$) = tr($\beta\alpha$) for all linear operators α, β. The last line of (5.2.2) can be rewritten as

$$\sum_{i=1}^{r} \text{tr}(\text{ad } z_i [d, \text{ad } y_i]). \tag{5.2.3}$$

Since d is a differentiation of L, (1.1.5) holds and from (5.2.2) and (5.2.3) we infer that

$$\text{tr}(\text{ad } x \cdot d) = \sum_{i=1}^{r} \text{tr}(\text{ad } z_i [d, \text{ad } y_i])$$

$$= \sum_{i=1}^{r} \text{tr}(\text{ad } z_i \text{ ad}(dy_i)) = \sum_{i=1}^{r} (z_i, dy_i). \tag{5.2.4}$$

By hypothesis, the Killing form is identically zero on the Lie algebra L. Hence the right side of (5.2.4) is zero; that is, tr(ad $x \cdot d$) = 0 for all $d \in$ Der (L) and so ad x is nilpotent in L for all $x \in L^{(1)}$. $\square$

The argument used in the proof of theorem 1 can be applied in a more general case.

I. *Let L be a Lie algebra over the field K. Let π be a homomorphism of L into the Lie algebra $gl(V)$. Let P be the set of $x \in L$ such that $B_\pi(x, y) = 0$ for all $y \in L$. Then P is an ideal in L and the operator $\pi(x)$ is nilpotent for all $x \in [P, L]$.*

Proof. Since the form B_π is invariant, we see that

$$B_\pi([x, y], z) + B_\pi(y, [x, z]) = 0 \tag{5.2.5}$$

for all $x, y, z \in L$. In particular, if $y \in P$ then $B_\pi(y, [x, z]) = 0$ for all $x, z \in L$. Thus (5.2.5) implies that $[x, y] \in P$ for all $x \in L$, i.e., P is an ideal. For $x \in [P, L]$, we write $x = \sum_{i=1}^r [y_i, z_i]$, where $y_i \in P$, $z_i \in L$. To prove that the operator $\pi(x)$ is nilpotent, we need only to show that $tr(\pi(x)R) = 0$ for every replica R of $\pi(x)$. If R is a replica of $\pi(x)$, ad R is a replica of ad $\pi(x)$ (see Proposition II in §4). Thus (see Proposition IV in §4), the operator ad R is a polynomial in ad $\pi(x)$, and hence $\pi(L)$ is invariant under ad R. Hence there are elements $u_i \in L$ such that $[\pi(z_i), R] = \pi(u_i)$ $(i = 1, \ldots, r)$. We then obtain

$$\begin{aligned}
tr(\pi(x)R) &= \sum_{i=1}^r tr([\pi(y_i), \pi(z_i)]R) \\
&= \sum_{i=1}^r tr(\pi(y_i)\pi(z_i)R - \pi(z_i)\pi(y_i)R) \\
&= \sum_{i=1}^r tr(\pi(y_i)\pi(z_i)R - \pi(y_i)R\pi(z_i)) \\
&= \sum_{i=1}^r tr(\pi(y_i)[\pi(z_i), R]) \\
&= \sum_{i=1}^r tr(\pi(y_i)\pi(u_i)) = \sum_{i=1}^r B_\pi(y_i, u_i) = 0,
\end{aligned}$$

since $y_i \in P$. Therefore $\pi(x)$ is nilpotent.   □

II. *Retain the hypotheses of Proposition* I. *If* π *is an exact homomorphism* (*that is,* $x = 0$ *if* $\pi(x) = 0$), *then* $[P, L]$ *is a nilpotent ideal in* L.

Proof. By Proposition I, the operator $\pi(x)$ is nilpotent for all $x \in [P, L]$ and $[P, L]$ is an ideal in L. Since π is an isomorphism of L onto $\pi(L)$, $\pi([P, L]) = [\pi(P), \pi(L)]$ is an ideal in $\pi(L)$, all elements of which are nilpotent operators. Proposition II in §3 implies that $\pi([P, L])$ is a nilpotent ideal in $\pi(L)$ and so $[P, L]$ is a nilpotent ideal in L.   □

III. *Under the hypotheses of Proposition* II, P *is a solvable ideal in* L.

Proof. The Lie algebra $[P, L]$ is nilpotent by Proposition II. Thus the Lie algebra $[P, P] \subset [P, L]$ is nilpotent and we need only apply corollary 1 in 2.3.   □

5.3. The Criterion for Semisimplicity

Theorem 2. *A Lie algebra L is semisimple if and only if the Killing form on L is nonsingular.*

Proof.

(a) Suppose that the Killing form on L is nonsingular. Let us assume that the radical R of L is nonzero. Let h be the height of R as a solvable algebra. In this case $A = R^{(h-1)}$ is a nonzero commutative ideal in L. For x, $y \in L$ and $a \in A$ we have $[a, y] \in A$, $[x, [a, y]] \in A$, $[a, [x, [a, y]]] = 0$, that is, ad a ad x ad $a = 0$. Thus we have (ad x ad $a)^2 = 0$, which implies that tr(ad x ad $a) = 0$; that is, $(x, a) = 0$ for all $x \in L$ and $a \in A$. Since $A \neq \{0\}$, this contradicts the hypothesis that the Killing form be a nonsingular.

(b) Let L be a semisimple Lie algebra. The elements $n \in L$ such that $(x, n) = 0$ for all $x \in L$ form a linear subspace N of L. Because the Killing form is invariant under the adjoint homomorphism, we obtain

$$(x, [y, n]) = (x, \text{ad } y(n)) = -(\text{ad } y(x), n) = 0$$

for all x, $y \in L$ and $n \in N$, so that $[y, n] \in N$. Thus N is an ideal in L. By the definition of N, we have $(n, n_1) = 0$ for $n, n_1 \in N$. Theorem 1 shows that the ideal N is solvable. Since L is semisimple, $N = \{0\}$, and hence the form (x, y) is nonsingular.

§6. The Universal Enveloping Algebra of a Lie Algebra

6.1. Tensor Algebras

Let V be a vector space over the field K. We define $J_0 = K$ and for $r = 1, 2, \ldots$, we define $J_r = V \otimes \cdots \otimes V$ (the tensor product of r copies of V). Elements of the space J_r are called (*homogeneous*) *tensors of order r*. Let J be the direct sum of all J_r ($r \geqslant 0$). We define a bilinear mapping $\{u, v\} \rightarrow u \otimes v$ of the direct product $J \times J$ into J by:

$$c \otimes v = cv = v \otimes c, \quad \text{for } c \in K = J_0 \text{ and } v \in J; \qquad (6.1.1a)$$

$$(x_1 \otimes \cdots \otimes x_r) \otimes (y_1 \otimes \cdots \otimes y_s) = x_1 \otimes \cdots \otimes x_r \otimes y_1 \otimes \cdots \otimes y_s \quad (6.1.1b)$$

for all natural numbers r, s and all $x_1, \ldots, x_r$ and $y_1, \ldots, y_s$ that belong to V. We extend this mapping by linearity over all sums of the expressions on the left sides of (6.1.1a) and (6.1.1b). The linear space J is an associative algebra over K under the multiplication $\otimes$. The identity element of this algebra is the identity element of the field $K = J_0$. The algebra J is called the *tensor algebra over V*. The space V can be identified with the subspace $J_1 \subset J$.

I. *The algebra J is generated (as an algebra) by the subspace V. If A is an associative algebra and f is a linear mapping of V into A, there is a unique homomorphism of J into A that extends f.*

This proposition is obvious.

6.2. The Definition and Construction of the Universal Enveloping Algebra of a Given Lie Algebra

Let L be a Lie algebra over the field K of real or complex numbers. Let J be the tensor algebra over L. For $x, y \subset L$, we write $u_{x,y}$ for the element $x \otimes y - y \otimes x - [x, y]$ of J. The linear subspace of J generated by all elements of the form $t \otimes u_{x,y} \otimes t' (t, t' \in J; x, y \in L)$ is denoted by M. That is,

$$M = \sum_{x,y \in L} J \otimes u_{x,y} \otimes J. \tag{6.2.1}$$

Since $u_{x,y}$ belongs to $J_1 + J_2$ for all $x, y \in L$, it is clear that $M \subset \sum_{m \geq 1} J_m$. Thus, M is a proper subspace of J. From (6.2.1) we see that M is a two-sided ideal in J. The factor algebra J/M is denoted by U and the canonical mapping of J onto U is written as γ. Because L and 1 generate J (see Proposition I in 6.1), $\gamma(L)$ and $\gamma(1)$ generate U. The identity element of the associative algebra U is denoted by 1. The product of elements $a, b \in U$ is written ab and the restriction of the mapping γ to the subspace $L = J_1$ is written as α.

Let L be a Lie algebra over K. Let A be an associative algebra with unit element 1 over K. Let ρ be a linear mapping of L into A. The algebra A, provided with the mapping ρ, is called a *universal enveloping algebra of the Lie algebra L* if the following conditions hold. (1) The algebra A is generated by 1 and $\rho(L)$. (2) We have $\rho([x, y]) = \rho(x)\rho(y) - \rho(y)\rho(x)$ for all $x, y \in L$. (3) Let B be an associative algebra with a unit element 1_B for which there is a linear mapping ξ of L into B such that $\xi([x, y]) = \xi(x)\xi(y) - \xi(y)\xi(x)$ for all $x, y \in L$. Then there is a homomorphism ξ' of A into B such that $\xi'(1) = 1_B$ and $\xi(x) = \xi'(\rho(x))$ for all $x \in L$.

I. *Let L be a Lie algebra over K. The associative algebra U, provided with the mapping $\alpha: L \to U$, is a universal enveloping algebra of L. If (U', α') is another universal enveloping algebra for L, there is a unique isomorphism ξ of U onto U' such that $\xi(1) = 1$ and $\xi(\alpha(x)) = \alpha'(x)$ for all $x \in L$.*

Proof. We already know that $\alpha(L) = \gamma(L)$ and 1 generate U. Since $\gamma(u_{x,y}) = 0$, we have

$$\gamma([x, y]) = \gamma(x)\gamma(y) - \gamma(y)\gamma(x)$$

for all $x, y \in L$, and so

$$\alpha([x, y]) = \alpha(x)\alpha(y) - \alpha(y)\alpha(x)$$

for all $x, y \in L$. Let A be an associative algebra and let ξ be a linear mapping of L into A such that $\xi([x, y]) = \xi(x)\xi(y) - \xi(y)\xi(x)$ for all $x, y \in L$. Let η be the homomorphism of the algebra J into A such that $\eta(x) = \xi(x)$ for all $x \in L$ (see Proposition I in 6.1). Since $\eta(u_{x,y}) = 0$ for all $x, y \in L$, η is identically 0 on M. Moving to the factor algebra U, we obtain a homomorphism ζ of the associative algebra U into A for which $\eta = \zeta \circ \gamma$. Therefore (U, α) is a universal enveloping algebra for L.

Let (U', α') be another universal enveloping algebra for L. Because (U, α) is also a universal enveloping algebra for L, there are homomorphisms $\zeta : U \to U'$ and $\zeta' : U' \to U$ such that $\zeta(\alpha(x)) = \alpha'(x)$ and $\zeta'(\alpha'(x)) = \alpha(x)$ for all $x \in L$. Therefore the mappings $\zeta \circ \zeta'$ and $\zeta' \circ \zeta$ are the identity mappings on $\alpha'(L)$ and $\alpha(L)$ respectively. We also have $\zeta(1) = 1$ and $\zeta'(1) = 1$. Since U and U' are generated by their units and $\alpha(L), \alpha'(L)$ respectively, $\zeta \circ \zeta'$ and $\zeta' \circ \zeta$ are the identity mappings. Thus ζ is an isomorphism of U onto U'. $\square$

6.3. The Poincaré-Birkhoff-Witt Theorem

Theorem (Poincaré-Birkhoff-Witt). *Let L be a Lie algebra over K, let $x_1, \ldots, x_n$ be a basis in L, and let $U(\alpha)$ be a universal enveloping algebra for L. The elements 1 and $\alpha(x_{i_1}) \cdots (x_{i_s})(s \geq 1, i_1 \leq \cdots \leq i_s)$ are a basis in the linear space U. The mapping α is an embedding of L into U; that is, the kernel of α is $\{0\}$.*

Proof. Fix an integer $p \geq 1$. The linear span of the elements of J having the form $x_{i_1} \otimes \cdots \otimes x_{i_p}$ $(i_1 \leq \cdots \leq i_p)$ id denoted by J_p^0. We write J_0 as J_0^0. It is clear that $J_1^0 = J_1$. We write J^0 for the direct sum of all the linear spaces J_p^0, $p \geq 0$. For $p \geq 2$, we break up J_p as follows. Let d be a positive integer and consider all sequences of distinct positive integers $(i_1, i_2, \ldots, i_p)$ displaying exactly d inversions (an inversion in the sequence is a pair (r, s) such that $1 \leq r < s \leq p$ and $i_r > i_s$). The linear span of all elements $x_{i_1} \otimes x_{i_2} \otimes \cdots \otimes x_{i_p}$ for which the sequence $(i_1, i_2, \ldots, i_p)$ displays exactly d inversions is denoted by J_p^d. Plainly J_p is the direct sum of the spaces J_p^d $(d \geq 0)$.

The present theorem is equivalent to the assertion that J is the direct sum of M and J^0, M being as in (6.2.1). We must show that $M + J^0 = J$ and $M \cap J^0 = \{0\}$.

To prove that $M + J^0 = J$, it suffices to show that for every $r \geq 0$

$$J_r \subset M + \sum_{q=0}^{r} J_q^0. \tag{6.3.1}$$

This inclusion is obvious for $r = 0$ and $r = 1$. We will prove it for $r \geq 2$ by induction on r. For $p \geq 2$, we suppose that (6.3.1) holds for $r = 0, 1, \ldots, p - 1$. Since $J_p = \sum_{d \geq 0} J_p^d$ it suffices to prove that

$$J_p^d \subset M + \sum_{q=0}^{p} J_q^0 \tag{6.3.2}$$

for all $d \geqslant 0$. We will prove (6.3.2) by induction on d. For $d = 0$, (6.3.2) is trivial. For $d \geqslant 1$, we suppose that $J_p^e \subset M + \sum_{q=0}^{p} J_q^0$ for $e = 0, 1, \ldots, d - 1$. Consider any $t = x_{i_1} \otimes \cdots \otimes x_{i_p} \in J_p^d$. Since $d \geqslant 1$, there exists a natural number $r \leqslant p - 1$ such that $i_r > i_{r+1}$. Let t' be the tensor $x_{j_1} \otimes \cdots \otimes x_{j_p}$, where $j_l = i_l$ for $l \neq r$, $l \neq r + 1$, $j_r = i_{r+1}$, $j_{r+1} = i_r$. We then have $t' \in J_p^{d-1}$ and by our second inductive hypothesis, t' belongs to $M + J^0$. We also have $x_{i_r} \otimes x_{i_{r+1}} - x_{i_{r+1}} \otimes x_r = [x_{i_r}, x_{i_{r+1}}] + \lambda$, where λ is in M. From this, we infer that $t - t' \in M + J_{p-1} \subset M + \sum_{q=0}^{p-1} J_q^0$, where we apply our first inductive hypothesis. Thus t belongs to $M + \sum_{q=0}^{p} J_q^0$, and we have completed our inductive proof of (6.3.2).

Now we will prove that $M \cap J^0 = \{0\}$. To do this, we construct a linear operator φ on J with the following properties. First,

$$\varphi(t) = t, \quad \text{for all } t \in J_p^0 \quad \text{and all } p \geqslant 0. \tag{6.3.3}$$

Second, for $p \geqslant 2$, $1 \leqslant s \leqslant p - 1$, and $i_s > i_{s+1}$, we have

$$\varphi(x_{i_1} \otimes \cdots \otimes x_{i_s} \otimes x_{i_{s+1}} \otimes \cdots \otimes x_{i_p})$$
$$= \varphi(\cdots \otimes x_{i_{s+1}} \otimes x_{i_s} \otimes \cdots) + \varphi(\cdots \otimes [x_{i_s}, x_{i_{s+1}}] \otimes \cdots). \tag{6.3.4}$$

If such a linear operator exists, (6.3.4) implies that $\varphi(t_1 \otimes u_{x_i, x_j} \otimes t_2) = 0$ for all $t_1, t_2 \in J$ and $1 \leqslant i \leqslant n$, $1 \leqslant j \leqslant n$. Therefore φ vanishes on M. Since φ is the identity on J^0, we have $M \cap J^0 = (0)$.

Let φ be the identity mapping on $J_0 + J_1$. Suppose that $r \geqslant 2$ and that a linear operator φ with properties (6.3.3) and (6.3.4) for all $p \leqslant r - 1$ has been constructed on the space $\sum_{q=0}^{r-1} J_q$. We will extend φ to a linear operator on the space $\sum_{q=0}^{r} J_q$ that satisfies (6.3.3) and (6.3.4) for all $p \leqslant r$. It suffices to define $\varphi(x_{i_1} \otimes \cdots \otimes x_{i_r})$ in such a way that (6.3.3) and (6.3.4) are satisfied for $p = r$. We construct the elements $\varphi(t)$ for $t = x_{i_1} \otimes \cdots \otimes x_{i_r}$ by induction on the number d of inversions in the sequence $(i_1, \ldots, i_r)$. If t belongs to J_r^0, we set $\varphi(t) = t$. Suppose that $d \geqslant 1$ and suppose that the mapping φ has been defined so that (6.3.3) and (6.3.4) are satisfied for all $t \in J_r^e$, where $e = 0$, $1, \ldots, d - 1$. Let q be a natural number $\leqslant r - 1$, such that $i_q > i_{q+1}$. We define

$$\varphi(x_{i_1} \otimes \cdots \otimes x_{i_q} \otimes x_{i_{q+1}} \otimes \cdots \otimes x_{i_r})$$
$$= \varphi(\cdots \otimes x_{i_{q+1}} \otimes x_{i_q} \otimes \cdots) + \varphi(\cdots \otimes [x_{i_q}, x_{i_{q+1}}] \otimes \cdots). \tag{6.3.5}$$

Since the number q is not defined uniquely, we need to prove that (6.3.5) defines φ unambiguously. Once this is proved, our induction on d and then on r will be completed and we will have a linear operator $\varphi : J \to J$ that satisfies (6.3.3) and (6.3.4). Accordingly, let l be another natural number $\leqslant r - 1$ such that $i_l > i_{l+1}$. We must show from our inductive hypothesis that

$$\varphi(\cdots \otimes x_{i_{l+1}} \otimes x_{i_l} \otimes \cdots) + \varphi(\cdots \otimes [x_{i_l}, x_{i_{l+1}}] \otimes)$$
$$= \varphi(\cdots \otimes x_{i_{q+1}} \otimes x_{i_q} \otimes \cdots) + \varphi(\cdots \otimes [x_{i_q}, x_{i_{q+1}}] \otimes \cdots). \tag{6.3.6}$$

If $|q - l| \geq 2$, we may suppose that $q \geq l + 2$, and so $p \geq 4$. Since the expressions on the two sides of (6.3.6) belong to $\sum_{p=0}^{r-1} J_p + \sum_{e=0}^{d-1} J_r^e$, we may apply our inductive hypothesis to both sides of (6.3.6). A simple computation then shows that both sides of (6.3.6) are equal to

$$\varphi(\cdots \otimes x_{i_{l+1}} \otimes x_{i_l} \otimes \cdots \otimes x_{i_{q+1}} \otimes x_{i_q} \otimes \cdots)$$
$$+ \varphi(\cdots \otimes [x_{i_l}, x_{i_{l+1}}] \otimes \cdots \otimes x_{i_{q+1}} \otimes x_{i_q} \otimes \cdots)$$
$$+ \varphi(\cdots \otimes x_{i_{l+1}} \otimes x_{i_l} \otimes \cdots \otimes [x_{i_q}, x_{i_{q+1}}] \otimes \cdots)$$
$$+ \varphi(\cdots \otimes [x_{i_l}, x_{i_{l+1}}] \otimes \cdots \otimes [x_{i_q}, x_{i_{q+1}}] \otimes \cdots). \qquad (6.3.7)$$

If $|q - l| = 1$, we may take $q = l + 1$. Thus we have $i_l > i_{l+1} > i_{l+2}$ and $p \geq 3$. Our inductive hypothesis shows that the left side of (6.3.6) is equal to

$$\varphi(\cdots \otimes x_{i_{l+2}} \otimes x_{i_{l+1}} \otimes x_{i_l} \otimes \cdots)$$
$$+ \varphi(\cdots \otimes x_{i_l} \otimes [x_{i_{l+1}}, x_{i_{l+2}}] \otimes \cdots)$$
$$+ \varphi(\cdots \otimes [x_{i_l}, x_{i_{l+2}}] \otimes x_{i_{l+1}} \otimes \cdots)$$
$$+ \varphi(\cdots \otimes x_{i_{l+2}} \otimes [x_{i_l}, x_{i_{l+1}}] \otimes \cdots), \qquad (6.3.8)$$

while the right side of (6.3.6) reduces to

$$\varphi(\cdots \otimes x_{i_{l+2}} \otimes x_{i_{l+1}} \otimes x_{i_l} \otimes \cdots)$$
$$+ \varphi(\cdots \otimes [x_{i_{l+1}}, x_{i_{l+2}}] \otimes x_{i_l} \otimes \cdots)$$
$$+ \varphi(\cdots \otimes x_{i_{l+1}} \otimes [x_{i_l}, x_{i_{l+2}}] \otimes \cdots)$$
$$+ \varphi(\cdots \otimes [x_{i_l}, x_{i_{l+1}}] \otimes x_{i_{l+2}} \otimes \cdots). \qquad (6.3.9)$$

Our inductive hypothesis also shows that

$$\varphi(t_1 \otimes x \otimes y \otimes t_2) - \varphi(t_1 \otimes y \otimes x \otimes t_2)$$
$$= \varphi(t_1 \otimes [x, y] \otimes t_2) \qquad (6.3.10)$$

for all x and y in L and t_1 in J_α, t_2 in J_β with $\alpha \geq 0$, $\beta \geq 0$ and $\alpha + \beta = p - 3$. Comparing (6.3.8) and (6.3.9) and applying (6.3.10), we see that (6.3.6) holds provided that the mapping φ vanishes on the element

$$s_1 \otimes [x_{i_l}, [x_{i_{l+1}}, x_{i_{l+2}}]] \otimes s_2 + s_1 \otimes [x_{i_{l+1}}, [x_{i_{l+2}}, x_{i_l}]] \otimes s_2$$
$$+ s_1 \otimes [x_{i_{l+2}}, [x_{i_l}, x_{i_{l+1}}]] \otimes s_2, \qquad (6.3.11)$$

which belongs to J_{r-2}. The element (6.3.11) itself vanishes, as is shown by Jacobi's identity for the Lie algebra L. Therefore φ is well defined and we have proved the theorem. $\square$

Remark. Let $x_1, \ldots, x_n$ be a basis in the Lie algebra L. Let c_{ijk} be the corresponding structural constants. The associative algebra U is generated as an algebra by elements x_i for which

$$x_i x_j - x_j x_i = \sum_{k=1}^{n} c_{ijk} x_k, \qquad (i, j = 1, \ldots, n). \qquad (6.3.12)$$

I. *Let L be a Lie algebra over the field K with universal enveloping algebra U. Let V be a vector space and let π be a homomorphism of L into the Lie algebra $gl(V)$. There exists a homomorphism π' of U into the associative algebra $L(V)$ of linear operators on V such that $\pi(x) = \pi'(x)$ for all $x \in L$. The homomorphism π' is defined uniquely by π.*

This proposition follows from the definition of a universal enveloping algebra. We take the algebra $L(V)$ as A and the mapping $x \to \pi(x)$ as ξ. We note an important special case.

II. *Let L be a Lie algebra and let π be a representation of L in the linear space L. There is a representation π' of the algebra U in the space L such that $\pi(x) = \pi'(x)$ for all $x \in L$. The representation π' is uniquely defined by π.*

6.4. The Casimir Element

Let π be a homomorphism of a Lie algebra L into the Lie algebra $gl(V)$, where V is a finite-dimensional vector space. We suppose that the space L admits a nonsingular symmetric bilinear form B_π, invariant under the homomorphism ad. We choose bases $\{e_i\}$, $\{f_j\}$ in the algebra L such that

$$B_\pi(e_i, f_j) = \delta_{ij}, \qquad (6.4.1)$$

where δ_{ij} is Kronecker's symbol. Let U be a universal enveloping algebra for L. Let b be the element of U defined by

$$b = \sum e_i f_i. \qquad (6.4.2)$$

The element b is called a *Casimir element corresponding to the form B*.

I. *The element b is in the center of U and is independent of the choice of bases $\{e_i\}$, $\{f_j\}$.*

Proof. Let Φ be the linear mapping of the tensor product of the Lie algebras $L \otimes L$ into the Lie algebra $gl(V)$ defined by

$$\Phi(x \otimes y)(z) = B_\pi(y, z)x \qquad (6.4.3)$$

for all $x, y, z \in L$. We will show that Φ is nonsingular. Let $\sum_{i=1}^{k} x_i \otimes y_i$ be an element of $L \otimes L$ lying in the kernel of the mapping Φ. We may suppose that the elements $x_1, \ldots, x_k$ are linearly independent in L. Thus the condition $\Phi(\sum_{i=1}^k x_i \otimes y_i) = 0$ is equivalent to the condition

$$\sum_{i=1}^k B_\pi(y_i, z)x_i = 0, \quad \text{for all } z \in L,$$

and also to

$$B_\pi(y_i, z) = 0, \quad \text{for all } z \in L \quad \text{and all } i = 1, \ldots, k.$$

Since B_π is nonsingular by hypothesis, we have $y_i = 0$ for all $i = 1, \ldots, k$; that is, $\sum_{i=1}^k x_i \otimes y_i = 0$. Thus the kernel of Φ is $\{0\}$. Because the dimensions of the finite-dimensional linear space $L \otimes L$ and $gl(L)$ are equal, the mapping Φ is an isomorphism of $L \otimes L$ onto $gl(L)$. It is easy to verify that the identity

$$\Phi((\text{ad } z)x \otimes y + x \otimes (\text{ad } z)y) = [\text{ad } z, \Phi(x \otimes y)] \qquad (6.4.4)$$

holds for all $x, y, z \in L$. Furthermore, (6.4.3) implies that

$$\Phi(\sum_i e_i \otimes f_i)(z) = \sum_i \Phi(e_i \otimes f_i)z = \sum_i B_\pi(f_i, z)e_i = z,$$

for all $z \in L$. That is, the mapping Φ carries the element $\sum e_i \otimes f_i$ into the identity operator in L. Hence the element $\sum e_i \otimes f_i \in L \otimes L$ is independent of the choice of bases $\{e_i\}, \{f_i\}$. Since $\Phi(\sum e_i \otimes f_i) = 1$, we have

$$[\text{ad } z, \Phi(\sum e_i \otimes f_i)] = 0 \quad \text{for all } z \in L.$$

Thus (6.4.4) implies that

$$\sum_i \{(\text{ad } z)e_i \otimes f_i + e_i \otimes (\text{ad } z)f_i\} = 0.$$

The element $b \in U$ is the image of the element

$$\sum_i e_i \otimes f_i \in L \otimes L$$

under the canonical mapping of the tensor algebra J onto U. Since the algebra U is generated by the subspace L, the element $b \in U$ is in the center of U. $\square$

By Proposition I in 6.3, the homomorphism π can be extended to a homomorphism of U into $L(V)$. The image $\pi(b)$ of the element b commutes with all operators $\pi(x)$ for $x \in L$, since b is in the center of U.

II. *If the form B_π is nonsingular and b is the Casimir element corresponding to B_π, we have $\pi(b) \neq 0$. If, in addition, $\pi(L)$ is an irreducible set of operators, $\pi(b)$ is a nonzero multiple of the identity operator.*

Proof. Note that $\operatorname{tr} \pi(b) = \sum \operatorname{tr}(\pi(e_i)\pi(f_i)) = \sum_i B_\pi(e_i, f_i) = \dim L \neq 0$. If the set of operators $\pi(L)$ is irreducible, Schur's Lemma implies that $\pi(b)$ is a multiple of the identity operator. Since $\pi(b) \neq 0$, $\pi(b)$ is a nonzero multiple of the identity operator. $\square$

6.5. The Universal Enveloping Algebra of a Lie Group

Let G be a Lie group with Lie algebra L. We recall that L is constructed with the aid of left invariant vector fields on G (see 3.2, chapter IX). A linear combination of finite products of left invariant vector fields on G is called a *left invariant differential operator* on G, where the product of vector fields is the operation of their consecutive application according to (1.6.7) in chapter IX. From (1.6.6) in chapter IX, we infer that any left invariant differential operator on G is written in every coordinate neighborhood as a linear differential expression with analytic coefficients. Such an operator commutes with left translations on G. The left invariant differential operators on G form an associative algebra under the usual linear operations and multiplication defined by composition of operators.

Let $U(L)$ be a universal enveloping algebra of L. The identity mapping π of L into the algebra of left invariant differential operators satisfies the identity $\pi([X, Y]) = XY - YX$, according to the definition of the Lie algebra of G. By proposition I in 6.2, this mapping has a unique extension to a homomorphism π' of the associative algebra $U(L)$ into the associative algebra of left invariant differential operators on G. It is easy to show that the kernel of π' consists of 0 alone and that the image of $U(L)$ is the algebra of all left invariant differential operators on G. Thus we obtain the following.

I. *The universal enveloping algebra $U(L)$ is isomorphic (as an associative algebra) to the algebra of all left invariant differential operators on G.*

The center $Z(L)$ of the enveloping algebra $U(L)$ plays an important role in many problems of representation theory. One of the elements of the center, Casimir's element, was introduced in 6.4. We will use Casimir's element below in 7.2 to prove Weyl's theorem on the complete reducibility of finite-dimensional representations of semisimple Lie algebras. We cite the description found in I. M. Gelfand [1*] of the elements of the center of a universal enveloping algebra of the Lie algebra L of a connected Lie group G.

Let $g \to \operatorname{Ad} g$ be the adjoint representation of G. The operator $\operatorname{Ad} g$ defines a Lie homomorphism of the Lie algebra L into itself. An element $a \in U(L)$ is called *homogeneous* if a belongs to the image of one of the subspaces J_r (see 6.1). Write m for $\dim L$. Let $e_1, \ldots, e_m$ be a basis in L and let

$(\widetilde{\operatorname{Ad} g})$ be the matrix of the linear operator $\operatorname{Ad} g$ in the basis $e_1, \ldots, e_m$. Let $(\widetilde{\operatorname{Ad} g})'$ be the transpose of the matrix $(\widetilde{\operatorname{Ad} g})$ and let $\varphi(z_1, \ldots, z_m) = \sum c_{i_1, \ldots, i_k} z_{i_1} \cdots z_{i_k}$ be a homogeneous polynomial of degree k in the variables $z_1, \ldots, z_m$. We set

$$a = \sum \frac{c_{i_1 \cdots i_k}}{k!} \sum_{\sigma \in S_k} e_{i_{\sigma(1)}} \cdots e_{i_{\sigma(k)}} \in U(L), \tag{6.5.1}$$

where S_k is the group of all permutations of the indices $1, \ldots, k$.

II. *Let G be a connected Lie group with Lie algebra L. Let a be a homogeneous element in the enveloping algebra $U(L)$. Then a belongs to $Z(L)$ if and only if a has the form* (6.5.1), *where the polynomial φ satisfies the identity $\varphi(z_1, \ldots, z_m) = \varphi((\widetilde{\operatorname{Ad} g})'z_1, \ldots, (\widetilde{\operatorname{Ad} g})'z_m)$ for all $g \in G$ and all $z_1, \ldots, z_m$.*

The proof can be found in Želobenko[2] and Kirillov[1].

Example

We set $G = SU(2)$. The Lie algebra L of G is $su(2)$, that is, the Lie algebra of complex skew-Hermitian matrices of order two with trace zero. Plainly L is isomorphic as a linear space to the space $\mathbf{R}^3$. The reader can easily verify that the adjoint representation of the group $SU(2)$ is equivalent to the representation of $SU(2)$ constructed in the last exercise of §1 in chapter IV. Thus the adjoint group of $GU(2)$ is the group $SO(3, \mathbf{R})$. A polynomial in three variables invariant under $SO(3, \mathbf{R})$ is constant on spheres and therefore is a polynomial in $x_1^2 + x_2^2 + x_3^2$. In the case of the group $SU(2)$, the center of the universal algebra is the algebra of polynomials in the Casimir operator constructed in 6.4. The ideals in the enveloping algebra of the Lie algebra $SU(2)$ are described in Kirillov [1]. The centers of the universal enveloping algebra for the groups $GL(n, C)$, $U(n)$, $SL(n, C)$ and $SU(n)$ are described in Želobenko[1].

§7. Semisimple Lie Algebras

In this section we study some elementary properties of semisimple Lie algebras and prove H. Weyl's theorem on the complete reducibility of finite-dimensional representations of these algebras.

7.1. Ideals in Semisimple Lie Algebras

Let L be a Lie algebra with a linear subspace A. The set of all $x \in L$ such that $(x, y) = 0$ for all $y \in A$ is called the *orthogonal complemeut* of A and is denoted by $A^{\perp}$.

I. *If A is an ideal in the Lie algebra L, $A^\perp$ is also an ideal in L.*

Proof. Plainly $A^\perp$ is a linear subspace in L. The invariance of the Killing form means that

$$([x, y], z) + (y, [x, z]) = 0$$

for all $x, y, z \in L$. If we take $y \in A$ and $z \in A^\perp$, we find $[x, y] \in A$ and $([x, y], z) = 0$. Hence $[x, z]$ belongs to $A^\perp$ for $x \in L$, $z \in A^\perp$.

II. *Let L be a semisimple Lie algebra with ideals A and B. The following conditions are equivalent:*

(a) $A \cap B = \{0\}$;
(b) $[A, B] = \{0\}$;
(c) $(A, B) = \{0\}$.

Proof. Because $[A, B]$ is contained in $A \cap B$, (a) implies (b). Suppose that (b) holds and choose $a \in A$, $b \in B$, and $x \in L$. We find $[b, x] \in B$ and $[a, [b, x]] \in [A, B] = \{0\}$. Therefore we have ad a ad $b = 0$ and $(a, b) = \text{tr } 0 = 0$. That is, (c) holds. If (c) holds, the Killing form vanishes throughout the ideal $A \cap B$. By Cartan's criterion, this ideal is solvable. Since L is semisimple, we have $A \cap B = \{0\}$, and so (a) is satisfied. Therefore we have (a) $\Rightarrow$ (b) $\Rightarrow$ (c) $\Rightarrow$ (a). $\square$

III. *Let M be an ideal in the semisimple Lie algebra L. Then we have $[M, M^\perp] = \{0\}$ and L is the direct sum of M and $M^\perp$. The Lie algebras M and L/M are semisimple.*

Proof. By definition, M and $M^\perp$ satisfy $(M, M^\perp) = \{0\}$. From Proposition II, we infer that $M \cap M^\perp = \{0\}$ and $[M, M^\perp] = \{0\}$. Since the Killing form on the semisimple Lie algebra is nonsingular, the definition of $M^\perp$ shows that $\dim L = \dim M + \dim M^\perp$. Combining this relation with the equality

$$M \cap M^\perp = \{0\},$$

we see that L is the direct sum of M and $M^\perp$. For $x \in M$, the restriction of the operator ad x to M defines the adjoint homomorphism of the Lie algebra M, and the restriction of ad x to $M^\perp$ vanishes. Hence the restriction of the Killing form on L to the ideal M is the Killing form on M. Assume that this form is singular on M, so that there is a nonzero element x_0 in M for which $(x_0, x) = 0$ for all $x \in M$. Since M and $M^\perp$ are orthogonal and $L = M + M^\perp$, we infer that $(x_0, x) = 0$ for all $x \in L$. Since L is semisimple, the form $(\cdot, \cdot)$ is is nonsingular, and we infer that $x_0 = 0$. Hence the Killing form on M is nonsingular and M is a semisimple Lie algebra. Similarly, $M^\perp$ is semisimple. Since L/M is isomorphic to $M^\perp$, L/M is a semisimple Lie algebra. $\square$

IV. *If L is semisimple, then L is equal to $[L, L]$.*

Proof. Let us assume $L \neq [L, L]$. Then $L/[L, L]$ is a nonzero abelian Lie algebra, which by Proposition III is semisimple. This is impossible. $\square$

V. *Let L be a semisimple Lie algebra with an ideal M. Let N be an ideal in M. Then N is an ideal in L. In particular, if M is a minimal ideal in L, M is simple.*

Proof. By Propositions III and II we have $L = M \dotplus M^{\perp}$ and $[M, M^{\perp}] = \{0\}$. This implies that $[N, L] = [N, M] \subset N$. Also, a minimal ideal in a semisimple algebra is not commutative, and so its dimension is greater than one. $\square$

VI. *A Lie algebra is semisimple if and only if it is the direct sum of simple Lie algebras L_i. In this case any ideal M of L is the direct sum of some of the L_i. This decomposition of a semisimple Lie algebra is unique.*

Proof. Suppose that L is semisimple. Since every ideal $H \subset L$ has a complementary ideal $H^{\perp}$, we see from Proposition V that L is the direct sum of certain minimal ideals L_i. Every such minimal ideal is a simple Lie algebra. Let M be any ideal in L. If N is an ideal in M, $N^{\perp} \cap M$ is also an ideal in M and the preceding argument shows that M is the direct sum of its minimal ideals M_i. According to Proposition V, every ideal M_i is an ideal in L. Thus it suffices to show that every nonzero minimal ideal A of L is one of the ideals L_i. Since the Killing form is nonsingular and the direct sum of the ideals L_i is all of the algebra L, there is an index i_0 such that A is not orthogonal to L_{i_0}. From Proposition II, we infer that $A \cap L_{i_0} \neq \{0\}$. The ideals A and L_{i_0} are minimal and we therefore have $A \cap L_{i_0} = A = L_{i_0}$. The decomposition $L = \sum \dotplus L_i$ is unique because the L_i's are the set of *all* minimal ideals of L. The set of ideals L_i is also uniquely defined in this manner.

Conversely, if L is the direct sum of simple algebras L_i, the ideals L_i are pairwise orthogonal under the Killing form and the Killing form is nonsingular on each of them. Therefore the Killing form is nonsingular on the Lie algebra L; that is, L is semisimple. $\square$

7.2. Complete Reducibility of Representations of Semisimple Lie Algebras

Theorem (H. Weyl). *Let L be a semisimple Lie algebra over the field K. Let π be a homomorphism of L into the Lie algebra $gl(V)$, where V is a finite-dimensional space over K. For any subspace V_1 of V invariant under $\pi(L)$, there is a subspace V_2 of V invariant under $\pi(L)$, such that $V_1 \cap V_2 = \{0\}$ and $V_1 + V_2 = V$.*

Proof. Let N be the kernel of the homomorphism π. We can consider π as a homomorphism of the Lie algebra L/N with kernel zero. Since L/N is semisimple by Proposition III, we may suppose in our proof that π has kernel zero.

Consider the set A of all elements x in L for which $B_\pi(x, y) = 0$ for all $y \in L$. By Proposition I in 5.2, the set A is an ideal in L. Proposition III of 5.2 shows that this ideal is solvable. Since L is semisimple, A is $\{0\}$. That is, the form B_π is nonsingular on L. Let $b \in U$ be the Casimir element corresponding to the form B_π (see 6.4). We extend the homomorphism π to the enveloping algebra U of the Lie algebra L (see Proposition I in 6.3). By Proposition II in 6.4, the operator $\pi(b)$ is nonzero, and if $\pi(L)$ is an irreducible set of operators in V, the operator $\pi(b)$ is invertible.

We proceed with the proof. Suppose first that V_1 is an invariant subspace in V of codimension 1. On the one-dimensional factor space $W = V/V_1$, we define the homomorphism $\tilde{\pi}$ of L into $gl(W)$. Since W is isomorphic to K, $gl(W)$ is a one-dimensional abelian Lie algebra. However, L is semisimple and therefore has no abelian Lie factor algebras (see Proposition III in 7.1). Thus we have $\tilde{\pi}(L) = \{0\}$. We will show that *the space V admits a one-dimensional subspace invariant under $\pi(L)$ and complementary to V_1.*

We may impose the added hypothesis that the subspace V_1 be irreducible under $\pi(L)$. Suppose that our last assertion holds when V_1 is irreducible under $\pi(L)$. Then our assertion is proved by induction on the dimension of the space V_1. If V_1 is reducible under $\pi(L)$, we have a $\pi(L)$-invariant subspace V' of V_1 such that $V' \neq \{0\}$, $V' \neq V_1$. Consider the homomorphism $\tilde{\pi}$ of L into $gl(V/V')$ defined by going to factor operators. The space V_1/V' is invariant under $\tilde{\pi}(L)$ and has codimension 1 in V/V'. By our inductive hypothesis, we have a line V''/V' complementary to V_1/V' and invariant under $\tilde{\pi}(L)$. Thus V'' is invariant under $\pi(L)$ and contains the invariant subspace V' of codimension 1 in V''. By again applying our inductive hypothesis, we obtain a $\pi(L)$-invariant line V_2, complementary to V' in V''. Then V_2 is complementary to V_1 in V. In fact, we have $V' \subset V_1$ and $V_2 + V' = V''$, so that $V_2 + V_1 = V_2 + V' + V_1 = V'' + V_1 = V$. On the other hand, if $V_2 \cap V_1 \neq \{0\}$, we have $V_2 \subset V_1$, because V_2 is one-dimensional. Thus we obtain $V' \subset V_1$ and $V_2 \subset V_1$, that is, $V'' = V' + V_2 \subset V_1$. However, V''/V' is complementary to V_1/V' in V/V' and therefore $V'' \not\subset V_1$. This contradiction shows that $V_2 \cap V_1 = \{0\}$, that is, V_2 is complementary to V_1 in V.

Thus *we may suppose that $\pi(L)$ is an irreducible set of operators in V_1.* As we proved above, the bilinear form B_π is nonsingular. If b is the Casimir element corresponding to the form B_π, we have $\pi(b)V \subset V_1$. For, the image of L in the Lie algebra $gl(V/V_1) \approx gl(K)$ is a one-dimensional Lie algebra and therefore abelian. That is, we have $\pi(x)V \subset V_1$ for all $x \in L$ and $\pi(b)V \subset V_1$. The set of operators $\pi(L)$ is irreducible in V_1 and therefore $\pi(b)$ is a nonzero multiple of the identity operator on the space V_1. Let V_2 be the kernel of the operator $\pi(b)$ in V. In this case V_2 is a one-dimensional subspace complementary to V_1. Since $\pi(b)$ commutes with the operators $\pi(x)$, $x \in L$, we have $\pi(b)(\pi(x)v) = \pi(x)(\pi(b)v) = \pi(x)0 = 0$ for all $v \in V_2$. That is, $\pi(x)v$ is in

V_2 and V_2 is invariant under $\pi(L)$. Note that

$$\pi(L)V_2 = \{0\}, \tag{7.2.1}$$

since $\pi(L)|_{V_2}$ is abelian, being a one-dimensional Lie algebra.

We now take up the general case. Let $\tilde{V}$ be the linear space of linear mappings of V into V_1 whose restrictions to V_1 are scalar multiples of the identity operator. Let $\tilde{V}_1$ be the set of linear operators from the space V into V_1 whose restrictions to V_1 are the zero operator. If V_1 is a nonzero subspace, the space $\tilde{V}_1$ has codimension 1 in $\tilde{V}$. We define a homomorphism $\tilde{\pi}$ of L into $gl(\tilde{V})$ by setting

$$\tilde{\pi}(x)\tilde{v} = [\pi(x), \tilde{v}], \quad \text{for all } x \in L, v \in V.$$

Plainly $\tilde{V}_1$ is invariant under $\tilde{\pi}(L)$. The statement we have just proved shows that there exists a one-dimensional subspace $\tilde{V}_2$ of $\tilde{V}$ invariant under $\tilde{\pi}(L)$ and complementary to $\tilde{V}_1$. Let ψ be a nonzero element in $\tilde{V}_2$. By the definition of $\tilde{V}$, the restriction of ψ to V_1 is a nonzero multiple of the identity operator. Multiplying ψ by a number, we may suppose that ψ is the identity operator on V_1. By (7.2.1), we have $\pi(L)\tilde{V}_2 = \{0\}$, and thus $\tilde{\pi}(x)\psi = [\pi(x), \psi] = 0$ for $x \in L$. That is, ψ is a mapping of V into V_1 equal to the identity on V_1 and commuting with all $\pi(x)$, $x \in L$. The kernel of the mapping ψ is the required subspace V_2. $\square$

7.3. Corollaries of H. Weyl's Theorem

I. *A Lie algebra L is semisimple if and only if, for every homomorphism π of L into $gl(V)$, the set of operators $\pi(L)$ is completely reducible. (That is, every $\pi(L)$-invariant subspace has a $\pi(L)$-invariant complementary subspace.)*

Proof. If L is semisimple, $\pi(L)$ is completely reducible by H. Weyl's Theorem. To prove the converse, suppose that all $\pi(L)$ are completely reducible. In particular, the adjoint homomorphism is completely reducible. Therefore every ideal in L admits a complementary ideal. Suppose that L is *not* semisimple, so that the radical R of L is nonzero. Let n be the height of R as a solvable algebra. Thus we have $n \geqslant 1$ and $A = R^{(n-1)}$ is a nonzero commutative ideal in L. Let B be an ideal of L complementary to A. The Lie algebra L is isomorphic to the direct sum of A and B and therefore any homomorphism of the Lie algebra A into $gl(V)$ is extensible to a homomorphism of L into $gl(V)$ that vanishes on B. The commutative Lie algebra A admits a homomorphism π into $gl(K^2)$ such that $\pi(A)$ is not completely reducible. To see this, let f be any nonzero linear functional on A. The homomorphism π can be defined, for example, by the formula

$$\pi(a) = \begin{Vmatrix} 0 & f(a) \\ 0 & 0 \end{Vmatrix}, \quad \text{for } a \in A.$$

Thus nonsemisimple Lie algebras admit noncompletely reducible representations. The proof is complete. $\square$

II. *Every differentiation d of a semisimple Lie algebra L has the form* ad x *for some $x \in L$.*

Proof. Let D be the Lie algebra of differentiations of L and let J be the set of differentiations of L of the form ad x, $x \in L$ (see Proposition I in 1.1). By Proposition I in 1.1, the set J is an ideal in D. The mapping $x \to$ ad x is an isomorphism of L onto J. To prove this, observe that the center of the semisimple Lie algebra L is a commutative ideal in L and so is $\{0\}$. The kernel of the mapping $x \to$ ad x is exactly the center of L. Thus, L *and* J *are isomorphic* and J is a semisimple Lie algebra. The restriction of the adjoint homomorphism of the Lie algebra D to the ideal J is a completely reducible set of operators in D. Thus D is the direct sum of the subspace $[D, J]$ and the subspace D_0 of elements orthogonal to $[D, J]$:

$$D = [D, J] \dotplus D_0. \tag{7.3.1}$$

For $a \in J, d \in D, d_0 \in D_0$, we have $([a, d], d_0) = 0$ and thus $(d, [a, d_0]) = 0$ for all $d \in D, a \in J, d_0 \in D_0$. We infer that $[a, d_0] = 0$. Conversely, if $[a, d_0] = 0$, then plainly d_0 belongs to D_0. Therefore D_0 consists of all elements $d_0 \in D$ that commute with all the elements of J. Thus if d_0 belongs to D_0, we have $\text{ad}(d_0 x) = [d_0, \text{ad } x] = 0$ for all $x \in L$. In this case we obtain $d_0 x = 0$ for all $x \in L$, so that $d_0 = 0$, and so $D_0 = \{0\}$ and $D = [D, J] \dotplus D_0 = [D, J]$ in view of (7.3.1). Since J is an ideal in D, we find $D = [J, D] \subset J$, so that $D = J$. $\square$

III. *Let L be a Lie algebra with an ideal M that is a semisimple Lie algebra. There is a unique ideal N in L such that L is the direct sum of M and N.*

Proof. Consider the restriction of the adjoint homomorphism of L to the ideal M. Plainly the subspace $M \subset L$ is invariant under all operators ad x, $x \in M$. According to H. Weyl's Theorem, L admits a subspace N that is complementary to M and invariant under the operators ad x, $x \in M$. We will show that $[M, N] = \{0\}$. Since M is an ideal, $[M, N]$ is contained in M. Since N is invariant under ad x, $x \in M$, $[M, N]$ is contained in N. Therefore we have $[M, N] \subset M \cap N = \{0\}$. We will show that $N = M'$, where M' is the set of elements $x \in L$ for which $[x, M] = \{0\}$. For $x \in L$, we write $x = m + n$ where $m \in M$ and $n \in N$. Thus we have $[M, x] = [M, m]$. If $[M, x] = \{0\}$, we also have $[M, m] = \{0\}$; that is, m lies in the center of M. Because M is semisimple, m must be 0. Hence, if x belongs to M', then $x \in N$; that is, M' is contained in N. Since $[M, N] = \{0\}$, it is clear that $M' \supset N$ and thus $M' = N$. Therefore N is uniquely defined (specifically, $N = M'$). Note that N is an ideal in L, by Proposition I in 7.1. $\square$

§8. Cartan Subalgebras

Let L be a complex Lie algebra with a nilpotent Lie subalgebra H. Let π be the restriction of the adjoint representation of the algebra L to H. By Proposition II in 2.4, the space L is the direct sum of the subspaces D^α, corresponding to the different weights α of the representation π. Any weight of π is called a *root of the Lie algebra L relative to the Lie subalgebra H.*

In other words, a linear functional α on H is called a *root of the Lie algebra L relative to H* if there exists a nonzero element x of L such that for all $h \in H$, we have

$$[h, x] = \alpha(h)x. \tag{8.1.1}$$

The space L^α is called the *root subspace corresponding to the root α.* We list some properties of the decomposition $L = \sum L^\alpha$.

I. *The Lie subalgebra H is contained in L^0. Furthermore, every nilpotent Lie subalgebra M of L containing H is contained in L^0.*

Proof. The adjoint representation of the Lie algebra M is a representation by nilpotent operators (see theorem 1 in 2.1). Therefore the restriction of this representation to H can have only zero weight and thus M is contained in L^0. $\square$

II. *The set $[L^\alpha, L^\beta]$ is contained in $L^{\alpha+\beta}$.*

Proof. Suppose that $x \in L^\alpha$ and $y \in L^\beta$. We will show that $[x, y] \in L^{\alpha+\beta}$. From the obvious identity

$$(\mathrm{ad}(h) - \alpha(h) - \beta(h))[x, y] = [(\mathrm{ad}(h) - \alpha(h))x, y] + [x, (\mathrm{ad}\,(h) - \beta(h))y]$$

we infer by induction an analogue of Leibniz's formula:

$$(\mathrm{ad}(h) - \alpha(h) - \beta(h))^n[x, y]$$

$$= \sum_{p=0}^{n} C_n^p[(\mathrm{ad}(h) - \alpha(h))^p x, (\mathrm{ad}(h) - \beta(h))^{n-p}y]. \tag{8.1.2}$$

If $n > \dim L^\alpha + \dim L^\beta$, all the summands in the right side of (8.1.2) are zero and thus we obtain $[x, y] \in L^{\alpha+\beta}$. $\square$

From Proposition II we see in particular that $[L^\alpha, L^\beta] = (0)$ if $\alpha + \beta$ is not a root.

III. *The set L^0 is a subalgebra of the Lie algebra L.*

Proof. Proposition II implies that $[L^0, L^0] \subset L^0$. $\square$

A nilpotent Lie subalgebra H of L is called a *Cartan subalgebra of L* if $L^0 = H$.

It is clear that a Cartan subalgebra is a maximal nilpotent subalgebra of L.

Let L be a Lie algebra and x an element of L. Let $L(\text{ad } x, 0)$ be the subspace of L consisting of the elements $y \in L$ such that $(\text{ad } x)^k y = 0$ for some $k \geqslant 0$. It is clear that $x \in L(\text{ad } x, 0)$. Thus we have $L(\text{ad } x, 0) \neq \{0\}$ for $x \neq 0$. The number $r = \min \dim L(\text{ad } x, 0)$ is called the *rank of L*. If x is in L and $\dim L(\text{ad } x, 0) = r$, x is called a *regular element of L*.

Every nonzero Lie algebra contains at least one regular element. To prove this, let x_1 be any nonzero element of L. If $\dim L(\text{ad } x_1, 0)$ is not minimal, there exists an element $x_2 \in L$, $x_2 \neq 0$, for which $\dim L(\text{ad } x_2, 0) < \dim L(\text{ad } x_1, 0)$. Since $\dim L$ is finite, we repeat this argument and arrive at an element x_k for which $\dim L(\text{ad } x_k, 0)$ has the least possible value.

IV. *Let H be a nilpotent Lie subalgebra of L containing a regular element h. In this case the Lie algebra L^0 is a Cartan subalgebra in L. Also we have $L^0 = L(\text{ad } h, 0)$.*

Proof. Let $L = L^0 + \sum_{\alpha \neq 0} L^\alpha$ be the decomposition of the space L relative to the nilpotent Lie subalgebra H. We set $\tilde{L} = \sum_{\alpha \neq 0} L^\alpha$. By Proposition II, the subspaces L^α are invariant under all operators ad x, $x \in L^0$. Thus the subspace $\tilde{L}$ is also invariant under these operators. For $x \in L^0$, let det (x) be the determinant of the transformation ad x restricted to the subspace $\tilde{L}$. The set of vectors $h \in H$ for which $\alpha(h) = 0$ for some $\alpha \neq 0$ is a finite union of hyperplanes in H. This union does not exhaust the space H. Hence there is a vector $h_0 \in H$ such that $\alpha(h_0) \neq 0$ for all $\alpha \neq 0$. Because H is contained in L^0, we have det $h = \prod_{\alpha \neq 0} \alpha(h) \neq 0$ for some $h \in L^0$. On the other hand, if $\det(x) \neq 0$ for some element $x \in L^0$, the restriction of the operator ad x to the subspace $\tilde{L}$ has only nonzero eigenvalues. Thus $L(\text{ad } x, 0)$ is contained in L^0. Let h be a regular element belonging to H. Since $h \in H$, we have $L(\text{ad } h, 0) \supset L^0$ and thus

$$L(\text{ad } x, 0) \subset L^0 \subset L(\text{ad } h, 0). \tag{8.1.3}$$

The dimension of $L(\text{ad } h, 0)$ is the least possible, so we conclude from (8.1.3) that

$$L(\text{ad } x, 0) = L^0 = L(\text{ad } h, 0). \tag{8.1.4}$$

We will prove that the Lie subalgebra L^0 is nilpotent. By (8.1.4), the operator $\text{ad}(x)$ is nilpotent on L^0 and so

$$\text{tr}(\text{ad}(x)|_{L^0})^p = 0, \quad \text{for all } p \geqslant 1. \tag{8.1.5}$$

The left side of (8.1.5) is a polynomial in the coordinates of the vector $x \in L^0$. We have already proved that (8.1.5) holds for all $x \in L^0$ such that $\det(x) \, w \, 0$. This set is nonvoid and open in L^0 and therefore (8.1.5) holds for all $x \in L^0$. Reducing $\mathrm{ad}(x)|_{L^0}$ to its Jordan canonical form, we see that (8.1.5) implies that all symmetric polynomials in eigenvalues of the operator $\mathrm{ad}(x)|_{L^0}$ are equal to zero. Hence the characteristic polynomial of the operator $\mathrm{ad}(x)|_{L^0}$ has only zero roots, that is, the operator $\mathrm{ad}(x)|_{L^0}$ is nilpotent. It follows that the Lie algebra L^0 is nilpotent. Finally, the Lie subalgebra H is contained in L^0 and $L(L^0, 0) \subset L(H, 0) = L^0$. This implies that $L(L^0, 0) = L^0$, that is, L^0 is a Cartan subalgebra of L.

V. *A Lie algebra L contains at least one Cartan subalgebra.*

Proof. Let h be a regular element in L and let H be the one-dimensional Lie subalgebra generated by h. Applying Proposition IV, we obtain a Cartan subalgebra L^0 in L.   □

We fix a Cartan subalgebra H in L. Then we have

$$L = H + \sum_{\alpha \neq 0} L^\alpha. \tag{8.1.6}$$

We define

$$v(\alpha) = \dim L^\alpha \tag{8.1.7}$$

for $\alpha \neq 0$.

VI. *The identity*

$$(h, h') = \sum_\alpha v(\alpha)\alpha(h)\alpha(h') \tag{8.1.8}$$

holds for all h and h' in H. Here $v(\alpha)$ is as in (8.1.7).

Proof. In view of the evident identity $(h, h') = (1/4)\{h + h', h + h') - (h - h', h - h')\}$, it suffices to prove (8.1.8) for $h = h'$. The operator $\mathrm{ad}\, h$ admits the unique eigenvalue $\alpha(h)$ in the subspace L^α; consequently we have $\mathrm{tr}((\mathrm{ad}\, h)^2) = \sum_\alpha v(\alpha)\alpha(h)^2$.   □

VII. *Let α and β be roots of the Lie algebra L with respect to H, and suppose that $\alpha \neq 0$. There exist a greatest nonpositive integer $p = p_{\beta\alpha}$ and a least nonnegative integer $q = q_{\beta\alpha}$ for which*

$$[L^{-\alpha}, L^{\beta + p\alpha}] = \{0\}, \quad and \quad [L^\alpha, L^{\beta + q\alpha}] = \{0\}. \tag{8.1.9}$$

Proof. Suppose that

$$[L^\alpha, L^\beta] \neq \{0\}, \quad [L^\alpha, L^{\beta + \alpha}] \neq \{0\}, \ldots, [L^\alpha, L^{\beta + q\alpha}] \neq \{0\}.$$

Proposition II implies that $\beta, \beta + \alpha, \ldots, \beta + (q + 1)\alpha$ are roots. Since the number of roots is finite, we have $[L^\alpha, L^{\beta+q\alpha}] = \{0\}$ for some $q \geqslant 0$. Let $q_{\beta\alpha}$ be the least of all these numbers q. We prove similarly that $p_{\beta\alpha}$ exists. $\square$

VIII. *Let* $\alpha, \beta, p = p_{\beta\alpha}, q = q_{\beta\alpha}$ *be as in* **VII.** *We define the rational number* $r_{\beta\alpha}$ *by*

$$r_{\beta\alpha} = \frac{\sum\limits_{k=p}^{q} k v(\beta + k\alpha)}{\sum\limits_{k=p}^{q} v(\beta + k\alpha)}. \tag{8.1.10}$$

Then we have $\beta(h) = r_{\beta\alpha}\alpha(h)$ *for all* $h \in [L^\alpha, L^{-\alpha}]$.

Proof. It suffices to verify **VIII** for elements $h \in H$ of the form $[x, y]$ with $x \in L^\alpha$ and $y \in L^{-\alpha}$. We write $V = \sum_{k=p}^{q} L^{\beta+k\alpha}$. The operator $\text{ad}(x)$ maps the space $L^{\beta+k\alpha}$ into the subspace $L^{\beta+(k+1)\alpha}$ and carries $L^{\beta+q\alpha}$ into zero, by our hypotheses. Therefore V is invariant under $\text{ad}(x)$. The proof that V is invariant under $\text{ad}(y)$ is similar. Since $\text{ad}(h) = [\text{ad}(x), \text{ad}(y)]$, V is invariant under $\text{ad}(h)$ and $\text{tr}(\text{ad}(h)) = 0$. The number $\beta(h) + k\alpha(h)$ is the unique eigenvalue of the operator $\text{ad}(h)$ in the subspace $L^{\beta+k\alpha}$. Thus we have

$$\sum_{k=p}^{q} v(\beta + k\alpha)(\beta(h) + k\alpha(h)) = 0,$$

and this proves the present proposition. $\square$

§9. The Structure of Semisimple Lie Algebras

9.1. Cartan Subalgebras of Semisimple Lie Algebras

Let L be a semisimple Lie algebra with a Cartan subalgebra H. We will use the notation and results of §8.

Recall that the Killing form is nonsingular on L.

I. *If* $\alpha + \beta \neq 0$, *the subspaces* L^α *and* L^β *are orthogonal.*

Proof. For $x \in L^\alpha$ and $y \in L^\beta$, the operator $\text{ad } x \, \text{ad } y$ carries each subspace L^γ into the subspace $L^{\gamma+\alpha+\beta}$. Suppose that $\alpha + \beta \neq 0$. Choose a basis of the linear space L consistent with the decomposition of L into the direct sum of subspaces L^γ. In this basis, all diagonal elements of the matrix of the operator $\text{ad } x \, \text{ad } y$ vanish. Thus we have $\text{tr}(\text{ad } x \, \text{ad } y) = 0$, that is, $(x, y) = 0$. $\square$

II. *The subspaces* L^α *and* $L^{-\alpha}$ *are dual under the Killing form (i.e., for each* $x \in L^\alpha$ *there is a* $y \in L^{-\alpha}$ *such that* $(x, y) \neq 0$ *and for each* $y \in L^{-\alpha}$ *there is an* $x \in L^\alpha$ *such that* $(x, y) \neq 0$).

Proof. If the element $x \in L^\alpha$ is orthogonal to the subspace $L^{-\alpha}$, Proposition I shows that x is orthogonal to $L = \sum_\alpha L^\alpha$. Since the Killing form is non-singular, we have $x = 0$.   □

III. *The equality* dim $L^\alpha =$ dim $L^{-\alpha}$ *holds.*

See Proposition II.

IV. *The restriction of the Killing form of the Lie algebra L to the subspace H is nonsingular.*

Proof. By Proposition II, $H = L^0$ is self-dual under the Killing form.   □

V. *The Lie algebra H is commutative. Write* $r =$ dim H. *There are r linearly independent roots of the Lie algebra L with respect to H.*

Proof. If the largest number of linearly independent roots is less than the dimension of the space H, H contains a nonzero element h such that $\alpha(h) = 0$ for all roots α. From (8.1.8) we infer that

$$(h, h') = \sum_\alpha v(\alpha)\alpha(h)\alpha(h') = 0 \qquad (9.1.1)$$

for all $h \in H$, which is impossible, as the Killing form is nonsingular on H.

We remark that every root α vanishes on $[H, H]$ (since $\alpha([h_1, h_2])h = [[h_1, h_2], h] = [h_1, [h_2, h]] - [h_2, [h_1, h]] = (\alpha(h_1)\alpha(h_2) - \alpha(h_2)\alpha(h_1))h = 0$ for all $h, h_1, h_2 \in H$). As we have just seen, the condition $\alpha(h) = 0$ for all roots α implies that $h = 0$. Hence $[H, H]$ is $\{0\}$ and H is commutative.   □

9.2. Properties of the Subspaces L^α

Since the form (x, y) is nonsingular on H, every linear functional λ on H has the form $\lambda(h) = (h, h'_\lambda)$ for all $h \in H$, h'_λ being a unique element of H. We introduce a scalar product in the space H^* of linear functionals on H by the equalities

$$(\lambda, \mu) = (h'_\lambda, h'_\mu) = \lambda(h'_\mu) = \mu(h'_\lambda) \qquad (9.2.1)$$

for all λ and μ in H^*.

In every subspace L^α we select a nonzero vector e_α that is a common eigenvector for all operators ad h, $h \in H$. We then have

$$[h, e_\alpha] = \alpha(h)e_\alpha \qquad (9.2.2)$$

for all $h \in H$.

I. *If α is a nonzero root and $x \in L^{-\alpha}$, then*

$$[e_\alpha, x] = (e_\alpha, x)h'_\alpha. \qquad (9.2.3)$$

Proof. Both sides of (9.2.3) belong to the Lie algebra H and for all $h \in H$, we have

$$(h, [e_\alpha, x]) = ([h, e_\alpha], x) = \alpha(h)(e_\alpha, x) = (e_\alpha, x)(h'_\alpha, h).$$

That is, the scalar products of both sides of (9.2.3) with every vector $h \in H$ are equal. Now apply Proposition IV of 9.1. $\square$

II. *If α is a nonzero root, we have $(\alpha, \alpha) \neq 0$.*

Proof. According to Proposition II in 9.1, $L^{-\alpha}$ contains an element x for which $(e_\alpha, x) = 1$. Applying (9.2.3), we obtain $[e_\alpha, x] = h'_\alpha$. Since $[e_\alpha, x]$ belongs to $[L^\alpha, L^{-\alpha}]$, Proposition VII in §8 implies that for any root β, there exists a rational number $r_{\beta\alpha}$ such that $\beta(h'_\alpha) = r_{\beta\alpha}\alpha(h'_\alpha)$. If $(\alpha, \alpha) = \alpha(h'_\alpha) = 0$, we have $\beta(h'_\alpha) = 0$ for all roots β. This implies that $h'_\alpha = 0$ (see the proof of Proposition V in 9.1). If $h'_\alpha = 0$, we cannot have $\alpha \neq 0$. $\square$

III. *If α is a nonzero root, the subspace L^α is one-dimensional.*

Proof. Choose x in $L^{-\alpha}$ so that $[e_\alpha, x] = h'_\alpha$ (see (9.2.3) and Proposition II in 9.1). Suppose that $y \in L^\alpha$. We will show that y is a scalar multiple of e_α. We set $y_k = (\text{ad } e_\alpha)^k y$. Since e_α belongs to L^α, Proposition II in §8 implies by induction on k that $y_k \in L^{(k+1)\alpha}$. Let us find $[x, y_k]$. According to Jacobi's identity, we have

$$[x, y_1] = [x, (\text{ad } e_\alpha)y] = -[e_\alpha, [y, x]] - [y, [x, e_\alpha]]. \tag{9.2.4}$$

We write $[x, y] = h$. The element h belongs to H according to Proposition II in §8. Since $[x, e_\alpha] = -h'_\alpha$, we infer from (9.2.4) that

$$[x, y_1] = -\alpha(h)e_\alpha - [h'_\alpha, y]. \tag{9.2.5}$$

Repeating this argument, we obtain by induction on k that

$$[x, y_k] = \frac{k(k-1)}{2}\alpha(h'_\alpha)y_{k-1} - k[h'_\alpha, y_{k-1}] \tag{9.2.6}$$

for $k > 1$. Since the number of roots of a Lie algebra is finite, there exists a least integer k_0 such that $y_k = 0$. (Recall that $y_k \in L^{(k+1)\alpha}$.) If $k_0 \geqslant 2$ and $y_{k_0-1} \neq 0$, (9.2.6) implies that for $k = k_0$, y_{k_0-1} is an eigenvector for the operator ad h'_α with eigenvalue $(k_0 - 1)\alpha(h'_\alpha)/2$. On the other hand, y_{k_0-1} belongs to $L^{k_0\alpha}$ and thus the eigenvalue of the operator ad(h'_α) corresponding to y_{k_0-1} is equal to $k_0\alpha(h'_\alpha)$. Since $\alpha(h'_\alpha) \neq 0$ by Proposition II and $(k_0 - 1)/2 \neq k_0$ for $k_0 \geqslant 2$, we have a contradiction. Thus, we have $y_{k_0-1} = 0$, and hence $y_1 = 0$. Therefore (9.2.5) implies that

$$[h'_\alpha, y] = -\alpha(h)e_\alpha. \tag{9.2.7}$$

We write $z = \alpha(h'_\alpha)y + \alpha(h)e_\alpha$. From (9.2.7) we infer that $[h'_\alpha, z] = 0$; that is, for $z \neq 0$, z is an eigenvector for the operator ad h'_α with eigenvalue 0. Because z belongs to L^α and $\alpha \neq 0$, this is impossible. We see that $z = 0$ and thus $\alpha(h'_\alpha)y + \alpha(h)e_\alpha = 0$. However, we have $\alpha(h'_\alpha) \neq 0$ and therefore y is a scalar multiple of e_α. $\square$

9.3. Series of Roots and Their Lie Subalgebras

Let α be a nonzero root of the Lie algebra L. If β is some root of L, the set S of all roots of the form $\beta + k\alpha$ (k an integer) is called an α-*series of roots*.

I. *The subspace* $L_S = \sum_{\alpha \in S} L^\alpha$ *is invariant under the operators of* ad L, ad $L^{-\alpha}$ *and* ad H.

See Proposition II of §8.

Proposition I allows us to contruct a Lie subalgebra of L isomorphic to $sl(2)$. We take up this construction.

II. *The subspace* $H^\alpha = [L^\alpha, L^{-\alpha}] \subset H$ *is one-dimensional for all nonzero roots* α.

See Proposition III in 9.2.

III. *There are vectors* $x_\alpha \in L^\alpha$, $y_\alpha \in L^\alpha$, $h_\alpha \in [L^\alpha, L^{-\alpha}] \subset H$ *such that*

$$[h_\alpha, x_\alpha] = 2x_\alpha, \; [h_\alpha, y_\alpha] = -2y_\alpha, \; [x_\alpha, y_\alpha] = h_\alpha. \tag{9.3.1}$$

Proof. By Proposition II in 9.2, we have $(\alpha, \alpha) = \alpha(h'_\alpha) \neq 0$ and by Proposition I in 9.2, h'_α belongs to $[L^\alpha, L^{-\alpha}]$. Multiplying h'_α by a suitable scalar, we obtain a vector $h_\alpha \in [L^\alpha, L^{-\alpha}]$ such that $\alpha(h_\alpha) = 2$. Since $[e_\alpha, e_{-\alpha}] = (e_\alpha, e_{-\alpha})h'_\alpha$ (see (9.2.3)) and $(e_\alpha, e_{-\alpha}) \neq 0$ by Proposition II in 9.1, we see that $[e_\alpha, e_{-\alpha}] \neq 0$ and is a scalar multiple of h_α. Multiplying $e_{-\alpha}$ by a suitable scalar k, we find $[e_\alpha, ke_{-\alpha}] = h_\alpha$. We write x_α for e_α and y_α for $ke_{-\alpha}$, so that $[x_\alpha, y_\alpha] = h_\alpha$. By applying (9.2.2) we see that $[h_\alpha, x_\alpha] = \alpha(h_\alpha)x_\alpha = 2x_\alpha$. Since y_α belongs to $L^{-\alpha}$, we find in like manner that $[h_\alpha, y_\alpha] = (-\alpha)(h_\alpha)(y_\alpha) = -2y_\alpha$. For future use, we note that

$$h_\alpha = 2h'_\alpha/(\alpha, \alpha) \tag{9.3.2}$$

since $\alpha(h_\alpha) = (2/(\alpha, \alpha))\alpha(h'_\alpha) = 2$. $\square$

IV. *Let* α *be a nonzero root. The linear span of the spaces* L^α, $L^{-\alpha}$ *and* H^α *is a Lie subalgebra* L_α *of* L. *The Lie algebra* L_α *is isomorphic to* $sl(2)$.

Proof. Proposition I shows that the subspace L_α is a subalgebra. Consider the Lie algebra $sl(2)$, which consists of all square matrices of order 2 with trace zero. The matrices

$$x = \begin{Vmatrix} 0 & 1 \\ 0 & 0 \end{Vmatrix}, \qquad h = \begin{Vmatrix} 1 & 0 \\ 0 & -1 \end{Vmatrix}, \qquad y = \begin{Vmatrix} 0 & 0 \\ 1 & 0 \end{Vmatrix} \qquad (9.3.3)$$

are a basis in $sl(2)$. We also have

$$[x, y] = h, \qquad [h, x] = 2x, \qquad [h, y] = -2y. \qquad (9.3.4)$$

The equalities (9.3.4) and (9.3.1) show that the mapping $\xi x + \eta y + \zeta h \to \xi x_\alpha + \eta y_\alpha + \zeta h_\alpha (\xi, \eta, \zeta \in \mathbf{C})$ is an isomorphism of the Lie algebra $sl(2)$ onto L_α. $\square$

9.4. Representations of the Lie Algebra $sl(2)$

We will need in the sequel certain information about the Lie algebra $sl(2)$ and its representations. We list them in Propositions I–III and in the theorem following.

I.

   (a) *The Lie algebra $sl(2)$ is simple.*
   (b) *The one-dimensional subspace H generated by the element h is a Cartan subalgebra of $sl(2)$.*

Proof. Let I be a nonzero ideal in the Lie algebra $sl(2)$. If z belongs to $I \backslash H$, (9.3.4) shows that the element $[h, z] \in I$ is a nonzero linear combination of the vectors x and y. Apply either ad x or ad y to this vector. We obtain a nonzero element of the ideal I, which is a multiple of $[x, y] = h$. Thus h belongs to I and $x = (1/2)[h, x]$ and $y = (-1/2)[h, y]$ belong to I. That is, $I = sl(2)$.

Since $[z, z] = 0$ for all z in a Lie algebra L, we have $r = \min \dim_{z \in L} L(\operatorname{ad} z, 0) \geqslant 1$. From (9.3.4) we infer that $L(\operatorname{ad} h, 0) = H$ so that h is a regular element of the Lie algebra $L = sl(2)$ and H is a Cartan sub-algebra of L (see Proposition IV in §8).

Let α denote the linear functional on H defined by the equality

$$\alpha(ch) = 2c \qquad (9.4.1)$$

for all $c \in \mathbf{C}$. Let π be a linear representation of $sl(2)$ in a finite dimensional space V. For a linear functional λ on H let V_λ denote the subspace of V consisting of the vectors $v \in V$ for which $\pi(h)v = \lambda(h)v$. If $v \in V_\lambda$, we have

$$\pi(h)\pi(x)v = \pi(x)\pi(h)v + 2\pi(x)v = (\lambda(h) + 2)\pi(x)v$$
$$= ((\lambda + \alpha)(h))\pi(x)v,$$

that is, $\pi(x)V_\lambda \subset V_{\lambda + \alpha}$. The inclusion $\pi(y)V_\lambda \subset V_{\lambda - \alpha}$ is proved similarly. $\square$

II. *The equality $V = \Sigma\, V_\lambda$ holds.*

Proof. Suppose first that the representation π is irreducible. The inclusions $\pi(x)V_\lambda \subset V_{\lambda+\alpha}$, $\pi(y)V_\lambda \subset V_{\lambda-\alpha}$ show that $\sum_\lambda V_\lambda$ is an invariant subspace in V. Since the operator $\pi(h)$ has at least one eigenvector, we see that $\sum_\lambda V_\lambda \neq \{0\}$. Because π is irreducible, we have $\sum_\lambda V_\lambda = V$. If π is not irreducible, the simplicity of $sl(2)$ and H. Weyl's Theorem show that π is the direct sum of certain irreducible representations, to each of which the foregoing applies. $\square$

Let π be an irreducible representation of the Lie algebra $sl(2)$. The number of weights of the restriction of π to H is finite. Hence there is a weight λ_0 such that $\lambda_0 + \alpha$ is not a weight of the restriction of π to H.

III. *The set of weights of the restriction of π to H is the set*

$$-\lambda_0, -\lambda_0 + \alpha, -\lambda_0 + 2\alpha, \ldots, \lambda_0 - \alpha, \lambda_0. \qquad (9.4.2)$$

For every weight λ the number $\lambda(h)$ is an integer.

Proof. Since $\lambda_0 + \alpha$ is not a weight, we have

$$\pi(x)V_{\lambda_0} \subset V_{\lambda_0+\alpha} = \{0\}. \qquad (9.4.3)$$

Let e_0 be a nonzero element of V_{λ_0} and write e_k for $\pi(y)^k e_0$. Also write $e_{-1} = 0$. We will show that the vector $\pi(x)e_{k+1}$ is a scalar multiple of e_k. For $k = -1$ the statement holds by virtue of (9.4.3). If $\pi(x)e_{k+1} = \mu_k e_k$ for all $k < r$, we have

$$\pi(x)e_{r+1} = \pi(x)\pi(y)e_r = \pi(y)\pi(x)e_r + \pi(h)e_r$$
$$= \mu_{r-1}\pi(y)e_{r-1} + (\lambda_0 - r\alpha)(h)e_r = \mu_r e_r,$$

where

$$\mu_r = \mu_{r-1} - (r\alpha - \lambda_0)(h), \qquad \mu_{-1} = 0.$$

Since $\alpha(h) = 2$ (see (9.4.1)) we obtain

$$\mu_k = -\sum_{r=0}^{k} (r\alpha - \lambda_0)(h) = -(k+1)(k\alpha(h)/2 - \lambda_0(h))$$
$$= (k+1)(\lambda_0(h) - k). \qquad (9.4.4)$$

Since the space V is finite-dimensional, there is a j such that $e_j \neq 0$, $e_{j+1} = 0$. (If no such j exists, we get $0 \neq e_k \in V_{\lambda_0-k\alpha}$ for all k. The spaces $V_{\lambda_0-k\alpha}$ being linearly independent, this violates the finite dimensionality of V.) We thus have $\mu_j e_j = \pi(x)e_{j+1} = 0$ and therefore $\mu_j = 0$. From (9.4.4) we infer that $\lambda_0(h) = j$ so that $\lambda_0(h)$ is a nonnegative integer. Hence (9.4.4) implies that

$$\mu_k = (k+1)(j-k). \tag{9.4.5}$$

Let us write

$$f_k = (j-k)!e_k, \qquad k = 0, 1, \ldots, j. \tag{9.4.6}$$

Then (9.4.5) yields

$$\left.\begin{array}{l} \pi(x)f_k = kf_{k-1}, \qquad \pi(y)f_k = (j-k)f_{k+1}, \\ \pi(h)f_k = (\lambda_0 - k\alpha)(h)f_k = (j-2k)f_k \quad \text{for } k = 0, 1, \ldots, j. \end{array}\right\} \tag{9.4.7}$$

The equalities (9.4.7) imply that the subspace spanned by the vectors $f_0, f_1, \ldots, f_j$ is invariant under the representation π. By hypothesis π is irreducible, and so the vectors $f_0, f_1, \ldots, f_j$ are a basis for V. Since $f_0 \in V_{\lambda_0}$ and $f_k \in V_{\lambda_0 - k\alpha}$, the set of weights connected with the representation π is $\{\lambda_0, \lambda_0 - \alpha, \ldots, \lambda_0 - j\alpha\}$. But we have $(\lambda_0 - j\alpha)(h) = \lambda_0(h) - 2j = j - 2j = -j = -\lambda_0(h)$ and therefore

$$\lambda_0 - j\alpha = -\lambda_0. \tag{9.4.8}$$

Finally, for any weight $\lambda = \lambda_0 - k\alpha$, the number $\lambda(h) = \lambda_0(h) - k\alpha(h) = j - 2k$ is an integer. $\square$

Theorem. *Let π be a linear representation of the Lie algebra $sl(2)$ in a finite-dimensional space V. We have:*

(1) $V = \sum_\lambda V_\lambda$ *and for any weight λ the number $\lambda(h)$ is an integer;*
(2) *for any weight λ there is a weight λ' such that $\lambda(h) = -\lambda'(h)$ and $\dim V_\lambda = \dim V_{\lambda'}$;*
(3) *if λ and $\lambda + \alpha$ are weights, then $\pi(x)V_\lambda \neq \{0\}$; if λ and $\lambda - \alpha$ are weights, then $\pi(y)V_\lambda \neq \{0\}$;*
(4) *if the representation π is irreducible, v belongs to V, s is the least nonnegative integer such that $\pi(x)^{s+1}v = 0$, and t is the least nonnegative integer such that $\pi(y)^{t+1}v = 0$, then we have*

$$\pi(x)\pi(y)v = (s+1)tv; \qquad \pi(y)\pi(x)v = (t+1)sv. \tag{9.4.9}$$

Proof. Since all representations of the Lie algebra $sl(2)$ are completely reducible, it is sufficient to prove the theorem for an irreducible representation. Statement (1) is implied by Propositions II and III. Statement (2) follows from Proposition III if we note that the subspaces V_λ are one-dimensional and that (9.4.8) implies that $-(\lambda_0 - k\alpha) = \lambda_0 - (j-k)\alpha$, $0 \leqslant k \leqslant j$. We will now prove statement (3). Let $\lambda = \lambda_0 - k\alpha$ be a weight of π. If $\lambda + \alpha$ is a weight, then $k \geqslant 1$ and $\pi(x)f_k \neq 0$ by Proposition III. Since f_k belongs to V_λ, $\pi(x)V_\lambda$ is not $\{0\}$. If $\lambda - \alpha$ is a weight, we have $k \leqslant j-1$ and $\pi(y)f_k \neq 0$.

Let us prove (4). It suffices to consider the case $v = f_k$. We then have $s = k$ and $t = j - k$, and (9.4.7) implies that

$$\pi(x)\pi(y)f_k = (j - k)\pi(x)f_{k+1} = (k + 1)(j - k)f_k = (s + 1)tf_k,$$
$$\pi(y)\pi(x)f_k = k\pi(y)f_{k-1} = k(j + 1 - k)f_k = s(t + 1)f_k.$$

This establishes the equalities (9.4.9). □

Remark. For every nonnegative integer j, the formulas (9.4.7) define a representation π of the Lie algebra $sl(2)$ in the space V with basis $f_0, f_1, \ldots, f_j$. This representation is irreducible. Indeed, if W is an invariant subspace of π, it is invariant in particular under the operator $\pi(h)$ and thus is generated by eigenvectors of the operator $\pi(h)$ contained in W. The eigenvectors of $\pi(h)$ are precisely $f_0, \ldots, f_j$. If f_k belongs to W, then $f_l \in W$ for all $l = 0, \ldots, j$, since f_l is a scalar multiple of $\pi(x)^{k-l}f_k$ for $k \geqslant l$ and f_l is a multiple of $\pi(y)^{l-k}f_k$ for $l \geqslant k$. Thus we have $W = V$, and so π is irreducible.

9.5. The Structure of the α-Series of Roots

We apply the foregoing results to the representation π of the Lie algebra L_α, isomorphic to $sl(2)$, where π is the restriction of the adjoint representation of the Lie algebra L to the subalgebra L_α and the subspace L_S (see Proposition I in 9.3).

I. *Let α, β be roots of the Lie algebra L with $\alpha \neq 0$. The α-series S containing β consists of all roots of the form $\beta + k\alpha$, where $p \leqslant k \leqslant q$. Furthermore we have*

$$-2(\beta, \alpha)/(\alpha, \alpha) = p + q. \tag{9.5.1}$$

Proof. First note that $\alpha(h'_\alpha) \neq 0$ and therefore distinct roots from the same α-series have distinct values on the element h'_α. Let H^α be the subspace of H generated by the vector h'_α. Each root γ of the Lie algebra L defines a weight of the Lie algebra $sl(2)$ under the restriction to H^α. Since the root subspaces L^α are one-dimensional for $\alpha \neq 0$, the representation π is irreducible or is the sum of an irreducible representation and the zero representation in some subspace W of $L_s \cap H$. (This is possible only when the α-series S contains the root zero. The form claimed for the α-series S is implied by Proposition III in 9.4. Since the restrictions of the roots $\beta + q\alpha$ and $\beta + p\alpha$ to H^α coincide with λ_0 and $-\lambda_0$ respectively, we find $(\beta + q\alpha)(h'_\alpha) = \lambda_0(h'_\alpha)$, $(\beta + p\alpha)(h'_\alpha) = -\lambda_0(h'_\alpha)$. Adding these equalities, we obtain

$$2\beta(h'_\alpha) + (p + q)(\alpha, \alpha) = 0,$$

which implies (9.5.1). □

II. *If α, β and $\alpha + \beta$ are nonzero roots, we have $[L^\alpha, L^\beta] = L^{\alpha+\beta}$.*

Proof. We already know that $[L^\alpha, L^\beta] \subset L^{\alpha+\beta}$ and that dim $L^{\alpha+\beta} = 1$. On the other hand, assertion (3) in the theorem in 9.4 shows that $\mathrm{ad}(e_\alpha)L^\beta \neq \{0\}$. $\square$

III. *The only roots that are scalar multiples of a nonzero root α are 0, α, $-\alpha$.*

Proof. Suppose that the set $\{p\alpha, (p+1)\alpha, \ldots, q\alpha\}$ is an α-series of roots containing α. Since $[L^\alpha, L^\alpha] = \{0\}$ and $[L^{-\alpha}, L^{-\alpha}] = \{0\}$, Proposition II implies that 2α and -2α are *not* roots. Since $(-\alpha)$ is a root (see Proposition II in 9.1), we have $p = -1$, $q = 1$. If $\beta = c\alpha$ is a root of our Lie algebra (c is a complex number), the number $2(\beta, \alpha)/(\alpha, \alpha) = 2c$ must be an integer by (9.5.1). Thus c is an integer or a half integer. According to (9.5.1), for an α-series of the root β, we have $p + q = -2k - 1$. The inequalities $p \leqslant 0 \leqslant q$ (p and q are integers) show that $p \leqslant -k - 1$, $q \geqslant -k$. Therefore $\alpha/2$ and $-\alpha/2$ are roots of the Lie algebra L. Then $\alpha = 2(\alpha/2)$ cannot be a root, as was shown above. This contradiction shows that c is an integer, which contradicts our hypothesis. $\square$

9.6. Positive and Simple Roots

The set of linear combinations of the vectors h'_α with *real* coefficients is denoted by H_0. This set H_0 is a vector space over the real field $\mathbf{R}$. Clearly H_0 is contained in H.

I. *The dimension of the space H_0 over $\mathbf{R}$ is equal to the dimension of H over $\mathbf{C}$. The restriction of the Killing form to H_0 is positive-definite and is real-valued. The Lie algebra H is the direct sum $H_0 + iH_0$.*

Proof. Let Δ be the set of nonzero roots of the Lie algebra L with respect to H. From Propositions VI in §8 and III in 9.1 we infer that

$$(h, h') = \sum_{\beta \in \Delta} \beta(h)\beta(h') \tag{9.6.1}$$

for all h, $h' \in H$, and in particular, $(h'_\alpha, h'_\alpha) = \sum_{\beta \in \Delta} (\beta(h'_\alpha))^2 = \sum_{\beta \in \Delta} r^2_{\beta\alpha}(\alpha, \alpha)^2$. On the other hand, we have $(h'_\alpha, h'_\alpha) = (\alpha, \alpha)$ and thus $(\alpha, \alpha) = (\sum_{\beta \in \Delta} r^2_{\beta\alpha})^{-1}$. Therefore (α, α) is a positive number. If $p_{\beta\alpha}$, $q_{\beta\alpha}$ are the integers defined in Proposition I of 9.5, we have $(\beta(h'_\alpha))^2 = ((p_{\beta\alpha} + q_{\beta\alpha})^2/4)(\alpha, \alpha)^2$ by (9.5.1). Substituting this expression in (9.6.1) we make some obvious computations and get $(\alpha, \alpha) = 4(\sum_{\beta \in \Delta} (p_{\beta\alpha} + q_{\beta\alpha})^2)^{-1}$. For $\beta \neq \alpha$, we have

$$(h'_\alpha, h'_\beta) = (\beta, \alpha) = -(p_{\beta\alpha} + q_{\beta\alpha})(\alpha, \alpha)/2.$$

This number is real because (α, α) is real. If x belongs to H_0, then

$$x = \sum_{\alpha \in \Delta} c_\alpha h'_\alpha,$$

where the c_α are real. Thus $\beta(x) = \sum_{\alpha \in \Delta} c_\alpha \beta(h'_\alpha) = \sum_{\alpha \in \Delta} c_\alpha(\beta, \alpha)$ is a real number, that is, all roots are real-valued on H_0. Furthermore, for $x \in H_0$, the number

$$(x, x) = \sum_{\beta \in \Delta} (\beta(x))^2$$

(see (9.6.1)) is nonnegative. If $(x, x) = 0$, then $\beta(x) = 0$ for all $\beta \in \Delta$ and so $x = 0$ (see Proposition IV in 9.1).

We will prove that $\dim_{\mathbf{R}} H_0 = \dim_{\mathbf{C}} H$. It suffices to show that any system $\{h'_{\alpha_i}\}$ linearly independent over $\mathbf{R}$ is linearly independent over $\mathbf{C}$. Assume that $\sum \lambda_i h'_{\alpha_i} = 0$ for complex numbers λ_i not all zero. We then have $\sum \lambda_i (h'_{\alpha_i}, h'_{\alpha_j}) = 0$. This is a system of homogeneous linear equations in the unknowns λ_i with real coefficients $(h'_{\alpha_i}, h'_{\alpha_j})$. If such a system has a complex nonzero solution, it has a real nonzero solution. This is impossible since $\{h'_{\alpha_i}\}$ is linearly independent over $\mathbf{R}$.

It is clear that $H_0 \cap iH_0 = \{0\}$, and also $\dim_{\mathbf{C}}(H_0 + iH_0) = \dim_{\mathbf{R}} H_0$. Since $\dim_{\mathbf{R}} H_0 = \dim_{\mathbf{C}} H$, we find $H_0 + iH_0 = H$. $\square$

An abstract set M is said to be *ordered by a relation* $<$ if: for any two elements a, b of this set exactly one of the relations $a < b$, $b < a$ or $a = b$ holds; and $a < b$ and $b < c$ imply that $a < c$. We define $b > a$ to mean $a < b$.

A finite-dimensional real vector space V is called an *ordered vector space* if V is an ordered set and the order $<$ satisfies the following conditions:

(a) if $x, y \in V$, $x > 0$, $y > 0$, then $x + y > 0$;
(b) if $x \in V$, $x > 0$ and a is a positive real number, then $ax > 0$.

An element x is called *positive* if $x > 0$.

Let $e_1, \ldots, e_n$ be a basis of the real vector space V. We set $x > y$ for $x, y \in V$, if $x - y = \sum_{i=1}^n a_i e_i$ and the first nonzero number in the sequence $a_1, \ldots, a_n$ is positive. This is called the *lexicographic ordering of the space V under the basis $e_1, \ldots, e_n$.*

II. *A lexicographically ordered vector space is an ordered vector space.*

We omit the proof.

Let V be a finite-dimensional real vector space. Let V^* be the space of linear functionals on V. Let $e_1, \ldots, e_n$ be a basis in V and $f_1, \ldots, f_n$ the basis in V^* biorthogonal to $e_1, \ldots, e_n$ (that is, $f_j(e_i) = \delta_{ij}$, where δ is Kronecker's symbol). The lexicographic ordering of the space V^* under the basis $f_1, \ldots, f_n$ is called the *lexicographic ordering under the basis* $e_1, \ldots, e_n$.

Thus for the lexicographic ordering of V^* under $e_1, \ldots, e_n$ we say that $f > g$ for $f, g \in V^*$, if $f(e_i) = g(e_i)$ for $i = 1, \ldots, k$ and $f(e_{k+1}) > g(e_{k+1})$

Let $h_1, \ldots, h_r$ be any basis in the space H_0. We order the space H_0^* of linear functionals on H_0 lexicographically under the basis $h_1, \ldots, h_r$.

A root is called *positive* if it is a positive element in H_0^*. A positive root is called *simple* if it is not the sum of two positive roots.

It is clear that $\alpha > 0$ if and only if $0 > -\alpha$. Therefore the set of all nonzero roots Δ is the union $\Delta_+ \cup (-\Delta_+)$, where Δ_+ is the set of all positive roots.

III. *If* $\dim H = r$, *we then have exactly* r *simple roots* $\alpha_1, \ldots, \alpha_r$. *These roots are a basis in the space* H_0^*. *Every root* β *is a sum* $\sum m_i \alpha_i$, *where* m_i *are integers of a single sign.*

Proof. We will show that the simple roots are linearly independent. First, if α_i, α_j are simple roots, their difference $\alpha_i - \alpha_j$ is not a root. For, if $\alpha_i - \alpha_j = \beta \in \Delta$, we have either $\beta > 0$ and then $\alpha_i = \alpha_j + \beta$ violates the simplicity of α_i or $\beta < 0$ and $\alpha_j = \alpha_i - \beta$ violates the simplicity of α_j. Proposition I in 9.5 implies that for $\beta = \alpha_i$, $\alpha = \alpha_j$,

$$-2(\alpha_i, \alpha_j)/(\alpha_j, \alpha_j) = q \geqslant 0, \qquad (9.6.2)$$

since $p = 0$, as just proved. Therefore we have $(\alpha_i, \alpha_j) \leqslant 0$.

Assume that the set of simple roots is linearly dependent. For certain nonnegative numbers a_i, b_j the relations $\sum (a_i - b_i)\alpha_i = 0$ and so $\sum_i a_i \alpha_i = \sum_j b_j \alpha_j > 0$ must hold. The roots appearing in the left and right sides of the last equality are distinct. We set $\gamma = \sum_i a_i \alpha_i$. We then have

$$(\gamma, \gamma) = \left(\sum_i a_i \alpha_i, \sum_j b_j \alpha_j \right) = \sum_{i,j} a_i b_j (\alpha_i, \alpha_j).$$

Since $(\gamma, \gamma) \geqslant 0$, $a_i b_j \geqslant 0$, $(\alpha_i, \alpha_j) \leqslant 0$, the equality can obtain only if $(\gamma, \gamma) = 0$ and thus $\gamma = 0$.

We will show that any root $\beta > 0$ is the sum of simple roots with nonnegative integral coefficients. This is obvious for the root that is the least element in the set Δ_+ under the above ordering. If $\beta > 0$ is not minimal, it is either simple, in which case our claim is obvious, or it is the sum $\beta = \gamma + \delta$, where γ, δ are positive roots. It is plain that $\gamma < \beta, \delta < \beta$. Supposing that all positive roots strictly smaller than β can be represented in the form $\beta = \sum m_i \alpha_i$ with nonnegative integers m_i, we complete the proof by induction. $\square$

Finally, positive roots generate the space H^* (by Proposition I, for example). Therefore simple roots also generate this space. Let Π denote the set $\alpha_1, \ldots, \alpha_r$ of simple roots. The integers

$$n_{ij} = -2(\alpha_i, \alpha_j)/(\alpha_i, \alpha_i) \tag{9.6.3}$$

are called *Cartan numbers.*

IV. *The set Δ_+ of positive roots can be constructed from the set Π of simple roots.*

Proof. Proposition III implies that any positive root β is simple or has the form $\beta = \gamma + \alpha_i$, where γ is a positive root and α_i is a simple root. We set $\beta = \sum m_i \alpha_i$, where the m_i are nonnegative integers. Since $(\beta, \beta) = \sum m_i(\beta, \alpha_i) > 0$ and $m_i \geqslant 0$, there is at least one i for which $(\beta, \alpha_i) > 0$. Proposition I in 9.5 implies that

$$p_{\beta \alpha_i} + q_{\beta \alpha_i} = -2(\beta, \alpha_i)/(\alpha_i, \alpha_i) < 0.$$

Thus we have $p_{\beta \alpha_i} < 0$, and $\gamma = \beta - \alpha_i$ is a root. In this case we have either $\gamma = 0$ and $\beta = \alpha_i$ is a simple root, or $\gamma > 0$ (since if $\gamma < 0$, $\alpha_i = \beta + (-\gamma)$ is not simple).

We will now identify the $\sum m_i \alpha_i$ (the m_i being nonnegative integers) that are roots. The natural number $m = \sum m_i$ is called the *order* of the linear combination $\sum m_i \alpha_i$. Linear combinations of order 1 are the roots α_i. Suppose that all the roots of order $\leqslant m$ have been found. Every root of order $m + 1$ has the form $\gamma + \alpha_i$, where γ is a root of order m. Thus it suffices to identify the sums $\gamma + \alpha_i$ that are roots. Suppose that $\gamma \neq \alpha_i$. We will show that the α_i-series containing γ consists only of positive roots. In fact, $\gamma = \sum_{j \neq i} m_j \alpha_j + m_i \alpha_i$, where at least one of the integers m_j is positive. Proposition III shows that the vector $\gamma + k\alpha_i = \sum_{j \neq i} m_j \alpha_j + (m_i + k)\alpha_i$ can be a root only if $m_i + k \geqslant 0$. We then have $\gamma + k\alpha_i > 0$. Furthermore, if $k \leqslant 0$, the sum $\gamma + k\alpha_i$ has order $m + k \leqslant m$. All these roots are known under our inductive hypothesis. Therefore we also know the number p from Proposition I in 9.5. The relation (9.5.1) takes the form

$$p + q = -2(\gamma, \alpha_i)/(\alpha_i, \alpha_i).$$

Thus we can also define q. The vector $\gamma + \alpha_i$ is a root if and only if $q \geqslant 1$. $\square$

A system of simple roots Π is said to *split* if $\Pi = \Pi' \cup \Pi''$, where the sets Π' and Π'' are nonvoid and the subspaces H' and H'' of H, generated by the vectors $h'_{\alpha'}$, $\alpha' \in \Pi'$ and $h'_{\alpha''}$, $\alpha'' \in \Pi''$, respectively, are orthogonal under the Killing form.

V. *Let L be a semisimple Lie algebra. This Lie algebra is simple if and only if the system of simple roots of L does not split.*

Proof. Suppose that $\Pi = \Pi' \cup \Pi''$ and $H' \perp H''$. Let Δ'_+ be the set of positive roots of the form $\sum m_i \alpha_i$, where m_i are nonnegative integers and $\alpha_i \in \Pi'$. We define Δ''_+ similarly. If α' belongs to Δ'_+, then $h'_{\alpha'} = \sum m_i h'_{\alpha_i} \in H'$. Similarly, if α'' belongs to Δ''_+, then $h'_{\alpha''} \in H''$. Therefore any two roots $\alpha' \in \Delta'_+$, $\alpha'' \in \Delta''_+$, are orthogonal. The difference $\alpha' - \alpha''$ cannot be a root. To see this, observe that $\alpha' - \alpha'' = \sum m_i \alpha_i - \sum n_j \beta_j$, where $\alpha_i \in \Delta'_+$, $\beta_j \in \Delta''_+$, $m_i \geq 0$, $n_j \geq 0$, while, like all roots, it must be a linear combination of simple roots with coefficients of one sign. On the other hand, Proposition I in 9.5 implies that for the α''-series containing α', we have $p + q = -2(\alpha', \alpha'')/(\alpha'', \alpha'') = 0$. As proved above, we have $p = 0$, so that $q = 0$. This means that $\alpha' + \alpha''$ is not a root. Therefore any positive root belongs either to Δ'_+ or to Δ''_+, i.e., $\Delta_+ = \Delta'_+ \cup \Delta''_+$.

We write $L' = H' + \sum_{\alpha \in \Delta'_+} (L^\alpha + L^{-\alpha})$. If α, β belong to Δ'_+, we have $\alpha \pm \beta \in \Delta'_+ \cup (-\Delta'_+)$. The inclusions $[H', L^\alpha] \subset L^\alpha$ and $[L^\alpha, L^{-\alpha}] = \{Ch'_\alpha\} \subset H'$ hold. Hence L' is a subalgebra of L. The subalgebra L'' is defined analogously. Since $\pm \alpha' \pm \alpha''$ is not a root, we have $[L^{\pm \alpha'}, L^{\pm \alpha''}] = \{0\}$. Furthermore the equality $(\alpha', \alpha'') = 0$ implies that $[h'_\alpha, e_{\alpha''}] = (\alpha', \alpha'') e_{\alpha''} = 0$. This implies that $[H', L^{\pm \alpha''}] = \{0\}$. Similarly we get $[H'', L^{\pm \alpha'}] = \{0\}$. Since H is commutative, we have $[H', H''] = \{0\}$. It follows that $[L', L''] = \{0\}$, that is, the Lie algebra L is the direct sum of the Lie subalgebras L' and L''. Hence L is not simple.

Suppose now that L is not simple, so that $L = L' + L''$, where L' and L'' are ideals in L. The vector $e_\alpha \in L^\alpha$ can be uniquely written in the form $e_\alpha = e'_\alpha + e''_\alpha$ where $e'_\alpha \in L'$ and $e''_\alpha \in L''$. If h belongs to H, we have

$$0 = [h, e_\alpha] - \alpha(h)e_\alpha = [h, e'_\alpha] - \alpha(h)e'_\alpha + [h, e''_\alpha] - \alpha(h)e''_\alpha.$$

Since $[h, e'_\alpha]$ belongs to L' and $[h, e''_\alpha]$ belongs to L'', we obtain $[h, e'_\alpha] - \alpha(h)e'_\alpha = 0$, $[h, e''_\alpha] - \alpha(h)e''_\alpha = 0$. This implies that e'_α and e''_α are in L^α. However, we have dim $L^\alpha = 1$ and the vectors e'_α, e''_α are orthogonal. Therefore we have either $e'_\alpha = 0$ or $e''_\alpha = 0$. If $e''_\alpha = 0$, L^α is contained in L'. Since L' is orthogonal to L'', $e_{-\alpha}$ is not in L'', by Proposition II in 9.1. Consequently $e_{-\alpha}$ is in L' and $L_{-\alpha} \subset L'$. We also find that $h'_\alpha = -[e_\alpha, e_{-\alpha}] \in L'$. Likewise, if $e'_\alpha = 0$, we have $L^\alpha \subset L''$, $L^{-\alpha} \subset L''$, $h'_\alpha \in L''$.

Let Δ' (Δ'') be the set of roots α such that $L^\alpha \subset L'$ ($L^\alpha \subset L''$). If $\alpha \in \Delta'$ and $\beta \in \Delta''$, we obtain $(\alpha, \beta) = (h'_\alpha, h'_\beta) = 0$, since $(L', L'') = 0$. We thus have $\Pi = (\Pi \cap \Delta') \cup (\Pi \cap \Delta'')$ and the subspaces $H' \subset L'$ and $H'' \subset L''$ are orthogonal. It is also clear that $H = (H \cap L') + (H \cap L'')$. $\square$

9.7. A Weyl Basis for a Semisimple Lie Algebra

Let $\{h_1, \ldots, h_r\}$ be a basis in H. If we add to this basis the vectors $e_\alpha \in L^\alpha$ for all nonzero roots α of the Lie algebra L, we obtain a basis for L. We may suppose that the vectors e_α are chosen so that $(e_\alpha, e_{-\alpha}) = -1$. Thus the following equalities hold:

$$[h, e_\alpha] = \alpha(h)e; \tag{9.7.1}$$

$$[e_\alpha, e_{-\alpha}] = -h'_\alpha; \tag{9.7.2}$$

$$[e_\alpha, e_\beta] = \begin{cases} 0, & \text{if } \alpha + \beta \neq 0 \text{ is not a root,} \\ N_{\alpha\beta} \, e_{\alpha+\beta}, & \text{if } \alpha + \beta \text{ is a nonzero root;} \end{cases} \tag{9.7.3}$$

$$(h, h'_\alpha) = \alpha(h). \tag{9.7.4}$$

We write $N_{\alpha\beta} = 0$ if $\alpha + \beta \neq 0$ is not a root. The equalities (9.7.1)–(9.7.4) show that the Lie algebra L is completely determined by the numbers $N_{\alpha\beta}$. We list some relations among these numbers.

I.

(a)   *For any roots α, β we have $N_{\beta\alpha} = -N_{\alpha\beta}$.*
(b)   *Let α, β, γ be nonzero roots, where $\alpha + \beta + \gamma = 0$. We then have*

$$N_{\alpha,\beta} = N_{\beta,\gamma} = N_{\gamma,\alpha}. \tag{9.7.5}$$

(c)   *Let α, β, γ, δ be nonzero roots, where $\alpha + \beta + \gamma + \delta = 0$, and the pairwise sums of α, β, γ, δ are nonzero. We then have*

$$N_{\alpha\beta}N_{\gamma\delta} + N_{\beta\gamma}N_{\alpha\delta} + N_{\gamma\alpha}N_{\beta\delta} = 0. \tag{9.7.6}$$

Proof. Since $[e_\alpha, e_\beta] = -[e_\beta, e_\alpha]$, we have $N_{\alpha\beta}e_{\alpha+\beta} = -N_{\beta\alpha}e_{\alpha+\beta}$. If $\alpha + \beta \neq 0$ is a root, the vector $e_{\alpha+\beta}$ is nonzero and $N_{\alpha\beta} = -N_{\beta\alpha}$. If $\alpha + \beta \neq 0$ is not a root, then $N_{\alpha\beta} = N_{\beta\alpha} = 0$. This proves (a).

Let α, β and γ be nonzero roots and let $\alpha + \beta + \gamma = 0$. This implies that $[e_\beta, e_\gamma] = N_{\beta\gamma}e_{\beta+\gamma} = N_{\beta\gamma}e_{-\alpha}$ and therefore $[e_\alpha, [e_\beta, e_\gamma]] = N_{\beta\gamma}[e_\alpha, e_{-\alpha}] = -N_{\beta\gamma}h'_\alpha$. From Jacobi's identity for e_α, e_β, e_γ, we infer that

$$N_{\beta\gamma}h'_\alpha + N_{\gamma\alpha}h'_\beta + N_{\alpha\beta}h'_\gamma = 0,$$

and thus

$$N_{\beta\gamma}\alpha + N_{\gamma\alpha}\beta + N_{\alpha\beta}\gamma = 0.$$

On the other hand, we have $\alpha + \beta + \gamma = 0$. If (9.7.5) fails, the roots α, β, γ are scalar multiples of one of them, say α. In this case, we have either $\beta = -\alpha$ or $\gamma = -\alpha$. One of the roots is therefore zero, a violation of hypothesis. Thus (9.7.5) holds.

Finally, let α, β, γ, δ be as in (c). We then have

$$[e_\alpha, [e_\beta, e_\gamma]] = N_{\beta\gamma}[e_\alpha, e_{\beta+\gamma}] = N_{\beta\gamma}N_{\alpha,\beta+\gamma}e_{-\delta},$$

since $\beta + \gamma \neq 0$ and $\alpha + \beta + \gamma = -\delta$. If $N_{\beta\gamma}$ is not zero, $\beta + \gamma$ is a root. Applying (9.7.5), we obtain $N_{\alpha,\beta+\gamma} = N_{\delta\alpha} = -N_{\alpha\delta}$. This implies that

$$[e_\alpha, [e_\beta, e_\gamma]] = -N_{\beta\gamma}N_{\alpha\delta}e_{-\delta}.$$

Jacobi's identity for e_α, e_β, e_γ now yields (9.7.6). If all pairwise sums of the roots α, β, γ, δ fail to be roots, (9.7.6) takes the form $0 = 0$. $\quad\square$

II. *Let α, β be nonzero roots, where $\alpha + \beta \neq 0$, $\alpha \neq \beta$. If the α-series containing β consists of all vectors of the form $\beta + k\alpha$, $p \leqslant k \leqslant q$, we have*

$$N_{\alpha\beta}N_{-\alpha,-\beta} = q(1 - p)(\alpha, \alpha)/2. \tag{9.7.7}$$

Proof. Since β is not a multiple of α, the subspaces $L^{\beta + k\alpha}$ are one-dimensional and the representation of the Lie subalgebra $L_\alpha = L^\alpha + L^{-\alpha} + H^\alpha$ in the space $\sum_{k=p}^q L^{\beta + k\alpha}$ is irreducible. We apply part (4) of the theorem in 9.4, that is, the identities (9.4.9). If we set $x_\alpha = e_\alpha$, $y_\alpha = -2e_{-\alpha}/(\alpha, \alpha)$, (9.7.2) and (9.3.2) imply that for these x_α and y_α, all of (9.3.1) holds. Thus from (9.4.9) we infer that

$$\begin{aligned}[e_{-\alpha}, [e_\alpha, e_\beta]] &= -(\alpha, \alpha)[y_\alpha, [x_\alpha, e_\beta]]/2 \\ &= -(\alpha, \alpha)(1 - p)q/2)e_\beta.\end{aligned} \tag{9.7.8}$$

We also have

$$[e_{-\alpha}, [e_\alpha, e_\beta]] = N_{\alpha\beta}[e_{-\alpha}, e_{\alpha+\beta}] = N_{\alpha\beta}N_{-\alpha, \alpha+\beta}e_\beta. \tag{9.7.9}$$

If $\alpha + \beta$ is not a root, we get $q = 0$ and $N_{\alpha\beta} = 0$ according to (9.7.3), so that (9.7.7) holds. If $\alpha + \beta$ is a root, we apply (9.7.5) to the roots $-\alpha$, $\alpha + \beta$, $-\beta$ and find

$$N_{-\alpha, \alpha+\beta} = N_{-\beta, -\alpha}. \tag{9.7.10}$$

Substituting (9.7.10) in (9.7.9) and comparing it with (9.7.8), we obtain

$$N_{\alpha\beta}N_{-\beta, -\alpha} = -(\alpha, \alpha)(1 - p)q/2. \tag{9.7.11}$$

Since $N_{-\beta, -\alpha} = -N_{-\alpha, -\beta}$, (9.7.11) implies (9.7.7). $\quad\square$

III. *Let L, L' be simple Lie algebras. Let H and H' be Cartan subalgebras of L and L' respectively. Let Δ (Δ') be the set of nonzero roots of the Lie algebra L (L'). Choose vectors $e_\alpha \in L^\alpha$ such that (9.7.1)–(9.7.4) hold with the given set of constants $N_{\alpha,\beta}$. Let φ be a one-to-one linear mapping of the space H_0 onto the space H'_0 such that the mapping φ^* conjugate to φ maps the system Δ' onto the system Δ. We set $\alpha' = (\varphi^*)^{-1}\alpha$ for all $\alpha \in \Delta$. There exist vectors $e'_{\alpha'} \in (L')^{\alpha'}$ that satisfy (9.7.1)–(9.7.4) with constants $N_{\alpha',\beta'} = N_{\alpha\beta}$. That is, we* *have*

$$[e'_{\alpha'}, e'_{\beta'}] = N_{\alpha\beta}e'_{\alpha'+\beta'} \qquad (9.7.12)$$

for all $\alpha', \beta' \in \Delta'$.

Proof. The restriction of the Killing forms of the Lie algebras L and L' to H_0 and H'_0 are positive-definite real bilinear forms. These forms allow the identification of H_0 with H_0^* and of H'_0 with $H'_0{}^*$. Therefore the Killing forms on L and L' define scalar products on the real spaces H_0^* and $H'_0{}^*$. We will show that the mapping φ^* of $H'_0{}^*$ onto H_0^* is an isometry under these scalar products. According to Proposition I in 9.5, we have $-2(\beta, \alpha)/(\alpha, \alpha) = p + q$, for all α, $\beta \in \Delta$, where the numbers p and q are defined by the condition that the α-series containing β consists of the vectors $\beta + k\alpha$ with $p \leqslant k \leqslant q$. Since φ^* is a one-to-one mapping of Δ' onto Δ and preserves linear operations, we see that $p = p'$, $q = q'$. Therefore we obtain $-2(\beta, \alpha)/(\alpha, \alpha) = -2(\beta', \alpha')/(\alpha', \alpha')$ for all $\alpha, \beta \in \Delta$ or

$$(\beta, \alpha)/(\alpha, \alpha) = (\beta', \alpha')/(\alpha', \alpha') \qquad (9.7.13)$$

for all $\alpha, \beta \in \Delta$. Switching α and β we obtain

$$(\alpha, \beta)/(\beta, \beta) = (\alpha', \beta')/(\beta', \beta'). \qquad (9.7.14)$$

From (9.7.13) and (9.7.14) we infer that

$$(\alpha, \alpha)/(\alpha', \alpha') = (\beta, \beta)/(\beta', \beta') \qquad (9.7.15)$$

if $(\alpha, \beta) \neq 0$. According to Proposition V in 9.6, for any two roots α, $\beta \in \Delta$, there exists a chain $\alpha = \alpha_1, \alpha_2, \ldots, \alpha_n = \beta$ of nonzero roots of the Lie algebra L such that $(\alpha_k, \alpha_{k+1}) \neq 0$ for all $k = 1, 2, \ldots, n - 1$. From this we infer (9.7.15) for all $\alpha, \beta \in \Delta$. We set $(\beta, \beta)/(\beta', \beta') = k$. For any $\alpha \in \Delta$ we have

$$(\alpha, \alpha) = k(\alpha', \alpha'), \qquad (9.7.16)$$

where the coefficient k does not depend upon α. Substituting (9.7.16) in (9.7.13), we obtain

$$(\alpha, \beta) = k(\alpha', \beta') \qquad (9.7.17)$$

for all $\alpha, \beta \in \Delta$, where the constant k does not depend upon α and β. On the other hand, (9.6.1) implies that

$$(\alpha, \beta) = (h'_\alpha, h'_\beta) = \sum_\gamma \gamma(h'_\alpha)\gamma(h'_\beta) = \sum_\gamma (\gamma, \alpha)(\gamma, \beta) \qquad (9.7.18)$$

and similarly

$$(\alpha', \beta') = \sum_{\gamma'} (\gamma', \alpha')(\gamma', \beta'). \tag{9.7.19}$$

Substituting (9.7.17) in (9.7.19), we obtain $(\alpha, \beta) = k^2(\alpha', \beta')$ for all $\alpha, \beta \in \Delta$. Thus we have $k = k^2$ and since $(\alpha, \alpha) \neq 0$ for $\alpha \neq 0$, k must be 1.

We will now prove the existence of vectors that satisfy (9.7.12). In order to construct these vectors by induction, we provide the space H_0^* with the lexicographic ordering defined by some fixed basis $h_1, \ldots, h_r$. Let ρ be a positive root of the Lie algebra L. Let Δ_ρ be the set of all nonzero roots α for which $-\rho < \alpha < \rho$. Suppose that the vectors $e'_{\alpha'} \in L'^{\alpha'}$ for all $\alpha \in \Delta_\rho$ are already constructed and that (9.7.12) holds for all $\alpha, \beta, \alpha + \beta \in \Delta_\rho$. We construct the vector $e'_{\rho'}$ as follows. If ρ is a simple root, we choose a nonzero $e'_{\rho'}$ arbitrarily. If ρ is not simple, it has the form $\rho = \alpha + \beta$, where $\alpha, \beta \in \Delta$. In this case we define the vector e'_ρ so that (9.7.12) holds for every decomposition $\rho = \alpha + \beta$. We then define the vector e'_ρ by the requirement that $(e'_{\rho'}, e'_{-\rho'}) = -1$.

If σ is a positive root such that $\rho < \sigma$ and there are no other positive roots between the roots ρ and σ, we have $\Delta_\sigma = \Delta_\rho \cup \{\rho, -\rho\}$. Thus the vectors $e'_{\rho'}$ and $e'_{-\rho'}$ having been defined, the required vectors $e'_{\alpha'} \in L'^{\alpha'}$ are defined for all $\alpha \in \Delta_\sigma$. Suppose that γ, δ and $\gamma + \delta$ are in Δ_σ. The number $N'_{\gamma', \delta'}$ is defined by the requirement $[e'_{\gamma'}, e'_{\delta'}] = N'_{\gamma', \delta} e'_{\gamma', \delta'}$. We will prove that $N_{\gamma, \delta} = N'_{\gamma', \delta'}$. This will show that (9.7.12) holds for all $\gamma, \delta, \gamma + \delta \in \Delta_\sigma$.

If γ, δ and $\gamma + \delta$ are in Δ_ρ, then $N_{\gamma, \delta} = N'_{\gamma'\delta'}$ by our inductive hypothesis. If γ, δ are in Δ_ρ and $\gamma + \delta = \rho$, we may suppose that the decomposition $\rho = \gamma + \delta$ does not coincide with the decomposition $\rho = \alpha + \beta$. Thus we have $\alpha + \beta + (-\gamma) + (-\delta) = 0$ and the pairwise sums of the roots α, β, $-\gamma$, $-\delta$ are not equal to zero. Part (c) of Proposition I implies (9.7.6). Therefore we have

$$N_{\alpha, \beta} N_{-\gamma, -\delta} = -N_{\beta, -\gamma} N_{\alpha, -\delta} - N_{-\gamma, \alpha} N_{\beta, -\delta}. \tag{9.7.20}$$

Similarly, we obtain the identity

$$N'_{\alpha', \beta'} N'_{-\gamma', -\delta'} = -N'_{\beta', -\gamma'} N'_{\alpha', -\delta'} - N'_{-\gamma', \alpha'} N'_{\beta', -\delta'} \tag{9.7.21}$$

in L'. Since the roots $\beta - \gamma$, $\alpha - \delta$, $\alpha - \gamma$, $\beta - \delta$ belong to Δ_ρ, our inductive hypothesis shows that the right sides of (9.7.20) and (9.7.21) are equal. Furthermore, the definition of the vector e'_ρ implies that $N_{\alpha, \beta} = N'_{\alpha', \beta'}$ and $N_{\alpha\beta} \neq 0$. From this we infer that $N_{-\gamma, -\delta} = N'_{-\gamma', -\delta'}$. We have already shown that the mapping φ^* is isometric. By applying this we see that $(\gamma, \gamma) = (\gamma', \gamma')$. Proposition II now shows that $N_{\gamma, \delta} N_{-\gamma, -\delta} = q(1 - p)(\gamma, \gamma)/2 = N'_{\gamma', \delta'} N'_{-\gamma', -\delta'}$, and thus $N_{\gamma, \delta} = N'_{\gamma', \delta'}$. If $\gamma + \delta = \rho$, we obtain $N'_{\gamma', \delta'} = N_{\gamma\delta}$ and $N'_{-\gamma', -\delta'} = N_{-\gamma, -\delta}$. If γ, δ belong to Δ_ρ and $\gamma + \delta = -\rho$, then $-\gamma$, $-\delta$ belong to Δ_ρ

and $(-\gamma) + (-\delta) = \rho$. This implies that $N'_{\gamma',\delta'} = N_{\gamma,\delta}$. Finally, if γ, δ and $-(\gamma + \delta)$ are in Δ_σ, only one of these roots is equal to $\pm\rho$. Applying part (b) of Proposition I, we reduce the proof that $N'_{\gamma',\delta'} = N_{\gamma,\delta}$ to the case $\gamma + \delta = \pm\rho$. Since there are only a finite number of positive roots, the foregoing argument completes our proof by induction.   □

IV. *Let L, L' be simple Lie algebras and let H (H') be Lie subalgebras in L (L'). Let $\Delta(\Delta')$ be the set of nonzero roots of $L(L')$. Let φ be a one-to-one linear mapping of the space H_0 onto the space H'_0 such that the mapping φ^* conjugate to φ maps the system Δ' onto the system Δ. There is an isomorphism f of L onto L' that extends φ.*

Proof. Choose $e_\alpha \in L^\alpha$ so that (9.7.1)–(9.7.4) hold. We will construct a system $e_{\alpha'}$ that satisfies the hypotheses of Proposition III. We write

$$f(h) = \varphi(h), \quad \text{for } h \in H_0, \qquad f(e_\alpha) = e_{\alpha'}. \tag{9.7.22}$$

Proposition I in 9.6 shows that there is a basis for the real vector space H_0 that is also a basis in the complex linear space H. Therefore the mapping f defined by (9.7.22) can be extended to a linear mapping of the complex vector space L into the complex vector space L'. Plainly f transforms a basis in L into a basis in L' and therefore is an isomorphism of the space L onto the space L'. We will show that f is a homomorphism of Lie algebras. We have:

$$f([h, h']) = f(0) = 0 = [f(h), f(h')];$$
$$f([h, e_\alpha]) = f(\alpha(h)e_\alpha) = \alpha(h)e'_{\alpha'} = \alpha'(\varphi(h))e'_{\alpha'}$$
$$= [\varphi(h), e'_{\alpha'}] = [f(h), f(e_\alpha)];$$
$$f([e_\alpha, e_\beta]) = f(N_{\alpha\beta}e_{\alpha+\beta}) = N_{\alpha\beta}e'_{\alpha'+\beta'} = [e'_{\alpha'}, e'_{\beta'}]$$
$$= [f(e_\alpha), f(e_\beta)], \qquad \text{if } \alpha + \beta \neq 0;$$
$$f([e_\alpha, e_{-\alpha}]) = f(-h'_\alpha) = [e'_{\alpha'}, e'_{-\alpha'}] = [f(e_\alpha), f(e_{-\alpha})].$$

This completes the proof.   □

We have the following corollary.

V. *A simple Lie algebra L is defined uniquely up to an isomorphism by the linear space H and the system of roots.*

VI. *The choice of vectors $e_\alpha \in L^\alpha$ can be made so that not only (9.7.1)–(9.7.4) but also the following identity hold:*

$$N_{\alpha,\beta} = N_{-\alpha,-\beta}. \tag{9.7.23}$$

Proof. The mapping $\varphi_0 : h \rightarrow -h$ of the space H_0 onto itself preserves the root system: the root α goes into the root $\alpha' = -\alpha$. Proposition IV shows the existence of an isomorphism Φ of the Lie algebra L onto itself that extends φ_0. We select vectors $f_\alpha \in L^\alpha$ such that $(f_\alpha, f_{-\alpha}) = -1$; suppose that $\Phi(f_\alpha) = f'_{-\alpha} = \rho_\alpha f_{-\alpha}$. We then have $(f'_{-\alpha}, f'_\alpha) = (f_\alpha, f_{-\alpha}) = -1$ and thus $\rho_\alpha \rho_{-\alpha} = 1$. Define e_α as $\rho_\alpha^{-1/2} f_\alpha$, where $\rho_\alpha^{-1/2}$ is either square root of ρ_α. We define $(\rho_{-\alpha})^{-1/2}$ as $(\rho_\alpha)^{1/2}$. From this we obtain $\Phi(e_\alpha) = \rho_\alpha^{-1/2} \rho_\alpha f_{-\alpha} = \rho_\alpha^{1/2} f_{-\alpha} = \rho_{-\alpha}^{-1/2} f_{-\alpha} = e_{-\alpha}$. Since Φ is an isomorphism of Lie algebras, we have $N_{\alpha,\beta} = N_{\alpha',\beta'} = N_{-\alpha,-\beta}$ for the vectors $e_\alpha \in L^\alpha$. $\square$

VII. *If* (9.7.23) *holds, we have* $N_{\alpha,\beta}^2 = q(1-p)(\alpha,\alpha)/2 \geqslant 0$. *Thus the* $N_{\alpha\beta}$ *are real numbers.*

Proof. Apply Propositions II and VI. Since $q \geqslant 0$, $p \leqslant 0$, the number $q(1-p)/2$ is nonnegative. $\square$

A basis of the space L obtained by extending an arbitrary basis $\{h_1, \ldots, h_r\}$ of the space H by vectors $e_\alpha \in L^\alpha$ such that $(e_\alpha, e_{-\alpha}) = -1$ and $N_{\alpha,\beta} = N_{-\alpha,-\beta}$ is called a *Weyl basis of the Lie algebra* L.

§10. Classification of Simple Lie Algebras

10.1. Dynkin Diagrams and Their Classification

Let L be a simple complex Lie algebra, and let H be a Cartan subalgebra of L. Let Π be the system of simple roots of L with respect to H. Let H_0^* be the real vector space of linear combinations of the elements of the system Π.

I.

   (a)   *The Killing form is real-valued and positive-definite in the space* H_0^*.
   (b)   *The set* Π *is linearly independent in* H_0^*.
   (c)   *If* α, β *are distinct elements of* Π, *the number* $n_{\alpha,\beta} = -2(\alpha,\beta)/(\alpha,\alpha)$ *is a nonnegative integer.*
   (d)   *The system* Π *is not the union of nonvoid subsets* Π' *and* Π'' *such that all vectors in* Π' *are orthogonal to all vectors in* Π''.

Proof. We infer (a) from Proposition I in 9.6, (b) from Proposition III in 9.6, (c) from (9.6.2) and (d) from Proposition V in 9.6. $\square$

II. *The system* Π *defines the Lie algebra* L *up to an isomorphism.*

Proof. According to Proposition IV in 9.6, the system Π defines the system Δ of nonzero roots of the Lie algebra L. By Proposition V in 9.6, the system Δ defines the Lie algebra L up to an isomorphism. $\square$

Propositions I and II reduce the problem of classification of simple complex Lie algebras to finding all systems Π of vectors in a finite-dimensional Euclidean space that satisfy conditions (b)–(d) of Proposition I. Note that when all the vectors of the system Π are multiplied by a nonzero number λ, these properties (b)–(d) are preserved. The system Π defines the system Δ of all roots of the Lie algebra L. For all roots α and $\beta \in \Delta$, the identity $(\alpha, \beta) = \sum_{\gamma \in \Delta} (\gamma, \alpha)(\gamma, \beta)$ holds (see (9.6.1)). Thus the "coefficient of proportionality" λ is determined by the condition that Π is the system of simple roots of a Lie algebra.

Let Π be a system of vectors in a finite-dimensional real Euclidean space that satisfies (b)–(d) of Proposition I. Suppose that α and β are in Π, that $\alpha \neq \pm\beta$, and that α is not orthogonal to β. We then have $n_{\alpha,\beta} \neq 0$, $n_{\beta,\alpha} \neq 0$ and (c) gives us

$$n_{\alpha,\beta} n_{\beta,\alpha} = 4(\alpha, \beta)^2 / \{(\alpha, \alpha)(\beta, \beta)\} = 4 \cos^2 \theta, \tag{10.1.1}$$

where θ is the angle between the vectors α and β. By hypothesis (α, β) is not zero and α is not proportional to β. The equality (10.1.1) gives

$$0 < n_{\alpha,\beta} n_{\beta,\alpha} = 4 \cos^2 \theta < 4,$$

and the product $n_{\alpha,\beta} n_{\beta,\alpha}$ is an integer. That is, $n_{\alpha,\beta} n_{\beta,\alpha}$ must be 1, 2, or 3. Therefore the smaller of the positive integers $n_{\alpha,\beta}$ and $n_{\beta,\alpha}$ is 1. For definiteness, we suppose that $(\alpha, \alpha) \geq (\beta, \beta)$. We then have $n_{\alpha,\beta} \leq n_{\beta,\alpha}$ and thus $n_{\alpha,\beta} = 1$, $n_{\beta,\alpha} = 4 \cos^2 \theta$, and

$$(\alpha, \alpha)/(\beta, \beta) = n_{\beta,\alpha}/n_{\alpha,\beta} = 4 \cos^2 \theta. \tag{10.1.2}$$

In the three cases $n_{\beta,\alpha} = 1, 2, 3$, we obtain $\cos \theta = -1/2, -\sqrt{2}/2, -\sqrt{3}/2$, and thus $\theta = 2\pi/3, 3\pi/4, 5\pi/6$, respectively.

For every system Π we define a *Dynkin diagram* S defined as follows. Every element $\alpha \in \Pi$ is represented by a point in the plane and is assigned the numerical coefficient (α, α). Two distinct points α, β are then joined, if α is not orthogonal to β, by one, two, or three line segments, depending on the number $4 \cos^2 \theta$: the number of connecting lines is equal to $4 \cos^2 \theta$. The points corresponding to α and β are not joined if α is orthogonal to β. We say that the diagram S *corresponds to the system of roots* Π. The number of elements of the system Π is called the *order* of the Dynkin diagram S. A Dynkin diagram is called *admissible* if it is constructed for a system Π that satisfies conditions (b) and (c) of Proposition I. *The system Π satisfies condition* (d) *if and only if the Dynkin diagram of Π is connected.*

III. *If a Dynkin diagram S is admissible, we obtain another Dynkin diagram by removing some of its points and the lines issuing from those points.*

Proof. Let the diagram S correspond to the system Π. The removal of some set of points from S corresponds to discarding certain elements from the system Π. Since (b) and (c) hold for all subsystems Π' of Π, the Dynkin diagram corresponding to Π' is admissible. $\square$

IV. *Every admissible connected Dynkin diagram of order* 2 *coincides with one of the following diagrams*:

$$
\overset{1}{\underset{\circ}{}}\!-\!\overset{1}{\underset{\circ}{}}\qquad
\overset{2}{\underset{\circ}{}}\!\equiv\!\overset{1}{\underset{\circ}{}}\qquad
\overset{3}{\underset{\circ}{}}\!\equiv\!\overset{1}{\underset{\circ}{}}.
$$

Proof. Since the diagram is connected, (α, β) is not zero and thus the restriction $(\alpha, \alpha) \geqslant (\beta, \beta)$ implies that $4 \cos^2 \theta = (\alpha, \alpha)/(\beta, \beta)$ is equal to 1, 2, or 3. The sketches above illustrate these three cases. $\square$

We will now study the properties of the connections in admissible Dynkin diagrams.

V. *In an admissible Dynkin diagram of order* n *there are no more than* $n - 1$ *connected pairs of points*.

Proof. We write x for the vector $\sum_{\alpha \in \Pi} \alpha/\sqrt{(\alpha, \alpha)}$. We know that $(\alpha, \beta) \leqslant 0$ if $\alpha \neq \beta$. Also, if $(\alpha, \beta) \neq 0$, then the number $4 \cos^2 \theta$ is not less than 1. Therefore we have $2(\alpha, \beta)/\sqrt{(\alpha, \alpha)(\beta, \beta)} \leqslant -1$. If there are more than $n - 1$ nonorthogonal pairs of vectors in the system Π, we have

$$
(x, x) \sum_{\alpha \in \Pi} (\alpha, \alpha)/(\alpha, \alpha) + 2 \sum_{\substack{\alpha, \beta \in \Pi \\ \alpha \neq \beta}} (\alpha, \beta)/\sqrt{(\alpha, \alpha)(\beta, \beta)} \leqslant n - n = 0,
$$

and thus $x = 0$. This contradicts the linear independence of Π. $\square$

VI. *An admissible Dynkin diagram contains no closed circuits*.

Proof. If S contains a closed circuit S', then S' is an admissible diagram by Proposition III. On the other hand, all the elements in S' are pairwise connected, which is impossible, as Proposition V shows. $\square$

VII. *If S' is a connected subdiagram of an admissible Dynkin diagram S, any element α not in S' is connected with at most one element in S'*.

Proof. Assume that an element α of $S \backslash S'$ is connected to elements β and γ of the subdiagram S'. Since S' is connected, there is a sequence $\beta = \beta_0$, $\beta_1, \ldots, \beta_n = \gamma$ of elements of S' in which every pair of adjacent elements are connected. The subdiagram of S consisting of the element α and the elements β_i $(i = 0, \ldots, n)$ contains $n + 2$ elements and no fewer than $n + 2$ connected pairs, which is impossible, as shown by Proposition V. $\quad\square$

VIII. *No more than three line segments emanate from each point of an admissible Dynkin diagram Π.*

Proof. Suppose that $\alpha \in \Pi$. Let $\beta_1, \ldots, \beta_n$ be the elements connected to α. If $i \neq j$, then β_i and β_j are *not* connected (otherwise the elements α, β_i, β_j would form a closed circuit, which is impossible according to Proposition VI).

Therefore, if $\beta_i \neq \beta_j$, β_i and β_j are orthogonal. Suppose that the vector γ is orthogonal to all β_j and also is in the subspace generated by α and the vectors β_j. Let θ_0 be the angle between α and γ and θ_j the angle between α and β_j. Since α can be expanded in the orthogonal basis formed by the vectors γ and β_j, we have $\cos^2 \theta_0 + \sum_i \cos^2 \theta_i = 1$. The set of vectors α and β_j is linearly independent, and hence the vector α is not orthogonal to γ. We infer that $\cos^2 \theta_0 \neq 0$ and so $\sum_j \cos^2 \theta_j < 1$. Therefore $\sum_j 4 \cos^2 \theta_j < 4$ and $4 \cos^2 \theta_j$ is the number of lines joining α and β_j. $\quad\square$

IX. *A connected admissible Dynkin diagram of order $n \geqslant 3$ admits no triple connections.*

Proof. Assume that α and β are connected by a triple line. This means (see Proposition VIII) that no other line can emanate from either α or β. Since the diagram is by hypothesis connected, any connected admissible diagram with a triple connection between α and β consists of elements α and β, i.e., has order 2. $\quad\square$

A sequence $\alpha_1, \ldots, \alpha_{n+1}$ of the points of a Dynkin diagram S, in which adjacent points α_k and α_{k+1} are connected for $k = 1, \ldots, n$, is called a *chain*. By Proposition VI, there are no other connections between the elements of the chain. A chain C is called *homogeneous* if all of its connections consist of a single line segment.

Every chain C is connected. As Proposition VII shows, each element $\beta \in S$ is connected with at most one element of C.

X. *Let S be an admissible Dynkin diagram and C a homogeneous chain in S. We form a new Dynkin diagram S' by replacing the chain C with a single point, which is connected to an element $\beta \notin C$ by a p-multiple connection ($p = 0, 1, 2$, or 3), if β is connected in S by a p-multiple connection to some element of the chain C. The diagram S' is admissible.*

Proof. Let Π' be the system of vectors formed by all vectors β in $\Pi \backslash C$ and the vector $\alpha = \sum_{k=1}^{n+1} \alpha_k$, where the chain C is $\{\alpha_1, \ldots, \alpha_{n+1}\}$. We will show that the diagram S' corresponds to the system Π'. If β does not belong to

C and β is connected to α_i, the vector β is orthogonal to all α_j, $j \neq i$. Thus we have $(\beta, \alpha) = (\beta, \alpha_i)$. On the other hand, we have $(\alpha, \alpha) = \sum_{i=1}^{n} ((\alpha_i, \alpha_i) + 2(\alpha_i, \alpha_{i+1})) + (\alpha_{n+1}, \alpha_{n+1}) = (\alpha_{n+1}, \alpha_{n+1})$. Since C is homogeneous, we have $n_{\alpha_i, \alpha_{i+1}} = 1$ for all $i = 1, \ldots, n$, but also $n_{\alpha_i, \alpha_{i+1}} = -2(\alpha_i, \alpha_{i+1})/(\alpha_i, \alpha_i)$ and thus $(\alpha_i, \alpha_i) + 2(\alpha_i, \alpha_{i+1}) = 0$ for all $i = 1, \ldots, n + 1$. Furthermore, by (10.1.2), the equality $n_{\alpha_i, \alpha_{i+1}} = 1$ implies that $(\alpha_i, \alpha_i) = (\alpha_{i+1}, \alpha_{i+1})$. Therefore we obtain $(\alpha_i, \alpha_i) = (\alpha_k, \alpha_k)$ for all $i, k = 1, \ldots, n + 1$ and $(\alpha, \alpha) = (\alpha_{n+1}, \alpha_{n+1}) = (\alpha_k, \alpha_k)$ for all $k = 1, \ldots, n + 1$. The connections between pairs of elements not contained in C do not change. Thus the diagram S' corresponds to the system Π' and Π' satisfies (c) in I. Clearly the vectors of Π' are linearly independent, that is, condition (b) of I holds. Hence the Dynkin diagram S' is admissible. □

XI. *The Dynkin diagrams*

(H),
(I),
(K)

are inadmissible.

Proof. Applying Proposition X to the chains $C = \{\alpha_1, \ldots, \alpha_{n+1}\}$ of diagrams (H), (I), and (K), we obtain the Dynkin diagrams

which by Proposition VIII are inadmissible. □

XII. *Every connected admissible Dynkin diagram of order $n \geqslant 3$ belongs to one of the classes*

(L),
(M),
(N).

Proof. By Proposition IX, our connected admissible Dynkin diagram S admits no triple connections. Suppose that S admits a double connection, *i.e.*, a subdiagram consisting of two circles joined by two lines. Embed this subdiagram in a chain C of maximum length. If C contains two double connections, it contains a subdiagram of the form (H), which is inadmissible. Suppose that $C \neq S$. Since S is connected, there is an element $\beta \notin C$ connected to some element of C (in view of Proposition VII exactly one). If β is connected to an end element of C, we can extend C by joining it with β. This contradicts the maximality of the chain C. Therefore β is connected to some middle element of C. That is, S contains a subdiagram (I), which is inadmissible. Proposition III shows that this is impossible. Thus C is all of S, *i.e.*, the Dynkin diagram S is a chain admitting one double connection. The diagram therefore has the form (L).

Suppose now that all connections in S are single. If every point of S is connected to no more than two other points, a chain C in S of maximal length coincides with S. To see this, assume that $C \neq S$. There is an element $\beta \notin C$ connected to exactly one element of C. If β is connected to an end element of C, we can extend C, which contradicts its maximality. However, β cannot be connected to a middle element of C since we suppose that each point of S is connected to no more than two other points. Therefore S has the form (M).

Finally, suppose that all connections in S are single and suppose that there is a point in S connected to three other points. In this case, S admits a subdiagram having the form of a "star" with three rays. We extend this star to a maximal subdiagram $\tilde{S} \subset S$ of the form (N). Assume that $\tilde{S} \neq S$. There exists an element $\beta \notin S$ connected to some element from $\tilde{S}$ (and by Proposition VII only one element). Since $\tilde{S}$ is a maximal subdiagram of the form (N), the element β cannot be connected to any end element in $\tilde{S}$. If β is connected to a middle element in $\tilde{S}$, S contains a subdiagram of the form (K), which is inadmissible. The resulting contradiction shows that $\tilde{S} = S$; that is, S has the form (N). $\square$

XIII. *Let the chain $C = \{\alpha_1, \ldots, \alpha_n\}$ be formed by vectors of identical length, i.e., $(\alpha_i, \alpha_i) = a$ for $i = 1, 2, \ldots, n$. We set $\alpha = \sum_{k=1}^{n} k\alpha_k$. Then we have $(\alpha, \alpha) = n(n + 1)a/2$.*

Proof. We compute:

$$(\alpha, \alpha) = \sum_{k=1}^{n} k^2(\alpha_k, \alpha_k)^2 + 2 \sum_{k=1}^{n-1} k(k + 1)(\alpha_k, \alpha_{k+1})$$

$$= n^2 a - \sum_{k=1}^{n-1} ka = n(n + 1)a/2. \quad \square$$

Theorem. *Every connected admissible Dynkin diagram coincides with one of the following diagrams* A_n, B_n, C_n, D_n, E_n, F_4, G_2 *(up to a common multiple of the numbers* (α, α) *labelling the elements):*

$$(A_n) \quad \overset{2}{\circ}\!-\!\overset{2}{\circ}\!-\!\cdots\!-\!\overset{2}{\circ} \qquad (n \geqslant 1);$$

$$(B_n) \quad \overset{2}{\circ}\!-\!\overset{2}{\circ}\!-\!\cdots\!-\!\overset{2}{\circ}\!=\!\overset{1}{\circ} \qquad (n \geqslant 2);$$

$$(C_n) \quad \overset{2}{\circ}\!-\!\overset{2}{\circ}\!-\!\cdots\!-\!\overset{2}{\circ}\!=\!\overset{4}{\circ} \qquad (n \geqslant 3);$$

$$(D_n) \quad \overset{2}{\circ}\!-\!\overset{2}{\circ}\!-\!\cdots\!-\!\overset{2}{\circ}\!-\!\overset{2}{\circ}\!-\!\overset{2}{\circ} \qquad (n \geqslant 4);$$
with a branch $\overset{2}{\circ}$

$$(E_n) \quad \overset{2}{\circ}\!-\!\overset{2}{\circ}\!-\!\cdots\!-\!\overset{2}{\circ}\!-\!\overset{2}{\circ}\!-\!\overset{2}{\circ}\!-\!\overset{2}{\circ} \qquad (n = 6,7,8);$$
with a branch $\overset{2}{\circ}$

$$(F_4) \quad \overset{2}{\circ}\!-\!\overset{2}{\circ}\!=\!\overset{1}{\circ}\!-\!\overset{1}{\circ};$$

$$(G_2) \quad \overset{2}{\circ}\!\equiv\!\overset{6}{\circ}.$$

(The subscripts n denote the number of elements in the Dynkin diagram.)

Proof. Consider a diagram of the form (L). Of the two elements with a double connection, the square of the length of one of the vectors is twice the square of the length of the other (see 10.1.2). Elements connected by a simple connection have the same length. Therefore, to within a common multiple of the labelling numbers, any diagram of the form (L) has the form

$$\underset{\alpha_1}{\overset{1}{\circ}}\!-\!\underset{\alpha_2}{\overset{1}{\circ}}\!-\!\cdots\!-\!\underset{\alpha_{p-1}}{\overset{1}{\circ}}\!-\!\underset{\alpha_p}{\overset{1}{\circ}}\!=\!\underset{\beta_q}{\overset{2}{\circ}}\!-\!\underset{\beta_{q-1}}{\overset{2}{\circ}}\!-\!\cdots\!-\!\underset{\beta_1}{\overset{2}{\circ}},$$

where the square of the length of the roots $\beta_1, \ldots, \beta_q$ is twice that of the roots $\alpha_1, \ldots, \alpha_p$. We set $\alpha = \sum_{k=1}^{p} k\alpha_k$, $\beta = \sum_{j=1}^{q} j\beta_j$. Proposition XIII gives $(\alpha, \alpha) = p(p + 1)/2$ and $(\beta, \beta) = q(q + 1)$. It is clear that $(\alpha, \beta) = pq(\alpha_p, \beta_q) = -pq$. Since the vectors α and β are not collinear, we have $(\alpha, \beta)^2 < (\alpha, \alpha)(\beta, \beta)$, that is, $p^2q^2 < pq(p + 1)(q + 1)/2$ or $2pq < (p + 1)(q + 1)$ and thus $(p - 1)(q - 1) < 2$. Therefore we have only three possibilities: $p = 1$ and q has any value; $q = 1$ and p has any value; or $p = q = 2$. The corresponding diagrams are (B_{q+1}), (C_{p+1}) and (F_4), respectively. The diagrams (B_2) and (C_2) differ only by a multiple.

The diagram (M) with n elements is denoted by (A_n).
Consider next a diagram (N):

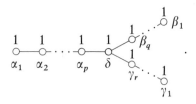

We set $\alpha = \sum_{i=1}^{p} i\alpha_i$, $\beta = \sum_{j=1}^{q} j\beta_j$, $\gamma = \sum_{k=1}^{r} k\gamma_k$. The vectors α, β, and γ are pairwise orthogonal and the vector δ is not a linear combination of α, β, and γ, since the entire system Π is linearly independent. Hence the sum of the squares of the cosines of the angles formed by the vector δ with the vectors α, β, and γ is less than 1. Let us compute this sum. Proposition XIII gives $(\alpha, \alpha) = p(p + 1)/2$. Note next that $(\delta, \delta) = 1$ and $(\alpha, \delta) = p(\alpha_p, \delta) = -p/2$. For the angle θ between α and δ, we get $\cos^2 \theta = (\alpha, \delta)^2/(\alpha, \alpha)(\delta, \delta) = (1 - 1/(p + 1))/2$. We compute similarly for the angles between β and δ and γ and δ. Adding, we obtain the inequalities

$$\left(1 - \frac{1}{p + 1} + 1 - \frac{1}{q + 1} + 1 - \frac{1}{r + 1}\right)\frac{1}{2} < 1,$$

or $(p + 1)^{-1} + (q + 1)^{-1} + (r + 1)^{-1} > 1$. Suppose that $p \geqslant q \geqslant r$, which implies that $(p + 1)^{-1} \leqslant (q + 1)^{-1} \leqslant (r + 1)^{-1}$ and $3 > r + 1$, i.e., $r = 1$. Hence we have $(p + 1)^{-1} + (q + 1)^{-1} > 1/2$, and $2/(q + 1) > 1/2$, which implies that $q < 3$. There are then two possibilities: either $q = 1$ and p is any number or $q = 2$ and $2 \leqslant p \leqslant 4$. The corresponding diagrams are denoted by $(D_{p+3})(p \geqslant 1)$ and $(E_{p+4})(2 \leqslant p \leqslant 4)$.

The diagram of order 2 with a triple connection is denoted by (G_2). The proof is complete. $\square$

The systems of the roots of Lie algebras of types $(A_n) - (G_2)$ are studied in detail in the book Bourbaki [1].

10.2. Some Auxiliary Propositions

In the sequel we will construct the simple Lie algebras, whose systems of simple roots correspond to Dynkin diagrams of the forms (A_n), (B_n), (C_n) and D_n. In this section we will prove two propositions that will be used to construct these algebras.

I. *Let L be a semisimple Lie algebra and let H be a Cartan subalgebra of dimension r. Suppose that a subsystem $\{\alpha_i\}$, $i = 1, \ldots, r$, of the system of all roots is such that every root α of L under H has the form $\alpha = \pm \sum m_i\alpha_i$, where the m_i are nonnegative integers. The set $\{\alpha_i\}$ is the system of all simple roots.*

Proof. The roots $\alpha_1, \ldots, \alpha_r$ generate the system of all roots of rank r, and hence they are a basis in the space H^*. We give the space H_0^* the lexicographic ordering that corresponds to the basis $\alpha_1, \ldots, \alpha_r$ (see 9.6). Under this ordering the roots $\alpha_1, \ldots, \alpha_r$ are positive. The positive roots are exactly the roots $\sum m_i \alpha_i$, where the m_i are nonnegative. Under this ordering, the roots $\alpha_1, \ldots, \alpha_r$ are simple. To see this, assume that $\alpha_i = \beta + \gamma$, where β and γ are positive roots. We write $\beta = \sum m_j \alpha_j$, $\gamma = \sum n_j \alpha_j$, where $m_j, n_j \geq 0$ and $\alpha_i = \sum_j (m_j + n_j)\alpha_j$. The vectors α_j being linearly independent, we find that $m_j + n_j = 0$ for $j \neq i$ and $m_i + n_i = 1$, that is, one of the roots β and γ is α and the other is zero. Thus, all roots α_i are simple. There are no other simple roots, since the number of simple roots is equal to r.   □

II. *Let L be a complex Lie algebra. Let H be an abelian subalgebra of L and Δ a finite set of nonzero linear functionals on H. For any linear functional λ on H, (not necessarily in Δ), the set of elements $\lambda \in L$ such that $[h, x] = \lambda(h)x$ for all $h \in H$ is denoted by L^λ. Suppose that the following hypotheses hold.*

(a)   *The linear span of the set Δ is the entire space H^* of linear functionals on H.*
(b)   *For all $\alpha \in \Delta$, $-\alpha$ belongs to Δ and $[L^\alpha, L^{-\alpha}] \neq \{0\}$.*
(c)   *We have*

$$L = H + \sum_{\alpha \in \Delta} L^\alpha. \tag{10.2.1}$$

Then L is a semisimple Lie algebra, H is a Cartan subalgebra of L and (10.2.1) is the decomposition of L with respect to H.

Proof. Plainly (10.2.1) is a decomposition of the Lie algebra L into a direct sum of subspaces. Condition (c) implies that $L^0 \subset H$. Since H is abelian, H is contained in L^0 and therefore is equal to L^0. The definition of L^λ and Jacobi's identity imply that $[L^\lambda, L^\mu] \subset L^{\lambda+\mu}$ for all $\lambda, \mu \in H^*$. (Compare Proposition II in §8.) In particular, we have $[L^\alpha, L^{-\alpha}] \subset H$ for all $\alpha \in \Delta$. Since $[L^\alpha, L^{-\alpha}] \neq \{0\}$, we have $L^\alpha \neq \{0\}$. Now fix $\alpha \in \Delta$ and choose elements $x'_\alpha \in L^\alpha$, $x'_{-\alpha} \in L^{-\alpha}$ so that $\bar{h}_\alpha = [x'_\alpha, x'_{-\alpha}]$ is nonzero.

Consider the set $\Delta_{\beta,\alpha}$ of elements in Δ that have the form $\beta + k\alpha$ for some integer k. Since the set Δ is finite, the set $\Delta_{\beta,\alpha}$ is finite. Consider the subspace $V = \sum_{\gamma \in \Delta_{\beta,\alpha}} L^\gamma$. The inclusion $[L^\lambda, L^\mu] \subset L^{\lambda+\mu}$ shows that $[x'_\alpha, V] \subset V$ and $[x'_{-\alpha}, V] \subset V$. Thus the restriction of the operator ad $\bar{h}_\alpha$ to the subspace V is a commutator of the restrictions of the operators ad x'_α and ad $x'_{-\alpha}$ to V. Therefore the trace of the operator ad $\bar{h}_\alpha$ in V is zero. Let d_k be the dimension of the space $L^{\beta+k\alpha}$ (we set $d_k = 0$ for $\beta + k\alpha \notin \Delta_{\beta,\alpha}$). In this case the trace of ad $\bar{h}_\alpha$ in $L^{\beta+k\alpha}$ is equal to $d_k(\beta + k\alpha)(\bar{h}_\alpha)$. Therefore we have

$$0 = \text{tr}(\text{ad } \bar{h}_\alpha|_V) = \sum_{k=-\infty}^{+\infty} d_k(\beta + k\alpha)(\bar{h}_\alpha). \tag{10.2.2}$$

(The sum is actually finite, since $d_k = 0$ for $\beta + k\alpha \notin \Delta_{\beta,\alpha}$.) Rewrite (10.2.2) to obtain

$$\beta(\bar{h}_\alpha) \sum_{k=-\infty}^{+\infty} d_k = -\alpha(\bar{h}_\alpha) \sum_{k=-\infty}^{+\infty} k d_k,$$

where $\sum_{k=-\infty}^{+\infty} d_k$ and $\sum_{k=-\infty}^{+\infty} k d_k$ are integers, the first of which is nonzero. Thus for all $\alpha, \beta \in \Delta$ there is a rational number $q_{\beta,\alpha}$ such that

$$\beta(\bar{h}_\alpha) = q_{\beta,\alpha}\alpha(\bar{h}_\alpha). \tag{10.2.3}$$

If $\alpha(\bar{h}_\alpha) = 0$, we obtain $\beta(\bar{h}_\alpha) = 0$ for all $\beta \in \Delta$ by (10.2.3). Thus hypothesis (a) implies that $\lambda(\bar{h}_\alpha) = 0$ for all $\lambda \in H^*$ and therefore $\bar{h}_\alpha = 0$, which violates the construction of $\bar{h}_\alpha$. We therefore have

$$\alpha(\bar{h}_\alpha) \neq 0 \quad \text{for all } \alpha \in \Delta. \tag{10.2.4}$$

We will show that $\dim(L^\alpha) = 1$ for all $\alpha \in \Delta$. Let

$$V' = \mathbf{C}x'_{-\alpha} + \mathbf{C}\bar{h}_\alpha + \sum_{k \geq 1} L^{k\alpha}$$

be a subspace of the space L (we define $L^{k\alpha}$ as $\{0\}$ if $k\alpha \notin \Delta$). One easily shows that the subspace V' is invariant under the operators ad x'_α and ad $x'_{-\alpha}$. Therefore the trace of the restriction of ad h'_α to V' is zero, since ad $\bar{h}_\alpha|_{V'} = [\text{ad } x'_\alpha|_{V'}, \text{ad } x'_{-\alpha}|_V]$. Let δ_k be the dimension of the space $L^{k\alpha}$. The trace of the operator ad h_α in $L^{k\alpha}$ is equal to $\delta_k(k\alpha)(\bar{h}_\alpha)$. Furthermore, we have (ad $\bar{h}_\alpha)(\bar{h}_\alpha) = 0$ and (ad $\bar{h}_\alpha)(x'_{-\alpha}) = -\alpha(\bar{h}_\alpha)x'_{-\alpha}$. Therefore we obtain

$$0 = \text{tr ad } \bar{h}_\alpha|_{V'} = -\alpha(\bar{h}_\alpha) + \sum_{k=1}^{\infty} k\delta_k\alpha(\bar{h}_\alpha),$$

or

$$0 = \alpha(\bar{h}_\alpha)(-1 + \delta_1 + 2\delta_2 + \cdots), \tag{10.2.5}$$

where $\delta_1 \geq 1$ (since $L^\alpha \neq \{0\}$) and $\delta_k \geq 0$ for $k \geq 2$. Since $\alpha(\bar{h}_\alpha) \neq 0$ by (10.2.4), we have $\delta_1 + 2\delta_2 + \cdots = 1$, where $\delta_1 \geq 1$ and $\delta_k \geq 0$. Substituting these relations in (10.2.5) we obtain $\delta_1 = 1$, that is, $\dim(L^\alpha) = 1$ for all $\alpha \in \Delta$.

We now define elements $h_\alpha \in H$, $x_\alpha \in L^\alpha$, $x_{-\alpha} \in L^{-\alpha}$ by

$$h_\alpha = 2\bar{h}_\alpha/\alpha(\bar{h}_\alpha), \qquad x_\alpha = x'_\alpha, \qquad x_{-\alpha} = 2x'_{-\alpha}/\alpha(h'_\alpha). \tag{10.2.6}$$

The identity $\bar{h}_\alpha = [x'_\alpha, x'_{-\alpha}]$ and the definition of the spaces L^α show that

$$[h_\alpha, x_\alpha] = 2x_\alpha, \qquad [h_\alpha, x_{-\alpha}] = -2x_{-\alpha}, \qquad [x_\alpha, x_{-\alpha}] = h_\alpha. \tag{10.2.7}$$

We will now prove that L is a semisimple Lie algebra. Let R be the radical of L. By Proposition IV in §3, R is invariant under the representation π of the Lie algebra H in the space L defined by $\pi(h) = \text{ad } h$. We will show that $R = (R \cap H) + \sum_{\alpha \in \varDelta} (R \cap L^\alpha)$. Let ρ be the representation of H defined by π in the factor space $\tilde{L} = L/R$. Let $\tilde{H}$, $\tilde{L}^\alpha$ be the images of the space H and L^α respectively in $\tilde{L}$. We see that $\rho(h)\tilde{h} = 0$, $\rho(h)\tilde{x}_\alpha = \alpha(h)\tilde{x}_\alpha$ for all $\alpha \in \varDelta$, $h \in H$, $\tilde{h} \in \tilde{H}$, $\tilde{x}_\alpha \in \tilde{L}^\alpha$. Let $\tilde{\varDelta}$ be the set of all $\alpha \in \varDelta$ such that $\tilde{L}^\alpha \neq \{0\}$. It is easy to check that the subspaces $\tilde{L}^\alpha$, $\alpha \in \tilde{\varDelta}$, are linearly independent and that if $\tilde{H} \neq \{0\}$, then the subspaces $\tilde{H}$ and $\tilde{L}^\alpha$, $\alpha \in \tilde{\varDelta}$, are linearly independent. If $h + \sum_{\alpha \in \varDelta} x_\alpha \in R$, where $h \in H$ and $x_\alpha \in L^\alpha$, we see that $h \in R$ and $x_\alpha \in R$; that is, $R \subset (R \cap H) + \sum_{\alpha \in \varDelta} (R \cap L^\alpha)$. The reverse inclusion is obvious. Thus we have

$$R = (R \cap H) + \sum_{\alpha \in \varDelta} (R \cap L^\alpha). \tag{10.2.8}$$

We will show that $R \cap L^\alpha = \{0\}$ for all $\alpha \in \varDelta$. Assume the contrary: $R \cap L^\alpha \neq \{0\}$ for some $\alpha \in \varDelta$. Since L^α is one-dimensional, we have $R \cap L^\alpha = L^\alpha$ and thus $x_\alpha \in R$. Since R is an ideal, (10.2.7) implies that $h_\alpha \in R$ and therefore $x_{-\alpha} \in R$. The ideal R is solvable, and from what we have already proved, it contains the elements h_α, x_α, $x_{-\alpha}$. The subspace $Ch_\alpha + L^\alpha + L^{-\alpha}$ generated by these elements is a semisimple Lie subalgebra of R. This contradicts the solvability of R. It follows that $R \cap L^\alpha = \{0\}$ for all $\alpha \in \varDelta$, and (10.2.8) implies that $R \subset H$. If R is nonzero and h is a nonzero element of R, there exists an element $\lambda \in H^*$ such that $\lambda(h) \neq 0$. Hypothesis (a) implies that there is an element $\alpha \in \varDelta$ such that $\alpha(h) \neq 0$. From $[h, x_\alpha] = \alpha(h)x_\alpha$, we obtain $x_\alpha = \alpha(h)^{-1}[h, x_\alpha] \in R$. This is impossible since $R \subset H$. Thus we have $R = \{0\}$ and L is a semisimple Lie algebra. From (10.2.1) we infer that $H = L^0$, i.e., H is a Cartan subalgebra of L. $\quad\square$

10.3. Lie Algebras of the Type (A_n) $(n \geqslant 1)$

Let L be the Lie algebra $sl(n + 1, \mathbf{C})$, that is, the Lie algebra of complex square matrices of order $n + 1$ with trace zero. Let H be the abelian Lie subalgebra consisting of all diagonal matrices. The diagonal matrix with diagonal elements $a_1, \ldots, a_{n+1}$, where $a_1, \ldots, a_{n+1}$ are complex numbers, is denoted by $\text{diag}(a_1, \ldots, a_{n+1})$. Let e_{ij} be the matrix of order $n + 1$ with entry 1 at the i-th row and j-th column and entries 0 everywhere else. Clearly the matrices $e_{ii} - e_{i+1, i+1}$ $(1 \leqslant i \leqslant n)$, e_{ij} $(i \neq j, 1 \leqslant i, j \leqslant n + 1)$ are a basis for L, and the matrices $e_{ii} - e_{i+1, i+1}$ $(i = 1, 2, \ldots, n)$ are a basis in H.

Let $\lambda_1, \ldots, \lambda_{n+1}$ be the linear functionals on H defined by $\lambda_i(\text{diag}(a_1, \ldots, a_{n+1})) = a_i$. It is plain that $\lambda_1 + \cdots + \lambda_{n+1} = 0$. A direct calculation shows that

$$[h, e_{ij}] = (\lambda_i - \lambda_j)(h)e_{ij} \tag{10.3.1}$$

for all $h \in H$. Let $\varDelta$ be the set of linear functionals on H of the form $\lambda_i - \lambda_j$, where $i \neq j$, $i, j = 1, \ldots, n + 1$. The identities (10.3.1) imply that

$$L^\alpha = L^{\lambda_i - \lambda_j} = Ce_{ij}, \qquad (i \neq j) \tag{10.3.2}$$

for all $\alpha = \lambda_i - \lambda_j$ in Δ. It follows that

$$L = H + \sum_{\alpha \in \Delta} L^{\alpha}. \tag{10.3.3}$$

Thus hypothesis (c) of Proposition II in 10.2 holds for the Lie algebras L and H and the set Δ. Clearly if $\alpha \in \Delta$, then $-\alpha \in \Delta$. Finally, if $h = \text{diag}(a_1, \ldots, a_{n+1})$ and $\alpha(h) = 0$ for all $\alpha \in \Delta$, it follows that $a_i = a_{i+1}$ for all $i = 1, \ldots, n$. This implies that $h = \text{diag}(a_1, \ldots, a_1)$ and $\text{tr}(h) = (n+1)a_1$. From $\text{tr}(h) = 0$, we have $a_1 = 0$ and so $h = 0$. Thus hypothesis (a) of Proposition II in 10.2 holds.

We will show that $[L^{\alpha}, L^{-\alpha}] \neq \{0\}$ for $\alpha \in \Delta$. For $\alpha = \lambda_i - \lambda_j$, $i \neq j$, we find that $e_{ij} \in L^{\alpha}$, $e_{ji} \in L^{-\alpha}$ and

$$[e_{ij}, e_{ji}] = e_{ii} - e_{jj}, \qquad (i \neq j). \tag{10.3.4}$$

This means that $[L^{\alpha}, L^{-\alpha}] \neq \{0\}$, and so all hypotheses of Proposition II in 10.2 hold. Therefore the Lie algebra $L = sl(n+1, \mathbf{C})$ is semisimple, H is a Cartan subalgebra of L, and (10.3.3) is the decomposition of L corresponding to H.

To find the Dynkin diagram for L, we first calculate the Killing form on L. If $h = \text{diag}(a_1, \ldots, a_{n+1})$ and $h' = \text{diag}(a'_1, \ldots, a'_{n+1})$ are elements of the Cartan subalgebra H, we apply (9.6.1) to obtain

$$
\begin{aligned}
(h, h') &= \sum_{\alpha \in \Delta} \alpha(h)\alpha(h') = \sum_{i \neq j} (a_i - a_j)(a'_i - a'_j) \\
&= 2 \sum_{1 \leqslant i < j \leqslant n+1} (a_i - a_j)(a'_i - a'_j) \\
&= 2 \sum_{1 \leqslant i < j \leqslant n+1} (a_i a'_i + a_j a'_j - a_j a'_i - a_i a'_j) \\
&= 2 \sum_{1 \leqslant i < j \leqslant n+1} (a_i a'_i + a_j a'_j) - 2 \sum_{i \neq j} a_i a'_j.
\end{aligned}
\tag{10.3.5}
$$

From $a_1 + \cdots + a_{n+1} = 0$ and $a'_1 + \cdots + a'_{n+1} = 0$ we have

$$(a_1 + \cdots + a_{n+1})(a'_1 + \cdots + a'_{n+1}) = \sum_{i=1}^{n+1} a_i a'_i + \sum_{i \neq j} a_i a'_j = 0. \tag{10.3.6}$$

Substituting (10.3.6) in (10.3.5), we find

$$
\begin{aligned}
(h, h') &= 2 \sum_{1 \leqslant i < j \leqslant n+1} (a_i a'_i + a_j a'_j) + 2 \sum_{i=1}^{n+1} a_i a'_i \\
&= 2(n+1) \sum_{i=1}^{n+1} a_i a'_i.
\end{aligned}
\tag{10.3.7}
$$

This formula allows us to calculate the Killing form of L for all elements of the Cartan subalgebra H. We set

$$h'_{\lambda_i - \lambda_j} = (2(n + 1))^{-1}(e_{ii} - e_{jj}) \tag{10.3.8}$$

for all $i \neq j$. From (10.3.7) we infer that

$$(h, h'_{\lambda_i - \lambda_j}) = (\lambda_i - \lambda_j)(h). \tag{10.3.9}$$

Comparing (10.3.4) and (10.3.8), we see that $h'_{\lambda_i - \lambda_j} \in [L^{\lambda_i - \lambda_j}, L^{\lambda_j - \lambda_i}]$ and therefore $h'_{\lambda_i - \lambda_j}$ is a scalar multiple of $\bar{h}_{\lambda_i - \lambda_j}$. The first identity in (10.2.6) implies that

$$h_{\lambda_i - \lambda_j} = 2((\lambda_i - \lambda_j)(h'_{\lambda_i - \lambda_j}))^{-1} h'_{\lambda_i - \lambda_j} = e_{ii} - e_{jj} \tag{10.3.10}$$

for all $i \neq j$.

Let α_i denote the root $\lambda_i - \lambda_{i+1}$ for $i = 1, \ldots, n$. We may write

$$\Delta = \{\pm(\alpha_i + \alpha_{i+1} + \cdots + \alpha_j), \quad 1 \leqslant i \leqslant j \leqslant n\}. \tag{10.3.11}$$

Proposition I in 10.2 shows that the system of roots

$$S = \{\alpha_1, \ldots, \alpha_n\} \tag{10.3.12}$$

is the system of simple roots of the Lie algebra L relative to H. The corresponding system of positive roots is the set of roots $\lambda_i - \lambda_j$, where $i < j$. Applying (9.2.1) and (10.3.8), we obtain from (10.3.7) the identities

$$\begin{aligned}(\alpha_i, \alpha_i) &= (h'_{\lambda_i - \lambda_{i+1}}, h'_{\lambda_i - \lambda_{i+1}}) \\ &= 2 \cdot 2(n + 1) \cdot 4^{-1}(n + 1)^{-2} = (n + 1)^{-1}\end{aligned} \tag{10.3.13}$$

for all $i = 1, \ldots, n$. Similarly, for $i \neq j$, we obtain

$$(\alpha_i, \alpha_j) = \begin{cases} 0, & \text{for } |i - j| > 1, \\ -2^{-1}(n + 1)^{-1}, & \text{for } |i - j| = 1. \end{cases} \tag{10.3.14}$$

From (10.3.13) and (10.3.14) we infer that

$$n_{\alpha_i, \alpha_{i+1}} = 1, \qquad n_{\alpha_i, \alpha_j} = 0, \quad \text{for } |i - j| > 1, \tag{10.3.15}$$

where n_{α_i, α_j} are the Cartan numbers defined in (9.6.3). Thus $sl(n + 1, \mathbf{C})$ is a Lie algebra of type (A_n) $(n \geqslant 1)$. In particular, $sl(n + 1, \mathbf{C})$ is simple.

10.4. Lie Algebras of Type (B_n) $(n \geqslant 1)$ and (D_n) $(n \geqslant 2)$

Let m be an integer greater than 2. Let L be the Lie algebra of skew-symmetric complex matrices of order m; i.e., the Lie algebra $so(m, \mathbf{C})$. The elements of

$L, a = (a_{pq})$ are the square matrices of order m such that $a_{pq} = -a_{qp}$ ($p, q = 1, \ldots, m$). The cases $m = 2n$ ($n \geqslant 2$) and $m = 2n + 1$ ($n \geqslant 1$) are completely different.

(a) Consider first the case $m = 2n + 1$, $n \geqslant 1$. It is convenient to write the subscripts p and q in our matrices (a_{pq}) as running through the numbers $-n, -n + 1, \ldots, -1, 0, 1, \ldots, n$.

I. *The Lie algebra $L = so(m, \mathbf{C})$ is isomorphic with the Lie algebra of all matrices $b = (b_{pq})$, $p, q = -n, -n + 1, \ldots, n$ for which*

$$b_{pq} = -b_{-q, -p}, \qquad p, q = -n, -n + 1, \ldots, n. \tag{10.4.1}$$

Proof. Suppose that $a_{pq} = -a_{qp}$ for all $p, q = -n, -n + 1, \ldots, n$. This means that

$$a' = -a, \tag{10.4.2}$$

where a' is the transpose of the matrix a. Define the matrix $\sigma = (\sigma_{pq})$, $p, q = -n, -n + 1, \ldots, n$ by

$$\sigma = \left\| \begin{matrix} (1+i)/2 & & 0 & & & (1-i)/2 \\ & \ddots & & & & \\ & & (1+i)/2 & (1-i)/2 & & \\ \mathbf{0} & & 1 & & \mathbf{0} \\ & & (1-i)/2 & (1+i)/2 & & \\ & & & & \ddots & \\ (1-i)/2 & & 0 & & & (1+i)/2 \end{matrix} \right\|. \tag{10.4.3}$$

Let $\tilde{L}$ be the Lie algebra of all matrices $b = \sigma a \sigma^{-1}$, $a \in L$. The Lie algebras L and $\tilde{L}$ are isomorphic, and $b \in \tilde{L}$ if and only if $a \in L$, which is to say

$$(\sigma^{-1} b \sigma)' = -\sigma^{-1} b \sigma. \tag{10.4.4}$$

A direct calculation shows that

$$\sigma^{-1} = \left\| \begin{matrix} (1-i)/2 & & & & (1+i)/2 \\ & \ddots & & & \\ \cdots & (1-i)/2 & (1+i)/2 & \cdots \\ & 1 & & \\ \cdots & (1+i)/2 & (1-i)/2 & \cdots \\ & & \ddots & \\ (1+i)/2 & \cdots\cdots\cdots\cdots\cdots & (1-i)/2 \end{matrix} \right\|, \qquad \sigma' = \sigma, \tag{10.4.5}$$

and

$$\sigma^2 = \left\| \begin{matrix} 0 & & & & 1 \\ & & & 1 & \\ & & \cdot\cdot & & \\ & 1 & & & \\ 1 & & & & 0 \end{matrix} \right\|. \tag{10.4.6}$$

From (10.4.4)–(10.4.5) we see that $b \in \tilde{L}$ if and only if $\sigma' b'(\sigma^{-1})' = -\sigma^{-1} b\sigma$, i.e., $\sigma b' \sigma^{-1} = -\sigma^{-1} b\sigma$. This yields

$$b' = -\sigma^{-2} b\sigma^2 = -\sigma^2 b\sigma^2. \tag{10.4.7}$$

From (10.4.6) we infer that (10.4.7) is equivalent to the relations (10.4.1). $\square$

Let H be the linear subspace of $\tilde{L}$ consisting of all diagonal matrices in $\tilde{L}$. The matrices $f_{pq} = e_{pq} - e_{-q,-p}$ $(-n \leqslant -q < p \leqslant n)$ are a basis in the Lie algebra $\tilde{L}$ and the matrices $e_{kk} - e_{-k,-k} = f_{kk}$, $k = 1, \ldots, n$ are a basis in the subspace H, so that the dimension of H is n. The identities

$$[e_{pq}, e_{kl}] = \begin{cases} 0, & \text{for } q \neq k \text{ or } p = k, q = l, \\ e_{pl}, & \text{for } q = k, p \neq l, \\ e_{pp} - e_{qq}, & \text{for } p = l, q = k, p \neq q \end{cases} \tag{10.4.8}$$

imply that

$$[f_{k,k}, f_{k_1,k_1}] = 0 \quad \text{for all } k, k_1, 1 \leqslant k, k_1 \leqslant n. \tag{10.4.9}$$

Thus the subspace H is an abelian Lie subalgebra of L. Furthermore, if we have $-n \leqslant -q < p \leqslant n$, $1 \leqslant k \leqslant n$, then (10.4.8) yields:

$$[f_{k,k}, f_{p,q}] = [e_{k,k} - e_{-k,-k}, e_{pq} - e_{-q,-p}]$$

$$= \begin{cases} 0 & \text{for } p \neq \pm k, q \neq \pm k \text{ or } p = q = \pm k; \\ (e_{kq} - e_{-q,-k}) & \text{for } p = k, q \neq k; \\ (e_{-k,q} - e_{-q,k}) & \text{for } p = -k, q \neq -k; \\ (e_{pk} - e_{-k,-p}) & \text{for } p \neq k, q = k; \\ (e_{p,-k} - e_{k,-p}) & \text{for } p \neq -k, q = -k. \end{cases} \tag{10.4.10}$$

The identities (10.4.10) imply that

$$[f_{k,k}, f_{p,q}] = c_{pqk} f_{pq}, \tag{10.4.11}$$

where

$$
c_{pqk} = \begin{cases} 0, & \text{for } p \neq \pm k, q \neq \pm k \quad \text{or } p = q = \pm k, \\ -1, & \text{for } p = -k, q \neq -k \quad \text{or } q = k, p \neq k, \\ 1, & \text{for } p = k, q \neq k \quad \text{or } p \neq -k, q = -k. \end{cases} \tag{10.4.12}
$$

For h in H we write $h = \sum_{k=1}^{n} h_k f_{k,k}$. Note that $h_0 = 0$, $h_{-k} = -h_k$, $k = 1, \ldots, n$. From (10.4.12) we infer that

$$
[h, f_{pq}] = \left(\sum_{k=1}^{n} c_{pqk} h_k \right) f_{pq} = (h_p - h_q) f_{pq} \tag{10.4.13}
$$

for all $h \in H$. We set $\lambda_k(h) = h_k$ ($k = -n, -n+1, \ldots, n$), which means that

$$
[h, f_{pq}] = (\lambda_p - \lambda_q)(h) f_{pq} \tag{10.4.14}
$$

for all $h \in H$. Let Δ be the set of linear functionals on H of the form $\lambda_p - \lambda_q$, $q \neq p$, $-p < q$. For $\lambda_p - \lambda_q = \alpha \in \Delta$, (10.4.14) gives us

$$
\tilde{L}^{\alpha} = \tilde{L}^{\lambda_p - \lambda_q} = \mathbf{C} f_{pq}, \tag{10.4.15}
$$

since no functional $\lambda_p - \lambda_q$, $q \neq p$, $-p < q$, is the zero functional on H, and these functionals are pairwise distinct. Since $\lambda_p - \lambda_q = -(\lambda_q - \lambda_p)$, Δ and $-\Delta$ are the same. If $(\lambda_q - \lambda_p)(h) = 0$ for all $p \neq q$, $-p < q$, we obtain $(\lambda_p - \lambda_0)(h) = h_p = 0$ for $p = 1, \ldots, n$, which means that $h = 0$. Finally, (10.4.15) implies that

$$
\tilde{L} = H + \sum_{\alpha \in \Delta} \tilde{L}^{\alpha}. \tag{10.4.16}
$$

From (10.4.8) we see that $[f_{pq}, f_{qp}] = 2(f_{pp} - f_{qq}) \neq 0$ for $p \neq q$, which means that $[\tilde{L}^{\alpha}, \tilde{L}^{-\alpha}] \neq \{0\}$. Thus all hypotheses of Proposition II in 10.2 hold. Therefore $\tilde{L}$ is a semisimple Lie algebra with a Cartan subalgebra H and Δ is the system of nonzero roots of $\tilde{L}$ relative to H.

Let us find the Dynkin diagram for $\tilde{L}$. Let $h = \sum_{k=1}^{n} h_k f_{kk}$, $h' = \sum_{k=1}^{n} h'_k f_{kk}$ be elements of H. From (9.6.1) we find

$$
(h, h') = \sum_{\alpha \in \Delta} \alpha(h) \alpha(h') = \sum_{\substack{p \neq q \\ p+q > 0}} (h_p - h_q)(h'_p - h'_q). \tag{10.4.17}
$$

Calculations similar to (10.3.5)–(10.3.7) show that

$$
(h, h') = (2n - 1) \sum_{k=1}^{n} h_k h'_k. \tag{10.4.18}
$$

Thus for $k \neq 0$ we have

$$\lambda_k(h) = (\lambda_k - \lambda_0)(h) = (h, h'_{\lambda_k}), \tag{10.4.19}$$

where

$$h'_{\lambda_k} = (2n - 1)^{-1} f_{kk}. \tag{10.4.20}$$

For $p \neq 0$, $q \neq 0$, we have

$$(\lambda_p - \lambda_q)(h) = (h, h'_{\lambda_p - \lambda_q}), \tag{10.4.21}$$

where

$$h'_{\lambda_p} = (2n - 1)^{-1}(f_{pp} - f_{qq}). \tag{10.4.22}$$

Write α_p for the root $\lambda_p - \lambda_{p+1}$ $(p = 1, \ldots, n - 1)$, and write α_n for λ_n. This means that

$$\begin{aligned}
\lambda_q &= \alpha_q + \alpha_{q+1} + \cdots + \alpha_n & \text{for } q > 0, \\
\lambda_q - \lambda_p &= \alpha_q + \alpha_{q+1} + \cdots + \alpha_{p-1} & \text{for } 0 < q < p, \\
\lambda_q &= -\lambda_{-q} & \text{for } q < 0, \\
\lambda_q - \lambda_p &= -(\lambda_{-q} + \lambda_p) & \text{for } q < 0, p > 0, \\
\lambda_q - \lambda_p &= -(\lambda_p - \lambda_q) & \text{for } q > p > 0, \\
\lambda_q - \lambda_p &= -(\lambda_{-q} - \lambda_{-p}) & \text{for } p < 0.
\end{aligned}$$

Proposition I in 10.2 then implies that $\alpha_1, \ldots, \alpha_n$ is the system of simple roots of the Lie algebra $\tilde{L}$. Applying (9.2.1), (10.4.20) and (10.4.22), we obtain

$$\begin{aligned}
(\alpha_p, \alpha_p) &= (h'_{\lambda_p - \lambda_{p+1}}, h'_{\lambda_p - \lambda_{p+1}}) \\
&= (2n - 1)^{-2} \cdot 2 \cdot (2n - 1) = 2(2n - 1)^{-1} \tag{10.4.23}
\end{aligned}$$

for $p = 1, \ldots, n - 1$. Similarly, we have

$$(\alpha_n, \alpha_n) = (2n - 1)^{-1}. \tag{10.4.24}$$

For $p \neq q$, we obtain

$$(\alpha_p, \alpha_q) = \begin{cases} 0, & \text{for } |p - q| > 1, \\ -(2n - 1)^{-1}, & \text{for } |p - q| = 1. \end{cases} \tag{10.4.25}$$

The equalities (10.4.23)–(10.4.25) show that the Dynkin diagram of the Lie algebra L has the form *i.e.*, L is a Lie algebra of type (B_n). For $n = 1$ we have $(B_1) = (A_1)$. Since the Dynkin diagram is connected, the Lie algebra $\tilde{L} \approx sl(2n + 1, \mathbf{C})$ is simple.

(b) Consider now the case $m = 2n$, $n \geqslant 2$. We will index the rows and columns of matrices in $so(2n, \mathbf{C})$ with the numbers $-n, -n + 1, \ldots, -1, 1, 2, \ldots, n$.

II. *The Lie algebra* $L = so(m, \mathbf{C})$ *is isomorphic with the Lie algebra of all matrices* $b = (b_{pq})$, $p, q = \pm 1, \ldots, \pm n$ *that satisfy the condition*

$$b_{pq} = -b_{-q, -p}, \qquad p, q = \pm 1, \ldots, \pm n. \tag{10.4.26}$$

Proof. The proof is like that of Proposition I. Define the matrix $\sigma = (\sigma_{pq})$, $p, q = \pm 1, \ldots, \pm n$, by

$$\sigma = \begin{Vmatrix} (1 + i)/2 & & 0 & & (1 - i)/2 \\ & \ddots & & \ddots & \\ 0 & & \ddots & & 0 \\ & \ddots & & \ddots & \\ (1 - i)/2 & & 0 & & (1 + i)/2 \end{Vmatrix}. \tag{10.4.27}$$

Let $\tilde{L}$ be the Lie algebra of all matrices of the form $b = \sigma a \sigma^{-1}$, $a \in L$. The Lie algebras L and $\tilde{L}$ are isomorphic. Note that $b \in \tilde{L}$ if and only if $a \in L$, which is to say

$$(\sigma^{-1} b \sigma)' = -\sigma^{-1} b \sigma. \tag{10.4.28}$$

Calculations like those leading from (10.4.4) to (10.4.7) show that (10.4.28) is equivalent to $b' = -\sigma^2 b \sigma^2$ and that

$$\sigma^2 = \begin{Vmatrix} 0 & & & & 1 \\ & & & 1 & \\ & & \ddots & & \\ & \ddots & & & \\ 1 & & & & 0 \end{Vmatrix}. \tag{10.4.29}$$

This implies the equivalence of (10.4.26) and (10.4.28).

Let H be the linear space of $\tilde{L}$ consisting of all diagonal matrices in $\tilde{L}$. Plainly the matrices $f_{pq} = e_{pq} - e_{-q, -p}$ ($p + q > 0$, $p, q = \pm 1, \ldots, \pm n$) are a basis in $\tilde{L}$ and the matrices $f_{kk} = e_{kk} - e_{-k, -k}$ ($k = 1, \ldots, n$) are a basis in H. Note that the dimension of H is n. Just as in case (a), we obtain from (10.4.8) the identities

$$[f_{k,k}, f_{k_1, k_1}] = 0, \quad \text{for all } k, k_1, 1 \leqslant k, k_1 \leqslant n \tag{10.4.30}$$

and

$$[f_{k,k}, f_{pq}] = c_{pqk} f_{pq}, \qquad (10.4.31)$$

where:

$$c_{pqk} = \begin{cases} 0, & \text{for } p \neq \pm k, q \neq \pm k \quad \text{or } p = q = \pm k; \\ -1, & \text{for } p = -k, q \neq -k \quad \text{or } q = k, p \neq k; \\ 1, & \text{for } p = k, q \neq k \qquad \text{or } p \neq -k, q = -k. \end{cases} \qquad (10.4.32)$$

The identities (10.4.30) show that H is an abelian Lie subalgebra of $\tilde{L}$. Let $h \in H$. We then have $h = \sum_{k=1}^{n} h_k f_{kk}$. We set $h_{-k} = -h_k$ for $k = 1, \ldots, n$. The linear functionals λ_k on H are defined by $\lambda_k(h) = h_k$, $k = \pm 1, \ldots, n$. From (10.4.31) and (10.4.32) we obtain

$$[h, f_{pq}] = (\lambda_p - \lambda_q)(h) f_{pq} \qquad (10.4.33)$$

for all $h \in H$. Let Δ be the set of linear functionals on H of the form $\lambda_p - \lambda_q$, $q \neq p$, $p + q > 0$. Thus for $\alpha \in \Delta$, $\alpha = \lambda_p - \lambda_q$, (10.4.33) gives

$$\tilde{L}^\alpha = \tilde{L}^{\lambda_p - \lambda_q} = \mathbf{C} f_{pq}, \qquad (10.4.34)$$

since the functionals $\lambda_p - \lambda_q$, $q \neq p$, $p + q > 0$, are pairwise distinct, and none of them is the zero functional. Since $(\lambda_p - \lambda_q) = -(\lambda_q - \lambda_p)$, Δ is equal to $-\Delta$. Suppose that h is such that $(\lambda_q - \lambda_p)(h) = 0$ for all $p \neq q$, $p + q > 0$. We get $(\lambda_q - \lambda_n)(h) = 0$, $(\lambda_{-q} - \lambda_n)(h) = 0$ for all $q \neq \pm n$ and therefore $2\lambda_q(h) = (\lambda_q - \lambda_{-q})(h) = 0$ for all $q \neq \pm n$. We also obtain $\lambda_n(h) = 0$; that is, $h_k = 0$ for all $k = 1, \ldots, n$, and so $h = 0$. Finally, (10.4.34) implies that $L = H + \sum_{\alpha \in \Delta} \tilde{L}^\alpha$. From (10.4.8) we infer that $[f_{pq}, f_{qp}] = 2(f_{pp} - f_{qq}) \neq 0$ for $p \neq q$ and thus we have $[L^\alpha, L^{-\alpha}] \neq \{0\}$. Proposition II in 10.2 implies that $\tilde{L}$ is a semisimple Lie algebra, H is a Cartan subalgebra of $\tilde{L}$, and Δ is the corresponding system of nonzero roots.

We will now find the Dynkin diagram for $\tilde{L}$. Consider $h, h' \in H$, and write $h = \sum_{k=1}^{n} h_k f_{kk}$, $h' = \sum_{k=1}^{n} h'_k f_{kk}$. From (9.6.1) we see that

$$(h, h') = \sum_{\alpha \in \Delta} \alpha(h)\alpha(h') = \sum_{\substack{p \neq q \\ p+q>0}} (h_p - h_q)(h'_p - h'_q). \qquad (10.4.35)$$

Calculations analogous to (10.3.5)–(10.3.7) show that

$$(h, h') = (2n - 2) \sum_{k=1}^{n} h_k h'_k. \qquad (10.4.36)$$

Thus we have $(\lambda_p - \lambda_q)(h) = (h, h'_{\lambda_p - \lambda_q})$, where

$$h'_{\lambda_p - \lambda_q} = (2n - 2)^{-1}(f_{pp} - f_{qq}). \qquad (10.4.37)$$

We write α_p for the root $\lambda_p - \lambda_{p+1}$, $p = 1, \ldots, n-1$, and α_n for $(\lambda_n - \lambda_{-(n-1)}) = \lambda_{n-1} + \lambda_n$. Thus $\alpha_1, \ldots, \alpha_n$ belong to Δ and for $i + j > 0$ we have:

$$\lambda_p - \lambda_q = \alpha_p + \cdots + \alpha_{q-1} \qquad\qquad \text{for } 0 < p < q;$$
$$\lambda_p - \lambda_q = -(\lambda_{-q} - \lambda_{-p}) = \lambda_{-p} - \lambda_{-q} \qquad \text{for } 0 < q < p,\, p < q < 0$$
$$\qquad\qquad\qquad\qquad\qquad\qquad\qquad\qquad\qquad \text{or } q < p < 0;$$
$$\lambda_p - \lambda_q = (\lambda_p - \lambda_n) + (\lambda_{-q} - \lambda_{n-1}) + \alpha_n \quad \text{for } p > 0 > q,$$
$$\lambda_p - \lambda_q = -(\lambda_q - \lambda_p) \qquad\qquad\qquad\quad \text{for } p < 0.$$

From Proposition I in 10.2, we now infer that $\alpha_1, \ldots, \alpha_n$ are the system of simple roots of the Lie algebra $\tilde{L}$. From (9.2.1) and (10.4.37) we obtain:

$$(\alpha_p, \alpha_q) = \begin{cases} 2(2n-2)^{-1} & \text{for } q = p; \\ -(2n-2)^{-1} & \text{for } |p-q| = 1,\, p < n,\, q < n; \\ -(2n-2)^{-1} & \text{for } p = n-2,\, q = n \\ & \text{or } p = n,\, q = n-2; \\ 0 & \text{in all other case.} \end{cases} \qquad (10.4.38)$$

According to (10.4.38), the Dynkin diagram of L has the form

which is a Dynkin diagram of type (D_n) $(n \geqslant 2)$. For $n = 2$ this diagram is disconnected and has the form

For $n = 3$ we obtain $(D_3) = (A_3)$. For $n > 3$ the Lie algebra $\tilde{L} \approx so(2n, \mathbf{C})$ is simple since its Dynkin diagram is connected.

10.5. Lie Algebras of Type (C_n) $(n \geqslant 2)$

Let $L = sp(2n, \mathbf{C})$, $n \geqslant 2$, be the Lie algebra of complex matrices x of order $2n$ having the form

$$x = \left\| \begin{matrix} a & b \\ c & -a^t \end{matrix} \right\|; \qquad\qquad (10.5.1)$$

here a, b, c are square complex matrices of order n for which b and c are symmetric and a^t is the transpose of a. We can verify that L is a Lie algebra by a direct calculation or by noting that a matrix x belongs to L if and only if $x^t s + sx = 0$, where

$$s = \left\| \begin{matrix} 0 & 1_n \\ -1_n & 0 \end{matrix} \right\|, \quad \text{and } 1_n \text{ is the identity matrix of order } n. \quad (10.5.2)$$

Let H be the linear subspace of L consisting of all diagonal matrices in L. We write $f_{ij} = e_{ij} - e_{j+n, i+n}$ $(i, j = 1, \ldots, n)$,

$$g_{ij} = e_{i+n, j} + e_{j+n, i}, \quad (i, j = 1, \ldots, n),$$
$$\tilde{g}_{ij} = e_{i, j+n} + e_{j, i+n}, \quad (i, j = 1, \ldots, n).$$

The set f_{ij}, $i, j = 1, \ldots, n$; g_{ij}, $\tilde{g}_{ij}$, $1 \leqslant i \leqslant j \leqslant n$, is a basis in L and the elements f_{ii}, $i = 1, \ldots, n$, are a basis in H. Therefore we have

$$L = H + \sum_{i \neq j} C f_{ij} + \sum_{i \leqslant j} C g_{ij} + \sum_{i \leqslant j} C \tilde{g}_{ij}. \quad (10.5.3)$$

For $h \in H$, let us write $h = \sum_{i=1}^n h_i f_{ii}$. Let $\lambda_i(h) = h_i$ $(i = 1, \ldots, n)$. From (10.4.8) we have:

$$[h, f_{ij}] = (\lambda_i - \lambda_j)(h) f_{ij}; \quad (10.5.4\text{a})$$
$$[h, g_{ij}] = (\lambda_i + \lambda_j)(h) g_{ij}; \quad (10.5.4\text{b})$$
$$[h, \tilde{g}_{ij}] = -(\lambda_i + \lambda_j)(h) \tilde{g}_{ij}. \quad (10.5.4\text{c})$$

Let Δ be the set of linear functionals on H of the form $\lambda_i - \lambda_j$ $(i \neq j, i, j = 1, \ldots, n)$ and $\pm(\lambda_i + \lambda_j)$ $(i, j = 1, \ldots, n)$. It is clear that $\Delta = -\Delta$. If h is such that $\lambda(h) = 0$ for all $\lambda \in \Delta$, then we have $(\lambda_i + \lambda_i)(h) = 2\lambda_i(h) = 0$ for all $i = 1, \ldots, n$, and so $h = 0$. From (10.5.4) we infer that

$$L^{\lambda_i - \lambda_j} = C f_{ij}, \quad L^{\lambda_i + \lambda_j} = C g_{ij}, \quad L^{-\lambda_i - \lambda_j} = C \tilde{g}_{ij}. \quad (10.5.5)$$

From the identities (10.4.8) we infer that

$$[f_{ij}, f_{ji}] = f_{ii} - f_{jj}, \quad [g_{ij}, \tilde{g}_{ij}] = f_{ii} + f_{jj}, \quad (10.5.6)$$

which shows that $[L^\alpha, L^{-\alpha}] \neq \{0\}$ for all $\alpha \in \Delta$. We can write (10.5.3) as $L = H + \sum_{\alpha \in \Delta} L^\alpha$. Thus all the hypotheses of Propostion II in 10.2 are satisfied. Hence L is semisimple, H is a Cartan subalgebra, and Δ is the set of nonzero roots of L relative to H.

We will now find the Dynkin diagram for L. For $h = \sum_{i=1}^{n} h_i f_{ii}$, $h' = \sum_{i=1}^{n} h'_i f_{ii}$, (9.6.1) and the definition of $\varDelta$ yield

$$
\begin{aligned}
(h, h') &= \sum_{\alpha \in \varDelta} \alpha(h)\alpha(h') \\
&= \sum_{i \neq j} (h_i - h_j)(h'_i - h'_j) + 2 \sum_{i \leqslant j} (h_i + h_j)(h'_i + h'_j) \\
&= 8 \sum_{i=1}^{n} h_i h'_i + \sum_{i \neq j} \{(h_i - h_j)(h'_i - h'_j) + (h_i + h_j)(h'_i + h'_j)\} \\
&= 8 \sum_{i=1}^{n} h_i h'_i + 2 \sum_{i \neq j} (h_i h'_i + h_j h'_j) = (4n + 4) \sum_{i=1}^{n} h_i h'_i. \quad (10.5.7)
\end{aligned}
$$

Thus we have $(\lambda_i - \lambda_j)(h) = (h, h'_{\lambda_i - \lambda_j})$, $(\lambda_i + \lambda_j)(h) = (h, h'_{\lambda_i + \lambda_j})$, where

$$
h'_{\lambda_i - \lambda_j} = (4(n + 1))^{-1}(f_{ii} - f_{jj}), \; h'_{\lambda_i + \lambda_j} = (4(n + 1))^{-1}(f_{ii} + f_{jj}). \quad (10.5.8)
$$

We write α_i for the root $\lambda_i - \lambda_{i+1}$ $(i = 1, \ldots, n - 1)$, α_n for the root $2\lambda_n$. Thus $\alpha_1, \ldots, \alpha_n$ belong to $\varDelta$ and we have:

$$
\begin{aligned}
\lambda_i - \lambda_j &= \alpha_i + \cdots + \alpha_{j-1} & \text{for } j > i; \\
\lambda_i - \lambda_j &= -(\lambda_j - \lambda_i) & \text{for } j < i; \quad (10.5.9) \\
\pm(\lambda_i + \lambda_j) &= \pm\{(\alpha_i + \cdots + \alpha_{n-1}) + (\alpha_j + \cdots + \alpha_n)\} & \text{for } i \leqslant j.
\end{aligned}
$$

From (10.5.9) we infer that the system $\alpha_1, \ldots, \alpha_n$ satisfies the hypotheses of Proposition I in 10.2 and therefore $\alpha_1, \ldots, \alpha_n$ is the system of simple roots of the Lie algebra L. From (9.2.1) and (10.5.8) we obtain:

$$
(\alpha_i, \alpha_j) = \begin{cases}
2/(4(n + 1)), & \text{for } 1 \leqslant i = j \leqslant n - 1; \\
4/(4(n + 1)), & \text{for } i = j = n; \\
-1/(4(n + 1)), & \text{for } |i - j| = 1, i < n, j < n; \\
-2/(4(n + 1)), & \text{for } i = n, j = n - 1, \\
& \quad \text{or } i = n - 1, j = n; \\
0, & \text{in all other cases.}
\end{cases} \quad (10.5.10)
$$

By (10.5.10) the Dynkin diagram of L has the form

$$
\begin{array}{ccccc}
\alpha_1 & \alpha_2 & & \alpha_{n-1} & \alpha_n \\
\circ\!\!-\!\!\!&\!\!\!-\!\!\circ & \cdots & \circ\!\!=\!\!\!&\!\!\!=\!\!\circ \\
1 & 1 & & 1 & 2
\end{array}.
$$

This is a Dynkin diagram of type (C_n) $(n \geqslant 2)$. For $n = 2$ we have $(C_2) = (B_2)$. Since the Dynkin diagram of L is connected, the Lie algebra $L = sp(2n, \mathbf{C})$ is simple.

Lie algebras of types $(A_n), (B_n), (C_n), (D_n)$ are called *classical*. Lie algebras of the types $(E_6), (E_7), (E_8), (F_4), (G_2)$ are called *exceptional*. They are constructed in the book Jacobson [1], among other places.

§11. The Weyl Group of a Semisimple Lie Algebra

Let L be a semisimple complex Lie algebra. Let H be a Cartan subalgebra of L and Δ the system of nonzero roots of L relative to H. For every $\alpha \in \Delta$ we introduce an element h'_α of H defined by the condition

$$\alpha(h) = (h, h'_\alpha) \tag{11.1.1}$$

for all $h \in H$. (The element h'_α exists and is uniquely defined for all $\alpha \in \Delta$ since the restriction of the Killing form to H is nonsingular by Proposition IV in 9.1.) Let H_0 be the real subspace of H spanned by the vectors $h'_\alpha, \alpha \in \Delta$ and let λ be a linear functional on H_0. The hyperplane in H defined by the identity $\lambda(h) = 0$ is denoted by P_λ. Let S_λ be the mapping of H_0 onto itself that maps every vector $h \in H_0$ to the vector symmetric to h with respect to the hyperplane P_λ. Let us find a formula for S_λ. By (11.1.1), the hyperplane P_λ is described by the identity $(h, h'_\lambda) = 0$. That is, the vector h'_λ is orthogonal to P_λ. Since we have $(h'_\lambda, h'_\lambda) = (\lambda, \lambda)$, S_λ is given by the formula

$$S_\lambda(h) = h - 2\lambda(h)(\lambda, \lambda)^{-1} h'_\lambda \quad (h \in H_0). \tag{11.1.2}$$

By the definition of S_λ, this is an orthogonal transformation of H_0 onto itself. The transformation S_λ^* of the space H_0^* conjugate to S_λ is defined by the relation

$$S_\lambda^*(\mu) = \mu - 2(\lambda, \mu)(\lambda, \lambda)^{-1}\lambda, \quad (\mu \in H_0^*). \tag{11.1.3}$$

For, (11.1.1) and (9.2.1) show that $\mu(h'_\lambda) = (\lambda, \mu)$, and so

$$(\mu - 2(\lambda, \mu)(\lambda, \lambda)^{-1}\lambda)(h) = \mu(h - 2\lambda(h)(\lambda, \lambda)^{-1}h'_\lambda). \tag{11.1.4}$$

Now choose λ as some α in Δ. By applying S_α^* to a root $\beta \in \Delta$, we obtain

$$S_\alpha^*(\beta) = \beta - 2(\alpha, \beta)(\alpha, \alpha)^{-1}\alpha \tag{11.1.5}$$

in consequence of (11.1.3). According to (9.5.1) we have

$$-2(\alpha, \beta)(\alpha, \alpha)^{-1} = p + q, \tag{11.1.6}$$

where $p \leqslant 0$, $q \geqslant 0$ and for all integers k such that $p \leqslant k \leqslant q$, the linear functional $\beta + k\alpha$ belongs to Δ. Since $p \leqslant p + q \leqslant q$, we may cite (11.1.5) and (11.1.6) to verify the following.

I. *For all α, β in Δ, the functional $S_\alpha^*(\beta)$ belongs to Δ.*

Let W be the group of orthogonal transformations of the space H_0 onto itself generated by the transformations S_α, $\alpha \in \Delta$.

II. *The group W is finite.*

Proof. By Proposition I, all transformations S_α^*, $\alpha \in \Delta$, of H_0 carry Δ into Δ. The same therefore holds for all w in W: the transformation w^* carries Δ into Δ. If $w \in W$ and $w_\alpha^* = \alpha$ for all $\alpha \in \Delta$, the transformation w^* of H_0^* is the identity mapping on the linear span of the set Δ. This linear span is all of H_0^*, by Proposition III in 9.6. Thus w^* is the identity mapping on H_0^* and so w is the identity mapping on H_0. Therefore the group W is isomorphic to a subgroup of the group of permutations of the finite set Δ, and so W is a finite group. $\square$

The group W is called the *Weyl group of the Lie algebra L.* The group W^*, formed by the operators conjugate to the elements of the group W, is sometimes called the Weyl group.

Let Q be the set of elements $h \in H_0$ such that $\alpha(h) \neq 0$ for all $\alpha \in \Delta$. The maximal convex subsets of Q are called *fundamental Weyl domains,* or *Weyl chambers.*

III. *Let $\alpha_1, \ldots, \alpha_r$ be the simple roots of the Lie algebra L relative to H. The set C_0 of elements $h \in H_0$ such that $\alpha_i(h) > 0$ for all $i = 1, \ldots, r$, is a Weyl chamber.*

Proof. For $h_1, h_2 \in C_0$, we have $\alpha_i(th_1 + (1-t)h_2) = t\alpha_i(h_1) + (1-t)\alpha_i(h_2) > 0$ for all $i = 1, \ldots, r$ and all $t \in [0,1]$. Hence C_0 is a convex set. We will show that C_0 is a subset of Q. If $C_0 \not\subset Q$, there is an $h \in C_0$ such that $\alpha(h) = 0$ for some $\alpha \in \Delta$. Any nonzero root α can be represented in the form $\alpha = \sum k_i \alpha_i$, where the integers k_i have the same sign and are not all zero. Since $\alpha_i(h) > 0$, we have $\alpha(h) \neq 0$, and so C_0 is contained in Q. Clearly, C_0 is a *maximal* convex subset of Q. $\square$

The Weyl chamber C_0 is called the *dominant Weyl chamber.*

Let W' be the subgroup of W generated by symmetries defined by simple roots, that is, by the elements $S_i = S_{\alpha_i}$ $(i = 1, \ldots, r)$.

IV. *The group W' (a fortiori the group W) is transitive on the collection of Weyl chambers: for every Weyl chamber C_1, there exists an element $w \in W'$ such that $C_0 = wC_1$.*

Proof. If C_1 is a Weyl chamber, wC_1 is convex for all $w \in W$. Since w^* carries Δ onto Δ, w maps Q into itself. Since $C_1 \subset Q$, we have $wC_1 \subset Q$. Thus wC_1

is a convex subset in Q and so is contained in some Weyl chamber. Given $w \in W'$, the inclusion $wC_1 \subset C_0$ means exactly that $wh_1 \in C_0$ for all $h_1 \in C_1$. Let us find such a w. Let h_0 be a fixed point of C_0. For fixed $h_1 \in C_1$, the set of points wh_1, $w \in W'$, is finite. Choose $w_0 \in W'$ so that $w_0 h_1$ is a point closest to h_0 among all of the points wh_1, $w \in W'$. If $w_0 h_1$ does not belong to C_0, we have $\alpha_i(w_0 h_1) < 0$ for some simple root α_i. Compare the distance between $w_0 h_1$ and h_0 with that between $S_{\alpha_i}(w_0 h_1)$ and h_0. The projections of the vectors $h_0 - w_0 h_1$ and $h_0 - S_{\alpha_i}(w_0 h_1)$ onto the hyperplane P_{α_i} are the same, but the projections of these vectors onto the line $\{th'_{\alpha_i}, t \in \mathbf{R}\}$ perpendicular to the plane P_{α_i} are different. We have in fact

$$\left|(h_0 - w_0 h_1, h'_{\alpha_i})\right| = \left|\alpha_i(h_0) - \alpha_i(w_0 h_1)\right| = \alpha_i(h_0) - \alpha_i(w_0 h_1), \quad (11.1.7)$$

since $\alpha_i(h_0) > 0$, $\alpha_i(w_0 h_1) < 0$. We further obtain

$$\begin{aligned}
\left|(h_0 - S_{\alpha_i}(w_0 h_1), h'_{\alpha_i})\right| &= \left|\alpha_i(h_0) - \alpha_i(S_{\alpha_i}(w_0 h_1))\right| \\
&= \left|\alpha_i(h_0) - \alpha_i(w_0 h_1 - 2\alpha_i(w_0 h_1)(\alpha_i, \alpha_i)^{-1} h'_{\alpha_i})\right| \quad (11.1.8) \\
&= \left|\alpha_i(h_0) + \alpha_i(w_0 h_1)\right|.
\end{aligned}$$

Comparing the right sides of the equalities (11.1.7) and (11.1.8), we see that $\left|\alpha_i(h_0) + \alpha_i(w_0 h_1)\right| < \alpha_i(h_0) - \alpha_i(w_0 h_1)$, and thus the point $S_{\alpha_i}(w_0 h_1)$ is strictly closer to h_0 than $w_0 h_1$ is. Therefore the point $w_0 h_1$ is not the closest to h_0. This contradiction shows that $w_0 h_1 \in C_0$ and thus $wC_1 \subset C_0$. An analogous argument shows that $w^{-1} C_0 \subset C_1$ and thus $C_0 = w(w^{-1} C_0) \subset wC_1$ and $C_0 = wC_1$. $\square$

V. *The group W' is all of W.*

Proof. We say that the hyperplane P_α *bounds* the Weyl chamber C if the boundary of C contains a nonvoid open subset of P_α. Plainly, the chamber C_0 is bounded by the hyperplanes P_{α_i}, $i = 1, \ldots, r$ and by no others. For $\alpha \in \Delta$, let C_1 be a Weyl chamber bounded by P_α. There exists a transformation $w \in W'$ such that $C_1 = wC_0$. Under this mapping P_α is the image of some hyperplane that bounds C_0, that is, some hyperplane P_{α_i}. The only roots for which P_α is the set of zeros are $\pm\alpha$. Thus we have $w\alpha_i = \pm\alpha$. If $w\alpha_i = -\alpha$, we obtain $wS_{\alpha_i}(\alpha_i) = w(-\alpha_i) = \alpha$ so that there is always an element $w \in W'$ and a simple root α_i, such that $w\alpha_i = \alpha$. It is clear that $S_\alpha = wS_{\alpha_i} w^{-1}$. Since $w \in W'$ and $S_{\alpha_i} \in W'$, we see that $S_\alpha \in W'$ for all $\alpha \in \Delta$ and thus $W = W'$. $\square$

Propositions IV and V yield the following.

VI. *The Weyl group W is generated by the symmetries S_i, $i = 1, \ldots, r$, defined by simple roots.*

Weyl groups of Lie algebras of types $(A_n) - (G_2)$ are studied in the book Bourbaki [1].

§12. Linear Representations of Semisimple Complex Lie Algebras

Let L be a semisimple complex Lie algebra with a Cartan subalgebra H. Let $\varDelta$ be the system of nonzero roots of L relative to H. Order the real vector space H_0^* lexicographically under some basis in H_0, so that Σ is the system of positive roots and $\varPi = \{\alpha_1, \ldots, \alpha_r\}$ is the system of simple roots.

I. *The linear spaces $N_+ = \sum_{\alpha > 0} L^\alpha$ and $N_- = \sum_{\alpha < 0} L^\alpha$ are nilpotent Lie algebras and the space L is the direct sum of the subspaces N_+, N_- and H.*

Proof. Since $[L^\alpha, L^\beta] \subset L^{\alpha+\beta}$ for all α, $\beta \in \varDelta$ (see Proposition II in §8), N_+ and N_- are Lie subalgebras of L. We will show that N_+ is nilpotent. Let $\alpha = \sum m_i \alpha_i$ be the decomposition of the root $\alpha \in \Sigma$ as a linear combination of simple roots (see Propostion III in 9.6). The number $m = \sum m_i$ is called the *order* of the root α. Let n be the largest order of the roots $\alpha \in \Sigma$. Since $[L^\alpha, L^\beta] \subset L^{\alpha+\beta}$, we infer that for all $x \in N_+$, the operator $(\text{ad } x)^{n+1}$ maps the subspaces L^α, $\alpha \in \Sigma$, into zero. Thus (see theorem 1 in 2.1) the Lie algebra N_+ is nilpotent. We prove that N_- is nilpotent by an analogous argument. The relation $L = H + \sum_{\alpha \neq 0} L^\alpha$ (see (8.1.6)) implies the equality $L = N_+ + H + N_-$. $\square$

The decomposition $L = N_- + H + N_+$ is called a *Cartan factorization of the Lie algebra* L.

II. *The Lie algebra N_+ (N_-) is generated by the subspaces L^{α_i} $(L^{-\alpha_i})$ $(i = 1, \ldots, r)$.*

Proof. Let N'_+ be the Lie subalgebra of N_+ generated by the subspaces L^{α_i}. Suppose that we already know that $L^\beta \subset N'_+$ for $0 < \beta < \alpha$. If α is a simple root, we have $L^\alpha \subset N'_+$ by construction. If $\alpha = \beta + \gamma$, where $\beta, \gamma > 0$, then $0 < \beta < \alpha$, $0 < \gamma < \alpha$. Applying Proposition II in 9.5, we see that $L^\alpha = [L^\beta, L^\gamma] \subset [N'_+, N'_+] \subset N'_+$. The proof for N_- is similar. $\square$

We define h_α as $2(\alpha, \alpha)^{-1} h'_\alpha$ and $h_i = h_{\alpha_i}$. We then have $\alpha(h_\alpha) = 2$ and $\alpha_i(h_j) = 2(\alpha_i, \alpha_j)(\alpha_j, \alpha_j)^{-1} = -n_{ji}$, where the n_{ji} are certain nonnegative integers (see Proposition III in 9.6 and (9.6.3)). If also $x \in L^\alpha$, $y \in L^{-\alpha}$, then $[x, y] = (x, y) h'_\alpha$ (see (9.2.3)). Therefore we can choose $x_i \in L^{\alpha_i}$ and $y_i \in L^{-\alpha_i}$ so that $[x_i, y_i] = h_i$. Since $\alpha_i - \alpha_j$ is not a root, we have $[x_i, y_j] = 0$ for $i \neq j$. Thus the Lie algebra L is generated by the elements x_i, y_i, h_i ($i =$

$1, 2, \ldots, r$). The following identities hold among these elements (in general other identities hold as well):

$$[x_i, y_j] = 0 \quad \text{for } i \neq j; \tag{12.1.1a}$$
$$[x_i, y_i] = h_i; \tag{12.1.1b}$$
$$[h_i, x_j] = -n_{ij}x_j; \tag{12.1.1c}$$
$$[h_i, y_j] = n_{ij}y_j; \tag{12.1.1d}$$
$$[h_i, h_j] = 0. \tag{12.1.1e}$$

Let π be a linear representation of L in a vector space V. Let λ be a linear functional on H. Let V_λ denote the set of vectors $v \in V$ such that $\pi(h)v = \lambda(h)v$ for all $h \in H$. Plainly V_λ is a linear subspace of V and $V_\lambda \subset V^\lambda$, where $V^\lambda = V(H, \lambda)$ is as defined in 2.4. If V_λ is not the zero subspace, the functional λ is called a *weight of the representation π of L*.

III. *The subspaces V_λ are linearly independent.*

See Proposition II (3) in 2.4 and note that $V_\lambda \subset V^\lambda$.

IV. *If W is an invariant subspace of V under the representation π, we have $W \cap (\sum_\lambda V_\lambda) = \sum_\lambda (W \cap V_\lambda)$.*

Proof. Consider the representation defined by π in the factor space V/W. By Proposition III, if $\sum_\lambda v_\lambda \in W$ (where $v_\lambda \in V_\lambda$), we must have $v_\lambda \in W$. The inclusion $W \cap (\sum_\lambda V_\lambda) \subset \sum_\lambda (W \cap V_\lambda)$ follows. The reversed inclusion $\sum_\lambda (W \cap V_\lambda) \subset W \cap (\sum_\lambda V_\lambda)$ is obvious. $\square$

V. *A finite-dimensional representation π of a semisimple Lie algebra L has at least one weight.*

Proof. The restriction of the representation π to a Cartan subalgebra H is a finite-dimensional representation of the Lie algebra H. Let V_1 be a nonzero subspace of V invariant under $\pi(H)$ such that the set of operators $\pi(H)|_{V_1}$ is irreducible. Since H is a commutative Lie algebra, $\pi(H)|_{V_1}$ is an irreducible set of pairwise commuting operators. Schur's lemma implies that V_1 is one-dimensional. If v is a nonzero vector in V_1, we have $(h)v \in V_1$ for all $h \in H$; that is, $(h)v = \lambda(h)v$ for some $\lambda(h)$. The function λ is a weight. $\square$

VI. *The inclusion $\pi(L^\alpha)V_\lambda \subset V_{\lambda+\alpha}$ holds.*

Proof. For $v \in V_\lambda$ and $x \in L^\alpha$, we have

$$\pi(h)\pi(x)v = \pi([h, x])v + \pi(x)\pi(h)v$$
$$= \pi(\alpha(h)x)v + \pi(x)(\lambda(h)v) = (\alpha(h) + \lambda(h))\pi(x)v. \tag{12.1.2}$$

This implies that $\pi(x)v \in V_{\lambda+\alpha}$. $\square$

VII. *The subspace $\sum_\lambda V_\lambda$ is invariant under the representation π.*

See Proposition VI.

The vector $v \in V$ is called a *highest vector of the representation π* if $v \in V_\lambda$ for some weight λ, $\pi(x_i)v = 0$ for all $i = 1, \ldots, r$ and the smallest subspace of V, invariant under all operators $\pi(x)$, $x \in L$, and containing the vector v, coincides with V. In this case the weight λ is called a *highest weight of the representation π.*[55]

Since the Lie subalgebra N_+ is generated by the elements x_i, $i = 1, \ldots, r$, we have $\pi(x)v = 0$ for all $x \in N_+$. Consider the subspace $B = H + N_+$ of L. Plainly B is a Lie subalgebra of L and N_+ is an ideal in B, since by (12.1.1c) $[H, N_+]$ is contained in N_+. Since $\pi(h)v = \lambda(h)v$ for $h \in H$ and $\pi(x)v = 0$ for $x \in N_+$, the one-dimensional subspace generated by the vector v is invariant under the operators $\pi(b)$, $b \in B$.

Let U, U_0, U_- be universal enveloping algebras of the Lie algebras L, B, N_-, respectively. Choose a basis in L of the form $n_1^-, \ldots, n_p^-, h_1, \ldots, h_r$, $n_1^+, \ldots, n_p^+$, where $n_i^\pm \in N_\pm$ and $h_i \in H$. By applying the Poincaré-Birkhoff-Witt Theorem to this basis (see the theorem in 6.3), we conclude that $\pi(U)v = \pi(U_-)v$, since $\pi(U_0)v$ is in the one-dimensional subspace $\mathbf{C}v$ generated by v.

Consider the subspace $\mathbf{C}v + \sum V_\mu$, where the sum is taken over all weights μ such that the difference $\lambda - \mu$ is a linear combination of the roots α_i with nonnegative integer coefficients. This subspace contains v and is invariant under N_-. In fact, for any positive root α we have $\pi(L^{-\alpha})V_\mu \subset V_{\mu-\alpha}$ and the functional $\lambda - (\mu - \alpha) = (\lambda - \mu) + \alpha$ has the form $\sum m_i\alpha_i$, where the m_i are nonnegative integers. Since $\pi(U_-)v = \pi(U)v$ and $\pi(U_-)v \subset \mathbf{C}v + \sum V_\mu$, the definition of a vector of highest weight shows that $\mathbf{C}v + \sum V_\mu = V$. We summarize as follows.

VIII. *If λ is the highest weight of the representation π, the subspace V_λ is one-dimensional and any weight μ of π has the form $\mu = \lambda - \sum m_i\alpha_i$, where the m_i are nonnegative integers. Furthermore, we have $V = \sum V_\mu = \pi(U_-)v$.*

We have a corollary to Proposition VIII.

IX. *The highest weight λ is uniquely defined by the representation π.*

X. *The weight subspaces V_μ are finite-dimensional.*

Proof. We know that $V = \pi(U_-)v$ and $y_i \in L^{-\alpha_i}$. Thus the subspace V_μ is spanned by vectors of the form $\pi(y_{i_1}) \cdots \pi(y_{i_p})v$, where $i_1, \ldots, i_p$ is a set such

---

[55] This definition is inconsistent with the definition of highest vector and highest weight given for representations of groups in chapters VI and VII. For, in the present definition, the highest vector must be a cyclic vector of the representation. For *irreducible* representations, cyclicity is automatic.

that $\lambda - \alpha_{i_1} - \cdots - \alpha_{i_p} = \mu$. There are only a finite number of sets $\{i_1, \ldots, i_p\}$ with this property and therefore V_μ is finite-dimensional. $\square$

We now turn to the classification of finite-dimensional representations of a semisimple complex Lie algebra L. According to the theorem of complete reducibility (see 7.2), we may limit our discussion to irreducible representations.

XI. *Any finite-dimensional irreducible representation π has a highest vector defined uniquely to within a scalar multiple.*

Proof. Proposition V shows that π has at least one weight. Since π is a finite-dimensional representation, Proposition III implies that the set of weights of π is finite. Thus there is a weight λ of π such that $\lambda + \alpha_i$ fails to be a weight for all $i = 1, \ldots, r$. From Proposition VI we see that $\pi(x_i)v = 0$ for $i = 1, \ldots, r$ for all $v \in V_\lambda$; that is, $\pi(N_+)v = \{0\}$. Since π is irreducible, any subspace of V invariant under the operators $\pi(x)$, $x \in L$ and containing a nonzero vector v coincides with V. Therefore any nonzero vector $v \in V_\lambda$ is a highest weight of π, and the subspace V_λ is one-dimensional by Proposition VIII. $\square$

Choose elements h_i of H so that $\alpha_i(h) = (h, h_i)$ for all $h \in H$ and all simple roots α_i.

Theorem 1. *The linear functional λ is the highest weight of a finite-dimensional irreducible representation π of the Lie algebra L if and only if all $\lambda(h_i)$, $i = 1, \ldots, r$, are nonnegative integers. Every finite-dimensional irreducible representation π of the Lie algebra L in a space V has a highest weight, and V is the direct sum of the subspaces V_μ. For every root α and every weight μ, the number $\mu(h_\alpha)$ is an integer. If μ and $\mu + \alpha$ are weights of π, we have $\pi(L^\alpha)V_\mu \neq \{0\}$. Let P be the set of all weights of the representation π. The set P is finite and invariant under the set of operators conjugate to the elements of the Weyl group W of L. If $\mu = \sigma^* v$ for some $\sigma \in W$, the equality $\dim V_\mu = \dim V_v$ holds.*

Proof. According to Proposition XI, π admits a highest vector v. Let λ be the highest weight of π. By Proposition VIII, every weight μ of π has the form $\mu = \lambda - \sum_{i=1}^r m_i \alpha_i$, where the m_i are nonnegative integers. Also V is the direct sum of the subspaces V_μ. Let μ be any weight of π and let α be a root of L. We write $\tilde{V} = \sum_j V_{\mu + j\alpha}$. Let L_α be the Lie subalgebra of L generated by the vectors $e_\alpha, e_{-\alpha}, h_\alpha$ (see Proposition IV in 9.3). The subspace $\tilde{V}$ is invariant under $\pi(L_\alpha)$ since $\pi(L^\alpha)V_\mu \subset V_{\mu + \alpha}$ for every weight μ and $\pi(H)V_\mu \subset V_\mu$. Since $\alpha(h_\alpha)$ is equal to 2, distinct weights $\mu + j\alpha$ assume distinct values at h_α. Apply the theorem in 9.4 to the representation $\tilde{\pi}$ of the Lie algebra L_α in the space $\tilde{V}$ defined by the formula $\tilde{\pi}(x)\tilde{v} = \pi(x)\tilde{v}$ for all $x \in L_\alpha$ and $\tilde{v} \in \tilde{V}$. By this theorem, the number $2\mu(h_\alpha)(\alpha(h_\alpha))^{-1}$ is an integer. Furthermore, there is a weight μ' of the representation $\tilde{\pi}$ of L_α in $\tilde{V}$ such that $\mu'(h_\alpha) = -\mu(h_\alpha)$. Therefore we have $(\mu + \mu')(h_\alpha) = 0$, that is, $(\mu + \mu', \alpha) = 0$. The weight μ', like all weights of $\tilde{\pi}$, has the form $\mu' = \mu + j\alpha$, where j is some integer. Substituting $\mu' = \mu + j\alpha$ in $(\mu + \mu', \alpha) = 0$, we see that $2(\mu, \alpha) + j(\alpha, \alpha) = 0$, that is, $j = -2(\mu, \alpha)(\alpha, \alpha)^{-1}$ and $\mu' = \mu + j\alpha = \mu - 2(\mu, \alpha)(\alpha, \alpha)^{-1} = S_\alpha^*(\mu)$. Therefore the set of weights of the representation π is invariant under the Weyl group W^* generated by the transformations S_α^*. The same theorem in 9.4 implies that dim $V_\mu = \dim V_{\mu'}$. Thus for any weight ν obtained from μ by a transformation in the Weyl group, we have dim $V_\mu = \dim V_\nu$. Finally, if $\mu + \alpha$ is a weight and $x \in L^\alpha$, $x \neq 0$, (9.4.7) shows that $\pi(x)V_\mu \neq \{0\}$.

We now apply these results to the highest weight λ. The functional $S_i^*(\lambda) = \lambda - \lambda(h_i)\alpha_i$ is a weight. Every weight has the form $\lambda - \sum m_k\alpha_k$, where the m_k are nonnegative integers. Therefore $\lambda(h_i)$ is a nonnegative integer. For every positive root $\alpha = \sum n_i\alpha_i$, the number $\lambda(h_\alpha) = (\lambda, \alpha) = (\lambda, \sum n_i\alpha_i) = \sum n_i\lambda(h_i)$ is a nonnegative integer.

Conversely, let λ be a linear functional on H. Let I_λ be the left ideal in the algebra U generated by the Lie algebra N_+ and elements of the form $h - \lambda(h)e$, where $h \in H$, and e is the unit of U. Consider the natural representation ρ of the Lie algebra L in the space U/I_λ, consisting of multiplication on the left by the elements of L. Let v be the image of the unit e of U in the space U/I_λ. We will show that v is a highest vector of the representation ρ with the weight λ. Since $h - \lambda(h)e \in I_\lambda$, we have $\rho(h - \lambda(h)e)v = 0$. Since $N_+ \subset I_\lambda$, we have $\rho(N_+)v = \{0\}$. Finally, every subspace of the space U/I_λ invariant under ρ is also invariant under all operators of left multiplication by elements of the algebra U. We have $Uv = U/I_\lambda$; that is, the smallest subspace of U/I_λ invariant under ρ and containing v coincides with U/I_λ. Thus v is a highest vector of the representation ρ with the weight λ.

We will show that $\rho \neq 0$, that is, $I_\lambda \neq U$. Let I_λ' be the left ideal in the algebra U_0 generated by the Lie algebra N_+ and elements of the form $h - \lambda(h)e$, $h \in H$. Let ρ_0 be the representation of U_0 that corresponds to the one-dimensional representation θ of the Lie algebra B for which $\theta(h) = \lambda(h)1$, $\theta(n) = 0$ for $h \in H$ and $n \in N_+$. (Since the linear functional θ vanishes on $[B, B] = N_+$, it is actually a representation of B.) The ideal I_λ' is contained in the kernel of ρ_0 and therefore $I_\lambda' \neq U_0$. Applying the Poincaré-Birkhoff-Witt theorem, we see that $U = U_- U_0$ does not coincide with $I_\lambda = U_- I_\lambda'$. Thus ρ is a nonzero (in general infinite-dimensional) representation of the Lie algebra L with highest weight λ.

Let T be an invariant subspace of the space $V' = U/I_\lambda$. By Proposition VIII, we have $V' = \sum_\mu V_\mu'$. By Proposition IV, T is equal to $\sum_\mu (T \cap V_\mu')$. The subspace V_λ' is one-dimensional according to VIII and thus $T \cap V_\lambda' = \{0\}$ or V_λ'. In the latter case, T contains a highest vector, that is, $T = V'$. Therefore, if $T \neq V'$, then $T \subset \sum_{\mu \neq \lambda} V_\mu'$. The sum V'' of all invariant subspaces contained in $\sum_{\mu \neq \lambda} V_\mu'$ is an invariant subspace distinct from V'. Plainly, the representation π of the Lie algebra L in the factor space $V = V'/V''$, defined by ρ, is an irreducible representation with highest weight λ.

Let ρ_1 be another irreducible representation of L with highest weight λ. Let I be the set of all $x \in U$ such that $\pi(x)v = 0$. It is clear that I contains N_+ and all elements of the form $h - \lambda(h)e$ for $h \in H$. Therefore I contains I_λ, and the representation ρ_1, isomorphic to the natural representation of L in U/I, is also isomorphic to some factor representation of ρ. Since ρ_1 is irreducible, I is a maximal left ideal of the algebra U containing I_λ and is uniquely defined. Thus ρ_1 is equivalent to π; that is, *an irreducible representation is uniquely defined by its highest weight.*

Suppose that $\lambda(h_i)$, $i = 1, \ldots, r$, are nonnegative integers. We will show that in this case the space V is finite-dimensional. Let L_i be the Lie subalgebra of L generated by the elements x_i, y_i, h_i. Let M_i be the linear span of the vector v and vectors of the form $\pi(z_1) \cdots \pi(z_n)v$, where n is an arbitrary natural number and $z_1, \ldots, z_n$ are any elements of L_i. The subspace M_i is invariant under L_i and has highest vector v. This subspace is the direct sum of weight spaces under L_i and any weight of L_i in M_i has the form $\lambda - k\alpha_i$, where k is a nonnegative integer. Thus the space M_i is generated by the vectors $(\pi(y_i))^k v$, where the k are nonnegative integers (see Proposition VIII). For $j \neq i$, we have $[x_j, y_i] = 0$, and thus $\pi(x_j)\pi(y_i)^k v = \pi(y_i)^k \pi(x_j)v$. We also have $x_j \in N_+$ and therefore $\pi(x_j)v = 0$ and $\pi(x_j)\pi(y_i)^k v = 0$. Hence the operator $\pi(x_j)$ vanishes on M_i. The space M_i is the direct sum of weight spaces and the space of the highest weight is one-dimensional (see Proposition VIII). Any proper subspace M_i' of M_i invariant under L_i is contained in the subspace $\sum_{\mu \neq \lambda} V_\mu$. Since $\pi(x_j) = 0$ on M_i' for $j \neq i$, we see that M_i' is also invariant under x_j for $j \neq i$ and thus under the Lie algebra N_+. Note that M_i' is invariant under the Cartan subalgebra H. Since $\alpha_i(h_i) = 2$, we infer that the various weights of the representation π (that is, weights of the form $(\lambda - k\alpha_i)$) in the space M_i correspond to distinct weights of the representation of L_i in M_i. Thus we have $M_i' = \sum_\mu (M_i' \cap V_\mu)$. The subspace M_i' is therefore invariant under the Lie algebra $B = H + N_+$ and M_i' is contained in $\sum_{\mu \neq \lambda} V_\mu$. We thus have $\pi(U)M_i' = \pi(U_-)\pi(U_0)M_i' \subset \pi(U_-)(\sum_{\mu \neq \lambda} V_\mu) \subset \sum_{\mu \neq \lambda} V_\mu$. Hence nonzero vectors in M_i' are not cyclic for the representation π, which contradicts the irreducibility of π if $\pi(U)M_i' \neq \{0\}$. We have proved that $\pi(U)M_i' = \{0\}$ and $M_i' = \{0\}$; that is, the representation of the Lie algebra L_i in M_i is irreducible. By Proposition III in 9.4 there is an irreducible finite-dimensional representation of L_i with highest weight λ. We already know that an irreducible representation is defined by its highest weight and thus M_i is finite-dimensional.

Let $\mathcal{T}_i$ be the set of all finite-dimensional subspaces of the space V that are invariant under L_i. If M and N are in $\mathcal{T}_i$, their sum $M + N$ is also in $\mathcal{T}_i$. Furthermore, if M belongs to $\mathcal{T}_i$, the subspace $\pi(L)M$ is finite-dimensional and we have

$$\pi(L_i)\pi(L)M \subset \pi([L, L_i])M + \pi(L)\pi(L_i)M \subset \pi(L)M, \qquad (12.1.3)$$

which shows that $\pi(L)M \in \mathcal{T}_i$. Let $\tilde{M}_i$ be the union of all subspaces in $\mathcal{T}_i$. Since $\pi(L)M \in \mathcal{T}_i$ for $M \in \mathcal{T}_i$, we infer that $\tilde{M}_i$ is invariant under L. We also have $M_i \in \mathcal{T}_i$ and $v \in M_i$, which implies that $v \in \tilde{M}_i$ and $\tilde{M}_i = V$.

Let μ be a weight of the representation π. Choose a vector v_μ in V_μ. The subspace $\sum_k V_{\mu+k\alpha_i}$ is invariant under L_i. By the foregoing, there is a finite-dimensional subspace M_i, invariant under L_i, that is contained in $\sum_k V_{\mu+k\alpha_i}$ and contains the vector v_μ. Just as for the representation π of the Lie algebra L_α in the space $\tilde V$, we conclude that $S_i^* \mu$ is a weight of π. Since the Weyl group W is generated by the transformations S_i (see Proposition VI in §11), the set P of weights of π is invariant under the Weyl group.

By Proposition X, all of the subspaces V_μ are finite-dimensional. We will show that the set P is finite, which implies that the space V is finite-dimensional. Let μ be an element of P. Choose a weight v from the finite set of weights conjugate to μ under the Weyl group, such that for all $\sigma \in W\backslash\{e\}$, $\sigma^* v - v$ is not a linear combination of the roots α_i with nonnegative integer coefficients.

Consider the weight $S_i^* v = v - v(h_i)\alpha_i$. The number $v(h_i)$ is an integer, in view of the equalities $v = \lambda - \sum m_i \alpha_i$ and $v(h_i) = \lambda(h_i) - \sum m_i \alpha_i(h_i)$, in which as we know $\lambda(h_i)$, the m_i and the $\alpha_i(h_i)$ are integers. Consequently $v(h_i)$ is nonnegative for all $i = 1, \ldots, r$. That is, every weight μ is conjugate to a weight v for which the $v(h_i)$ are nonnegative integers for $i = 1, \ldots, r$. We write $v = \lambda - \sum m_i \alpha_i$ and $\beta = \sum m_i \alpha_i$, obtaining

$$(\lambda, \lambda) = (v, v) + (\beta, \beta) + 2(v, \beta), \tag{12.1.4}$$

where

$$(v, \beta) = \sum m_i(v, \alpha_i) = (1/2) \sum m_i v(h_i)(\alpha_i, \alpha_i) \geq 0. \tag{12.1.5}$$

Therefore (12.1.4) implies that

$$(\lambda, \lambda) \geq (v, v). \tag{12.1.6}$$

The weight v is an element of the integral lattice of functionals $\lambda - \sum m_j \alpha_j$ in the space H_0^*, that belong to the ball with radius $\sqrt{(\lambda, \lambda)}$ and center at the origin. Thus the number of weights v is finite. Any weight is conjugate under the Weyl group (which is of course finite) to one of the weights v. Hence P is finite. $\square$

§13. Characters of Finite-Dimensional Irreducible Representations of a Semisimple Lie Algebra

13.1. The Definition of a Character and Its Simplest Properties

Let π be an irreducible linear representation of a Lie algebra L in a finite-dimensional complex linear space V. The representation π determines a representation, also denoted by π, of the enveloping algebra U of the Lie algebra L in the same space V (see Proposition II in 6.3). We define

$$\chi(x) = (\dim V)^{-1} \operatorname{tr} \pi(x) \tag{13.1.1}$$

for all $x \in U$. Clearly (13.1.1) defines a linear function χ on the algebra U. This function χ is called *the character*[56] *of the representation.*

I. *The representation π is uniquely defined by its character to within equivalence.*

Proof. By Burnside's Theorem (see Naĭmark [2]), the operators $\pi(x)$, $x \in U$, are the algebra of all linear operators on the space V. Let I be the set of all $a \in U$ such that $\chi(ax) = 0$ for all $x \in U$. For $a \in I$, we have tr $\pi(ax) = \mathrm{tr}(\pi(a)\pi(x))$ $= 0$ for all $x \in U$. This implies that

$$\mathrm{tr}(\pi(a)T) = 0 \qquad (13.1.2)$$

for all linear operators T in V. It follows that

$$\pi(a) = 0. \qquad (13.1.3)$$

This equality implies that the factor algebra U/I of U by the ideal I is isomorphic to the algebra $L(V)$ of all linear operators on the space V.

Let π and π' be representations of the Lie algebra L in the spaces V and V', with equal characters. In this case the algebras $L(V)$ and $L(V')$ are isomorphic, since they are both isomorphic to the algebra U/I. It is known (see for example Naĭmark [2]) that every isomorphism between $L(V)$ and $L(V')$ is generated by an isomorphism between the spaces V and V'. This isomorphism also realizes the equivalence of the representations π and π'. $\qquad \square$

13.2. Some Properties of the Enveloping Algebra

We will show that a character χ is determined by its restriction to the subalgebra $U(H) \subset U(L)$, that is, to the universal enveloping algebra of the Cartan subalgebra H of L.

I. *Every element $x \in U$ has the form $h + \sum_{i=1}^{m} [x_i, y_i]$, where $h \in U(H)$, $x_i \in L$, $y_i \in U$. Thus the linear space U is the sum of its subspaces $U(H)$ and $[L, U]$. It is also true that $U = U(H) + [U, U]$.*

Proof. Consider the symmetric algebra S over the space L. This is the universal enveloping algebra of the abelian Lie algebra for which $[x, y]_1 = 0$ for all x, y in the linear space L. The algebra S is the factor algebra of the tensor algebra T over L by the ideal J which is generated by elements $x \otimes y - y \otimes x$ $(x, y \in L)$. Let M be the space of symmetric tensors on L. Applying the operation of symmetrization to all elements of the tensor algebra T, we see that the space T is the direct sum of the spaces J and M. Thus the canonical mapping of T onto $T/J = S$ produces an isomorphism f of the space M onto S. We identify the spaces M and S by means of the isomorphism f.

---

[56] Sometimes this function $\chi(x)$ is called a *normalized character of the representation* π and the function $\tilde{\chi}(x) = \mathrm{tr}\,\pi(x)$, $x \in V$ the *character of the representation* π.

We extend the adjoint representation of the Lie algebra L to the representation of L in the space T by defining

$$\text{ad } x(x_1 \otimes \cdots \otimes x_n) = \sum_{i=1}^{n} x_1 \otimes \cdots \otimes [x, x_i] \otimes \cdots \otimes x_n. \quad (13.2.1)$$

The operator ad x obviously leaves the subspaces J and M invariant. Thus the restriction of the operators ad x to the subspace M defines a representation ρ of L in M. The isomorphism f of the space M onto S defines a representation σ of L in the space S equivalent to ρ: $\sigma = f\sigma f^{-1}$.

Let I be the ideal in the tensor algebra T generated by the elements $\varphi_{x,y} = x \otimes y - y \otimes x - [x, y]$. From (13.2.1) we see that

$$
\begin{aligned}
(\text{ad } z)(\varphi_{x,y}) &= [z, x] \otimes y + x \otimes [z, y] - [z, y] \otimes x \\
&\quad - y \otimes [z, x] - [z, [x, y]] \\
&= \varphi_{(\text{ad } z)(x), y} + \varphi_{x, (\text{ad } z)(y)}.
\end{aligned}
\quad (13.2.2)
$$

These identities show that I is invariant under all operators ad x, $x \in L$. Let τ denote the representation of L in the factor algebra $U = T/I$ defined by the representation $z \to \text{ad } z$ of L in T. The Poincaré-Birkhoff-Witt Theorem shows that the space T is the direct sum of the subspace M and the ideal I. The canonical mapping of T onto $U = T/I$ defines an isomorphism of M and U. Plainly this isomorphism realizes the equivalence of the representations ρ and τ of L in the spaces M and U respectively. Hence the representations σ and τ are also equivalent. The isomorphism between U and S that realizes the equivalence of τ and σ obviously carries the subspace $U(H) \subset U$ onto the subspace $S(H) \subset S$, where $S(H)$ is the symmetric algebra over H.

Let R be the smallest subspace of S that contains $S(H)$ and is invariant under the representation σ of L. Let h be an element of H such that $\alpha(h) \neq 0$ for all roots α. Let $\alpha_1, \ldots, \alpha_p$ be a certain set of roots. Let e_{α_i} be in L^{α_i}. Define the numbers n_i by the equalities $[e_{\alpha_1}, e_{\alpha_i}] = n_i e_{\alpha_1 + \alpha_i}$. The equality (13.2.1) yields the identity

$$
\begin{aligned}
(\text{ad } e_{\alpha_1})(e_{\alpha_2} \cdots e_{\alpha_p} h^{n-p+1}) &= \sum n_i e_{\alpha_2} \cdots e_{\alpha_1 + \alpha_i} \cdots e_{\alpha_p} h^{n-p+1} \\
&\quad - (-p + n + 1)\alpha_1(h) e_{\alpha_1} \cdots e_{\alpha_p} h^{n-p}. \quad (13.2.3)
\end{aligned}
$$

The subspace R contains $S(H)$; in particular, h^{n+1} belongs to R. From (13.2.3) and induction on p, we see that the subspace R (which is invariant under ad L) contains all elements $e_{\alpha_1}, \ldots, e_{\alpha_p} h^{n-p}$, $p = 1, \ldots, n$. (Recall that $\alpha(h) \neq 0$ for all roots α.) Every element h of H is the difference of elements h_1 and h_2 such that $\alpha(h_1) \neq 0$ and $\alpha(h_2) \neq 0$ for all nonzero roots α. (To prove this, observe that the set of $h \in H$ such that $\alpha(h) = 0$ for some nonzero root α is a finite union of hyperplanes in H.) The space $S(H)$ is generated by powers of the elements $h \in H$ such that $\alpha(h) \neq 0$ for nonzero roots α. Hence the subspace R contains all elements $e_{\alpha_1} \cdots e_{\alpha_p} h^k$ for *all* h in H. Hence R and S coincide. By the definition of R, every element of R is the sum of an element of $S(H)$ and a linear combination of elements of the subspaces of the form $\sigma(x_1)(\sigma(x_2)(\cdots(\sigma(x_k)(S(H)))\cdots), x_1, \ldots, x_k \in L$. Thus we have $S = R \subset S(H) + \sum_{k \geqslant 1}(\sigma(L))^k(S(H))$. Applying the isomorphism between S and U, we see that $U \subset U(H) + \sum_{k \geqslant 1}(\text{ad } L)^k(U(H))$, that is, $U \subset U(H) + [L, U] \subset U(H) + [U, U] \subset U$ and $U = U(H) + [L, U] = U(H) + [U, U]$. $\square$

II. *A character χ is determined by its restriction to the subalgebra $U(H)$.*

Proof. This proposition follows from Proposition I. The commutator of two linear operators has trace zero. For all $x \in U$, we thus have

$$\chi(x) = (\dim V)^{-1} \operatorname{tr} \pi(x) = (\dim V)^{-1} \left(\operatorname{tr} \pi\left(h + \sum_{i=1}^{m} [x_i, y_i] \right) \right)$$

$$= (\dim V)^{-1} \operatorname{tr}\left(\pi(h) + \sum_{i=1}^{m} [\pi(x_i), \pi(y_i)] \right) = (\dim V)^{-1} \operatorname{tr} \pi(h) = \chi(h),$$

if

$$x = h + \sum_{i=1}^{m} [x_i, y_i]. \quad \square$$

13.3. Some Auxiliary Propositions

Let V and V' be finite-dimensional vector spaces dual to each other via a bilinear form (v, v'). Let $S^m(V)$ be the linear span of formal products $x_1 x_2 \cdots x_m$, where $x_i \in V$ and the order of the elements $x_1, \ldots, x_m$ in $x_1 x_2 \cdots x_m$ plays no role. We define

$$(v_1 \cdots v_m, v_1' \cdots v_m') = \sum_{\sigma \in S_m} \prod_{i=1}^{m} (v_i, v_{\sigma(i)}'), \tag{13.3.1}$$

where S_m is the group of all permutations of the indices $1, 2, \ldots, m$. Since the right side of (13.3.1) is polylinear and symmetric separately for v_i and v_i', (13.3.1) defines a bilinear form on $S^m(V) \times S^m(V')$. A special case of (13.3.1) is

$$(v_1 \cdots v_m, v'^m) = m! \prod_{i=1}^{m} (v_i, v') \tag{13.3.2}$$

for all $v_1, \ldots, v_m \in V$ and $v' \in V$. For every $v' \in V'$, we define a linear functional $e^{v'}$ on $S(V)$ (the direct sum of the spaces $S^m(V)$) by setting

$$(v_1 \cdots v_p, e^{v'}) = \prod_{i=1}^{p} (v_i, v') \tag{13.3.3}$$

for all positive integers p and all $v_1, \ldots, v_p \in V$. From (13.3.2) we obtain the formal equality

$$e^{v'} = \sum_{m \geq 0} (m!)^{-1} v'^m, \qquad (13.3.4)$$

under the hypothesis that v'^m vanishes on $S^n(V)$ for $n \neq m$. From (13.3.3) we conclude that the mapping $x \to (x, e^{v'})$ of the algebra $S(V)$ into the field of complex numbers is a homomorphism that maps 1 into 1 and v into (v, v'). Furthermore, (13.3.4) and the binomial formula give

$$e^{v_1'} e^{v_2'} = e^{v_1' + v_2'}, \qquad (13.3.5)$$

where the product on the left side is the product of homomorphisms of $S(V)$ onto the field of complex numbers; that is, $(e^{v_1'} e^{v_2'})(x) = e^{v_1'}(x) e^{v_2'}(x)$ for all $x \in S(V)$.

Let $\alpha_1, \ldots, \alpha_r$ be the system of simple roots of the Lie algebra L. Suppose that $\alpha_i(h) = (h, h_i)$ for all $h \in H$. Let P be the additive group of linear functionals λ on H such that the values $\lambda(h_i)$ are integers. Let P_+ be the set of all $\lambda \in P$ such that $\lambda(h_i) \geq 0$ for all $i = 1, \ldots, r$. Let A be the algebra of complex functions on P vanishing off of finite sets, in which multiplication is defined as convolution:

$$(f * g)(a) = \sum_{b+c=a} f(b) g(c). \qquad (13.3.6)$$

Let e^λ be the function on P equal to one at λ and zero elsewhere. The set $\{e^\lambda, \lambda \in P\}$ is a basis in A and we have

$$e^\lambda * e^\mu = e^{\lambda + \mu}. \qquad (13.3.7)$$

This identity allows us to identify the element $e^\lambda \in A$ with the functional e^λ on $S(H)$ defined by (13.3.4).

Let B be the subalgebra of A consisting of linear combinations of the elements e^λ, where $\lambda \in P_+$. We set $x_i = e^{\lambda_i}$, where λ_i is a functional such that $\lambda_i(h_j) = 1$ for $i = j$ and $\lambda_i(h_j) = 0$ for $i \neq j$.

The equality (13.3.7) shows that every element e^λ with $\lambda \in P_+$ has the form $e^\lambda = (e^{\lambda_1})^{\lambda(h_1)} \cdots (e^{\lambda_r})^{\lambda(h_r)} = x_1^{\lambda(h_1)} \cdots x_r^{\lambda(h_r)}$, where the $\lambda(h_i)$ are nonnegative integers. Since the elements e^λ, $\lambda \in P_+$ are a basis in B, the algebra B is isomorphic to the algebra of all polynomials in the variables $x_1, \ldots, x_r$ with complex coefficients. Similarly, the algebra A can be identified with the algebra of all complex polynomials in the variables $x_1, \ldots, x_r, x_1^{-1}, \ldots, x_r^{-1}$.

Let W be the Weyl group of the Lie algebra L, and W^* the set of transformations conjugate to the transformations of the group W. For $w \in W^*$, we define

$$\tilde{w}(e^\lambda) = e^{w(\lambda)}, \qquad \lambda \in P \qquad (13.3.8)$$

and extend the transformation $\tilde{w}$ linearly over the entire algebra A. The Weyl group consists of orthogonal transformations. Therefore the determinant of every transformation w in W^* is ± 1. An element a of A is called *symmetric* if $\tilde{w}(a) = a$ for all $w \in W^*$ and *skew-symmetric* if $\tilde{w}(a) = (\det w)a$ for all $w \in W^*$.

An element $a \in A$ is symmetric (skew-symmetric) if and only if $S_i^* a = a$ ($S_i^* a = -a$) for all $S_i^* = S_{\alpha_i}^*$, $i = 1, \ldots, r$. This is implied by the fact that $\det S_i^*$ is -1, since the elements S_i generate the entire Weyl group W (see §11).

We introduce the *operation of alternation* in the algebra A, setting

$$\mathrm{Alt}(a) = \sum_{w \in W^*} (\det w)\tilde{w}(a). \tag{13.3.9}$$

I. *The element* $\mathrm{Alt}(a)$ *is skew-symmetric for all* $a \in A$.

Proof. For $w_0 \in W$, we have

$$\tilde{w}_0(\mathrm{Alt}(a)) = \sum_{w \in W^*} (\det w)\tilde{w}_0\tilde{w}(a) = \sum_{w \in W^*} \det(w_0^{-1}w)\tilde{w}(a)$$

$$= \det(w_0^{-1}) \sum_{w \in W^*} (\det w)\tilde{w}(a) = \det(w_0^{-1})\,\mathrm{Alt}(a) = \det(w_0)\,\mathrm{Alt}(a). \qquad \square$$

If a is a skew-symmetric element, (13.3.9) implies that $\mathrm{Alt}(a) = |W|a$, where $|W|$ is the order of the Weyl group. Therefore the mapping $a \to |W|^{-1}\,\mathrm{Alt}(a)$ is a mapping of the algebra A onto the set C of skew-symmetric elements of A. Plainly C is a linear subspace of A. Every element $x \in C$ is a linear combination of the elements $\mathrm{Alt}(e^\lambda) = \sum_{w \in W^*} (\det w)e^{w(\lambda)}$. It is clear that $\mathrm{Alt}(e^{w(\lambda)}) = (\det w)\,\mathrm{Alt}(e^\lambda)$ for all $\lambda \in P$ and $w \in W^*$. In constructing a basis in the space C of elements $\mathrm{Alt}(e^\lambda)$, we can restrict ourselves to elements e^λ with linear functionals λ such that $\lambda \geq w(\lambda)$ for all $w \in W^*$. We prove this. For every $\lambda \in P$, the finite set $w(\lambda)$, $w \in W^*$, contains the largest functional λ_0 under the lexicographic ordering, and $\mathrm{Alt}(e^{\lambda_0})$ is a scalar multiple of $\mathrm{Alt}(e^\lambda)$.

We will study properties of the functionals $\lambda \in P$ for which $w(\lambda) \leq \lambda$ for all $w \in W^*$.

II. *The symmetry* S_i^* *permutes positive roots distinct from* α_i *among themselves. Also we have* $S_i^*(\alpha_i) = -\alpha_i$.

Proof. Let $\alpha = \sum_{k=1}^r m_k \alpha_k$ be positive. We then have

$$S_i^* \alpha = \alpha - 2(\alpha, \alpha_i)(\alpha_i, \alpha_i)^{-1}\alpha_i = (m_i - 2(\alpha, \alpha_i)(\alpha_i, \alpha_i)^{-1})\alpha_i + \sum_{k \neq i} m_k \alpha_k.$$

Since $\{\alpha_k\}$ is the system of simple roots, all coefficients in the expansion of $S_i^*\alpha$ in this system must be of the same sign. If $m_i < 2(\alpha, \alpha_i)(\alpha_i, \alpha_i)^{-1}$, it follows that $m_k \leqslant 0$ for $k \neq i$. But α is a positive root and therefore $m_k \geqslant 0$. Thus we have $m_k = 0$ for $k \neq i$, and so α is a scalar multiple of α_i. Proposition III in 9.5 implies that $\alpha = \alpha_i$ and thus $S_i^*(\alpha_i) = \alpha_i - 2(\alpha_i, \alpha_i)(\alpha_i, \alpha_i)^{-1}\alpha_i = -\alpha_i$. If on the other hand $m_i \geqslant 2(\alpha, \alpha_i)(\alpha_i, \alpha_i)^{-1}$, all coefficients in the expansion of $S_i^*(\alpha)$ in the system $\{\alpha_k\}$ are nonnegative and $S_i^*(\alpha)$ is a positive root. $\square$

III. *If we have $\lambda \geqslant S_i^*(\lambda)$, then $\lambda(h_i) \geqslant 0$.*

Proof. We have $S_i^*(\lambda) = \lambda - 2(\lambda, \alpha_i)(\alpha_i, \alpha_i)^{-1}\alpha_i$ and therefore the condition $\lambda \geqslant S_i^*(\lambda)$ is equivalent to $(\lambda, \alpha_i) \geqslant 0.$, which is to say $\lambda(h_i) \geqslant 0$. $\square$

IV. *Let λ be in P. The condition $w(\lambda) < \lambda$ holds for all $w \in W^*$ different from 1 if and only if $\lambda(h_i)$ is positive for $i = 1, \ldots, r$.*

Proof. If $\lambda(h_i) \leqslant 0$, we find $S_i^*(\lambda) = \lambda - \lambda(h_i)\alpha_i \geqslant \lambda$. Conversely, suppose that $\lambda(h_i) > 0$ for all $i = 1, \ldots, r$. We will show that $w(\lambda) < \lambda$ for all $w \in W^*$ different from 1. This inequality holds for $w = S_i^*$, since $\lambda > S_i^*(\lambda) = \lambda - \lambda(h_i)\alpha_i$. Suppose that $p > 1$ and that the inequality $\lambda > w(\lambda)$ holds for all $w \in W^*$ that are products of fewer than p symmetries S_i^*. Consider an element $w = S_{i_1}^* \cdots S_{i_p}^* = w_0 S_{i_p}^*$, where $w_0 = S_{i_1}^* \cdots S_{i_{p-1}}^*$. We have $w(\lambda) = w_0(S_{i_p}^*\lambda) = w_0(\lambda) - \lambda(h_{i_p})w_0(\alpha_{i_p})$. If $w_0(\alpha_{i_p})$ is a positive root, we get $w(\lambda) < w_0(\lambda) < \lambda$. If $w_0(\alpha_{i_p})$ is a negative root, let k be the smallest number with the property that $S_{i_l}^* \cdots S_{i_{p-1}}^*(\alpha_{i_p}) > 0$ for all $l \geqslant k$. Since $w_0(\alpha_{i_p}) < 0$, the number k exceeds 1. We may suppose that $S_{i_{p-1}}^* \neq S_{i_p}^*$. (In the opposite case we can shorten the expression for w, using the identity $S_{i_p}^{*2} = 1$. Proposition II implies that $S_{i_{p-1}}^*(\alpha_{i_p}) > 0$; that is, $k \leqslant p - 1$. By our definition, we have $S_{i_k}^* \cdots S_{i_{p-1}}^*(\alpha_{i_p}) > 0$ and $S_{i_{k-1}}^*(S_{i_k}^* \cdots S_{i_{p-1}}^*(\alpha_{i_p})) < 0$. Proposition II shows that $S_{i_k}^* \cdots S_{i_{p-1}}^*(\alpha_{i_p}) = \alpha_{i_{k-1}}$. Setting $w_0' = S_{i_1}^* \cdots S_{i_{k-2}}^*$ and $w_0'' = S_{i_k}^* \cdots S_{i_{p-1}}^*$, we see that $w_0 = w_0' S_{i_{k-1}}^* w_0''$, where $w_0''(\alpha_{i_p}) = \alpha_{i_{k-1}}$. It is clear that $w_0'' S_\alpha^*(w_0'')^{-1} = S_{w_0''(\alpha)}^*$ and thus $w_0'' S_{i_p}^*(w_0'')^{-1} = S_{i_{k-1}}^*$, $w_0'' S_{i_p}^* = S_{i_{k-1}}^* w_0''$ and $w = w_0 S_{i_p}^* = w_0' S_{i_{k-1}}^* w_0'' S_{i_p}^* = w_0' S_{i_{k-1}}^{*2} w_0'' = w_0' w_0''$. Hence w is the product of $p - 2$ symmetries S_i^*, and the inequality $w(\lambda) < \lambda$ follows from our inductive hypothesis. $\square$

V. *If $\lambda(h_i) = 0$ for some $i = 1, \ldots, r$, then $\mathrm{Alt}(e^\lambda)$ is zero.*

Proof. Our hypothesis implies that $S_i^*(\lambda) = \lambda$. Let W' be a subset of W containing exactly one element of every coset of the subgroup $H = \{1, S_i\}$. We find

$$\mathrm{Alt}(e^\lambda) = \sum_{w \in (W')^*} \{(\det w)e^{w(\lambda)} + \det(wS_i^*)e^{(wS_i^*)(\lambda)}\} = 0,$$

since $\det S_i^* = -1$. $\square$

VI. *Every skew-symmetric element of the algebra* A *is a linear combination of elements* $\mathrm{Alt}(e^\lambda)$ *such that* $\lambda(h_i) > 0$ *for all* $i = 1, \ldots, r$.

Proof. Every a in C is a linear combination of elements $\mathrm{Alt}(e^\lambda)$. We may suppose that $\lambda \geqslant w(\lambda)$ for $w \in W^*$. To complete the proof, we need only apply Propositions III, IV, and V. □

VII. *Let* α *be a nonzero root and* a *an element of* C *such that* $\widetilde{S}^*_\alpha a = -a$. *Then* a *has the form* $(1 - e^{-\alpha})b$ *for some* $b \in A$.

Proof. We have $(1 - \widetilde{S}^*_\alpha)a = 2a$. Setting $a = \sum a_\lambda e^\lambda$ $(a_\lambda \in C)$ and applying (13.3.8) and (13.3.5), we obtain

$$a = (1/2)(1 - \widetilde{S}^*_\alpha)a = (1/2) \sum a_\lambda (1 - \widetilde{S}^*_\alpha)e^\lambda$$
$$= \sum (a^\lambda/2)(e^\lambda - e^{\lambda - \lambda(h_\alpha)\alpha}) = \sum (a_\lambda/2)e^\lambda \{1 - (e^{-\alpha})^{\lambda(h_\alpha)}\},$$

the numbers $\lambda(h_\alpha)$ being integers. Since $1 - (e^{-\alpha})^{\lambda(h_\alpha)} = (1 - e^{-\alpha})b_\lambda$ for certain elements $b_\lambda \in A$, it follows that $a = (1 - e^{-\alpha}) \sum (a_\lambda/2)e^\lambda b_\lambda$. □

The partial sum of all positive roots of the Lie algebra L is denoted by ρ.

VIII. *If* $a \in C$, *then* $a = Db$, *where* $b \in A$ *and*

$$D = e^\rho \prod_{\alpha \geqslant 0} (1 - e^{-\alpha}) = \prod_{\alpha > 0} (e^{\alpha/2} - e^{-\alpha/2}). \qquad (13.3.10)$$

Proof. Recall that the algebra A is isomorphic to the algebra of polynomials in the variables $x_1, \ldots, x_r, x_1^{-1}, \ldots, x_r^{-1}$. The elements e^α correspond to certain monomials in A. The elements $x_i - 1$, $i = 1, \ldots, r$, are polynomials in $x_1, \ldots, x_r$ and clearly are irreducible elements[57] of the algebra of all such polynomials. For any $\alpha > 0$, the element e^α goes into e^{α_i} under some transformation w of the Weyl group. Elements $e^\alpha - 1$ of the subalgebra $B \subset A$ are irreducible elements of A. We prove this as follows. If $e^\alpha - 1 = a_1 a_2$, where $a_1, a_2 \in A$, then $e^{\alpha_i} - 1 = b_1 b_2$, where $b_1 = \tilde{w}(a_1)$ and $b_2 = \tilde{w}(a_2)$. Thus we have $x_1 - 1 = f_1(x)f_2(x)$, where f_1 and f_2 are polynomials in $x_1, \ldots, x_r, x_1^{-1}, \ldots, x_r^{-1}$. Multiplying the last equality by a large power of the monomial $x_1 \cdots x_r$, we see that $x_1^{2N} \cdots x_r^{2N}(x_i - 1) = P_1(x)P_2(x)$, where $P_i = x_1^N \cdots x_r^N f_i(x)$, $i = 1, 2$, are polynomials in $x_1, \ldots, x_r$. The unique factorization theorem for the ring of polynomials in n variables (see the book [1] of A. G. Kuroš, §51, page 315) implies that one of the polynomials P_1 and P_2 is a monomial in $x_1, \ldots, x_r$. Therefore either f_1

---

[57] An element x of a commutative algebra A is called *irreducible* or *prime*, if given a factorization $x = x_1 x_2$, $x_1 \in A$, $x_2 \in A$, at least one of the elements x_1 and x_2 admits an inverse in A.

or f_2 is a monomial in $x_1, \ldots, x_r, x_1^{-1}, \ldots, x_r^{-1}$. Hence it is an invertible element of A.

Suppose that $a \in C$. Since $\det S_\alpha^* = -1$, we have $S_\alpha^* a = -a$ for all $\alpha > 0$. Proposition VII shows that for all $\alpha > 0$, a is a product $(1 - e^{-\alpha})b_\alpha$, where $b_\alpha \in A$. The elements $(1 - e^{-\alpha}) = e^{-\alpha}(e^\alpha - 1)$ are relatively prime for distinct $\alpha > 0$. Again we apply unique factorization in a polynomial ring to represent a in the form $a = \prod_{\alpha>0} (1 - e^{-\alpha})b_1 = e^\rho \prod_{\alpha \geqslant 0} (1 - e^{-\alpha})(e^{-\rho}b_1) = Db$, where b_1 and b belong to A. $\quad\square$

Let us compute $S_i^* D$. Factors of D distinct from $e^{\alpha_i/2} - e^{-\alpha_i/2}$ are permuted among themselves (Proposition II). The element $e^{\alpha_i/2} - e^{-\alpha_i/2}$ goes into $e^{-\alpha_i/2} - e^{\alpha_i/2} = -(e^{\alpha_i/2} - e^{-\alpha_i/2})$. Therefore we have $S_i^* D = -D$ for all $i = 1, \ldots, r$, and so D belongs to C.

IX. *We have* $D = \mathrm{Alt}(e^\rho)$.

Proof. Definition (13.3.10) shows that D is a linear combination of elements e^λ with exponents λ such that $\rho \geqslant \lambda \geqslant -\rho$. Thus if we write $D = \sum c_\lambda \mathrm{Alt}(e^\lambda)$, we may extend the sum on the right side only over λ such that $\rho \geqslant w(\lambda) \geqslant -\rho$ for all $w \in W^*$. For, assume that there are $\lambda \in P$ and $w \in W^*$ such that $c_\lambda \neq 0$ holds but $\rho \geqslant w(\lambda) \geqslant -\rho$ does not. The equality

$$\mathrm{Alt}(e^\lambda) = \sum_{w \in W^*}(\det w)e^{w(\lambda)}$$

and the definition of the numbers c_λ imply that there is an element e^μ with exponent $\mu = w(\lambda)$ in the expansion of D in the basis $\{e^\lambda\}$ that fails to satisfy $\rho \geqslant \mu \geqslant -\rho$. On the other hand, every element $\mathrm{Alt}(e^\lambda)$ is skew-symmetric and Proposition VIII implies that $\mathrm{Alt}(e^\lambda) = Db_\lambda$, where $b_\lambda \in A$.

Compare the highest and lowest terms of the left and right sides of the equality $\mathrm{Alt}(e^\lambda) = Db_\lambda$. The highest term e^μ of the product Db_λ is equal to the product of the highest terms in D and in b_λ; that is, $\mu \geqslant \rho$. Similarly, the lowest term e^ν of the product Db_λ is equal to the product of the lowest terms in D and in b_λ; that is $\nu \leqslant -\rho$. By hypothesis, the highest and lowest terms of $\mathrm{Alt}(e^\lambda)$ have exponents between ρ and $-\rho$. Thus the highest and lowest terms in b_λ are constants; that is, b_λ is a constant, and λ is obtained from ρ by a transformation of the Weyl group. Hence we may suppose that $\lambda = \rho$. Thus D is a scalar multiple of $\mathrm{Alt}(e^\rho)$. Since the highest terms of D and $\mathrm{Alt}(e^\rho)$ are equal to e^ρ, the equality $D = \mathrm{Alt}(e^\rho)$ follows. $\quad\square$

X. *We have* $\rho(h_i) = 1$.

Proof. We have

$$S_i^* \rho = \rho - \rho(h_i)\alpha_i. \tag{13.3.11}$$

On the other hand, Proposition II implies that

$$S_i^* \rho = S_i^*((1/2) \sum \alpha_j) = (1/2)S_i^*(\sum \alpha_j)$$
$$= (1/2)(\sum \alpha_j - 2\alpha_i) = \rho - \alpha_i. \qquad (13.3.12)$$

Compare the right sides of (13.3.11) and (13.3.12) to see that $\rho(h_i) = 1$. $\quad\square$

XI. *A symmetric element a of the algebra A is a linear combination of elements of the form* $\mathrm{Alt}(e^{\lambda+\rho})/\mathrm{Alt}(e^{\rho})$, *where* λ *belongs to* P_+.

Proof. The element aD is skew-symmetric. Proposition VI gives it a representation $aD = \sum c_\mu \, \mathrm{Alt}(e^\mu)$, where the $\mu(h_i)$ are positive integers for all $i = 1, \ldots, r$. Define λ by $\mu = \lambda + \rho$. Proposition X implies that $\lambda(h_i) = \mu(h_i) - \rho(h_i) = \mu(h_i) - 1 \geqslant 0$. Therefore λ is in P_+ and $aD = \sum c_{\lambda+\rho} \, \mathrm{Alt}(e^{\lambda+\rho})$, where $\lambda \in P_+$. Every element $\mathrm{Alt}(e^{\lambda+\rho})$ is skew-symmetric, by Proposition I. Propositions VIII and IX show that $\mathrm{Alt}(e^{\lambda+\rho}) = a_\lambda D = a_\lambda \, \mathrm{Alt}(e^\rho)$, where a_λ is an element of A. This implies that $aD = \sum c_{\lambda+\rho} a_\lambda D$, that is, $a = \sum c_{\lambda+\rho} a_\lambda$, where $a_\lambda = \mathrm{Alt}(e^{\lambda+\rho})/\mathrm{Alt}(e^\rho)$. $\quad\square$

13.4. Weyl's Character Formula

We now turn to the calculation of the character of an irreducible linear representation π of the Lie algebra L in the finite-dimensional vector space V. Plainly $\pi([x, y]) = [\pi(x), \pi(y)]$ has trace zero for all $x, y \in U(L) = U$. Proposition II in 13.2 shows that the character of π is determined by the quantities $\mathrm{tr}\,\pi(h)$ for $h \in U(H)$.

Let λ be the highest weight of the representation π. Let n_μ be the multiplicity of the weight μ in π, i.e., the dimension of the space V_μ. Consider an element $h \in U(H) \subset U(L)$. The operator $\pi(h)$ is a scalar operator on the subspace V_μ. The eigenvalue of $\pi(h)$ on V_μ defines a homomorphism of the algebra $U(H)$ onto the field $\mathbf{C}$. Since H is an abelian Lie algebra, $U(H)$ and $S(H)$ are canonically isomorphic. The homomorphism of the algebra $S(H)$ into $\mathbf{C}$ that carries 1 to 1 and elements $h \in H$ to $\mu(h)$ is the homomorphism $e^\mu = \sum_{m \geqslant 0} (m!)^{-1} \mu^m$. Thus the eigenvalue of the operator $\pi(h)$ in the subspace V_μ is (h, e^μ), and we have

$$\mathrm{tr}\,\pi(h) = \sum_\mu n_\mu(h, e^\mu) = \left(h, \sum_\mu n_\mu e^\mu \right); \qquad (13.4.1)$$

note that $\dim V = \sum_\mu n_\mu$.
Consider the element

$$\varphi_\lambda = \sum_\mu n_\mu e^\mu \qquad (13.4.2)$$

of the algebra A. If μ_1 and μ_2 are conjugate under the Weyl group, we have $n_{\mu_1} = n_{\mu_2}$, by theorem 1 in §12. Therefore φ_λ is symmetric. Proposition XI in 13.3 shows that φ_λ is a linear combination of elements $\mathrm{Alt}(e^{\mu+\rho})/\mathrm{Alt}(e^\rho)$.

I. *For all $\alpha \neq 0$, choose $e_\alpha \in L^\alpha$. Let $t(\mu, \alpha)$ be the trace of the restriction of the operator $\pi(e_\alpha)\pi(e_{-\alpha})$ to the subspace V_μ. If $(e_\alpha, e_{-\alpha}) = 1$, we have*

$$t(\mu, \alpha) - t(\mu + \alpha, \alpha) = n_\mu(\mu, \alpha). \tag{13.4.3}$$

Proof. Let E be the sum of the subspaces V_λ and $V_{\lambda+\alpha}$. We define linear operators P_+, P_- on E by $P_+x = \pi(e_\alpha)x$, $P_+y = 0$; $P_-x = 0$, $P_-y = \pi(e_{-\alpha})y$ for all $x \in V_\lambda$ and $y \in V_{\lambda+\alpha}$. We then obtain

$$[P_+, P_-]x = -\pi(e_{-\alpha})\pi(e_\alpha)x = (\lambda, \alpha)x - \pi(e_\alpha)\pi(e_{-\alpha})x;$$
$$[P_+, P_-]y = \pi(e_\alpha)\pi(e_{-\alpha})y.$$

The trace of the operator $[P_+, P_-]$ is zero and is also equal to $n_\mu(\lambda, \alpha) - t(\mu, \alpha) + t(\mu + \alpha, \alpha)$, which proves (13.4.3). $\square$

Choose $k_i \in H$ so that $(h_j, k_i) = \delta_{ij}$, where $\delta_{ij} = 1$ is Kronecker's symbol. The definition (6.4.2) shows that the element $z = \sum_\alpha e_\alpha e_{-\alpha} + \sum_i h_i k_i$ is the Casimir element corresponding to the adjoint representation of the Lie algebra L. The element z is in the center of the algebra $U = U(L)$. Thus the irreducibility of π and Schur's lemma imply that the operator $\pi(z) = c \cdot 1$. Computing the trace of the operator $\pi(z)$ in the subspace V_μ, we obtain

$$cn_\mu = \mathrm{tr}_{v_\mu}\pi(z) = \sum \mathrm{tr}_{V_\mu}(\pi(e_\alpha)\pi(e_{-\alpha})) + \sum \mathrm{tr}_{V_\mu}(\pi(h_i)\pi(k_i))$$
$$= \sum_\alpha t(\mu, \alpha) + \sum_i n_\mu\mu(h_i)\mu(k_i). \tag{13.4.4}$$

From the definition $(h_i, k_j) = \delta_{ij}$, we see that $\sum_i \mu(h_i)\mu(k_i) = (\mu, \mu)$ and thus (13.4.4) implies that

$$cn_\mu = \sum_\alpha t(\mu, \alpha) + n_\mu(\mu, \mu),$$

i.e.,

$$\sum_\alpha t(\mu, \alpha) = n_\mu(c - (\mu, \mu)). \tag{13.4.5}$$

Let Q be the tensor product $A \otimes H^*$. We define in $Q \times Q$ a "scalar product" with values in A by

$$(a \otimes \lambda, b \otimes \mu) = ab(\lambda, \mu), \tag{13.4.6}$$

and extend it linearly over all of Q. We also define

$$a(b \otimes \lambda) = (ab) \otimes \lambda. \tag{13.4.7}$$

This rule, extended linearly over Q, defines a representation of the algebra A in the space Q. We define

$$\varDelta(e^\lambda) = (\lambda, \lambda)e^\lambda \tag{13.4.8}$$

and

$$g(e^\lambda) = e^\lambda \otimes \lambda \tag{13.4.9}$$

for all $e^\lambda \in A$. We extend the mappings $\varDelta$ and g linearly to linear operators from the space A to the spaces A and Q respectively. It is easy to verify that

$$g(ab) = bg(a) + ag(b), \tag{13.4.10}$$
$$\varDelta(ab) = a\,\varDelta(b) + 2(g(a), g(b)) + \varDelta(a)b. \tag{13.4.11}$$

(It suffices to verify these equalities for $a = e^\lambda$, $b = e^\mu$; then (13.4.10) follows from (13.4.7) and (13.4.9), while (13.4.11) uses the equality $(\lambda + \mu, \lambda + \mu) = (\lambda, \lambda) + 2(\lambda, \mu) + (\mu, \mu)$.)

Consider next the element

$$R = \prod_{\alpha \neq 0} (e^\alpha - 1) = \prod_{\alpha > 0} (e^\alpha - 1)(e^{-\alpha} - 1) = -D^2. \tag{13.4.12}$$

We compute the product

$$R(c\varphi_\lambda - \varDelta\varphi_\lambda) = T. \tag{13.4.13}$$

By (13.4.2) we have

$$T = \prod_{\beta \neq 0} (e^\beta - 1)(c \sum n_\mu e^\mu - \varDelta(\sum n_\mu e^\mu))$$

$$= \prod_{\beta \neq 0} (e^\beta - 1) \sum_\mu n_\mu e^\mu(c - (\mu, \mu)). \tag{13.4.14}$$

From (13.4.5) we infer that

$$T = \prod_{\beta \neq 0} (e^\beta - 1) \sum_\mu e^\mu \sum_\alpha t(\mu, \alpha)$$

$$= \sum_\alpha \left(\prod_{\substack{\beta \neq \alpha \\ \beta \neq 0}} (e^\beta - 1) \right)(e^\alpha - 1) \sum_\mu t(\mu, \alpha)e^\mu$$

$$= \sum_\alpha \left(\prod_{\substack{\beta \neq \alpha \\ \beta \neq 0}} (e^\beta - 1) \right) \sum_\mu t(\mu, \alpha)(e^{\mu+\alpha} - e^\mu). \tag{13.4.15}$$

Changing the order of summation, we obtain

$$T = \sum_{\substack{\alpha \\ \beta \neq \alpha \\ \beta \neq 0}} \prod (e^\beta - 1) \sum_\mu e^{\mu + \alpha}(t(\mu, \alpha) - t(\mu + \alpha, \alpha)).$$

Proposition I and (13.4.6) imply that

$$T = \sum_\alpha \prod_{\beta \neq \alpha, \beta \neq 0} (e^\beta - 1) \sum_\mu n_\mu(\mu, \alpha)e^{\mu + \alpha}$$

$$= \left(\sum_\alpha \prod_{\beta \neq \alpha, \beta \neq 0} (e^\beta - 1)(e^\alpha \otimes \alpha), \sum_\mu n_\mu(e^\mu \otimes \mu) \right). \qquad (13.4.16)$$

By (13.4.12) we have $R = \prod_{\alpha \neq 0} (e^\alpha - 1)$ and therefore (13.4.10) and (13.4.9) yield

$$g(R) = \sum_{\alpha \neq 0} \left(\prod_{\beta \neq \alpha, \beta \neq 0} (e^\beta - 1) \right)(e^\alpha \otimes \alpha). \qquad (13.4.17)$$

From (13.4.2) we have $\sum_\mu n_\mu(e^\mu \otimes \mu) = g(\sum n_\mu e^\mu) = g(\varphi_\lambda)$. Substituting (13.4.7) in (13.4.16), we obtain

$$T = (g(R), g(\varphi_\lambda)). \qquad (13.4.18)$$

Substituting (13.4.18) in (13.4.13), we see that $R(c\varphi_\lambda - \varDelta\varphi_\lambda) = (g(R), g(\varphi_\lambda))$. Divide this equality by $(-D)$ and observe from (13.4.10) that $g(R) = g(-D^2) = -2Dg(D)$. This yields

$$D(c\varphi_\lambda - \varDelta\varphi_\lambda) = 2(g(D), g(\varphi_\lambda)). \qquad (13.4.19)$$

Rewrite this to find

$$c(D\varphi_\lambda) = D\,\varDelta(\varphi_\lambda) + 2(g(D), g(\varphi_\lambda)).$$

Using (13.4.11), we obtain

$$c(D\varphi_\lambda) = \varDelta(D\varphi_\lambda) - \varDelta(D)\varphi_\lambda. \qquad (13.4.20)$$

Consider the expression $\varDelta(D) = \varDelta(\text{Alt}(e^\rho))$. For every $w \in W^*$, we have $(w(\rho), w(\rho)) = (\rho, \rho)$, since the transformation w is orthogonal. Thus (13.4.8) gives $\varDelta(D) = (\rho, \rho)D$, and so (13.4.20) yields

$$\varDelta(D\varphi_\lambda) = (c + (\rho, \rho))(D\varphi_\lambda). \qquad (13.4.21)$$

Thus $D\varphi_\lambda$ is an eigenvector of the operator $\varDelta$. We will find the corresponding eigenvalue. Note that the highest term of the product $D\varphi_\lambda$ is $e^\rho e^\lambda = e^{\rho + \lambda}$. Under the operator $\varDelta$, this term is multiplied by $(\lambda + \rho, \lambda + \rho)$ and so we obtain

$$\Delta(D\varphi_\lambda) = (\lambda + \rho, \lambda + \rho)(D\varphi_\lambda). \tag{13.4.22}$$

The element $D\varphi_\lambda$ is skew-symmetric, which implies that

$$|W|D\varphi_\lambda = \text{Alt}(D\varphi_\lambda) = \sum_{w \in W^*} \sum_\mu (\det w)n_\mu \, \text{Alt}(e^{\mu + w\rho}). \tag{13.4.23}$$

The term of this sum that corresponds to $w \in W^*$ is an eigenvector of the operator Δ with eigenvalue $(\mu + w\rho, \mu + w\rho)$. Therefore we may restrict ourselves to summands with $(\mu + w\rho, \mu + w\rho) = (\lambda + \rho, \lambda + \rho)$. We will identify these summands.

II. *If μ is a weight of the representation π different from λ, we have* $(\mu + w\rho, \mu + w\rho) < (\lambda + \rho, \lambda + \rho)$.

Proof. By theorem 1 in §12 (see (12.1.6)) every weight μ is conjugate under the Weyl group to a weight ν such that $\nu(h_i) \geq 0$, $(\nu, \nu) \leq (\lambda, \lambda)$, and $(\mu, \mu) = (\nu, \nu)$. On the other hand, $w^{-1}(\mu)$ is a weight of the representation π and thus $w^{-1}(\mu) = \lambda - \sum m_i \alpha_i$, where the m_i are nonnegative integers. Therefore we have

$$(\mu, w\rho) = (w^{-1}(\mu), \rho) = (\lambda, \rho) - \sum_i m_i(\rho, \alpha_i) \tag{13.4.24}$$

and

$$(\lambda + \rho, \lambda + \rho) - (\mu + w\rho, \mu + w\rho) = (\lambda, \lambda) - (\mu, \mu) + 2\sum_i m_i(\rho, \alpha_i). \tag{13.4.25}$$

Proposition IX in 13.3 shows that $\rho(h_i) = 2(\rho, \alpha_i)(\alpha_i, \alpha_i)^{-1} = 1$ so that

$$(\lambda + \rho, \lambda + \rho) - (\mu + w\rho, \mu + w\rho)$$
$$= (\lambda, \lambda) - (\mu, \mu) + \sum_i m_i(\alpha_i, \alpha_i) \geq 0. \tag{13.4.26}$$

This equality can obtain only if all of the m_i are zero, i.e., $w^{-1}(\mu) = \lambda$. $\quad\square$

Proposition II implies that the sum over μ in the right side of (13.4.23) admits only the term with $\mu = \nu$ as nonzero. Therefore we have

$$D\varphi_\lambda = |W|^{-1} \sum_{w \in W^*} (\det w)^2 n_\lambda \, \text{Alt}(e^{\lambda + \rho}) = n_\lambda \, \text{Alt}(e^{\lambda + \rho}),$$

i.e.,

$$\varphi_\lambda = \text{Alt}(e^{\lambda + \rho})/\text{Alt}(e^\rho), \tag{13.4.27}$$

(recall that $n_\lambda = 1$).

We summarize (13.4.1), (13.4.2) and (13.4.27).

Theorem 1. *The character of an irreducible finite-dimensional representation π of a semisimple complex Lie algebra L is given by*

$$\operatorname{tr}\pi(h) = \left(h, \frac{\sum\limits_{w \in W^*} (\det w)e^{w(\lambda + \rho)}}{\sum\limits_{w \in W^*} (\det w)e^{w(\rho)}} \right) \tag{13.4.28}$$

for all $h \in U(H)$, where λ is the highest weight of the representation π, H is a Cartan subalgebra of L, W^* is the corresponding Weyl group and ρ is the partial sum of the positive roots.

This formula is called *Weyl's character formula.*
We will now find the dimension of the representation π.

Theorem 2. *The dimension of the irreducible finite-dimensional representation π of the semisimple complex Lie algebra L is given by*

$$\dim \pi = \frac{\prod\limits_{\alpha > 0} (\lambda + \rho, \alpha)}{\prod\limits_{\alpha > 0} (\rho, \alpha)}. \tag{13.4.29}$$

Proof. Consider the homomorphism of the algebra A onto the field $\mathbf{C}$ under which all elements e^λ, $\lambda \in P$, map into 1. This homomorphism is the composition of two homomorphisms. First we map e^λ into the power series $e^{x(\lambda, \rho)}$ in the variable x and then we read off the constant term of this power series. Let ψ_μ be the homomorphism that carries e^λ into $e^{x(\lambda, \mu)}$. We get

$$\psi_\mu(\operatorname{Alt}(e^\lambda)) = \sum_w (\det w)e^{x(w\lambda, \mu)}$$

$$= \sum_w (\det w)e^{x(\lambda, w^{-1}\mu)} = \psi_\lambda(\operatorname{Alt}(e^\mu)), \tag{13.4.30}$$

and hence

$$\psi_\rho(\operatorname{Alt}(e^\lambda)) = \psi_\lambda(\operatorname{Alt}(e^\rho)) = \prod_{\alpha > 0} (e^{x(\alpha, \lambda)/2} - e^{-x(\alpha, \lambda)/2}). \tag{13.4.31}$$

The lowest term of this series is $\prod_{\alpha > 0} x(\alpha, \lambda)$. Thus the constant term of the series $\psi_\rho(\operatorname{Alt}(e^{\lambda + \rho}))/\psi_\rho(\operatorname{Alt}(e^\rho))$ is $\prod_{\alpha > 0}(\lambda + \rho, \alpha)/\prod_{\alpha > 0}(\rho, \alpha)$. This means that $\dim \pi = \prod_{\alpha > 0}(\lambda + \rho, \alpha)/\prod_{\alpha > 0}(\rho, \alpha)$, and we have proved (13.4.29). $\quad\square$

There are a number of corollaries of Weyl's character formula.
Multiplying both sides of the formula $\varphi_\lambda = \operatorname{Alt}(e^{\lambda + \rho})/D$ by the denominator D, we obtain

$$\sum_w \sum_\mu n_\mu(\det w)e^{\mu + w\rho} = \sum_w (\det w)e^{w(\lambda + \rho)}. \tag{13.4.32}$$

Let μ be a weight different from λ. Proposition II shows that $(\mu + \rho, \mu + \rho) < (\lambda + \rho, \lambda + \rho)$. Therefore $\mu + \rho$ is different from all functionals $w(\lambda + \rho)$. Thus the total coefficient of $e^{\mu + \rho}$ in the left side of (13.4.32) is zero. To get this coefficient, we sum over the weights μ' such that $\mu' + w\rho = \mu + \rho$, i.e.,

$$\sum_{w \in W} n_{\mu + \rho - w\rho}(\det w) = 0. \tag{13.4.33}$$

Propositions IV and X in 13.3 show that $\rho - w\rho > 0$ for $w \neq 1$. Therefore (13.4.33) is a recurrence formula for the multiplicities n_μ: we rewrite (13.4.33) as

$$n_\mu = -\sum_{w \neq 1} (\det w) n_{\mu + \rho - w\rho}. \tag{13.4.34}$$

We offer without proofs two other recurrence formulas for the multiplicity of a weight[58]:

(1) Freudenthal's formula:

$$n_\mu = 2((\lambda + \rho, \lambda + \rho) - (\mu + \rho, \mu + \rho))^{-1} \sum_{\alpha > 0} \sum_{k=1}^{\infty} n_{\mu + k\alpha}(\mu + k\alpha, \alpha).$$

$$\tag{13.4.35}$$

(2) Kostant's formula:

$$n_\mu = \sum_{w \in W*} (\det w) P(w(\lambda + \rho) - (\mu + \rho)), \tag{13.4.36}$$

where $P(0) = 0$ and for $\mu \neq 0$, the number $P(\mu)$ is equal to the number of distinct representations of the vector μ as a sum of positive roots.

The proofs for these formulas can be found in D. P. Želobenko's book [1] and in Cartier's article [1*].

13.5. Representations of the Lie Algebra $sl(n + 1, \mathbf{C})$

Let L be a simple Lie algebra of type (A_n); for example, $L = sl(n + 1, \mathbf{C})$ (see §10, 10.3, from which we take our notation). Let H be the Cartan subalgebra of L consisting of all diagonal matrices in L. We define the linear functional λ_i on H by the formula

$$\lambda_i(h) = h_i, \quad \text{for } h = \begin{Vmatrix} h_1 & & 0 \\ & \ddots & \\ 0 & & h_{n+1} \end{Vmatrix}. \tag{13.5.1}$$

---

[58] These formulas do not exhaust all the known formulas for the multiplicity of a weight. See, for example, Klimyk [1*].

It is clear that

$$\lambda_1 + \cdots + \lambda_{n+1} = 0, \tag{13.5.2}$$

and that the functionals $\lambda_1, \ldots, \lambda_n$ are a basis in the conjugate space H^* of H.

The set Δ of the roots of L relative to H consists of the linear functionals

$$\omega_{ij} = \lambda_i - \lambda_j, \qquad i, j = 1, \ldots, n + 1. \tag{13.5.3}$$

We introduce in the subspace H_0 of real diagonal matrices the natural lexicographic ordering (see Proposition II in 9.6 and 10.3). The set Δ^+ of positive roots of L relative to H is the set of roots

$$\omega_{ij} = \lambda_i - \lambda_j, \qquad 1 \leqslant i < j \leqslant n + 1. \tag{13.5.4}$$

The roots

$$\omega_{i, i+1} = \lambda_i - \lambda_{i+1}, \qquad i = 1, \ldots, n, \tag{13.5.5}$$

are the simple roots. The nilpotent Lie algebra $\sum_{\alpha > 0} L^\alpha$ is the Lie algebra N^+ of L consisting of all upper triangular matrices with zeros on the main diagonal. Similarly, $N^- = \sum_{\alpha < 0} L^\alpha$ is the Lie subalgebra of lower triangular nilpotent matrices.

We define $h'_{\omega_{ij}}$ in H_0 as $h'_{\omega_{ij}} = (2(n + 1))^{-1}(e_{ii} - e_{jj})$, where e_{ij} is the matrix whose only nonzero element is 1 and is located in the i-th row and j-th column. We then have

$$\omega_{ij}(h) = (h, h'_{\omega_{ij}}) \tag{13.5.6}$$

for all $h \in H$, where $(\cdot, \cdot)$ is the Killing form on H:

$$(h, \tilde{h}) = (2n + 1) \sum_{i=1}^{n+1} h_i \tilde{h}_i. \tag{13.5.7}$$

We define

$$h_{\omega_{ij}} = 2(\omega_{ij}(h'_{\omega_{ij}}))^{-1} h'_{\omega_{ij}} = e_{ii} - e_{jj}. \tag{13.5.8}$$

We will find the irreducible finite-dimensional representations of the Lie algebra $L = sl(n + 1, \mathbf{C})$ and their characters.

I. *A linear functional λ on H is the highest weight of an irreducible representation π of the Lie algebra L if and only if*

$$\lambda = m_1 \lambda_1 + \cdots + m_n \lambda_n, \tag{13.5.9}$$

where the m_i are nonnegative integers and $m_k \geqslant m_{k+1}$ for all $k = 1, \ldots, n - 1$.

Proof. Every element $\lambda \in H^*$ has the form (13.5.9). From (13.5.8) we infer that

$$\lambda(h_{\omega_{i,\,i+1}}) = \lambda(e_{ii} - e_{i+1,\,i+1}) = m_i - m_{i+1}, \qquad (13.5.10)$$

where m_{n+1} is defined as 0. By theorem 1 in §12, the functional λ is the highest weight of a finite-dimensional irreducible representation if and only if the numbers $\lambda(h_{\omega_{i,\,i+1}})$ are nonnegative integers for all $i = 1, \ldots, n$. Thus (13.5.10) implies the present proposition. $\square$

We will construct a representation with highest weight $\lambda = m_1\lambda_1 + \cdots + m_n\lambda_n$. Let π be the identity representation of the Lie algebra L in the space $V = \mathbf{C}^{n+1}$. The weights of π are $\lambda_1, \ldots, \lambda_{n+1}$ ($\lambda_{n+1} = -\lambda_1 - \cdots - \lambda_n$) and therefore the highest weight of π is λ_1.

II. *Let k be one of the numbers $1, 2, \ldots, n$. Let V_k denote the exterior product of k copies of the space V. Let $\bigotimes_{i=1}^{k} \pi$ be the tensor product of k copies of the representation π of L. The space V_k is invariant under the representation $\bigotimes_{i=1}^{k} \pi$. The representation π_k defined by the restriction of $\bigotimes_{i=1}^{k} \pi$ to V_k is irreducible. The highest weight of the representation π_k is equal to $\lambda_1 + \cdots + \lambda_k$.*

Proof. The space V_k is the linear span of elements of the form $f_{i_1} \wedge \cdots \wedge f_{i_k}$. Here we take $1 \leqslant i_1 < \cdots < i_k \leqslant n+1$, $f_1, \ldots, f_{n+1}$ is the canonical basis in $V = \mathbf{C}^{n+1}$, and $f_{i_1} \wedge \cdots \wedge f_{i_k} = \sum_{\sigma \in S_k} (\operatorname{sgn} \sigma) f_{i_{\sigma(1)}} \otimes \cdots \otimes f_{i_{\sigma(k)}}$ where S_k is the group of all permutations and $\operatorname{sgn} \sigma = 1$ for even permutations σ and $\operatorname{sgn} \sigma = -1$ for odd permutations σ. Using (1.2.2), we see that

$$\left(\bigotimes_{i=1}^{k} \pi\right)(x)(f_{i_1} \wedge \cdots \wedge f_{i_k}) = \pi(x)f_{i_1} \wedge \cdots \wedge f_{i_k} + \cdots + f_{i_1} \wedge \cdots \wedge \pi(x)f_{i_k}.$$

$$(13.5.11)$$

This implies first of all that the subspace $V_k \subset \bigotimes_{i=1}^{n} V$ is invariant under the representation $\bigotimes_{i=1}^{k} \pi$. Let π_k denote the restriction of the representation $\bigotimes_{i=1}^{k} \pi$ to V_k. Formula (13.5.11) shows that

$$\pi_k(h)(f_{i_1} \wedge \cdots \wedge f_{i_k}) = \pi(h)f_{i_1} \wedge \cdots \wedge f_{i_k} + \cdots + f_{i_1} \wedge \cdots \wedge \pi(h)f_{i_k}$$
$$= (\lambda_{i_1} + \cdots + \lambda_{i_k})(h)(f_{i_1} \wedge \cdots \wedge f_{i_k}). \qquad (13.5.12)$$

Formula (13.5.12) in turn shows that the linear functionals $\lambda_{i_1} + \cdots + \lambda_{i_k}$ are the complete set of weights of the representation π_k.

We will show that the vector $f_1 \wedge \cdots \wedge f_k$ is a highest vector of π_k. For every $i = 1, \ldots, n+1$, the subspace $\pi(N^+)f_i$ is contained in the linear span of the vectors $f_1, \ldots, f_{i-1}$ and so (13.5.11) implies that $\pi_k(N^+)(f_1 \wedge \cdots \wedge f_k) = \{0\}$. To prove that $f_1 \wedge \cdots \wedge f_k$ is the highest vector, it remains to show that any $\pi_k(N^-)$-invariant subspace of V_k containing the vector $f_1 \wedge \cdots \wedge f_k$

coincides with V_k. We note that

$$\pi(e_{i+1,i})f_k = \begin{cases} 0, & \text{for } k \neq i, \\ f_{k+1}, & \text{for } k = i. \end{cases} \tag{13.5.13}$$

From (13.5.11) and (13.5.13) we obtain

$$\pi_k(e_{i_j+1,i_j})(f_{i_1} \wedge \cdots \wedge f_{i_k}) = f_{i_1} \wedge \cdots \wedge f_{i_j+1} \wedge \cdots \wedge f_{i_k}. \tag{13.5.14}$$

By induction (13.5.14) implies that a $\pi_k(N^-)$-invariant subspace containing $f_1 \wedge \cdots \wedge f_k$ contains all vectors $f_{i_1} \wedge \cdots \wedge f_{i_k}$, $1 \leqslant i_1 < \cdots < i_k \leqslant n+1$; that is, it coincides with V_k. Thus, $f_1 \wedge \cdots \wedge f_k$ is a highest vector of the representation π_k and the corresponding highest weight is equal to $\lambda_1 + \cdots + \lambda_k$. Since the representation π_k has a highest vector, it is irreducible. $\square$

The following proposition if valid for all complex semisimple Lie algebras.

III. *Let ρ_1, ρ_2 be irreducible representations of the Lie algebra L in the spaces E_1, E_2. Let μ_1, μ_2 be their highest weights and f_1, f_2 their highest vectors. Let ρ be the restriction of the representation $\rho_1 \otimes \rho_2$ to E, the smallest $\rho_1 \otimes \rho_2$-invariant subspace of the space $E_1 \otimes E_2$ that contains the vector $f_1 \otimes f_2$. Then ρ is an irreducible representation of the Lie algebra L, the highest vector of ρ is $f_1 \otimes f_2$ and the highest weight is $\mu_1 + \mu_2$.*

Proof. By definition we have

$$(\rho_1 \otimes \rho_2)(x)(f_1 \otimes f_2) = \rho_1(x)f_1 \otimes f_2 + f_1 \otimes \rho_2(x)f_2 \tag{13.5.15}$$

for all $x \in L$. It follows that $(\rho_1 \otimes \rho_2)(L^\alpha)(f_1 \otimes f_2) = \{0\}$ for $\alpha > 0$ and $f_1 \otimes f_2$ is an eigenvector for the operators $(\rho_1 \otimes \rho_2)(h)$, $h \in H$, with the eigenvalue $\mu_1(h) + \mu_2(h)$. Hence E is the linear span of all vectors $(\rho_1 \otimes \rho_2)(x_1) \cdots (\rho_1 \otimes \rho_2)(x_k)(f_1 \otimes f_2)$, where $x_1, \ldots, x_k \in \sum_{\alpha < 0} L^\alpha$. The preceding argument shows that ρ is a representation with highest vector $f_1 \otimes f_2$ and so ρ is irreducible. The highest weight of the representation ρ is $\mu_1 + \mu_2$. $\square$

IV. *Every irreducible representation π of the Lie algebra $L = sl(n+1, \mathbf{C})$ is a subrepresentation of a tensor power of the identity representation of L.*

Proof. Let $\lambda = m_1\lambda_1 + \cdots + m_n\lambda_n$ be the highest weight of the representation π (see Proposition I). We have

$$\lambda = (m_1 - m_2)\lambda_1 + (m_2 - m_3)(\lambda_1 + \lambda_2) + \cdots + m_n(\lambda_1 + \cdots + \lambda_n), \tag{13.5.16}$$

where $m_1 - m_2, \ldots, m_n$ are nonnegative integers. The weights $\lambda_1, \ldots, \lambda_1 + \cdots + \lambda_n$ are the highest weights of the representations $\pi_1, \ldots, \pi_n$ (see Proposition II), which are subrepresentations of the tensor powers of the identity representation. The present proposition follows from (13.5.16) and Proposition III. $\square$

We will now find the character of the irreducible representation π of L with highest weight $\lambda = m_1\lambda_1 + \cdots + m_n\lambda_n$. To apply (13.4.28), we must find the partial sum of the positive roots ρ and also describe the Weyl group W^*.

First note that (13.5.2) and (13.5.4) imply that

$$2\rho = (n + 1)\lambda_1 + ((n + 1) - 2)\lambda_2 + \cdots + (-(n + 1))\lambda_{n+1}$$
$$= 2(n + 1)\lambda_1 + 2n\lambda_2 + \cdots + 2\lambda_{n+1},$$

so that

$$\rho = (n + 1)\lambda_1 + n\lambda_2 + \cdots + \lambda_{n+1}$$
$$= n\lambda_1 + (n - 1)\lambda_2 + \cdots + \lambda_n. \qquad (13.5.17)$$

Furthermore, for any simple root $\omega_{i,i+1}$, the corresponding symmetry $S_i^* \in W^*$ has the form

$$S_i^*(\mu) = \mu - 2(\mu, \omega_{i,i+1})(\omega_{i,i+1}, \omega_{i,i+1})^{-1}\omega_{i,i+1}$$
$$= \mu - \mu(e_{ii} - e_{i+1,i+1})\omega_{i,i+1} \qquad (13.5.18)$$

for all $\mu \in H_0^*$. Replacing μ in (13.5.18) by the root ω_{kl}, $k \neq l$, we see that an application of S_i^* reduces to the transposition of i and $i + 1$ in the indices of the roots ω_{kl} (indices not equal to i or $i + 1$ remain fixed). Hence the Weyl group W^* is isomorphic to the group S_{n+1} of all permutations of the indices $(1, \ldots, n + 1)$. Applying (13.4.28) we see that for all $h \in U(H)$

$$\text{tr } \pi(h) = \left(h, \frac{\sum\limits_{\sigma \in S_{n+1}} (\text{sgn } \sigma)e^{(m_1+n)\lambda_{\sigma(1)} + \cdots + (m_n+1)\lambda_{\sigma(n)}}}{\sum\limits_{\sigma \in S_{n+1}} (\text{sgn } \sigma)e^{n\lambda_{\sigma(1)} + \cdots + \lambda_{\sigma(n)}}} \right). \qquad (13.5.19)$$

As shown above, (13.5.19) can be rewritten as

$$\text{tr } \pi(h) = \frac{\begin{vmatrix} e^{(m_1+n)\lambda_1}(h) & \cdots & e^{(m_1+n)\lambda_{n+1}}(h) \\ \vdots & & \\ e^{(m_n+1)\lambda_1}(h) & \cdots & e^{(m_n+1)\lambda_{n+1}}(h) \\ 1 & \cdots & 1 \end{vmatrix}}{\begin{vmatrix} e^{n\lambda_1}(h) & \cdots & e^{n\lambda_{n+1}}(h) \\ \vdots & & \\ e^{\lambda_1}(h) & \cdots & e^{\lambda_{n+1}}(h) \\ 1 & \cdots & 1 \end{vmatrix}} \qquad (13.5.20)$$

for all $h \in U(H)$.

We will now find the dimension of the representation π. By (13.4.29) we have

$$\dim \pi = \frac{\prod\limits_{\alpha > 0} (\lambda + \rho, \alpha)}{\prod\limits_{\alpha > 0} (\rho, \alpha)} = \frac{\prod\limits_{i < j} (\lambda + \rho, \omega_{ij})}{\prod\limits_{i < j} (\rho, \omega_{ij})} = \frac{\prod\limits_{i < j} (m_i - m_j + j - i)}{\prod\limits_{i < j} (j - i)}. \qquad (13.5.21)$$

A useful corollary of (13.5.21) is the following. The formula (13.5.21) shows that dim π is a strictly increasing function of $m_i - m_j$, $i < j$. Thus if $m_k - m_{k+1} > 0$, the dimension of the representation π is not less than the dimension of the representation π_k. Suppose that dim $\pi = n + 1$. Since dim $\pi_k = C^k_{n+1}$, we find that dim $\pi_k > n + 1$ for $1 < k < n$. This implies the following.

If π is an irreducible representation of the Lie algebra $L = sl(n + 1, \mathbf{C})$ of dimension $n + 1$, we have either $\pi \approx \pi_1$, or $\pi \approx \pi_n$.

It is clear that the mapping

$$\tilde{\pi}(x) = -x' \qquad (13.5.22)$$

(x' is the transpose of the matrix x) is an $(n + 1)$-dimensional irreducible representation of the Lie algebra L with weights $-\lambda_1, \ldots, -\lambda_n, -\lambda_{n+1} = \lambda_1 + \cdots + \lambda_n$. Proposition V gives us the following.

VI. *The representations $\tilde{\pi}$ and π_n are equivalent.*

It is of interest to compare these results with those of 3.3–3.4 in chapter VI.

§14. Real Forms of Semisimple Complex Lie Algebras

14.1. Real Forms

Let L be a complex Lie algebra. Let L_0 be a real Lie subalgebra of L (L is considered as a real Lie algebra). The Lie algebra L_0 is called a *real form* of the Lie algebra L if the canonical embedding of the complexification $(L_0)_{\mathbf{C}}$ of L_0 into L (defined as the extension to $(L_0)_{\mathbf{C}}$ of the embedding homomorphism of L_0 into L) is an isomorphism of the complex Lie algebras $(L_0)_{\mathbf{C}}$ and L. It is clear in this case that $\dim_{\mathbf{R}} L_0 = \dim_{\mathbf{C}} L$.

If L_0 is a real form of the Lie algebra L, every element $z \in L$ can be uniquely represented in the form

$$z = x + iy; \quad x, y \in L_0. \qquad (14.1.1)$$

Consider the mapping θ of L onto itself defined by

$$\theta(x + iy) = x - iy, \quad \text{for all } x, y \in L_0. \qquad (14.1.2)$$

It is clear that

$$\theta(\theta(z)) = z, \quad \text{for all } z \in L. \tag{14.1.3}$$

Thus θ is an *involution of L onto L*. The identities

$$\theta(z + w) = \theta(z) + \theta(w), \tag{14.1.4}$$
$$\theta(\lambda z) = \bar{\lambda}\theta(z), \tag{14.1.5}$$
$$\theta([z, w]) = [\theta(z), \theta(w)] \tag{14.1.6}$$

for all $z, w \in L$ and all $\lambda \in \mathbf{C}$, *are evident*. The mapping θ is determined by the Lie subalgebra L_0. We call θ the *involution defined by the real form L_0 of the Lie algebra L*.

Next, suppose that we have a mapping θ of L into itself that satisfies (14.1.3)–(14.1.6). Let L_0 be the set of elements $z \in L$ such that $\theta(z) = z$. For $z, w \in L_0$ and real numbers λ, the equalities (14.1.4)–(14.1.6) imply that $z + w$, λz, and $[z, w]$ are in L_0. That is, L_0 is a real Lie algebra.

I. *Every element $z \in L$ can be written uniquely as $z = x + iy$, with $x, y \in L_0$.*

Proof. From (14.1.3) and (14.1.5) we get $z + \theta(z) \in L_0$ and $(-i)(z - \theta(z)) \in L_0$. We also have

$$z = (1/2)(z + \theta(z)) + i(-i/2)(z - \theta(z)) = x + iy, \tag{14.1.7}$$

where $x = (z + \theta(z))/2 \in L_0$, $y = (z - \theta(z))/2i \in L_0$. Furthermore, if $z = x + iy$ where $x, y \in L_0$, we have $\theta(z) = x - iy$. We then have $x = (1/2)(z + \theta(z))$, $y = (1/2i)(z - \theta(z))$, so that the representation (14.1.7) of $z \in L$ is unique. $\square$

Proposition I shows that L_0 is a real form of the Lie algebra L. There is a one-to-one correspondence between the real forms of a given complex Lie algebra L and involutions in L that satisfy (14.1.4)–(14.1.6).

II. *The Killing form of the Lie algebra L_0 is the restriction to L_0 of the Killing form of the Lie algebra L.*

Proof. Suppose that x, y are in L_0. Consider the linear operator $\text{ad } x \cdot \text{ad } y$ in the space L. Since L_0 is a real Lie algebra, we have $(\text{ad } x \cdot \text{ad } y)(z) = [x, [y, z]] \in L_0$ for $z \in L_0$. That is, the real subspace L_0 of L is invariant under the operator $\text{ad } x \cdot \text{ad } y$. Let $e_1, \ldots, e_n$ be a basis for the real linear space L_0. Proposition I shows that $e_1, \ldots, e_n$ is a basis for the complex linear space L. Let $(\text{ad } x \cdot \text{ad } y)e_k = \sum_{i=1}^{n} c_{ik}e_i$ be the expansion of the element $(\text{ad } x \cdot \text{ad } y)e_k$ in this basis of the space L_0. The same formula yields the expansion of the element $(\text{ad } x \cdot \text{ad } y)e_k$ in the basis $e_1, \ldots, e_n$ for the complex space L. Thus the trace of the operator $\text{ad } x \text{ ad } y$ in the real space L_0 is equal to the trace of the operator ad x ad y in the complex linear space L. In symbols, we have

$$(x, y)_{L_0} = \mathrm{tr}_{L_0}(\mathrm{ad}\ x\ \mathrm{ad}\ y) = \mathrm{tr}_L(\mathrm{ad}\ x\ \mathrm{ad}\ y) = (x, y)_L. \qquad (14.1.8)$$

$\square$

III. *The form* $(x, y)_L$ *is real-valued on the Lie algebra* L_0.

Proof. In (14.1.8), observe that the second term is real-valued. $\square$

IV. *We have*

$$(x, y)_L = \overline{(\theta(x), \theta(y))_L} \qquad (14.1.9)$$

for all $x, y \in L$.

Proof. For $x, y \in L_0$, we have $\theta(x) = x$, $\theta(y) = y$ and the number $(x, y)_L$ is real (Proposition III). Thus (14.1.9) holds for all $x, y \in L_0$. The identities (14.1.4) and (14.1.5) show that $\overline{(\theta(x), \theta(y))}$ is a complex bilinear form on L. Proposition I proves that complex bilinear forms coinciding on L_0 coincide throughout L. $\square$

A real Lie algebra M is said to be *compact* if the Killing form on M is negative-definite; that is, if $(x, x)_M < 0$ for all nonzero x in M. Let L_0 be a real form of the complex Lie algebra L. If the real Lie algebra L_0 is compact, it is said to be a *compact real form* of the Lie algebra L.

V. *A real form* L_0 *of the complex Lie algebra* L *is compact if and only if the Hermitian form* $\{x, y\} = (x, \theta(y))$ *on* L *is negative-definite.*

Proof. Let L_0 be a compact real form of the Lie algebra L. For $z \in L$, we write $z = x + iy$ with $x, y \in L_0$. For $z \neq 0$ at least one of the elements x and y is nonzero. It is clear that

$$(z, \theta(z)) = (x + iy, x - iy) = (x, x) + (y, y) < 0.$$

Conversely, if the Hermitian form $(x, \theta(y))$ is negative-definite and $x \in L_0\backslash\{0\}$, we have $\theta(x) = x$ and $(x, x) = (x, \theta(x)) < 0$. $\square$

We now study involutions θ in a *complex semisimple* Lie algebra L that satisfy conditions (14.1.4)–(14.1.6); that is, we will study real forms L_0 of L. We need two general auxiliary propositions.

VI. *Let H be a complex linear subspace of the Lie algebra L. The space H is invariant under an involution θ if an only if every element $z \in H$ has the form $z = x + iy$, where $x, y \in H_0 = L_0 \cap H$.*

Proof. Suppose that $\theta(H) = H$. If $z \in H$, then $x = (1/2)(z + \theta(z))$ belongs to H and to L_0. Similarly, we have $y = (1/2i)(z - \theta(z)) \in L_0 \cap H = H_0$ and of course $z = x + iy$. Conversely, suppose that every element $z \in H$ has the form $z = x + iy$, where $x, y \in H_0 = L_0 \cap H$. We then have $\theta(z) = x - iy \in H$ and so $\theta(H) \subset H$. Since $\theta^2 = 1$, we get $\theta(H) = H$. $\square$

VII. *Let τ, θ be involutions in the Lie algebra L that satisfy* (14.1.4)–(14.1.6) *and suppose that $\tau\theta = \theta\tau$. Let L_u be the set of fixed points for the involution τ. The set of points $z \in L_u$ such that $\theta(z) = z$ ($\theta(z) = -z$) is denoted by L_u^+ (L_u^-). The space L_u is the direct sum of L_u^+ and L_u^- and the space L_0 of fixed points for the involution θ is the direct sum of the spaces L_u^+ and iL_u^-.*

Proof. For z in L_u, we have $\tau\theta(z) = \theta(\tau z) = \theta(z)$ and therefore $\theta(z) \in L_u$ for all $z \in L_u$: the operator θ leaves the space L_u invariant. The restriction σ of the operator θ to the subspace L_u is a real linear mapping of L_u onto itself for which $\sigma^2 = 1$. It is clear that $L_u^+ \cap L_u^- = \{0\}$. Furthermore, we have $z + \sigma(z) \in L_u^+$ and $z - \sigma(z) \in L_u^-$ for all $z \in L_u$, so that from

$$z = (1/2)(z + \sigma(z)) + (1/2)(z - \sigma(z))$$

we infer that $L_u = L_u^+ \dotplus L_u^-$.

The space L is the direct sum of the spaces L_u and iL_u, which means that $L = L_u^+ \dotplus L_u^- \dotplus iL_u^+ \dotplus iL_u^-$. The mapping θ is the identity mapping on L_u^+ and is multiplication by -1 on the subspace L_u^-. Since we have $\theta(iz) = -i\theta(z)$ for all $z \in L$, the mapping θ is multiplication by -1 on the subspace iL_u^+ and is the identity mapping on iL_u^-. $\square$

VIII. *Let L_0 be the real form of the complex semisimple Lie algebra L for the involution θ in L. There is a Cartan subalgebra H of L invariant under θ. Write H_0 for the real subspace of H generated by the vectors h'_α, $\alpha \in \Delta$. We have $\theta(H_0) = H_0$. Let θ^* be the mapping in the space H^* that is conjugate to θ. We have $\theta^* \Delta = \Delta$.*

Proof. Let M_x be the subspace of L consisting of all $y \in L$ such that $(\mathrm{ad}\ x)^q y = 0$ for some natural number q. Plainly M_x is invariant under $\mathrm{ad}\ x$. Reducing the operator $\mathrm{ad}\ x$ in L to its Jordan canonical form, we see that M_x is the root subspace of $\mathrm{ad}\ x$ that corresponds to the root zero. Thus the dimension of the space M_x is equal to the multiplicity of the zero root of the characteristic polynomial of $\mathrm{ad}\ x$. We write

$$\det(\mathrm{ad}\ x - \lambda 1) = (-\lambda)^n + (-\lambda)^{n-1}\varphi_{n-1}(x) + \cdots + (-\lambda)^k\varphi_k(x).$$

If $\varphi_k(x) \not\equiv 0$ on L, k is the smallest possible dimension of the subspace M_x as x runs through L. Therefore if x_0 is such that $\varphi_k(x_0) \neq 0$, x_0 is a regular element of L.

Consider a positive integer k such that $\varphi_k(x) \not\equiv 0$ on L. The function $\varphi_k(x)$ is a polynomial in the entries of the matrix of the operator ad x in some basis. Using Taylor's formula for polynomials, we see that if $\varphi_k(x)$ vanishes throughout L_0, then $\varphi_k(x)$ vanishes throughout L. Therefore there is an element $x_0 \in L_0$ such that $\varphi_k(x_0) \neq 0$, and so the Lie algebra L_0 contains a regular element, x_0. By Proposition IV in §8, the set H of elements $y \in L$ such that $[y, x_0] = 0$ is a Cartan subalgebra of L. For $h \in H$, (14.1.6) implies that

$$[\theta(h), x_0] = \theta([h, \theta(x_0)]) = \theta([h, x_0]) = \theta(0) = 0. \qquad (14.1.10)$$

Hence $\theta(h)$ belongs to H: the subspace H is invariant under θ. Since $\theta^2 = 1$, we have in fact $\theta H = H$.

Let $\alpha \neq 0$ be a root of the Lie algebra L relative to the Cartan subalgebra H. Let $\bar{\alpha}$ be the linear functional on H defined by

$$\bar{\alpha}(h) = \overline{\alpha(\theta h)} \quad \text{for all } h \in H. \qquad (14.1.11)$$

For $y \in L^\alpha$, we have $[h, y] = \alpha(h)y$ and $[\theta(h), \theta(y)] = \theta([h, y]) = \theta(\alpha(h)y) = \overline{\alpha(h)}\theta(y) = \bar{\alpha}(\theta h)\theta(y)$. Since $\theta H = H$, we infer that

$$[h, \theta(y)] = \bar{\alpha}(h)\theta(y) \qquad (14.1.12)$$

for all $h \in H$. This means that $\bar{\alpha}$ is a root of the Lie algebra L relative to H. That is, we have $\theta^* \Delta \subset \Delta$ and $\theta(y) \in L^{\bar{\alpha}}$ for all $y \in L^\alpha$, which implies that $\theta L^\alpha \subset L^{\bar{\alpha}}$. Proposition IV also shows that

$$(\theta(h'_\alpha), h) = (\overline{h'_\alpha, \theta(h)}) = \overline{\alpha(\theta(h))} = \bar{\alpha}(h) = (h'_{\bar{\alpha}}, h). \qquad (14.1.13)$$

Therefore we have $\theta h'_\alpha = h'_{\bar{\alpha}}$ and so θ carries the real subspace H_0 of H onto itself.

We thus obtain $\theta H \subset H$, $\theta^* \Delta \subset \Delta$, $\theta H_0 \subset H_0$. Since $\theta^2 = 1$, we have $\theta H = H$. $\square$

IX. *Let L be a complex semisimple Lie algebra with the real form L_0 for an involution θ. Let H be the Cartan subalgebra of L defined in Proposition VIII. We choose vectors $e_\alpha \in L^\alpha$, $\alpha \in \Delta$, such that $(e_\alpha, e_{-\alpha}) = -1$. There exist complex numbers C_α such that $\theta e_\alpha = C_\alpha e_{\bar{\alpha}}$ and*

$$\bar{C}_\alpha C_{\bar{\alpha}} = 1, \qquad C_\alpha C_{-\alpha} = 1, \qquad C_{\alpha+\beta}\bar{N}_{\alpha,\beta} = C_\alpha C_\beta N_{\bar{\alpha},\bar{\beta}}{}^{59} \qquad (14.1.14)$$

for all α, $\beta \in \Delta$.

---

[59] Editor's note. For the definition of $N_{\alpha,\beta}$ see (9.7.3).

Proof. By (14.1.12), we have $\theta(e_\alpha) \in L^{\bar\alpha}$. Since all the spaces L^α, $\alpha \in \Delta$, are one-dimensional, we have $\theta(e_\alpha) = C_\alpha e_{\bar\alpha}$ for certain complex numbers C_α.

Since $\theta^2 h = h$ for $h \in H_0$ and $\theta^2 e_\alpha = \theta(C_\alpha e_{\bar\alpha}) = \bar C_\alpha \theta e_{\bar\alpha} = \bar C_\alpha C_{\bar\alpha} e_{\bar{\bar\alpha}}$, we obtain $\bar C_\alpha C_{\bar\alpha} e_{\bar{\bar\alpha}} = e_\alpha$. From (14.1.11) we infer that $\bar{\bar\alpha} = \alpha$ and thus $\bar C_\alpha C_{\bar\alpha} = 1$, which proves the first equality in (14.1.14).

We now use (14.1.5). Since $\theta([e_\alpha, e_{-\alpha}]) = [\theta(e_\alpha), \theta(e_{-\alpha})]$ and $\theta h'_\alpha = h'_{\bar\alpha}$, we find that $\theta(-h'_\alpha) = [C_\alpha e_{\bar\alpha}, C_{-\alpha} e_{\bar\alpha}] = C_\alpha C_{-\alpha}(-h'_{\bar\alpha})$. Since $\theta h'_\alpha = h'_{\bar\alpha}$, the identity $C_\alpha C_{-\alpha} = 1$ for all $\alpha \in \Delta$ follows.

Finally, the identity (14.1.6) gives $\theta([e_\alpha, e_\beta]) = [\theta(e_\alpha), (e_\beta)]$ and this implies that $\bar N_{\alpha,\beta} C_{\alpha+\beta} e_{\bar\alpha+\bar\beta} = C_\alpha C_\beta N_{\bar\alpha,\bar\beta} e_{\bar\alpha+\bar\beta}$. This proves the last equality in (14.1.14). $\qquad\square$

X. *Let L be a complex semisimple Lie algebra with a Cartan subalgebra H. Let H_0 be the real linear subspace of H generated by the vectors h'_α, where α runs through the set of nonzero roots of L relative to H. Let φ be a real linear isomorphism of the space H_0 onto itself such that $\varphi^2 = 1$; and, for all $\alpha \in \Delta$, the complex linear functional $\bar\alpha$ on H defined by $\bar\alpha(h) = \alpha(\varphi(h))$ for all $h \in H_0$, is a root of L relative to H. Suppose that $(e_\alpha, e_{-\alpha}) = -1$ for all $\alpha \in \Delta$. The mapping φ can be extended to an involution τ of L that satisfies (14.1.4)–(14.1.6) and $\tau(e_\alpha) = C_\alpha e_{\bar\alpha}$ if and only if the complex numbers C_α satisfy (14.1.14).*

Proof. The identities (14.1.13) (with θ replaced by φ) hold in view of the definition of $\bar\alpha$, and thus we have $\varphi(h'_\alpha) = h'_{\bar\alpha}$ for all $\alpha \in \Delta$. Since the space H_0 and the vectors e_α, $\alpha \in \Delta$, generate the complex linear space L, there is a unique linear mapping τ of L into itself such that $\tau(h_0) = \varphi(h_0)$ for $h_0 \in H_0$, $\tau(e_\alpha) = C_\alpha e_{\bar\alpha}$, and for which (14.1.14) and (14.1.15) hold (τ replacing θ, of course).

We now determine properties of the numbers C_α under which the mapping τ satisfies (14.1.3) and (14.1.6). From the definition of the root $\bar\alpha$ and from the condition $\varphi^2 = 1$, we infer that $\bar{\bar\alpha} = \alpha$. We also have $\tau(h'_\alpha) = \varphi(h'_\alpha) = h'_{\bar\alpha}$, $\tau(ih'_\alpha) = -\tau(h'_\alpha) = -h'_{\bar\alpha}$ and thus $\tau^2 = 1$ on the Cartan subalgebra H. Furthermore, we have $\tau^2(e_\alpha) = \tau(C_\alpha e_{\bar\alpha}) = \bar C_\alpha C_{\bar\alpha} e_{\bar{\bar\alpha}} = \bar C_\alpha C_{\bar\alpha} e_\alpha$. The first of the conditions in (14.1.14) is therefore necessary and sufficient for τ to satisfy (14.1.3).

We now consider (14.1.6), which is equivalent to the three identities

$$\begin{aligned}
\tau([e_\alpha, e_{-\alpha}]) &= [\tau(e_\alpha), \tau(e_{-\alpha})], \\
\tau([h, e_\alpha]) &= [\tau(h), \tau(e_\alpha)], \\
\tau([e_\alpha, e_\beta]) &= [\tau(e_\alpha), \tau(e_\beta)]
\end{aligned} \qquad (14.1.15)$$

for all α, $\beta \in \Delta$ and $h \in H$. The condition $\tau([h_1, h_2]) = [\tau(h_1), \tau(h_2)]$ holds automatically (the Lie algebra H is abelian). Retracing the proof of Proposition IX, we see that the first and third identities in (14.1.15) are equivalent to the second and third identities in (14.1.14). The second identity in (14.1.15) is equivalent to the identity $\tau(\alpha(h)e_\alpha) = [\tau(h), C_\alpha e_\alpha] = C_\alpha \bar\alpha(\tau(h))e_{\bar\alpha}$ and hence to the identity $\overline{\alpha(h)}C_\alpha e_{\bar\alpha} = C_\alpha \bar\alpha(\tau(h))e_{\bar\alpha}$. This in turn follows from the definition of the functional $\bar\alpha$. □

XI. *Retain the hypotheses of Proposition* VIII. *If L_0 is a compact real form of the Lie algebra L, we have $\theta(h) = -h$ for all $h \in H_0$ and there exists a Weyl basis of the Lie algebra L such that $\theta(e_\alpha) = e_{-\alpha}$.*

Proof. Proposition V and compactness of L_0 imply that $(x, \theta(x)) < 0$ for all nonzero x in L. Since $\theta H_0 = H_0$ by Proposition VIII, the restriction of θ to H_0 is a real linear mapping of H_0 into itself (we denote it by σ) such that $\sigma^2 = 1$. Thus the space H_0 is the direct sum of subspaces H_0^+ and H_0^-, which consist of the vectors $h \in H_0$ for which $\sigma h = h$ and $\sigma h = -h$, respectively. For all $h \in H_0^+$, we have $(h, \theta(h)) = (h, \sigma(h)) = (h, h) = \sum_{\alpha \in \Delta}(\alpha(h))^2 \geqslant 0$. If $h \neq 0$, we have $(h, \theta(h)) < 0$ and so h must be 0. This means that $H_0^+ = \{0\}$ and $H_0^- = H_0$. Thus on H_0 the operator θ is multiplication by -1. In particular, we have

$$\theta(h'_\alpha) = -h'_\alpha. \tag{14.1.16}$$

We conclude that $h'_{\bar\alpha} = -h'_\alpha$ and $\bar\alpha = -\alpha$. Choose a Weyl basis in L. We then have

$$N_{\alpha,\beta} = \bar N_{\alpha,\beta} = N_{\bar\alpha,\bar\beta}. \tag{14.1.17}$$

If $\theta e_\alpha = C_\alpha e_{-\alpha}$, the identities (14.1.14) must hold for the numbers C_α. Combined with (14.1.16) and (14.1.17), these relations become

$$\bar C_\alpha C_{-\alpha} = C_\alpha C_{-\alpha} = 1, \qquad C_{\alpha+\beta} = C_\alpha C_\beta. \tag{14.1.18}$$

Note that $(e_\alpha, \theta(e_\alpha)) = C_\alpha(e_\alpha, e_{-\alpha}) = -C_\alpha < 0$, so that $C_\alpha > 0$. We write $\delta_\alpha = (C_\alpha)^{-1/2}$ and $e'_\alpha = \delta_\alpha e_\alpha$. Thus (14.1.18) yields $(e'_\alpha, e'_{-\alpha}) = -\delta_\alpha \delta_{-\alpha} = -1$ and $\theta(e'_\alpha) = C_\alpha \delta_\alpha e_{-\alpha} = (\delta_\alpha)^{-1}e_{-\alpha} = \delta_{-\alpha}e_{-\alpha} = e'_{-\alpha}$. Finally, we define the numbers $N'_{\alpha,\beta}$ by $[e'_\alpha, e'_\beta] = N'_{\alpha\beta}e'_{\alpha+\beta}$. From (14.1.16) and (14.1.17) we get $N'_{\alpha,\beta} = N'_{-\alpha,-\beta}$. Thus $\{e'_\alpha\}$ is the desired Weyl basis. □

XII. *Let L_0 be a real form of the complex semisimple Lie algebra L. Let $\{e_\alpha, \alpha \in \Delta\}$ be a Weyl basis in L. Let θ be an antilinear mapping of L into itself (uniquely) defined by the conditions*

$$\theta(h) = -h, \quad \text{for all } h \in H_0, \tag{14.1.19}$$

$$\theta(e_\alpha) = e_{-\alpha}, \quad \text{for all } \alpha \in \Delta. \tag{14.1.20}$$

In this case θ is an involution in L that satisfies (14.1.4)–(14.1.6). The real form of L associated with θ is L_0, and L_0 is compact.

Proof. Suppose that $\bar{\alpha} = -\alpha$ and $C_\alpha = 1$. The involution θ defined by the formula

$$\theta(h_0 + i\tilde{h}_0 + \sum_{\alpha \in \Delta} \xi_\alpha e_\alpha) = -h_0 + i\tilde{h}_0 + \sum_{\alpha \in \Delta} \bar{\xi}_\alpha e_{-\alpha},$$

for $h_0, \tilde{h}_0 \in H_0$, $\xi_\alpha \in \mathbf{C}$, is an antilinear mapping of L into itself that satisfies (14.1.3)–(14.1.5) and (14.1.19)–(14.1.20). Let us verify that $[\theta(x), \theta(y)] = \theta([x, y])$ for all $x, y \in L$. It suffices to check the following equalities:

$$\theta([e_\alpha, e_{-\alpha}]) = [\theta(e_\alpha), \theta(e_{-\alpha})],$$
$$\theta([h, e_\alpha]) = [\theta(h), \theta(e_\alpha)],$$
$$\theta([e_\alpha, e_\beta]) = [\theta(e_\alpha), \theta(e_\beta)]$$

for all $\alpha, \beta \in \Delta$, $h \in H_0$. Note that $\theta([h_1, h_2]) = 0 = [\theta(h_1), \theta(h_2)]$, since the Lie algebra H is abelian. Proposition X shows that the equality $\theta([e_\alpha, e_{-\alpha}]) = [\theta(e_\alpha), \theta(e_{-\alpha})]$ is equivalent to $C_\alpha C_{-\alpha} = 1$, which holds here because $C_\alpha = 1$. The equality $\theta([e_\alpha, e_\beta]) = [\theta(e_\alpha), \theta(e_\beta)]$ is equivalent to the equality $C_{\alpha + \beta} N_{\alpha\beta} = C_\alpha C_\beta N_{\bar{\alpha}, \bar{\beta}}$, which holds because $C_\alpha = 1$ and the numbers $N_{\alpha, \beta} = N_{-\alpha, -\beta}$ are real and $\bar{\alpha} = -\alpha$. The equality $\theta([h, e_\alpha]) = [\theta(h), \theta(e_\alpha)]$ is equivalent to the equality $C_\alpha \bar{\alpha}(\theta(h))e_{\bar{\alpha}} = \overline{\alpha(h)}\theta(e_\alpha)$, which follows from the definition of the functional $\bar{\alpha}$. We have proved that $\theta([x, y]) = [\theta(x), \theta(y)]$ for all $x, y \in L$. We will show that $(x, \theta(x)) < 0$ for all nonzero x. Write x as

$$h_0 + i\tilde{h}_0 + \sum_{\alpha \in \Delta} \xi_\alpha e_\alpha.$$

We get $\theta(x) = h_0 + i\tilde{h}_0 + \sum_{\alpha \in \Delta} \bar{\xi}_\alpha e_\alpha$, and (9.6.1) implies that

$$(x, \theta(x)) = (h_0 + i\tilde{h}_0, -h_0 + i\tilde{h}_0) + \sum_\alpha \xi_\alpha \bar{\xi}_\alpha (e_\alpha, e_{-\alpha})$$

$$= \sum_\alpha \{\alpha(h)\alpha(\theta(h)) - \xi_\alpha \bar{\xi}_\alpha\}, \tag{14.1.21}$$

where $h = h_0 + i\tilde{h}_0$. Note that $\alpha(\theta(h)) = \overline{\bar{\alpha}(h)} = -\alpha(h)$, in view of (14.1.11) and (14.1.20). Thus (14.1.21) yields

$$(x, \theta(x)) = -\sum_\alpha (|\alpha(h)|^2 + |\xi_\alpha|^2),$$

and therefore $(x, \theta(x)) < 0$ for all nonzero x. $\square$

XIII. *Let L be a semisimple complex Lie algebra with a real form L_0. Let θ be the involution in L defined by L_0. There exists a compact real form L_u of L that is invariant under θ.*

Proof. According to Proposition VII, it suffices to prove that there is an involution τ in L satisfying (14.1.4)–(14.1.6), commuting with θ, and for which $(x, \tau(x)) < 0$ for all nonzero x. The desired Lie algebra L_u can be taken as the set of elements $x \in L$ for which $\tau(x) = x$.

By Proposition VIII, L contains a Cartan subalgebra H that is invariant under the involution θ. According to Proposition IX, there is a Weyl basis in L such that $\theta(e_\alpha) = C_\alpha e_{\bar\alpha}$ for all $\alpha \in \Delta$. We define

$$\tau(e_\alpha) = |C_\alpha| e_{-\alpha}, \qquad \tau(h'_\alpha) = h'_{-\alpha}, \qquad (14.1.22)$$

and extend the mapping τ to an involution on L that satisfies (14.1.4) and (14.1.5). Proposition X shows that τ satisfies (14.1.6) if and only if the numbers $|C_\alpha|$ satisfy (14.1.14). The equality $|C_\alpha| |C_{-\alpha}| = 1$ holds because $C_\alpha C_{-\alpha} = 1$. Thus the first two relations in (14.1.14) hold for the numbers $|C_\alpha|$. Since $N_{\alpha,\beta} = N_{-\alpha,-\beta}$ the third equality in (14.1.14) for the numbers $|C_\alpha|$ is $|C_{\alpha+\beta}| = |C_\alpha| |C_\beta|$. We check this equality. First note that $(\bar\alpha, \bar\beta) = \bar\beta(h'_\alpha) = \overline{\beta(\theta(h'_\alpha))} = \overline{\beta(h'_\alpha)} = \overline{(\alpha, \beta)}$. Since (α, β) is real for all $\alpha, \beta \in \Delta$, we find that $(\bar\alpha, \bar\beta) = (\alpha, \beta)$ for all $\alpha, \beta \in \Delta$. Proposition II in 9.7 shows that $N^2_{\alpha,\beta} = (1/2)(\alpha, \alpha)q_{\alpha,\beta}(1 - p_{\alpha,\beta})$. Since the mapping $\alpha \to \bar\alpha$ is an additive involution of the system of roots Δ onto itself, the linear functional $\beta + k\alpha$ is a root if and only if $\bar\beta + k\bar\alpha \in \Delta$. Therefore we have $q_{\bar\alpha,\bar\beta} = q_{\alpha,\beta}, p_{\bar\alpha,\bar\beta} = p_{\alpha,\beta}$ and $N^2_{\bar\alpha,\bar\beta} = (1/2)(\bar\alpha, \bar\alpha)q_{\bar\alpha,\bar\beta}(1 - p_{\bar\alpha,\bar\beta}) = (1/2)(\alpha, \alpha)q_{\alpha,\beta}(1 - p_{\alpha,\beta}) = N^2_{\alpha,\beta}$ or $|N_{\alpha,\beta}| = |N_{\bar\alpha,\bar\beta}|$. Take absolute values in the last equality in (14.1.14) to obtain $|C_{\alpha+\beta}| = |C_\alpha| |C_\beta|$. Proposition X implies that the involution τ in L satisfies (14.1.4)–(14.1.6).

Let us show that τ and θ commute. Since $\tau\theta(h'_\alpha) = \tau(h'_{\bar\alpha}) = h'_{-\bar\alpha}$ and $\theta\tau(h'_\alpha) = \theta(h'_{-\alpha}) = h'_{-\bar\alpha}$, we see that $\tau\theta$ and $\theta\tau$ coincide on H_0 and thus $\tau\theta = \theta\tau$ on H. It remains to show that $\theta\tau(e_\alpha) = \tau\theta(e_\alpha)$. We have

$$\tau\theta(e_\alpha) = \tau(C_\alpha e_{\bar\alpha}) = \bar{C}_\alpha |C_{\bar\alpha}| e_{-\bar\alpha},$$
$$\theta\tau(e_\alpha) = |C_\alpha|\theta(e_{-\alpha}) = |C_\alpha| C_{-\alpha} e_{-\bar\alpha},$$

and $\bar{C}_\alpha |C_{\bar\alpha}| = |C_\alpha| C_{-\alpha}$, since the first two equalities in (14.1.14) imply that

$$C_\alpha(\bar{C}_\alpha |C_{\bar\alpha}|) = |C_\alpha|^2 |C_\alpha| = |C_\alpha|(|C_\alpha| |C_\alpha|) = |C_\alpha| = |C_\alpha| C_\alpha C_{-\alpha},$$

and $C_\alpha \neq 0$. $\square$

XIV. *Retain the hypotheses of Proposition XIII. Let L_u be as in Proposition XIII. Define $L_u^+ = L_u \cap L_0$ and $L_u^- = L_u \cap iL_0$. Let H_u^- be a maximal commutative Lie subalgebra of L_u^-. Let H_u be a maximal commutative Lie subalgebra of L_u containing H_u^- and define H_u^+ as $H_u \cap L_u^+$. The subspace $H = H_u + iH_u$ is a Cartan subalgebra of L and*

$$H_u = H_u^+ + H_u^-, \qquad H \cap L_0 = H_u^+ + iH_u^-.$$

Proof. Consider any x in L_u. The definitions of L_u^+ and L_u^- show that $x \in L_u^+ (L_u^-)$ if and only if $\theta x = x$ ($\theta x = -x$). Proposition VII shows that

$L_u = L_u^+ + L_u^-$. For all $h \in H_u$ and $x, y \in L_u$, we have

$$((\mathrm{ad}\ h)(x),\ y) + (x, (\mathrm{ad}\ h)(y)) = ([h, x], y) + (x, [h, y])$$
$$= \mathrm{tr}([\mathrm{ad}\ h, \mathrm{ad}\ x]\ \mathrm{ad}\ y + \mathrm{ad}\ x[\mathrm{ad}\ h, \mathrm{ad}\ y])$$
$$= \mathrm{tr}[\mathrm{ad}\ h, \mathrm{ad}\ x\ \mathrm{ad}\ y] = 0.$$

That is, the operator ad h on the space L_u is skew-symmetric with respect to the negative-definite form (x, y). The operator ad h is therefore completely reducible; i.e., every invariant subspace in L_u admits an invariant orthogonal complement. Since H_u is a commutative Lie algebra, the subspace H_u of L_u is invariant under all operators ad h, $h \in H_u$. Thus the orthogonal complement $H_u^\perp$ of the space H_u in L_u is also invariant under all these operators. Since H_u is a maximal commutative Lie subalgebra of L_u, the condition $(\mathrm{ad}\ h)y = 0$ for a given $y \in L_u$ and all $h \in H_u$ implies that $y \in H_u$. Thus the space H_u contains no nonzero root subspaces that correspond to the zero root of the Lie algebra L_u relative to H_u. That is, H_u is the root space that corresponds to the zero root of L_u relative to H_u. Thus H_u is a Cartan subalgebra of L_u (see §8).

We will show that $H = H_u + iH_u$ is a Cartan subalgebra of the Lie algebra L. Choose any $z \in L^0$. Thus for every $h \in H$ there exists a natural number k such that $(\mathrm{ad}\ h)^k z = 0$. We will show that $z \in H$. For every $h \in H_u$ there exists a natural number k such that $(\mathrm{ad}\ h)^k z = 0$. Since L_u is a real form of L, we have $z = x + iy$ where $x, y \in L_u$. The space L_u is invariant under all the operators ad h, $h \in H_u$. Therefore we have $0 = (\mathrm{ad}\ h)^k z = (\mathrm{ad}\ h)^k x + i(\mathrm{ad}\ h)^k y$; that is, $(\mathrm{ad}\ h)^k x = 0$ and $(\mathrm{ad}\ h)^k y = 0$. This means that $x, y \in L_u^0 = H_u$ and thus $z = x + iy \in H_u + iH_u = H$. From this we obtain $L^0 = H$, and so H is a Cartan subalgebra of L.

We will show that the Cartan subalgebra H_u of L_u is invariant under θ. Suppose that $h \in H_u$ and $h_u^- \in H_u^-$. Since H_u^- is contained in H_u and H_u is commutative, we have $[h, h^-] = 0$, which implies that $[\theta(h), \theta(h^-)] = \theta([h, h^-]) = \theta(0) = 0$. We also have $h^- \in H_u^- \subset L_u^-$ and thus $\theta(h^-) = -h^-$. Therefore $\theta(h)$ commutes with all elements of H_u^- and $h - \theta(h)$ also commutes with all elements of H_u^-. Since $h - \theta(h) \in L_u$ and $\theta(h - \theta(h)) = -(h - \theta(h))$, $h - \theta(h)$ belongs to L_u^-. Since H_u^- is a maximal commutative Lie subalgebra of L_u^- and $h - \theta(h)$ commutes with all of its elements, $h - \theta(h)$ belongs to H_u^-. Since $h \in H_u$, $\theta(h)$ belongs to H_u. Thus H_u is invariant under the involution θ, which implies that $H_u = H_u^+ + H_u^-$.

Proposition VII shows that $L_0 = L_u^+ + iL_u^-$. The equalities $H_u = H_u^+ + H_u^-$ and $H = H_u + iH_u$ imply that $H = H_u^+ + H_u^- + iH_u^+ + iH_u^-$. Observe as well that $H_u^+ \subset L_u^+$, $iH_u^- \subset iL_u^-$, and that $H_u^- \cap L_0 = iH_u^+ \cap L_0 = \{0\}$. Thus we obtain $H \cap L_0 = H_u^+ + iH_u^-$. $\square$

XV. *Retain the hypotheses of Proposition XIV. Let H_0 be the set of real linear combinations of the elements h'_α. We then have $H_0 = H_0^+ + H_0^-$, where $H_0^\pm = \{x \in H, \theta x = \pm x\}$. The system Δ of nonzero roots of the Lie algebra*

L relative to H splits into three parts, $\Delta = \Sigma \cup \Delta' \cup (-\Sigma)$, where

$$\Sigma = \{\alpha \in \Delta, \alpha > 0, \bar{\alpha} > 0\},$$
$$\Delta' = \{\alpha \in \Delta, \bar{\alpha} = -\alpha\} = \{\alpha \in \Delta, \alpha(H_0^+) = \{0\}\}.$$

Proof. The space H_0 is invariant under θ and therefore $H_0 = H_0^+ + H_0^-$. Let $h_1^+, \ldots, h_m^+$ be a basis in H_0^+ and let $h_1^-, \ldots, h_n^-$ be a basis in H_0^-. We order H_0 lexicographically, using the basis $h_1^+, \ldots, h_m^+, h_1^-, \ldots, h_n^-$.

We will show that $\bar{\alpha} = -\alpha$ if and only if $\alpha(H_0^+) = \{0\}$. The real linear space H_0^+ is generated by the vectors $h + \theta(h)$, $h \in H_0$. Since α is real-valued on H_0, we see that $\bar{\alpha}(h) = \overline{\alpha(\theta(h))} = \alpha(\theta(h))$ for $h \in H_0$. Thus we have $\bar{\alpha} = -\alpha$ if and only if $\alpha(h + \theta(h)) = 0$ for all $h \in H_0$.

Suppose that $\alpha > 0$ and $\alpha \notin \Delta'$, i.e., $\bar{\alpha} \neq -\alpha$. We then have $\alpha(H_0^+) \neq \{0\}$ and there is a nonnegative integer i such that $\alpha(h_1^+) = \cdots = \alpha(h_i^+) = 0$, $\alpha(h_{i+1}^+) > 0$. Since $h_j^+ \in H_0^+$, we have $\theta(h_j^+) = h_j^+$ and thus $\bar{\alpha}(h_j^+) = \alpha(\theta(h_j^+)) = \alpha(h_j^+)$ for $j = 1, \ldots, m$. It follows that $\bar{\alpha}(h_j^+) = 0$ for $j \leq i$, $\bar{\alpha}(h_{i+1}^+) = \alpha(h_{i+1}^+) > 0$. We have proved that $\bar{\alpha} > 0$ and $\alpha \in \Sigma$. $\square$

XVI. *Let L be a semisimple complex Lie algebra with a real form L_0. Let L_u be a compact real form of L. Let θ and τ be the involutions on L defined by the real forms L_0 and L_u respectively. We define L_u^+ as $L_u \cap L_0$. Suppose that θ and τ commute and that for the root $\alpha \in \Delta$ we have $\bar{\alpha} = -\alpha$. Then θ and τ are equal on L^α, and in particular $\theta(e_\alpha) = e_{-\alpha}$. Furthermore, $x + \theta(x)$ belongs to L_u^+ for all x in L^α.*

Proof. From (14.1.12) we see that $\theta(x) \in L^\alpha$ for all $x \in L^\alpha$. If $\bar{\alpha} = -\alpha$, we then have $\theta(x) \in L^{-\alpha}$ for $x \in L^\alpha$. Proposition X implies that $\tau(x) \in L^{-\alpha}$ for $x \in L^\alpha$; thus the element $y = \theta(x) - \tau(x)$ belongs to the space $L^{-\alpha}$. We will show that y commutes with the subspace H_0^+. For $h \in H_0^+$, we have $h = \theta(h)$. Proposition XV shows that $\alpha(H_0^+) = \{0\}$ and Proposition XI shows that $\tau(h) = -h$ for all $h \in H_0$. Therefore we have $[h, \theta(x)] = \theta([\theta(h), x]) = \theta([h, x]) = \theta(\alpha(h)x) = \overline{\alpha(h)}\theta(x) = 0$ for all $h \in H_0^+$ and similarly

$$[h, \tau(x)] = \tau([\tau(h), x]) = -\tau([h, x]) = -\overline{\alpha(h)}\tau(x) = 0.$$

Thus we obtain $[h, y] = 0$. Note that the real space H_u is generated by the vectors ih'_α and that $iH_0^+ \subset H_u^-$, $iH_0^- \subset H_u^+$. Therefore the real space H_u^- is generated by the set iH_0^+. Since y commutes with H_0^+, it also commutes with H_u^-. We know that the subspace H_u^- is invariant under θ, which implies that $\theta(y)$ commutes with H_u^- and thus $y - \theta(y)$ also commutes with H_u^-. The element $y - \theta(y)$ belongs to L_u^- and H_u^- is a maximal commutative subalgebra of L_u^-. Therefore $y - \theta(y)$ belongs to H_u^-. We also have $y \in L^{-\alpha}$, $\theta(y) \in L^\alpha$ and the sum of the subspaces L^α, $L^{-\alpha}$ and H_u^- is direct. From this we infer that $y = \theta(y) = 0$, or $\theta(x) = \tau(x)$ for all $x \in L^\alpha$. Finally we conclude that $x + \theta(x) = x + \tau(x) \in L_0 \cap L_u = L_u^+$. $\square$

XVII. *Retain the hypotheses of Proposition* XIII. *We write* $L_u^+ = L_u \cap L_0$, $L_u^- = L_u \cap iL_0$. *Then* L_u *is the direct sum of* L_u^+ *and* iL_u^-, *and* L_0 *is the direct sum of the subalgebra* L_u^+ *and a certain solvable subalgebra* M_0.

Proof. We write

$$N = \sum_{\alpha \in \Sigma} L^\alpha, \qquad N' = \sum_{\alpha \in \Sigma} L^{-\alpha}. \tag{14.1.23}$$

If $\alpha \in \Sigma$, then α and $\bar{\alpha}$ are positive roots, which means that $\bar{\alpha}$ and $\bar{\bar{\alpha}} = \alpha$ are positive roots. Thus we have $\bar{\alpha} \in \Sigma$ and the subspaces N and N' are invariant under θ. From (14.1.22) we infer that $\tau(L^\alpha) = L^{-\alpha}$ and thus $\tau(N) = N'$.

Since N is invariant under θ, N is the complex span of the set $N_0 = N \cap L_0$. We will prove that

$$L_0 = L_u^+ \dotplus iH_u^- \dotplus N_0. \tag{14.1.24}$$

The sum in the right side of (14.1.24) is direct. To see this, set $l \in L_u^+$, $h \in iH_u^-$ and $n \in N_0$ and suppose that $l + h + n = 0$. Applying τ to this equality we obtain $l - h + \tau(n) = 0$. Subtracting, we get $2h + n - \tau(n) = 0$. Note that $h \in H$, $n \in N$, $\tau(n) \in N'$ and that the sum of the subspaces H, N and N' is direct. Thus we have $h = n = 0$ and $l = 0$ as well. Therefore the sum of L_u^+, iH_u^- and N_0 is direct. Note that the real space L_0 is generated by the subspace $H \cap L_0 = H_u^+ + iH_u^- \subset L_u^+ + iH_u^-$ and by elements $y = x + \theta(x)$ for $x \in L^\alpha$. If $\bar{\alpha} = -\alpha$, Proposition XVI implies that the element $x + \theta(x)$ is in L_u^+. If $\alpha \in \Sigma$, we find that $\bar{\alpha} > 0$ and $y \in L^\alpha + L^{\bar{\alpha}} \subset N$. Since $y \in L_0$, we see that $y \in N \cap L_0 = N_0$. If $\alpha \in -\Sigma$, we find that $\tau(y) \in L^{-\alpha} + L^{-\bar{\alpha}} \subset N$ and $y + \tau(y) \in L_u^+$. Therefore we have $y \in N_0 + L_u^+$. Hence the real space L_0 is generated by the spaces L_u^+, iH_u^-, and N_0.

Since $[L^\alpha, L^\beta] \subset L^{\alpha+\beta}$ and since $\alpha + \beta \in \Sigma$ if $\alpha, \beta \in \Sigma$, we see that N and N' are Lie subalgebras of L. Since Σ is finite, the inclusion $[L^\alpha, L^\beta] \subset L^{\alpha+\beta}$ implies that N and N' are nilpotent Lie algebras. We write M_0 for $iH_u^- + N_0$. From (14.1.24) we infer that $L_0 = L_u^+ + M_0$. We also have $[N_0, N_0] \subset N_0$, $[iH_u^-, iH_u^-] = \{0\}$, and $[iH_u^-, N_0] \subset N_0$. Therefore M_0 is a Lie subalgebra of L_0, with the property that $[M_0, M_0] \subset N_0$. Since N_0 is nilpotent, M_0 is a solvable Lie algebra. $\square$

14.2. Linear Representations of Real Forms

I. *Let* L *be a complex Lie algebra with a real form* L_0. *Let* π *be a representation of the Lie algebra* L_0 *in the finite-dimensional complex linear space* V. *The formula*

$$\tilde{\pi}(x + iy) = \pi(x) + i\pi(y), \qquad x, y \in L_0, \tag{14.2.1}$$

defines a representation $\tilde{\pi}$ of L in V. The representation π is irreducible if and only if $\tilde{\pi}$ is irreducible. If π_1 is a representation of the Lie algebra L_0 in the space V_1, then π and π_1 are equivalent if and only if $\tilde{\pi}$ and $\tilde{\pi}_1$ are equivalent.

Proof. Proposition I in 14.1 shows that (14.2.1) defines a complex linear mapping of the Lie algebra L into the Lie algebra $gl(V)$. Since

$$[x + iy, x_1 + iy_1] = [x, x_1] - [y, y_1] + i([x, y_1] + [y, x_1]),$$

the condition $\pi([x, y]) = [\pi(x), \pi(y)]$ ensures that $\tilde{\pi}$ is a representation of the Lie algebra L in V. Furthermore (14.2.1) shows that any complex linear subspace $V_1 \subset V$ invariant under the operators $\pi(x)$, $x \in L_0$, is invariant under the operators $\tilde{\pi}(z)$, $z \in L$. Finally, if we have $\pi_1(x) = A\pi(x)A^{-1}$ for all $x \in L_0$, where A is an invertible linear operator carrying V onto V_1, (14.2.1) implies that $\tilde{\pi}_1(z) = A\tilde{\pi}(z)A^{-1}$ for all $z \in L$.   □

II. *Let L be a complex Lie algebra with a real form L_0. Let π be an irreducible representation of L and let ρ be its restriction to L_0. The representation ρ is irreducible. As π runs through a set of all pairwise inequivalent irreducible representations of L, ρ runs through a set of all pairwise inequivalent representations of L_0.*

Proof. Proposition I in 14.1 and (14.2.1) show that the representation π is equivalent to the representation $\tilde{\rho}$. Apply Proposition I to prove the present proposition.   □

14.3. Classification of Real Forms of Simple Complex Lie Algebras

We will list the real forms of complex semisimple Lie algebras of types $(A_n) - (G_2)$, whose root systems are listed in §11. This will give us a list of all real simple Lie algebras, since the complexification of a real simple Lie algebra is a complex simple Lie algebra.

First we introduce some notation. Let 1_m be the identity matrix of order m. We write

$$I_{p,q} = \left\| \begin{matrix} -1_p & 0 \\ 0 & 1_q \end{matrix} \right\|, \quad J_n = \left\| \begin{matrix} 0 & 1_n \\ -1_n & 0 \end{matrix} \right\|, \quad K_{p,q} = \left\| \begin{matrix} I_{p,q} & 0 \\ 0 & I_{p,q} \end{matrix} \right\|. \quad (14.3.1)$$

For a matrix $x \in gl(n, \mathbf{C})$, we denote by x^* the matrix conjugate to x. The transpose of x is denoted by x^+ and the matrix obtained from x by replacing all entries by their complex conjugates is denoted by $\bar{x}$.

We will examine the Lie algebras $u(p, q)$, $su(p, q)$, $su^*(2n)$, $so(p, q)$, $so^*(2n)$, $sp(p, q)$ defined as follows.

The Lie algebra of all matrices $x \in gl(n, \mathbf{C})$ for which $x = -I_{p,q}x^*I_{p,q}$ is written $u(p, q)$.

The Lie subalgebra of $u(p,q)$ consisting of all matrices $x \in u(p,q)$ with trace 0 is written $su(p,q)$.

The Lie algebra of all matrices $x \in gl(n, \mathbf{C})$ for which $\operatorname{tr} x = 0$ and $xJ_n = J_n \bar{x}$, is denoted by $su^*(2n)$.

The Lie algebra of all matrices $x \in gl(n, \mathbf{R})$ for which $\operatorname{tr} x = 0$ and $x = -I_{p,q} x^+ I_{p,q}$ is denoted by $so(p,q)$.

The Lie algebra of all matrices $x \in su^*(2n)$ for which $x = -x^+$ is denoted by $so^*(2n)$.

The Lie algebra of all matrices $x \in gl(2n, \mathbf{C})$ such that $xJ_n + J_n x^+ = 0$ and $x = -K_{p,q} x^* K_{p,q}$ is denoted by $sp(2p, 2q)$ $(p + q = n)$. For the Lie algebra $sp(2n, 0)$ we write $sp(2n)$.

We now list the real forms of the Lie algebras of types $(A_n) - (D_n)$.

Theorem 1.

(1) *The real forms of the complex semisimple Lie algebra $sl(n + 1, \mathbf{C})$ of type (A_n) are the following real Lie algebra: $sl(n + 1, \mathbf{R})$; $su(p,q)$, where $p \geqslant q$ and $p + q = n + 1$ (for $q = 0$ we obtain the compact real form $su(n + 1)$), and if n is odd, the Lie algebra $su^*(n + 1)$.*

(2) *The real forms of the complex semisimple Lie algebra $so(2n + 1, \mathbf{C})$, of type (B_n), are the real Lie algebras $so(p,q)$, where $p + q = 2n + 1$, $p > q$ (for $q = 0$ we obtain the compact real form $so(2n + 1, \mathbf{R})$).*

(3) *The real forms of the complex semisimple Lie algebra $sp(2n, \mathbf{C})$ of type (C_n) are: the real Lie algebras $sp(2n, \mathbf{R})$; $sp(2p, 2q)$, where $p \geqslant q$, $p + q = n$ (for $q = 0$ we obtain the compact real form $sp(2n)$).*

(4) *The real forms of the complex semisimple Lie algebra $so(2n, \mathbf{C})$ of type (D_n) are the real Lie algebras $so(p,q)$, where $p + q = 2n$, $p \geqslant q$ (for $q = 0$ we obtain the compact real form $so(2n, \mathbf{R})$; and if n is even, the Lie algebra $so^*(2n)$.*

The proof of this theorem is rather tedious. We will give the proof only for the Lie algebra $L = sl(n + 1, \mathbf{C})$ of type (A_n).

Let L_0 be a real form of the Lie algebra L with corresponding involution θ. Let L_u be a compact real form of L invariant under θ (see Proposition XIII in 14.1). Let τ be the involution in L corresponding to the form L_u. This means that $\theta\tau = \tau\theta$: we write σ for $\theta\tau$. The mapping σ is an automorphism of L for which $\sigma^2 = (\theta\tau)^2 = 1_L$. The compact real form L_u is invariant under σ. The mapping $\pi: x \to \sigma(x)$ is an $(n + 1)$-dimensional representation of the Lie algebra $L = sl(n + 1, \mathbf{C})$. Since σ is an automorphism, the representation π is irreducible. Propositions V and VI in 13.5 show that π is equivalent either to the identity representation or to the representation $x \to -x^+$. We will consider these two possibilities.

(a) If π is equivalent to the identity representation of the Lie algebra L, there is an invertible matrix A such that $\sigma(x) = AxA^{-1}$ for all $x \in sl(n + 1, \mathbf{C})$. Since $\sigma(\sigma(x)) = x$, we see that $A^2 x A^{-2} = x$ for all $x \in sl(n + 1, \mathbf{C})$; that is, $A^2 = \lambda 1_{n+1}$, with $\lambda \neq 0$. Multiplying A by $\lambda^{-1/2}$, we may suppose that

$A^2 = 1_{n+1}$. Thus the matrix A is similar to the matrix $I_{p,q}$ (see (14.3.1)) for some p, q such that $p + q = n + 1$. Therefore the automorphism σ is similar to the automorphism $\tilde{\sigma}$ defined by

$$\tilde{\sigma}(x) = I_{p,q} x I_{p,q}. \qquad (14.3.2)$$

We write $\tilde{\sigma} = \alpha \sigma \alpha^{-1}$, where α is an automorphism of the Lie algebra $sl(n + 1, \mathbf{C})$. Plainly, a compact real form of $su(n + 1)$ is invariant under $\tilde{\sigma}$. We may suppose that the compact form L_u has been chosen so that $\alpha(su(n + 1)) = L_u$. The mapping $\tilde{\tau} = \alpha \tau \alpha^{-1}$ is the involution in L associated with $su(n + 1)$; that is, the involution $\tilde{\tau}$ is defined by $\tilde{\tau}(x) = -x^*$. We write $\tilde{\theta} = \alpha \theta \alpha^{-1}$. We then have $\tilde{\theta} = \tilde{\sigma}\tilde{\tau}$ and thus

$$\tilde{\theta}(x) = -I_{p,q} x^* I_{p,q}. \qquad (14.3.3)$$

Clearly the set $\tilde{L}_0$ of fixed points for the involution $\tilde{\theta}$ is exactly $su(p, q)$. Since $\alpha(\tilde{L}_0) = L_0$, we see that L_0 is isomorphic to $su(p, q)$.

   (b) If π is equivalent to the representation $x \to -x^+$, there exists an invertible matrix A such that $\sigma(x) = Ax^+ A^{-1}$ for all $x \in sl(n + 1, \mathbf{C})$. Since $\sigma(\sigma(x)) \equiv x$, we have $A(A^{-1})^+ x A^+ A^{-1} = x$ for all $x \in sl(n + 1, \mathbf{C})$. This means that $A^+ A^{-1} = \lambda 1_{n+1}$ and so $A^+ = \lambda A$. We next write $A = (A^+)^+ = \lambda^2 A$ and infer that $\lambda = \pm 1$.

   Suppose first that $\lambda = 1$, which implies that $A^+ = A$ and that A is a symmetric nonsingular matrix. Let $p(\lambda)$ be a polynomial in λ such that $(p(A))^2 = A$ (see the book Gantmaher [1], chapter V). Write B for $p(A)$. We then have $B^+ = p(A^+) = p(A) = B$ and $B^2 = BB^+ = A$. Consider the automorphism α of the Lie algebra L defined by $\alpha(x) = B^{-1}xB$. A direct computation shows that $\alpha \sigma \alpha^{-1}(x) = -x^+$, that is, the automorphism σ is similar to the automorphism $\tilde{\sigma}$ defined by

$$\tilde{\sigma}(x) = -x^+. \qquad (14.3.4)$$

If $\tilde{\tau}$ is the involution corresponding to the compact form $su(n + 1)$, the mapping $\tilde{\theta} = \tilde{\sigma}\tilde{\tau}$ has the form

$$\tilde{\theta}(x) = -(-x^*)^+ = \bar{x}. \qquad (14.3.5)$$

The set $\tilde{L}_0$ of fixed points for the involution $\tilde{\theta}$ is the set of all real matrices in $sl(n + 1, \mathbf{C})$, i.e., $\tilde{L}_0 = sl(n + 1, \mathbf{R})$, and $L_0 \approx \tilde{L}_0 = sl(n + 1, \mathbf{R})$.

   Suppose next that $\lambda = -1$. In this case we have $A^* = -A$ and A is a skew-symmetric nonsingular matrix. This is possible only if $n + 1$ is an even number, say $n + 1 = 2m$. Consider the automorphism α of the Lie algebra L defined by $\alpha(x) = BxB^{-1}$, where B is a certain nonsingular matrix. A direct computation shows that the automorphism $\tilde{\sigma} = \alpha \sigma \alpha^{-1}$ of L has the form

$$\tilde{\sigma}(x) = -\tilde{A}x^+ \tilde{A}^{-1}, \qquad x \in sl(n + 1, \mathbf{C}), \qquad (14.3.6)$$

where

$$\tilde{A} = BAB^+. \tag{14.3.7}$$

The following is proved in the treatise Bourbaki [1], chapter IX, §5, 1. For any nonsingular skew-symmetric matrix A there exists a nonsingular matrix B such that the matrix $\tilde{A}$ in (14.3.7) is J_m. Thus the automorphism $\tilde{\sigma}$ in (14.3.6) has the form

$$\tilde{\sigma}(x) = -J_m x^+ J_m^{-1}. \tag{14.3.8}$$

If we define $\tilde{\tau}(x) = -x^*$ and $\tilde{\theta} = \tilde{\sigma}\tilde{\tau}$, we find

$$\tilde{\theta}(x) = -J_m(-x^*)^+ J_m^{-1} = J_m \bar{x} J_m^{-1}. \tag{14.3.9}$$

The equalities (14.3.9) show that the set of fixed points for the involution $\tilde{\theta}$ is $su^*(n + 1)$, that is, $L_0 \approx su^*(n + 1)$.   □

We now list the real forms of the exceptional complex semisimple Lie algebras of types $(E_6), (E_7), (E_8), (F_4), (G_2)$.

The Lie algebra L of type (E_6) has a compact real form L_u and four non-compact real forms L_1, L_2, L_3, L_4. For these we have $L_1 \cap L_u \approx sp(8)$, $L_2 \cap L_u \approx su(6)$, $L_3 \cap L_u \approx so(10, \mathbf{R}) + \mathbf{R}$ and $L_4 \cap L_u$ is isomorphic to a compact real form of the complex Lie algebra of type (F_4).

The Lie algebra L of type (E_7) has a compact real form L_u and three noncompact real forms L_1, L_2, L_3, where $L_1 \cap L_u \approx su(8)$; $L_2 \cap L_u \approx so(12, \mathbf{R}) + su(2)$; $L_3 \cap L_u \approx \tilde{L} + \mathbf{R}$, $\tilde{L}$ being a compact real form of the complex Lie algebra of type (E_6).

The Lie algebra L of type (E_8) has a compact real form L_u and two non-compact real forms L_1, L_2, where $L_1 \cap L_u \approx so(16, \mathbf{R})$, $L_2 \cap L_u \approx \tilde{L} + su(2)$, $\tilde{L}$ being a compact real form of the complex Lie algebra of type (E_7).

The Lie algebra L of type (F_4) has a compact real form L_u and two noncompact real forms L_1, L_2, where $L_1 \cap L_u \approx sp(6) + su(2)$, $L_2 \cap L_u \approx so(9, \mathbf{R})$.

The Lie algebra L of type (G_2) has a compact real form L_u and a non-compact real form L_0, where $L_0 \cap L_u \approx su(2) + su(2)$.

The proof of these statements as well as a description of the corresponding involutions in terms of roots is found in the article É. Cartan [1*]. See also the article by Sirota and Solodovnikov [1*].

§15. General Theorems on Lie Algebras

I. *Let L be a Lie algebra. Let π be a homomorphism of L into the Lie algebra $gl(V)$ such that $\operatorname{Ker} \pi = \{0\}$. Let A be an abelian ideal in L. Suppose that a nonzero element v of V has the following properties: (1) $\pi(L)v = \pi(A)v$; (2) the* *correspondence* $a \to \pi(a)v$ *is a one-to-one mapping of* A *onto* $\pi(A)v$. *Let* J *be the set of* $x \in L$ *such that* $\pi(x)v = 0$. *Then* J *is a Lie subalgebra of* L *and* L *is the direct sum of the vector spaces* A *and* J.

Proof. Since $\pi([x, y]) = [\pi(x), \pi(y)]$, the definition of J shows that J is a Lie subalgebra of L. By hypothesis $J \cap A$ consists only of zero. If x belongs to L and a is an element of the ideal A such that $\pi(x)v = \pi(a)v$, then we have $\pi(x - a)v = 0$ and so $x - a \in J$. $\square$

II. *Let* L *be a Lie algebra with an abelian ideal* A. *We write* S *for* L/A. *Suppose that* S *is semisimple and that the restrictions of the set of operators* $\mathrm{ad}(x)$, $x \in L$, *to the subspace* $A \subset L$ *is a nonzero, irreducible set of operators on* A. *Then* L *contains a subalgebra* J *that is complementary to* A *and isomorphic to* S.

Proof. We define V as $gl(L)$; that is, V is the linear space of all linear operators on L. We write $\sigma(x)\varphi = [\mathrm{ad}\ x, \varphi]$ for all $x \in L$ and $\varphi \in V$. Thus σ is a homomorphism of L into $gl(V)$. Define subspaces P, Q, R of V as follows: P is the set of all operators $\mathrm{ad}\ a$ for $a \in A$; Q is the set of all $\varphi \in V$ such that $\varphi(L) \subset A$ and $\varphi(A) = \{0\}$; R is the set of all $\varphi \in V$ such that $\varphi(L) \subset A$ and $\varphi|_A$ is a multiple of the identity operator. It is clear that $P \subset Q \subset R$, that all three subspaces are invariant under all $\sigma(x)$, $x \in L$, and that Q has codimension 1 in R. Consider $\varphi \in R$, and suppose that $\varphi|_A$ is multiplication by the number λ. For $a \in A$, we get

$$\sigma(a)\varphi = \mathrm{ad}\ a \circ \varphi - \varphi \circ \mathrm{ad}\ a = -\lambda\ \mathrm{ad}\ a.$$

That is, $\varphi(a)R$ is contained in P for $a \in A$. Thus we may consider the factor spaces Q/P and R/P and the homomorphisms π_1, π of the Lie algebra $S = L/A$ into the Lie algebras $gl(Q/P)$ and $gl(R/P)$ respectively, which are induced by the homomorphism σ. The homomorphism π_1 is the restriction of π to the invariant subspace R/P, which contains the invariant subspace Q/P of codimension 1 in R/P. Since the algebra S is semisimple by hypothesis, the set $\pi_1(S)$ is completely reducible on R/P. Thus there is an invariant complement of the subspace Q/P, and so R/P contains an element $\bar{v}$ invariant under π (that is, $\pi(S)\bar{v} = \{0\}$. We may suppose that the restriction to A of the representatives $v \in R$ of the element $\bar{v}$ is the identity mapping. Let $v \in R$ be any inverse image of the element $\bar{v}$. We will show that v satisfies the hypotheses of Proposition I. For $a \in A$, we have $\sigma(a)v = -\mathrm{ad}\ a$ (since the restriction of the mapping v to A is the identity map). If $\sigma(a)v = 0$, then $\mathrm{ad}\ a$ is 0, which is to say that $[a, x] = 0$ for all $x \in L$. This implies that $(\mathrm{ad}\ x)(A) = \{0\}$ for all $x \in L$. If $a \neq 0$, the set of elements $b \in A$ such that $(\mathrm{ad}\ x)(b) = 0$ is a nonzero subspace of A. This contradicts the condition that the restriction of the set of operators $\mathrm{ad}(x)$, $x \in L$, to the ideal A is a nonzero irreducible set. Thus $a \to \sigma(a)v$ is a one-to-one mapping of the ideal A onto $\sigma(A)v$. Consider any x in L. We must show that $\sigma(x)v$ has the form $\sigma(a)v$ for some a in A. We have $\sigma(a)v = -\operatorname{ad} a$, and so we must show that $\sigma(x)v \in P$. The element $\bar{v}$ is invariant under $\pi(S)$, which implies that $\pi(S)\bar{v} = \pi(L/A)\bar{v} = \{0\}$, that is, $\sigma(L)v \subset P$. Apply Proposition I to the representation σ and the vector v to find that the set J of elements $x \in L$ such that $[\operatorname{ad} x, v] = 0$ is a Lie subalgebra of L and that L is the direct sum of the vector spaces A and J. The canonical mapping of L onto $L/A = S$ is also an isomorphism of the Lie subalgebra J onto S. □

The Levi-Mal'cev Theorem. *Let φ be a homomorphism of the Lie algebra L onto the semisimple Lie algebra S. Let A be the kernel of φ. The radical of L is contained in A and L contains a Lie subalgebra J that is complementary to A and isomorphic to S.*

Proof.

(a) If $A = \{0\}$, the theorem is trivial and therefore we will suppose that $A \neq \{0\}$. We suppose initially that the set B of restrictions of the operators ad x, $x \in L$, to the ideal A is an irreducible set of operators. Let R be the radical of the Lie algebra L (see §3). The Lie algebra $\varphi(R)$ is isomorphic to some factor algebra of the algebra R and $\varphi(R)$ is thus a solvable Lie algebra (see Proposition IV in 1.6). Since R is an ideal in L, $\varphi(R)$ is an ideal in $\varphi(L) = S$. The Lie algebra S is semisimple by hypothesis and therefore the solvable ideal $\varphi(R)$ in S is zero, which is to say that $R \subset A$. Since the ideal R is invariant under all operators ad x, $x \in L$ (see Proposition VI in §3) and the ideal A is irreducible under the set B, we have either $R = \{0\}$ or $R = A$. If $R = \{0\}$, then L is a semisimple Lie algebra and we can take J to be $A^{\perp}$ (see Proposition III in 7.1). If $R = A \neq \{0\}$, A is a solvable ideal. Thus we obtain $A \neq [A, A]$ and the subspace $[A, A]$ is invariant under B. By hypothesis, the ideal A is irreducible under the set B, which means that $[A, A] = \{0\}$, *i.e.*, A is an abelian ideal. If B consists of the zero operator alone, we have $[x, a] = 0$ for all $a \in A$, $x \in L$. Then the ideal A lies in the kernel of the adjoint homomorphism of the Lie algebra L and the set of operators ad x, $x \in L$, can be considered as the image under the homomorphism π of $S = L/A$ in $gl(V)$, defined by the formula $\pi(s) = \operatorname{ad} x$ for $x \in s \in S$, $x \in L$. The Lie algebra S is semisimple and therefore the set of operators $\{\operatorname{ad} x, x \in L\} = \{\pi(s), s \in S\}$ is completely reducible, by Weyl's Theorem. Therefore the Lie algebra L admits a subspace J that is complementary to A and invariant under the set of operators ad x, $x \in L$. In this case J is an ideal in L complementary to A. If B is not the zero operator alone, Proposition II implies the existence of a complementary subalgebra J that satisfies the requirements of the theorem.

(b) We will now prove the Levi-Mal'cev Theorem by induction on the dimension of the ideal A. If dim $A = 1$, then A is an abelian ideal in L and any set of operators in the (one-dimensional) space A is irreducible. By part (a) of the proof, the Levi-Mal'cev Theorem holds. Suppose that the theorem holds for dim $A \leqslant n - 1$, where $n > 1$ and let dim $A = n$. If the set B is irreducible, then again by part (a), the theorem holds. If the set B is reducible, then A admits a proper subspace A_1 that is invariant under B. Thus A_1 is an ideal in L and is contained in A. Consider the homomorphism $\tilde{\varphi}$ of the Lie algebra L/A_1 onto the Lie algebra $L/A = S$, defined by the canonical mapping $\varphi : L \rightarrow L/A$ upon going to the factor algebra L/A_1. Plainly the kernel of the homomorphism $\tilde{\varphi}$ is A/A_1. Since $\dim(A/A_1) <$ dim $A = n$, our inductive hypothesis shows that the Lie algebra L/A_1 contains a Lie subalgebra L_1/A_1 that is isomorphic to S and complementary to A/A_1 in L/A_1. By this we have defined a homomorphism of L/A_1 onto S. The kernel of this homomorphism is the ideal A_1, the dimension of which is strictly less than n. Using the inductive hypothesis once more, we see that the Lie algebra L_1 contains a subalgebra J_1 that is isomorphic to S and complementary to A_1 in J_1. The composite mapping $L \rightarrow L/A_1 \rightarrow L/A$, restricted to L_1, is an isomorphism of J_1 onto L/A. Thus J_1 is the complement of the ideal A in the Lie algebra L. $\quad \square$

III. *Let L be a Lie algebra with radical R. The algebra L contains a semi-simple Lie subalgebra S such that $L = S \dotplus R$.*

Proof. Let R_1 be the radical of the Lie algebra L/R. Write $\tilde{R}$ for the inverse image of R_1 in L. Note that $\tilde{R}$ is a solvable ideal in L, since R and $\tilde{R}/R$ are solvable (see Proposition II in 2.2). Every solvable ideal in L is contained in R (see §3) and therefore we have $\tilde{R} \subset R$, $\tilde{R} = R$ and $R_1 = \{0\}$ so that the Lie algebra L/R is semisimple. By applying the Levi-Mal'cev Theorem to the canonical homomorphism of L onto the semisimple Lie algebra L/R, we see that L contains a semisimple subalgebra S that is isomorphic to L/R and complementary to R. $\quad \square$

We quote without proof a theorem that allows us to reduce the study of Lie algebras to the study of Lie subalgebras of the Lie algebra $gl(n, \mathbf{R})$ or the Lie algebra $gl(n, \mathbf{C})$.

Ado's Theorem. *Every real (or complex) Lie algebra has a faithful finite-dimensional linear representation.*

The proof of this theorem can be found in the books Chevalley [1], Serre [1] and in the Proceedings of the Seminar "Sophus Lie" [1].

Chapter XI

Lie Groups

§1. The Campbell-Hausdorff Formula

1.1. Statement of the Problem

Let G be a Lie group with Lie algebra L. For $x, y \in L$, consider the product $\exp(x) \exp(y)$. In a certain neighborhood of the point $0 \in L$ the exponential mapping is a one-to-one analytic mapping of L into G. Hence there are a neighborhood U of the point 0 in L and an analytic mapping $f : U \times U \to L$ that satisfies

$$\exp(x) \exp(y) = \exp f(x, y) \tag{1.1.1}$$

for all $x, y \in U$. In this section we will find a formula for the mapping f.

We introduce a norm in L that makes L a finite-dimensional normed space. (For example, if $e_1, \ldots, e_m$ is a basis in L, we may define

$$\|x\| = \sum_{i=1}^{m} |x_i|, \quad \text{for } x = \sum_{i=1}^{m} x_i e_i.)$$

Suppose that f is as in (1.1.1) and that $f(0, 0) = 0$. Let L_r denote the set of elements $x \in L$ such that $\|x\| < r$, $r > 0$. Choose a number $\varepsilon > 0$ such that the exponential mapping is one-to-one on the set L_ε. We also select a number $\delta > 0$ such that $\delta < \varepsilon$ and $\exp(L_\delta) \exp(L_\delta) \subset \exp(L_\varepsilon)$. Thus the mapping f defines an analytic mapping of $L_\delta \times L_\delta$ into L_ε. We set

$$\varphi(u, v) = f(ux, vy), \tag{1.1.2}$$

where u and v are numbers, real or complex according as G is a real or complex Lie group. The function $\varphi(u, v)$ is analytic in some neighborhood of the origin in $\mathbf{R}^2$ (or $\mathbf{C}^2$) for example in the neighborhood defined by the conditions $\|ux\| < \delta$, $\|vy\| < \delta$. From (1.1.1) and (1.1.2) we infer that

$$\exp(ux) \exp(vy) = \exp \varphi(u, v) \tag{1.1.3}$$

for all (u, v) in this neighborhood.

We now define

$$\psi(t) = \psi(t, x, y) = \varphi(t, t). \tag{1.1.4}$$

The function ψ is analytic in some neighborhood of the point $t = 0$ (for example, for $\|tx\| < \delta$ and $\|ty\| < \delta$). We define

$$c_n(x, y) = \frac{1}{n!} ((d^n/dt^n)\psi(t, x, y))\Big|_{t=0}$$ (1.1.5)

for all $n \geqslant 0$. Since the function ψ is analytic, we have

$$\psi(t, x, y) = \sum_{n=0}^{\infty} t^n c_n(x, y)$$ (1.1.6)

for all sufficiently small t, and furthermore the series

$$\sum_{n=0}^{\infty} |t|^n \|c_n(x, y)\|$$ (1.1.7)

converges. To define the function $\psi(t, x, y)$ it thus suffices to find the coefficients $c_n(x, y)$. The function ψ defines the function f in view of the identity

$$f(x, y) = \psi(1, x, y)$$ (1.1.8)

(cf. (1.1.1), (1.1.3) and (1.1.4)).

1.2. A Recursion Formula for the Coefficients c_n

By (3.5.15) in chapter IX we have $c_0(x, y) = 0$, $c_1(x, y) = x + y$. We will find a recursion formula for $c_n(x, y)$. We define $g(z) = z^{-1}(1 - e^{-z}) = \sum_{n=0}^{\infty} (-1)^n((n + 1)!)^{-1} z^n$, $z \neq 0$; $g(0) = 1$. Plainly g is an entire function. We next define $h(z) = (g(z))^{-1}$, so that h is analytic in a neighborhood of the point $z = 0$ and $h(0) = 1$. A direct computation shows that $h(-z) = h(z) - z$. Next define $k(z) = h(z) - z/2 = z(1 - e^{-z})^{-1} - z/2$. Thus k is an even function, analytic in a neighborhood of the point $z = 0$ and $k(0) = 1$. We write $k(z) = 1 + \sum_{p=1}^{\infty} k_{2p} z^{2p}$.[60] We define linear operators $k(\mathrm{ad}\, x)$, $g(\mathrm{ad}\, x)$ in the space L by

$$g(\mathrm{ad}\, x) = \sum_{n=0}^{\infty} \frac{(-1)^n}{(n + 1)!} (\mathrm{ad}\, x)^n,$$ (1.2.1)

and

$$k(\mathrm{ad}\, x) = 1 + \sum_{p=1}^{\infty} k_{2p}(\mathrm{ad}\, x)^{2p}$$ (1.2.2)

---

[60] The numbers $k_{2p} \cdot (2p)!$ are called the *Bernoulli numbers*; see, for example, Markuševič [1], chapter 3, §7. The numbers k_{2p} are rational.

for all $x \in L_\delta$. Since the series $\sum_{n=0}^{\infty}((n+1)!)^{-1}\|\operatorname{ad} x\|^n$ and $\sum_{p=0}^{\infty} k_{2p}\|\operatorname{ad} x\|^{2p}$ converge for $x \in L_\delta$, (1.2.1) and (1.2.2) define linear operators $g(\operatorname{ad} x)$, $k(\operatorname{ad} x)$. We can also carry out arithmetic operations with the series (1.2.1) and (1.2.2) just as with complex-valued power series. In particular, the identity $g(z)(k(z) + z/2) = 1$ implies that $g(\operatorname{ad} x)(k(\operatorname{ad} x) + \frac{1}{2}\operatorname{ad} x) = (k(\operatorname{ad} x) + \frac{1}{2}\operatorname{ad} x)g(\operatorname{ad} x) = 1$. Hence the operator $g(\operatorname{ad} x)$ is invertible and

$$(g(\operatorname{ad} x))^{-1} = k(\operatorname{ad} x) + (\operatorname{ad} x)/2. \tag{1.2.3}$$

By (3.6.6) in chapter IX, the operator $g(\operatorname{ad} x)$ for all $x \in L_\delta$ coincides with the differential of the exponential mapping:

$$(d \exp)_x = g(\operatorname{ad} x). \tag{1.2.4}$$

We construct a differential equation satisfied by the function $\psi(t, x, y)$.

I. *Let x, y be elements of L and suppose that the function $\psi(t, x, y)$ is defined by (1.1.3)–(1.1.4). We choose $a > 0$ such that $a\|x\| < \delta$, $a\|y\| < \delta$. In the domain $|t| < a$, the function ψ satisfies the differential equation*

$$d\psi/dt = k(\operatorname{ad} \psi)(x + y) + (1/2)[x - y, \psi], \tag{1.2.5}$$

and the initial condition

$$\psi(0, x, y) = 0. \tag{1.2.6}$$

Proof. The identity (1.1.3) holds for all u, v such that $|u| < a$, $|v| < a$. We will find the differentials of the mappings $\alpha : v \to \exp(ux)\exp(vy)$ and $\beta : v \to \exp \varphi(u, v)$ of the manifold $\{v : |v| < a\}$ into the group G. Let w be the vector field on the manifold $\{v : v < a\}$ defined by $w(v)z = 1$ for all v, $|v| < a$, z being the function defined by $z(v) \equiv v$, $|v| < a$. Let $y(g)$ be the tangent vector defined by the vector field y at the point $g \in G$. The mapping α is a composition of the one-parameter subgroup $\gamma : v \to \exp(vy)$ followed by multiplication by $\exp(ux)$. Hence the definition of the mapping γ, the identity (3.4.18) in Proposition II of 3.4, chapter IX and definition (1.5.1) in chapter IX imply that

$$(d\alpha)_v(\omega(v)) = \exp(ux)(d\gamma)_v(\omega(v))$$
$$= \exp(ux)\exp(vy)y(e) = y(\exp(ux)\exp(vy)). \tag{1.2.7}$$

We also have

$$(d\beta)_v(\omega(v)) = (d \exp)_{\varphi(u,v)}\left(\frac{\partial\varphi}{\partial v}\right). \tag{1.2.8}$$

Thus (1.1.3), (1.2.7) and (1.2.8) show that

$$y(\varphi(u,v)) = (d \exp)_{\varphi(u,v)}\left(\frac{\partial\varphi}{\partial v}\right). \tag{1.2.9}$$

Apply (1.2.4) to the right side of (1.2.9) to find that

$$y(\varphi(u,v)) = g(\text{ad } \varphi(u,v))\left(\frac{\partial\varphi}{\partial v}\right). \tag{1.2.10}$$

The operator $g(\text{ad } \varphi(u,v))$ is invertible. Multiply (1.2.9) by $g(\text{ad } \varphi(u,v))^{-1}$ and apply (1.2.3) to obtain

$$(\partial\varphi/\partial v) = k(\text{ad } \varphi)y + (1/2)[\varphi, y] \tag{1.2.11}$$

for all $|u| < a$, $|v| < a$.
   Consider now the equality

$$\exp(-vy)\exp(-ux) = \exp(-\varphi(u,v)) \tag{1.2.12}$$

(which is an immediate corollary of (1.1.3)). Differentiating the left and right sides of (1.2.12) by u, we obtain

$$-x(\exp(-vy)\exp(-ux)) = (d \exp)_{-\varphi(u,v)}\left(-\frac{\partial\varphi}{\partial u}\right)$$

$$= g(-\text{ad } \varphi)\left(-\frac{\partial\varphi}{\partial u}\right), \tag{1.2.13}$$

which in turn yields

$$\frac{\partial\varphi}{\partial u} = g(-\text{ad } \varphi)^{-1}x \tag{1.2.14}$$

for all $|u| < a$, $|v| < a$. Since k is an even function, we cite (1.2.3) to infer that

$$\frac{\partial\varphi}{\partial u} = k(\text{ad } \varphi)x - (1/2)[\varphi, x] \tag{1.2.15}$$

for all $|u| < a$, $|v| < a$. We also have

$$\frac{d\psi}{dt} = \left(\frac{\partial\varphi}{\partial u} + \frac{\partial\varphi}{\partial v}\right)_{u=v=t}, \tag{1.2.16}$$

and thus (1.2.5) follows from (1.2.16), (1.2.11) and (1.2.15). The identities (1.1.5) and (3.5.15) of chapter IX give (1.2.6). $\square$

II. *For elements* x, y *in* L, *let the coefficients* $c_n(x, y)$ *be defined by* (1.1.5). *Then they are uniquely defined by the recursion formula*

$$(n + 1)c_{n+1}(x, y) = [x - y, c_n(x, y)]/2 + \sum_{p \geqslant 1,\, 2p \leqslant n} k_{2p}$$

$$\times \sum_{\substack{m_1, \ldots, m_{2p} > 0 \\ m_1 + \cdots + m_{2p} = n}} [c_{m_1}(x, y), [\ldots, [c_{m_{2p}}(x, y), x + y] \cdots]]$$

$$\tag{1.2.17}$$

for $n \geqslant 1$ *and by the equality*

$$c_1(x, y) = x + y. \tag{1.2.18}$$

Proof. The equality (1.2.18) follows from (3.5.15) in chapter IX. We will prove (1.2.17). Fix an $n \geqslant 1$. Differentiating the equality (1.1.6), we see that

$$d\psi/dt = c_1(x, y) + 2tc_2(x, y) + \cdots + (n + 1)t^n c_{n+1}(x, y) + o(t^n). \tag{1.2.19}$$

Since the adjoint representation is analytic, (1.1.6) gives us

$$\mathrm{ad}\,\psi(t) = t\,\mathrm{ad}\,c_1(x, y) + \cdots + t^n\,\mathrm{ad}\,c_n(x, y) + o(t^n). \tag{1.2.20}$$

By raising (1.2.20) to a power, we obtain the equality

$$(\mathrm{ad}\,\psi(t))^{2p} = \sum_{2p \leqslant q \leqslant n} t^q \sum_{\substack{m_1 > 0, \ldots, m_{2p} > 0 \\ m_1 + \cdots + m_{2p} = q}} \mathrm{ad}\,c_{m_1}(x, y) \cdots \mathrm{ad}\,c_{m_{2p}}(x, y) + o(t^n)$$

$$\tag{1.2.21}$$

for all p such that $1 \leqslant p \leqslant \frac{1}{2}n$. Also (1.2.20) implies that $\mathrm{ad}\,\psi(t) = t\,\mathrm{ad}\,c_1(x, y) + o(t)$. We may thus substitute (1.2.21) in (1.2.2) with $x = \psi(t)$ and obtain

$$k(\mathrm{ad}\,\psi(t)) = 1 + \sum_{p \geqslant 1,\, 2p \leqslant n} k_{2p}(\mathrm{ad}\,\psi(t))^{2p} + o(t^n)$$

$$= 1 + \sum_{1 \leqslant q \leqslant n} t^q \sum_{p \geqslant 1,\, 2p \leqslant q} k_{2p}$$

$$\times \sum_{\substack{m_1, \ldots, m_{2p} \geqslant 1 \\ m_1 + \cdots + m_{2p} = q}} \mathrm{ad}\,c_{m_1}(x, y) \cdots \mathrm{ad}\,c_{m_{2p}}(x, y) + o(t^n). \tag{1.2.22}$$

Substituting (1.2.19), (1.2.20) and (1.2.22) in (1.2.5) and equating coefficients of t^n on the two sides of the equality, we obtain (1.2.17).

Clearly the coefficients $c_n(x, y)$ are uniquely defined by (1.2.17)–(1.2.18) for all $n \geqslant 1$. $\square$

Note that $c_2(x, y) = (1/2)[x, y]$. From (1.1.1), (1.1.8), (1.1.6), (1.2.17) and (1.2.18), we find

$$f(x, y) = \sum_{n=1}^{\infty} c_n(x, y) \qquad (1.2.23)$$

for $\|x\| < a$, $\|y\| < a$. The coefficients $c_n(x, y)$ are defined by (1.2.17) and (1.2.18). The relations (1.2.23), (1.2.17) and (1.2.18) are called the *Campbell-Hausdorff formula* and the series on the right side of (1.2.23) is called the *Campbell-Hausdorff series*.

1.3. Convergence of the Campbell-Hausdorff Series

Let $q(z)$ be the function of the complex variable z defined by

$$q(z) = 1 + \sum_{p=1}^{\infty} |k_{2p}| z^{2p}. \qquad (1.3.1)$$

By the Cauchy-Hadamard formula, the radius of convergence of the series $1 + \sum_{p=1}^{\infty} k_{2p} z^{2p}$ for the function $k(z)$ is equal to the radius of convergence of the series $q(z)$. The singularities of the function $k(z) = z(1 - e^{-z})^{-1} - z/2$ closest to zero are the points $\pm 2\pi i$. Therefore the radius of convergence of the series for $k(z)$ and also of the series (1.3.1) is 2π.

Consider the differential equation

$$dy/dz = (y/2) + q(y). \qquad (1.3.2)$$

According to the general theory of differential equations, there is a positive number b, $b < 2\pi$, such that (1.3.2) has an analytic solution $y(z)$ in the disk $\{z: |z| < b\}$ for which

$$y(0) = 0. \qquad (1.3.3)$$

We fix this positive number b. Let G be a Lie group with Lie algebra L. Let L be a normed linear space under some norm $\|x\|$, $x \in L$. Let $M \geqslant 1$ be a number such that

$$\|[x, y]\| \leqslant M \|x\| \, \|y\| \qquad (1.3.4)$$

for all $x, y \in L$. Let U be the set of elements $x \in L$ such that $\|x\| < b/2M$, that is,

$$U = L_{b/2M}. \qquad (1.3.5)$$

Note that $b/2M < \pi$, since $b < 2\pi$ and $2M \geqslant 2$.

I. *Let $c_1(x, y) = x + y$ and let the functions $c_n(x, y)$ be defined by the recursion formulas (1.2.17) for all $n > 1$. The series $\sum_{n=1}^{\infty} \|c_n(x, y)\|$ converges for all $x, y \in U$. We write*

$$\sum_{n=1}^{\infty} c_n(x, y) = F(x, y), \qquad x, y \in U. \tag{1.3.6}$$

The function $F(x, y)$ defines an analytic mapping of the manifold $U \times U$ into L, for which

$$\exp(x)\exp(y) = \exp F(x, y) \tag{1.3.7}$$

for all $x, y \in U$.

Proof. For $x, y \in L$, let $r = \max(\|x\|, \|y\|)$. Since $c_1(x, y) = x + y$, we have $\|c_1(x, y)\| \leqslant 2r$. From (1.2.17) we obtain

$$(n + 1)\|c_{n+1}(x, y)\| \leqslant Mr\|c_n(x, y)\| + 2r \sum_{p \geqslant 1,\, 2p \leqslant n} |k_{2p}|M^{2p}$$
$$\times \sum_{\substack{m_1 > 0, \ldots, m_{2p} > 0 \\ m_1 + \cdots + m_{2p} = n}} \|c_{m_1}(x, y)\| \cdots \|c_{m_{2p}}(x, y)\| \tag{1.3.8}$$

for all $n \geqslant 1$. Let $y = y(z)$ be the analytic function in the disk $\{z : |z| < b\}$ that solves Cauchy's problem (1.3.2)–(1.3.3). We write

$$y(z) = \sum_{n=1}^{\infty} \rho_n z^n \tag{1.3.9}$$

for $|z| < b$. Substituting (1.3.9) in (1.3.2) and equating coefficients of z^n on the two sides of the resulting equality, we see that the coefficients ρ_n satisfy the recursion formulas

$$(n + 1)\rho_{n+1}$$
$$= \rho_n/2 + \sum_{p \geqslant 1,\, 2p \leqslant n} |k_{2p}| \sum_{\substack{m_1 > 0, \ldots, m_{2p} > 0 \\ m_1 + \cdots + m_{2p} = n}} \rho_{m_1} \cdots \rho_{m_{2p}}, \tag{1.3.10}$$

along with

$$\rho_1 = 1. \tag{1.3.11}$$

From (1.3.10)–(1.3.11) we see that $\rho_n > 0$ for all $n \geqslant 1$.

By induction on n we will show that

$$\|c_n(x, y)\| \leqslant M^{n-1}(2r)^n \rho_n \tag{1.3.12}$$

for $n \geqslant 1$. Since we have $\|c_1(x, y)\| \leqslant 2r$, (1.3.12) holds for $n = 1$. Next suppose that (1.3.12) holds for all $n = 1, 2, \ldots, m$. Then (1.3.8) and (1.3.10) imply that

$$(m + 1)\|c_{m+1}(x, y)\| \leqslant MrM^{m-1}(2r)^m \rho_m$$

$$+ 2r \sum_{p \geqslant 1, \, 2p \leqslant m} |k_{2p}| M^{2p}$$

$$\times \sum_{\substack{m_1 > 0, \, \ldots, \, m_{2p} > 0 \\ m_1 + \cdots + m_{2p} = m}} M^{m - 2p}(2r)^m \rho_{m_1} \cdots \rho_{m_{2p}}$$

$$= M^m(2r)^{m+1}(m + 1)\rho_{m+1}.$$

That is, (1.3.12) holds for all $n \geqslant 1$. Since the series $\sum_{n=1}^{\infty} \rho_n |z|^n$ converges for $|z| < b$, (1.3.12) implies that the series $\sum_{n \geqslant 1} \|c_n(x, y)\|$ converges for $2Mr < b$, and hence for $x, y \in U$.

For fixed $x, y \in L$ and any t, we have

$$c_n(tx, ty) = t^n c_n(x, y). \tag{1.3.13}$$

(This identity is easy to prove by induction with the aid of (1.2.17).) Thus (1.1.6), (1.1.8), (1.3.6) and (1.3.13) imply that

$$F(tx, ty) = f(tx, ty) \tag{1.3.14}$$

for all sufficiently small t. Since the mappings F and f are analytic in a neighborhood of the point $(0, 0) \in L \times L$, (1.3.14) implies that these functions coincide in some neighborhood of the point $(0, 0) \in L \times L$. In particular, we have $\exp(x) \exp(y) = \exp F(x, y)$ in some neighborhood of $(0, 0) \in L \times L$. An analytic function on a connected set is uniquely defined by its values on any nonvoid open subset; and so we obtain $\exp(x) \exp(y) = \exp F(x, y)$ for all $x, y \in U$. $\quad\square$

1.4. The Connection Between Homomorphisms of Lie Groups and Lie Algebras

I. *Let G, H be connected Lie groups with Lie algebras L, M respectively. Let π_1, π_2 be analytic homomorphisms of G into H. If the homomorphisms $d\pi_1$ and $d\pi_2$ of L into M coincide, then π_1 and π_2 also coincide.*

Proof. Let W be a canonical neighborhood of the element $e \in G$ (see 3.4 in chapter IX). From (3.4.25) in chapter IX, we infer that $\pi_1(g) = \pi_2(g)$ on W. Since G is connected, we have $G = \bigcup_{n \geqslant 1} W^n$ (see Proposition V in 1.2, chapter V) and thus $\pi_1(g) = \pi_2(g)$ for all $g \in G$. $\quad\square$

Therefore a homomorphism of a Lie group is determined by the corresponding homomorphism of its Lie algebra. In particular, an analytic representation of a Lie group G (which is an analytic homomorphism of G into the Lie group G_E, where E is a finite-dimensional linear space) is determined by the corresponding representation of the Lie algebra of G.

Suppose next that we have a representation ρ of the Lie algebra of G in a finite-dimensional complex linear space E. Generally speaking, the representation ρ *cannot* be represented in the form $\rho = d\pi$, where π is a representation of G in the space E.

Example

Let $G = \Gamma^1$ be the unit circle. The Lie algebra L of G is isomorphic to $\mathbf{R}^1$ (see example 2 in 3.2 of chapter IX). Let X be the left invariant vector field on Γ^1 defined by $X(1)(\sin \varphi) = 1$. We then have $X(1)(\cos \varphi) = X(1)(\sqrt{1 - \sin^2 \varphi}) = X(1)(1 + \sin^2 \varphi f(\varphi))$, where $f(\varphi)$ is an analytic function in a neighborhood of the point $\varphi = 0$. Since $X(1)(1) = 0$, the identities (1.4.1) and (1.4.2) in chapter IX show that $X(1)(\cos \varphi) = 0$. Since the vector field X is left invariant, we obtain

$$X(e^{i\theta})(\sin \varphi) = X(1)(\sin(\varphi + \theta)) = X(1)(\sin \varphi \cos \theta + \sin \theta \cos \varphi) = \cos \theta.$$

The identity $X(e^{i\theta})(\cos \varphi) = -\sin \theta$ is proved similarly. A simple induction shows that $X(e^{i\theta})(\cos n\varphi) = -n \sin n\theta$, $X(e^{i\theta})(\sin n\varphi) = n \cos n\theta$ for all integers n. The element X is a basis for the Lie algebra L. Let $\tilde{L}$ be the Lie algebra of the group $\mathbf{C}$ and let Y be the vector field on $\mathbf{C}$ defined by the relation $Yx = 1$, where x is the function $x(z) = z$ on $\mathbf{C}$. Thus Y is a basis in the complex linear space $\tilde{L}$. For any complex number a, the formula $\rho(\lambda x) = \lambda a Y$, $\lambda \in \mathbf{R}$, defines a representation ρ of L in the space $\mathbf{C}$. We already know all the continuous representations of the group Γ^1 (see (d) in 3.3, chapter III). Let $\tilde{\pi}_n$, $n \in \mathbf{Z}$ be the representation of the group Γ^1 defined by $\pi_n(e^{i\varphi}) = e^{in\varphi}$, $0 \leqslant \varphi \leqslant 2\pi$. We then have $d\pi_n(X)(0) = (d\pi_n)_1 X(1)$ (see (3.3.1)) and therefore $d\pi_n(X)(0)(x) = X(1)(x \circ \pi_n) = X(1)(e^{in\varphi}) = X(1)(\cos n\varphi + i \sin \varphi) = (-n \sin 0) + in \cos 0 = in$, that is, $d\pi_n(X) = in Y$. For $a \neq in$, the representation ρ does not have the form $\rho = d\pi$.

In general, therefore, a homomorphism of Lie algebras need correspond to no homomorphism of Lie groups. However, for simply connected Lie groups the situation is as follows.

II. *Let G, H be Lie groups with Lie algebras L, M respectively. Let ρ be a homomorphism of L into M. If G is simply connected, there is an analytic homomorphism π of G into H for which $d\pi = \rho$.*

Proof. Let W be a canonical neighborhood of the identity element in the group G (see 3.4 of chapter IX). Let V be a neighborhood of the zero element in the Lie algebra L which is mapped homeomorphically onto W by the exponential mapping (see V in 3.4, chapter IX). We write

$$\pi(\exp(x)) = \exp(\rho(x)) \tag{1.4.1}$$

for all $x \in V$. We will show that (1.4.1) defines a local homomorphism of G into H. Consider the set $U \subset L$ defined by (1.3.5). The set $U \cap V$ is a neighborhood of zero in L and $\exp(U \cap V)$ is a neighborhood of the identity element e in G. Let U_0 be a neighborhood of e in G such that $U_0^2 \subset \exp(U \cap V)$. For all $g_1, g_2 \in U_0$, we have

$$g_1 = \exp x, \qquad g_2 = \exp y \tag{1.4.2}$$

for uniquely determined $x, y \in U \cap V$. By (1.3.7) we have

$$\exp x \exp y = \exp F(x, y) \tag{1.4.3}$$

where $\exp x \exp y \in \exp(U \cap V)$ and thus $F(x, y) \in U \cap V$ in (1.4.3). From (1.4.3) and (1.4.1) we infer that

$$\pi(g_1 g_2) = \pi(\exp x \exp y) = \pi(\exp F(x, y)) = \exp \rho(F(x, y)). \tag{1.4.4}$$

The representation ρ is linear and therefore continuous. Applying ρ to both sides of (1.3.6), we obtain

$$\rho(F(x, y)) = \sum_{n=1}^{\infty} \rho(c_n(x, y)). \tag{1.4.5}$$

From (1.2.17) and (1.2.18) an easy induction shows that

$$\rho(c_n(x, y)) = c_n(\rho(x), \rho(y)) \tag{1.4.6}$$

for all $n \geqslant 1$. Substituting (1.4.5) and (1.4.6) in (1.4.4) we obtain

$$\pi(g_1 g_2) = \exp \rho(F(x, y)) = \exp F(\rho(x), \rho(y)). \tag{1.4.7}$$

From (1.4.7) and (1.3.6) we then have

$$\pi(g_1 g_2) = \exp F(\rho(x), \rho(y)) = \exp \rho(x) \exp \rho(y)$$
$$= \pi(\exp(x) \exp(y)) = \pi(g_1)\pi(g_2) \tag{1.4.8}$$

for all $g_1, g_2 \in U_0$.

Therefore the mapping π is a local homomorphism of G into H. By hypothesis, G is simply connected. Proposition I in 2.3, chapter VIII, implies that π can be extended to a continuous homomorphism of the group G into the group H, which we also denote by π. From (1.4.1) we conclude that π is an analytic mapping in some neighborhood of the identity element of G. Thus π is an analytic homomorphism of G into H.

Let $d\pi$ be the homomorphism of the Lie algebra L into the Lie algebra M defined by the homomorphism π. By (3.4.25) of chapter IX, we have $\pi(\exp x) = \exp(d\pi(x))$ for all $x \in L$. Comparing this with (1.4.1), we see that

$$\exp(\rho(x)) = \exp(d\pi(x)) \tag{1.4.9}$$

for all $x \in V$. Since the exponential mapping is a homeomorphism in some neighborhood of zero in M, we have $\rho(x) = d\pi(x)$ in this neighborhood, according to (1.4.9). This means that $\rho = d\pi$. $\square$

Theorem. *Let G be a simply connected Lie group with Lie algebra L. Let ρ be a representation of L in the finite-dimensional complex linear space E. There is an analytic representation π of G in E such that $\rho = d\pi$.*

Apply Proposition II to the case $H = G_E$, $M = gl(E)$.

§2. Cartan's Theorem

Let G be a connected Lie group with Lie algebra L. We will suppose that L has been given a norm $\|\cdot\|$. Let U be the neighborhood of zero in L defined by (1.3.5). Let $V \subset U$ be a symmetric neighborhood of zero in L on which the exponential mapping into G is a homeomorphism. We set $W = \exp(V)$. Thus W is a symmetric neighborhood of the identity in G. Let H be a closed subgroup of G. The intersection $W \cap H$ is closed in W. Let F be the subset of V such that $\exp F = W \cap H$. Since $\exp$ is a homeomorphism on V, F is closed in V. Plainly the subset $W \cap H$ has the following properties.

(a) If $g_1, \ldots, g_m \in W \cap H$ and $g_1 \cdots g_m \in W$, then $g_1 \cdots g_m$ belongs to $H \cap W$.
(b) If $g \in H \cap W$, then g^{-1} belongs to $H \cap W$.

Now let M be the set of elements $x \in L$ such that for some $\varepsilon > 0$, we have $\exp(tx) \in H \cap W$ for all $|t| < \varepsilon$; that is, $tx \in F$ for all $|t| < \varepsilon$.

I. *If x belongs to M, then $\lambda x \in M$ for all real λ.*

Proof. For $\lambda \neq 0$, we have $t(\lambda x) \in F$ for $|t| < \varepsilon\lambda^{-1}$. $\square$

II. *If $x \in V \cap M$, then x belongs to F.*

Proof. Since x belongs to M, there is some $\varepsilon > 0$ such that $tx \in F$ for $|t| < \varepsilon$; that is, $\exp(tx) \in H \cap W$ for $|t| < \varepsilon$. We also have $x \in V$, so that $\exp(x) = (\exp(xn^{-1}))^n \in H \cap W$ by property (a). $\square$

III. *Suppose that* $x_n \in F \backslash \{0\}$ *for* $n = 1, 2, \ldots$. *Suppose that* $x_n \to 0$ *and that* $x_n / \|x_n\| \to y$ *in* L. *Then* y *belongs to* M.

Proof. Let L_ε be a ball of radius ε with center at 0 that is contained in V. Let m be a natural number and let $\varepsilon_m = \varepsilon m^{-1}$. We write

$$S_k = \{x \in L : (k-1)\varepsilon_m \leqslant \|x\| \leqslant k\varepsilon_m\} \tag{2.1.1}$$

(note that $S_1 = L_{\varepsilon_m}$). There is a number N_m such that $x_n \in S_1$ for all $n \geqslant N_m$. We choose k so that $1 < k \leqslant m$. Proposition I and the definition of S_k show that for every integer $n \geqslant N_m$, there exists an element $y_n^{(k)}$ of the form jx_n (j is a natural number) that belongs to S_k. Since $x_n / \|x_n\|$ converges to y (note that $\|y\| = 1$) and all S_k are compact, some subsequence of points $y_{n_k}^{(k)}$ converges to a point of the form λy, where $(k-1)\varepsilon_m \leqslant |\lambda| \leqslant k\varepsilon_m$. Since $x_n \in S_1 = L_{\varepsilon_m}$, we see that $jx_n \in L_\varepsilon \subset V$ for all natural numbers j such that $jx_n \in S_k$. This implies that $y_n^{(k)} \in V$. We have $x_n = j^{-1} y_n^{(k)} \in M$ and Proposition II shows that $y_n^{(k)} \in F$. Since $y_n^{(k)} \to \lambda y$, where $\lambda y \in S_m \subset L_\varepsilon$ and the set F is closed in V, we infer that $\lambda y \in F$. Thus we have proved that *for any natural numbers* m *and* k *such that* $1 < k \leqslant m$, *there exists an element* $\lambda y \in F$ *for which* $(k-1)\varepsilon_m \leqslant |\lambda| \leqslant k\varepsilon_m$.

Property (b) shows that the set D of real numbers λ such that $\lambda y \in F$ is symmetric around zero. This implies that D is everywhere dense in the interval $(-\varepsilon, \varepsilon)$. Since the set F is closed, D contains the entire interval $(-\varepsilon, \varepsilon)$ and therefore $y \in M$. $\square$

IV. *For* $x, y \in V$, *we have*

$$\lim_{m \to +\infty} \left(\exp \frac{x}{m} \exp \frac{y}{m} \right)^m = \exp(x+y), \tag{2.1.2}$$

$$\lim_{m \to +\infty} \left[\exp \frac{x}{m}, \exp \frac{y}{m} \right]^{m^2} = \exp[x, y]. \tag{2.1.3}$$

Proof. By hypothesis, the exponential mapping is a homeomorphism on V. Thus it suffices to show that if $\exp(x) \exp(y) = \exp f(x, y)$, where f is defined as in (1.2.17), (1.2.18) and (1.2.23), then

$$\lim_{m \to +\infty} mf \left(\frac{x}{m}, \frac{y}{m} \right) = x + y, \tag{2.1.4}$$

and

$$\lim_{m \to +\infty} m^2 f \left(f \left(\frac{x}{m}, \frac{y}{m} \right), f \left(-\frac{x}{m}, -\frac{y}{m} \right) \right) = [x, y]. \tag{2.1.5}$$

From (1.2.17)–(1.2.18) and induction on n we find

$$c_n\left(\frac{x}{m}, \frac{y}{m}\right) = \frac{1}{m^n} c_n(x, y). \tag{2.1.6}$$

Thus (2.1.4) follows from (2.1.6) and (1.2.23). Furthermore, (2.1.6) and (1.2.23) also imply that

$$f\left(f\left(\frac{x}{m}, \frac{y}{m}\right), f\left(-\frac{x}{m}, -\frac{y}{m}\right)\right) = f\left(\frac{x}{m}, \frac{y}{m}\right) + f\left(-\frac{x}{m}, -\frac{y}{m}\right)$$
$$+ \frac{\left[f\left(\frac{x}{m}, \frac{y}{m}\right), f\left(-\frac{x}{m}, -\frac{y}{m}\right)\right]}{2} + o\left(\frac{1}{m^2}\right), \tag{2.1.7}$$

where we have

$$f\left(\frac{x}{m}, \frac{y}{m}\right) = \frac{x+y}{2} + \frac{1}{2m^2}[x, y] + o\left(\frac{1}{m^2}\right). \tag{2.1.8}$$

Substitute (2.1.8) in (2.1.7) to obtain

$$f\left(f\left(\frac{x}{m}, \frac{y}{m}\right), f\left(-\frac{x}{m}, -\frac{y}{m}\right)\right) = \frac{x+y}{m} + \frac{1}{2m^2}[x, y] + \frac{x+y}{m}$$
$$+ \frac{1}{2m^2}[x, y] + \frac{[x+y, x+y]}{2m^2} + o\left(\frac{1}{m^2}\right)$$
$$= \frac{[x, y]}{m^2} + o\left(\frac{1}{m^2}\right). \tag{2.1.9}$$

$\square$

V. *The set M is a Lie subalgebra of L.*

Proof. If $x, y \in M$, then $\exp(tx)$ and $\exp(ty)$ are in $H \cap W$ for all sufficiently small t. Since H is a subgroup, the elements $\exp(tx)\exp(ty)$ and $[\exp(tx), \exp(ty)]$ also belong to $H \cap W$ for sufficiently small t. We find that $f(tx, ty) \in F$ for sufficiently small t and $f(x/n, y/n) \in F$ for sufficiently large n. Therefore (2.1.2) and Proposition III imply that $x + y \in M$. Similarly, we have $[\exp(x/n), \exp(y/n)] \in F$. Thus (2.1.3) and Proposition III imply that $[x, y] \in M$. Since $\lambda x \in M$ for $x \in M$ (Proposition I), M is a Lie subalgebra of L. $\square$

Theorem 1. *A closed subgroup of a real Lie group is a Lie group.*

Proof. Let G be a Lie group with a closed subgroup H. The intersection of H with the component G_0 of the identity element in G is both an open and closed subgroup (and even a normal subgroup) in H. Thus if $H \cap G_0$ is a Lie group, H is also a Lie group. It is sufficient to prove that the intersection $H \cap G_0$ is a Lie group. Hence we may suppose that G is connected. Let L be the Lie algebra of G and let M be the Lie subalgebra that was defined before Proposition I (see also Proposition V). Let $\tilde{H}$ be the analytic subgroup of G corresponding to M. Consider any g in $W \cap \tilde{H}$. Since $\tilde{H}$ is connected, we have $\tilde{H} = \bigcup_{m=1}^{\infty} (\exp(M \cap V))^m$, and thus $g = \exp x_1 \cdots \exp x_k$, where $x_1, \ldots, x_k$ belong to $M \cap V$. Since $x_1 \in M \cap V \subset M$, we have $tx_i \in F$ for $|t| < \varepsilon_i$ for some $\varepsilon_i > 0$; in particular, x_i/m_i belongs to F for some natural number m_i. This implies that $g = (\exp(x_1/m_1))^{m_1} \cdots (\exp(x_k/m_k))^{m_k}$, where the factors $\exp(x_i/m_i)$ are in $\exp F = H \cap W$. Property (a) and the relation $g \in W$ show that $g \in H \cap W$. We have shown that $\tilde{H} \cap W \subset H \cap W = \exp F$. We will prove that F is contained in a certain neighborhood of zero in the space M. Assume the contrary. We then have a sequence of points $x_n \in F \backslash M$ such that $x_n \to 0$ for $n \to \infty$. Let N be the subspace of L complementary to M. By construction, the mapping of the space $(N \cap V) \times (M \cap V)$ into V defined by $(y, z) \to f(y, z)$, $y \in N \cap V$, $z \in M \cap V$, is a local isomorphism at the point 0 (since the differential of this mapping is defined by the formula $\{y, z\} \to y + z$ and is therefore an isomorphism of tangent spaces). Thus for n sufficiently large, the element x_n can be uniquely represented in the form $x_n = f(y_n, z_n)$, where $y_n \in N \cap V$, $z_n \in M \cap V$. Note that $y_n \to 0$ and $z_n \to 0$ as $n \to \infty$. For n sufficiently large, we therefore have $z_n \in F$ (since $\exp F \supset W \cap \tilde{H}$), and then we have $y_n \in F$ in view of (a) and (b). We may therefore suppose that the sequence x_n lies in N. Since the unit sphere is compact, a subsequence $x_{n_k}/\|x_{n_k}\|$ converges to a nonzero element $y_0 \in N$. We also have $y_0 \in M$, which contradicts the fact that N is the complement of M in L. Thus we have $\exp(F) \subset H$ and the Lie group $\tilde{H}$ is the component of the identity element in H. $\square$

Theorem 2. *Every continuous homomorphism $\varphi: G_1 \to G_2$ of Lie groups is (real) analytic.*

Proof. Let H be the set of all elements in $G_1 \times G_2$ of the form $(g_1, \varphi(g_1))$, $g_1 \in G_1$. Thus H is a closed subgroup of the group G. By theorem 1, H is a Lie subgroup of $G_1 \times G_2$. The mapping $\varphi: (g_1, g_2) \to g_1$ is an analytic homomorphism of the group $G_1 \times G_2$ onto G_1 and its restriction to H is a one-to-one analytic mapping. Clearly it defines an isomorphism of the Lie groups H and G_1. The inverse mapping $g_1 \to (g_1, \varphi(g_1))$ is analytic. Since the mapping $(g_1, g_2) \to g_2$ is also analytic, the composite mapping $g_1 \to (g_1, \varphi(g_1)) \to \varphi(g_1)$ is analytic. $\square$

VI. *Every continuous finite-dimensional representation of a (real or complex) Lie group is real analytic.*

Proof. A continuous representation π of the Lie group G in the finite-dimensional space V is a continuous homomorphism of G into the Lie group G_V. By theorem 2 this homomorphism is real analytic. □

Example

The proof of theorem 1 allows us to determine the Lie algebra of any closed subgroup of a Lie group. Consider a particular case. Define a closed subgroup G of the Lie group $GL(n, \mathbf{R})$ by a set of conditions of the form $F_\alpha(g) = 0$, $\alpha \in A$, where the F_α are analytic functions on G. Identify the Lie algebra of the group $GL(n, \mathbf{R})$ canonically with $gl(n, \mathbf{R})$ (see Proposition VI in 3.2 of chapter IX). The Lie algebra of the Lie group G can then be identified with the Lie subalgebra $M \subset gl(n, \mathbf{R})$ that consists of all $x \in gl(n, \mathbf{R})$ such that $(x^n F_\alpha)(e) = 0$ for all $\alpha \in A$ and all natural numbers n. In the notation of Proposition I, we know that the set M is the Lie algebra of the Lie group G, M is the set of all $x \in gl(n, \mathbf{R})$ such that $\exp(tx) \in G$ for all sufficiently small t; that is, the set of $x \in gl(n, \mathbf{R})$ such that $F_\alpha(\exp(tx)) \equiv 0$ for all sufficiently small t. Since $F_\alpha(\exp(tx))$ is an analytic function, we have $F_\alpha(\exp(tx)) \equiv 0$ if and only if $(d^n/dt^n)F_\alpha(\exp(tx))|_{t=0} = 0$ for all $n \geq 0$; that is, $(x^n F_\alpha)(e) = 0$ for all natural numbers n.

We suppose first that one of the functions F_α has the form $F_{\alpha_0}(g) = \det g - 1$. It is clear that $F_{\alpha_0}(g) = 0$ if and only if $g \in SL(n, \mathbf{R})$. The group $SL(n, \mathbf{R})$ is connected (see Proposition XVI in 1.2, chapter V) and is the kernel of the analytic homomorphism φ of the Lie group $GL(n, \mathbf{R})$ into the Lie group $\mathbf{R}^* = GL(1, \mathbf{R})$ defined by the formula $\varphi(g) = \det g$. Proposition IV in 3.3 shows that the group $SL(n, \mathbf{R})$ is the analytic subgroup that corresponds to the kernel of the homomorphism $d\varphi$ of the Lie algebra $gl(n, \mathbf{R})$. Formula (3.4.25) in chapter VIII shows that $\varphi(\exp(tx)) = \exp(d\varphi(tx)) = \exp(t(d\varphi(x)))$ for all $x \in gl(n, \mathbf{R})$. Reducing the matrix x to its Jordan canonical form, we see that $\varphi(\exp(tx)) = \det(e^{tx}) = e^{t \cdot \operatorname{tr}(x)}$, where $\operatorname{tr}(x)$ is the trace of the matrix x. We thus obtain $e^{t \operatorname{tr}(x)} = e^{t \, d\varphi(x)}$ for all $x \in gl(n, \mathbf{R})$; that is, $d\varphi(x) = \operatorname{tr}(x)$. The Lie algebra of $SL(n, \mathbf{R})$ is then the Lie subalgebra $L \subset gl(n, \mathbf{R})$ of all matrices with trace zero; that is, $L = sl(n, \mathbf{R})$ (see example 2 in 2.1, chapter IX).

We consider yet another important special case. We set $F_\alpha(g) = g^* B g - B$, where B is a nonsingular matrix. The identity $F_\alpha(\exp(tx)) \equiv 0$ is equivalent to $\exp(tx^*)B \exp(tx) - B \equiv 0$ and to $\exp(tx^*)B - B \exp(-tx) = 0$. Differentiating by t and then setting $t = 0$, we see that the matrix x must satisfy the identity $x^* B + Bx = 0$, or $x^* B = -Bx$. Conversely, suppose that $x^* B = -Bx$, x being a matrix in $gl(n, \mathbf{R})$. By induction on n we obtain $(x^*)^n B = (-1)^n B x^n$ for all natural numbers n. This in turn implies that $(d^n/dt^n)(\exp(tx^*)B - B \exp(-tx)|_{t=0} = (x^{*n}B + (-1)^{n+1}Bx^n) = 0$ for all natural numbers n; that is, $\exp(tx^*)B = B\exp(-x)$ for all t, and we have

$F_\alpha(\exp(tx)) \equiv 0$. Similarly, if we have $F_\alpha(g) = g'Bg - B$, then the identity $F_\alpha(\exp(tx)) \equiv 0$ holds if and only if $x'B = -Bx$. Similar constructions for closed subgroups of $GL(n, \mathbf{C})$ can be carried out.

Applying this result to the groups $U(n)$, $SU(n)$, $o(n, \mathbf{R})$, $SO(n, \mathbf{R})$, $Sp(2n)$, $O(n, \mathbf{C})$, $SO(n, \mathbf{C})$, $Sp(2n, \mathbf{C})$, we see that their Lie algebras are $u(n)$, $su(n)$, $o(n, \mathbf{R})$, $so(n, \mathbf{R})$, $sp(2n)$, $o(n, \mathbf{C})$, $so(n, \mathbf{C})$, $sp(2n, \mathbf{C})$, respectively.

§3. Lie's Third Theorem

3.1. Semidirect Products of Lie Groups

Let G, H be connected Lie groups. Suppose that we have a mapping α that associates with each $h \in H$ an analytic isomorphism α_h of the Lie group G onto itself. Suppose further that the following conditions hold.

(1) For all $h_1, h_2 \in H$, we have

$$\alpha_{h_1 h_2} = \alpha_{h_1} \alpha_{h_2}. \tag{3.1.1}$$

(2) The mapping of the product $G \times H$ into G defined by

$$(g, h) \to \alpha_h(g), \qquad g \in G, h \in H, \tag{3.1.2}$$

is an analytic mapping of manifolds.

We define multiplication in the topological space $G \times H$ by

$$(g, h)(g_1, h_1) = (g\alpha_h(g_1), hh_1) \tag{3.1.3}$$

for all $g, g_1 \in G$ and $h, h_1 \in H$. Let e_G, e_H be the identity elements of the groups G and H respectively. It is easy to check that the multiplication defined in (3.1.3) and the product topology induce the structure of a topological group on $G \times H$. The identity element of this group is (e_G, e_H) and inverses are computed by

$$(g, h)^{-1} = (\alpha_{h^{-1}}(g^{-1}), h^{-1}) \tag{3.1.4}$$

for all $g, h \in G$. Since the mapping (3.1.2) is analytic, (3.1.3) and (3.1.4) imply that $G \times H$ is a Lie group under (3.1.3) and (3.1.4). This Lie group is denoted by $G \times_\alpha H$ and is called the *semidirect product of the Lie groups G and H under α*. If α_h is the identity mapping of the group G onto itself for all $h \in H$, then $G \times_\alpha H$ is the ordinary direct product of G and H.

We write

$$g' = (g, e_H), \qquad h' = (e_G, h) \tag{3.1.5}$$

for all $g \in G$, $h \in H$, and also

$$G' = G \times \{e_H\}, \qquad H' = \{e_G\} \times H. \tag{3.1.6}$$

Plainly G' and H' are closed Lie subgroups of $G \times_\alpha H$. From (3.1.3) and (3.1.4) we infer that

$$(g, h)(g_1, h_1)(g, h)^{-1} = (g\alpha_h(g_1)\alpha_{hh_1h^{-1}}(g^{-1}), hh_1h^{-1}) \tag{3.1.7}$$

for all $g, g_1 \in G$, $h, h_1 \in H$. In particular we find

$$(g, h)g'_1(g, h)^{-1} = (g\alpha_h(g_1)g^{-1})', \tag{3.1.8}$$

and

$$h'g'h'^{-1} = \alpha_h(g)'. \tag{3.1.9}$$

From (3.1.8) we see that G' is a normal subgroup of the group $G \times_\alpha H$.

3.2. Semidirect Products of Lie Algebras

We preserve the notation of 3.1. Let L, M be Lie algebras of the groups G and H respectively. Let β_h be the differential of the mapping $\alpha_h : G \to G$. Thus β_h is an automorphism of L for all $h \in H$ (see 3.5 of chapter IX). By (3.4.25) in chapter IX, we have

$$\alpha_h(\exp(x)) = \exp \beta_h(x) \tag{3.2.1}$$

for all $x \in L$. From (3.1.1) we infer that

$$\beta_{h_1 h_2} = \beta_{h_1} \beta_{h_2} \tag{3.2.2}$$

for all $h_1, h_2 \in H$, and (3.2.1) implies that the mapping β is an analytic mapping of the group H into the group G_L. Comparing this with (3.2.2), we see that β is an analytic homomorphism of H into G_L. Let $\gamma = d\beta$ be the corresponding homomorphism of the Lie algebra M of H (see Proposition III in 3.3, chapter IX). This proposition shows that the mapping γ is a homomorphism of M into the Lie algebra $gl(L)$. Note that

$$\beta_h([x, y]) = [\beta_h(x), \beta_h(y)] \tag{3.2.3}$$

for all $x, y \in L$ and $h \in H$, since β_h is an automorphism of L. From (3.2.3) we obtain

$$\gamma(z)[x, y] = [\gamma(z)x, y] + [x, \gamma(z)y] \tag{3.2.4}$$

for all $x, y \in L$ and $z \in M$. That is, γ is a homomorphism of the Lie algebra M into the Lie algebra of differentiations of L.

Let L, M be Lie algebras. Let δ be a homomorphism of M into the Lie algebra $\mathrm{Der}(L)$. In the linear space $L \times M$, we define

$$[(x, y), (x_1, y_1)] = ([x, x_1] + \delta(y)x_1 - \delta(y_1)x, [y, y_1]) \qquad (3.2.5)$$

for all $x, x_1 \in L$ and $y, y_1 \in M$. It is simple to check that $L \times M$ is a Lie algebra under the operation (3.2.5). This Lie algebra is denoted by $L \times_\delta M$ and is called the *semidirect product of the Lie algebras L and M under δ*. We write

$$L' = L \times \{0\}, \qquad M' = \{0\} \times M. \qquad (3.2.6)$$

From (3.2.5) and (3.2.6) we infer that L' is an ideal in $L \times_\delta M$ and that M' is a subalgebra of $L \times_\delta M$. Note too that

$$L' + M' = L \underset{\delta}{\times} M, \qquad L' \cap M' = \{0\}. \qquad (3.2.7)$$

I. *The Lie algebra of the Lie group $G \times_\alpha H$ is isomorphic to the Lie algebra $L \times_\gamma M$.*

Proof. Let A be the Lie algebra of $G \times_\alpha H$. Let L', M' be the Lie subalgebras of A that correspond to the subgroups $G', H' \subset G \times_\alpha H$ (see (3.1.6)). Since G' is a normal subgroup of $G \times_\alpha H$, L' is an ideal in A (see Proposition V in 3.5, chapter IX). From (3.1.6) we also see that $G \times_\alpha H = G'H'$ and $G' \cap H' = \{e\}$. Thus Propositions I and II in 3.3 and (3.5.15) in chapter IX imply that

$$L' + M' = A, \qquad L' \cap M' = \{0\}. \qquad (3.2.8)$$

Let $x \to x'(y \to y')$ be the isomorphism of the Lie algebra L onto the Lie algebra L' (M onto M') defined by the isomorphism $g \to g'(h \to h')$ of the Lie group G onto G' (H onto H'). From (3.1.9) and (3.4.25) in chapter IX, we obtain

$$(\exp \beta_h(x))' = h'(\exp x')h'^{-1} \qquad (3.2.9)$$

for all $h \in H$ and $x \in L$. It follows that

$$\beta_h(x)' = \mathrm{Ad}_{G \times_\alpha H}(h')(x'). \qquad (3.2.10)$$

Differentiating both sides of (3.2.10) as mappings of the group H into L', we see that

$$(\gamma(y)x)' = [y', x'] \qquad (3.2.11)$$

for all $y \in M$ and $x \in L$. From (3.2.11) we infer that

$$
\begin{aligned}
\left[x' + y', x'_1 + y'_1\right] &= \left[x', x'_1\right] + \left[y', x'_1\right] - \left[y'_1, x'\right] + \left[y', y'_1\right] \\
&= \left[x', x'_1\right] + (\gamma(y)x_1)' - (\gamma(y_1)x)' + \left[y', y'_1\right] \quad (3.2.12)
\end{aligned}
$$

for all $x, x_1 \in L$ and $y, y_1 \in M$. Comparing (3.2.12) with (3.2.5) and applying (3.2.7) and (3.2.8) we see that the mapping

$$
(x, y) \to x' + y'
$$

is an isomorphism of the Lie algebras $L \times_\gamma M$ and A. □

II. *Let G, H be simply connected Lie groups, with Lie algebras L, M respectively. Let δ be a homomorphism of M into the Lie algebra $\mathrm{Der}(L)$. There exists exactly one mapping α of the Lie group H into the set of analytic automorphisms of G onto itself that satisfies (1) and (2) in 3.1 and such that the mapping $(x, y) \to x' + y'$ is an isomorphism of the Lie algebra of the group $G \times_\alpha H$ and the Lie algebra $L \times_\delta M$. The group $G \times_\alpha H$ is simply connected.*

Proof. Since H is simply connected, we can use the homomorphism δ to construct an analytic homomorphism β of H into the Lie group $gl(L)$. These homomorphisms are connected by the identity $d\beta = \delta$ (see Proposition II in 1.4). We will show that

$$
\beta(h)[x_1, x_2] = [\beta(h)x_1, \beta(h)x_2] \quad (3.2.13)
$$

for all $h \in H$, $x_1, x_2 \in L$. The group H being connected, it suffices to prove (3.2.13) for all h in some neighborhood of the identity in H. Thus we need only show that the function

$$
\varphi(t) = \beta(\exp(ty))[x_1, x_2] - [\beta(\exp(ty))x_1, \beta(\exp(ty))x_2] \quad (3.2.14)
$$

vanishes identically in t for all $y \in M$ and $x_1, x_2 \in L$. Using (3.4.25) in chapter IX, we may write (3.2.14) as

$$
\varphi(t) = \exp(t\delta(y))[x_1, x_2] - [\exp(t\delta(y))x_1, \exp(t\delta(y))x_2]. \quad (3.2.15)
$$

Differentiating (3.2.15) with respect to t and then setting $t = 0$, we obtain

$$
\left(\frac{d^n \varphi(t)}{dt^n}\right)\bigg|_{t=0} = \delta(y)^n[x_1, x_2] - \sum_{k=0}^{n} C_n^k[\delta(y)^k x_1, \delta(y)^{n-k} x_2]. \quad (3.2.16)
$$

By hypothesis, $\delta(y)$ is a differentiation and thus we have

$$
\delta(y)[x_1, x_2] = [\delta(y)x_1, x_2] + [x_1, \delta(y)x_2] \quad (3.2.17)
$$

for all $y \in M$ and $x_1, x_2 \in L$. Therefore we have $(d\varphi/dt)|_{t=0} = 0$. By induction on n, we infer from (3.2.16) that $(d^n\varphi(t)/dt^n)|_{t=0} = 0$ for all $n \geqslant 1$. Thus we obtain $\varphi(t) = \varphi(0) = 0$ and we have proved (3.2.13).

In other words, then, $\beta(h)$ is an automorphism of the Lie algebra L of the group G. Since G is simply connected, there is exactly one analytic auto-morphism α_h of G onto itself such that $d\alpha_h = \beta(h)$ (see Proposition II in 1.4). Since α_h is determined by the automorphism $\beta(h)$ and $\beta(h_1)\beta(h_2) = \beta(h_1 h_2)$, we have $\alpha_{h_1}\alpha_{h_2} = \alpha_{h_1 h_2}$ for all $h_1, h_2 \in H$.

The relation $d\beta = \delta$ shows that $\beta(\exp(y)) = \exp(\delta(y))$ for all $y \in L$. From this and the equality $d\alpha_{\exp(y)} = \beta(\exp(y))$, we conclude that

$$\alpha_{\exp(y)}(\exp(x)) = \exp(\beta(\exp y)x) = \exp(\exp \delta(y))x) \qquad (3.2.18)$$

for all $x \in L$ and $y \in M$. From (3.2.18) we see that the mapping $(g, h) \rightarrow \alpha_h(g)$ is an analytic mapping of the manifold $G \times H$ into G. The mapping α satisfies conditions (1) and (2) in 3.1. We will now construct the group $G \times_\alpha H$. Since $\delta = d\beta$, where $\beta(h) = d\alpha_h$, Proposition I implies that the Lie algebra of $G \times_\alpha H$ is isomorphic to the Lie algebra $L \times_\delta M$. Our isomorphism is the mapping $(x, y) \rightarrow x' + y'$. The group $G \times_\alpha H$ is homeomorphic to the product $G \times H$ (see 3.1). Since G and H are simply connected, $G \times H$ and $G \times_\alpha H$ are simply connected (see Proposition II in 2.1, chapter IX). $\square$

3.3. Lie's Third Theorem

Theorem. *Let L be a Lie algebra. There exists a simply connected Lie group G whose Lie algebra is isomorphic to L.*

Proof. We suppose first that L is solvable. In this case we prove the theorem by induction on the dimension of the Lie algebra L. If $\dim L = 1$, then we have $L = \mathbf{R}$ or $L = \mathbf{C}$ and so we can take $G = \mathbf{R}$ or $G = \mathbf{C}$. Suppose that $\dim L = n > 1$ and that the theorem has been proved for all solvable Lie algebras of dimension less than n. Since L is solvable, we have $[L, L] \neq L$ (see 1.6 of chapter X). Let I be a linear subspace of L containing $[L, L]$, such that $\dim L/I = 1$. Let J be a one-dimensional subspace in L that is com-plementary to I. Since we have $[I, L] \subset [L, L] \subset I$, I is an ideal in L. Clearly J is a commutative Lie subalgebra of L and we have

$$I + J = L, \qquad I \cap J = \{0\}. \qquad (3.3.1)$$

For $x \in I$ and $y \in J$, we write

$$\delta(y)x = [y, x]. \qquad (3.3.2)$$

Plainly $\delta(y)$ is a differentiation of the Lie algebra I for all $y \in J$ and the map-ping δ is a homomorphism of the Lie algebra J into the Lie algebra $\mathrm{Der}(I)$.

It is easy to check that L is isomorphic to the Lie algebra $I \times_\delta J$. By our inductive hypothesis, there are simply connected Lie groups S and T whose Lie algebras are isomorphic to I and J, respectively. Proposition II in 3.2 shows that there is a simply connected Lie group whose Lie algebra is isomorphic to L.

We now deal with semisimple Lie algebras L. The center of L is zero. Thus the adjoint representation of L is one-to-one and the Lie subalgebra L' of $gl(L)$ consisting of all operators ad x, $x \in L$, is isomorphic to L. Proposition II in 3.3, chapter IX, shows that G_L contains an analytic subgroup G' whose Lie algebra is L'. Let G be the universal covering group of the group G. Thus G is a simply connected Lie group whose Lie algebra is isomorphic to L.

Finally, let L be an arbitrary Lie algebra. Let R be the radical of L. Proposition III in §15, chapter X shows that there is a semisimple Lie subalgebra Q of L such that

$$L = Q + R, \qquad Q \cap R = 0.$$

For $x \in R$ and $y \in Q$ we write

$$\delta(y)x = [y, x].$$

It is simple to show that the Lie algebra L is isomorphic to the Lie algebra $R \times_\delta Q$. Since R is solvable and Q is semisimple, there exist simply connected Lie groups S and T whose Lie algebras are isomorphic to R and Q respectively. Use Proposition II in 3.2 to construct a simply connected Lie group whose Lie algebra is isomorphic to $R \times_\delta Q$, that is, to L. □

§4. Some Properties of Lie Groups in the Large

4.1. Decompositions of Simply Connected Lie Groups

I. *Let G be a simply connected Lie group with Lie algebra L. Let M be an ideal in L and H the analytic subgroup of G corresponding to M. The subgroup H is a closed normal subgroup of G.*

Proof. In view of Proposition V in 3.5, chapter IX, it suffices to show that H is closed. Let S be a Lie group whose Lie algebra is isomorphic to L/M (see the theorem in 3.3). The canonical mapping ρ of the Lie algebra L onto the Lie algebra L/M is a homomorphism. Since the group G is simply connected, there is an analytic homomorphism π of G into S such that $d\pi = \rho$ (see Proposition II in 1.4). Proposition IV in 3.3, chapter IX, shows that the homomorphism π maps G onto the entire group S. Thus H is the component of the identity element in the kernel of the homomorphism π, and so H is closed. □

II. *Let T be a simply connected Lie group with Lie algebra S. Let L be an ideal in S and M a Lie subalgebra of L such that*

$$L + M = S, \qquad L \cap M = \{0\}. \tag{4.1.1}$$

Let G, H be analytic subgroups of the group T corresponding to the Lie subalgebras L and M in S. We define

$$\alpha_h(g) = hgh^{-1} \tag{4.1.2}$$

for all $g \in G$, $h \in H$. The mapping

$$(g, h) \to gh \tag{4.1.3}$$

($g \in G$ and $h \in H$) is an isomorphism of the Lie groups $G \times_\alpha H$ and T. The groups G and H are closed and simply connected and we have

$$GH = T, \qquad G \cap H = \{e\}. \tag{4.1.4}$$

Proof. For $y \in M$, let $\delta(y)$ be the operator on L obtained by restricting the operator ad y in the Lie algebra S to L. It is easy to verify that the mapping ρ defined by

$$\rho(x, y) = x + y \tag{4.1.5}$$

is an isomorphism of the Lie algebra $L \times_\delta M$ onto the Lie algebra S. Let G', H' be simply connected Lie groups with Lie algebras L and M respectively (see the theorem in 3.3). Let $G' \times_{\alpha'} H'$ be the semidirect product of the Lie groups G' and H', whose Lie algebra is isomorphic to the Lie algebra $L \times_\delta M$ (see Proposition II in 3.2). Let τ be the corresponding isomorphism of the Lie algebra of the group $G' \times_{\alpha'} H'$ onto the Lie algebra $L \times_\delta M$. The mapping $\rho \circ \tau$ is an isomorphism of the Lie algebra of the group $G' \times_{\alpha'} H'$ onto the Lie algebra of the group T. Since the groups T and $G' \times_{\alpha'} H'$ are simply connected, there is an analytic isomorphism π of the Lie group $G' \times_{\alpha'} H'$ onto the group T such that $d\pi = \rho \circ \tau$. From our definitions we infer that $\rho \circ \tau(L \times \{0\}) = L$, $\rho \circ \tau(\{0\} \times M) = M$ and thus $\pi(G' \times \{e\}) = G$, $\pi(\{e\} \times H') = H$. $\square$

We now generalize Proposition II.

III. *Let G be a simply connected Lie group with Lie algebra L. Let J, $I_1, \ldots, I_r$ be Lie subalgebras of L such that:*

(1) *L is the direct sum of the linear spaces $J, I_1, \ldots, I_r$;*
(2) *with the notation $Q_0 = J$ and $Q_i = J + I_1 + \cdots + I_i$ for $i \geqslant 1$, the Q_i are Lie subalgebras of L and Q_i is an ideal in Q_{i+1}, $i = 1, \ldots, r - 1$.*

Let $T, S_1, \ldots, S_r$ be the analytic subgroups of G defined by the Lie subalgebras $J, I_1, \ldots, I_r$ respectively. The groups $T, S_1, \ldots, S_r$ are closed and simply connected and the mapping

$$(t, s_1, \ldots, s_r) \to t s_1 \ldots s_r, \tag{4.1.6}$$

$t \in T, s_i \in S_i, i = 1, \ldots, r$, is an analytic isomorphism of the manifold $T \times S_1 \times \cdots \times S_r$ onto G.

Proof. The case $r = 1$ is Proposition II. For $m \geqslant 2$, we suppose that Proposition III is already proved for $r = 1, \ldots, m - 1$. Let H_{m-1} be the analytic subgroup in G defined by the Lie subalgebra Q_{m-1}. Proposition II shows that the subgroups H_{m-1} and S_m are closed and simply connected in G and that the mapping $(h, s_m) \to h s_m$ is an analytic isomorphism of the manifold $H_{m-1} \times S_m$ onto G. To complete the proof, apply the inductive hypothesis to H_{m-1}. $\square$

To apply Propositions II and III, we require a lemma.

IV. *Let L be a Lie algebra. Let I be a maximal ideal in L (that is, I admits no proper superideals). There is a Lie subalgebra M of L such that*

$$I + M = L, \qquad I \cap M = \{0\}. \tag{4.1.7}$$

Proof. Since the Lie algebra L/I has no nontrivial ideals, either L/I is simple or $\dim(L/I) = 1$. If $\dim(L/I) = 1$, then we can take M to be a one-dimensional subspace of L generated by an element $x \in L \backslash I$. If L/I is a simple Lie algebra, it is certainly semisimple and we may apply the Levi-Mal'cev Theorem (§15, chapter X). $\square$

V. *Let G be a simply connected Lie group. Let H be an analytic normal subgroup of G. Then H is closed and the groups H and G/H are simply connected. Let $\pi : G \to G/H$ be the canonical analytic homomorphism of the group G onto G/H. There is an analytic mapping $\rho : G/H \to G$ such that $\pi \circ \rho$ is the identity mapping of G/H onto itself.*

Proof. Let L be the Lie algebra of G. Let I be the ideal in L corresponding to the analytic subgroup H. Using the Jordan-Hölder series for the Lie algebra L/I (see 1.5 of chapter X), we find a sequence of Lie subalgebras $L = L_0 \supset L_1 \supset \cdots \supset L_r = I$ such that L_i is a maximal ideal in L_{i-1} for all $i = 1, \ldots, r$. We write M_i for L_{r-i}. Proposition IV shows the existence of Lie subalgebras $I_i, i = 1, \ldots, r$, such that $M_i = M_{i-1} + I_i, I_i \cap M_{i-1} = \{0\}$. Let S_i be the analytic subgroups of G that correspond to $I_i, i = 1, \ldots, r$. Proposition III shows that the analytic subgroup H is closed and simply connected and that the mapping $(s_1, \ldots, s_r) \to H s_1 \cdots s_r$ is an analytic isomorphism of the manifold $S_1 \times \cdots \times S_r$ onto the group G/H. Since the groups S_i are simply connected, G/H is also simply connected. Finally, for the mapping ρ, we take $\rho(Hs_1 \cdots s_r) = s_1 \cdots s_r$. $\square$

4.2. Commutators in Connected Lie Groups[61]

Let G be a group and g_1, g_2 elements of G. We write

$$[g_1, g_2] = g_1 g_2 g_1^{-1} g_2^{-1}. \tag{4.2.1}$$

If H_1, H_2 are subgroups of G, We write $[H_1, H_2]$ for the subgroup of G generated by all elements $[h_1, h_2]$ for $h_1 \in H_1$ and $h_2 \in H_2$.

I. *The subgroup* $[H_1, H_2]$ *is the set of all elements*

$$[a_1, b_1][b_2, a_2][a_3, b_3] \cdots [a_{2r-1}, b_{2r-1}][b_{2r}, a_{2r}], \tag{4.2.2}$$

where $a_i \in H_1, b_i \in H_2, i = 1, \ldots, 2r, r \geqslant 1$.

Proof. All elements (4.2.2) belong to H_1, H_2. To see this, it suffices to check that $b, [b,a] \in [H_1, H_2]$ for $a \in H_1, b \in H_2$. Since

$$[b, a] = bab^{-1}a^{-1} = (aba^{-1}b^{-1})^{-1} = [a, b]^{-1}, \tag{4.2.3}$$

$[b, a]$ belongs to $[H_1, H_2]$. Products of elements (4.2.2) evidently have the same form, and by (4.2.3), so do inverses of such elements. $\square$

II. *Let* H_1, H_2 *be connected subgroups of a connected topological group* G. *In this case,* $[H_1, H_2]$ *is connected.*

For the proof, see Proposition II in 2.3 of chapter V.

Now let L be a Lie algebra with Lie subalgebras M_1 and M_2. Let $[M_1, M_2]$ denote the linear span of all elements of the form $[x, y]$ with $x \in M_1$ and $y \in M_2$.

III. *Let* G *be a connected Lie group with Lie algebra* L. *Let* M_1, M_2, *and* $N = [M_1, M_2]$ *be Lie subalgebras of* L *such that* $[M_1, N] \subset N$ *and* $[M_2, N] \subset N$. *Let* H_1, H_2 *and* K *be the analytic subgroups of* G *corresponding to the Lie subalgebras* M_1, M_2, *and* N, *respectively. We then have* $K = [H_1, H_2]$.

Proof. Proposition IV in 3.5, chapter IX implies that

$$h_1 K h_1^{-1} \subset K \quad \text{and} \quad h_2 K h_2^{-1} \subset K, \quad \text{for all } h_1 \in H_1, h_2 \in H_2. \tag{4.2.4}$$

---

[61] Editor's note. Professor Naĭmark here repeats, with altered notation, certain material already set forth in 2.3 of chapter V. We have decided to present the present section as he wrote it: the reader may want to refer to 2.3 of chapter V as well.

We will show that the vector $\text{Ad}(h_1)y - y$ belongs to N for all $y \in M_2$ and $h_1 \in H_1$. Applying (3.4.25) in chapter IX to the adjoint representation, we obtain

$$\text{Ad}(\exp(tx))(y) - y = \sum_{n=1}^{\infty} \frac{t^n}{n!} (\text{ad } x)^n(y) \qquad (4.2.5)$$

for all $x \in M_1$, $y \in M_2$, and $t \in \mathbf{R}$. Since $(\text{ad } x)^n(y)$ belongs to N for $n \geqslant 1$, (4.2.5) implies that $\text{Ad}(\exp(tx))(y) - y \in N$. On the other hand, the set F of the elements $h_1 \in H_1$ such that $\text{Ad}(h_1)y - y \in N$ for all $y \in M_2$, is a subgroup in H_1. For, if h_1, h_2 belong to F, (4.2.4) shows that $\text{Ad}(h_1 h_2)y - y = \text{Ad } h_1((\text{Ad } h_2)y - y) + ((\text{Ad } h_1)y - y) \in N$ and

$$\text{Ad}(h_1^{-1})y - y = \text{Ad}(h_1^{-1})(y - \text{Ad}(h_1)y) \in N,$$

so that $h_1 h_2 \in F$ and $h_1^{-1} \in F$. The subgroup F contains all elements $\exp(tx)$ for $x \in M_1$, that is, it contains a neighborhood of the identity. Since H_1 is a connected group, we have $F = H_1$, and so $\text{Ad}(h_1)y - y \in N$ for all $h_1 \in H_1$.

We will show that elements $\text{Ad}(h_1)y - y$, $y \in M_2$, $h_1 \in H_1$, generate the entire Lie algebra N. Let λ be a linear functional on N. It suffices to show that if $\lambda(\text{Ad}(h_1)y - y) = 0$ for all $y \in M_2$, $h_1 \in H_1$, then λ is the zero functional. From (4.2.5) we infer that

$$\frac{d}{dt} \lambda(\text{Ad}(\exp(tx))y - y)\big|_{t=0} = \lambda([x, y]). \qquad (4.2.6)$$

Hence if $\lambda(\text{Ad}(h_1)y - y) = 0$ for all $y \in M_2$ and $h_1 \in H_1$, we conclude that $\lambda([x, y]) = 0$ for all $x \in M_1$ and $y \in M_2$. Since $[M_1, M_2] = N$ by hypothesis, we find $\lambda(N) = \{0\}$.

Let us show that $[H_1, H_2] \subset K$. Let W be the symmetric neighborhood of zero in L that is defined in (1.3.5). For $x, y \in W$, the series on the right side of the Campbell-Hausdorff formula converges and defines a function $F(x, y)$ for which

$$\exp x \exp y = \exp F(x, y) \qquad (4.2.7)$$

for all $x, y \in W$. Let $V \subset W$ be a symmetric neighborhood of zero in L such that $\text{Ad}(\exp x)(y) \in W$ for $x, y \in V$. In particular, for $x \in M_1 \cap V$, $y \in M_2 \cap V$, we have $\text{Ad}(\exp x)(y) \in W$ so that (4.2.7) yields

$$\begin{aligned}
[\exp x, \exp y] &= \exp x \exp y(\exp x)^{-1}(\exp y)^{-1} \\
&= (\text{Ad}(\exp x) \exp y)(\exp(-y)) \\
&= \exp(\text{Ad}(\exp x)y) \exp(-y) \\
&= \exp F(\text{Ad}(\exp x)y, -y). \qquad (4.2.8)
\end{aligned}$$

We write $x' = \text{Ad}(\exp x)y$ and $y' = -y$. We use the formulas (1.2.17)–(1.2.18) for the coefficients $c_n(x, y)$. Since $\text{Ad}(\exp x)y - y$ belongs to N, we have $c_1(x', y') = x' + y' = \text{Ad}(\exp x)y - y \in N$.

We will show by induction that $c_n(x', y') \in N$ for all n. Suppose that $c_m(x', y') \in N$ for $m = 1, \ldots, n$. Since $x' + y' \in N$ and $y' \in M_2$, the hypothesis $c_n(x', y') \in N$ yields

$$[x' - y', c_n(x', y')] = [x' + y', c_n(x', y')] - 2[y', c_n(x', y')] \in N.$$

(Recall that $[M_2, N] \subset N$ by hypothesis.) From (1.2.17) we conclude that $c_{n+1}(x', y') \in N$, and so we have proved that $c_n(x', y') \in N$ for all natural numbers n. Thus (4.2.8) implies that $[\exp x, \exp y] \in \exp N \subset K$. We can therefore find neighborhoods U_1, U_2 of the identity element in the groups H_1 and H_2 respectively such that $[h_1, h_2] \in K$ for $h_1 \in U_1$ and $h_2 \in U_2$. We now note that

$$[h_1, h'_1 h''_2] = [h_1, h'_2](h'_2[h_1, h''_2]h_2'^{-1}) \tag{4.2.9}$$

for all $h_1 \in H_1$ and $h'_2, h''_2 \in H_2$. From (4.2.9) and the inclusion $h'_2 K h_2'^{-1} \subset K$ we infer that $[h_1, h_2] \in K$ for all $h_1 \in U_1$ and $h_2 \in U_2^n$, where n is an arbitrary natural number. Since H_2 is a connected group, we have $\bigcup_{n \geq 1} U_2^n = H_2$, that is, $[h_1, h_2] \in K$ for all $h_1 \in U_1$ and $h_2 \in H_2$. Reversing the roles of h_1 and h_2, we obtain $[H_1, H_2] \subset K$.

We will show finally that the group $[H_1, H_2]$ contains a neighborhood of the identity element in K. Choose elements $y_1, \ldots, y_m \in M_2$ and $h_1^{(1)}, \ldots, h_1^{(m)} \in H_1$ such that the vectors $\text{Ad}(h_1^{(k)})y_k - y_k$, $k = 1, \ldots, m$, generate the space N. Consider the analytic mapping ψ of the direct sum of m copies of the manifold $H_1 \times H_2$ into G, defined by

$$\psi((h_1^{(1)}, h_2^{(1)}), \ldots, (h_1^{(m)}, h_2^{(m)})) = [h_1^{(1)}, h_2^{(1)}] \cdots [h_1^{(m)}, h_2^{(m)}].$$

Proposition II implies that the range of ψ is contained in $[H_1, H_2]$. The inclusion $[H_1, H_2] \subset K$ shows that the range of ψ is contained in K. That is, ψ is an analytic mapping of the product of m copies of the manifold $H_1 \times H_2$ into K. Note that $\psi((h_1^{(1)}, e), \ldots, (h_1^{(m)}, e)) = e$. Proposition VIII in 1.5 of chapter IX shows that the image of ψ contains a neighborhood of the identity in K, provided that $d\psi$ in the point $z = ((h_1^{(1)}, e), \ldots, (h_1^{(m)}, e))$ is a mapping onto the tangent space $\tilde{T}_e(K)$. Apply (3.4.25) of chapter IX to $\pi = \alpha_{h_1^{(k)}}$ for all $k = 1, \ldots, m$ to obtain

$$\frac{d}{dt}\psi((h_1^{(1)}, e), \ldots, (h_1^{(k)}, \exp(ty_k)), \ldots, (h_1^{(m)}, e))\bigg|_{t=0}$$

$$= \frac{d}{dt}(\exp(t\,\text{Ad}(h_1^{(k)})y_k)\exp(-ty_k))\bigg|_{t=0}$$

$$= \text{Ad}(h_1^{(k)})y_k - y_k.$$

Thus the image under the mapping $(d\psi)(z)$ contains all vectors $\mathrm{ad}(h_1^{(k)})y_k - y_k$, and so $(d\psi)(z)$ is a mapping onto $\tilde{T}_e(K)$. $\quad\square$

IV. *Let G be a connected Lie group with Lie algebra L. We write $G^{(1)} = [G, G]$, $G^{(n)} = [G^{(n-1)}, G^{(n-1)}]$, $G_{(1)} = [G, G]$, $G_{(n)} = [G, G_{(n-1)}]$ $(n > 1)$. For every $n \geqslant 1$, the group $G^{(n)}(G_{(n)})$ is the analytic subgroup of the Lie group G corresponding to the Lie subalgebra $L^{(n)}(L_{(n)})$. All groups $G^{(n)}$ and $G_{(n)}$ are normal subgroups of G. If G is simply connected, all of the groups $G^{(n)}$ and $G_{(n)}$ are closed and simply connected.*

See Propositions III and V in 4.1.

V. *A connected Lie group is solvable (nilpotent) if and only if its Lie algebra is solvable (nilpotent).*

See Proposition IV.

VI. *If G is a connected Lie group whose Lie algebra is semisimple, then $G = [G, G]$.*

See Propositions III and IV in 7.1, chapter X.

4.3. The Structure of Solvable Lie Groups

I. *Let L be a solvable Lie algebra of dimension n. There is a basis $e_1, \ldots, e_n$ in L that enjoys the following properties.*

(1) *The linear span L_k of the elements $e_1, \ldots, e_k$ is a Lie subalgebra of L for $k - 1, \ldots, n - 1$.*
(2) *For $k = 1, \ldots, n - 1$, L_k is an ideal in L_{k+1} (note that $L_n = L$).*

Proof. Since L is solvable, $[L, L]$ is not all of L. Let L_{n-1} be a subspace of codimension 1 in L that contains $[L, L]$. We then have $[L, L_{n-1}] \subset [L, L] \subset L_{n-1}$. Thus L_{n-1} is an ideal in L, and so a Lie subalgebra of L.

Suppose that we have constructed Lie subalgebras $L_{n-1}, L_{n-2}, \ldots, L_{k+1}$ of L such that L_m is an ideal in L_{m+1} for $m = k + 1, \ldots, n - 1$ ($L_n = L$). Proposition III in 1.6, chapter X, shows that the Lie subalgebra L_{k+1} is solvable. Let L_k be a subspace of codimension 1 in L_{k+1} containing $[L_{k+1}, L_{k+1}]$. Thus L_k is an ideal in L_{k+1} and also a Lie subalgebra of L. Thus we have constructed the subalgebras L_k by induction. Let e_1 be any nonzero vector in L_1. Suppose that vectors $e_1, \ldots, e_k$ have been constructed so that the linear span of the vectors $e_1, \ldots, e_j$ is L_j for all $j = 1, \ldots, k$. Take any vector in $L_{k+1}\backslash L_k$ as the vector e_{k+1}. Thus the basis $e_1, \ldots, e_n$ is constructed by induction.

II. *We retain the notation of Proposition I. Let G be a simply connected solvable real (complex) Lie group with Lie algebra L. The mapping φ of the* *space* $\mathbf{R}^n(\mathbf{C}^n)$ *into G defined by*

$$\varphi(t_1, \ldots, t_n) = \exp(t_1 e_1) \cdots \exp(t_n e_n), \tag{4.3.1}$$

is an analytic isomorphism of the manifold $\mathbf{R}^n(\mathbf{C}^n)$ *onto G.*

Proof. Let M_k be the one-dimensional Lie subalgebra of L generated by the vector e_k. Let H_k be the analytic subgroup of the group G corresponding to M_k. Apply Proposition III in 4.1 (with $I_k = M_k$ and $J = \{0\}$) to conclude that the H_k are closed one-dimensional subgroups of G and the mapping $(h_1, \ldots, h_n) \rightarrow h_1 h_2 \cdots h_m$ is an analytic isomorphism of the manifold $H_1 \times \cdots \times H_n$ onto G. Since the subgroups H_k are one-dimensional and simply connected, the mapping $t \rightarrow \exp(te_k)$ is an analytic isomorphism of the group $\mathbf{R}(\mathbf{C})$ onto H_k. $\square$

III. *Let G be a simply connected solvable Lie group. Every analytic subgroup H of G is closed and simply connected.*

Proof. Let L be the Lie algebra of the group G. Let M be the Lie subalgebra of L corresponding to the subgroup H. Choose a basis $e_1, \ldots, e_n$ in L as described in Proposition I. Write m for dim M. Choose indices $k_1, \ldots, k_m$ $(1 \leqslant k_1 < k_2 < \cdots < k_m \leqslant n)$ for which the following conditions hold. Writing d_j for $\dim(M \cap L_j)$, $j = 1, \ldots, n$, we have $d_j = 0$ for $j < k_1$, $d_j = i$ for $k_i \leqslant j < k_{i+1}$; $d_j = m$ for $j \geqslant k_m$. Let $f_1, \ldots, f_m$ be a basis in M such that the linear span of the vectors $f_1, \ldots, f_i$ is $M \cap L_{k_i}$ for all $i = 1, \ldots, m$. In the basis $e_1, \ldots, e_m$, replace the vectors e_{k_i} by f_i. This replacement does not alter the subspaces L_k. We may thus suppose that $e_{k_i} = f_i$ for all $i = 1, \ldots, m$. Let M_i be the linear span of the vectors $f_1, \ldots, f_i$. We then have $M_i = L_{k_i} \cap M$ for all $i = 1, \ldots, m$, and so the M_i are Lie subalgebras of M. For $1 \leqslant i \leqslant m - 1$ and $k_i \leqslant k < k_{i+1}$ we have $L_k \cap M = L_{k_i} \cap M$ and therefore

$$\begin{aligned}[M_{i+1}, M_i] &= [L_{k_{i+1}} \cap M, L_{k_{i+1}-1} \cap M] \\ &\subset [L_{k_{i+1}}, L_{k_{i+1}-1}] \cap M \subseteq L_{k_{i+1}-1} \cap M = M_i.\end{aligned} \tag{4.3.2}$$

The relations (4.3.2) show that M_i is an ideal in M_{i+1}.

Let $\tilde{H}$ be the simply connected Lie group that is the universal covering group for H under a homomorphism π (see 3.2, chapter VIII). The Lie algebra M can be considered as the Lie algebra of the group $\tilde{H}$, since the mapping $d\pi$ is an isomorphism of the Lie algebras of $\tilde{H}$ and H (see Proposition VII of 3.3, chapter IX). The properties established for the basis $f_1, \ldots, f_m$ in the Lie algebra M allow us to apply Proposition II to the group H. Thus the mapping $(u_1, \ldots, u_m) \rightarrow \exp_{\tilde{H}}(u_1 f_1) \cdots \exp_{\tilde{H}}(u_m f_m)$ is an analytic isomorphism of an m-dimensional vector space onto $\tilde{H}$. Since $\pi(\tilde{H}) = H$, the mapping

$$(u_1, \ldots, u_m) \rightarrow \exp(u_1 f_1) \cdots \exp(u_m f_m) \tag{4.3.3}$$

is an analytic mapping of an m-dimensional vector space onto the entire group H. On the other hand, the mapping (4.3.3) is the restriction of the mapping φ (defined in (4.3.1)) to the set of points $(t_1, \ldots, t_m)$ such that $t_k = 0$ for $k \neq k_1, \ldots, k \neq k_m$. Since φ is an analytic isomorphism (Proposition II), the image of a linear subspace is closed and simply connected. That is, H is closed and simply connected. $\square$

4.4. Semisimple Lie Algebras: the Levi-Mal'cev Theorem

I. *Let G be a Lie group with Lie algebra L. Let R be the radical of L. Let H be the analytic subgroup of G corresponding to the Lie subalgebra $R \subset L$. The subgroup H is a closed, connected, solvable, normal subgroup of G.*

Proof. By Proposition V in 4.2, we need only to show that H is closed. Let $\bar{H}$ be the closure of H in G; $\bar{H}$ is a closed connected subgroup of G and so is a connected Lie subgroup of G (see Cartan's Theorem in §2). Let M be the Lie subalgebra of L corresponding to $\bar{H}$. We will show that $\bar{H} = H$. We need only show that $M \subset R$. Since R is the radical of L, it suffices to show that M is a solvable Lie algebra. By Proposition V in 4.2, we need only show that the group $\bar{H}$ is solvable. Proposition I in 4.2 implies that the subgroup $[\bar{H}_1, \bar{H}_2]$ is contained in the closure of the group $[H_1, H_2]$ for all subgroups $H_1, H_2 \subset G$. Let n be the least natural number for which $H^{(n)} = \{e\}$. If n is 1, then $[\bar{H}, \bar{H}] \subset \overline{[H, H]} = \overline{H^{(1)}} = \overline{\{e\}} = \{e\}$, which is to say that $\bar{H}$ is solvable. Suppose that $n > 1$ and that the solvability of the group H has been proved for all subgroups H such that $H^{(n-1)} = \{e\}$. Thus we have $[H, H] \subset \overline{[H, H]} = \overline{H^{(1)}}$, where $(H^{(1)})^{n-1} = \{e\}$. By our inductive hypothesis, $(\bar{H})^{(1)}$ is a solvable group. The preceding argument implies that $\bar{H} = H$. Since R is an ideal in L, H is a normal subgroup of G (see Proposition V in 3.5, chapter IX). $\square$

Recall that a topological group G is called *semisimple* if $\{e\}$ is its sole closed connected solvable normal subgroup.

II. *A Lie group is semisimple if and only if its Lie algebra is semisimple.*

Proof. Let G be a semisimple Lie group with Lie algebra L. Let R be the radical of L. If H is the analytic subgroup in G corresponding to R, H is a closed connected solvable normal subgroup of G. If R is not $\{0\}$, then $H \neq \{e\}$, which violates the semisimplicity of G. Thus $R = \{0\}$ and L is semisimple.

Conversely, let L be a semisimple Lie algebra and let N be a connected solvable normal subgroup of G. Let M be the ideal in L corresponding to N. If $N \neq \{0\}$, then $M \neq \{0\}$ and M is solvable by Proposition V in 4.2. This is a contradiction. Therefore $N = \{e\}$ and G is semisimple. $\square$

III (The Levi-Mal'cev Theorem). *Let G be a Lie group with Lie algebra L. Let R be the radical of L. Let S be a semisimple Lie subalgebra of L such that*

$L = S + R$ (see Proposition III, §15, chapter X). *Let H be the analytic sub-group of G corresponding to R. We then have:*

(a) *H is a connected closed solvable normal subgroup of G and the factor group G/H is a semisimple Lie group whose Lie algebra is isomorphic to S;*

(b) *if G is simply connected and T is the analytic subgroup of G corresponding to S, then T is a simply connected closed semisimple subgroup, H is simply connected and the mapping $(t, h) \to th$, $t \in T$, $h \in H$, is an analytic isomorphism of the manifold $T \times H$ onto G.*

Proof. We infer (a) from Propositions II, I, IV and VIII in 3.3, chapter IX. Assertion (b) follows from Proposition II in 4.1. □

IV. *The center Z of a semisimple Lie group is discrete.*

Proof. Let G be a semisimple Lie group with Lie algebra L. Since L is a semisimple Lie algebra (see Proposition II), the center of L is zero. The kernel of the adjoint homomorphism of L is the zero element of L. Thus the mapping $x \to \operatorname{ad} x$, $x \in L$, is an isomorphism of the Lie algebra L onto the Lie algebra $\operatorname{ad}(L)$ of the Lie group $\operatorname{Ad}(G)$. From (3.4.25) of chapter IX, we infer that the adjoint representation of the Lie group G is a local isomorphism. Since the kernel of the adjoint representation of G is the center Z of G, and since the canonical mapping of G onto $\operatorname{Ad}(G) \approx G/Z$ is a local isomorphism, Z is a discrete subgroup of G. □

V. *Let G be a connected semisimple Lie group with Lie algebra L. The adjoint representation of G is a closed subgroup of the group G_L.*

Proof. Let $\operatorname{Ad}: G \to G_L$ be the adjoint representation of the Lie group G. Let $\operatorname{Aut}(L)$ be the set of elements $h \in G_L$ such that $\alpha_h(y) = h(y)$, $y \in L$, is an automorphism α_h of the Lie algebra L. By 3.5 of chapter IX, every element of G_L of the form $\operatorname{Ad} g$, $g \in G$, is in $\operatorname{Aut}(L)$. Plainly $\operatorname{Aut}(L)$ is a closed subgroup of G_L. By §2 and Proposition III in 3.5, chapter IX, the Lie algebra of the group $\operatorname{Aut}(L)$ is all linear operators X in the space L for which $\exp(tX) \in \operatorname{Aut}(L)$ for all $t \in \mathbf{R}$. This condition is equivalent to $X \in \operatorname{Der}(L)$ (compare the proof of Proposition VI, §3, chapter X and the example of §2). On the other hand, the Lie algebra of the adjoint group $\operatorname{Ad} G$ is the image of the Lie algebra L of G under the adjoint homomorphism ad (see Proposition IV in 3.3, chapter IX). However, $\operatorname{ad}(L)$ is the ideal of inner differentiations of the Lie algebra L (see Proposition I in 1.1, chapter X); and every differentiation of a semisimple Lie algebra is inner. Thus the Lie algebra $\operatorname{Der}(L)$ is its ideal of inner differentiations. Hence the Lie algebra of the group $\operatorname{Ad} G$ coincides with the Lie algebra of the closed subgroup $\operatorname{Aut}(L)$ of G_L. The group $\operatorname{Ad} G$ is connected as a continuous image of the connected group G. Hence the group $\operatorname{Ad} G$ is the connected component of the identity element in $\operatorname{Aut}(L)$. Since the component of the identity element is a closed subgroup of a topological group (see Proposition III in 1.2, chapter V) Ad G is closed in $\text{Aut}(L)$. Since $\text{Aut}(L)$ is closed in G_L, Ad G is closed in G_L.   □

§5. Gauss's Decomposition

5.1. Gauss's Decomposition in a Linear Semisimple Complex Lie Group

Let G be an analytic subgroup of the group $GL(n, \mathbf{C})$ and Let L be the Lie algebra of G. We suppose that the Lie subalgebra H of L formed by diagonal matrices in L is a Cartan subalgebra of L. Let N_+ and N_- denote the nilpotent Lie subalgebras of L formed by upper and lower triangular (nilpotent) matrices, respectively. We suppose further that

$$L = N_- + H + N_+ \tag{5.1.1}$$

Let D_0 be the analytic subgroup of G with Lie algebra H. Let Z_+, Z_- be the analytic subgroups of G corresponding to N_+, N_-, respectively (see Propositions I and II in 3.3, chapter IX).

Consider the mapping φ of the product $Z_- \times D_0 \times Z_+$ into the group G defined by

$$\varphi(z_-, d, z_+) = z_- d z_+ \tag{5.1.2}$$

for all $z_- \in Z_-$, $d \in D_0$, $z_+ \in Z_+$. Plainly φ is an analytic mapping of manifolds.

I. *The mapping φ is a homeomorphism of a certain neighborhood of the point $(e, e, e) \in Z_- \times D_0 \times Z_+$ onto a neighborhood U of the identity element in G.*

Proof. From (3.5.15), chapter IX, (5.1.2), and (5.1.1), we infer that the differential of the mapping φ at the point (e, e, e) is a mapping *onto* the tangent space $T_G(e)$. On the other hand, $\dim(Z_- \times D_0 \times Z_+) = \dim G$, and therefore the mapping $d\varphi_{(e,e,e)}$ is an isomorphism of tangent spaces (specifically, φ is regular at the point (e, e, e)). Now apply Proposition VIII of 1.5, chapter IX.

Let $D(n)$ be the subgroup of diagonal matrices in the group $GL(n, \mathbf{C})$ Let $Z_+(n), Z_-(n)$ be the subgroups of upper and lower triangular matrices in $GL(n, \mathbf{C})$ with entries 1 on the main diagonal, respectively. Thus we have $Z_+ \subset Z_+(n)$, $Z_- \subset Z_-(n)$, $D_0 \subset D(n)$. We set $D = D(n) \cap G$. Plainly, H is the Lie algebra of the group D.   □

II. *The groups Z_+ and Z_- are simply connected and $Z_+ = Z_+(n) \cap G$, $Z_- = Z_-(n) \cap G$.*

Proof. The Lie algebras of the groups Z_+ and $Z_+(n) \cap G$ coincide. By Proposition III in 4.3, it suffices to show that the group $Z_+(n)$ is simply connected. Consider $z_+ \in Z_+(n)$, so that $z_+ = 1 + w$, where 1 is the identity matrix and w is a nilpotent matrix. Hence z_+ can be uniquely represented as $\exp(x)$, where x is an upper nilpotent matrix. The exponential mapping of the Lie algebra $N_+(n)$ of the group $Z_+(n)$ into the group $Z_+(n)$ is therefore one-to-one and onto the entire group $N_+(n)$, and so $Z_+(n)$ is homeomorphic to $N_+(n)$. Being homeomorphic to a finite-dimensional linear space, $N_+(n)$ is simply connected. (See chapter VIII, Proposition I in 4.1 and Proposition II in 2.1.) □

Let G_{reg} be the set of elements $g \in G$ admitting a Gauss decomposition in the group $GL(n, \mathbf{C})$, that is, elements g having the form

$$g = z_- \, dz_+, \tag{5.1.3}$$

where $z_- \in Z_-(n)$, $z_+ \in Z_+(n)$, $d \in D(n)$.

III. *The set G_{reg} is connected, open and dense in G.*

Proof. A matrix $g \in G$ is in G_{reg} if and only if all principal minors of the matrix g are distinct from zero, that is, $g \in G \backslash G_{\mathrm{reg}}$ if and only if

$$\Delta_k(g) = 0$$

for at least one k, $k = 1, \ldots, n-1$. The function $\Delta_k(g)$ is a polynomial in the entries of the matrix, and so $\Delta_k(g)$ is an analytic function on the group G. If $\Delta_k(g)$ vanishes on some open set in G, then $\Delta_k(g) \equiv 0$ on G, since G is connected. But $\Delta_k(e)$ is not zero. Therefore the closed set of zeros of the function $\Delta_k(g)$ has void interior and $G \backslash G_{\mathrm{reg}}$ is the union of a finite number of these sets.

Thus the set G_{reg} is the complement in G of the set F of zeros of the analytic function $f(g) = \Delta_1(g) \cdots \Delta_n(g)$, which does not vanish identically on G. Recall that the manifold G is connected. Let g_1, g_2 be elements of G_{reg} and let $\{g(t)\}, t \in [0, 1]$, be a path in G joining g_1 and g_2; that is, $g(0) = g_1, g(1) = g_2$ and $t \to g(t)$ is a continuous mapping of the closed interval $[0, 1]$ into G. The definition of a manifold implies that the set of points $g_2 \in G$ that can be joined by a path $\{g(t)\}$ to a given point $g_1 \in G$ is nonvoid, open, and closed in G. Since G is connected, any two elements g_1, g_2 G can be joined by a path. Cover every point of the path $\{g(t)\}$ with a connected coordinate neighborhood and choose a finite subcovering $U_1, \ldots, U_p$. In each of the connected coordinate neighborhoods U_k, $k = 1, \ldots, p$, the set of zeros of the function $f(g)$ is homeomorphic to the set of zeros of an ordinary analytic function in an open connected subset of the space $\mathbf{C}^n$. If $\Phi(z_1, \ldots, z_n)$ is an analytic function on an open connected subset Ω of the space $\mathbf{C}^n$ not vanishing identically, the complement in Ω of the space N of zeros of the function Φ

is connected (see Gunning and Rossi [1], chapter I). Thus the points g_1 and g_2 are contained in a finite union of connected sets $U_k \backslash F$. Since $\bigcup_{k=1}^{p} U_k$ covers the path $\{g(t), t \in [0, 1]\}$, we may suppose that the sets U_k and U_{k+1} intersect $(k = 1, 2, \ldots, p - 1)$. Hence the sets $U_k \backslash F$, $U_{k+1} \backslash F$ also intersect for all $k = 1, \ldots, p - 1$. Since all of the sets $U_k \backslash F$, $k = 1, \ldots, p$, are connected, their union is also connected. Therefore any two elements g_1, g_2 of the set G_{reg} lie in the connected subset $\bigcup_{k=1}^{p} (U_k \backslash F)$ of G_{reg} and thus G_{reg} is connected. $\square$

IV. *For any* $g \in G_{\text{reg}}$, *the element* $z_+ \in Z_+(n)$ *is uniquely defined by* (5.1.3). *It belongs to* Z_+, *and the mapping* $g \to z_+$ *is analytic on* G_{reg}.

Proof. Let M be the set of points $g \in G_{\text{reg}}$ such that $z_+ \in Z_+$. We know already that $U \subset M$ (see Proposition I). Let M_1 be the closure of the interior of the set M. Since M contains a neighborhood of the identity element in G, the set M_1 is nonvoid. On the other hand, the entries of the matrix z_+ are rational functions of the entries of the matrix g. Hence the mapping $g \to z_+$ is analytic on G_{reg}, and in particular z_+ is a continuous function of g on G_{reg}. Since Z_+ is closed in $GL(n, \mathbf{C})$ and z_+ depends continuously on g, the set M is closed. Hence M_1 is contained in M. We now show that M_1 is open. Consider any g_0 in M_1. Let V be a coordinate neighborhood of the point g_0 in the group G contained in G_{reg}. (Such a V exists, since G_{reg} is open by Proposition III.) Choose a neighborhood W of g_0 in $GL(n, \mathbf{C})$ and a system of analytic coordinates $(y_1, \ldots, y_{n^2})$ in W such that $W \cap G \subset V$ and the submanifold $G \subset GL(n, \mathbf{C})$ is defined by the conditions $y_{m+1} = \cdots = y_{n^2} = 0$, where $m = \dim G$ (see 1.7, chapter IX). Since the mapping $g \to z_+$ is analytic, $y_k(z_+)$ is an analytic function on G_{reg}. On the other hand, the intersection $V \cap W \cap M_1$ contains a subset open in G_{reg} on which $z_+ \in G$. That is, the functions $y_k(z_+)$ vanish on $V \cap W \cap M$ for $k = m + 1$, $\ldots, n^2$. Thus we have $y_k(z_+) = 0$ for $k = m + 1, \ldots, n^2$ on the entire intersection $V \cap W$; i.e., z_+ belongs to G_{reg} for $g \in V \cap W$. Hence M_1 is open in G_{reg}. Since G_{reg} is connected (Proposition III), M_1 and G_{reg} coincide. $\square$

We prove similarly that the element z_- depends analytically on g and that $z_- \in Z_-$ for all $g \in G_{\text{reg}}$. Thus we arrive at the following.

V. *For every* $g \in G_{\text{reg}}$, *the element* $d \in D(n)$ *defined by* (5.1.3) *belongs to* G_{reg}. *That is,* d *belongs to* D *and the mapping* $g \to d$ *is analytic.*

VI. *The group* D *is connected and coincides with the subgroup* D_0.

Proof. Consider the mapping ψ of the manifold $Z_-(n) \times D(n) \times Z_+(n)$ into the group $GL(n, \mathbf{C})$ defined by $\psi(z_-, d, z_+) = z_- d z_+$ for all $z_- \in Z_-(n)$, $d \in D(n), z_+ \in Z_+(n)$. Plainly ψ is an analytic mapping of manifolds. Furthermore, ψ is a homeomorphism of the manifold $M = Z_-(n) \times D(n) \times Z_+(n)$

onto the image $\psi(M)$ of the manifold M in $GL(n, \mathbf{C})$. Thus the restriction of ψ to $Z_- \times D \times Z_+$ is a homeomorphism of $Z_- \times D \times Z_+$ onto G_{reg}. Since G_{reg} is connected, $Z_- \times D \times Z_+$ is connected, and so D is connected. Therefore the group D coincides with the analytic subgroup D_0. $\square$

We have proved the following theorem.

Theorem. *Let G be an analytic subgroup of the group $GL(n, \mathbf{C})$. Let L be the Lie algebra of G. Suppose that the Lie subalgebra $H \subset L$ consisting of the diagonal matrices in L is a Cartan subalgebra of L. Let N_+ and N_- be the nilpotent Lie subalgebras of L consisting of upper and lower triangular nilpotent matrices, respectively. Suppose that $L = N_+ + H + N_-$. Let D, Z_+, Z_- be the analytic subgroups of G corresponding to the Lie subalgebras H, N_+ and N_-, respectively. The group G contains a connected open dense set G_{reg} such that every element $g \in G_{\text{reg}}$ can be written uniquely as*

$$g = z_- d z_+, \tag{5.1.4}$$

where $z_- \in Z_-$, $d \in D$, $z_+ \in Z_+$. The elements z_-, d, z_+ depend analytically on $g \in G_{\text{reg}}$. The groups Z_+ and Z_- are simply connected and $Z_+ = Z_+(n) \cap G$, $Z_- = Z_-(n) \cap G$, $D = D(n) \cap G$. The set G_{reg} contains a neighborhood of the identity element in G.

The relation (5.1.4) is called *Gauss's decomposition of the element $g \in G$.*

5.2. Gauss's Decomposition in a Connected Semisimple Lie Group

Let G be a connected semisimple complex Lie group. Let L be the Lie algebra of G and H a Cartan subalgebra of L. Let Δ be the system of roots of L relative to H. Let H_0 be the real linear span of the system of vectors h'_α, $\alpha \in \Delta \backslash \{0\}$ (see 9.2 and 9.6 of chapter X). Let Δ_+ be the system of positive roots of L under some lexicographic ordering of the space H_0. We set

$$N_+ = \sum_{\alpha \in \Delta_+} L^\alpha, \qquad N_- = \sum_{\alpha \in -\Delta_+} L^\alpha, \tag{5.2.1}$$

where L^α are the root subspaces of L (see §8, chapter X). By Proposition I in §12 of chapter X, the subspaces N_+ and N_- are nilpotent Lie algebras and we have

$$L = N_- + H + N_+. \tag{5.2.2}$$

I. *There is a basis in the Lie algebra L such that the operators ad h, $h \in H$, have diagonal matrices and the operators ad x, $x \in N_+$ (ad $x \in N_-$) are upper (lower) triangular matrices with zeros on the main diagonal.*

Proof. Let $\{h_\alpha, e_\alpha\}$ be the Weyl basis in the Lie algebra L defined by a Cartan subalgebra H. Write r for dim H. We order the elements of the Weyl basis as follows:

$$\{e_\alpha, \alpha \in -\varDelta_+\,;\, h_1, \ldots, h_r\,;\, e_\alpha, \alpha \in \varDelta_+\}. \tag{5.2.3}$$

We will suppose that the vectors e_α are ordered in such a way that the vectors $h'_\alpha \in H_0$ are in increasing order under our lexicographic ordering in H_0. The inclusions $[L^\alpha, L^\beta] \in L^{\alpha+\beta}$ for $\alpha, \beta \in \varDelta$ (see Proposition II in §8 of chapter X) and the definitions (5.2.1) show that in the basis (5.2.3), the operators ad h, $h \in H$ are diagonal matrices and the operators ad x, $x \in N_+$ ($x \in N_-$) are upper (lower) nilpotent matrices. $\square$

Let $\tilde{G}$ be the adjoint group of the group G. Let $\tilde{L}$ be the Lie algebra of G and $\tilde{H}$, $\tilde{N}_+$, $\tilde{N}_-$ the images of the Lie subalgebras H, N_+, N_- under the isomorphism of the Lie algebra L onto $\tilde{L}$ defined by the adjoint homomorphism $x \to$ ad x, $x \in L$. Using the basis (5.2.3), we identify the group $\tilde{G}$ with a Lie subgroup of the group $GL(n, \mathbf{C})$ and the Lie algebra $\tilde{L}$ with a Lie subalgebra of the Lie algebra $gl(n, \mathbf{C})$, where $n = $ dim $G = $ dim L. Proposition I shows that the theorem in 5.1 is applicable to the group $\tilde{G}$. Let Z_- and Z_+ be analytic subgroups of G with Lie algebras N_- and N_+, respectively. Let $\tilde{D}$, $\tilde{Z}_+$, $\tilde{Z}_-$ be analytic subgroups of $\tilde{G}$ corresponding to the Lie subalgebras $\tilde{H}$, $\tilde{N}_+$, $\tilde{N}_-$, respectively. Let C be the center of the group G. Let π be the homomorphism of G onto $\tilde{G}$ defined by the adjoint representation. The kernel of the homomorphism π coincides with C (see Proposition VI in 3.5 of chapter IX) and C is a discrete group (see Proposition IV in 4.4). Hence G is a covering group for $\tilde{G}$ with respect to the mapping π.

II. *The groups Z_+ and Z_- are the components of the identity element in the groups $\pi^{-1}(\tilde{Z}_+)$ and $\pi^{-1}(\tilde{Z}_-)$ respectively.*

Proof. This follows from the obvious equalities $\pi(Z_+) = \tilde{Z}_+$, $\pi(Z_-) = \tilde{Z}_-$, since Z_+ and Z_- are connected. $\square$

Let D be the inverse image of $\tilde{D}$ under the homomorphism π.

III. *The group D is commutative.*

Proof. Let D_0 be the component of the identity in D. Since D_0 is an analytic subgroup corresponding to the commutative Lie subalgebra $H \subset L$, D_0 is commutative. On the other hand, D is generated by the subgroups D_0 and C. Since D_0 and C are commutative and their elements commute in pairs, the group D is commutative. $\square$

IV. *The restrictions of the mapping π to the groups Z_+ and Z_- define isomorphisms of Z_+ and Z_- onto the groups $\tilde{Z}_+$ and $\tilde{Z}_-$ respectively. The groups Z_+ and Z_- are simply connected.*

Proof. Proposition II show that we may apply Proposition VII of §1 in chapter VIII. Thus Z_+ and Z_- are covering groups for $\tilde{Z}_+$ and $\tilde{Z}_-$, respectively, under the restrictions of the mapping π. The groups $\tilde{Z}_+$ and $\tilde{Z}_-$ are simply connected according to Proposition II of 5.1. Hence Z_+ and $\tilde{Z}_+$ (Z_- and $\tilde{Z}_-$) are isomorphic under the restriction of π to Z_+ (Z_-). □

V. *The group G contains an open everywhere dense set G_{reg} such that every element g of G_{reg} can be written uniquely as*

$$g = z_- \, d z_+,$$

where $z_- \in Z_-$, $d \in D$, $z_+ \in Z_+$. The mappings $g \to z_-$, $g \to d$ and $g \to z_+$ are analytic mappings of the manifold G_{reg} into the Lie groups Z_-, D, and Z_+, respectively.

Proof. Let $\tilde{G}_{\text{reg}}$ be the set of elements $\tilde{g} \in \tilde{G}$ that allow Gauss's decomposition

$$\tilde{g} = \tilde{z}_- \, \tilde{d} \tilde{z}_+,$$

where $\tilde{z}_- \in \tilde{Z}_-$, $\tilde{d} \in \tilde{D}$, $\tilde{z}_+ \in \tilde{Z}_+$. By Proposition III in 5.1, the set $\tilde{G}_{\text{reg}}$ is open and dense in $\tilde{G}$. Let G_{reg} be the inverse image of $\tilde{G}_{\text{reg}}$ in G. If g belongs to G_{reg}, then $\pi(g) = \tilde{z}_- \, \tilde{d} \tilde{z}_+$ for certain $\tilde{z}_-$, $\tilde{d}$, $\tilde{z}_+$. Let $z_- \in Z_-$, $z_+ \in Z_+$ be the uniquely defined elements for which $\pi(z_-) = \tilde{z}_-$, $\pi(z_+) = \tilde{z}_+$ (see Proposition IV). Let d' be an inverse image in D of the element $\tilde{d} \in \tilde{D}$. We then have $\pi(g) = \pi(z_- d' z_+)$, i.e., $\pi(g^{-1} z_- d' z_+) = e$. Thus we obtain

$$g^{-1} z_- d' z_+ = c,$$

where c belongs to the center C. This gives us

$$g = z_- d' z_+ c^{-1} = z_- (d' c^{-1}) z_+, \quad \text{where } d' c^{-1} \in D.$$

Setting $d = d' c^{-1}$, we obtain

$$g = z_- \, d z_+, \tag{5.2.4}$$

where $z_- \in Z_-$, $z_+ \in Z_+$, $d \in D$. Since the elements z_- and z_+ are uniquely defined by $g \in G_{\text{reg}}$, the element $d \in D$ is also uniquely defined by $g \in G_{\text{reg}}$. The mapping $g \to z_+$ is the composition of analytic mappings $g \to \pi(g) \to \tilde{z}_+ \to z_+$ (see Proposition IV in 5.1 of chapter IV). Thus the mapping $g \to z_+$ is analytic. Similarly, the mapping $g \to z_-$ is analytic and so the mapping $g \to d = (z_-)^{-1} g (z_+)^{-1}$ is analytic as well. □

The decomposition (5.2.4) is called *Gauss's decomposition of the element* $g \in G$.

§6. Iwasawa's Decomposition

6.1. Iwasawa's Decomposition in an Adjoint Group

Let L be a complex semisimple Lie algebra with a real form L_0. Let σ be the corresponding involution in L. Let L_u be a compact real form of L invariant under σ (see Proposition XIII in 14.1 of chapter X). Let τ be the involution in L corresponding to the real form L_u. The adjoint representation $x \to \text{ad } x$, $x \in L$, is an isomorphism of L into the Lie algebra $gl(L)$. Let G, G_0, G_u be analytic subgroups of the real Lie group G_L corresponding to the Lie subalgebras $ad(L)$, $ad(L_0)$, $ad(L_u)$. Since σ and τ are involutive linear operators in L, the formulas

$$s(g) = \sigma g \sigma^{-1}, \qquad t(g) = \tau g \tau^{-1} \qquad (6.1.1)$$

define involutive automorphisms of the Lie group G.

I. *The set $G_0(G_u)$ is a connected component of the set of fixed points of the automorphism s (t).*

Proof. Let H be the connected component that contains e of the set of fixed points of the automorphism s. Note that H is a Lie subgroup of G (see theorem 1 of §2). Let M be the Lie subalgebra of L corresponding to H. From (6.1.1) we infer that $x \in M$ if and only if

$$\exp(tx)(y) = s(\exp(tx))(y) = \sigma(\exp tx)\sigma^{-1}(y) \qquad (6.1.2)$$

for all $y \in L$. Differentiate (6.1.2) with respect to t and set $t = 0$. Using the fact that σ is an automorphism of L, we obtain

$$[x, y] = \sigma[x, \sigma^{-1}y] = [\sigma x, y]. \qquad (6.1.3)$$

That is, $[x - \sigma x, y] = 0$ for all $y \in L$. Since the Lie algebra L is semisimple, this yields $x - \sigma x = 0$, which is to say that $\sigma x = x$ and $x \in L_0$. Conversely, if x belongs to L_0, we have $\sigma x = x$ and so

$$\sigma(\exp tx)\sigma^{-1}(y) = \sigma\left(\sum_{n=0}^{\infty} \frac{t^n}{n!}(\text{ad } x)^n(\sigma^{-1}y)\right)$$

$$= \sum_{n=0}^{\infty} \frac{t^n}{n!}(\text{ad}(\sigma x))^n(y) = \sum_{n=0}^{\infty} \frac{t^n}{n!}(\text{ad } x)^n(y) = \exp(tx)y$$

for all $y \in L$. Thus (6.1.2) holds. Therefore we have $L_0 = M$ and $G_0 = H$. The analogous assertion for the automorphism t is proved similarly. □

II. *The groups G_0 and G_u are closed subgroups of G.*

Proof. The sets of fixed points of the automorphisms s and t are closed. According to Proposition I, G_0 and G_u are connected components of these sets and so are also closed in G. □

The group G is closed in G_L (see Proposition V in 4.4). Hence G, G_0 and G_u are closed subgroups of G_L.

III. *The group G_u is compact.*

Proof. Recall that $(\tau y, \tau z) = (y, z)$ for all $y, z \in L$, where $(\cdot, \cdot)$ is the Cartan-Killing form. If $x \in L_u$ and $y, z \in L$, use the invariance of the Killing form to see that

$$((\text{ad } x)y, \tau(z)) = (\tau\tau(\text{ad } x)y, \tau(z))$$
$$= (\tau([x, y]), z) = ([\tau x, \tau y], z) = ([x, \tau y], z)$$
$$= (\text{ad } x(\tau y), z) = -(\tau y, \text{ad } x(z)) = -(y, \tau(\text{ad } x)(z)). \quad (6.1.4)$$

Let $\{y, z\}$ be the bilinear form on L defined by $\{y, z\} = -(y, \tau z)$. The equality (6.1.4) shows that

$$\{(\text{ad } x)y, z\} = -\{y, (\text{ad } x)z\} \quad (6.1.5)$$

for all $x \in L_u$ and $y, z \in L$.
  Thus for all $t \in \mathbf{R}$, $y, z \in L$ and $x \in L_u$ (6.1.5) gives us

$$\frac{d}{dt}\{\exp(tx)y, \exp(tx)z\} = \{(\text{ad } x)\exp(tx)y, \exp(tx)z\}$$
$$+ \{\exp(tx)y, (\text{ad } x)\exp(tx)z\} = 0. \quad (6.1.6)$$

This implies that

$$\{\exp(tx)y, \exp(tx)z\} = \{\exp(0x)y, \exp(0x)z\} = \{y, z\} \quad (6.1.7)$$

for all $t \in \mathbf{R}$, $y, z \in L$ and $x \in L_u$. Hence we have

$$\{gy, gz\} = \{y, z\} \quad (6.1.8)$$

for all $y, z \in L$ and all g in a certain neighborhood of the identity in G_u. Since G_u is connected, (6.1.8) holds for all $g \in G_u$. Thus G_u is contained in the set of linear transformations of the space L that preserve the bilinear form

$\{x, y\}$. According to Proposition V in 14.1 of chapter X, the form $\{x, x\} = -\{x, \tau x\}$ is negative-definite on L. Thus G_u is a closed (by Proposition II) subgroup of the group U of transformations of L that are unitary under the form $\{x, x\}$. Since U is compact, G_u is compact. $\square$

Let K be the connected component of the identity in the group $G_0 \cap G_u$.

IV. *K is a compact Lie group whose Lie algebra coincides with $L_u \cap L_0$.*

Proof. The group $G_0 \cap G_u$ is a closed subgroup of the compact group G_u, and so $G_0 \cap G_u$ and K are compact. The Lie algebra of the subgroup K is plainly the intersection of the Lie algebras of the groups G_0 and G_u. $\square$

Let H_u be the Cartan subalgebra of the Lie algebra L_u constructed as in Proposition XIV in 14.1 of chapter X. Let H_u^-, N_0, H_0 and H_0^+ have the same meanings as in 14.1 of chapter X. Let R be the Lie subgroup of G_0 whose Lie algebra is the solvable Lie subalgebra $M_0 = iH_u^- + N_0 \subset L$ (see Proposition XVII in 14.1 of chapter X).

Note that the real vector space iH_u^- is generated by the set H_0^+ of elements h of the Cartan subalgebra $H_0 \subset L_0$ for which $\sigma(h) = h$. Choose the Weyl basis $\{e_{-\alpha_k}, \ldots, e_{-\alpha_1}, h_1, \ldots, h_r, e_{\alpha_1}, \ldots, e_{\alpha_k}\}$, where $e_{\pm \alpha_k} \in L^{\pm \alpha_k}$ ($\alpha_1, \ldots, \alpha_k$ is the set of all positive roots and $\alpha_i < \alpha_{i+1}$ under a lexicographic ordering) and $h_1, \ldots, h_r$ is an orthonormal basis in H_0 under the Killing form. Since $\alpha(H_0^+)$ is contained in $\mathbf{R}$ for all roots α, the transformations in $\mathrm{ad}(iH_u^-)$ have diagonal matrices with a real diagonal in this Weyl basis. Since transformations in $\mathrm{ad}(N_0)$ have triangular matrices in this basis with zeros on the main diagonal, we arrive at the following.

V. *Transformations in $\mathrm{ad}(M_0)$ have triangular matrices in the given Weyl basis with real entries on the main diagonal.*

VI. *The group R is a simply connected solvable group. The intersection $R \cap K$ is the identity element. The exponential mapping is a homeomorphism of $M_0 = iH_u^- + N_0$ onto R.*

Proof. The group R is solvable, since its Lie algebra M_0 is solvable (see Proposition V in 4.2). Since G is a linear group, the mapping exp coincides with ordinary matrix exponentiation (see the example in 3.4 of chapter IX). If x is a triangular matrix with real entries on the diagonal, e^x is a triangular matrix with positive entries on the diagonal. Every triangular matrix y with positive entries on the diagonal has the form e^x for exactly one triangular matrix x with real diagonal. (To see this, reduce the transformation y to its Jordon canonical form.) Hence if M is a real Lie subalgebra of $gl(L)$ whose elements are represented in our Weyl basis by triangular matrices with real diagonal, exp is a one-to-one mapping of M onto the subgroup H of G_L

formed by triangular matrices with positive diagonal entries. Plainly the subgroup H is homeomorphic to $\mathbf{R}^n_+ \times \mathbf{C}^{n(n-1)/2}$ and is therefore simply connected. Proposition III in 4.3 thus implies that R is simply connected. Furthermore, the fact that exp is a one-to-one mapping onto M means that exp is also a one-to-one mapping of M_0 into R.

We will prove that exp maps M_0 onto R homeomorphically. Since the eigenvalues of operators in $\mathrm{ad}(M_0)$ are real, exp is a local homeomorphism (see Proposition II in 3.6 of chapter IX and Proposition VIII in 1.5). It remains to prove that the image of exp coincides with R. The mapping $x \to e^x$, $x \in M$, is also a local homeomorphism and $x \to e^x$ is a one-to-one mapping of M onto H. Since exp is the restriction of the mapping $x \to e^x$ to M_0, the image of exp is closed in R. Since exp is a local homeomorphism, the image under exp is open in R. Since R is connected, we find that $\exp(M_0) = R$.

Finally, we will prove that $R \cap K = \{e\}$. By (14.1.21) of chapter X, if we have $x = \sum_{j=1}^r t_j h_j + \sum_{\alpha \neq 0} \lambda_\alpha e_\alpha = h + \sum_{\alpha \neq 0} \lambda_\alpha e_\alpha$, then $\{x, x\} = -(x, \tau(x)) = \sum_{\alpha \neq 0} (|\alpha(h)|^2 + |\lambda_\alpha|^2)$. Therefore the chosen Weyl basis is orthonormal under the form $\{x, x\}$. The identity (6.1.8) implies that transformations of the group K have unitary matrices in the given Weyl basis. A matrix that is both unitary and triangular is diagonal. If only positive real numbers occupy the diagonal, the matrix is the identity. We have proved that $K \cap R = \{e\}$. $\quad\square$

VII. *The group G_0 is equal to $K \cdot R$; i.e., an element $g \in G_0$ can be uniquely written as $g = kr$, where $k \in K$ and $r \in R$. The mappings $g \to k$ and $g \to r$ are analytic.*

Proof. Since $\tilde{T}_e(R) = M_0$ and $\tilde{T}_e(K) = L_u \cap L_0$, the tangent space to the manifold $K \times R$ at the point (e, e) can be identified with $(L_u \cap L_0) + M_0 = (L_u \cap L_0) + iH_u^- + N_0$ (see 1.5 of chapter IX). Consider the mapping of $G_0 \times G_0$ onto G_0 defined by $(g_1, g_2) \to g_1 g_2$. The differential of this mapping at (e, e) is the mapping of $L_0 + L_0$ onto L_0 defined by $(x_1, x_2) \to x_1 + x_2$. Hence the mapping $\pi: (k, r) \to kr$ of the manifold $K \times R$ into the group G_0 induces an isomorphism of tangent spaces at (e, e). By Proposition VIII in 1.5 of chapter IX, there exist neighborhoods U and V of the identities in K and R, respectively, such that the restriction of π to $U \times V$ is an analytic homeomorphism of $U \times V$ into G_0. Since multiplication is analytic, there exist neighborhoods $U_1 \subset U$, $V_1 \subset V$ and analytic mappings $k^0: U_1 \times V_1 \to U$ and $r^0: U_1 \times V_1 \to V$, such that

$$rk = k^0(r, k) r^0(r, k) \qquad (6.1.9)$$

for all $k \in U_1$, $r \in V$.

Let $S = \{k_1, \ldots, k_p\}$ be a set of elements in U_1. Construct a neighborhood $V(S) \subset V_1$ and analytic mappings r^S and k^S such that

$$rk(S)k = k^S(r, k) r^S(r, k) \qquad (6.1.10)$$

for all $k \in U_1$ and $r \in V(S)$, where $k(S)$ is defined as $k_1 k_2 \cdots k_p$. For $p = 0$ $(S = \varnothing)$ we define $k^S = k^0, r^S = r^0$ (as in (6.1.9)). We will make a construction by induction on p. Let $S' = \{k_0\} \cup S$, and suppose that the neighborhood $V(S)$ and the mapping r^S and k^S are already constructed for the set S. We have $k(S') = k_0 k(S)$ and (1.6.10) and (1.6.9) yield

$$rk(S')k = rk_0k(S)k = k^0(r, k_0)r^0(r, k_0)k(S)k$$

for all $r \in V_1$. Since (6.1.9) is unique, we have $r^0(e, k_0) = e$. Thus there is a neighborhood $V(S') \subset V_1$ such that $r^0(r, k_0) \in V(S)$ for $r \in V(S')$. We then have

$$rk(S')k = k^0(r, k_0)k^S(r^0(r, k_0), k)r^S(r^0(r, k_0), k)$$

for all $r \in V(S')$. This proves (6.1.10).

Recall that the group K is compact. For every U_1, there exist a finite number of sets $S_j, j = 1, \ldots, m$, such that $K = \bigcup_{j=1}^{m} k(S_j)U_1$. We write V_2 for the set $\bigcap_{j=1}^{m} V(S_j)$. For $r \in V_2$ and $k \in K$ (6.1.10) shows that $rk \in KR$.

Next suppose that $T = \{r_1, \ldots, r_q\}$ is a set of elements from V_2 and $r(T) = r_1 \cdots r_q$. We will show that $r(T)k \in KR$. For $q = 1$ this is already proved. If we take $T' = \{r_0\} \cup T$, then $r(T') = r_0 r(T)$ and $r(T')k = r_0 r(T)k \in r_0 KR$. Since $r_0 \in V_2$, $r_0 K$ is contained in KR, and so $r(T')k \in KR$. Since the connected group R is generated by the neighborhood V_2, we find that $RK \subset KR$. Hence KR is a subgroup of G_0. The group KR contains the set $U_1 V_1$, which is a neighborhood of the identity in the connected group G_0. By Proposition VI in 1.2 of chapter V, we have $G_0 = KR$. Uniqueness of the decomposition follows from the equality $K \cap R = \{e\}$. For, if we have $g = k_1 r_1 = k_2 r_2$, then $k_2^{-1} k_1 = r_2 r_1^{-1} \in K \cap R$ and thus $k_2^{-1} k_1 = r_2 r_1^{-1} = e$ and $k_1 = k_2, r_1 = r_2$. The mapping of the product $K \times R$ into G_0 defined by $(k, r) \to kr$ is therefore an analytic isomorphism of the manifolds $K \times R$ and G_0. The inverse mapping $kr \to (k, r)$ is also analytic; i.e., the mappings $g \to k$ and $g \to r$ are analytic. $\square$

6.2. Iwasawa's Decomposition in an Arbitrary Connected Semisimple Lie Group

Let G' be a connected semisimple Lie group, that is, a connected Lie group whose Lie algebra is semisimple. Let L_0 be the Lie algebra of G' and L_C its complexification. Let G_0 be the analytic subgroup of the group G_{L_C} corresponding to the Lie subalgebra $\mathrm{ad}(L_0) \subset \mathrm{ad}(L_C)$. The adjoint homomorphism $\rho: g \to \mathrm{Ad}(g)$ of G' into G_{L_0} admits a unique extension to a homomorphism $\pi: G' \to G_{L_C}$, where $\pi(g)$ is defined for all $g \in G'$ by

$$\pi(g)(x + iy) = \rho(g)x + i\rho(g)y, \qquad x, y \in L_0.$$

The image $\pi(G')$ is the analytic subgroup of G_{L_C} that corresponds to the Lie subalgebra $\mathrm{ad}(L_0)$ (see Proposition IV in 3.3 of chapter IX). Hence we have $\pi(G') = G_0$.

Plainly the conditions $\pi(g) = 1$ and $\rho(g) = 1$ are equivalent. Since the kernel of the adjoint representation ρ is the center Z of G, the group $\pi(G') = G_0$ is isomorphic to G'/Z. Since the center of the Lie algebra L_0 is $\{0\}$, the group Z is discrete. Hence the mapping π is a covering. Let K and R be the analytic subgroups of G_0 constructed as in 6.1 (see Propositions IV–VII of 6.1). Let $\tilde{R}$ be the inverse image of R in G'. Let R' be the component of the identity in R. The group R' covers the group R with respect to the mapping π since $\pi(R') = R$ and the kernel of π is discrete. The group R is simply connected (see Proposition VI of 6.1). Thus π defines an analytic isomorphism of R' onto R. In particular, $R' \cap Z$ is $\{e\}$. Clearly R' is closed, since it is the component of the identity in the inverse image of the closed (see Proposition VI of 6.1) subgroup R of G_0. Thus R' is the simply connected closed solvable Lie subgroup of the group G' corresponding to the Lie subalgebra $M_0 = iH_u^- + N_0 \subset L_0$(see 6.1).

We write K' for $\pi^{-1}(K)$; thus K' contains Z. Since K is closed in G_0, K' is closed in G'. If g belongs to $K' \cap R'$, then $\pi(g) \in K \cap R = \{e\}$ and g is in Z. But $Z \cap R'$ is $\{e\}$ and therefore $g = e$ and $K \cap R = \{e\}$. Furthermore, if g belongs to G', then $\pi(g) = kr$ for certain $k \in K$ and $r \in R$. Since $K' = \pi^{-1}(K)$ and $\pi(R') = R$, we see that there are elements $k'' \in \pi^{-1}(k) \subset K$ and $r' \in \pi^{-1}(r) \cap R'$ such that $\pi(g) = \pi(k'')\pi(r') = \pi(k''r')$. That is, there is a $z \in Z$ such that $g = z(k''r') = (zk'')r'$. Setting $zk'' = k'$ and using the inclusion $K \supset Z$, we see that $k' \in K'$ and $g = k'r'$. If we have $g = k_1'r_1'$ and $k_1' \in K'$, $r_1' \in R'$, then $k'r' = k_1'r_1'$, $k_1'^{-1}k' = r_1'r'^{-1} \in K' \cap R' = \{e\}$, so that $k' = k_1'$, $r' = r_1'$ and the decomposition $g = k'r'$, $k' \in K'$, $r' \in R'$ is unique.

Hence the mapping $\psi:(k',r') \to k'r'$ of the product manifold $K' \times R'$ into G' is a (plainly analytic) one-to-one mapping *onto* the entire group G'. To prove that ψ is an isomorphism of the analytic manifolds $K' \times R'$ and G', it thus suffices to show that at every point (k',r'), the mapping $d\psi(k',r')$ is an isomorphism of the tangent spaces $T_{(k',r')}(K' \times R')$ and $T_{k'r'}(G')$. Let x, y be elements of the Lie algebras of the groups K' and R' respectively. Every tangent vector in $T_{(k',r')}(K' \times R')$ has the form $(x(k'), y(r'))$ for certain x, y. Thus for every function f that is analytic in some neighborhood of the point $g' = k'r'$, we have

$$d\psi(k',r')(x(k'), y(r'))f = (x(k'), y(r'))(f \circ \psi)$$
$$= x(k')(f \circ \psi) + y(r')(f \circ \psi). \tag{6.2.1}$$

Using the relation

$$f(k'r') = f(r'r'^{-1}k'r'),$$

we see that

$$x(k')(f \circ \psi) = \mathrm{Ad}(r'^{-1})x(k')(f \circ \varphi_{r'}) \tag{6.2.2}$$

and that

$$y(r')(f \circ \psi) = y(r')(f \circ \varphi_{k'}). \tag{6.2.3}$$

Substituting (6.2.2) and (6.2.3) in (6.2.1), and applying (3.2.1) of chapter IX, we obtain

$$d\psi(k',r')(x(k'),y(r'))f = (d\varphi_{(k',r')})_e(\mathrm{Ad}(r'^{-1})x + y)(e)f$$
$$= (\mathrm{Ad}(r'^{-1})x + y)(k'r')f.$$

If we have $d\psi(k',r')(x(k'),y(r')) = 0$, then $\mathrm{Ad}(r'^{-1})x + y = 0$, so that $x + \mathrm{Ad}(r')y = 0$. But x is in the Lie algebra of the group K' and y and $\mathrm{Ad}(r')y$ are in the Lie algebra of the group R'. These Lie algebras are complementary in L_0 and thus $x = \mathrm{Ad}(r')y = 0$, which is to say that $x = y = 0$. Therefore $d\psi(k',r')$ is an isomorphic mapping into the tangent space $T_{k'r'}(G')$. The dimensions of the manifolds $K' \times R'$ and G' match and therefore $d\psi$ is an isomorphism of tangent spaces at every point $(k',r') \in K' \times R'$. Hence the mapping ψ is an isomorphism of analytic manifolds. The group K' is connected because G' is connected. The group K' is the analytic subgroup of G' that corresponds to the Lie subalgebra $L_0 \cap L_u$ (see 6.1).

We have proved the following theorem.

Theorem. *Let G' be a connected semisimple Lie group. Let K' and R' be analytic subgroups of G' corresponding to the Lie subalgebras $L_0 \cap L_u$ and $iH_u^- + N_0$ of the Lie algebra L_0 of G (see Propositions I and II in 3.3 of chapter IX). The group K' contains the center of G' and the image of K' under the adjoint representation of G' is a compact group. The group R' is a simply connected solvable Lie group. The groups K' and R' are closed. The mapping $(k',r') \to k'r'$ is an analytic isomorphism of the manifold $K' \times R'$ onto G'.*

We will now find the formula for Haar measure on G'.

I. *Haar measure on a connected semisimple Lie group G' is two-sided invariant.*

Proof. Let Δ be the modular function of the group G'. Plainly Δ is an analytic (see theorem 2 of §2) homomorphism of G' into the group $\mathbf{R}^+$. Accordingly, $d\Delta$ is a homomorphism of the semisimple Lie algebra of G' into a one-dimensional commutative Lie algebra. On the other hand, Proposition III of §7 in chapter X implies that any factor algebra of a semisimple Lie algebra is semisimple. If $d\Delta \neq 0$, we have a contradiction, since the image of a Lie algebra under a homomorphism is isomorphic to some factor algebra. Therefore $d\Delta$ is 0, and Δ maps the group G' onto the number 1. $\square$

II. *Under a proper normalization of left invariant Haar measures dx', dr', dg' on the groups K', R', G', we have*

$$\int f(g')\,dg' = \iint f(k'r'^{-1})\,dk'\,dr' \tag{6.2.4}$$

for all continuous functions f on G' having compact supports.

Proof. Consider functions f on G' of the form $f(g') = f_1(k')f_2(r')$, where $g' = k'r'$. For a fixed function $f_1 \geq 0$, the integral $\int f(g')\,dg'$ is a positive right invariant linear functional on the set of all continuous functions f_2 on R' with compact supports. Therefore we have $\int f(g')\,dg' = c(f_1)\int f_2(r'^{-1})\,dr'$. Fixing $f_2 \geq 0$, we see that $c(f_1)$ is a left invariant positive linear functional of f_1. Thus we can normalize the measure dk' to obtain $c(f_1) = \int f_1(k')\,dk'$. Hence (6.2.4) holds for all continuous functions f with compact supports having the form $f_1(k')f_2(r')$. Apply Stone's Theorem (see 1.4 of chapter IV) to compact subsets of the group G' to see that any continuous function f on G' with compact support is the uniform limit on G' of a sequence of (finite) linear combinations of functions of the form $f_1(k')f_2(r')$, where f_1, f_2 are continuous functions with compact supports on K', R'. Going to the limit in (6.2.4), we prove the present proposition. $\square$

§7. The Universal Covering Group of a Semisimple Compact Lie Group

7.1. Some Properties of Homomorphisms of Compact Groups

Let $\tilde{G}$ be a connected locally compact group. Let C be a discrete subgroup of $\tilde{G}$ contained in the center of $\tilde{G}$. Then C is a closed normal subgroup of $\tilde{G}$. Let G be the factor group $\tilde{G}/C$ and let π be the canonical homomorphism of $\tilde{G}$ onto G.

I. *Suppose that the group G is compact. The group C is then finitely generated. In particular, if C is infinite, there is a nontrivial homomorphism of C into the additive group of real numbers.*

Proof. Let U be a neighborhood of the identity element $\tilde{e}$ of the group $\tilde{G}$ with compact closure $\bar{U}$. Then $\pi(gU)$ is a neighborhood of the element $\pi(g) \in G$. Since G is compact, there exists a finite set of elements $g_1, \ldots, g_n \in \tilde{G}$ such that $\bigcup_{k=1}^{n} \pi(g_k U)$ covers G. Let $D = \bigcup_{k=1}^{n} g_k \bar{U}$. The set D is a compact subset of $\tilde{G}$ whose interior D_0 contains the union $\bigcup_{k=1}^{n} g_k U$. Therefore we have $\pi(D_0) = G$, so that $G = CD_0$. Enlarging the set D if necessary, we may suppose that $\tilde{e} \in D$ and that $D = D^{-1}$. The set $D \cdot D^{-1}$ is compact. On the other hand, since $D \cdot D^{-1} \subset \tilde{G} = \bigcup_{c \in C} cD_0$, the family of sets of the form cD_0, $c \in C$, is an open covering of the set $D \cdot D^{-1}$. Thus there exist elements $c_1, \ldots, c_m \in C$ such that $D \cdot D^{-1} \subset \bigcup_{i=1}^{m} c_i D_0$. Let C_1 be the subgroup of C generated by the elements $c_1, \ldots, c_m$. Since C is contained in the center of $\tilde{G}$, C_1 is a closed normal subgroup of $\tilde{G}$. Let E be the image of the set D in the factor group $\tilde{G}/C_1$. The set E contains the identity element; we also have $E = E^{-1}$ and $E \cdot E^{-1} \subset E$, i.e., E is a subgroup of $\tilde{G}/C_1$. Since D_0 is a neighborhood of the identity in $\tilde{G}$, the set E contains a neighborhood of the identity in $\tilde{G}/C_1$. Since $\tilde{G}$ is connected, the group $\tilde{G}/C_1$ is also connected. Hence the subgroup E (containing as it does a neighborhood of the identity in $\tilde{G}/C_1$) coincides with $\tilde{G}/C_1$. This means that $G = C_1 D$. Since C is a discrete group, $D \cap C$ is finite. Furthermore, if $c \in C$ and $c = c_1 d$, where $c_1 \in C_1 \subset C$, we have $d = cc_1^{-1} \in C$ so that $C \subset C_1 \cdot (D \cap C)$. Since C_1 is finitely generated and $D \cap C$ is finite, the group C is finitely generated.

If C is infinite, it is isomorphic to a group of the form $C_0 \times \mathbf{Z}^q$, where C_0 is a finite group and $q \geqslant 1$. Hence there exists a nontrivial homomorphism of the group C into $\mathbf{R}$. □

II. *Let φ be a homomorphism of C into $\mathbf{R}$. There exists a continuous real-valued function ψ on $\tilde{G}$ with the following properties:* $\psi(xc) = \psi(x) + \varphi(c)$ *for all $x \in \tilde{G}$, $c \in C$, and $\psi(\tilde{e}) = 0$.*

Proof. Let D be a compact subset of $\tilde{G}$ such that $\tilde{G} = CD$. Let f be a non-negative continuous function on $\tilde{G}$ vanishing outside a certain compact set Q and equal to 1 on the set D. We write

$$h_1(x) = \sum_{c \in C} f(xc)e^{-\varphi(c)} \tag{7.1.1}$$

for all $x \in \tilde{G}$. The sum in the right side of (7.1.1) is finite for all $x \in \tilde{G}$, since the set $xC \cap Q$ is finite for all $x \in \tilde{G}$ (Q is compact and xC is discrete). Therefore the function $h_1(x)$ is well defined and is continuous on $\tilde{G}$. If $x \in \tilde{G}$, then x belongs to CD and there is an element $c \in C$ such that $xc \in D$. We then have $f(xc) = 1$. Hence $h_1(x)$ is positive everywhere. We write $h(x)$ for the function $(h_1(\tilde{e}))^{-1}h_1(x)$ on $\tilde{G}$: note that h is continuous and positive everywhere. For $c_0 \in C$ we have

$$h(xc_0) = (h_1(\tilde{e}))^{-1}h_1(xc_0) = (h_1(\tilde{e}))^{-1} \sum_{c \in C} f(xc_0 c)e^{-\varphi(c)}$$

$$= (h_1(\tilde{e}))^{-1} \sum_{c_1 \in C} f(xc_1)e^{-\varphi(c_1 c_0^{-1})} = (h_1(\tilde{e}))^{-1} \sum_{c_1 \in C} f(xc_1)e^{-\varphi(c_1)+\varphi(c_0)}$$

$$= e^{\varphi(c_0)}(h_1(\tilde{e}))^{-1}h_1(x) = e^{\varphi(c_0)}h(x). \tag{7.1.2}$$

Finally we define $\psi(x) = \ln h(x)$. Since $h(\tilde{e}) = 1$, $\psi(\tilde{e}) = 0$. Take the logarithm of both ends of (7.1.2) to see that $\psi(xc_0) = \psi(x) + \varphi(c_0)$ for all $c_0 \in C$. □

III. *Let G be a connected compact group. Let F be a continuous real-valued function on $G \times G$ such that*

$$F(e, e) = 0, \tag{7.1.3}$$

$$F(xy, z) + F(x, y) = F(x, yz) + F(y, z) \tag{7.1.4}$$

for all x, y, z *in* G. *There is a continuous real-valued function* f *on* G *such that*

$$f(e) = 0, \qquad\qquad\qquad (7.1.5a)$$
$$F(x, y) = f(xy) - f(x) - f(y) \qquad\qquad (7.1.5b)$$

for all x, y *in* G.

Proof. Let dg be the invariant measure on the group G for which $\int_G dg = 1$. We define

$$f(x) = -\int_G F(x, g)\, dg \qquad\qquad (7.1.6)$$

for all $x \in G$. Apply (7.1.4) to obtain

$$
\begin{aligned}
f(xy) - f(x) - f(y) &= -\int_G F(xy, g)\, dg + \int_G F(x, g)\, dg + \int_G F(y, g)\, dg \\
&= \int_G F(x, g)\, dg + \int_G F(x, y)\, dg - \int_G F(x, yg)\, dg \\
&= F(x, y) + \int_G F(x, g)\, dg - \int_G F(x, yg)\, dg. \qquad (7.1.7)
\end{aligned}
$$

The identity (7.1.5b) follows from (7.1.7) and the invariance of dg. Setting $x = y = e$ in (7.1.4) and applying (7.1.3), we see that $F(e, z) = F(e, z) + F(e, z)$, that is, $F(e, z) = 0$ for all $z \in G$. From this and (7.1.6) we infer that $f(e) = 0$ and thus (7.1.5a) holds. $\square$

IV. *Retain the hypotheses and notation of Proposition* I. *For every homomorphism* φ *of the group* C *into* **R** *there exists a continuous homorphism* χ *of the group* $\tilde{G}$ *into* **R** *such that* $\chi(c) = \varphi(c)$ *for all* $c \in C$.

Proof. Let ψ be the continuous real-valued function on G constructed in Proposition II. We set

$$\Phi(\tilde{x}, \tilde{y}) = \psi(\tilde{x}\tilde{y}) - \psi(\tilde{x}) - \psi(\tilde{y}) \qquad\qquad (7.1.8)$$

for all $\tilde{x}, \tilde{y} \in \tilde{G}$. The identity $\psi(\tilde{x}c) = \psi(\tilde{x}) + \varphi(c)$ and (7.1.8) show that

$$\Phi(\tilde{x}c_1, \tilde{y}c_2) = \Phi(\tilde{x}, \tilde{y}) \qquad\qquad (7.1.9)$$

for all $c_1, c_2 \in C$. Thus the formula

$$F(x, y) = \Phi(\tilde{x}, \tilde{y}), \qquad x, y \in G, \quad \tilde{x} \in x, \quad \tilde{y} \in y, \qquad (7.1.10)$$

unambiguously defines a continuous real-valued function F on $G \times G$. Since $\psi(\tilde{e}) = 0$, $F(e, e)$ is also zero, and (7.1.8) and (7.1.10) imply that the function F satisfies (7.1.4) for all x, y, $z \in G$. By Proposition III there exists a real-valued function f that satisfies (7.1.5a) and (7.1.5b). We define $\chi(\tilde{x}) = \psi(x) - f(\pi(\tilde{x}))$ for all $\tilde{x} \in \tilde{G}$. Then χ is a continuous real-valued function on G and (7.1.8)–(7.1.10) and (7.1.5) give us

$$\chi(\tilde{x}\tilde{y}) - \chi(\tilde{x}) - \chi(\tilde{y}) = \Phi(\tilde{x}, \tilde{y}) - F(\pi(\tilde{x}), \pi(\tilde{y})) = 0$$

for all $\tilde{x}, \tilde{y} \in \tilde{G}$. That is, χ is a continuous homomorphism of the group $\tilde{G}$ into $\mathbf{R}$.

We will show that $\chi(c) = \varphi(c)$ for all $c \in C$. Set $\tilde{x} = \tilde{e}$ in the identity $\psi(\tilde{x}c) = \psi(\tilde{x}) + \varphi(c)$ and recall that $\psi(\tilde{e}) = 0$. This gives $\psi(c) = \varphi(c)$ for all $c \in C$. Since $f(\pi(c)) = f(e) = 0$ for all $c \in C$, we have $\chi(c) = \psi(c) - f(\pi(c)) = \varphi(c)$ for all $c \in C$. $\square$

V. *Let $\tilde{G}$ be a connected locally compact group. Let C be a discrete subgroup of $\tilde{G}$ contained in the center of $\tilde{G}$. Suppose that the group $G = \tilde{G}/C$ is compact and that the group $\tilde{G}$ admits no nontrivial continuous homomorphisms into the group $\mathbf{R}$. Then the group $\tilde{G}$ is compact.*

Proof. If the group C is infinite, there exists a nontrivial homomorphism of the group C into $\mathbf{R}$ (see Proposition I). Hence there exists a nontrivial continuous homomorphism of the group $\tilde{G}$ into $\mathbf{R}$ (Proposition IV), which violates our hypothesis. Therefore the group C is finite and compactness of the group G implies compactness of the group $\tilde{G}$. $\square$

VI. *Let $\tilde{G}$ be a connected locally compact group. Let C be a discrete subgroup of $\tilde{G}$ contained in the center of $\tilde{G}$. If the group $[\tilde{G}, \tilde{G}]$ is dense in $\tilde{G}$ and the group $G = \tilde{G}/C$ is compact, then $\tilde{G}$ is also compact.*

Proof. The subgroup $[\tilde{G}, \tilde{G}]$ is contained in the kernel of every homomorphism of $\tilde{G}$ into $\mathbf{R}$. Therefore the only continuous homomorphism of $\tilde{G}$ into $\mathbf{R}$ is trivial. Now apply Proposition V. $\square$

7.2. The Universal Covering Group of a Compact Lie Group

Recall from the definition of a manifold that every manifold is locally connected and locally simply connected. Therefore any connected Lie group has a universal covering group (see chapter VIII, Proposition IV of 3.1 and Proposition I of 3.2). According to Proposition VII of 3.3, chapter IX, this universal covering group is also a Lie group.

I (Weyl's Theorem). *Suppose that G is a connected compact semisimple real Lie group. Its universal covering group is also compact.*

Proof. Let $\tilde{G}$ be the universal covering group of G. We may suppose that $G = \tilde{G}/C$, where C is a discrete subgroup of $\tilde{G}$ contained in the center of

$\tilde{G}$. Since the Lie algebras of the Lie groups G and $\tilde{G}$ are isomorphic, $\tilde{G}$ is also semisimple. Hence $\tilde{G} = [\tilde{G}, \tilde{G}]$ (see Proposition VI of 4.2). To complete the proof, apply Proposition VI of 7.1. □

We know that every representation of the Lie algebra of a simply connected Lie group is the differential of some representation of the group (see Proposition II of 1.4). The converse holds for connected compact semisimple Lie groups.

II. *Let G be a compact semisimple connected Lie group with Lie algebra L. If every representation of L is the differential of some representation of G, then G is simply connected.*

Proof. Let $\tilde{G}$ be the universal covering group of G, with homomorphism π. By Proposition I, $\tilde{G}$ is compact. Let C be the kernel of the homomorphism π. We identify the Lie algebra of $\tilde{G}$ with the Lie algebra L in such a way that the mapping $d\pi$ is the identity mapping. Assume that C contains an element $c \neq \tilde{e}$. By theorem 4 of 2.4, chapter IV (cf. the proof of Proposition I of 8.3), there exists a representation $\tilde{\rho}$ of $\tilde{G}$ such that $\tilde{\rho}(c)$ is an operator in the space of the representation ρ different from the identity operator. Let $d\tilde{\rho}$ be the differential of the representation $\tilde{\rho}$. By hypothesis there exists a representation ρ of G such that $d\rho = d\tilde{\rho}$. Hence the representations $\tilde{\rho}$ and $\rho \circ \pi$ of $\tilde{G}$ have equal differentials. Proposition I of 1.4 shows that $\tilde{\rho} = \rho \circ \pi$; in particular we have $\tilde{\rho}(c) = \rho(\pi(c)) = \rho(e) = 1$, which is contrary to hypothesis. Therefore C reduces to the identity element of $\tilde{G}$. Hence $\tilde{G}$ and G are isomorphic, which is to say that G is simply connected. □

III. *The group $SO(n, \mathbf{R})$ is not simply connected.*

Proof. The entries in the matrix of the identity representation π of the compact group $G = SO(n, \mathbf{R})$ are real. Therefore the set of matrix entries of all possible tensor powers of the representation π satisfies the conditions of Stone's Theorem. By the theorem of 1.4, chapter IV, the irreducible components of the tensor powers of the representation are a complete system of irreducible unitary representations of G. Let L be the Lie algebra of G. The representation of the Lie algebra corresponding to a tensor power of π is a tensor power of the representation $d\pi$ of the Lie algebra L corresponding to the identity representation π of L. The reader can easily verify that every weight of the representation $\bigotimes_{k=1}^{n} d\pi$ is a sum of weights of $d\pi$. Formulas (10.4.19), (10.4.21) and (10.4.37) of chapter X imply that the numbers

$$2(\lambda, \alpha_k)/(\alpha_k, \alpha_k) = \lambda(h_k)$$

are even for every weight λ of the representation of the Lie algebra $so(n, \mathbf{C})$ corresponding to the identity representation of the group $SO(n, \mathbf{C})$. Therefore not every irreducible representation of the Lie algebra $so(n, \mathbf{R})$ can be extended to a representation of the group $SO(n, \mathbf{R})$. (In particular, a representation of the Lie algebra $so(2n, \mathbf{R})$ or $so(2n + \mathbf{R})$ corresponding to the highest weight $(1/2)(\lambda_1 + \cdots + \lambda_n)$ of the corresponding complex Lie algebra is not extensible to a representation of the group.) Therefore $SO(n, \mathbf{R})$ is not simply connected.   □

§8. Complex Semisimple Lie Groups and Their Real Forms

8.1. Compact Real Forms

The definition of a real form of a complex Lie algebra is given in §14 of chapter IX. Recall that any complex semisimple Lie algebra has at least one real form—in particular, one compact real form (see Proposition XII in §14 of chapter X).

We will now define a real form of a complex Lie group. Let G be a complex Lie group with Lie algebra L. Let H be a (real) analytic subgroup of the Lie group G, considered as a real Lie group. Let L_0 be the real Lie subalgebra of L corresponding to H. The Lie subgroup H is called a *real form* of the complex Lie group G if L_0 is a real form of the Lie algebra L. A real form H is called *compact* if it is compact as a topological group.

I. *Let G be a connected complex semisimple Lie group with Lie algebra L. Then G has a compact real form. Indeed, a real analytic Lie subgroup $H \subset G$ corresponding to a compact real form $L_u \subset L$ is a compact real form of the Lie group G.*

Proof. It suffices to prove that the group H is compact, which is done by analogy with the proof of Proposition III in 6.1.   □

II. *Any finite-dimensional representation of a semisimple Lie algebra is completely reducible.*

Proof. In view of 14.2 of chapter X, it suffices to prove this assertion for a complex semisimple Lie algebra L. Let G be a simply connected complex Lie group with Lie algebra L (see the theorem of 3.3). Let L_u and H be as in Proposition I. Let π be a finite-dimensional representation of L. Then there exists a representation ρ of the group G such that $\pi = d\rho$ (see Proposition II of 1.4). Let σ be the restriction of ρ to H. Then $d\sigma$ is the restriction of π to L_u. Since every finite-dimensional representation of a compact group is completely reducible (see theorem 2 in 2.2, chapter IV), the representation σ is completely reducible. Hence $d\sigma$ is completely reducible (see the theorem in 3.4, chapter IX). The representation $d\sigma = \pi|_{L_u}$ is a representation of the real form L_u of L and thus the complete reducibility of $d\sigma$ implies the complete reducibility of π (see 14.2, chapter X).   □

8.2. Weyl's "Unitary Method"

I. *Let G be a connected compact semisimple Lie group. Let G_0 be a real form of G and N a connected neighborhood of the set G_0 in G. If F is a complex-analytic function on N whose restriction to G_0 is zero, then $F = 0$.*

Proof. Let L be the (complex) Lie algebra of the group G. Let L_0 be the real Lie subalgebra of L corresponding to the subgroup G_0. We will consider the elements of the universal enveloping algebra U of G as left invariant complex-analytic differential operators on G. The elements $x(g_0)$, $x \in L_0$, $g_0 \in G_0$, are tangent vectors to G_0. Since F vanishes on G_0, we have $xF(g_0) = 0$ for all $x \in L_0$ and $g_0 \in G_0$. Since the function F is complex-analytic and $L = L_0 + iL_0$, we see that $xF(g_0) = 0$ for all $g_0 \in G_0$ and $x \in L$. Thus xF vanishes on G_0 for all $x \in L$. Apply the preceding argument to xF and use induction on the degree of the element $y \in U$ to see that yF vanishes on G_0 for all $y \in U$, and so $yF(e) = 0$ for all $y \in U$. Thus the function F vanishes identically in some neighborhood of the identity element e. Since N is connected, F vanishes throughout N. □

II. *Let G be a simply connected semisimple complex Lie group with a real form G_u. Then G_u is simply connected.*

Proof. Let L be the Lie algebra of the group G. Let L_u be the compact real form corresponding to the subgroup G_u. Let π be a representation of the Lie algebra L_u and let π_C be its complexification (see (14.2.1) in chapter X). Since G is simply connected, there exists a complex-analytic representation ρ of G for which $d\rho = \pi_C$. Plainly the restriction σ of ρ to G_u has the property that $d\sigma = \pi$. Therefore G_u satisfies the hypotheses of Proposition VIII in 8.1, and so G_u is simply connected. □

III. *Every compact connected semisimple Lie group K can be embedded as a compact real form in some semisimple connected complex Lie group.*

Proof. Let L_K be the Lie algebra of K, and let L be the complexification of L_K. Both L_K and L are semisimple Lie algebras. Let G be the simply connected complex Lie group corresponding to the Lie algebra L (see the theorem in 3.3). Let G_K be the real analytic subgroup of G corresponding to the Lie subalgebra $L_K \subset L$. By Proposition II, the group G_K is simply connected. Since G_K and K have isomorphic Lie algebras, there exists a discrete subgroup $C \subset G_K$ such that G_K/C is isomorphic to K. Since C is discrete, it is contained in the center of G. Hence K is isomorphic to the compact real form G_K/C of the complex group G/C. □

IV. *Let G be a simply connected complex semisimple Lie group with a real form G_0. If π is an irreducible complex-analytic representation of G, the* *restriction of π to G_0 is irreducible and the equivalence class of this restriction is uniquely defined by the equivalence class of π. Also every irreducible real-analytic representation of G_0 is the restriction to G_0 of exactly one complex-analytic representation of G.*

Proof. Let ρ be an irreducible real-analytic representation of the group G_0. Let $d\rho$ be the corresponding representation of the real form L_0 of the Lie algebra L of the group G. Let $(d\rho)_\mathbf{C}$ be the complexification of $d\rho$. Since G is simply connected, there exists a complex-analytic representation π of the group G such that $d\pi = (d\rho)_\mathbf{C}$ (see Proposition II of 1.4). Plainly the differential of the restriction of π to G_0 coincides with $d\rho$. From Proposition I in 1.4, we infer that $\pi|_{G_0} = \rho$. On the other hand, if $\pi|_{G_0} = \rho$, then $d|_{L_0} = d\rho$ and so the representation $d\pi$ is determined by its restriction to L_0 (note that $L = L_0 + iL_0$). Since the representation π is determined by its differential, the representation ρ is the restriction to G_0 of exactly one complex-analytic representation of the group G.

Plainly, if ρ is irreducible, then π is irreducible, and if π_1 and π_2 are equivalent, then ρ_1 and ρ_2 are equivalent (where $\rho_i = \pi_i|_{G_0}$ for $i = 1, 2$). Let π be a complex-analytic representation of G in a finite-dimensional complex vector space V. Let ρ be the restriction of π to the subgroup G_0. Let $E(\pi)$, $E(\rho)$ be the complex linear spans of all operators of the form $\pi(g)$, $g \in G$, and $\rho(g_0)$, $g_0 \in G_0$, respectively. Clearly $E(\rho)$ is contained in $E(\pi)$. We will show that $E(\rho) = E(\pi)$. If $E(\rho) \neq E(\pi)$, there is a nonzero linear functional f on the linear space $E(\pi)$ that vanishes on $E(\rho)$. Let $F(g) = f(\pi(g))$ for $g \in G$. Then we have $F(g) \not\equiv 0$, but $F(g_0) = 0$ for all $g_0 \in G_0$. Proposition I implies that $F(g) \equiv 0$. This contradiction shows that $E(\rho) = E(\pi)$. If π is irreducible, then $E(\pi)$ consists of all linear operators on V ("Burnside's Theorem": see Naĭmark [2]). Since $E(\rho) = E(\pi)$, the irreducibility of π implies the irreducibility of ρ.

Now let ρ_1 and ρ_2 be representations of the Lie group G_0. Let π_1 and π_2 be complex-analytic representations of the group G such that $\pi_i|_{G_0} = \rho_i$, $i = 1, 2$. Suppose that ρ_1 and ρ_2 are equivalent, so that there is a linear isomorphism T of the space V_1 of ρ_1 onto the space V_2 of ρ_2 such that $\rho_2(g_0) = T\rho_1(g_0)T^{-1}$ for all $g_0 \in G$. If $\pi_2(g) \neq T\pi_1(g)T^{-1}$ for some $g \in G$, there exist a linear functional f on V_2 and an element $v \in V_2$ for which $f((\pi_2(g) - T\pi_1(g)T^{-1})v) \neq 0$. The function $F(g) = f((\pi_2(g) - T\pi_1(g)T^{-1})v)$ satisfies the hypotheses of Proposition I and therefore $F(g) \equiv 0$. This contradiction shows that $\pi_2(g) = T\pi_1(g)T^{-1}$ for all $g \in G$. □

Formulas for characters of and dimensions of irreducible continuous unitary representation of compact connected semisimple Lie groups were found by Weyl. They are closely related to the corresponding formulas for characters and representations of complex semisimple Lie algebras found in 13.4, chapter X. For an exposition of this matter, see Varadarajan [1] and Seminaire "Sophus Lie" [1].

8.3. Linearity of Compact Lie Groups

I. *A connected compact Lie group has a faithful real-analytic linear representation in a finite-dimensional vector space.*

Proof. Theorem 4 of 2.4, chapter IV implies that for any element $g \in G$ there exists a finite-dimensional representation T such that $T(g) \neq T(e)$. In the opposite case, all matrix entries of irreducible unitary representations of the group G would assume equal values in the points g and e and a continuous function on G equal to 0 at g and 1 at e would not allow a uniform approximation by linear combinations of these matrix entries. Therefore, for all $g \in G$, $g \neq e$, there exists a finite-dimensional representation T such that g is not contained in the kernel of T; that is, the open sets $G \backslash \mathrm{Ker}\, T$ cover $G \backslash \{e\}$. Let V be a neighborhood of e admitting canonical coordinates. Let L be the Lie algebra of the group G. We choose a ball L_ε in L of radius ε with its center at the origin of the coordinates so that the exponential mapping maps the ball $L_{2\varepsilon}$ homeomorphically into V. Let $W = \exp(L_\varepsilon)$. The sets $G \backslash \mathrm{Ker}\, T$ cover $G \backslash W$. The compactness of G implies that there exists a finite set $T_1, \ldots, T_n$ of finite-dimensional representations of G such that $\bigcap_{i=1}^{n} \mathrm{Ker}\, T_i \subset W$. The intersection $\bigcap_{i=1}^{n} \mathrm{Ker}\, T_i$ is a subgroup of G and the construction of W and the properties of canonical coordinates imply that the neighborhood W contains no subgroups of G other than $\{e\}$. Therefore $\bigcap_{i=1}^{n} \mathrm{Ker}\, T_i$ is $\{e\}$, and the direct sum $T_i + \cdots + T_n$ is a faithful representation. □

8.4. The Complex Hull of a Complex Semisimple Lie Algebra

Let L be a complex semisimple Lie algebra. Let $L_{\mathbf{R}}$ be the Lie algebra L considered as a real Lie algebra and let $L_{\mathbf{C}}$ be the complexification of $L_{\mathbf{R}}$. Since $L_{\mathbf{R}}$ is semisimple, $L_{\mathbf{C}}$ is semisimple. Let S be the linear operator in L defined by

$$Sx = ix$$

for all $x \in L$. We define $S\{x, y\} = \{Sx, Sy\}$ to obtain a complex-linear operator S on $L_{\mathbf{C}}$. Since $S^2 = -1$ in L and $[Sx, y] = S[x, y]$ for all $x, y \in L$, we have

$$S^2 = -1, \qquad [Sx, y] = S[x, y]$$

in the space $L_{\mathbf{C}}$. Let L' be the set of all elements $z \in L_{\mathbf{C}}$ for which $Sz = iz$ and L'' the set of all elements $z \in L_{\mathbf{C}}$ for which $Sz = -iz$. Note that $S^2 = -1$ and $(iS)^2 = 1$. Therefore z is in L' (L'') if and only if $z = ((iS + 1)/2)z$ ($z = ((1 - iS)/2)z$). Observe that $(1 + iS)/2$ and $(1 - iS)/2$ are projections onto L' and L'' respectively. Hence $L_{\mathbf{C}}$ is the direct sum of the subspaces L' and L''. If $x \in L'$ and $y \in L_{\mathbf{C}}$, then we have $S[x, y] = [Sx, y] = [ix, y] = i[x, y]$; that is, L' is an ideal in $L_{\mathbf{C}}$. The set L'' is also an ideal in $L_{\mathbf{C}}$.

Let $\tilde{L}$ be the Lie subalgebra of $L_{\mathbf{C}}$ consisting of all elements $\tilde{x} = \{x, 0\}$ where $x \in L$. The Lie algebra $\tilde{L} \subset L_{\mathbf{C}}$ intersects L' and L'' in zero alone since for $\tilde{x} \in \tilde{L}$, we have $i\tilde{x} = J\{x, 0\} = \{0, x\}$ (see 1.4 in chapter X). The relation $S\tilde{x} = \pm i\tilde{x}$ means that $\{ix, 0\} = \pm\{0, x\}$, i.e., $x = 0$. The projections of $\tilde{L}$ onto L' and L'' are therefore one-to-one. Since $Sz = iz$ for $z \in L'$, the projection of $\tilde{L}$ onto L' is an isomorphism of the complex Lie algebras $\tilde{L}$ and L'. Since $Sz = -iz$ for $z \in L''$, the projection of $\tilde{L}$ onto L'' is an *anti-isomorphism* of the complex Lie algebras $\tilde{L}$ and L''. (This is a one-to-one mapping $\varphi : \tilde{L} \to L''$ such that $\varphi(\alpha x + \beta y) = \bar{\alpha}\varphi(x) + \bar{\beta}\varphi(y)$, $\varphi([x, y]) = [\varphi(x), \varphi(y)]$ for all $x, y \in \tilde{L}$, $\alpha, \beta \in \mathbf{C}$.)

Let L_u be a compact real form of the complex semisimple Lie algebra L. Let L'_u and L''_u be its images in L', L'' under the projections defined above. Plainly $L'_u + L''_u$ is a compact real form of the complex semisimple Lie algebra $L_{\mathbf{C}} = L' + L''$. The intersection of the Lie algebra $\tilde{L} \subset L_{\mathbf{C}}$ and $L'_u + L''_u$ is equal to $\tilde{L}_u = \{(x, 0), x \in L_u\}$.

8.5. Iwasawa's Decomposition of a Complex Semisimple Lie Group

I. *Let G be a complex semisimple Lie group with Lie algebra L. Let L_u be a compact real form of L. There exists a (real) solvable Lie subalgebra M of L such that $L_u \dotplus M = L$. Let K and R be the analytic subgroups of G corresponding to the Lie subalgebras L_u and M. Then K is a compact Lie group and R is a solvable Lie group. The center of the group G is contained in K, R is simply connected and the mapping $(k, r) \to kr$, $k \in K$, $r \in R$, is an analytic isomorphism of the manifold $K \times R$ onto G.*

Proof. Let $\bar{G}$ be the complex Lie group defined as follows. The group $\bar{G}$ is isomorphic to the group G as a topological group. If $\pi : g \to \bar{g}$ is a topological isomorphism of G onto $\bar{G}$, for any open set $\bar{U} \subset \bar{G}$ the algebra $D(\bar{U})$ consists exactly of the functions f on $\bar{U}$ for which the function $\overline{f(\pi(g))}$ belongs to $D(\pi^{-1}(\bar{U}))$. We will use the results of 8.4. We write the Lie algebra $L_{\mathbf{C}}$ as $L' \dotplus L''$, where L' is isomorphic to L and L'' is anti-isomorphic to L. The Lie algebra of $G \times \bar{G}$ is isomorphic to $L_{\mathbf{C}}$. Therefore the complex Lie group G can be considered as a real form of the complex Lie group $G \times \bar{G}$. Now apply Proposition I of 8.1, the results of 8.4 and the theorem in §6 to the group $G \subset G \times \bar{G}$. $\square$

II. *A complex semisimple Lie group G is simply connected if and only if its compact real form G_u is simply connected.*

Proof. If G is simply connected, G_u is simply connected according to Proposition II of 8.2. Conversely, if G_u is simply connected, G is homeomorphic to the direct product of the simply connected spaces $K = G_u$ and R (see Proposition I) and so is simply connected. $\square$

We infer that the groups $SL(n, \mathbf{C})$ and $Sp(2n, \mathbf{C})$ fail to be simply connected. The group $SO(n, \mathbf{C})$ is not simply connected, since its compact real form $SO(n, \mathbf{R})$ is not simply connected (Proposition III of 7.2).

8.6. Representations of Complex Semisimple Lie Groups

Let T be a real-analytic representation of the complex Lie group G in a finite-dimensional linear space V. The representation T is called *complex-analytic* if for all $v \in V$ and $f \in V^*$ the function $\varphi(g) = f(T(g)v)$ belongs to $D(G)$. The representation T is called *complex-antianalytic* if the function $\overline{\varphi(g)} = \overline{f(T(g)v)}$ belongs to $D(G)$ for all $v \in V$ and $f \in V^*$.

I. *A connected semisimple complex Lie group G admits a faithful complex-analytic linear representation in a finite-dimensional complex vector space.*

Proof. Let $\tilde{G}$ be the universal covering group of the group G with homomorphism π. Let G_u be a compact real form of G. By Proposition I of 8.3, there exists a faithful real-analytic representation T of the group G_u in some finite-dimensional complex vector space V. We can consider T as a representation of the corresponding real form $\tilde{G}_u$ of $\tilde{G}$. Let ρ be a complex-analytic representation of $\tilde{G}$ whose restriction to $\tilde{G}_u$ coincides with T (see Proposition IV of 8.2). Since the representation T has a discrete kernel, dT is faithful, and so $d\rho$ is a faithful representation of the Lie algebra of $\tilde{G}$. Therefore the kernel of the representation ρ is discrete. Proposition VI in 1.2 of chapter V shows that the kernel of ρ is contained in the center of $\tilde{G}$. By Proposition I in 8.5, the center of $\tilde{G}$ is contained in $\tilde{G}_u$; i.e., the kernel of ρ is contained in $\tilde{G}_u$. We see that $\ker \rho = \operatorname{Ker} T$ and since $\pi(\tilde{G}_u) = G_u$ and the center of $\tilde{G}$ is contained in $\tilde{G}_u$, $\operatorname{Ker} T$ is equal to $\operatorname{Ker} \pi$. Thus the kernel of ρ also coincides with $\operatorname{Ker} \pi$. Hence ρ can be considered as a faithful representation of the group $G \approx \tilde{G}/\operatorname{Ker} \pi$. $\square$

II. *Let G be a complex semisimple Lie group with Lie algebra L. A representation T of G is complex-analytic (complex-antianalytic) if and only if $dT(ix) = i\,dT(x)\,(dT(ix) = -i\,dT(x))$ for all $x \in L$.*

Proof. The representation T is complex-analytic if and only if dT is a representation not only of the real, but also the complex Lie algebra L, i.e., if and only if $dT(ix) = i\,dT(x)$ for all $x \in L$. The assertion for antianalytic representations is proved similarly. $\square$

III. *An irreducible real-analytic linear representation of a complex Lie group in a complex finite-dimensional linear space is equivalent to the tensor product of a complex-analytic irreducible representation and a complex-antianalytic irreducible representation.*

Proof. Proposition IV of 8.2 and the proof of Proposition I of 8.5 show that an irreducible real-analytic representation T is the restriction to the group G of an irreducible complex-analytic representation S of the group $G \times \bar{G}$. By 2.7 of chapter I, the representation S is equivalent to the tensor product of irreducible representations T_1 and $\bar{T}_2$ of the groups G and $\bar{G}$, respectively. The representations T_1 and $\bar{T}_2$ are subrepresentations of the restriction of S to $G \times \{e\}$ and $\{e\} \times G$ respectively. Therefore T_1 and $\bar{T}_2$ are analytic representations of the groups G and $\bar{G}$, respectively. Let $\pi: G \to \bar{G}$ be the mapping constructed in the proof of Proposition I of 8.4. Then the representation T_2 defined by $T_2(g) = \bar{T}_2(\pi(g))$, $g \in G$, is an antianalytic representation of G, as follows from the definition of $\bar{G}$. The relation $S \approx T_1 \otimes \bar{T}_2$ and the definition of the tensor product of representations show that $T \approx T_1 \otimes \bar{T}_2$. $\square$

Chapter XII

Finite-Dimensional Irreducible Representations of Semisimple Lie Groups

The task of describing (up to equivalence) all finite-dimensional irreducible representations of a connected complex semisimple Lie group G was reduced in chapter IX to a similar problem for the Lie algebra L of G. The latter was solved in chapter X. Thus we obtained a solution of the original problem. We call it the *classical* solution, due to E. Cartan $[1^*]$–$[3^*]$ and H. Weyl [1]. Actually, in the classical solution we describe not the representation T of G but their differentials dT. In the present chapter, we supply another solution based on Gauss's decomposition, in which explicit formulas for the representation T appear.

§1. Representations of Complex Semisimple Lie Groups

1.1. Description of Representations with the Aid of Highest Weights and Vectors of Highest Weight

Let G be a connected semisimple complex Lie group with Gauss's decomposition

$$G_{\text{reg}} = KZ^+, \qquad K = Z^- D.^{62} \tag{1.1.1}$$

Let $T : g \to T(g)$ be a representation of the group G in a finite-dimensional space V. A vector $v \in V$ is called a *weight vector of* T if

$$T(\delta)v = v(\delta)v, \quad \text{for all } \delta \in D, \tag{1.1.2}$$

where $v(\delta)$ is a complex-valued function on D. Plainly $v(\delta)$ is a character of the group D. It is called *the weight of the vector v in the representation T.* A weight vector $v \in V$ is called a *vector of highest weight* if

$$T(z)v = v, \quad \text{for all } v \in Z^+ \tag{1.1.3}$$

---

[62] The group K is called a *Borel subgroup of G.*

and a *vector of lowest weight* if

$$T(\zeta)v = v, \quad \text{for all } \zeta \in Z^-. \tag{1.1.4}$$

A weight $\alpha(\delta)$ of a vector of highest weight is called a *highest weight* of the representation T. In the sequel, $\alpha(\delta)$ will denote a highest weight of the representation T.

Repeating in essence the arguments in 2.1, chapter VI (see also 2.1 of chapter VII), we arrive at the following.

Theorem 1. *In the space of a finite-dimensional representation T there exists a vector of highest weight and a vector of lowest weight. If T is irreducible, then the space V admits only one vector of highest weight (up to a scalar multiple) and T is determined (up to equivalence) by its highest weight $\alpha(\delta)$.*

This theorem implies the following.

I. *Two irreducible finite-dimensional representations of a group G are equivalent if and only if their highest weights agree.*

We write

$$\alpha(g) = \alpha(\delta), \quad \text{for } g = \zeta \, \delta z, g \in G_{\text{reg}}. \tag{1.1.5}$$

A character α of a group D is called *inductive under G* if:

(1) the function $\alpha(g)$ (defined on G_{reg}) can be extended to a continuous function on the entire group G;
(2) the linear span (denoted by Φ_α) of all functions $\alpha(gg_0)$, $g_0 \in G$, is finite-dimensional.

Let α be inductive under G. We define a representation $T_\alpha : g \to T_\alpha(g)$ of G in Φ_α as follows:

$$T_\alpha(g_0)f(g) = f(gg_0), \quad \text{for } f \in \Phi_\alpha. \tag{1.1.6}$$

Repeating the arguments in §2 of chapter VI, we conclude the following.

Theorem 2. *A character α of the group D defines an irreducible finite-dimensional representation of the group G with highest weight α if and only if α is inductive under G. In this case, T_α is a finite-dimensional irreducible representation with highest weight α and every finite-dimensional irreducible representation of G with highest weight α is equivalent to the representation T_α.*

The representation T_α is called the *canonical model of the irreducible finite-dimensional representation*. The function $\alpha(g)$ corresponding to an inductive character $\alpha(\delta)$ is called the *generating function of this representation*.

1.2. Realization of Representations in a Space of Functions on Z^+

One can easily verify that a generating function $\alpha(g)$ satisfies the condition $\alpha(kg) = \alpha(k)\alpha(g)$, $k \in K$. Hence for all translates $\alpha(gg_0)$ and for all of their linear combinations $f(g) \in \Phi_\alpha$, we have

$$f(kg) = \alpha(k)f(g), \quad \text{for } k \in K. \tag{1.2.1}$$

For all functions $f \in \Phi_\alpha$, we therefore have

$$f(g) = f(kz) = \alpha(k)f(z), \quad \text{for } k \in K,\, z \in Z^+,\, g \in G_{\text{reg}}. \tag{1.2.2}$$

Arguing as in 2.3 of chapter V, we conclude that the correspondence $f(g) \to f(z)$ of (1.2.2) is linear and one-to-one. Let F_α be the image of the space Φ_α under the mapping $f(g) \to f(z)$ of (1.2.2). Under this mapping the operators $T_\alpha(g)$ turn into operators on F_α; we write them as $\dot{T}_\alpha(g)$. We also have

$$\dot{T}_\alpha(g_0)f(z) = \alpha(zg_0)f(z\bar{g}_0), \tag{1.2.3}$$

where $z\bar{g}_0$ is the element of the group Z^+ defined by Gauss's decomposition

$$zg_0 = k_1 z\bar{g}_0, \quad k_1 \in K, \quad z\bar{g}_0 \in Z^+. \tag{1.2.4}$$

Formula (1.2.3) shows that the function $f_0(z) \equiv 1$ is a vector of highest weight and thus F_α is the linear span of all functions $\dot{T}_\alpha(g)f_0 = \alpha(zg)$.

Hence we have the following theorem.

Theorem 3. *A finite-dimensional irreducible representation T of a connected semisimple complex Lie group G is equivalent to a representation $\dot{T}_\alpha$, where α is the highest weight of T, defined as follows. The space F_α of the representation T_α is the linear span of all $\alpha(zg)$, $g \in G$ and the operators of the representation are defined by the formula*

$$\dot{T}_\alpha(g)f(z) = \alpha(zg)f(z\bar{g}),$$

where $z_1 = z\bar{g}$ is defined by the condition $zg = k_1 z_1$.

1.3. Realization of Representations in a Space of Functions on U

We now use Iwasawa's decomposition (see 8.5 of chapter XI) in place of Gauss's decomposition. If $f \in \Phi_\alpha$, then we have

$$f(g) = f(ku) = \alpha(k)f(u) \tag{1.3.1}$$

and the correspondence

$$f(g) \to f(u) \tag{1.3.2}$$

of (1.3.1) is one-to-one and linear. Let $\dot{F}_\alpha$ be the image of Φ_α under the mapping (1.3.2). Under this mapping the representation T_α becomes an equivalent representation in $\dot{F}_\alpha$, which we also write as $\dot{T}_\alpha$. Repeating the argument in 2.6 of chapter VI, we find the following.

I. *The operators $\dot{T}_\alpha(g)$ of the representation $\dot{T}_\alpha$ are given by the formula*

$$\dot{T}_\alpha(g_0)f(u) = \alpha(k)f(u_{g_0}), \quad \text{for } ug_0 = ku_{g_0}. \tag{1.3.3}$$

Furthermore, by (1.3.1) we obtain the following.

II. *The functions $f \in \dot{F}_\alpha$ satisfy*

$$f(\gamma u) = \alpha(\gamma)f(u), \quad \text{for } \gamma \in \Gamma = K \cap U. \tag{1.3.4}$$

In view of (1.3.4), the nonuniqueness of the decomposition $ug_0 = ku_{g_0}$ in (1.3.3) plays no role. For, any other such decomposition has the form $ug_0 = k\gamma^{-1}\gamma u_{g_0}$. In (1.3.3) we obtain $k\gamma^{-1}$ instead of k and γu_{g_0} instead of u_{g_0}. We also have $\alpha(k\gamma^{-1})f(\gamma u_{g_0}) = \alpha(k)\alpha(\gamma^{-1})\alpha(\gamma)f(u_{g_0})$. To make u_{g_0} unique, we write $ug_0 = \zeta \varepsilon u_{g_0}$. Then (1.3.3) takes the form

$$T_\alpha(g_0)f(u) = \alpha(\varepsilon)f(u_{g_0}), \quad \text{for } ug_0 = \zeta \varepsilon u_{g_0}. \tag{1.3.5}$$

1.4. A Test for Inductivity

Let G be a simply connected semisimple Lie group with Lie algebra L. Let

$$L = N^- + H + N^+ \tag{1.4.1}$$

be the Cartan decomposition of L corresponding to Gauss's decomposition (1.1.1). Then N^-, H, and N^+ are the Lie algebras of the groups Z^-, D, and Z^+, respectively. Write r for the rank of the group L. We choose a basis in $H : h_k = 2h_{\alpha k}/(h'_{\alpha_k}, h'_{\alpha_k})$, $k = 1, \ldots, r$, where α_k are the simple roots of the algebra L with respect to H. Each element $h \in H$ can be written as

$$h = \sum_{k=1}^{r} t_k h_k. \tag{1.4.2}$$

We choose $\lambda_k = \exp t_k$ as coordinates in the group D. Plainly D is isomorphic to the direct product $\mathbf{C}_0^k$. Therefore a character $\alpha(\delta)$ of D must have the form

$$\alpha(\delta) = \prod_{k=1}^{r} \lambda_k^{p_k} \bar{\lambda}_k^{q_k}, \quad \text{for } \delta = \exp h, \tag{1.4.3}$$

p_k and q_k being complex numbers such that $p_k - q_k$ are integers. This last condition is necessary and sufficient for the function $\alpha(\delta)$ on D to be well defined. From (1.4.3) we conclude that

$$\alpha(\delta) = \exp\left(\sum_{k=1}^{r} t_k p_k + \sum_{k=1}^{r} \bar{t}_k q_k \right), \quad \text{for } \delta = \exp h.$$

This and (1.4.2) imply that

$$\alpha(\delta) = \exp[(p, h) + (q, \bar{h})], \quad \text{for } \delta = \exp h, \tag{1.4.4}$$

where p and q are vectors in H with coordinates p_k and q_k, respectively.

Let T be an irreducible finite-dimensional representation of the group G in V and let $\pi = dT$ be the differential of T. Then π is an irreducible finite-dimensional representation of the Lie algebra L (see the theorem of 3.4 of chapter IX) in V.

Let v_0 be a vector of highest weight of the representation T, so that

$$T(\delta)v_0 = \alpha(\delta)v_0, \quad \text{for all } \delta \in D, \tag{1.4.5}$$

$\alpha(\delta)$ being the highest weight of T and

$$T(z)v_0 = v_0, \quad \text{for all } z \in Z^+. \tag{1.4.6}$$

Turn now to π and take account of (1.4.4)–(1.4.6). We obtain

$$\pi(h)v_0 = [(p, h) + (q, \bar{h})]v_0, \quad \text{for all } h \in H, \tag{1.4.7}$$

$$\pi(n)v_0 = 0, \quad \text{for all } n \in N^+. \tag{1.4.8}$$

Since π is irreducible, the smallest subspace of V invariant under $\pi(L)$ and containing v_0 is V itself. Thus (1.4.7) and (1.4.8) imply that v_0 is a vector of highest weight of the representation π and $(p, h) + (q, \bar{h})$ is the highest weight of this representation. Then (see the theorem of §12 in chapter X) for every simple root α of L with respect to H, the numbers

$$p_\alpha = 2(p, \alpha)/(\alpha, \alpha), \qquad q = 2(q, \alpha)/(\alpha, \alpha) \tag{1.4.9}$$

must be nonnegative integers.

Conversely, suppose that the numbers (1.4.9) are nonnegative integers for every simple root α. Then there exists an irreducible representation π of L in a finite-dimensional space V with highest weight $(p, h) + (q, \bar{h})$ (see Proposition III of 8.6, chapter XI). According to Proposition II in 1.4 of chapter XI this representation π is the differential of some irreducible representation T of the group G in V.

Let $\alpha'(\delta)$ be the highest weight of the representation T. Then $\alpha'(\delta)$ is inductive under G by theorem 2 in 1.1 and $\alpha'(\delta) = \exp((p, h) + (q, \bar{h})) = \alpha(\delta)$, i.e., $\alpha(\delta)$ is inductive under G. We have proved the following theorem.

Theorem 4. *A character*

$$\alpha(\delta) = \exp[(p,h) + (q,\overline{h})], \qquad h, p, q \in H,$$

of the group D is inductive under G if and only if all of the numbers

$$p_\alpha = 2(p,\alpha)/(\alpha,\alpha), \qquad q = 2(q,\alpha)/(\alpha,\alpha)$$

are nonnegative integers for each simple root α.

Remark. If the character $\alpha(\delta)$ in (1.4.4) is inductive under G, then p_α, q_α are nonnegative integers for *all* roots α. The system of roots can in fact be ordered in such wise that α becomes simple (see Propositions IV and V of §11 in chapter X).

Combining theorems 1–4, we arrive at the following.

Theorem 5. *Every finite-dimensional irreducible representation of a simply connected semisimple complex Lie group G is determined by a system of nonnegative integers (the signature) p_j, q_j, $j = 1, \ldots, r$, where r is the rank of G. Every such system p_j, q_j, $j = 1, \ldots, r$, describes an irreducible finite-dimensional representation of G. The representation T corresponding to a given signature p_j, q_j, $j = 1, \ldots, r$, is defined by its generating function*

$$\alpha(\zeta\,\delta z) = \alpha(\delta) = \exp[(p,h) + (q,\overline{h})], \qquad p, q, h \in H, \qquad \delta = \exp h,$$

where

$$2(p,\alpha_j)/(\alpha_j,\alpha_j) = p_j, \qquad 2(q,\alpha_j)/(\alpha_j,\alpha_j) = q_j, \qquad j = 1, \ldots, r;$$

the α_j are simple roots of the Lie algebra L of G with respect to a Cartan subalgebra H of L.

The space F of the representation is the linear span of all functions $\alpha(zg)$, $g \in G$. The operators $T(g)$ are defined by

$$T(g)f(z) = \alpha(zg)f(z\overline{g}), \quad \text{for } zg = k \cdot z\overline{g}, z\overline{g} \in Z^+.$$

Example

Suppose that $G = SL(n, \mathbf{C})$. Then Z^- and Z^+ coincide with the groups introduced in chapter VI and D is the group of diagonal matrices

$$\delta = \begin{Vmatrix} \lambda_1 & & 0 \\ & \ddots & \\ 0 & & \lambda_n \end{Vmatrix}$$

for which $\lambda_1 \cdots \lambda_n = 1$. The algebra H is the algebra of all diagonal matrices

$$h = \left\| \begin{array}{ccc} h_1 & & \\ & \ddots & \\ 0 & & h_n \end{array} \right\|$$

with trace $h_1 + \cdots + h_n = 0$. Simple roots have the form

$$\alpha_1 = h_1 - h_2, \qquad \alpha_2 = h_2 - h_3, \ldots, \alpha_{n-1} = h_{n-1} - h_n, \qquad r = n - 1.$$

An easy computation shows that $p_{\alpha_k} = p_k - p_{k+1}$, $q_{\alpha_k} = q_k - q_{k+1}$ (see (10.3.9) in chapter X). Therefore for inductivity of a character with respect to $SL(n, \mathbf{C})$ it is necessary and sufficient that the numbers $p_1 - p_2$, $p_2 - p_3, \ldots, p_{n-1} - p_n$, $q_1 - q_2$, $q_2 - q_3, \ldots, q_{n-1} - q_n$ be nonnegative integers. We have obtained the condition for inductivity for the group $SL(n, \mathbf{C})$ obtained by a different method in chapters VI and VII.

Still another derivation of the condition of inductivity of a character, not using the classical results for representations of a semisimple Lie algebra, is found in Želobenko [1].

1.5. Description of the Space of an Irreducible Finite-Dimensional Representation of the Group G

Consider a representation T_α of G realized by functions on a group Z^+. As we have seen, the space F_α of the representation is the linear span of all functions $\alpha(zg)$, $g \in G$. This description of the space is insufficiently precise. The following description of the space F_α is due to D. P. Želobenko [1].

Theorem 6. *Let D_j and $\bar{D}_j$ be the analytic and antianalytic infinitesimal operators of left translations on Z, corresponding to a root vector $e_{\alpha_j}, j = 1, \ldots, r$, where the α_j are all of the simple roots of the Lie algebra L of G. Then F_α consists of all solutions of the system of equations*

$$D_j^{p_j + 1} f(z) = 0, \qquad \bar{D}_j^{q_j + 1} f(z) = 0, \qquad j = 1, 2, \ldots, r, \qquad (1.5.1)$$

where

$$p_j = 2(p, \alpha_j)/(\alpha_j, \alpha_j), \qquad q_j = 2(q, \alpha_j)/(\alpha_j, \alpha_j), \qquad (1.5.2)$$

and

$$\alpha(\delta) = \exp[(h, p) + (\bar{h}, q)], \qquad \delta = \exp h. \qquad (1.5.3)$$

For a proof, see D. P. Želobenko [1], chapter XVI.

§2. Representations of Real Semisimple Lie Groups

Let $\tilde{G}$ be a connected real semisimple Lie group. Suppose that $\tilde{G}$ is the real form of some complex semisimple Lie group G.[63] We use theorem 5 of 1.4. Plainly a finite-dimensional irreducible representation T of $\tilde{G}$ is analytic if and only if all of the numbers $q_j, j = 1, \ldots, r$, are zero. Apply Proposition IV of 8.2, chapter XI to obtain the following theorem.

Theorem. *Every finite-dimensional irreducible representation T of a connected semisimple real Lie group G that is the real form of a complex semisimple Lie group G is described by a system of nonnegative integers (the signature of T), $p_j, j = 1, \ldots, r$, where r is the rank of G. Every such system describes an irreducible finite-dimensional representation of $\tilde{G}$. The representation $\tilde{T}$ corresponding to a given signature $p_j, j = 1, \ldots, r$, is determined by its generating function $\alpha(\zeta \, \delta z) = \alpha(\delta) = \exp(p, h)$, $p, q, h \in H$, $\delta = \exp h$, where $2(p, \alpha_j)/(\alpha_j, \alpha_j) = p_j, j = 1, \ldots, r$ (the α_j are the simple roots of the Lie algebra L of G with respect to a Cartan subalgebra H of L). The space F of $\tilde{T}$ is the linear span of all functions of the form $\alpha(zg), g \in G$. The operators of the representation $\tilde{T}$ are defined by the formula $\tilde{T}(\tilde{g})f(z) = \alpha(z\tilde{g})f(z\overline{\tilde{g}})$, where $\tilde{g} \in \tilde{G}$, $z\tilde{g} = k \cdot z\overline{\tilde{g}}, z\overline{\tilde{g}} \in Z^{+}$.*

Example

Let $\tilde{G}$ be the group $SU(n)$ or the group $SL(n, \mathbf{R})$. Then $\tilde{G}$ is the real form of the complex semisimple Lie group $G = SL(n, \mathbf{C})$. According to the theorem just proved, every finite-dimensional irreducible representation of $\tilde{G}$ is the restriction to $\tilde{G}$ of an irreducible analytic representation of the group G; i.e., of a representation with signature of the form $(p_1, \ldots, p_n, 0, \ldots, 0)$, where all of the numbers $p_1 - p_2, \ldots, p_{n-1} - p_n$ are nonnegative integers.

---

[63] Not every real semisimple Lie group is the real form of a complex semisimple Lie group. For example, the universal covering group of the group $SL(2, \mathbf{R})$ has no complex hull. For more details, see Želobenko [1].

Bibliography

Part A: Monographs and Textbooks

Bourbaki, N.: [1] *Algèbre. Nouv. éd.* Paris: Hermann & Cie., 1970. English translation *Algebra.* Reading, Mass.: Addison-Wesley Pub. Co., 1974.
— [2] *Groupes et algèbres de Lie.* Paris: Hermann & Cie., 1975. English translation, *Lie groups and Lie algebras.* Reading, Mass.: Addison-Wesley Pub. Co., 1975.
— [3] *Topologie Générale.* Nouv. éd. Paris: Hermann & Cie., 1971. English translation of 1st ed., *General Topology.* Reading, Mass.: Addison-Wesley Pub. Co., 1966.
Burnside, W.: [1] *Theory of Groups of Finite Order.* 2nd ed. Cambridge, England, Cambridge University Press, 1911.
Chevalley, C.: [1] *Theory of Lie Groups. Vol. I*, Princeton, N.J.: Princeton University Press, 1946. Vols. II, III (in French), Paris: Hermann & Cie., 1951, 1954.
Curtis, C. W., and Reiner, I.: [1] *Representation Theory of Finite Groups and Associative Algebras.* New York, N.Y.: Interscience Publishers, 1962.
Eyring, H., Walter, J., and Kimball, G.E.: [1] *Quantum Chemistry.* New York, N.Y.: John Wiley & Sons, 1944.
Fihtengol'c, G. M.: [1] *A Course of Differential and Integral Calculus* (in Russian), 3 vols., rev. ed. Moscow: Nauka, 1970, 1971. German translation of 1959 ed., *Differential- und Integralrechnung*, 3 vols. Berlin: VEB Deutscher Verlag der Wiss., 1964.
Frobenius, G.: [1] *The Theory of Characters and Group Representations* (in Russian). Har'kov, USSR: ONTI, 1937.
Gantmaher, F. R.: [1] *Theory of Matrices* (in Russian), rev. ed. Moscow: Nauka, 1967. English translation of 1st ed. New York, N.Y.: Chelsea Pub. Co., 1959.
Gel'fand, I. M., Minlos, R. A., and Šapiro, Z. Ja.: [1] *Representations of the Group of Rotations and of the Lorentz Group* (in Russian). Moscow, Fizmatgiz, 1958. English translation, *Representations of the rotation and Lorentz groups and their applications.* Oxford-New York: Pergamon Press, 1963.
—, and Naĭmark, M. A.: [1] *Unitary Representations of the Classical Groups* (in Russian). Moscow: Trudy Mat. Inst. Akad. Nauk SSSR, 1950. German translation, *Unitäre Darstellungen der klassischen Gruppen.* Berlin: Akademie-Verlag, 1957.
Gunning R. C., and Rossi, H.: [1] *Analytic Functions of Several Complex Variables.* Englewood Cliffs, N.J.: Prentice-Hall, 1965.
Hall, M.: [1] *The Theory of Groups*, 2nd ed. New York, N.Y.: MacMillan Co., 1968. Reprint, Bronx, N.Y.: Chelsea Pub. Co., 1976.
Hamermesh, M.: [1] *Group Theory and Its Application to Physical Problems.* Reading, Mass.: Addison-Wesley Pub. Co., 1962.
Helgason, S.: [1] *Differential Geometry and Symmetric Spaces.* New York, N.Y.: Academic Press, 1962.
Jacobson, N.: [1] *Lie Algebras.* New York, N.Y.: Interscience Publishers, 1962.
Kaplansky, I.: [1] *Lie Algebras and Locally Compact Groups.* Chicago, Ill.: University of Chicago Press, 1974.
Kirillov, A. A.: [1] *Elements of The Theory of Representations* (in Russian). Moscow, Nauka, 1972. English translation, same title. Heidelberg-Berlin-New York: Springer-Verlag, 1976.
Kokkedee, J. J. J.: [1] *The Quark Model.* New York, N.Y.: W. A. Benjamin, 1969.
Kolmogorov, A. N., and Fomin, S. V.: [1] *Elements of the Theory of Functions and Functional analysis* (in Russian), rev. ed. Moscow: Nauka, 1972. English translation of previous ed., *Introductory Real Analysis.* Englewood Cliffs, N.J.: Prentice-Hall, 1970.

Kuroš, A. G.: [1] *A Course of Higher Algebra* (in Russian). Moscow: Nauka, 1965.
—[2] *Theory of groups*, 3rd ed. Moscow: Nauka, 1967. English translation of 2nd ed, same title. New York, N.Y.: Chelsa Pub. Co., 1960.
Lie, S., and Engel, F.: [1] *Theorie der Transformationsgruppen*. 3 vols. Leipzig: B. G. Teubner Verlag, 1888–1893.
Littlewood, D. E.: [1] *The Theory of Group Characters and Matrix Representations of Groups*, 2nd ed. Oxford: Clarendon Press, 1958.
Ljubarskij, T. Ja.: [1] *The Theory of Groups and its Application in Physics* (in Russian). Moscow: Fizmatgiz, 1957.
Markuševič, A. I.: [1] *The Theory of Analytic Functions* (in Russian). Moscow-Leningrad: Gostehizdat, 1950. English translation, same title. Englewood Cliffs, N.J.: Prentice-Hall, 1965.
Montgomery, D., and Zippin, L.: [1] *Topological Transformation Groups*. New York, N.Y.: Interscience Publishers, 1955.
Murnaghan, F. D.: [1] *The Theory of Group Representations*. Baltimore, Md.: The Johns Hopkins Press, 1938.
Naĭmark, M. A.: [1] *Normed Rings* (in Russian), rev. ed. Moscow: Nauka, 1968. English translation of 1st ed., same title. Groningen: P. Noordhoff, 1964. German translation of 1st ed., *Normierte Algebren*. Berlin: VEB Deutscher Verlag der Wiss., 1959.
Pontrjagin, L. S.: [1] *Continuous Groups* (in Russian), 3rd ed. Moscow: Nauka, 1973. English translation of 2nd ed., *Topological groups*. New York, N.Y.: Gordon & Breach, 1966. German translation of 2nd ed., *Topologische Gruppen*, 2 vols. Leipzig: Teubner Verlagsgesellschaft, 1957, 1958.
Séminaire "Sophus Lie"; [1] Paris, École Normale Supérieure, 1954/1955, 1955/1956.
Serre, J.-P.: [1] *Lie Algebras and Lie Groups*. New York, N.Y.: W. A. Benjamin, 1965.
— [2] *Représentations Linéaires des Groupes Finis*, 2ᵉ éd. Paris: Hermann & Cie., 1971. English translation, *Linear Representations of Finite Groups*. Berlin-Heidelberg-New York: Springer-Verlag, 1977.
—[3] *Cours d'Arithmétique*. Paris: Presses universitaires de France, 1970. English translation, *A Course in Arithmetic*. Berlin-Heidelberg-New York: Springer-Verlag, 1973.
Šilov, G. E.: [1] *Mathematical Analysis, A Special Course* (in Russian). Moscow: Fizmatgiz, 1961. English translations, same title. Oxford-New York: Pergamon Press, 1965. Rev. ed. Cambridge, Mass.: MIT Press, 1973.
Varadarajan, V. S. [1] *Lie groups, Lie Algebras, and Their Representations*. Englewood Cliffs, N.J.: Prentice-Hall, 1974.
Vilenkin, N. Ja.: [1] *Special Functions and the Theory of Representations of Groups* (in Russian). Moscow: Nauka, 1965.
Waerden, B. L. van der: [1] *Algebra*, 7th ed. Berlin-Heidelberg-New York: Springer-Verlag, 1966. English translation, 2 vols. New York, N.Y.: Ungar Pub. Co., 1970.
— [2] *Group Theory and Quantum Mechanics*. Berlin-Heidelberg-New York: Springer-Verlag, 1974.
Weil, A.: [1] *L'intégration dans les Groupes Topologiques et ses Applications*. Paris: Hermann & Cie., 1941, 1951.
Weyl, H.: [1] *The Classical Groups, their Invariants and Representations*. Princeton, N.J.: Princeton University Press, 1939.
—[2] *Gruppentheorie und Quantenmechanik*, 2. Auflage. Leipzig: F. Hirzel, 1931. English translation, *The Theory of Groups and Quantum Mechanics*. New York, N.Y.: E. P. Dutton & Co., 1931, reprinted by Dover Publ., New York, 1949.
Zamansky, M.: [1] *Introduction à l'Algèbre et l'Analyse Modernes*. Paris: Dunod, 1967.
Želobenko, D. P. [1] *Compact Lie Groups and Their Representations* (in Russian). Moscow: Nauka, 1970. English translation, same title. Providence, R. I., Amer. Math. Soc., 1973.
— [2] *Lectures on the Theory of Lie Groups* (in Russian). Dubna, 1965.
— [3] *Harmonic Analysis on Semisimple Complex Lie Groups* (in Russian). Moscow: Nauka, 1974.

Part B: Journal Articles

Ado, I. D.: [1*] On the representation of finite continuous groups by linear substitutions. Izv. Fiz.-Mat. Obšč. Kazan' 7 (1934/1935), 3–43.
Berezin, F. A.: [1*] Some remarks about the associative envelope of a Lie algebra. *Funct. Anal. Appl.* **1** (1967), 91–102.

Bernšteĭn, I. N., Gel'fand, I. M., and Gel'fand, S. I.: [1*] Structure of representations generated by vectors of highest weight. *Funct. Anal. Appl.* **5** (1971), 1–8.

Cartan, É. J.: [1*] Sur la structure des groupes de transformations finis et continus. Thèse. Paris: Nony, 1894. 2e éd. Paris: Vuibert, 1933. Also in *Oeuvres Complètes*. Paris: Gauthier-Villars, 1952, Vol. I, 143–225.

— [2*] Les groupes réels simples, finis et continus. *Ann. Ec. N.S.* (3) **31** (1914), 263–355. Also in *Oeuvres Complètes*, Vol. I, 399–491.

— [3*] Les tenseurs irréductibles et les groupes linéaires simples et semi-simples. *Bull. Sci. Math.* (2) **49** (1925), 2e 130–152. Also in *Oeuvres Complètes*. Vol. I, 531–553.

Cartier, P.: [1*] On H. Weyl's character formula. *Bull. Amer Math. Soc.* **67** (1961), 228–230.

Curtis, C. W. [1*] Representations of finite groups of Lie type. *Bull. Amer. Math. Soc.* (N.S.) **1** (1979), 721–757.

Dynkin, E. B. [1*] The structure of semisimple Lie algebras (in Russian). *Uspehi Mat. Nauk* **2**, No. 4 (1947), 59–127.

— [2*] On the representation of the series $\log(e^x e^y)$ of noncommuting x, y by commutators (in Russian). *Mat. Sb. (N.S.)* **25** (1949), 155–162.

— and Oniščik, A. L.: [1*] Compact Lie groups in the large (in Russian). *Uspehi Mat. Nauk* **10**, No. 4 (1955), 3–74.

Gel'fand, I.M.: [1*] The center of an infinitesimal group ring (in Russian). *Mat. Sb. (N.S.)* **26** (1950), 103–112.

— [2*] On one-parametrical groups of operators in a normed space. *Dokl. Akad. Nauk SSSR* **25** (1939), 713–718.

— and Graev, M. I.: [1*] Finite-dimensional irreducible representations of the unitary and the full linear group and the special functions connected with them (in Russian). *Izv. Akad. Nauk SSSR Ser. Mat.* **29** (1965), 1329–1356.

Gel'fand, S. I.: [1*] Representations of the full linear group over a finite field (in Russian). *Mat. Sb (N.S.)* **83** (1970), 15–41.

Godement, R.: [1*] A theory of spherical functions. *Trans. Amer. Math. Soc.* **73** (1952), 496–556.

Green, J. A.: [1*] The characters of the finite general linear groups. *Trans. Amer. Math. Soc.* **80** (1955), 402–447.

Haar, A.: [1*] Der Massbegriff in der Theorie der kontinuierlichen Gruppen. *Ann. of Math.* (2) **34** (1933), 147–169.

Harish-Chandra: [1*] On some applications of the universal enveloping algebra of a semisimple Lie algebra. *Trans. Amer. Math. Soc.* **70** (1951), 28–96.

Iwasawa, K.: [1*] On some types of topological groups. *Ann. of Math.* (2) **50** (1949), 507–558.

Klimyk, A. U.: [1*] On the multiplicities of the weights of representations and multiplicities of representations of semisimple Lie groups (in Russian). *Dokl. Akad. Nauk SSSR* **177** (1967), 1001–1004.

Kostant, B.: [1*] A formula for the multiplicity of a weight. *Trans. Amer. Math. Soc.* **93** (1959), 53–73.

Kreĭn, M. G.: [1*] The principle of duality for a bicompact group and a quadratic block-algebra (in Russian). *Dokl. Akad. Nauk SSSR* **69** (1949), 725–728.

Mackey, G. W.: [1*] Infinite-dimensional group representations. *Bull. Amer. Math. Soc.* **69** (1963), 628–686.

Mal'cev, A. I.: [1*] On the representation of an algebra as a direct sum of the radical and a semi-simple subalgebra. *Dokl. Akad. Nauk SSSR* **36** (1942), 42–45.

Molčanov, V. F.: [1*] On matrix elements of irreducible representations of a symmetric group (in Russian). *Vestnik Moskov. Univ. Ser. I Mat. Meh.* **21** (1966), No. 1, 52–57.

Neumann, J. von: [1*] Zum Haarschen Mass in topologischen Gruppen. *Compositio Math.* **1** (1934), 106–114.

Raševskij, P. K.: [1*] On some fundamental theorems of the theory of Lie groups (in Russian). *Uspehi Mat. Nauk* **8**, No. 1 (1953), 3–20.

Schur, I.: [1*] Arithmetische Untersuchungen über endliche Gruppen linearer Substitutionen. *Sitzungsber. Preuss. Akad. Wiss. Berlin* 1906, 164–184. Also in *Gesammelte Abhandlungen*, Berlin-Heidelberg-New York: Springer-Verlag, 1973, Vol. I, 177–197.

Schwartz, L.: [1*] Sur une propriété de synthèse spectrale dans les groupes non compacts. *C.R. Acad. Sci. Paris* **227** (1948), 424–426.

Sirota, A. I., and Solodovnikov, A. S.: [1*] Noncompact semisimple Lie groups (in Russian). *Uspehi Mat. Nauk* **18**, No. 3 (1963), 87–144.

Stone, M. H. [1*]: Applications of the theory of Boolean rings to general topology. *Trans. Amer. Math. Soc.* **41** (1937), 375–481.

Tannaka, T.: [1*] Über den Dualitätssatz der nichtkommutativen topologischen Gruppen. *Tôhoku Math. J.* **45** (1939), 1–12.

Weyl, H.: [1*] Theorie der Darstellung kontinuierlicher halb-einfacher Gruppen durch lineare Transformationen. I, II, III, Nachtrag. *Math Z.* **23** (1925), 271–309; **24** (1926), 328–376, 377–395, 789–791. Also in *Gesammelte Abhandlungen*, Bd. II, No. 68, 543–647. Berlin-Heidelberg-New York: Springer-Verlag, 1968.

— and Peter, F.: [1*] Die Vollständigkeit der primitiven Darstellungen einer geschlossenen kontinuierlichen Gruppen. *Math. Ann.* **97** (1927), 737–755. Also in *Gesammelte Abhandlungen* Bd. III, No. 73, 58–75.

Young, A.: [1*] On quantitative substitutional analysis. *Proc. London Math. Soc.* **33** (1900–1901), 97–146; **34** (1901–1902), 361–397.

Želobenko, D. P.: [1*] Description of all irreducible representations of an arbitrary connected Lie group (in Russian). *Dokl. Akad. Nauk SSSR* **139** (1961), 1291–1294.

— [2*] The classical groups. Spectral analysis of finite-dimensional representations (in Russian). *Uspehi Mat. Nauk* **17**, No. 1 (1962), 27–120.

— [3*] On the theory of linear representations of complex and real Lie groups (in Russian). *Trudy Moskov. Mat. Obšč.* **12** (1963), 53–98.

Index

Grundlehren der mathematischen Wissenschaften

A Series of Comprehensive Studies in Mathematics

A Selection